W9-DFB-219

2ND EDITION

FUNDAMENTALS OF MOLECULAR

VIROLOGY

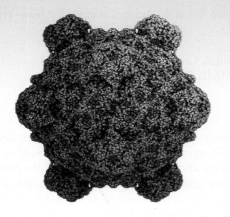

NICHOLAS H. ACHESON

WILEY

John Wiley & Sons, Inc.

Vice President and Publisher	Kaye Pace
Acquisitions Editor	Kevin Witt
Associate Editor	Michael Palumbo/Lauren Morris
Assistant Editor	Jenna Paleski
Marketing Manager	Clay Stone
Senior Media Editor	Linda Muriello
Media Specialist	Daniela DiMaggio
Production Manager	Janis Soo
Senior Production Editor	Joyce Poh
Designer	Maddy Lesure/Seng Ping Ngieng

Cover images: Enterobacteria Phage Phi X174, Human Rhinovirus 3, Simian Virus 40. Images created by Jean-Yves Sgro, University of Wisconsin, Madison, with software Qutemol and VMD.

This book was set in 10/12 Janson Text Roman by MPS Limited, a Macmillan Company, and printed and bound by Markono Print Media Pte Ltd. The cover was printed by Markono Print Media Pte Ltd.

This book is printed on acid free paper.

Founded in 1807, John Wiley & Sons, Inc. has been a valued source of knowledge and understanding for more than 200 years, helping people around the world meet their needs and fulfill their aspirations. Our company is built on a foundation of principles that include responsibility to the communities we serve and where we live and work. In 2008, we launched a Corporate Citizenship Initiative, a global effort to address the environmental, social, economic, and ethical challenges we face in our business. Among the issues we are addressing are carbon impact, paper specifications and procurement, ethical conduct within our business and among our vendors, and community and charitable support. For more information, please visit our website: www.wiley.com/go/citizenship.

Copyright © 2011, 2007 John Wiley & Sons, Inc. All rights reserved. No part of this publication may be reproduced, stored in a retrieval system or transmitted in any form or by any means, electronic, mechanical, photocopying, recording, scanning or otherwise, except as permitted under Sections 107 or 108 of the 1976 United States Copyright Act, without either the prior written permission of the Publisher, or authorization through payment of the appropriate per-copy fee to the Copyright Clearance Center, Inc. 222 Rosewood Drive, Danvers, MA 01923, website www.copyright.com. Requests to the Publisher for permission should be addressed to the Permissions Department, John Wiley & Sons, Inc., 111 River Street, Hoboken, NJ 07030-5774, (201)748-6011, fax (201)748-6008, website http://www.wiley.com/go/permissions.

Evaluation copies are provided to qualified academics and professionals for review purposes only, for use in their courses during the next academic year. These copies are licensed and may not be sold or transferred to a third party. Upon completion of the review period, please return the evaluation copy to Wiley. Return instructions and a free of charge return shipping label are available at www.wiley.com/go/returnlabel. Outside of the United States, please contact your local representative.

Library of Congress Cataloging-in-Publication Data
Acheson, N. H.
    Fundamentals of molecular virology / Nicholas H. Acheson.—2nd ed.
        p. ; cm.
    Includes bibliographical references and index.
    ISBN 978-0-470-90059-8 (pbk. : alk. paper)
        1. Molecular virology.   I. Title.
    [DNLM: 1. Viruses.   2. Virus Physiological Phenomena.   3. Viruses—genetics.   QW 160]
    QR389.A24   2011
    616.9'101—dc22

                                                                        2011002024

Printed in Asia
10 9 8 7 6 5 4 3 2

I dedicate this book to four mentors whose enthusiasm for virology stimulated my interest when I was a student, and who encouraged me to follow my own path.

Johns Hopkins III
James D. Watson
Igor Tamm
Purnell Choppin

BRIEF CONTENTS

CONTENTS

SECTION III: POSITIVE-STRAND RNA VIRUSES OF EUKARYOTES

14. Coronaviruses   159

SECTION IV: NEGATIVE-STRAND AND DOUBLE-STRANDED RNA VIRUSES OF EUKARYOTES

15. Paramyxoviruses and Rhabdoviruses   175

16. Filoviruses   188

27. Viruses of Algae and Mimivirus 325

CHLOROVIRUSES 327

COCCOLITHOVIRUSES 331

PRASINOVIRUSES 334

PHAEOVIRUSES 335

PRYMNESIOVIRUSES AND RAPHIDOVIRUSES 335

MIMIVIRUS 336

SECTION VII: VIRUSES THAT USE A REVERSE TRANSCRIPTASE

28. Retroviruses 342

29. Human Immunodeficiency Virus 354

PREFACE

This book is written for students who are learning about viruses for the first time in a university course at the undergraduate or graduate level. As the title implies, it concentrates on the molecular mechanisms of virus replication, and on the interactions between viruses and the cells in which they replicate. The book approaches learning about virology by presenting a set of chapters each of which covers a specific virus family, using one or two well-studied viruses as examples. These chapters are each designed to tell a story about the viruses being considered, and to portray the "personality" of those viruses, with the idea that this will help students to learn about and remember each virus group.

This organizational scheme has been used in a number of successful virology textbooks, including Salvador Luria's classic 1953 book, *General Virology*. Luria was one of the founding members of the "phage group", a coalition of physicists, biologists and chemists who chose during the 1940s to study bacteriophages in order to understand the molecular basis of life, and who invented the field of molecular biology. Their approach was to study how the proteins and nucleic acids of viruses interact with cellular molecules and organelles, transforming the cell into a factory that can produce many new progeny virus particles. Their underlying hope, largely achieved, was to use viruses as a tool to help understand how cells work.

The amount of knowledge that has accumulated about viruses has expanded enormously in recent years, as in many other areas of biomedical sciences. *Fields Virology* has become the classic reference book for knowledge about human and animal viruses during the past 25 years; that book is also organized in chapters that cover specific virus families. My own teaching experience and conversations with numerous colleagues convinced me that there is a real need for a concise, up-to-date textbook organized around the concept of virus families and designed specifically for teaching university students.

The problem was to make such a book accessible for beginning students but not to over-simplify the material. My approach was to ask a number of prominent virology researchers and teachers to write chapters on viruses that they knew well, using a set of criteria that I provided. I then edited and sometimes rewrote these chapters into a common style, and in many cases I created or redesigned the illustrations.

No individual can possibly write knowledgeably about the large spectrum of viruses that a virology course should cover, so a collaborative approach was necessary. However, a textbook that is an effective learning tool must have a coherent organization and a clear and consistent style of writing and illustration. My job has been to craft the original chapters that I received into what I hope are readable and easily understood units.

The emphasis of this textbook is on virus replication strategies; it is directed towards university students studying microbiology, cell and molecular biology, and the biomedical sciences. It does not go deeply into pathogenesis, epidemiology, or disease symptoms. However, substantial information and stories about medical and historical aspects of virology are included, particularly in introductory sections of each chapter. Students who understand what diseases are caused by particular viruses, and the importance of these diseases in human history, may be motivated to learn more about those viruses.

What Is New in the Second Edition

The first edition of this book was well-received and was adopted as a text by over 100 university-based virology courses in North America and overseas. When we considered creating a second edition, my editor and I solicited reviews and suggestions for improvements from a number of university teachers. We also set out to improve the graphic qualities of the book, by introducing full-color figures and by incorporating the impressive computer-generated figures of viruses created by Philippe Lemercier, of the Swiss Institute of Bioinformatics, Swiss-Prot Group, University of Geneva. These virion figures and many others can be found on the web at Viralzone: http://ca.expasy.org/viralzone/.

The second edition includes five new chapters: two survey chapters, "Viruses of Archaea" and "Viruses of Algae and Mimivirus"; a chapter on a well-studied plant virus, "Cucumber Mosaic Virus"; and two chapters on the host response, "Cellular Defenses Against Virus Infection" and "Innate and Adaptive Immune Responses to Virus Infection". To make room for these chapters, a chapter on human T-cell leukemia virus was removed, but it is available for book users on the text's companion website (www.wiley.com/college/acheson). Additionally, parts of the chapter on Interferons were incorporated into the new chapter on Cellular Defenses. Furthermore, all but one of the remaining chapters in the first edition were revised and updated by the original contributors or, in several cases, by other contributors recruited for that purpose. For example, the original chapter on herpes simplex virus now is entitled "Herpesviruses", and includes a substantial section on Epstein–Barr virus.

How To Use This Book

This textbook is designed to be used in a modular fashion. No course would be expected to use all the chapters in the book, nor necessarily in same order in which they appear. The organization of the book gives wide latitude to course coordinators to make their own choices of which virus groups will be covered. Chapters are designed to accompany a 50-minute lecture on the subject, or in some cases, two or three such lectures. It should be possible to read each chapter in 30–60 minutes, including examination of figures and tables. Lecturers might want to supplement material given in the text with experimental methods or results, which are not covered because of lack of space.

The book is organized into ten sections and 37 chapters. Four introductory chapters in Section I cover the history of virology and the virus life cycle, virus structure, virus classification, and the entry of viruses into animal cells. Four chapters in Section II cover well-studied bacteriophages. These are included because bacteriophages are among the best-known viruses, and because much of our knowledge of molecular biology and virology began with their study. Furthermore, bacteriophages are the source of many tools commonly used in modern molecular and cell biology laboratories. A final chapter in Section II covers exciting new knowledge about the sometimes bizarre viruses that infect archaea, members of the third domain of life alongside bacteria and eukaryotes.

Sections III through VII cover viruses of eukaryotes, with some emphasis on viruses that infect humans, although included are chapters on viruses that infect plants, insects, and algae. The division into sections is based on the nature of the virus genome and virus replication strategies: positive-strand RNA viruses (Section III), negative-strand and double-stranded RNA viruses (Section IV), DNA viruses (Sections V and VI), and viruses that use a reverse transcriptase (Section VII). Within a section, smaller and simpler viruses are discussed first, then larger and more complex viruses. In this way, concepts that are learned about simpler viruses can be applied when more complex viruses are encountered.

Section VIII covers small infectious entities that are not viruses: viroids, which are virus-like nucleic acids that replicate but code for no proteins; and prions, which are infectious proteins that contain no detectable nucleic acid. Section IX includes the two new chapters on host responses to virus infection, with important new information on detection of virus infection, intrinsic cellular responses to virus infection, and innate and adaptive immune responses. Finally, Section X finishes the book by reviewing some important applications in virology: antiviral vaccines, antiviral chemotherapy, and virus vectors.

Each chapter begins with an outline. For chapters that cover virus families, these outlines are "thumbnail sketches" that contain some basic information about virion structure, genome organization, replication strategies, diseases caused, and distinctive characteristics shared by viruses in that family. These outlines are designed to serve as study aids that will help students understand and remember common features of the viruses they study.

Subheadings within each chapter are explanatory phrases, telling the reader what will be discussed in the next several paragraphs. These subheadings (collected in the Table of Contents) can also be read separately to provide an overview of the material presented in the chapter, and to follow the steps of the virus replication cycle. Figures concentrate on individual well-studied steps in virus replication. Most figures are designed to be simple and easily understood while one is reading the accompanying text, rather than comprehensive (and sometimes complicated!) descriptions of the entire replication cycle. Figure legends are kept to a minimum.

Specialized terms that may be unfamiliar to students are presented in bold type at their first appearance in each chapter. These Key Terms are collected at the end of each chapter as a review aid, and definitions are given in a combined glossary at the end of the book. Many chapters have text boxes that cover intriguing applications or recent developments in research. Each chapter finishes with a list of Fundamental Concepts, statements outlining the most important facts or conclusions that the reader should have learned. Finally, a set of Review Questions is included as a further review tool and to alert the student to the kinds of knowledge that might be expected in test questions.

Answers to Review Questions are available to course instructors at the Instructor Companion Site of Wiley Higher Education at: www.wiley.com/college/acheson. The full text and figures of the chapter on Human T-cell Leukemia Virus Type I that appeared in the first edition but was not included in the second edition are also available at that site.

Key Features of This Book

- A concise, up-to-date textbook designed for university-level virology courses for students in biomedical sciences and microbiology
- Written in a simple and clear style for students with a background in cell and molecular biology

- Explains replication mechanisms of viruses representing many of the major virus families
- Many full-color figures complement the text and illustrate virus structure, genome organization and individual steps in virus replication
- Each chapter is designed to tell a story about a specific virus family and to portray the "personality" of the virus covered
- Chapter introductions give historical background and information about viral diseases
- Includes study aids such as thumbnail sketches of each virus group, informative chapter subheadings, text boxes outlining recent research and applications, a list of fundamental concepts after each chapter, sample test questions, and a comprehensive glossary with definitions of numerous terms
- An introductory section provides basic information about the history of virology, virus replication, virus structure, classification of viruses, and virus entry into cells
- A section on viruses of bacteria and archaea covers four of the best-known bacteriophages: single-stranded RNA phages, φX174, T7 and lambda; as well as a survey of the known viruses of archaea
- Five sections containing 21 chapters cover a wide variety of viruses that infect animals, plants, algae and insects, with emphasis on viruses that cause human disease
- Includes chapters that cover important human pathogens such as Ebola virus, hepatitis B and C viruses, herpes viruses, human immunodeficiency virus, influenza viruses, measles virus, poliovirus, SARS coronavirus, smallpox virus, West Nile virus and others
- A chapter on viroids: small infectious nucleic acids that do not code for proteins but cause important plant diseases
- A chapter on prions: infectious proteins that cause mad cow disease and Creutzfeld–Jacob disease in humans
- A section on host defenses, with discussion of intrinsic cellular responses, innate and adaptive immune responses to virus infections
- A concluding section with chapters on antiviral vaccines, antiviral chemotherapy, and virus vectors

ACKNOWLEDGMENTS

This textbook is the outgrowth of an undergraduate science course in virology taught by myself and colleagues in the Department of Microbiology and Immunology at McGill University for 25 years. I am grateful to Professors Dal Briedis, Mike DuBow, and John Hassell, with whom I collaborated in designing and offering this course. Their high academic standards and constant effort ensured its success. Among other colleagues who contributed significantly to this course during recent years are Alan Cochrane, Matthias Gotte, John Hiscott, Arnim Pause, and Mark Wainberg.

David Harris, then acquisitions editor at John Wiley and Sons, enthusiastically endorsed and welcomed my book project when I first proposed it. During its gestation, I was ably helped by a succession of editors at Wiley: Joe Hefta, Keri Witman, Patrick Fitzgerald, and finally Kevin Witt, under whose tutelage the book first saw the light of day. Kevin also launched the present second edition, and both Associate Editor Michael Palumbo and Senior Production Editor Joyce Poh have been of constant and uwavering help throughout this process.

A number of university and college teachers of virology reviewed the concept of the book, or parts of the manuscript at various stages, and offered helpful suggestions and comments. On behalf of all of my colleagues who contributed chapters to this book, I would like to thank the following reviewers:

Lawrence Aaronson, *Utica College*

John R. Battista, *Louisiana State University*

Karen Beemon, *Johns Hopkins University*

Martha Brown, *University of Toronto*

Craig E. Cameron, *Pennsylvania State University*

Howard Ceri, *University of Calgary*

Jeffrey DeStefano, *University of Maryland, College Park*

Rebecca Ferrell, *Metropolitan State College of Denver*

Lori Frappier, *University of Toronto*

Eric Gillock, *Fort Hays State University*

Michael Graves, *University of Massachusetts, Lowell*

Sidney Grossberg, *Medical College of Wisconsin*

Tarek Hamouda, *University of Michigan Medical Center*

Richard W. Hardy, *Indiana University*

Hans Heidner, *University of Texas at San Antonio*

Richard Kuhn, *Purdue University*

Alexander C. K. Lai, *Oklahoma State University*

Lorie LaPierre, *Ohio University*

Maria MacWilliams, *University of Wisconsin, Parkside*

Phillip Marcus, *University of Connecticut*

Nancy McQueen, *California State University, Los Angeles*

Joseph Mester, *Northern Kentucky University*

Thomas Jack Morris, *University of Nebraska*

Brian Olson, *Saint Cloud State University*

Arnim Pause, *McGill University*

Marie Pizzorno, *Bucknell University*

Sharon Roberts, *Auburn University*

Michael Roner, *University of Texas at Arlington*

Miroslav Sarac, *Our Lady of the Lake College*

David A. Sanders, *Purdue University*

Robert Sample, *Mississippi State University*

Jeff Sands, *Lehigh University*

Ann M. Sheehy, *College of the Holy Cross*

Kenneth Stedman, *Portland State University*

Carol St. Angelo, *Hofstra University*

Suresh Subramani, *University of California, San Diego*

William Tapprich, *University of Nebraska, Omaha*

Milton Taylor, *Indiana University*

Michael N. Teng, *Pennsylvania State University*

Loy Volkman, *University of California, Berkeley*

Darlene Walro, *Walsh University*

Jeannine Williams, *College of Marin*

During the preparation of the first edition of this book, preliminary versions of a number of chapters were made available to students taking our virology course at McGill, and many of those students gave precious feedback that improved the book. Furthermore, a number of chapters were read and reviewed in detail by the following McGill undergraduate students, who contributed insightful comments and suggestions: Jonathan Bertram, Yasmin D'Souza, Eric Fox, Caroline Lambert, Kathryn Leccese, Edward Lee, Alex Singer, Brian Smilovici, and Claire Trottier. Claire Trottier helped organize these student reviews.

Thanks to the following McGill University students who worked with me on chapter summaries, permissions and editing in the final phases of preparation of the book: Meoin Hagege, Jennifer LeHuquet, Melany Piette, Pooja Raut, and Emilie Mony. Thanks also to Joan Longo and Mei Lee of the Department office for their help.

Michael Roner kindly agreed to write review questions that are placed at the end of each chapter in this edition.

Work on this book began during a sabbatical year I spent in the laboratory of Steve Harrison at Harvard University. Thanks to McGill for approving my sabbatical leave, and to Steve and the members of the Harrison and Wiley laboratories for their stimulation and support.

This book is the result of an enjoyable and fruitful collaboration between myself and 49 other virologists from around the world who contributed or revised chapters. Their expertise, energy and enthusiasm made this book possible. Thank you, all.

Finally, I would like to thank my wife, Françoise, for enduring the seemingly endless task of writing, editing, and correcting the text and figures for this book, through two editions.

Nicholas H. Acheson
Montreal, September 2010

SCHEMATIC DIAGRAMS OF VIRUSES COVERED IN THIS BOOK

The diagrams on the following two pages illustrate most of the viruses discussed in detail in this book. Virions are shown as cross-sections, revealing the capsids or nucleic acid genomes within. Capsid subunits are shown in green; capsids with icosahedral symmetry are shown as circles or polygons, and capsids with helical symmetry are shown as chains or coils. Envelopes are shown as light blue membrane bilayers, and envelope proteins are shown as yellow spikes inserted in the membrane. DNA or RNA genomes are shown as coils or double helices.

The diagrams on the first page show virions at a scale of **50 nanometers (nm)** per inch. The smallest virion illustrated, a single-stranded RNA bacteriophage, has a diameter of 26 nm; the largest virions illustrated, retroviruses and influenza virus, have diameters of 100 nm.

The diagrams on the second page show virions at a scale of **200 nm per inch**, in order to be able to accommodate all the larger virions on a single page. These virions would therefore appear four times larger if they were shown at the same scale as the first page. To illustrate the scale change, the same diagram of a retrovirus shown at the top left of the second page is also shown, four times larger, at the bottom left of the first page. The largest virion illustrated, mimivirus, is shown both as a cross-section and as an intact virion. Mimivirus has a capsid diameter of 450 nm and a total diameter including fibers of 700 nm. Some filamentous virions, not shown here in their entirety, are 1000 nm or more in length.

These diagrams, and the figures illustrating the opening pages of each chapter in this book, were drawn by Philippe Lemercier, Swiss Institute of Bioinformatics, Swiss-Prot Group, University of Geneva. These virion figures and many others can be found at Viralzone (http://www.expasy.org/viralzone/all_by_protein/230.html). This resource has basic information on many viruses and facilitates entry into protein and nucleic acid databases relevant to each virus family or species.

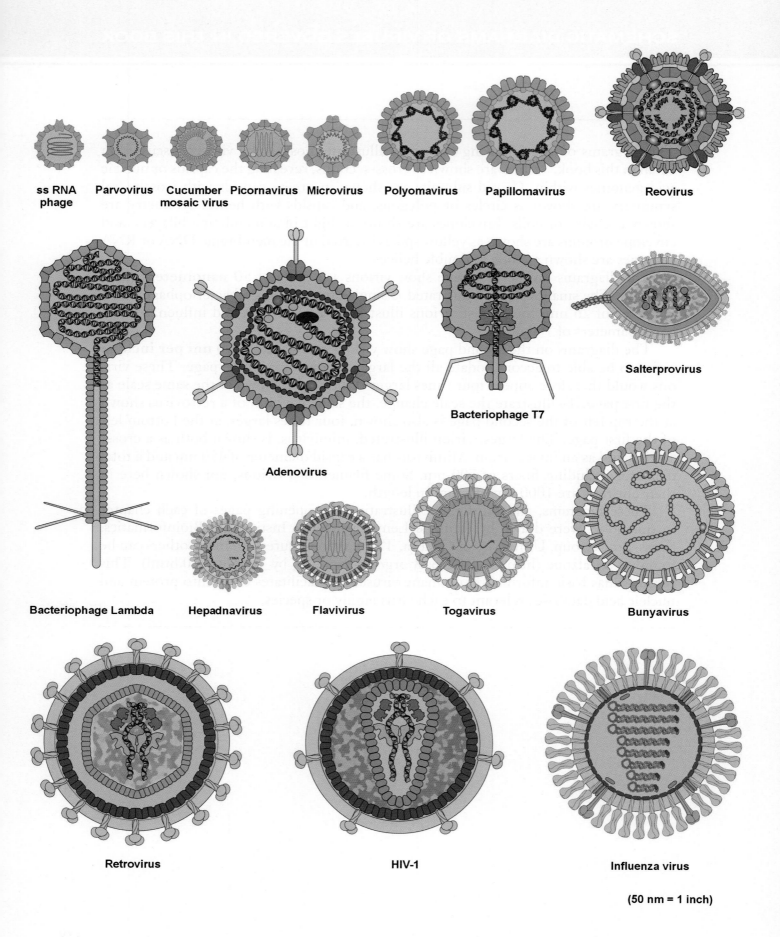

ss RNA phage

Parvovirus

Cucumber mosaic virus

Picornavirus

Microvirus

Polyomavirus

Papillomavirus

Reovirus

Salterprovirus

Bacteriophage T7

Adenovirus

Bacteriophage Lambda

Hepadnavirus

Flavivirus

Togavirus

Bunyavirus

Retrovirus

HIV-1

Influenza virus

(50 nm = 1 inch)

Retrovirus

Fusellovirus

Globulovirus

Guttavirus

Rhabdovirus

Ampullavirus

Alphalipothrixvirus

Rudivirus

Bicaudavirus

Betalipothrixvirus

Coronavirus

Paramyxovirus

Herpesvirus

Phycodnavirus

Poxvirus

Baculovirus

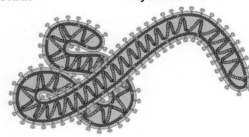

Filovirus

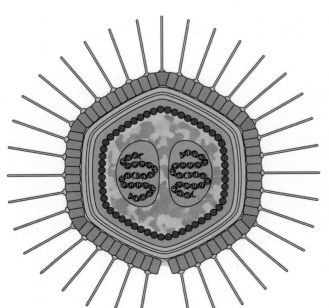

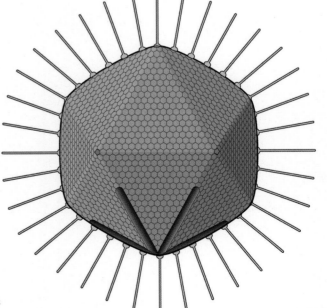

Mimivirus

(200 nm = 1 inch)

INTRODUCTION TO VIROLOGY

Virus particles, or **virions**, consist of an RNA or DNA genome packaged within a protein coat, and in some cases a lipid envelope. Viruses can only reproduce themselves by infecting living cells. Cells provide molecular building blocks such as nucleotides and amino acids, a source of chemical energy, the cellular protein-synthesizing machinery, and the controlled intracellular environment needed to carry out life processes. Without these, a virus is just a package of genes; once inside a cell, a virus organizes a "factory" that produces progeny virus particles and sends them on their way to infect other cells.

This section of the book introduces the student to the study of viruses, their structure and classification, and how they enter cells and begin their replication cycles. Chapter 1 outlines the properties of viruses and how they replicate. Chapter 2 examines the unique features of the intricately constructed symmetrical capsids that most viruses use to package their genomes, and discusses how lipid envelopes are formed. Chapter 3 explores the wide variety of viruses that exist on earth; probably all living organisms can be infected by at least one species of virus.

How and when did viruses first appear during evolution? Chapter 3 concludes with some speculation on their origin, perhaps billions of years ago. Chapter 4 describes how viruses that infect eukaryotic organisms bind to and enter host cells. Enveloped viruses usually fuse their envelope with a cellular membrane, releasing the genome or capsid into the cell. Non-enveloped viruses interact with cellular membranes, leading to the penetration of either the capsid or the genome into the cell, beginning the replication cycle.

Introduction to Virology

Nicholas H. Acheson

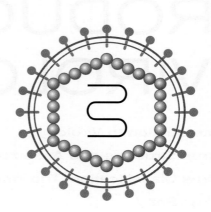

THE NATURE OF VIRUSES

Virus particles contain:
- *A nucleic acid genome (either DNA or RNA)*
- *A protein coat (**capsid**) that encloses the genome*
- *In some cases, a lipid membrane (**envelope**)*

The infectious virus particle is called a **virion**.

Virus particles are very small: between 20 and 500 nanometers (nm) in diameter.

Viruses are **obligatory intracellular parasites**.

Viruses multiply inside cells by expressing and replicating their genomes.

Viruses need the following machinery provided by cells:
- *Enzyme systems that synthesize amino acids, nucleotides, carbohydrates, and lipids*
- *Enzyme systems that generate useable chemical energy in the form of ATP*
- *Ribosomes, tRNAs, and enzymes used in protein synthesis*
- *Membranes that concentrate cellular macromolecules, small molecules, and ions*

WHY STUDY VIRUSES?

Viruses are important disease-causing agents.

Probably all different forms of life can be infected by viruses.

Viruses can transfer genes between organisms.

Viruses are important players in the regulation of the Earth's ecology.

Viruses can be engineered to prevent and cure disease.

Study of viruses reveals basic mechanisms of gene expression, cell physiology, and intracellular signaling pathways.

A BRIEF HISTORY OF VIROLOGY

Viruses were first distinguished from other microorganisms by their small size and their ability to pass through fine filters that retain bacteria.

Viruses can be crystallized: they lie on the edge between chemical compounds and life.

Study of bacterial viruses (bacteriophages) by the "phage group" led to understanding of the nature of genes and helped found molecular biology.

In vitro culture of eukaryotic cells led to rapid advances in the study of viruses and in vaccine production.

Study of tumor viruses led to the discovery of viral and cellular **oncogenes**.

DETECTION AND MEASUREMENT OF VIRUSES

The **plaque assay** is widely used to measure virus infectivity.

Hemagglutination is a cheap and rapid method for detection of virus particles.

Virus particles can be seen and counted by electron microscopy.

The ratio of physical particles to infectious particles is greater than 1 for many viruses.

VIRUS REPLICATION CYCLE

1. The virion binds to cell surface receptors.
2. The virion or viral genome enters the cell; the viral genome is uncoated.
3. Early viral genes are expressed (Baltimore classification scheme).

4. Early viral proteins direct replication of the viral genome.
5. Late viral genes are expressed from newly replicated viral genomes.
6. Late viral proteins package genomes and assemble progeny virus particles.
7. Virions are released from the host cell.

THE NATURE OF VIRUSES

Viruses consist of a nucleic acid genome packaged in a protein coat

Viruses are the smallest and simplest forms of life on Earth. They consist of a set of nucleic acid genes enclosed in a protein coat, called a **capsid,** which in some cases is surrounded by or encloses a lipid membrane, called an **envelope** (Figure 1.1). The viral genome encodes proteins that enable it to replicate and to be transmitted from one cell to another, and from one organism to another. The complete, infectious virus particle is called a **virion.**

Viruses are dependent on living cells for their replication

Viruses can replicate only within living cells. Another way of saying this is that viruses are **obligatory intracellular parasites.** Viruses depend on cells for their replication because they lack the following basic elements required for growth and replication, which are present in all living cells:

- Enzyme systems that produce the basic chemical building blocks of life: nucleotides, amino acids, carbohydrates, and lipids

- Enzyme systems that generate useable chemical energy, usually in the form of adenosine triphosphate (ATP), by photosynthesis or by metabolism of sugars and other small molecules
- Ribosomes, transfer RNAs, and the associated enzymatic machinery that directs protein synthesis
- Membranes that localize and concentrate in a defined space these cellular macromolecules, the small organic molecules involved in growth and metabolism, and specific inorganic ions

Virus particles break down and release their genomes inside the cell

Viruses are not the only obligatory intracellular parasites known. A number of small unicellular organisms including chlamydiae and rickettsiae, certain other bacterial species, and some protozoa can multiply only inside other host cells. However, viruses replicate by a pathway that is very different from the mode of replication of these other intracellular parasites.

Virus replication begins with at least partial disintegration of the virus particle, and release (**uncoating**) of the viral genome within the cell. Once uncoated, the viral genome can be used as a template for synthesis of

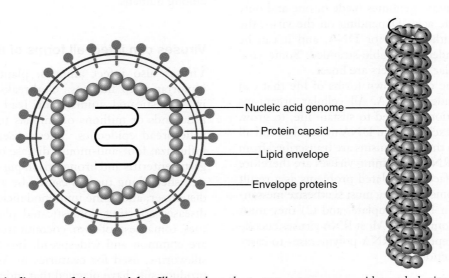

Nucleic acid genome
Protein capsid
Lipid envelope
Envelope proteins

Figure 1.1 Schematic diagram of virus particles. Illustrated are the two most common capsid morphologies: a roughly spherical shell (left) and a tubular rod (right). Some virus particles have an envelope (left) and some do not (right). Nucleic acid genomes are shown as black curved lines, capsid proteins as green spheres, and envelope proteins as orange knobbed spikes.

messenger RNAs, which in turn synthesize viral proteins using the enzyme systems, energy, ribosomes, and molecular building blocks that are present in the cell. These viral proteins then direct replication of the viral genome. Viral structural proteins encapsidate the newly replicated genomes to form progeny virus particles.

In contrast, unicellular organisms that replicate inside other cells invariably remain intact and retain their genomes within their own cellular membranes. They replicate not by disintegration and reassembly, but by growth and division into daughter cells. Such cellular parasites always contain their own ribosomes and protein synthetic machinery, and their genes code for enzymes that direct many of the basic metabolic pathways.

In summary, viruses in their simplest form contain a nucleic acid genome, packaged in a protein coat. To replicate, a virus must transport its genome into a host cell, where the genome directs synthesis of viral proteins, is replicated, and is packaged. The host cell provides the virus with all of the other biological molecules required for its reproduction.

Virus genomes are either RNA or DNA, but not both

There are many different viruses in the world, and probably all organisms on Earth can be infected by at least one virus. Viruses have a variety of distinct morphologies, genome and particle sizes, and mechanisms of replication. The smallest known viruses are 20 nanometers[1] (nm) in diameter; their genomes contain fewer than 2000 nucleotides, and they code for as few as 2 proteins. The largest known viruses are some 500 nm in diameter; their genomes are as large as 1.2 million nucleotides, and they code for over 1200 proteins. An overview of the variety of known viruses is given in Chapter 3.

All viruses contain genomes made of one and only one type of nucleic acid. Depending on the virus, the genome can be either RNA or DNA, and it can be either single-stranded or double-stranded. Some viral genomes are circular and others are linear.

Viruses are the only known forms of life that can have genomes made of RNA. All cellular organisms store the information required to sustain life, to grow, and to reproduce exclusively in DNA molecules, and all RNA molecules in these organisms are transcribed from DNA sequences. RNA-containing viruses are therefore unique, and they face two related problems as a result of their RNA genomes: (1) they must synthesize messenger RNAs from an RNA template, and (2) they must replicate their genome RNA. Most RNA viruses encode their own RNA-dependent RNA polymerases to carry out both these functions.

WHY STUDY VIRUSES?

Viruses are important disease-causing agents

As living organisms arose and evolved during the past 4 billion years on Earth, they were probably always accompanied by viruses that could replicate within cells and pass from cell to cell. Some of these viruses interfere with normal cellular processes and cause disease (although many other viruses infect their host organisms without causing overt disease). Some of the most feared, widespread, and devastating human diseases are caused by viruses (see Table 1.1). These include smallpox, influenza, **poliomyelitis**, yellow fever, measles, and AIDS (acquired immunodeficiency syndrome). Viruses are responsible for many cases of human **encephalitis**, meningitis, pneumonia, **hepatitis**, and cervical cancer, as well as warts and the common cold. Viruses causing respiratory infections, **gastroenteritis**, and diarrhea in young children lead to millions of deaths each year in less-developed countries.

A number of newly emerging human diseases are caused by viruses. In addition to the worldwide AIDS epidemic that started in the early 1980s, there have been localized outbreaks in Africa of the highly fatal Marburg and Ebola **hemorrhagic fevers** during the past 30 years, a short-lived epidemic in southern Asia and Canada of severe acute respiratory syndrome (SARS) in 2003, and spread of acute and chronic hepatitis via both hepatitis B and C viruses. An invasion of North America by West Nile virus, transmitted by mosquitoes, began in 1999 and fortunately has only caused disease and death in a limited number of victims. There are fears that a new deadly pandemic of human influenza could occur if a recently emerged, highly pathogenic strain of avian influenza virus mutates to a form that is easily transmitted among humans.

Viruses can infect all forms of life

Viruses also infect animals, plants, and insects of importance to humans. Outbreaks of virus diseases in domesticated animals can lead to destruction of thousands or millions of animals to avoid even more widespread epidemics. These diseases include avian influenza; foot-and-mouth disease of cattle; infectious gastroenteritis and bronchitis in pigs, cattle, and chickens; sheep lung tumors caused by a retrovirus; canine distemper; and feline immunodeficiency disease. Virus diseases affecting domesticated plants such as potatoes, tomatoes, tobacco, coconut trees, and citrus trees are common and widespread. Insect viruses that kill silkworms, used for centuries in Asia and Europe to produce silk, have plagued that industry over the ages. Viruses can also infect and kill bacteria, archaea, algae, fungi, and protozoa.

---

[1] 1 nanometer = 10^{-9} meter or 10^{-6} millimeter.

Table 1.1 Some human diseases caused by viruses

Disease	Virus	Family
Acquired immunodeficiency syndrome (AIDS)	HIV-1	Retrovirus
Cervical carcinoma	Human papillomavirus types 16, 18, 31	Papillomavirus
Chickenpox	Varicella virus	Herpesvirus
"Cold sores"	Herpes simplex virus type 1	Herpesvirus
Common cold	Adenoviruses	Adenovirus
	Coronaviruses	Coronavirus
	Rhinoviruses	Picornavirus
Diarrhea	Norwalk virus	Calicivirus
	Rotaviruses	Reovirus
Genital herpes	Herpes simplex virus type 2	Herpesvirus
Hemorrhagic fevers	Dengue virus	Flavivirus
	Ebola and Marburg viruses	Filovirus
	Lassa fever virus	Arenavirus
Hepatitis	Hepatitis A virus	Picornavirus
	Hepatitis B virus	Hepadnavirus
	Hepatitis C virus	Flavivirus
Influenza	Influenza A and B virus	Othomyxovirus
Measles	Measles virus	Paramyxovirus
Mononucleosis	Epstein–Barr virus	Herpesvirus
	Cytomegalovirus	Herpesvirus
Mumps	Mumps virus	Paramyxovirus
Poliomyelitis	Poliovirus types 1, 2, and 3	Picornavirus
Rabies encephalitis	Rabies virus	Rhabdovirus
Severe acute respiratory syndrome (SARS)	SARS coronavirus	Coronavirus
Smallpox	Variola virus	Poxvirus
Warts	Human papillomavirus types 1, 2, 4	Papillomavirus
Yellow fever	Yellow fever virus	Flavivirus

Viruses are the most abundant form of life on Earth

Recent studies of soil and seawater have revealed that bacterial viruses, also called bacteriophages, are much more numerous than previously imagined. There are 10–50 million bacteriophages on average per mL of seawater, and even more in many soils. Given the enormous volume of the oceans, scientists have calculated that there may be as many as 10^{31} bacteriophages in the world. This is about 10-fold greater than the estimated number of bacteria. In terms of mass, this many phages would weigh about 100 million tons, or the equivalent of 1 million blue whales (the largest animal on Earth). More astonishing, these 10^{31} phages, if lined up head-to-tail, would stretch some 200 million light years into space—that is, far into the universe beyond many of our known neighboring galaxies (see text box on page 6)!

More important is the ecological role played by bacteriophages and viruses that infect unicellular eukaryotic organisms such as algae and cyanobacteria. From 95 to 98% of the biomass in the oceans is microbial (the remaining 2–5% being made up of all other forms of life, including fish, marine invertebrates, marine mammals, birds, and plants), and roughly half of the oxygen in the Earth's atmosphere is generated by photosynthetic activity of marine microbes. It has been estimated that 20% of the microbes in the Earth's oceans are destroyed each day by virus infections. Therefore, these viruses play a major role in the carbon and oxygen cycles that regulate our atmosphere and help feed the world's population.

The study of viruses has led to numerous discoveries in molecular and cell biology

Because viruses replicate within cells but express a limited number of viral genes, they are ideal tools for understanding the biology of cellular processes. The intensive study of bacteriophages led to discovery of some of the fundamental principles of molecular biology and genetics. Research on animal, insect, and plant viruses has shed light on the functioning of these organisms, their diseases, and molecular mechanisms of replication, cell division, and signaling pathways. For example:

Phages lined up through the universe

Scientists estimate that there are approximately 10^{31} tailed bacteriophages on Earth. Each phage measures approximately 200 nm (0.2 μm) in length from top of head to base of tail. Aligned head to tail, these phages would therefore cover the following distance:

$$10^{31} \times 0.2\ \mu m = 0.2 \times 10^{25}\ \text{meters} = 2 \times 10^{24}\ \text{meters}$$
$$= \mathbf{2 \times 10^{21}\ kilometers}.$$

Because 1 light year (the distance traveled by light in one year) = 10^{13} kilometers,

$$2 \times 10^{21}\ \text{kilometers} = 2 \times 10^{21}/10^{13}\ \text{light years}$$
$$= \mathbf{2 \times 10^8\ light\ years}\ (200\ \text{million light years}).$$

Note that our Milky Way galaxy measures approximately 100,000 light years edge to edge, and the furthest visible galaxies in the universe are approximately 10 billion (10×10^9) light years distant.

- Study of gene expression in small DNA viruses led to the identification of promoters for eukaryotic RNA polymerases.
- Research on the replication of bacteriophage and animal virus DNAs laid the foundations for understanding the enzymes involved in cellular DNA replication.
- RNA splicing in eukaryotic cells was first discovered by studying messenger RNAs of DNA viruses.
- Study of cancer-producing viruses led to the isolation of numerous cellular **oncogenes** and the understanding that cancer is caused by their mutation or unregulated expression.

Given this track record, the study of viruses will undoubtedly continue to shed light on many important aspects of cell and molecular biology.

A BRIEF HISTORY OF VIROLOGY: THE STUDY OF VIRUSES

The scientific study of viruses is very recent

Although viruses were probably present among the first forms of life and have evolved over several billion years, humans only began to understand the nature of viruses near the end of the nineteenth century (see Table 1.2). It had been appreciated for some time that infectious diseases were transmitted by air, water, food, or close contact with sick individuals. Many diseases were considered to be caused by mysterious elements in fluids termed *virus* ("poison" in Latin), but the distinction between what are now called viruses and cellular microorganisms was not clear. Scientists had begun to use light microscopes to discover and describe fungi and bacteria in the first half of the nineteenth century. Louis Pasteur and Robert Koch firmly established the science of bacteriology in the latter part of that century by isolating and growing a variety of bacteria, some of which were shown to cause disease (e.g., tuberculosis). Even though effective vaccines against smallpox (Edward Jenner, 1798) and rabies (Louis Pasteur, 1885) were developed, there was no understanding of the nature of these disease agents, which we now know to be viruses.

Viruses were first distinguished from other microorganisms by filtration

In the last decade of the nineteenth century, Russian scientist Dimitrii Ivanovski and Dutch scientist Martinus Beijerinck independently showed that the agent that causes tobacco mosaic disease could pass through fine earth or porcelain filters, which retain bacteria. Shortly afterward, similar experiments were carried out on agents that cause foot-and-mouth disease in cattle, and yellow fever in humans. These landmark experiments established that certain infectious agents are much smaller than bacteria, and they were called **filterable viruses**. For some time, it was not clear whether viruses were a form of soluble small molecules ("infectious living fluid"), or alternatively a particle like bacteria but too small to be retained by these filters.

Using filtration as a diagnostic tool, numerous viruses infecting humans, other vertebrate animals, plants, insects, and bacteria were described during the first half of the twentieth century. A tumor virus, Rous Sarcoma virus, was isolated from sarcomas of chickens in 1911, and was only many years later recognized as a representative of the important retrovirus family, of which human immunodeficiency virus (HIV) is a member. Scientists working in England and France discovered in 1915–1917 that bacterial cultures could be lysed by filterable agents, the first known bacterial viruses. Vaccines against the important human pathogens responsible for influenza and yellow fever were developed during the 1930s.

The crystallization of tobacco mosaic virus challenged conventional notions about genes and the nature of living organisms

Wendell Stanley found in the mid-1930s that highly purified tobacco mosaic virus could form crystals. This discovery shook the scientific world, because it placed viruses at the edge between living organisms and simple chemical compounds like sodium chloride. It posed the question: Are viruses living or inanimate? We now know that viruses are inanimate when their genomes are

Table 1.2 **Some milestones in virology research**

Discovery	Date	Scientists	Nobel prize awarded
Smallpox vaccine	1798	Edward Jenner	
Rabies vaccine	1885	Louis Pasteur	
Filterable viruses:			
Tobacco mosaic virus	1892	Dimitrii Ivanovski	
	1898	Martinus Beijerinck	
Foot-and-mouth disease (cattle)	1898	Friedrich Loeffler and Paul Frosch	
Yellow fever (humans: transmitted by mosquitoes)	1900	Carlos Finlay and Walter Reed	
Discovery of Rous Sarcoma virus	1911	Peyton Rous	1966
Discovery of bacteriophages and the plaque assay	1915 1917	Frederick Twort Felix d'Herelle	
Vaccine against yellow fever	1930s	Max Theiler	1951
Crystallization of tobacco mosaic virus	1935	Wendell Stanley and John Northrup	1946
Studies with bacteriophages	1940s	Max Delbruck and Salvador Luria	1969
Growth of poliovirus in cultured cells	1949	John Enders, Frederick Robbins, and Thomas Weller	1954
Bacteriophage lambda and lysogeny	1950s	Andre Lwoff	1965
Bacteriophage genes are DNA	1952	Alfred Hershey and Martha Chase	1969
Discovery of interferon	1957	Alick Isaacs and Jean Lindenmann	
Poliovirus vaccines:			
killed	1955	Jonas Salk	
live	1960	Albert Sabin	
Studies on polyomavirus: a tumor virus	1960s	Renato Dulbecco	1975
Kuru is caused by an infectious agent	1965	D. Carleton Gajdusek	1976
Discovery of hepatitis B virus	1968	Baruch Blumberg	1976
Reverse transcriptase in retroviruses	1971	Howard Temin and David Baltimore	1975
Virus vectors and genetic engineering	1970s	Paul Berg	1980
Cellular oncogene is part of a retrovirus genome	1976	Michael Bishop and Harold Varmus	1989
RNA splicing in adenovirus	1977	Phillip Sharp and Richard Roberts	1993
Prions: infectious proteins	1975–1990	Stanley Prusiner	1997
Human papillomaviruses cause cervical cancer	1972–1984	Harald zur Hausen	2008
Discovery of AIDS virus (HIV-1)	1983	Luc Montagnier and Françoise Barré-Sinoussi	2008

packaged in virions, but they share many attributes of life, including the ability to mutate, evolve, and reproduce themselves, when they enter cells that can support their replication.

Studies by Stanley and others showed that viruses contain both protein and nucleic acids. At that time, most scientists believed that genes were made of proteins, not nucleic acids, because only proteins were believed sufficiently complex to encode genetic information. The development of the electron microscope in the late 1930s allowed scientists for the first time to actually see viruses—tobacco mosaic virus (a long rod-shaped virus),

bacteriophages with their polygonal heads and tubular tails, and vaccinia virus, one of the largest animal viruses (Figure 1.2).

The "phage group" stimulated studies of bacteriophages and helped establish the field of molecular biology

During the late 1930s and early 1940s, a group of scientists led by German physicist Max Delbruck, American biologist Emory Ellis, and Italian biologist Salvador Luria decided that the study of bacterial viruses

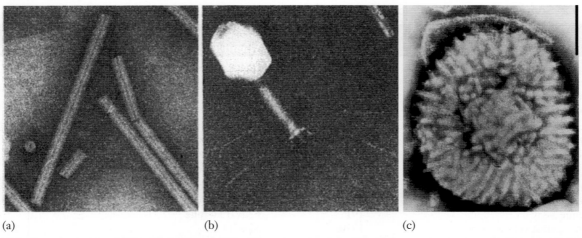

(a)   (b)   (c)

Figure 1.2 Electron micrographs of some representative virus particles. (a) Tobacco mosaic virus. (b) Bacteriophage T4. (c) Vaccinia virus.

(bacteriophages) could lead to understanding of the basic processes of life at a molecular level. They reasoned that bacteriophages show heritable traits, and therefore must contain and express genes as do all other organisms. Because bacteriophages are small and simple and can be propagated easily in bacterial cultures, they would be a fertile terrain for scientific discovery.

These scientists formed an informal network called the "phage group", which stimulated studies of bacteriophages and their host bacteria by numerous physicists, chemists, and biologists. These studies led to the isolation and analysis of phage genomes, the mapping of phage and bacterial genes by genetic crosses, and the elucidation of phage replication cycles. The phage group helped to found the field of molecular biology, which developed rapidly during the 1950s and 1960s.

One outcome of the phage group's work was the demonstration, by Alfred Hershey and Martha Chase in 1952, that the DNA of a bacteriophage is injected into the host cell and the protein coat remains outside the cell. This strongly backed up the chemical and enzymatic data published eight years earlier by Oswald Avery, Maclyn McCarty, and Colin MacLeod showing that a bacterial gene was made of DNA. Thus studies of a bacteriophage were important in proving the chemical nature of genes. Hershey and Chase's paper was followed a year later by the proposal by James Watson (a student of Luria) and Francis Crick of the double-helical model of DNA, which galvanized thinking and research throughout biology, but particularly in virology and molecular biology.

Bacteriophages can be used as targeted antibiotics against bacterial diseases

Bacteriophages may be called upon to play an important role in the fight against bacterial diseases in humans and animals. Because bacteriophages can attack and kill specific bacteria, they have long been considered as possible alternatives to antibiotics in treating disease. Bacteriophages specific to a variety of pathogenic bacteria have been isolated and characterized. There is a long history of their medical use, particularly in the Republic of Georgia, in an institute cofounded by one of the discoverers of bacteriophages, Felix d'Herelle. However, their use as antibacterial agents in humans is not accepted in most countries; there are significant unsolved problems, not the least of which is the induction of immune reactions to the bacteriophage. Future research and development may well reveal situations in which their use will be able to control otherwise runaway infections by bacteria that have developed resistance to many commonly used antibiotics.

Study of tumor viruses led to discoveries in molecular biology and understanding of the nature of cancer

Virus research underwent an explosive development in the second half of the twentieth century that led to the discovery of many new viruses and basic concepts in cell and molecular virology. Among the most important were the discovery and intensive study of DNA tumor viruses (polyomaviruses, papillomaviruses, and adenoviruses; see Chapters 21–23) and RNA tumor viruses (retroviruses; see Chapters 28–29). Research on DNA tumor viruses led to the discovery of viral oncogenes, whose protein products (tumor antigens) interact with numerous cell signaling pathways to stimulate cell growth and division.

Research on RNA tumor viruses led to the discovery of **reverse transcriptase**, an enzyme that can make a DNA copy of an RNA molecule, upsetting the one-way central dogma that "DNA makes RNA makes proteins". Numerous cellular oncogenes were discovered incorporated into retrovirus genomes. These oncogenes are

Virus vectors can replace defective genes, serve as vaccines, and combat cancer

Advances in molecular cloning have allowed the construction of numerous virus vectors (see Chapter 37). These are viruses in which some or all viral genes are removed by genetic engineering and are replaced by foreign genes. Because viruses can efficiently target specific cell types and express their genes at high levels, they can be used as vectors for expression of a variety of genes. Introduction of virus vectors into host cells or organisms can correct genetic diseases in which specific gene products are missing or defective. Vectors can also be used as vaccines that generate immune responses against a variety of unrelated pathogens. In a new twist, virus vectors have recently been used to combat cancers by expressing proteins that specifically kill tumor cells.

normal cellular regulatory genes whose mutation and/or over-expression can lead to the development of cancer; their protein products are involved in a variety of cellular signaling pathways. The study of viral and cellular oncogenes has led to major advances in the detection and treatment of cancer.

DETECTION AND TITRATION OF VIRUSES

Most viruses were first detected and studied by infection of intact organisms

Many viruses cause disease in the host organism, and this is how scientists and medical doctors usually become aware of their existence. The original methods for the study of such viruses relied on inoculation of animals or plants with filtered extracts from infected individuals, and their observation for the effects of virus infection. However, this is expensive and time-consuming work, and in most cases is no longer ethically acceptable when applied to humans. Experimental laboratory animals such as suckling mice, in which many animal viruses are able to replicate, were adopted for use because they are relatively easy and inexpensive to raise. A number of animal viruses have also been adapted to grow in embryonated chicken eggs, which are readily available from farms; this reduced both the expense and the time of virus assays.

The plaque assay arose from work with bacteriophages

Bacterial viruses can be easily studied by inoculating bacteria grown in tubes or on Petri dishes at the lab bench. Intact bacteria diffract visible light and therefore dense liquid cultures appear cloudy. Many bacteriophages lyse their host cell, and this **lysis** causes a loss in light diffraction leading to clearing of the bacterial culture; "clear lysis" serves as an indicator of phage replication.

A simpler and more quantitative application of cell lysis is to spread bacteria on the surface of nutrient agar in a Petri dish and to apply dilutions of a phage suspension. Wherever a phage binds to a bacterial cell and replicates, that cell releases the progeny phage particles, which are then taken up by neighboring cells and further replicated. After several such cycles all the cells in a circular area surrounding the original infected cell are lysed. The lysis area can be seen as a clear "plaque" against the cloudy background of the uninfected cells, which grow in multiple layers on the surface of the agar in the Petri dish. Use of this **plaque assay** (Figure 1.3) allows scientists to count the number of infectious virus particles in a suspension with a high degree of precision and reproducibility. The chance observation of such plaques played an important role in the discovery of bacteriophages.

Eukaryotic cells cultured *in vitro* have been adapted for plaque assays

In vitro culture of human, animal, insect, and plant cells was achieved in the mid-twentieth century, and allowed more convenient and cheaper growth and titration of many viruses. In the case of animal viruses, cells of many different tissues (especially from embryos or newborn animals) can be induced to grow in a monolayer on a glass or plastic surface underneath liquid media. Cultured cells also facilitated production of numerous antiviral vaccines (see Chapter 35), starting with the vaccines against poliovirus that were developed in the 1950s.

Plaque assays were subsequently developed for animal, plant, and insect viruses using cells cultured *in vitro*. When cells growing in a monolayer are infected, the progeny virus is released into the medium and can travel to distant sites, infecting other cells. To restrict diffusion of the progeny virus, the infected cells are overlaid with nutrient medium in melted agar, which solidifies on cooling. In gelled agar, released virus can infect only nearby cells on the monolayer, forming a local area of dead cells or a plaque.

Because cell monolayers are too thin to diffract light well, plaques in cultured animal or insect cells are usually visualized by staining the cells. When cells die and/or lyse, they do not stain well; therefore plaques are seen as clear, unstained circular areas on the background of the stained cell monolayer. Virus present in individual plaques can be isolated by sampling with a needle or Pasteur pipette, allowing the "cloning" of progeny virus derived from a single virion that initiated the infection leading to the plaque.

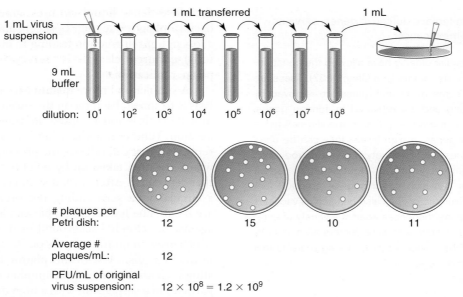

Figure 1.3 Plaque assay: an example. A virus suspension was subjected to 10-fold serial dilutions by adding 1 mL of the original suspension to 9 mL of a dilution buffer. After mixing, 1 mL of that dilution was added to 9 mL of fresh dilution buffer and mixed; these steps were repeated a total of eight times. Each successive tube contains a 10-fold dilution of the contents of the previous tube. The eighth tube therefore is diluted by a factor of 10^8 compared with the original virus suspension. One-mL aliquots from this 10^8-fold dilution were applied to four different Petri dishes of susceptible cells and plaques (bottom) were allowed to develop. Plaque-forming units (PFU) per mL are calculated as shown.

When a plaque assay is used to measure the infectious titer of a virus suspension, the results are usually expressed as **plaque-forming units** (PFU) per mL of suspension. To determine the titer, the number of plaques on a plate is multiplied by the factor by which the original virus suspension was diluted before an aliquot was applied to the plate. For an example, see Figure 1.3.

Hemagglutination is a convenient and rapid assay for many viruses

A number of animal viruses bind to **sialic acid** residues or other carbohydrates on cell surface proteins and lipids. Red blood cells have carbohydrate-containing receptors on their surface, and have the advantage of being visible because of their color. Furthermore, they can be isolated easily from the blood of a variety of animals, are sturdy during manipulation, and can be stored for days or weeks. This makes red blood cells an ideal substrate for assaying viruses.

Virus particles have multiple copies of receptor-binding proteins on their surface, and red blood cells contain many copies of surface receptors. Binding between an excess of virus particles and an aliquot of red blood cells forms an interlaced network of cells, held together by virus particles that form bridges between adjacent cells. These "agglutinated" red blood cells, when allowed to settle, form a light pink hemispherical shell in the bottom of a tube or plastic well. In contrast,

individual red blood cells slide to the bottom of the tube and form a compact, dark red pellet (Figure 1.4). This is the basis of **hemagglutination assays** for viruses.

Virus suspensions are diluted, usually in twofold steps, and the dilutions are added to aliquots of red blood cells in a buffer and mixed in tubes or multiple-well plates. After allowing the cells to settle, the tubes or plates are examined; the highest dilution that will agglutinate the aliquot of cells is considered to have one hemagglutinating unit (HAU) of virus. Such assays are sensitive to pH, temperature, and buffer composition, and some viruses will agglutinate only cells of a particular mammalian or avian species.

Because many (approximately 10^5) red blood cells are used in each tube, one hemagglutinating unit represents 10^5 or more virus particles. Therefore, hemagglutination assays are much less sensitive than plaque assays, but they are rapid and cheap. They can also be used to detect antibodies that bind to viral surface antigens, because addition of such antibodies will inhibit hemagglutination.

Virus particles can be seen and counted by electron microscopy

A variety of staining or shadowing methods can be used to detect virus particles by electron microscopy. One of the simplest methods is to mix a virus suspension with an electron-dense stain, usually phosphotungstate or uranyl acetate, and to spread the mixture on a grid for

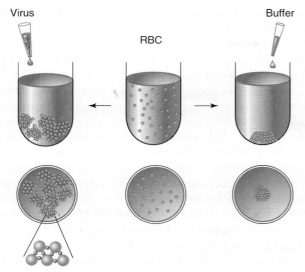

Figure 1.4 Hemagglutination assay. Red blood cells (RBC; small orange circles in central tube) are mixed with virus (small green spheres), or with buffer, and are allowed to settle. Individual red blood cells settle to form a compact pellet in the bottom of the tube (right), but when agglutinated, form a thin shell on sides and bottom of tube (left). The lower set of images show what is seen when tubes are viewed from below. An enlargement at left shows red blood cells bound together by virus particles.

examination in the electron microscope. The stain tends to form electron-dense pools around virus particles; virus particles exclude the stain and therefore show up as light images against a dark background, and much fine surface detail can be seen (Figure 1.2). This technique is called "negative staining". Measured aliquots of dilutions of virus suspensions can be applied to grids, and the number of virus particles in a given area can be counted. Standard suspensions of tiny latex spheres or other small uniform objects are often added to the virus suspension to help establish absolute numbers of virus particles per unit volume.

The ratio of physical virus particles to infectious particles can be much greater than 1

Measurement of the number of infectious virus particles by use of plaque assays, and of the number of physical virus particles in the same virus suspension by electron microscopy, allows calculation of the ratio of physical particles to infectious particles. Naively, one would expect that most intact virus particles are infectious. This is true for some bacteriophages and for a small number of animal viruses. However, for many viruses the ratio of physical particles to infectious particles can be 10, 100, or even 1000! There are several possible reasons for the low infectivity of virus preparations:

- *Not all virus particles may be intact*. For example, virus envelopes are fragile and can be disrupted, rendering the particle non-infectious. Some viral surface protein molecules can be denatured and therefore unable to bind to the cell receptor. However, virions contain numerous copies of receptor-binding proteins on their surface, so the loss or denaturation of a few protein molecules should not lead to loss of infectivity.

- *Some virus particles may contain defective genomes*. Mutations, including deletions in viral genes, occur frequently during genome replication, and such defective genomes are often incorporated into virus particles. In extreme cases, 90% or more of a virus preparation consists of particles with defective genomes.

- *"Empty" capsids that contain no viral genome can be made in large numbers*. Some viruses can form capsids in the absence of the viral genome. Others incorporate cellular DNA or RNA instead of the viral genome into the capsid. However, many viruses have specific packaging signals that ensure incorporation only of viral nucleic acid into virions.

- *Cells have antiviral defense mechanisms*. Many virus preparations consist of fully intact virions that contain infectious genomes. However, cells have a variety of defense mechanisms that can interfere with many steps in virus infection. Therefore, even though a cell takes up an intact and potentially infectious virion, it may not produce any progeny virus. This can be a major cause of the high ratio of physical particles to infectious particles of some viruses.

THE VIRUS REPLICATION CYCLE: AN OVERVIEW

The single-cycle virus replication experiment

Use of the plaque assay enables the quantitative study of the kinetics of virus replication. To understand the time course of events taking place during the replication cycle, scientists usually study cultures containing thousands to millions of infected cells, because only then can sufficient viral nucleic acid or proteins be isolated and analyzed. All cells must be infected simultaneously, and with some luck the events of the virus replication cycle will be synchronized so that similar steps will be taking place at the same time in all cells. To ensure simultaneous infection, cell cultures are infected with a sufficient number of virus particles such that each cell receives at least one infectious particle.

These considerations led to the concept of **multiplicity of infection (m.o.i.)**. This is defined as the number of infectious virus particles added per susceptible cell. An m.o.i. of 10 to 100 plaque-forming units per cell is often used in studies of bacterial or animal viruses.

Nearly all cells in a culture are infected simultaneously and there remain very few uninfected cells.

In practice, there are limitations in this method for distinguishing the time course of the various steps in a viral growth cycle; different steps in the cycle invariably overlap somewhat. However, the single-cycle approach does simplify the study of virus replication and is nearly universally used. Some bacteriophages complete their replication cycles in as little as 20 minutes; some animal viruses can take several days to complete one replication cycle. Certain viruses do not always undergo a productive growth cycle, but instead lodge one or more "silent" copies of their genome in the host cell until conditions are appropriate for a lytic cycle (lysogeny or **latent infection**; see Chapters 8, 9, 24, and 28).

An example of a virus replication cycle: mouse polyomavirus

An idealized example of a single-cycle growth curve for mouse polyomavirus (see Chapter 21), a small DNA virus that replicates in cultured baby mouse kidney cells, is shown in Figure 1.5. Each 6-cm diameter Petri dish contains approximately 5 million cells, and 0.5 mL of a suspension containing 100 million (10^8) PFU/mL of polyomavirus is used to infect the cells (m.o.i. = 10 PFU/cell).

After 1 hour, fresh medium is added to each culture and the cells are incubated at 37°C in a humidified chamber with a 5% CO_2 atmosphere. At intervals, samples of the infected cells or of the medium are harvested and analyzed for infectious virus, viral mRNA, viral proteins, or viral DNA. The results are expressed as PFU/cell for virus (logarithmic scale on left) or as a percent of the maximum amount of the viral macromolecule (linear scale on right).

During the first hour or two after adding virus, most of the infecting virions are taken up into the cells or are subsequently washed away by changing the medium. This leads to an initial loss in the titer of virus detected in the medium. Eventually, many of the virus particles taken up into the cells are uncoated, rendering intracellular virus non-infectious, and thus the titer of infectious intracellular virus also drops. This phase has been called the "latent phase" of infection because the infecting virus has disappeared and no new progeny virus has yet been made.

Some time later (in this case, starting at 18–20 hours after infection), new progeny virus begins to appear. Polyomavirus particles are assembled in the cell nucleus and are not efficiently released from the cell until after cell death. This means that most progeny virus can be detected only by lysing the cells after harvesting them. A much lower virus titer is detected in the medium surrounding the cells (not shown). This is true of many non-enveloped viruses that replicate and assemble in the nucleus or in the cytoplasm of eukaryotic cells, and of most bacteriophages. In contrast, many enveloped viruses form virions at the cell surface, where they acquire a lipid envelope, and therefore newly assembled virus particles are immediately released into the medium.

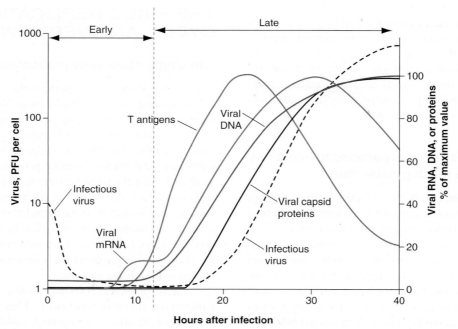

Figure 1.5 Replication cycle of mouse polyomavirus. The time course of appearance of infectious virus particles, viral mRNA, DNA, and proteins during a typical replication cycle of mouse polyomavirus in baby mouse kidney epithelial cells.

There is often a long period during which progeny virus is being produced, but at some point virus replication stops, often because the host cell dies. At this point no further virus is made, but release can take some time as cells slowly disintegrate after death.

Analysis of viral macromolecules reveals the detailed pathways of virus replication

This analysis of the virus growth cycle is very elementary and gives only an overview of the beginning and the end of steps in virus replication. Many important events of the virus replication cycle take place during the latent period. These steps are studied by extracting from the infected cell the macromolecules that constitute viral genomes, messenger RNAs, and proteins. Their interactions with cellular macromolecules and organelles can be analyzed, and their size, number per cell, and rates of synthesis and turnover can be studied.

This is often done by using radioactive compounds that are incorporated into viral or cellular DNA, RNA, protein, lipid, or carbohydrates. Antibodies against viral or cellular proteins can be used to detect specific molecules with fluorescent dyes or other assays. Molecular hybridization with oligonucleotide probes, or the polymerase chain reaction (PCR), can be used to detect specific viral DNA or RNA molecules. Macromolecules are often separated from each other by electrophoresis on polyacrylamide or agarose gels.

Staining of macromolecules can be observed in fixed or living cells by microscopy, especially by using computerized methods that allow visualization of horizontal layers of thickness 1 micrometer (μm) or less within individual cells. These views, revealed by **confocal microscopes** using focused laser light sources, show the localization of viral macromolecules within the nucleus, or in association with organelles such as the endoplasmic reticulum, lysosomes, or the plasma membrane. Other localization methods use the higher-resolution electron microscope coupled with specific staining: for example, using colloidal gold particles linked to antibodies.

Study of virus replication cycles by these and other methods has led to a general understanding of the various steps in virus replication. In the case of polyomavirus, Figure 1.5 shows that small amounts of "early" viral messenger RNAs are made beginning about 8–10 hours after infection, and early viral proteins, known as T antigens, begin to appear shortly afterward. This is followed by the beginning of viral DNA replication at 12–15 hours after infection. Once viral DNA replication has begun, much larger amounts of viral messenger RNAs can be detected, and these "late" messenger RNAs are copied from a different set of viral genes. Mature progeny virions begin to appear by 18–20 hours after infection, and virus titers increase slowly for the next 24 hours, by which time most of the infected mouse cells are dead.

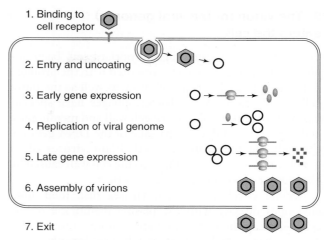

1. Binding to cell receptor
2. Entry and uncoating
3. Early gene expression
4. Replication of viral genome
5. Late gene expression
6. Assembly of virions
7. Exit

Figure 1.6 Steps in virus replication cycle. A schematic diagram showing a simplified virus replication cycle in seven steps, as described in text. The host cell is shown as a rounded box. Capsids are represented as green icosahedra, genomes as black circles, messenger RNAs as blue lines with translating ribosomes, viral replication proteins as orange ovals, and viral structural proteins as green squares.

Virus replication cycles vary greatly, depending on the host cell, genome type, and complexity of the virus. However, there is logic to the progression of steps in virus replication. The following section gives an outline and analysis of the replication pathway starting with the initial infection of a cell and progressing to the release of progeny virions (Figure 1.6). Designation and naming of these steps is somewhat arbitrary but it is useful to help understand the important elements of the replication cycles of all viruses.

STEPS IN THE VIRUS REPLICATION CYCLE

1. Virions bind to receptors on the cell surface

Viruses must first recognize and bind to cells that they infect. Virus-coded proteins on the surface of the virion bind to specific proteins, carbohydrates, or lipids on the cell surface. Structural surface proteins of many viruses bind to carbohydrate residues found in surface **glycoproteins** and **glycolipids** that are widely distributed on many cell types. Other viruses bind to cell surface proteins that are found only on specific cell types, and are often species-specific, thus limiting the **tropism** of the virus to a particular tissue and organism. Many viruses first bind to a relatively non-specific primary receptor such as a carbohydrate, and subsequently bind to a specific cell surface protein that serves as a secondary receptor.

2. The virion (or the viral genome) enters the cell

Once they have located an appropriate cell, bacteriophages and plant viruses are presented with the problem of passing through a rigid cell wall, as well as the outer cell membrane(s). Many bacteriophages have specialized tails that drill holes in the cell wall and membranes, and serve as a conduit for passage of the DNA genome through the hole into the cell. Plant viruses often penetrate as a result of damage to the cell wall caused by abrasion or a wound caused by an insect.

Some enveloped animal viruses fuse their lipid envelopes directly with the plasma membrane of the cell, releasing the viral capsid and genome into the cell. Many other animal viruses are taken up into the cytoplasm in vesicles formed at the plasma membrane. These vesicles then release the virion or its genome into the cytoplasm, and in some cases the genome is transported to the nucleus. Release can happen by disintegration of the vesicle membrane or by fusion of the viral envelope with the vesicle membrane. In cases where the capsid is released into the cell, a variety of pathways lead to disintegration of the capsid and release of the genome, called uncoating. Entry of animal viruses into cells is discussed in detail in Chapter 4.

3. Early viral genes are expressed: the Baltimore classification of viruses

Once in the appropriate compartment of the cell, the viral genome must direct expression of "early" proteins that will enable genome replication. The molecular pathways that lead to synthesis of early viral proteins from viral messenger RNAs depend on the chemical nature and strandedness of the genome. David Baltimore first recognized that all viruses can be divided into six (now seven) groups based on the pathways leading to mRNA and protein synthesis; this has come to be known as the *Baltimore classification system* (see Figure 1.7). One advantage of this system is that it draws our attention to the distinct kinds of RNA polymerases needed by viruses in each group, and indicates whether these enzymes are available in the cell or must be provided by the virus.

Viruses with single-stranded RNA genomes fall into two categories. Some viruses package the "sense" RNA strand, which contains coding regions that can be directly translated into viral proteins. By convention, this is labeled the **positive** (or **plus**) **strand**. Other viruses package the "antisense" strand, which cannot be translated because it does not contain meaningful coding regions; only its complementary copy codes for viral proteins. This is labeled the **negative** (or **minus**) **strand**. Viruses in these two groups differ fundamentally in how their replication cycle begins.

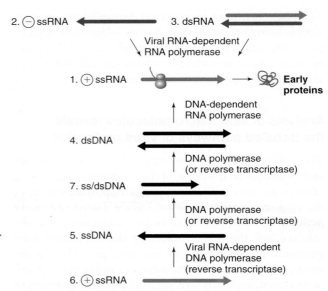

Figure 1.7 Baltimore classification of viruses. Seven categories of viruses are distinguished based on the nature of the viral genome and the pathway leading to synthesis of early messenger RNAs. Numbering is arbitrary. RNA and DNA polymerases that carry out different steps are shown. DNAs are shown as black lines and RNAs as light blue (positive-strand) or dark blue (negative-strand) lines. ss = single-stranded; ds = double-stranded; + = positive-strand RNA; − = negative-strand RNA. Messenger RNA is shown being translated by a ribosome.

The seven groups in the Baltimore classification system

1. *Viruses with positive-strand RNA genomes* bring their messenger RNA directly into the cell in the form of the genome. This RNA can bind to ribosomes and is translated into viral proteins as the first step in virus replication. These proteins will then direct replication of the viral genome.

2. *Viruses with negative-strand RNA genomes* must first synthesize complementary positive-strand copies of the genome, which serve as viral messenger RNAs. However, the cell does not produce an enzyme able to carry out this step, so the virus itself must supply this enzyme. Negative-strand RNA viruses package a virus-coded **RNA-dependent RNA polymerase** within the virion, and this enzyme accompanies the negative-strand RNA genome as it enters the host cell. The first intracellular step in their replication cycle is therefore the synthesis of positive-sense viral mRNAs by this viral RNA-dependent RNA polymerase, using the negative-sense genome as a template.

3. *Viruses with double-stranded RNA genomes* are faced with the same problem as those with negative-strand RNA genomes: the cell does not produce an enzyme that can generate an mRNA by transcribing

one of the RNA strands in the double-stranded genome. It is theoretically possible that the double-stranded genome RNA could be denatured within the cell, allowing the positive-strand RNA to be translated; however, this is energetically unlikely and does not occur. These viruses therefore also package a virus-coded RNA-dependent RNA polymerase in the virion and bring this enzyme into the host cell along with the genome. This RNA polymerase can specifically recognize the double-stranded RNA genome and transcribe its negative-strand into positive-strand messenger RNAs.

4. *Viruses with double-stranded DNA genomes* must first transcribe their genomes into messenger RNAs, using a **DNA-dependent RNA polymerase**. Cellular messenger RNAs are synthesized in exactly the same way, and therefore all cells contain such enzymes. Most DNA viruses use a host cell RNA polymerase for this purpose, but some of the larger DNA viruses bring their own RNA polymerase into the cell along with their genome, or make an additional RNA polymerase later in the cycle to transcribe a subset of their genes.

5. *Viruses with single-stranded DNA genomes* must first convert their DNA into a double-stranded form, as there are no known enzymes that will transcribe single-stranded DNA directly into RNA. This conversion is carried out by a cellular DNA polymerase. The resulting double-stranded DNA is then transcribed by a cellular RNA polymerase to generate viral mRNAs. Note that either of the complementary DNA strands can be packaged in virions. Most viruses preferentially package only one of the two DNA strands, but some viruses package both strands, making two kinds of virions that are identical except for the polarity of their single-stranded DNA genomes. In all cases the single-stranded DNA is converted into a double-stranded form before being transcribed, so the polarity of the packaged strand is of little importance.

6. *Retroviruses have a single-stranded, positive-sense RNA genome*, but they follow a pathway different from other positive-strand RNA viruses. Retroviruses package a virus-coded *reverse transcriptase* (an **RNA-dependent DNA polymerase**) within the virion, and use this enzyme to produce a double-stranded DNA copy of their RNA genome after they enter the cell. Once the double-stranded DNA is made (and integrated into the cellular genome: see Chapter 28 for details), it is transcribed into viral messenger RNAs by host cell RNA polymerase II, as for most DNA viruses.

7. There remains a maverick category of viruses in the Baltimore classification system, of which the best known representatives are the hepadnaviruses (see Chapter 30). These viruses package a circular DNA genome that is partly double-stranded, with a single-stranded gap of variable length on one strand. Hepadnaviruses can be seen as a special case of the category of viruses that have single-stranded DNA genomes (#5 above), because like those viruses, their genomes are made fully double-stranded by cellular DNA polymerases after entering the cell. This double-stranded DNA is subsequently transcribed into messenger RNAs by cellular RNA polymerase, as for other DNA viruses.

However, like retroviruses and unlike other DNA viruses, hepadnaviruses code for a reverse transcriptase. In the case of the hepadnaviruses, this reverse transcriptase is used to synthesize the partially single-stranded DNA genome that is packaged in virions. This DNA is made by reverse transcribing a genome-length RNA molecule that is one of the transcription products of the fully double-stranded DNA. See Chapter 28 for a complete description of how this occurs.

4. Early viral proteins direct replication of viral genomes

Once early viral proteins are made, they act to promote replication of the viral genome. All RNA viruses (except retroviruses) must synthesize an RNA-dependent RNA polymerase to replicate their genomes, as this enzyme is not present in host cells. A number of other early proteins, as well as a variety of cellular proteins, help form RNA replication complexes in the cell. These replication complexes are often associated with cellular membranes.

The early proteins of many DNA viruses induce the production of a number of cellular enzymes that are involved in the synthesis of DNA and of its building blocks, deoxyribonucleoside triphosphates. This is often achieved by interaction of early viral proteins with cellular signaling pathways that affect the cell cycle and direct the cell to enter the DNA synthesis (S) phase. Small DNA viruses usually use host cell DNA polymerases to replicate their genomes; larger DNA viruses often code for their own DNA polymerases as well as other enzymes involved in DNA replication.

Hundreds to tens of thousands of copies of the viral genome can be made in each cell. These progeny genomes can be used as templates for synthesis of more viral messenger RNAs or for further genome replication. At this stage, the cell has become a factory for the expression and replication of viral genomes.

5. Late messenger RNAs are made from newly replicated genomes

Many viruses synthesize a distinct set of "late" messenger RNAs after genome replication has begun. Because of genome replication, there can be many templates for late messenger RNA synthesis, and these mRNAs may therefore be abundant. The mechanisms that control the switch between early and late messenger RNA synthesis have been extensively studied as paradigms for understanding how gene expression is controlled in the host cell.

6. Late viral proteins package viral genomes and assemble virions

Structural proteins used to package viral genomes and to assemble the capsid are usually the most abundant viral proteins made in an infected cell. The simplest virus capsids consist of one protein that forms either a closed shell or a helical tube within which the viral genome is packaged (Figure 1.1). Larger and more complex viruses may have numerous capsid proteins, and some viruses make **scaffolding proteins** that are involved in virion assembly but are subsequently discarded and do not form part of the mature virion. Many bacteriophages have both a polygonal head that contains the genome and a tubular tail involved in attachment to the cell and delivery of the genome. Enveloped viruses code for glycoproteins that are inserted into lipid membranes and that direct formation of the viral envelope by a process called **budding**. For a detailed discussion of virion structure and assembly, see Chapter 2.

7. Progeny virions are released from the host cell

Once formed, virions leave the cell to find and infect new host cells and reinitiate the replication cycle. For viruses that infect unicellular organisms, this usually involves death and lysis of the host cell, and many viruses code for specialized late proteins that lead to cell death. However, some viruses of bacteria and archaea are released by extrusion from the cell membrane and do not kill their host cells.

For viruses that infect multicellular organisms, the problem becomes more complex. Viruses can spread from one cell to another within an organism, or they can move from one organism to another; virions that follow these two pathways are not necessarily identical. For example, many viruses of higher plants spread from cell to cell throughout a plant by passage through intracellular channels (**plasmodesmata**), using specific viral proteins to enhance passage of unencapsidated viral genomes. However, spread from plant to plant is often carried out by insects that feed on plant juices, ingesting virions and reinjecting them into other plants.

Some insect viruses spread as individual virions from cell to cell within the host organism, but wrap one or more virions in a specialized protein coat that is released when the insect dies. This protein coat protects virions in the environment and is ingested by other insects, in which it dissolves and releases the infectious virions to start another cycle (see Chapter 25).

Release of animal viruses can occur by a number of pathways. Some viruses accumulate within host cells and are released only on cell death. Certain viral proteins function to retard cell death, prolonging the period of virus replication, and other viral proteins can kill the host cell, leading to virus release. Many enveloped viruses are assembled by budding at the plasma membrane, and can be released with little or no effect on the host cell. Others bud at internal cell membranes or form their envelopes *de novo;* many of these viruses use cellular transport vesicles to reach the cell membrane and to exit the cell.

KEY TERMS

Budding	Obligatory intracellular
Capsid	parasite
Confocal microscope	Oncogenes
DNA-dependent RNA	Plaque assay
polymerase	Plaque-forming
Encephalitis	units (PFU)
Envelope	Plasmodesmata
Filterable virus	Poliomyelitis
Gastroenteritis	Positive (or plus) strand
Glycolipid	Reverse transcriptase
Glycoprotein	RNA-dependent DNA
Hemagglutination assay	polymerase
Hemorrhagic fever	RNA-dependent RNA
Hepatitis	polymerase
Latent infection	Scaffolding protein
Lysis	Sialic acid
Multiplicity of infection	Tropism
(m.o.i.)	Uncoating
Negative (or minus) strand	Virion

FUNDAMENTAL CONCEPTS

- A virus particle, or virion, consists of a nucleic acid genome enclosed in a protein coat and in some cases a lipid-containing envelope
- Viruses depend on cells for molecular building blocks, energy, and the machinery for protein synthesis

- Viruses can have genomes made of either RNA or DNA
- Viruses replicate by uncoating and expressing the viral genome inside the cell, directing synthesis of viral proteins, replicating the genome, and encapsidating it to form new progeny virus particles
- Viruses can be detected and enumerated by plaque assays, hemagglutination assays, and electron microscopy
- The sequence of steps during virus replication can be studied by infecting all cells in a culture at a high multiplicity of infection
- Seven different classes of virus can be distinguished based on the pathway to synthesis of early mRNAs; this is known as the Baltimore classification system.

REVIEW QUESTIONS

1. State two problems faced by viruses with RNA genomes, and explain how these viruses address these challenges.

2. How were viruses first distinguished from other microorganisms, and what did this ultimately reveal about viruses?

3. Viruses can replicate only within living cells. What four elements that are required for virus replication are provided by host cells?

4. Are viruses living or inanimate?

CHAPTER 2

Virus Structure and Assembly

Stephen C. Harrison

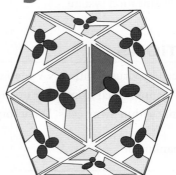
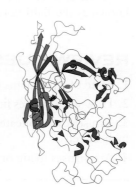

CHARACTERISTICS OF CAPSIDS

Virus capsids are assembled from many copies of a
small number of viral proteins: principle of **genetic
economy**.

Capsids made from multiple identical subunits have
regular, symmetrical shapes.

Symmetrical capsids are strong, stable, and can
self-assemble.

Most capsids are constructed as either:

- *Closed spherical shells with* **icosahedral symmetry** *or*
- *Elongated tubes with* **helical symmetry**.

The simplest icosahedral capsids contain 60 $\times$ T protein subunits.

- *T, the* **triangulation number**, *is equal to $h^2 + hk + k^2$, where h and k are integers.*

The polypeptide chain of many capsid proteins is folded as a **jelly-roll β-barrel**.

Some viruses include internal or **core proteins** bound to the nucleic acid.

VIRAL ENVELOPES

Many viruses enclose the nucleocapsid in a lipid membrane, called an **envelope**.

Viral envelopes contain viral **glycoproteins** inserted into the lipid membrane by a transmembrane anchor.

Most envelopes are formed by **budding** from preexisting cellular membranes.

Some virions contain a layer of **matrix protein** between the nucleocapsid and the envelope.

Some viruses have internal lipid membranes that lie underneath an external capsid.

VIRION ASSEMBLY

Many capsids self-assemble.

- *Some use viral* **scaffolding proteins** *to assemble virions and package genomes.*
- *Others use cellular machinery to assist assembly or to facilitate budding.*

Specific packaging signals on virus genomes are recognized and bound by viral proteins.

Virion proteins often undergo proteolytic cleavage during or after assembly, preparing the virion for uptake by
target cells.

BASIC CONCEPTS OF VIRUS STRUCTURE

Virus particles are molecular structures that package virus genomes in infected cells and transmit them to new host cells. The word **virion** is used to denote a *complete, infectious virus particle*. The structures of virions have evolved to carry out the task of transmitting the RNA or DNA genome of a virus from one cell to another and from one host organism to another. Virions must be assembled, exit the cell in which they are made, withstand the extracellular environment, attach to and enter another cell, and release the viral genome. These different functions determine the structural characteristics of virions.

What do we mean by the term *structure?* In everyday life, there can be packages with different structural characteristics. A cardboard box is rigid, symmetrical, and fixed in volume and shape; a plastic bag is flexible, variable in volume and shape, and roughly spherical when stuffed full. At the molecular level, the outer shell of a virion can be a rigid, symmetrical container called a **capsid** (derived from the Latin *capsa* = box). When the viral nucleic acid genome is packaged within the capsid, the resulting structure is called a **nucleocapsid**.

Many viruses enclose their nucleocapsids in lipid-bilayer membranes called **envelopes**. The nucleocapsids enclosed in this way are sometimes flexible, helical rods and sometimes more rigid, roughly spherical, closed shells. Enveloped virions are usually more flexible than non-enveloped virions, and they can have a variety of shapes. Certain enveloped viruses of bacteria, archaea, and algae have a rigid capsid surrounding an internal lipid bilayer, which is stuffed full of genome DNA or RNA.

Virus structure is studied by electron microscopy and X-ray diffraction

Most virus particles are too small to be visible by light microscopy. The higher resolution of the electron microscope reveals the general outlines of a virus particle, and shows the bumps and projections formed by oligomers of viral structural proteins on the surface of the virion (Figure 2.1). Virions can be visualized by **negative staining**, which reveals the structure of the virion by the exclusion of electron-absorbing molecules such as uranyl acetate (Figures 2.1a–g).

Alternatively, proteins, nucleic acids, and lipids are revealed by **positive staining** in plasticized thin sections of infected cells (e.g., human immunodeficiency virus; Figure 2.1j). However, both of these methods can distort or blur structural details of virions because of staining artifacts and sample damage by the electron beam.

In **cryoelectron microscopy**, samples are flash-frozen in liquid nitrogen and kept at that temperature in the microscope. This method improves preservation of structure, but image contrast is low. Data from images of many different virions are combined by computer programs to build an overall averaged image of the structure at a higher resolution than is possible by direct imaging of individual particles (e.g., Semliki Forest virus, Figure 2.1i).

When the particle can be crystallized, as is the case for many non-enveloped virions, a very detailed model can be obtained from **x-ray diffraction** data (e.g., simian virus 40; Figure 2.1h). The resolution of structures determined by x-ray diffraction is much greater than the resolution obtained by electron microscopy, allowing precise determination of the position of every atom in the virion. X-rays cannot be focused, however, and direct x-ray imaging of the virion is not possible;

instead, diffraction patterns are captured and analyzed. High-intensity x-rays are produced in giant **synchrotrons**, and the diffraction data are analyzed by sophisticated computer programs to yield the three-dimensional structure of the virion.

In contrast to many highly symmetrical non-enveloped virions, many enveloped viruses such as influenza virus (Figure 2.1g) have virions of variable shape and size. The lipid bilayer of the envelope ensures impermeability to ions or nucleases that might damage the viral genome located inside the envelope. The viral proteins that are inserted into this membrane do not need to form a perfectly tight barrier as in a capsid, and therefore are not usually organized in a symmetrical fashion. Because of their structural variability, the virions of most enveloped viruses cannot form crystals, and therefore high-resolution structures of complete virions cannot be obtained by x-ray diffraction. Individual viral membrane proteins such as the influenza viruses hemagglutinin and neuraminidase can be purified and crystallized, allowing determination of their molecular structures.

Many viruses come in simple, symmetrical packages

The simplest virus particles assemble spontaneously from their protein subunits and nucleic acid genomes; this process is called **self-assembly**. The information that determines their three-dimensional structures is built into the way in which their protein subunits bind to each other and to the packaged RNA or DNA.

The principle of **genetic economy** is an important factor that affects the evolution of designs for virus particles. Consider a 5000-nucleotide single-stranded RNA or DNA genome, which can code for proteins totaling about 1500 amino acid residues (1 amino acid per 3 nucleotides). The genome can be condensed into a compact sphere like a ball of string, about 18 nm in diameter. A protein shell 3 nm thick surrounding such a sphere will have about 30,000 amino acid residues. It is clearly impossible for the viral genome to code for a single structural protein of that size. The protein shell must therefore be made of a large number of identical protein subunits, in order not to exhaust the coding capacity of the genome.

Genetic economy leads to a requirement for symmetry of capsid structures. The large number of identical protein subunits will not form a tight and enclosed package if they are bound to each other by random contacts. Instead, the interactions between neighboring protein subunits must be well defined and specific. If all protein subunits on the surface of a capsid are identical, then the contacts will be identical (or nearly so), generating a symmetrical object.

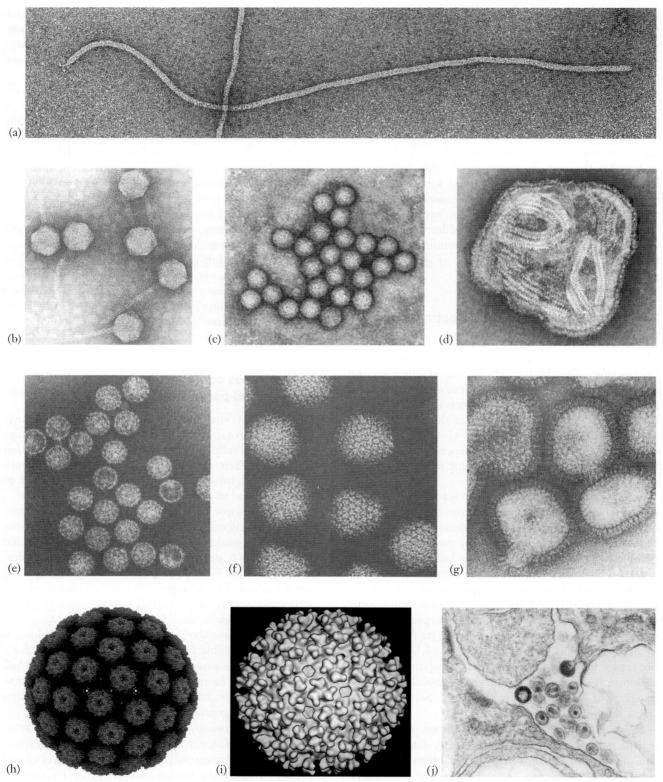

Figure 2.1 Structures of some typical virus particles. (a) Bacteriophage fd. (b) Bacteriophage lambda. (c) Hepatitis A virus (a picorna-virus). (d) Human parainfluenza virus (a paramyxovirus; in this image, the helical nucleocapsid is revealed by penetration of stain inside the particle). (e) Human papillomavirus. (f) Adenovirus. (g) Influenza virus (an orthomyxovirus). (h) Simian virus 40 (a polyomavirus). (i) Semliki Forest virus (a togavirus). (j) Human immunodeficiency virus (a retrovirus). Parts (a–g) are electron micrographs of negatively stained virions, Part (h) is a computer-generated image derived from x-ray diffraction, Part (i) is a computer-generated image derived from cryoelectron microscopy, and part (j) is an electron micrograph of a thin section of infected cells. Scales are different for each image.

The rigorous definition of symmetry involves spatial operations such as rotations, that when applied to an object give a result that appears identical to the original object. The symmetry of the object is defined by the collection of all such operations that apply to it. A symmetry axis that involves rotation by 180° is called a **twofold axis**, because two such rotations return the object to its initial position; one that involves rotation by 120° is called a threefold axis (Figure 2.2a), etc. There is an important distinction between shape and symmetry: the *shape* of an object refers to the geometry of its outline (spherical, cubic, tubular, etc.), whereas its *symmetry* refers to the rotational and translational operations that describe it.

CAPSIDS WITH ICOSAHEDRAL SYMMETRY

Geometric principles originally discovered by mathematicians in ancient Greece demonstrate that closed shells made of identical subunits can have one of only three kinds of symmetry—tetrahedral (a tetrahedron is a polygon with 4 identical triangular faces), cubic (a cube has 6 identical square faces), or **icosahedral** (a regular icosahedron has 20 triangular faces; from Greek *eikosi* = 20). Shells with icosahedral symmetry can accommodate exactly 60 structural subunits on their surface (Figure 2.2b, c). The capsids of many viruses are constructed as shells with icosahedral symmetry; tetrahedral and cubic symmetries have not been observed.

An object with icosahedral symmetry has fivefold, threefold, and twofold rotation axes. Figures 2.2b and c show side-by-side drawings of an icosahedron and a sphere onto which 60 commas are distributed with icosahedral symmetry. The fivefold rotation axes (pentagons shown at the vertices of the icosahedron) pass from the center of the sphere through the positions where the tails of five commas converge. An object with icosahedral symmetry has exactly 12 fivefold rotation axes, equivalent to the number of vertices on an icosahedron. The threefold axes (triangles at the center of each face of the icosahedron) pass through the centers of the spherical triangles outlined by the dashed lines on the image in Figure 2.2c. Objects with icosahedral symmetry have 20 threefold rotation axes, the number of triangular faces on an icosahedron. The twofold axes (ovals at the center of each edge of the icosahedron) pass through the midpoints of the edges of those triangles, where the heads of two commas lie opposite each other. Objects with icosahedral symmetry have 30 twofold rotation axes.

Note that the sphere shown in Figure 2.2c has icosahedral *symmetry*, but it does not have the *shape* of an icosahedron (Figure 2.2b). Most small viruses with icosahedral symmetry are roughly spherical, but many larger viruses (adenoviruses, phycodnaviruses, many tailed bacteriophages) have capsids with the shape of a regular icosahedron.

Some examples of virions with icosahedral symmetry

Parvoviruses are among the smallest viruses, with a 5 kb single-stranded DNA genome. Their capsids contain 60 copies of a protein of about 520 amino acid residues. The parvovirus capsid protein therefore uses up about one-third of the coding capacity of the viral genome. Parvovirus capsid protein subunits are arranged on the surface of the capsid with icosahedral symmetry (Figure 2.3a; compare with Figures 2.2b and c). This is the simplest type of capsid that can be constructed with icosahedral symmetry.

The folded structure of the parvovirus capsid protein is based on a domain known as a **jelly-roll β-barrel** (Figure 2.3b). The polypeptide chain is folded into a series of β strands that interact with each other to form a rigid blocklike structure. Loops that extend from the β strands of each subunit tie together the mature virion by binding to neighboring subunits. The same kind of protein fold is found in icosahedral capsids of many other viruses.

Poliovirus (Figure 2.4) illustrates one way in which viruses have evolved to produce larger icosahedral shells and thus to package larger genomes. The poliovirus capsid has an internal cavity about 19 nm in diameter, which contains the single-stranded RNA genome of about 8 kb. The genome encodes three structural protein subunits, known as VP1, VP2, and VP3, which are processed by cleavage from a single precursor polypeptide. Again, about one-third of the genome is used to encode the virion structural proteins. As with parvoviruses, the three poliovirus capsid proteins are all folded as jelly-roll β-barrels. The capsids of poliovirus and other members of the picornavirus family therefore contain three sets of 60 similarly folded protein subunits, for a total of 180 subunits.

The concept of quasi-equivalence

A more economical way to build a capsid with 180 subunits is to use 180 identical protein subunits, rather than 60 copies each of three different proteins. These subunits will not all be in identical positions on the surface of the capsid, and therefore they will have distinct binding interactions with their nearest neighbors, corresponding to their distinct environments. If a capsid contains certain multiples of 60 subunits, all these subunits can interact with their neighbors in *nearly* equivalent ways.

This concept is illustrated in Figure 2.5a (compare with Figure 2.2c). All 180 commas on the surface of this

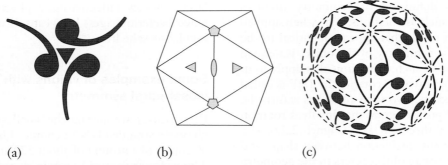

(a)   (b)   (c)

Figure 2.2 Elements of icosahedral symmetry. (a) Three (asymmetrical) commas, arranged with threefold symmetry. The threefold axis passes perpendicularly to the figure through the center of the triangle. (b) Outline of an icosahedron, showing fivefold (pentagons), threefold (triangles), and twofold (oval) symmetry axes. (c) Sixty commas distributed with icosahedral symmetry on the surface of a sphere. Tails meet in 12 groups of five on the surface, and heads meet in 20 groups of three.

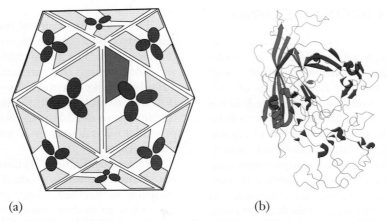

(a)   (b)

Figure 2.3 A simple capsid with icosahedral symmetry. (a) Distribution of 60 protein subunits in the shell of a parvovirus capsid. The trapezoids represent a jelly-roll β-barrel domain, and the ovals represent loops that form contacts about the threefold axes. Compare this image with that of the sphere in Figure 2.2c, oriented in the same manner. (b) Diagram showing the polypeptide chain of the canine parvovirus capsid protein in the same orientation as the red trapezoid in (a), with a jelly-roll β-barrel in red (upper left) and loop structures in blue (lower right).

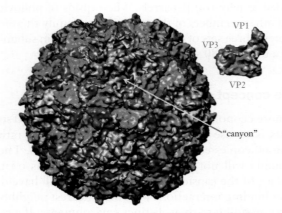

Figure 2.4 Picornavirus capsids. Model of poliovirus type I based on x-ray crystallography. The capsid is formed of 60 trimers (shown to the right) of VP1 (blue), VP2 (green), and VP3 (red), grouped in 12 pentamers, with 5-fold symmetry axes in the centers of the blue regions. VP4 is not visible as it is in the interior of the virion.

sphere have head-to-head, neck-to-neck, and tail-to-tail interactions with their neighbors. Sixty of them have tail-to-tail interactions in clusters of five; for example, the five commas whose tails meet at a point immediately above the equator, near the tail of the comma marked "A". The remaining 120 commas have tail-to-tail interactions in clusters of six; for example, the clusters on the equator including the commas marked "B" and "C".

Provided that the proteins can evolve to tolerate the distortions required by these slightly different contacts, they can use the same domains to make stable contacts at all 180 positions on the capsid surface. In other words, if the capsid corresponds to one of these **quasi-equivalent** designs, the protein subunits do not need to evolve three alternative binding sites for interactions with each other, as in the case of poliovirus; instead, they need to have built-in *flexibility*. The protein subunits that make up the capsids of many simple viruses exhibit this property. An example is shown in Figure 2.5b.

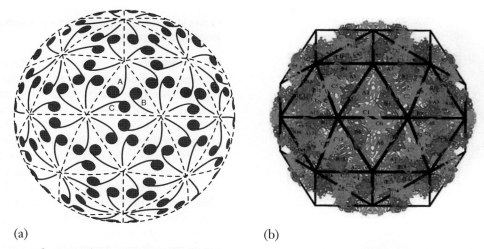

(a)   (b)

Figure 2.5 More complex capsid protein arrangements. (a) Quasi-equivalent arrangement of 180 commas in a T = 3 icosahedral surface lattice on a sphere. Compare with Figure 2.2c, a T = 1 arrangement. (b) Capsid of Norwalk virus, a human pathogen with a T = 3 structure, with a superimposed surface lattice in black, corresponding to the triangles on (a). The capsid is formed from 180 identical subunits, shown as 60 groups of three subunits labeled A, B, and C.

How many subunits can be accommodated on the capsid surface?

The theory of quasi-equivance was originally proposed by Donald Caspar and Aaron Klug to explain the observed distribution of subunits on the surfaces of a variety of viruses with icosahedral symmetry. They showed that the multiples of 60 that lead to quasi-equivalent structures are given by the **triangulation number** series, T = h² + hk + k², where h and k are any two integers, including zero. The total number of protein subunits in a capsid is equal to 60 × T. The integers h and k can be considered to represent the number of "jumps" required along two axes on the surface of a capsid to get from one cluster of 5 subunits to the next, touching down on clusters of 6 subunits along the way. On the simplest (T = 1) capsids, there are only clusters of 5 subunits, and therefore only one jump is required between such clusters: h = 1, k = 0. For the examples shown in Figures 2.5a and b, it takes two single jumps in two different directions to get from the fivefold axis above the equator to the fivefold axis below the equator. Therefore h = 1 and k = 1, and T = 1² + 1² + 1² = 3. Examples of T = 1, 3, 4, 7, 13, 16, 25 and greater are known. Capsids with larger T numbers require additional "glue" proteins to stabilize the interactions between subunits.

Larger viruses come in more complex packages

Quasi-equivalent icosahedral designs are neat and economical, but accurate assembly of larger particles is hard to achieve based solely on this principle. A protein subunit must take on the appropriate configuration when it enters a growing icosahedral capsid, and the more choices it has available, the more likely it is to make a mistake—which could result in a fatal flaw in the particle. As a result, the structures of larger and more complex virions depart in one way or another from the simple designs described above.

The polyomaviruses (Figure 2.6a) and papillomaviruses have capsids that assemble from 72 pentamers of the major capsid protein, VP1. These pentamers are assembled in a T = 7 icosahedral shell, which contains only 5 × 72 = 360 protein subunits, not 420 as would be predicted by the theory of quasi-equivalence. Sixty of the 72 pentamers have six nearest neighbors rather than five! Therefore, the pentamers at these positions have to fit into a "sixfold hole" on the capsid surface, even though they have fivefold symmetry. Apparently, this is easier to accomplish than forming pentamers for the fivefold positions and hexamers for the sixfold positions. A pattern of interlocking arms from the C-termini of the capsid subunits holds the structure together (Figure 2.6b).

The adenoviruses build their capsids with two kinds of structural subunits. **Hexons** lie at the 240 sixfold positions of a T = 25 icosahedral capsid, and **pentons** lie at the 12 fivefold positions (Figure 2.7). The hexon is actually constructed as a trimer of the adenovirus hexon protein, but each protein subunit contains two very similar subdomains, so that the assembled hexon has pseudo-sixfold symmetry.

These examples show that defined *subassemblies* of protein subunits confer significant advantage. The major capsid protein of polyomaviruses forms stable pentamers, and mechanisms have evolved for specific interactions with either five or six neighboring pentamers. Likewise, the adenovirus hexon is a defined trimer subassembly. The high-resolution crystal structures of polyomavirus capsid proteins and of the adenovirus hexon protein reveal that these proteins, like the parvovirus and picornavirus capsid proteins, contain jelly-roll β-barrels.

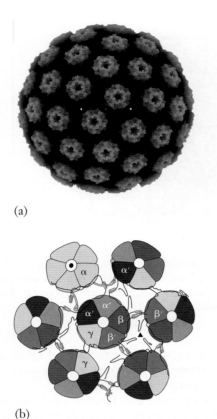

(a)

(b)

Figure 2.6 Structure of the capsid of simian virus 40.
(a) Seventy-two pentamers of VP1 are centered on lattice points of a $T = 7$ icosahedral surface lattice. Only 12 of these pentamers are located at centers of fivefold symmetry; the other 60 pentamers are each surrounded by 6 neighboring pentamers. (b) The arrangement of VP1 subunits in a 6-coordinated group. C-terminal arms of VP1 link adjacent pentamers together.

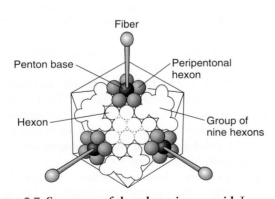

Figure 2.7 Structure of the adenovirus capsid. Larger and more complex capsids are made from more than one kind of protein subunit. For adenovirus, the penton base is formed from 5 copies of the penton protein, and hexons are formed from trimers of the hexon protein. Hexons are clustered in groups of 5 surrounding the twelve penton bases, and in groups of 9 on the 20 faces of the capsid, for a total of 240 per virion. Other proteins (not shown) link together hexons and pentons.

Other structures, large and small, display icosahedral symmetry

The principles of construction of closed shells using identical subunits are universal. The architect Buckminster Fuller exploited these principles to construct "geodesic domes" of various sizes from steel bars clustered in groups of 5 or 6 on their surface; a good example is the dome built for Expo '67 in Montreal (Figure 2.8a). Many soccer balls are made from 12 pentagonal and 20 hexagonal subunits, and display icosahedral symmetry (Figure 2.8b). A molecule in the shape of a stable closed shell composed of 60 carbon atoms was discovered in 1985 and named **buckminsterfullerene**. This molecule has icosahedral symmetry; the 60 carbon atoms lie at the vertices of 12 pentamers and 20 hexamers on the surface of the shell (Figure 2.8c). These objects range in size from tens of meters (geodesic domes) to 0.7 nm in diameter (C_{60}), and all share unusual strength, stability, and simplicity of construction based on their use of identical subunits joined in a symmetrical pattern.

(a)

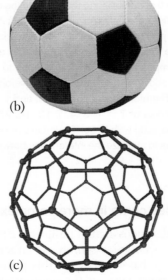

(b)

(c)

Figure 2.8 Structures that share symmetry elements with icosahedral viruses. (a) The geodesic dome, designed by Buckminster Fuller, erected for Expo '67 in Montreal. (b) A soccer ball. (c) C_{60}, known as buckminsterfullerene.

CAPSIDS WITH HELICAL SYMMETRY

An alternative to a closed shell for packaging viral genomes is an elongated tube. The same arguments for genetic economy, simplicity, and strength apply to these structures: they must be built from identical subunits and are therefore symmetric. In this case, the symmetry is **helical**.

Helical structures extend along a single helix axis. The symmetry of a helical assembly of subunits is defined by two parameters: u, the number of subunits per turn, and p, the displacement along the helix axis between one subunit and the next. P, the pitch of the helix, is the distance along the helix axis that corresponds to exactly one turn of the helix; $P = u \times p$. Figure 2.9 shows a diagram of the tobacco mosaic virus particle, in which many copies of a single protein subunit bind to the viral RNA and form a tube with helical symmetry. Here u is 16.33 (there are $3 \times 16.33 = 49$ subunits in three turns of the helix), p is 0.14 nm, and P is 2.3 nm.

In tobacco mosaic virus, as well as in the nucleocapsids of negative-strand RNA viruses like measles or influenza, the RNA winds along a groove that follows the helical path of the protein subunits. Each protein subunit binds to a fixed number of nucleotides in the RNA: 3 in the case of tobacco mosaic virus; 6 in the case of paramyxoviruses. By contrast, the protein coat of the filamentous bacteriophage fd forms a sleeve that surrounds the circular, single-stranded DNA, without a specific interaction between each subunit and a particular number of nucleotides in the DNA.

One advantage of helical nucleocapsids is that they can accommodate a variety of genome lengths; the longer the genome, the longer the helical nucleocapsid that assembles on the genome. This is particularly advantageous for viruses with a number of different genome fragments of different lengths, such as bunyaviruses and influenza viruses (see Chapters 17 and 18).

Virions of a number of bacteriophages contain both elements with icosahedral symmetry (the capsid) and elements with helical symmetry (the tail) (Figure 2.10).

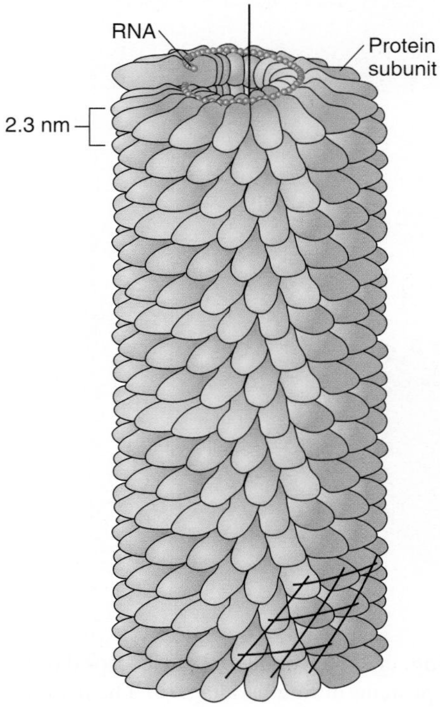

Figure 2.9 The helical nucleocapsid of tobacco mosaic virus. At the top is a cutaway view to reveal the RNA, protected within a groove, in the interior of the capsid. Each protein subunit binds to 3 nucleotides of the single-stranded viral RNA. Each turn of the helix is separated by 2.3 nm.

Figure 2.10 Bacteriophage φKZ virion. Many bacteriophages are constructed from a head with icosahedral symmetry and a tail with helical symmetry. The major capsid protein of bacteriophage φKZ is colored blue, the vertices are colored gray, the tail helices are in various colors, and the baseplate is colored gray. Below the baseplate, the device that punctures the cells is colored red. (Scale bar: 30 nm).

Phage tails are specialized structures that allow binding to cellular receptors and help to introduce the viral genome in the bacterial cell by piercing the cell wall and membranes. These particles elegantly illustrate the above principles of construction of molecular packages from a limited number of proteins.

VIRAL ENVELOPES

Viral envelopes are made from lipid bilayer membranes

Most viral envelopes are acquired at cellular membranes by a process of budding (Figure 2.11). During budding, the viral nucleocapsid is wrapped in a membrane into which viral glycoproteins have been inserted. Viral envelopes contain a lipid bilayer, whose lipid content reflects the composition of the cellular membrane from which the envelope was constructed. Viruses that bud outward from the cell surface, like influenza virus, contain phospholipids and cholesterol in the proportions characteristic of the plasma membrane. Flaviviruses, which bud into the endoplasmic reticulum, have much lower amounts of cholesterol, as does the endoplasmic reticulum membrane itself.

The lipid bilayer provides a permeability barrier that protects the viral genome from the environment, so that enveloped viruses do not need to assemble their proteins as perfectly as do viruses that lack envelopes. Nonetheless, some of the smallest enveloped viruses, such as flaviviruses (Figure 12.2) and togaviruses

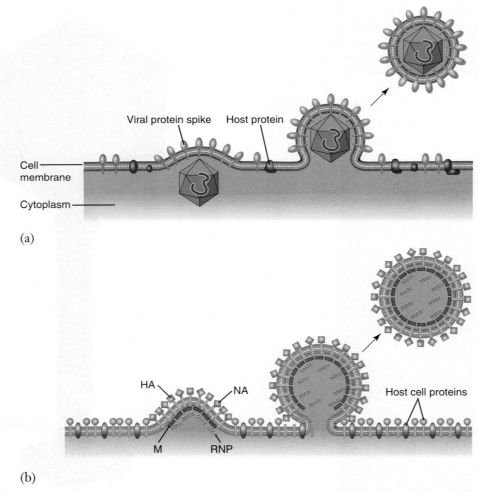

(a)

(b)

Figure 2.11 Release of enveloped virions from the cell surface by budding. (a) The icosahedral nucleocapsid of Semliki Forest virus interacts directly with the cytoplasmic tail of virus envelope glycoproteins during budding. (b) The matrix (M) protein of influenza virus interacts both with the cytoplasmic tail of surface glycoproteins (HA and NA) and with helical nucleocapsids (RNP) during budding.

(Figure 13.1), have virions with essentially perfect icosahedral symmetry. For togaviruses, each viral glycoprotein molecule in the envelope contacts a capsid protein molecule lying underneath the membrane, with the result that the viral glycoproteins are distributed on the envelope with the same symmetry as the proteins in the underlying capsid. Virions of larger enveloped viruses are often somewhat irregular in shape, and their envelopes have no obvious symmetry. Influenza virus (Figure 2.1g), which contains helical nucleocapsids, can adopt a variety of shapes and sizes from roughly spherical to elongated and filamentous (see Figure 18.2).

Viral glycoproteins are inserted into the lipid membrane to form the envelope

Most of the proteins found in viral envelopes have a relatively large glycosylated external domain (**ectodomain**), a hydrophobic **transmembrane anchor** of about 20 amino acid residues, and a short internal tail. A well-studied example is the influenza hemagglutinin (Figure 2.12). The tail is sometimes described as "cytoplasmic", since it faces the cytoplasm of the cell before the virus buds. The 20 non-polar amino acids of the transmembrane anchor form an alpha helix that neatly spans the 3-nm-thick hydrophobic part of a lipid bilayer membrane. Glycosylation of the ectodomain ensures that the external surface of virus particles is hydrated, reducing protein/protein interactions that could aggregate the virions.

Envelope proteins are synthesized on ribosomes bound to the endoplasmic reticulum and are inserted into the cell membrane by the standard route for cell-surface integral membrane proteins. Modifications such as glycosylation of the ectodomain and **palmitoylation** of cysteines near the inner margin of the lipid bilayer occur in the endoplasmic reticulum and in the Golgi complex.

Most viral envelope proteins are **type I integral membrane proteins**, with the N-terminus facing outward and the transmembrane anchor nearest the C-terminus of the polypeptide. However, some are **type II integral membrane proteins** of the opposite polarity, with the C-terminus facing outward and the anchor near the inward-facing N-terminus. Type I membrane proteins have a **signal sequence** that is cleaved from their N-terminus by a peptidase as they are inserted in the endoplasmic reticulum during synthesis. In type II proteins, the transmembrane anchor itself serves as the signal for membrane insertion.

A number of viruses have internal lipid membranes that lie beneath an external capsid with icosahedral symmetry. Several of these are viruses of

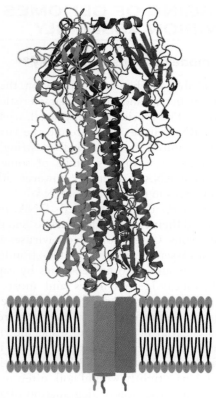

Figure 2.12 A typical viral envelope protein, influenza virus hemagglutinin. Ribbon diagram showing the amino acid chains of the hemagglutinin trimer, with each monomer in a distinct color. Alpha helices are shown as spirals, and beta sheets as flat segments. The external domain (above) is responsible for binding to cell receptors and for mediating fusion between the viral envelope and cell membranes. The transmembrane helices (cylinders) span the membrane, and a short internal domain interacts with the matrix protein underneath the viral envelope.

bacteria and archaea; the presence of a rigid cell wall would seem to rule out budding through the cytoplasmic membrane as a mechanism of formation of these internal envelopes. How they are formed is not well understood. It is known that certain capsid proteins of these viruses are integral membrane proteins that anchor the capsid in the underlying lipid membrane. Seen in this light, external capsids are simply a more elaborate and structured version of the envelope glycoproteins that coat the membrane of many viruses. The envelope proteins of both flaviviruses and alphaviruses (Chapters 12 and 13) are arranged with icosahedral symmetry, and these viruses could therefore also be considered to have an external "capsid" embedded in an internal lipid membrane.

PACKAGING OF GENOMES AND VIRION ASSEMBLY

Multiple modes of capsid assembly

Capsid assembly mechanisms depend on the nature of the final structure and on the characteristics of the viral genome. The capsids of many small, icosahedral, plus-strand RNA viruses grow around the viral RNA, condensing it into a compact "ball of string" as the assembly proceeds. Capsid subunits of some viruses require genome RNA to assemble; however, others can form empty shells in the absence of RNA.

Viruses with double-stranded DNA or RNA genomes wind their much stiffer nucleic acid contents into regular coils. Capsids of polyomaviruses and papillomaviruses assemble around double-stranded DNA genomes pre-wound into nucleosomes by capture of cellular histones. Herpesviruses and most double-stranded DNA bacteriophages insert a naked DNA genome into a preformed shell, in an energy-requiring process carried out by one or more specific viral proteins ("packaging motors"), sometimes with the help of specific RNAs. As the DNA is spooled into the capsid, it forms a tightly wound coil, often in coordination with the expulsion or degradation of an internal **scaffolding protein**. Scaffolding proteins help promote formation of a precursor particle ("procapsid") by coassembly with capsid proteins, but are not retained in the fully completed virion. Dismantling of the scaffold generally leads to rearrangement of the capsid subunits into a "mature" form of the shell. For examples of virion assembly using scaffolding proteins, see Chapters 6 (bacteriophage φX174) and 8 (bacteriophage lambda).

Specific packaging signals direct incorporation of viral genomes into virions

A variety of mechanisms have evolved to ensure specific genome packaging. Small, icosahedral, positive-strand RNA viruses, including some plant viruses and the togaviruses, have a positively charged N-terminal arm on their capsid proteins. This arm can interact specifically with a **packaging sequence** on the genome RNA, or nonspecifically with other segments of the RNA. The specific interaction often involves an RNA sequence element containing a stem-loop structure (the folded packaging sequence). During capsid assembly, one or a few capsid protein subunits bind to this packaging sequence, initiating both genome packaging and capsid assembly. The majority of the capsid subunits bind to RNA non-specifically, perhaps just through neutralization of the negative charge on the phosphate groups. The encapsidated RNA is highly condensed, with much of its stem-loop secondary structure retained during assembly.

In helical capsids, such as tobacco mosaic virus (Figure 2.9) or the nucleocapsids of many negative-strand RNA viruses, the RNA extends along a groove in the helical array of protein subunits. The RNA packaging sequence fits particularly well into a short length of this groove, giving the RNA genome an assembly advantage over any cellular RNAs that might be present during packaging. Once this packaging sequence is bound to the capsid proteins, the helical structure can form like a zipper around the remainder of the RNA molecule. For many negative-strand RNA viruses, packaging occurs as the RNA chain is being synthesized.

The machinery that spools the DNA of herpesviruses and most double-stranded DNA bacteriophages into the procapsid recognizes specific packaging sequences on viral DNA, and often involves cleavage of the DNA genome from a larger replicative intermediate (see Chapters 6, 8, 24). The insertion of DNA requires energy derived from ATP hydrolysis by the packaging motor.

Core proteins may accompany the viral genome inside the capsid

Some DNA viruses package not just the viral genome but a nucleoprotein complex. **Core proteins** condense viral DNA by neutralizing the negative charges of the phosphate groups on DNA. Polyomaviruses and papillomaviruses use cellular histones to condense their DNA genomes; the resulting DNA-protein complex, which resembles cellular chromatin, is called a "minichromosome" (see Figure 21.2). In the absence of viral genomes, these viruses can package segments of histone-associated, non-viral DNA to form "**pseudovirions**". Other DNA viruses such as adenoviruses and herpesviruses code for their own histone-like core proteins. Some viruses incorporate cellular polyamines such as spermidine into their capsids to neutralize the charge of phosphate groups and condense the nucleic acid genome.

Formation of viral envelopes by budding is driven by interactions between viral proteins

Viral envelope proteins inserted into a cell membrane tend to aggregate in local two-dimensional pools from which most of the cellular membrane proteins are excluded. This lateral association is possible because membrane proteins can move in the plane of the fluid membrane. The viral envelope proteins may have regular side-by-side interactions or may be clustered by virtue of interacting through their cytosolic "tails" with a multivalent internal viral protein or protein assembly. Regions of the membrane containing clusters of envelope proteins will eventually form the viral envelope by a process of budding (Figure 2.11).

Budding involves curvature of the membrane as it is wrapped around the viral nucleocapsids. There are several different patterns of budding. Icosahedral togavirus nucleocapsids assembled in the cytoplasm approach the plasma membrane and bind to the cytoplasmic tails of viral envelope proteins (Figure 2.11a). The capsid effectively wraps itself in the membrane, and progressive formation of contacts between the surface of the capsid and the cytoplasmic tails of the viral glycoproteins drives pinching-off of the bilayer by membrane fusion.

Many enveloped viruses with helical nucleocapsids have an internal **matrix protein** that mediates the association between the viral glycoproteins inserted into the plasma membrane and the nucleocapsids in the cytoplasm. Matrix proteins located at the cytoplasmic face of the membrane bind to the nucleocapsids and cooperate with the glycoprotein layer to drive the budding process. This is shown in Figure 2.11b for influenza virus.

A third variation on budding is carried out by some retroviruses, such as HIV-1. Instead of being assembled in the cytoplasm and migrating to the membrane, the nucleocapsid is assembled from the Gag precursor protein directly at the membrane during budding (Figure 2.1j; see Chapter 28). The matrix and nucleocapsid proteins are generated by cleavage of the Gag precursor during and after virion formation.

Budding is driven primarily by the interactions between envelope glycoproteins and matrix or nucleocapsid proteins. However, some viruses can generate empty envelopes that bud in the absence of nucleocapsids; in this case, the envelope glycoproteins themselves drive budding. In contrast, some retroviruses that have lost their envelope glycoprotein genes can form "bald" particles that contain nucleocapsids wrapped in membranes lacking envelope proteins. The layer of Gag proteins at the inner surface of the plasma membrane interacts sufficiently strongly with the lipid bilayer (in many cases through an N-terminal covalently bound fatty acid, **myristate**) that its assembly into a nucleocapsid can drive budding. Many viruses need to recruit cellular proteins to complete the final pinching-off of the lipid bilayer and release a completed particle.

DISASSEMBLY OF VIRIONS: THE DELIVERY OF VIRAL GENOMES TO THE HOST CELL

Virions are primed to enter cells and release their genome

Virus particles are more than passive packages that simply enclose the viral genome. They are vehicles evolved to deliver their cargo (the viral genome) to a host cell according to a programmed set of instructions (see Chapter 4, Virus Entry). Assembly and disassembly are not simply reverse processes. In many cases, the assembled virion is in an energetically metastable state, primed to dissociate and release its cargo upon binding to or entering the host cell. This primed state has been likened to a "Jack-in-the-box", awaiting activation by some triggering step. This step usually involves interaction with a specific cellular receptor on the cell surface or in the cytoplasm.

Proteolytic cleavage is the most common way in which an assembled virion is "primed" for the delivery program. For example, self-cleavage of the capsid protein VP0 in the picornaviruses (Chapter 11) creates a small internal peptide, VP4, which emerges and helps mediate penetration when receptor binding forces the virus particle to expand or dissociate (see Figure 11.3).

In the reoviruses, a protector protein called σ3 is cleaved by acid-activated proteases within the intestine or in lysosomes to expose another protein, σ1, that is responsible for penetration of the virus core into the cell (see Figures 19.4 and 19.5). Similarly, the enveloped flavivirus PrM protein is cleaved during infection and dissociates to reveal the E protein hidden underneath; the E protein can then enable virus entry by fusion of the viral envelope with the plasma membrane of the target cell. Cleavage of fusion proteins such as influenza virus hemagglutinin or human immunodeficiency virus gp160 primes them for a dramatic, fusion-inducing conformational rearrangement triggered by low pH (influenza hemagglutinin, see Figure 4.4) or by receptor and coreceptor binding (HIV-1 gp160; Chapter 29).

The "Jack-in-a-box" metaphor is particularly apt in the case of herpesviruses and certain bacteriophages, in which DNA is coiled in a spring-like fashion into the virion by the action of an energy-requiring protein spooling machine during virion assembly. Release of a latch that seals the particle leads to release of the genome. This occurs at the cell surface for bacteriophages, and at the nuclear pore for herpesviruses. The DNA of other bacteriophages (e.g., bacteriophage T7) is not spring-loaded sufficiently to drive complete injection into the cell; rather, after partial entry by injection, the remainder of the genome is actively spooled into the cell by the transcriptional machinery (Figure 7.2).

Plant viruses, which are usually injected into host cells by their insect vectors, do not require a spring-loading mechanism. They appear to undergo a reversible structural transition when exposed to the cytoplasm, allowing ribosomes or initiation factors access to their RNA. The ribosomal translocation process then drives uncoating. A related process may mediate uncoating of some positive-strand RNA animal viruses.

KEY TERMS

Buckminsterfullerene

Budding

Capsid

Core proteins

Cryoelectron microscopy

Ectodomain

Envelope

Genetic economy

Glycoprotein

Helical symmetry

Hexon

Icosahedral symmetry

Jelly-roll β-barrel

Matrix protein

Myristate

Negative staining

Nucleocapsid

Packaging sequence

Palmitoylation

Penton

Positive staining

Pseudovirion

Quasi-equivalent

Scaffolding proteins

Self-assembly

Signal sequence

Synchrotron

Transmembrane anchor

Triangulation number

Twofold axis

Type I integral membrane protein

Type II intregal membrane protein

Virion

X-ray diffraction

FUNDAMENTAL CONCEPTS

• Viruses enclose their nucleic acid genomes in protein capsids; some viruses also have an internal or external lipid-containing envelope.

• Capsids of small viruses are constructed from one or a small number of identical subunits; they are therefore highly symmetrical.

• Capsids that are closed, quasi-spherical shells have icosahedral symmetry, with fivefold, threefold, and twofold symmetry axes.

• The smallest capsids with icosahedral symmetry are constructed from 60 identical subunits.

• Larger capsids with icosahedral symmetry usually are made from $60 \times T$ subunits, where T is the triangulation number; larger capsids may also have additional capsid proteins that strenghten the structure.

• Many capsid proteins have a common three-dimensional fold called a jelly-roll β-barrel, which provides a rigid, brick-like structure.

• Tubular capsids have helical symmetry, and their length depends on the length of the enclosed nucleic acid genome.

• Many, but not all, viral envelopes are derived by budding from preformed cellular membranes.

• Viral envelopes usually incorporate viral glycoproteins with a transmembrane anchor.

• Some capsids are assembled around the nucleic acid genome; others are pre-assembled and the genome is stuffed into the capsid.

• Scaffolding proteins help to assemble capsids but are subsequently ejected and do not form part of the mature capsid.

• Many viral genomes contain specific packaging sequences that are recognized by capsid proteins, ensuring that only intact genomes are encapsidated.

REVIEW QUESTIONS

1. In simple terms, how do virions transmit the genome of a virus from one cell to another?

2. Why is x-ray diffraction unsuitable for the study of the structure of enveloped viruses?

3. How do Type I and Type II integral membrane proteins differ?

4. How are the virions of some of the simplest viruses assembled?

5. What is the role of the core proteins that some DNA viruses package along with their genomes?

Virus Classification: The World of Viruses

Nicholas H. Acheson

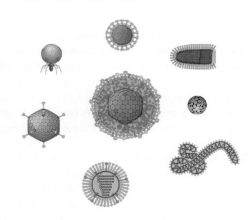

VIRUS CLASSIFICATION

Viruses are classified into related groups based on:
- *Genome composition (DNA or RNA, single- or double-stranded)*
- *Genome topology (linear or circular; single or multiple segments)*
- *Capsid symmetry*
- *Presence or absence of an envelope*
- *Genetic relatedness (nucleotide and amino acid sequence similarity)*
- *Mechanisms for expressing messenger RNAs and replication of genomes*
- *Host organisms*

THE MAJOR VIRUS GROUPS

Viruses are classified into species, genera, families, and (in some cases) orders, within the following categories:
- *Viruses with single-stranded DNA genomes*
- *Viruses with double-stranded DNA genomes*
- *Viruses with positive-strand RNA genomes*
- *Viruses with negative-strand RNA genomes*
- *Viruses with double-stranded RNA genomes*
- *Viruses that use a reverse transcriptase*
- *Satellite viruses, satellite nucleic acids, and viroids*

THE EVOLUTIONARY ORIGIN OF VIRUSES

RNA viruses may be relicts of the prebiotic RNA world.

Small DNA viruses could have evolved from escaped, self-replicating fragments of cellular genome DNA.

Some large DNA viruses could have evolved from degenerate intracellular parasites.

VIRUS CLASSIFICATION

Many different viruses infecting a wide variety of organisms have been discovered

Several thousand different viruses have been isolated and characterized during the approximately 100 years since the science of virology was established. These viruses infect numerous organisms including bacteria, archaea, fungi, algae, higher plants, invertebrates, and vertebrates. Several hundred viruses have been studied in some detail. To make sense out of this large collection of viruses, scientists have attempted to classify viruses into related groups.

Classification of organisms that are genetically and evolutionarily related to each other is useful because detailed knowledge about one organism can often be generalized and applied to related organisms. The classification of viruses into **species** and into a hierarchy of related groups of species (**genera**, **families**, and **orders**) helps virologists to compare viruses and to analyze and understand many aspects of virus structure, replication pathways, interactions with host cells, and disease transmission. The science of classification is called **taxonomy**. This chapter presents an overview of the different virus groups that have been recognized and named.

Virus classification is based on molecular architecture, genetic relatedness, and host organism

A rational classification of viruses must take into account the nature of the viral genome and the structure of the virus particle, because these are fundamental, inherited characteristics of viruses. The following molecular criteria are used to classify viruses:

- Type of nucleic acid genome (DNA or RNA)
- Strandedness of nucleic acid (single- or double-stranded)
- Topology of nucleic acid (linear or circular)
- Symmetry of capsid (icosahedral, helical, or no symmetry)
- Presence or absence of an envelope

Genetic relatedness among viruses can be determined by comparison of viral genome sequences and amino acid sequences of viral proteins. Other characteristics such as the order of genes on viral genomes, the mechanism of viral messenger RNA synthesis, the presence of viral genes coding for DNA or RNA polymerases, and the use of reverse transcriptase during replication, are also useful criteria for virus classification.

Many viruses have been associated with their host organisms during millions to billions of years of evolution. Viruses must be able to enter and replicate in their host organism, and must then be released and transmitted to another individual. Viruses therefore adapt to the host organisms in which they replicate, and evolve as these organisms evolve. Some examples of these adaptations are:

- Specific mechanisms for transmission between individual cells or organisms (via direct contact, diffusion through liquid media, dispersion via the circulatory system, respiratory droplets, stool or urine contamination of food or water, insect vectors, etc.)
- Entry pathways into cells (binding to specific cell surface receptors, passage through cell walls, entry by membrane fusion, uptake via vesicles, etc.)
- Interaction with cellular RNA or DNA polymerases, RNA processing enzymes, protein synthesizing systems, membrane systems, cellular signaling proteins, and other cellular proteins used by the virus to express and replicate its genes

As a result of these adaptations, viruses that are genetically and evolutionarily related tend to infect related organisms. Virus taxonomy takes into account the ability of viruses to infect different categories of organisms. For this purpose, it is convenient to divide all living organisms into the following six broad categories:

1. Bacteria
2. Archaea
3. Lower eukaryotes (fungi, protozoa, algae)
4. Plants
5. Invertebrates
6. Vertebrates (including humans)

Note that these categories do not correspond to those in the current "six-kingdom" model of classification of organisms; instead, this is a pragmatic grouping of life forms that is useful for the present discussion of virus classification.

Viruses that are evolutionarily and genetically related tend to infect organisms within only one of the above six categories. However, this is not always the case. For example, a number of viruses of vertebrates are transmitted from one animal to another by blood-sucking invertebrate arthropods (mosquitoes, ticks, sandflies), in which they can also replicate. Many plant viruses are also transmitted by insects. This means that such viruses can evolve and spread in both vertebrates (or plants) and invertebrates. As a result, a number of virus families include viruses that can infect organisms in two or more of the above categories.

Viruses are grouped into species, genera, and families

This chapter outlines information about the major virus groups presently identified, with particular emphasis on viruses that infect bacteria and vertebrates, the subject of most of the remainder of this book. This description of virus groups is not meant to be exhaustive; the periodical reports on virus taxonomy published by the International Committee on Virus Taxonomy (ICTV) are the definitive source on known viruses and their classification. Much of the information in this chapter is derived from the ICTV's 8th Report (2005) and recent updates (see http://www.ictvonline.org/index.asp). Following their rules, names of virus families are italicized and end in the Latin suffix -*viridae*, and names of genera end in the suffix -*virus*. However, in this book, virus family names are usually referred to in their English version for sake of simplicity (e.g., "picornaviruses" instead of "*Picornaviridae*").

Different virus isolates are generally considered to be members of a virus *species* if they share a high degree of nucleic acid sequence identity and most of their proteins have highly similar amino acid sequences and common antigenic properties. Members of a species usually infect a limited number of organisms or specific target cells and tissues. This implies that they are in genetic contact with each other and are therefore members of a common genetic **lineage**.

Virus species are grouped into *genera* by virtue of shared characteristics such as genome organization and size, virion structure, and replication strategies. Although they are related by evolution, species that are members of a common genus may show significant divergence in nucleotide and amino acid sequences. Different members of a genus may infect different organisms or different tissues in the same organism, and therefore may not usually come in contact with each other.

Genera are in turn grouped into virus *families*. Members of a virus family share overall genome organization, virion structure, and replication mechanisms. Because of this, they are presumed to be evolutionarily related. However, different members of a virus family may vary significantly in virion size and genome length, and may have a variety of unique genes not shared by other family members. Because members of a virus family may have evolved separately over considerable periods of time in distinct host organisms, they often share only limited nucleotide or amino acid sequence homology in a small number of genes. Presently the ICTV recognizes 87 virus families, and 13 genera that are not yet assigned to families.

Distinct naming conventions and classification schemes have developed in different domains of virology

Research on viruses has historically been carried out by scientists who study viruses that infect only one category of organism; for example, bacteria, fungi, algae, plants, insects, fish, domesticated animals, or humans. As a result, distinct domains such as bacterial virology, plant virology, veterinary virology, and medical virology have arisen.

Although there has been some convergence among these fields of study, their different histories have led to adoption of distinct conventions for naming viruses and for virus classification. For example, bacteriophages are given names that include the name of the host bacterial genus and an arbitrary number or Greek letter (e.g., Enterobacteria phage T4, Bacillus phage SP01, Enterobacteria phage Mu). In practice, these official names are often shortened to "bacteriophage" or "phage" followed by the number.

Naming of plant viruses usually makes reference to the plant species in which the virus was first isolated and to the disease symptoms caused by the virus (e.g., tobacco mosaic virus, tomato bushy stunt virus, maize streak virus). Many insect viruses are named using the Latin name of the insect species from which the virus was isolated, followed by the virus genus name (e.g., *Aedes aegypti* densovirus, *Autographica californica* multiple nucleopolyhedrovirus, *Chironomus luridus* entomopoxvirus).

A variety of naming conventions are used for viruses of vertebrates. These include the host species of origin (e.g., bovine papillomavirus, simian virus 40, porcine parvovirus), the locality where the virus was first isolated (e.g., Semliki Forest virus, Ebola virus, Norwalk virus), or the disease caused by the virus (e.g., measles virus, influenza A virus, hepatitis B virus).

Because most virology research has been carried out in the fields of bacterial, plant, and vertebrate viruses, the majority of viruses that have been discovered infect organisms in one of these groups. There undoubtedly remain many viruses yet to be discovered, particularly in poorly studied organisms such as archaea, protists, small invertebrates, and marine organisms.

Scientists who classify viruses that infect distinct groups of organisms tend to use different criteria. For example, bacteriophages are currently classified into families based on the nature of their genomes and on their capsid and tail morphology. Bacteriophages can mutate rapidly because of their rapid growth cycles, and they can readily exchange genes with each other by recombination or by integration into (and excision from) the genomes of their host bacteria. This means that bacteriophages tend to be genetic mosaics, and determining evolutionary or genetic relatedness among different phages is difficult if not impossible.

Plant viruses are presently grouped into 24 different families, but a large number of genera (most of which contain viruses with positive-strand RNA genomes) have not yet been assigned to families. Viruses of vertebrates and invertebrates are more evenly represented in all genome categories; they are currently grouped into 28 distinct families and only 3 unassigned genera.

MAJOR VIRUS GROUPS

Study of the major groups of viruses leads to understanding of shared characteristics and replication pathways

For convenience and simplicity, many of the major known virus groups (families and unassigned genera) are listed below in seven separate tables, based on the nature of their genomes and their replication strategy. Within each table, these lists are divided into sections based on host organism, starting with the simplest (prokaryotes) and ending with the most complex (vertebrates). Within each section, virus groups are listed in order of increasing genome length and therefore complexity. Virus groups that are covered in detail in this book are highlighted in green.

The attentive reader will note that some interesting observations can be made about viruses grouped in these different tables. For example:

- Viruses with single-stranded DNA genomes tend to be very small.
- Most plant viruses have positive-strand RNA genomes.
- Many fungal viruses have double-stranded RNA genomes.
- Most bacteriophages are tailed and have double-stranded DNA genomes.
- Viruses with very large double-stranded DNA genomes (400 to 1200 kb) are limited to free-living unicellular algae and protozoa.

Understanding these and other conclusions that are derived from our knowledge of virus classification will be an interesting future challenge for virologists.

Viruses with single-stranded DNA genomes are small and have few genes

Given the great variations in size and shape of known viruses (Figure 3.1), it is striking that viruses with single-stranded DNA genomes (Table 3.1) lie within a very narrow size range. Their genomes are among the smallest of all viral genomes, ranging from less than 2 kb up to 9 kb. The genomes of six of the seven families are circular, and those of the seventh family, *Parvoviridae*, have hairpin ends. These characteristics may reflect the fragility of single-stranded DNA compared with double-stranded DNA; small size and circularity or hairpin ends ensure that the genome DNA cannot be easily attacked by nucleases.

None of these viruses has a lipid envelope, and viruses in six out of the seven families have icosahedral

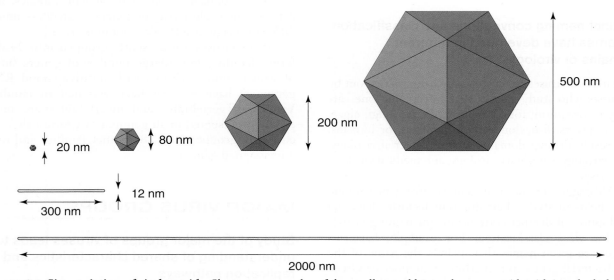

Figure 3.1 Size variation of viral capsids. Shown are examples of the smallest and largest known capsids with icosahedral symmetry, and two intermediate-size capsids. From left to right: a parvovirus (beak and feather disease virus), diameter 20 nm; adenovirus, diameter 80 nm; a phycodnavirus (Paramecium bursaria Chlorella virus 1), diameter 200 nm; and mimivirus, diameter 500 nm. The ratios of the volumes of these capsids is 1:64:1000:15,000. Below are examples of helical capsids: tobacco mosaic virus, 12 × 300 nm; and pseudomonas phage Pf1, 7 × 2000 nm.

Table 3.1 Viruses with single-stranded DNA genomes

Virus group	Hosts	Example	Genome segments	Genome size (kb) and topology		Capsid symmetry	Envelope
Microviridae	Bacteria	Bacteriophage φX174	1	4–6	Circular	Icosahedral	No
Inoviridae	Bacteria	Bacteriophage M13	1	5–9	Circular	Helical	No
Geminiviridae	Plants	Maize streak virus	1–2	3–6	Circular	Icosahedral	No
Nanoviridae	Plants	Banana bunchy top virus	6–8	6–9	Circular	Icosahedral	No
Circoviridae	Vertebrates	Beak and feather disease virus	1	2–4	Circular	Icosahedral	No
Parvoviridae	Vertebrates & invertebrates	Adeno-associated virus	1	4–6	Linear	Icosahedral	No
Anelloviridae	Vertebrates	Torque teno virus 1	1	4	Circular	Icosahedral	No

How small are viruses?

Most viruses are too small to be detected by light microscopy, but can be readily seen with the electron microscope. The smallest spherical virions have a diameter of approximately 20 nanometers (nm) (Figure 3.1); 50,000 such virions lined up in a row would cover a length of 1 millimeter. The largest known virions (mimivirus) have a diameter of approximately 500 nm, and these can be barely seen with a light microscope. It would take only 2000 such virions lined up in a row to cover a length of 1 millimeter. These giant viruses are as large as some of the smallest free-living cellular organisms, including mycoplasma and some archaea, and unlike most viruses, are retained by small-pore filters. Viruses with filamentous or rod-shaped capsids have diameters between 7 and 30 nm but can be as long as several thousand nm (several micrometers, mm), which approaches the diameter of many mammalian cells.

genes are contained in a single DNA molecule. Some of these DNA genomes are circular; others are linear. The large genome size of some of these viruses is probably a result of the superior qualities of double-stranded DNA as a gene storage facility compared with other forms of nucleic acids: increased resistance to degradation, a low error rate during DNA replication, and the ability to recognize and repair mutations. Among the families that infect vertebrates, those with smaller genomes (*Polyomaviridae*, *Papillomaviridae*, and *Adenoviridae*) do not have enveloped virions, but those with larger genomes all have enveloped virions.

Given the variety and number of these viruses, it is striking that virtually no known double-stranded DNA viruses (for the exception, see reverse transcriptase viruses below) infect higher plants. In contrast, the only viruses known to infect algae are double-stranded DNA viruses (*Phycodnaviridae*), and these are some of the largest known viruses (many of their genomes are 300 kb or larger). A virus that infects amoebae, called mimivirus, has by far the largest known viral genome; recently sequenced, it contains nearly 1.2 million base pairs, larger than the genomes of many bacteria and archaea.

Some 95% of the known bacteriophages are tailed phages with double-stranded DNA genomes. These are grouped into three families based on the tail structure: *Myoviridae* have a long contractile tail, *Siphoviridae* have a long non-contractile tail, and *Podoviridae* have a short non-contractile tail. Within each of these families there is a wide range of genome sizes, gene organization, and replication mechanisms. Because it is based primarily on virion morphology, this classification system is somewhat arbitrary. However, the members of individual genera within each family share significant biological characteristics.

Virions of three additional families of double-stranded DNA phages contain lipid membranes either outside or inside their protein capsids. Each of these families includes only one or a small number of different phage isolates. A number of viruses that infect archaea have been recently described; all but one of these contain double-stranded DNA genomes, and in many cases their virions have distinct and unusual morphologies.

capsids. It is tempting to speculate that most viruses with single-stranded DNA genomes may have evolved from a common ancestor (see Chapter 6).

Viruses with double-stranded DNA genomes include the largest known viruses

Viruses with double-stranded DNA genomes (Table 3.2) show a wide variation in genome size (between 5 and 1180 kb; Figure 3.2). Without exception, all of these viruses have unfragmented genomes; that is, all viral

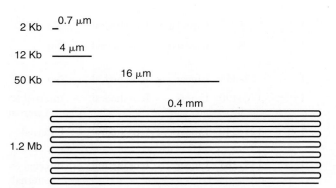

Figure 3.2 Size variation of viral genomes. Lines are drawn in proportion to genome lengths. Genome sizes in kb (thousands of nucleotides) are shown on left, and genome lengths in micrometers (μm) or mm are shown above each line. From top to bottom: circoviruses have genomes of less than 2 kb; many positive-strand RNA viruses have genomes of approximately 12 kb; medium-size bacteriophages such as lambda have genomes of 50 kb; the largest known viral genome, that of mimivirus, is 1.2 Mb long.

Most plant viruses and many viruses of vertebrates have positive-strand RNA genomes

Members of nine different virus families with positive-strand RNA genomes infect vertebrates, and positive-strand RNA viruses classified into twelve different families and seven unassigned genera infect plants (Table 3.3). The majority of known plant viruses have positive-strand RNA genomes. There is only one family of bacteriophages with positive-strand RNA genomes; no archaeal viruses with RNA genomes have

Table 3.2 Viruses with double-stranded DNA genomes

Virus group	Hosts	Example	Genome segments	Genome size (kb) and topology		Capsid symmetry	Envelope
Corticoviridae	Bacteria	Bacteriophage PM2	1	10	Circular	Icosahedral	Internal
Plasmaviridae	Mycoplasma	Bacteriophage L2	1	12	Circular	Asymmetric	External
Tectiviridae	Bacteria	Bacteriophage PRD1	1	15	Linear	Icosahedral	Internal
Podoviridae	Bacteria	Bacteriophage T7	1	16–70	Linear	Icosahedral, tailed	No
Siphoviridae	Bacteria & archaea	Bacteriophage lambda	1	22–121	Linear	Icosahedral, tailed	No
Myoviridae	Bacteria & archaea	Bacteriophage T4	1	34–497	Linear	Icosahedral, tailed	No
Fuselloviridae	Archaea	*Sulfolobus* spindle-shaped virus 1	1	15–24	Circular	Asymmetric	External
Lipothrixviridae	Archaea	*Sulfolobus islandicus* filamentous virus	1	16–56	Linear	Asymmetric	External
Guttaviridae	Archaea	*Sulfolobus neozealandicus* droplet-shaped virus	1	20	Circular	Asymmetric	External
Ampullaviridae	Archaea	*Acidianus* bottle-shaped virus	1	24	Linear	Asymmetric	External
Globulaviridae	Archaea	*Pyrobaculum* spherical virus	1	28	Linear	Asymmetric	External
Rudiviridae	Archaea	*Sulfolobus islandicus* rod-shaped virus 2	1	32–35	Linear	Asymmetric	No
Bicaudaviridae	Archaea	*Acidianus* two-tailed virus	1	63	Circular	Asymmetric	External
Phycodnaviridae	Algae	*Paramecium bursaria* chlorella virus 1	1	100–560	Linear or circular	Icosahedral	Internal &/or external
Mimiviridae	Protozoa (amoeba)	*Acanthamoeba polyphaga* mimivirus	1	1180	Linear	Icosahedral	Internal
Baculoviridae	Invertebrates	*A. californica* multiple nucleopolyhedrovirus	1	80–180	Circular	Helical	External
Polyomaviridae	Vertebrates	Mouse polyomavirus	1	5	Circular	Icosahedral	No
Papillomaviridae	Vertebrates	Human papillomavirus 16	1	8	Circular	Icosahedral	No
Adenoviridae	Vertebrates	Human adenovirus 5	1	25–45	Linear	Icosahedral	No
Asfarviridae	Vertebrates	African swine fever virus	1	170–190	Linear	Icosahedral	Internal & external
Herpesviridae	Vertebrates	Herpes simplex virus 1	1	125–240	Linear	Icosahedral	External
Iridoviridae	Vertebrates & invertebrates	Frog virus 3	1	135–303	Linear	Icosahedral	Internal & external
Poxviridae	Vertebrates & invertebrates	Variola virus (smallpox)	1	130–375	Linear	Complex	External

The great disparity of viral genome size and complexity

Viral genomes (Figure 3.2) range from slightly less than 2000 nucleotides (coding for 2–3 proteins) up to nearly 1.2 million nucleotides (coding for over 1200 proteins). In comparison, the smallest cellular genomes contain approximately 400,000 nucleotides, and the genomes of mammals contain approximately 3 billion nucleotides. Viral genomes are very compact and economical; most nucleotide sequences are used for protein coding, and a good yardstick is one protein coded per 1000 nucleotides. In contrast, genomes of complex plants and animals contain mostly noncoding sequences; for example, the human genome codes for about 30,000 proteins, or one protein per 100,000 nucleotides.

yet been discovered. All known RNA viruses have linear genomes. However, some very small satellite viruses and viroids (see below) have circular RNAs.

Most of these viruses have genomes in the size range of 5 to 19 kb; an exception is the *Coronaviridae*, whose genomes are as long as 31 kb (the largest known RNA virus genomes). As with single-stranded DNA genomes, single-stranded RNA genomes are susceptible to degradation by nucleases or by divalent cation attack, and this may be a factor that limits their size. A number of plant viruses have fragmented positive-strand RNA genomes that are packaged separately into distinct particles. This may be another means of reducing the likelihood of inactivation by genome degradation, by reducing the size of each genome fragment; however, these viruses can only infect their hosts

Table 3.3 Viruses with positive-strand RNA genomes

Virus group	Hosts	Example	Genome segments	Genome size (kb) and topology		Capsid symmetry	Envelope
Leviviridae	Bacteria	Bacteriophage Qβ	1	3–4	Linear	Icosahedral	No
Marnaviridae	Algae	Heterosigma akawisho RNA virus	1	9	Linear	Icosahedral	No
Barnaviridae	Fungi	Mushroom bacilliform virus	1	4	Linear	Icosahedral	No
Bromoviridae	Plants	Cucumber mosaic virus	3	8	Linear	Icosahedral	No
2 other families and 8 genera	Plants	Tomato bushy stunt virus	1–3	4–15	Linear	Icosahedral	No
Closteroviridae	Plants	Beet yellows virus	1–2	15–19	Linear	Helical	No
5 other families and 5 genera	Plants	Tobacco mosaic virus	1–5	6–19	Linear	Helical	No
2 families and 1 genus	Invertebrates	Cricket paralysis virus	1–2	6–10	Linear	Icosahedral	No
Roniviridae	Invertebrates	Gill-associated virus	1	26	Linear	Helical	Yes
Nodaviridae	Vertebrates & invertebrates	Flock house virus	2	4–5	Linear	Icosahedral	No
Astroviridae	Vertebrates	Human astrovirus	1	6–7	Linear	Icosahedral	No
Picornaviridae	Vertebrates	Poliovirus	1	7–9	Linear	Icosahedral	No
Caliciviridae	Vertebrates	Norwalk virus	1	7–8	Linear	Icosahedral	No
Flaviviridae	Vertebrates & invertebrates	Yellow fever virus	1	10–12	Linear	Icosahedral	Yes
Togaviridae	Vertebrates & invertebrates	Semliki Forest virus	1	10–12	Linear	Icosahedral	Yes
Arteriviridae	Vertebrates	Equine arteritis virus	1	13–16	Linear	Icosahedral	Yes
Coronaviridae	Vertebrates	SARS coronavirus	1	28–31	Linear	Helical	Yes

if at least one virus particle carrying each of the genome fragments gets into the same cell.

Most viruses with RNA genomes smaller than 10 kb have non-enveloped capsids, while the capsids of many viruses with larger genomes (*Flaviviridae*, *Togaviridae*, *Arteriviridae*, *Coronaviridae*, and *Roniviridae*) are enveloped; envelopes may afford additional protection to the viral genome. Viruses with the largest RNA genomes (*Coronaviridae*, *Roniviridae*), as well as many plant viruses, have helical nucleocapsids. The close association between capsid proteins and the single-stranded RNA in helical nucleocapsids may afford additional protection against degradation, particularly inside the cell.

Viruses with negative-strand RNA genomes have helical nucleocapsids; some have fragmented genomes

Some of the most widespread and/or deadly human diseases (measles, influenza, rabies, Ebola hemorrhagic fever, Lassa fever) are caused by viruses with negative-strand RNA genomes (Table 3.4). All known viruses in this group have helical nucleocapsids. This may be related to their mechanism of replication; the genome RNA must first be transcribed within the cell to generate complementary, positive-strand messenger RNAs, and the helical nucleocapsid may provide stability and structure to the genome during this process. All negative-strand RNA viruses that infect vertebrates are enveloped. It is notable that no negative-strand RNA viruses of prokaryotes (bacteria and archaea) have yet been discovered.

Viruses in this group have genomes with a limited size range of 9–19 kb. Members of four families (*Bornaviridae*, *Paramyxoviridae*, *Rhabdoviridae*, and *Filoviridae*) share several distinctive characteristics: a single genome segment with a common gene order, transcribed in series from a single promoter at the 3' end of the genome. These four families have been grouped into the order *Mononegavirales* (for "single segment, negative-strand RNA viruses").

All other families in this group have fragmented genomes (as many as eight genome segments). In most cases, each negative-strand genome segment codes for a single messenger RNA and one or two viral proteins. Viruses with fragmented genomes face the problem of ensuring that at least one copy of each genome segment is packaged into the virion. However, in some plant viruses, the different genome segments are separately packaged into distinct virions, as is the case with some positive-strand RNA plant viruses.

Viruses with double-stranded RNA genomes have fragmented genomes and capsids with icosahedral symmetry

Most viruses with double-stranded RNA genomes (Table 3.5) contain multiple genome segments, each of which codes for a single messenger RNA and usually a single viral protein. All of these viruses have capsids with icosahedral symmetry, in contrast to the helical nucleocapsids of viruses with negative-strand RNA genomes. This may be related to their mechanism of transcription. The capsid or a subviral particle remains

Table 3.4 Viruses with negative-strand RNA genomes

Virus group	Hosts	Example	Genome segments	Genome size (kb) and topology		Capsid symmetry	Envelope
Ophiovirus	Plants	Citrus psorosis virus	3–4	11–12	Linear	Helical	No
Tenuivirus	Plants & invertebrates	Rice stripe virus	4–6	17–18	Linear	Helical	No
Bornaviridae	Vertebrates	Borna disease virus	1	9	Linear	Helical	Yes
Arenaviridae	Vertebrates	Lassa fever virus	2	11	Linear	Helical	Yes
Orthomyxoviridae	Vertebrates & invertebrates	Influenza A virus	6–8	10–15	Linear	Helical	Yes
Rhabdoviridae	Vertebrates, invertebrates, & plants	Rabies virus	1	11–15	Linear	Helical	Yes
Paramyxoviridae	Vertebrates	Measles virus	1	13–18	Linear	Helical	Yes
Filoviridae	Vertebrates	Ebola virus	1	19	Linear	Helical	Yes
Bunyaviridae	Vertebrates, invertebrates, & plants	California encephalitis virus	3	11–19	Linear	Helical	Yes

Table 3.5 Viruses with double-stranded RNA genomes

Virus group	Hosts	Example	Genome segments	Genome size (kb) and topology		Capsid symmetry	Envelope
Cystoviridae	Bacteria	Bacteriophage phi6	3	13	Linear	Icosahedral	Yes
Totiviridae	Fungi & protozoa	Saccharomyces cerevisiae virus L-A	1	5–7	Linear	Icosahedral	No
Chrysoviridae	Fungi	Penicillium chrysogenum virus	4	13	Linear	Icosahedral	No
Partitiviridae	Fungi & plants	Penicillium stoloniferum virus S	2	3–4	Linear	Icosahedral	No
Birnaviridae	Invertebrates & vertebrates	Infectious bursal disease virus	2	5–6	Linear	Icosahedral	No
Reoviridae	Fungi, plants, invertebrates, & vertebrates	Rotavirus A	10–12	19–31	Linear	Icosahedral	No

intact in the cell while messenger RNAs are transcribed from each genome segment by viral RNA polymerase molecules packaged with the genome RNAs; the resulting messenger RNA is extruded into the cytoplasm. The capsid therefore provides a structure that positions each RNA polymerase molecule and the various genome segments, and this allows reiterated transcription of each segment and directional transfer of the mRNAs to the cytoplasm.

Different members of the *Reoviridae* family can infect vertebrates, invertebrates, plants, or fungi; this is one of the few virus families (along with the negative-strand *Rhabdoviridae* and *Bunyaviridae*) whose members can replicate in such a wide range of host organisms. Viruses with negative-strand or double-stranded RNA genomes import their RNA polymerases into the host cell, and the capsids serve as tiny intracellular machines for synthesizing viral messenger RNAs. Because of this lifestyle, and because they make their own RNA replication machinery, these viruses may be less dependent on the specific cellular environment where they replicate, giving them freer range to adapt to a wide variety of organisms.

Viruses with a reverse transcription step in their replication cycle can have either RNA or DNA genomes

Three virus families are grouped separately from other viruses because they have a reverse transcriptase that makes a DNA copy of an RNA during their replication cycle (Table 3.6). Therefore, unlike all other viruses, genome sequences can occur as both DNA and RNA at different stages of replication. *Retroviridae* package the RNA copy of the genome, while *Hepadnaviridae* and *Caulimoviridae* package the DNA copy of the genome. Viruses that use reverse transcriptase package this enzyme in the virion, as viruses with negative-strand or

The length of viral and cellular genomes

One turn of the DNA double helix contains 10 base pairs and is 3.4 nanometers (nm) in length. This means that a double-stranded viral DNA genome of 2000 nucleotides measures approximately 700 nm (0.7 micrometers, μm) from end-to-end (Figure 3.2). A viral genome containing 1.2 million base pairs is 600 times longer, or 420 μm (0.4 mm). Thus, just three molecules of the DNA genome of mimivirus, the largest known virus, stretched out end-to-end, would cover a length of more than 1 millimeter. In comparison, the total length of the genome contained within the 23 pairs of chromosomes in a single human cell, 3 billion base pairs, would be approximately 1 meter if stretched out end-to-end. Because there are approximately 10^{13} cells in a single human being, the aggregate length of all of the DNA in a single individual is 10^{13} meters (10^{10} kilometers), or 50 times the distance between the Earth and the sun!

double-stranded RNA genomes package their respective RNA polymerases in the virion.

All retroviruses package two identical copies of the viral genome. This is unique to retroviruses, and it may be important in the reverse transcription step that makes a full-length DNA copy of the genome within the capsid once it has entered the cell (see Chapter 28). During reverse transcription, the growing DNA chain is transposed ("jumps") from one end of the RNA template to the other end; chain transfer can alternatively occur to the second packaged copy of the RNA genome. Having a second copy may be a backup mechanism in case one of the RNA molecules is damaged. Retroviruses are found only in vertebrates, but a variety of retrovirus-like elements, some of which make virus-like particles, have been found in plants, invertebrates, and fungi.

Table 3.6 Viruses with a reverse transcriptase

Virus group	Hosts	Example	Genome segments	Genome size (kb) and topology		Capsid symmetry	Envelope
Caulimoviridae	Plants	Cauliflower mosaic virus	1	7–8	Circular ds DNA	Icosahedral	No
Hepadnaviridae	Vertebrates	Hepatitis B virus	1	3–4	Circular ss/ds DNA	Icosahedral	Yes
Retroviridae	Vertebrates	Human immuno-deficiency virus type 1	1 (dimer)	7–13	Linear + RNA	Icosahedral	Yes

Hepadnaviruses have an unusual DNA genome: it is circular and mostly double-stranded, with a single-stranded region of variable length. This results from interruption of synthesis of the second DNA strand by the viral reverse transcriptase (DNA polymerase) during virion maturation (see Chapter 30). Caulimoviruses have a circular double-stranded DNA genome that contains strand breaks at specific sites. Viruses in these two families carry out the reverse transcription step at the end of the replication cycle and package the resulting DNA molecule in the virion, whereas retroviruses carry out the reverse transcription step at the beginning of the replication cycle.

Satellite viruses and satellite nucleic acids require a helper virus to replicate

A number of very small virus-like genomes replicate only in association with other viruses. These genomes can replicate only if present in a cell that is simultaneously infected with a **helper virus**; therefore they have been called **satellite viruses** and **satellite nucleic acids** (Table 3.7). Genomes of satellite viruses encode their own capsid proteins, while genomes of satellite nucleic acids either contain no coding regions or encode only non-capsid proteins; their genomes are encapsidated by structural proteins of the helper virus. Satellite nucleic acids are presently classified based on the nature, length, and topology of their genomes.

Most known satellite viruses and nucleic acids replicate in plants, but an insect satellite virus (chronic bee paralysis virus), several double-stranded RNA satellite nucleic acids of fungi and protozoa, and a human satellite virus (hepatitis delta virus) are known. The genomes of all known satellite viruses are less than 2 kb in length, and those that contain no protein-coding regions are less than 700 nucleotides in length. Strictly speaking, adeno-associated viruses, which are single-stranded DNA viruses whose replication depends on coinfection with adenoviruses or herpesviruses, should be included in this category, but those viruses are closely related to numerous other parvoviruses that replicate independently of helper viruses.

Viroids do not code for proteins, but replicate independently of other viruses

Viroids (Table 3.7) are small, virus-like RNAs that are similar to satellite nucleic acids, but viroids are able to replicate and cause disease in a variety of plants in the absence of helper viruses. Viroid genomes are single-stranded, circular RNAs from 250 to 400 nucleotides long that do not code for any proteins and are not encapsidated. Viroids use cellular DNA-dependent RNA polymerases to replicate their RNA genomes. These genomes have a high degree of internal base pairing and therefore resemble double-stranded RNAs over much of their length.

Some viroids, as well as hepatitis delta virus RNA and the small, circular single-stranded RNA satellites, have the ability to self-cleave and self-ligate. These enzymatic activities play important roles in viroid replication, during which RNA oligomers are synthesized. These oligomers can cleave themselves at specific sites to generate linear monomers, which then self-ligate into circular monomers. These enzymatic activities link viroids and circular RNA satellites to the ancient RNA world (see below), in which it is postulated that RNAs were able to replicate themselves in the absence of proteins by serving as enzymes (**ribozymes**).

THE EVOLUTIONARY ORIGIN OF VIRUSES

Because viruses need living cells for their replication, they must have evolved hand-in-hand with their hosts as these organisms evolved. But how and when did viruses first appear? Because viruses leave no fossil record, we can only speculate on their first appearance based on what we know about their genomes, their proteins, and their replication strategies.

The first steps in the development of life on Earth: the RNA world

Within perhaps 300 million years after the formation of the Earth, once the oceans had formed and cooled, macromolecules may have arisen by spontaneous

Table 3.7 Satellite viruses, satellite nucleic acids, and viroids

Virus group	Hosts	Example	Genome	Genome segments	Genome size (kb) and topology		Function of encoded protein
Satellite viruses	Plants & invertebrates	Tobacco necrosis satellite virus	ss RNA	1–3	1–3	Linear	Capsid protein
Single-stranded DNA satellites	Plants	Tomato leaf curl virus satellite DNA	ss DNA	1	0.7–1.3	Circular	Unknown function
Double-stranded RNA satellites	Fungi & protozoa	M Satellites of saccharomyces cerevisiae L-A virus	ds RNA	1	0.5–1.8	Linear	Toxin
Circular single-stranded RNA satellites	Plants	Tobacco ringspot virus satellite RNA	ss RNA	1	0.35	Circular	No encoded proteins
Small linear single-stranded RNA satellites	Plants	Cucumber mosaic virus satellite RNA	ss RNA	1	0.7	Linear	No encoded proteins
Large single-stranded RNA satellites	Plants	Bamboo mosaic virus satellite RNA	ss RNA	1	0.8–1.5	Linear	Replication
Deltavirus	Humans	Hepatitis delta virus	ss RNA	1	1.7	Circular	Replication and packaging
Viroids	Plants	Avocado sunblotch viroid	ss RNA	1	0.25–0.4	Circular	No encoded proteins

polymerization of simple organic molecules such as ribonucleotides and amino acids. Some primitive RNAs may have by chance contained nucleotide sequences that enabled them to bind to other ribonucleotides and form covalent bonds. This would enable such RNAs to extend themselves, and by using base-pairing, replicate themselves in the absence of any protein-based enzymes. These RNAs may have represented the first self-replicating macromolecules in the **RNA world** (see Figure 3.3).

As these RNAs developed and evolved, some of them may have acquired the property of binding covalently to a limited variety of amino acids, a role presently carried out by transfer RNAs and amino acid-activating enzymes. Simple polypeptides could have been assembled by such adaptor RNAs on templates of self-replicating RNAs, and those polypeptides that helped to protect the RNAs and increase their rate of replication would have enhanced selection and further multiplication of their template RNAs. Thus a self-replicating RNA linked to a primitive protein-synthesizing system developed. This could have eventually led to the selection of RNAs that coded for a polypeptide with RNA polymerase activity, increasing the rate and efficiency of RNA replication beyond that of a purely RNA-based enzymatic activity.

At some point, these RNAs and their associated polypeptides may have become localized and concentrated within lipid droplets. Polypeptides coded by these RNA genomes could be inserted into the lipid membranes and control the passage of small molecules and ions, forming the first primitive cells. These primitive cells would have contained small RNA genomes that coded for only a few polypeptides, whose main functions were to replicate the genome RNA and to provide for the entry of substrate nucleotides and amino acids.

Viroids and RNA viruses may have originated in the RNA world

Viroids and small single-stranded RNA satellites share some unique characteristics that indicate that they may be relics of the RNA world that survive to this day. These small, circular RNAs are able to catalyze some steps (cleavage and ligation) in their replication in the absence of proteins, but depend on RNA polymerases coded by cellular genes or the genes of helper viruses. They could be descended from self-replicating RNA molecules that arose in the RNA world.

Viruses with RNA genomes may also have arisen from primitive RNA-based life forms. They are the only known present-day biological entities with RNA genomes. To replicate, these viruses must encode their own RNA-dependent RNA polymerases, because such enzymes are not made by their host organisms. These enzymes may well have been inherited from the RNA world.

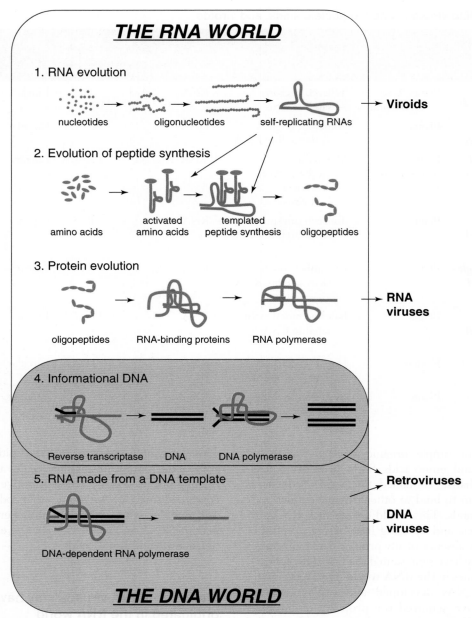

Figure 3.3 Possible steps in the evolution of life and the emergence of viruses. This figure shows a hypothetical series of steps that could have led from pools of complex organic molecules such as nucleotides and amino acids to the generation of self-replicating RNAs, which over time could have evolved to organize templated protein synthesis and eventually self-sustaining life forms. At some point these macromolecules would have been confined within lipid membranes to form cells. Overlapping rounded boxes enclose molecules that would have been part of the RNA world and the present DNA world. At the right are shown possible origins of different kinds of viruses. See text for details.

The transition to the DNA-based world

At some point during the evolution of the RNA world, polypeptides produced by self-replicating RNAs could have acquired the abilities to (1) catalyze the production of deoxyribonucleotides from ribonucleotides and (2) use these deoxyribonucleotides to make DNA copies of the RNA genomes (i.e., the first reverse transcriptase). In this way, the presently known DNA world could

have emerged. Because DNA molecules are more stable than RNAs, they would have a tendency to accumulate within primitive RNA-based cells even before they could be replicated.

One of the polypeptide products made by a self-replicating RNA, perhaps a mutant of the RNA-dependent RNA polymerase or the reverse transcriptase, could then have acquired the ability to act as a primitive

DNA-dependent RNA polymerase. This would allow DNAs to take over the function of storing the information for protein synthesis, while the RNA copies could continue to serve as messengers and amino-acid-binding adaptors for templating protein synthesis.

A final development would have been the generation of a polypeptide with DNA polymerase activity, enabling the replication of the DNA genomes. The fundamental properties of cellular life as we know it would have then been established. Self-replicating RNAs could have coexisted for some time alongside the DNA genomes in these primitive cells.

The steps postulated to have taken place in the evolution from simple self-replicating RNAs to DNA-based cellular life forms imply the existence of nucleic acid polymerases of four types:

1. RNA-dependent RNA polymerases, to help replicate RNAs in the RNA world
2. RNA-dependent DNA polymerases (reverse transcriptases), to transfer the information in these RNAs to DNA and to begin the DNA world
3. DNA-dependent RNA polymerases, to express the information in the DNA via messenger RNAs
4. DNA-dependent DNA polymerases, to replicate the DNA genomes

Retroviruses could have originated during the transition to DNA-based cells

At this stage of evolution, retroviruses and other viruses that use a reverse transcriptase could have emerged. A copy of the RNA template for reverse transcriptase could have been packaged along with reverse transcriptase, allowing transmission of this gene from cell to cell by a virus. This could have also have hastened the transformation of cells from RNA-based to DNA-based genes. As DNA-based cells evolved, they would no longer need to maintain genes for reverse transcriptase activities, and they therefore could have lost genes that code for these enzymes. However, viruses that use reverse transcriptases to replicate would have continued to exist and replicate in these DNA-based cells, using their DNA-dependent RNA polymerases to transcribe the DNA form of the retrovirus genome.

Small- and medium-sized DNA viruses could have arisen as independently replicating genetic elements in cells

A number of small DNA viruses replicate their genomes and express their messenger RNAs by using host cell DNA and RNA polymerases. Their limited number of viral genes code for (1) virion structural proteins, (2) proteins that stimulate production of cellular enzymes involved in DNA replication, and (3) proteins that recognize and bind to the viral DNA and assemble the cellular DNA replication machinery (for examples, see bacteriophage φX174, parvoviruses, polyomaviruses, and papillomaviruses, in Chapters 6, 20–22).

It is possible that these small DNA viruses originated from fragments of cellular DNA that broke away from the cell genome and were able to replicate independently within the cell as plasmids. Such DNA fragments must have contained a nucleotide sequence that was recognized by the cellular DNA replication machinery as a replication origin.

As they evolved, these independently replicating DNAs could have acquired genes coding for proteins that bind to the DNA, providing a package for its extracellular protection and delivery. Such DNAs would therefore have become primitive viruses, able to jump from one cell to the next. As they evolved, they would have undergone selection for the production of a stronger capsid that could better package and protect the genome DNA.

Capsid proteins that recognize and bind to receptors on the surface of susceptible cells would have been selected. As these DNA viruses replicated and evolved, virus-coded proteins that specifically bind to the viral origin of replication and attract the cellular DNA replication machinery would ensure preferential replication of viral DNA. Some of the larger DNA viruses could have evolved from these small viruses through recombination and gene duplication events.

Large DNA viruses could have evolved from cellular forms that became obligatory intracellular parasites

An alternative origin has been suggested for some of the largest DNA viruses. These viruses contain hundreds of genes that supply many functions usually provided by cellular genomes. It is possible that they began as independently replicating cells, which at some point parasitized other cell types and became obligatory intracellular parasites. With time, some of the progeny of these intracellular parasites would lose genes that were no longer needed, and other genes that can package the viral genome and transmit it to susceptible cells could have evolved.

This argument makes most sense for the giant mimivirus (Chapter 27) and the poxviruses (Chapter 26). These viruses package within the virion core a large number of enzymes needed to express and process viral early messenger RNAs. These RNAs are synthesized within the cores after they enter the host cell, and at least for poxviruses, mRNAs are transported through the core membrane into the cytoplasm. This strongly resembles the mechanisms by which cellular RNAs are expressed, modified, and transported from the cell nucleus to the cytoplasm.

After the production of poxvirus early mRNAs, the poxvirus core membrane is broken down, allowing viral DNA replication to be directed by early viral proteins.

This resembles the breakdown of the nuclear membrane during mitosis. At the end of the virus life cycle, lipid membranes reform around the progeny DNA and associated proteins to make the virus core, just as nuclei are reformed by enclosure of chromatin within a lipid membrane after mitosis.

An interesting twist on this hypothesis has been proposed. Instead of being derived from a degenerate intracellular parasite, a large DNA virus could have been the precursor to the nucleus in the evolution of the first eukaryotic cells. This argument suggests that a large, enveloped DNA-containing virus that had evolved to code for most of the functions required for its replication could have established itself persistently in a prokaryotic cell. Eventually, the resident viral genes, localized within an internal membrane system, could have taken over the functions of the prokaryotic genome, which would have subsequently been lost. In this way, a nucleus would be established, producing the first eukaryotic cell.

These arguments about the origin of viruses are only speculations

Because there are very few hard data on which to base theories of the origin of viruses (and the origin of life itself!), the above ideas are really only informed speculation. However, it is striking that:

- Many viruses have genomes not found within any other extant life form (single-stranded DNA, positive- and negative-strand RNA, double-stranded RNA).
- Some viruses code for replication proteins not normally found in other organisms (RNA-dependent RNA polymerases, reverse transcriptase).
- The smallest viruses are almost totally dependent on their host cells for the enzymatic machinery to express and replicate their genomes, while the largest viruses require only substrates, energy, and protein synthesis machinery.

The great variation in genome design and replication strategies among known viruses suggests that there could very well have been multiple origins of viruses during the evolution of life. Could new viruses still be arising by some of these mechanisms?

KEY TERMS

Family	RNA world
Genus (plural: genera)	Satellite nucleic acid
Helper virus	Satellite virus
Lineage	Species
Order	Taxonomy
Ribozyme	Viroid

FUNDAMENTAL CONCEPTS

- Viruses are classified into families, genera, and species based on characteristics of their genomes, capsids, and replication strategies.
- Members of most but not all virus families infect only organisms within one of the six major divisions of life.
- Viruses with single-stranded DNA genomes are uniformly small and none has a lipid envelope.
- Viruses with double-stranded DNA genomes are widespread among all organisms except plants, and they can have genomes ranging from 5000 to more than a million base pairs.
- Viruses with positive-stand RNA genomes are predominant in the plant kingdom and common among vertebrates.
- Viruses with negative-strand RNA genomes are among the major infectious disease agents of humans, but are not found in prokaryotes.
- Viruses with double-stranded RNA genomes can have up to 12 different genome fragments; they are transcribed by a packaged RNA polymerase.
- Viruses with a reverse transcriptase are known only from vertebrates and plants.
- Satellite viruses and viroids have very small genomes of from 250 to 1800 nucleotides, and are highly dependent on host cells or other viruses for packaging as well as replication.
- Viruses with different genome types and replication strategies may have originated independently at different times during the evolution of life.

REVIEW QUESTIONS

1. What molecular criteria are used to classify viruses?

2. Genetic relatedness is used to classify viruses. What other characteristics are used for virus classification?

3. Name three ways viruses adapt to the host organisms in which they replicate.

Virus Entry

Ari Helenius

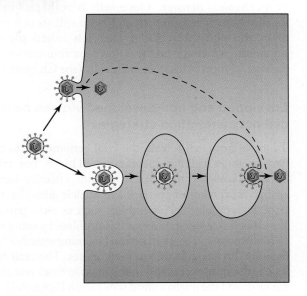

STEPS IN VIRUS ENTRY

Virions bind to specific receptors on the cell surface.

Virions enter the cell either in specialized vesicles or by direct penetration through the plasma membrane.

Virions or nucleocapsids are transported within the cell to sites of transcription and replication (cytoplasm or nucleus).

Viral genomes are released into the cell by passage through the vesicle membrane or by disintegration of capsids within the cell (uncoating).

PENETRATION THROUGH CELLULAR MEMBRANES

Membrane penetration is mediated by conformational changes of virion proteins.

These changes are triggered by:

- *interaction of viral surface proteins with cellular receptors in membranes, or*
- *exposure to low pH inside endosomes*

Enveloped viruses penetrate by fusion of the virus envelope with either:

- *the plasma membrane, or*
- *membranes of intracellular vesicles*

Non-enveloped viruses enter the cytosol by either:

- *forming transmembrane channels through which the viral nucleocapsid or genome can pass, or*
- *inducing lysis of endocytic vesicles*

INTRACELLULAR TRANSPORT

Virions and virus nucleocapsids are transported within the cell along microtubules using the cellular transport motor dynein.

Viruses that replicate in the nucleus transport their genomes into the nucleus via the nuclear pore complex.

How do virions get into cells?

Virus particles have a single mission: to mediate the transport of the viral genome from an infected cell to an uninfected cell, thus allowing the spread of infection from cell to cell and from organism to organism. The transfer process begins with the assembly of virions in the infected cell, followed by their release, transmission within the organism or in the environment to uninfected cells, and finally, entry into a new host cell.

This chapter focuses on the final stage of the journey for viruses of vertebrates: entry of the virion into a host cell, its transport to the proper intracellular location, and disassembly of the virion to release the viral genome. This complex process involves several steps and exploits a variety of cellular activities ranging from endocytosis to nuclear import. In many ways, viruses work like Trojan horses; they can only infect a cell that provides extensive assistance during entry. Viruses follow the molecular rules and regulations that control events inside the cell, but manipulate these pathways to serve their own ends. In this way, viruses can take advantage of the cell while remaining virtually invisible to the antiviral defenses of the organism.

Viruses that infect bacteria, archaea, algae, and higher plants use mechanisms of entry quite distinct from those of animal viruses, because access to cells is restricted by the presence of cell walls. Bacteriophages have evolved complex mechanisms for puncturing this barrier and penetrating the membranes of bacteria. Some of these are described in detail in Chapters 5 through 8. To enter plant tissues and access cells, plant viruses rely on insect bites, grafting, and other forms of mechanical damage. Once cells have been infected locally, viruses can then spread from cell to cell in the plant through intercellular channels called **plasmodesmata**, which normally allow communication and passage of metabolites between cells (see Chapter 10).

Enveloped and non-enveloped viruses have distinct penetration strategies

To move from cell to cell, the viral genome has to cross two or more cellular membranes. This is not trivial, because biological membranes provide effective barriers against movement of proteins and nucleic acids, let alone large macromolecular assemblies such as virus particles. Enveloped viruses circumvent this problem by using **vesicle transport** and coupled **fusion** ("coming together") and **fission** ("breaking apart") of membranes. The viral envelope is the transport vessel in the passage from one cell to the next, and the nucleocapsid is its cargo (Figure 4.1). The viral envelope forms around the nucleocapsid and other internal components such as matrix proteins during **budding** at a cellular membrane. Often budding takes place at the (external) plasma membrane, but some viruses use the membranes of the nucleus, the endoplasmic reticulum, the intermediate compartment, endosomes, or the Golgi complex. The enveloped virion is then released into the extracellular space directly, or after transport through the secretory pathway.

Delivery of the nucleocapsid cargo into the **cytosol** of the target cell occurs when the viral envelope fuses with a cellular membrane (Figure 4.1, right). Fusion can take place either at the plasma membrane or at the membranes of early or late **endosomes**. Endosomes are membranous organelles in which the virion is enclosed after undergoing **receptor-mediated endocytosis** (see below). The membrane fusion pathway has the advantage that viral macromolecules do not need to pass directly across the hydrophobic interior of any cellular membrane. In fact, in a topological sense, the capsid never leaves the cytoplasmic compartment. Given the simplicity and elegance of this strategy, it is not surprising that the entry of enveloped viruses such as parainfluenzaviruses or togaviruses is often very efficient.

Non-enveloped viruses must use other mechanisms to penetrate membrane barriers because they have no lipid membrane. To exit from the infected cell, non-enveloped viruses typically rely on rupture of cellular membranes. Cell death is usually necessary to release virus particles trapped in the nucleus or the cytoplasm. These viruses enter host cells in a variety of ways. Adenoviruses, for example, are taken up into endosomes, where they proceed to rupture the endosomal membrane, releasing the virion and other endosomal contents into the cytosol. Picornaviruses, on the other hand, undergo conformational changes during binding to surface receptors or after entering endosomes, so that viral proteins form membrane channels that allow the genome direct access to the cytosol.

Some viruses can pass directly from cell to cell

The release of virions from an infected cell is not always necessary for transmission of infection to a neighboring cell. Particles still associated with the surface of the

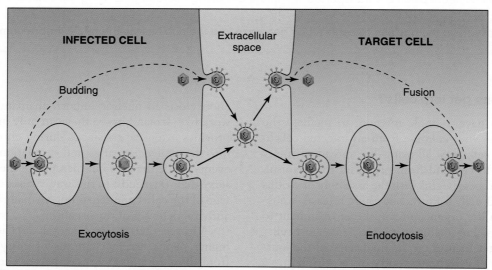

Figure 4.1 Exit and entry of enveloped viruses occurs by vesicle transport. Left: formation of a simple enveloped virus particle by budding at the plasma membrane (above) or at an internal cellular membrane (below), followed by release via exocytosis. Center: virions in the extracellular space. Right: uptake by fusion at the plasma membrane, or within a vesicle after endocytosis, releases the viral nucleocapsid into the cytosol.

infected cell may reach neighboring cells through contacts established via **filopodia** and other actin-based extensions. In such cases, the infected host cells play an active role in transmitting the infection, with minimal exposure of the virus to the extracellular environment. This can also help to avoid recognition of the virus by the host immune system. Some retroviruses have been shown to generate a special region where viruses are preferentially assembled and transmitted—a **virological synapse**—in the contact area between the virus-producing cell and its neighbor.

In a few cases, infection can spread from one cell to the next without formation of a complete virus particle. Measles and some other enveloped viruses induce fusion of infected cells with neighboring uninfected cells. This results in the formation of multinucleated cells called **syncytia**. The same viral proteins that serve as fusion agents in the virion envelope are responsible for cell–cell fusion when they are incorporated into the plasma membrane of the infected cell. Transmission of infection in this way is obviously limited to cells located next to each other; spread of infection between organisms cannot occur by this mechanism unless there is exchange of cells.

In plants, the cell wall prevents diffusion of virus particles between cells. Instead, infection spreads directly through special channels, called plasmodesmata, which connect the cytosolic compartments of neighboring cells. These channels traverse the cell wall and provide a conduit for transport of viral genomes and other macromolecules. In some cases, only the viral genome is transported in plasmodesmata; in other cases, assembled virions or nucleocapsids are transported. For some viruses, the plasmodesmata are too narrow to allow passage; virus-encoded "movement proteins" are expressed that cause expansion of plasmodesmata into wider channels.

A variety of cell surface proteins can serve as specific virus receptors

The first step in infection of a target cell is binding of the virion to the cell surface. Many different molecules located in the plasma membrane can serve as **attachment factors** and **receptors** for viruses (Figure 4.2). These are proteins, lipids, or carbohydrates that have their own cellular functions, and are only fortuitously exploited by viruses; therefore they tend to be highly conserved, and are unlikely to undergo mutations that could impede virus uptake. Since virus binding is a prerequisite for infection, the presence or absence of an appropriate receptor is one of the most important factors determining which species and which cell types can be infected by a given virus.

Cell surface components involved solely in binding of virions to cells are called attachment factors. For many viruses these include the carbohydrate moities of **proteoglycans** or **glycoproteins**. Receptors also bind to virions, but they play a more active role by inducing conformational changes in the particles, triggering cell signaling pathways, or mediating endocytosis.

Cell surface proteins exploited by viruses as receptors include members of the **immunoglobulin family** (CD4, PVR, CAR, CD46); the low-density lipoprotein (LDL) receptor family; a variety of multimembrane-spanning transport proteins; **integrins**; and others. Some viruses bind to specific carbohydrate groups present on glycoproteins, proteoglycans, and **glycolipids**. For example, influenza viruses and paramyxoviruses bind to **sialic acid** residues present on glycoproteins and glycolipids. While this allows them to attach to virtually any cell, binding may also occur to cells or membrane fragments that cannot support productive infection. To avoid such

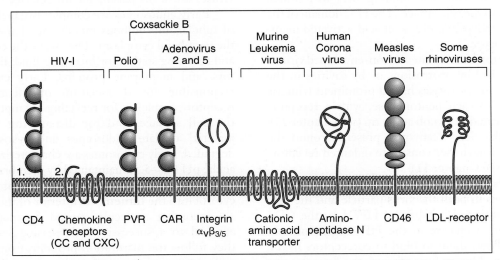

Figure 4.2 Cell surface glycoproteins used as virus receptors. Schematic views of several known virus receptors, shown protruding from the plasma membrane at the cell surface. Members of the immunoglobulin family have typical domains shown as green spheres. PVR = poliovirus receptor; CAR = Coxsackie virus and adenovirus receptor; LDL = low density lipoprotein.

non-productive binding, these viruses carry **neuraminidases** in their envelopes. Whenever necessary, these "receptor-destroying enzymes" release the virions by cleaving the sialic acid residues to which they are bound.

A number of viruses (e.g., HIV-1 and adenovirus) require more than one type of receptor (Figure 4.2). The primary receptor is usually sufficient for attachment and inducing conformational changes, the secondary receptor (or "coreceptor") may be needed to trigger endocytosis or penetration. Association with the receptors can occur sequentially because the binding sites for the coreceptor are exposed only after interaction of the virion with the primary receptor. For example, the envelope glycoproteins of HIV-1 and other lentiviruses first bind to CD4, a glycoprotein of the immunoglobulin family. Only after CD4 binding do these viruses engage the chemokine receptors CCR4 or CXCR5, which allow penetration by membrane fusion (see Figure 4.2 and Chapter 29).

Receptors interact with viral glycoproteins, surface protrusions, or "canyons" on the surface of the virion

Binding of viruses to their receptors occurs via specific binding sites exposed on the surface of the virion. Because each virus particle carries many such sites, there are usually multiple receptor-virus contacts. Attachment can therefore be exceptionally tight even though the affinity of individual interactions may be rather low. Enveloped viruses have specific surface spike glycoproteins that serve as the binding sites for receptors. Some viruses devote a glycoprotein exclusively to attachment. More often, the same glycoproteins mediate both attachment and membrane fusion. For example, the HA_1 domain of the trimeric influenza virus hemagglutinin (HA) protein contains the sialic acid binding sites with which the virus binds to the cell surface. The HA_2 domain of the protein serves independently as an acid-activated membrane fusion factor.

The receptor-binding sites in non-enveloped viruses are formed either by protrusions or by cavities in the capsid surface. Adenoviruses have a prominent trimeric fiber attached to each penton base, with a receptor-binding site located in a knob at the tip (see Chapter 23). In contrast, the deep "canyons" present around the pentons in picornaviruses constitute indented receptor-binding sites (see Chapter 11).

The interaction between receptor and virus often leads to changes in both the virus particle and the cell. As already mentioned, binding of HIV-1 to the CD4 protein triggers a change in the HIV-1 glycoprotein trimers that allows them to bind to coreceptors in the cell membrane. This in turn triggers the fusion-active conformation in the spikes and leads to penetration. Attachment of the adenovirus fibers to the primary receptor protein, CAR, triggers fiber dissociation. This exposes specific amino acid sequences in the penton base proteins that are recognized by binding sites in integrins. Interaction with integrins is needed for internalization by endocytosis and subsequent penetration of the adenovirus virion into the cytosol.

Attachment of adenovirus to the $\alpha_v\beta_3$ integrins also activates a signaling cascade inside the cell that temporarily increases endocytic activity and prepares the cell for the incoming virus. The activation of cellular signaling pathways is an essential process for the entry or replication of many viruses. Signaling pathways promote changes in cell architecture and induce macropinocytosis and caveolar/lipid raft-dependent endocytosis, enabling crucial steps in the entry of certain viruses (see below).

Many viruses enter the cell via receptor-mediated endocytosis

After binding to cell receptors, some viruses are able to penetrate directly into the cytosol by membrane fusion and other mechanisms. However, the majority of viruses depend on endocytosis to be first carried into cytoplasmic vacuoles, from which they can then enter the cytosol. Endocytosis occurs frequently via **clathrin-coated pits** and vesicles (Figure 4.3). These vesicles ferry virus particles through the cortical network of cytoskeletal fibers, which could otherwise constitute a major obstacle, and transport them away from the cell surface and toward the cell nucleus. A few minutes after leaving the cell surface, virions are delivered into **early endosomes**, as are other substances taken up by clathrin-coated vesicles. After internalization, the virus particle is still separated from the cytosol by the barrier of the endosomal membrane, which it can overcome by the same strategies that other viruses use at the plasma membrane (see below).

Early endosomes are complex organelles composed of tubular and bulbous membrane elements, found in the peripheral cytoplasm. They serve the cell as a sorting and recycling station for incoming fluid, **ligands**, receptors, and membrane proteins. Early endosomes are responsible for dissociation of incoming ligand-receptor complexes, for recycling of empty receptors to the cell surface, and for dispatching non-recycled material to late endosomes and **lysosomes** to be degraded. Early endosomes are characterized by a mildly acidic pH (6.3–5.5) maintained by membrane-associated vacuolar-type ATPases that pump protons from the cytosol into the **lumen** (the interior of the vesicle).

Incoming molecules sorted for degradation are exposed to a successively decreasing level of pH as they follow the maturation program from early to late endosomes, and ultimately to lysosomes. In the process, late endosomes are formed that have the characteristics of multivesicular bodies; they contain intralumenal

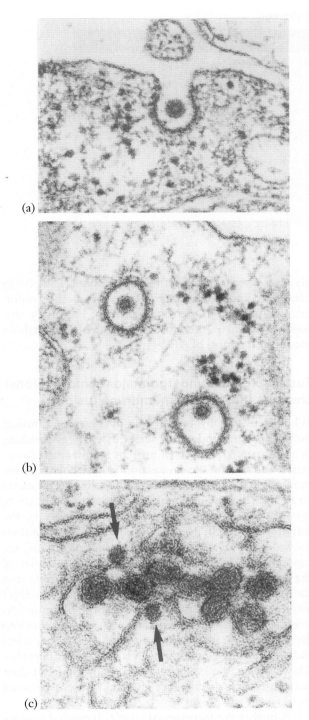

(a)

(b)

(c)

Figure 4.3 Receptor-mediated endocytosis and acid-activated penetration of Semliki Forest virus. (a) Virus particles bound to the cell surface via their spike glycoproteins are trapped in clathrin-coated pits. The clathrin coat is seen in these thin-section electron micrographs as a densely stained layer on the cytosolic side of the plasma membrane. (b) The coated vesicles detach from the plasma membrane and the clathrin coat dissociates. (c) The vesicle fuses with early endosomes, from which the nucleocapsid is extruded into the cytosol (see arrows) upon fusion of the viral envelope with the endosomal membrane at a pH below 6.2.

vesicles, they move by microtubule-mediated transport into the perinuclear region of the cell, and they fuse with lysosomes. For viruses such as influenza virus and some rhinoviruses, which require transport to late endosomes before penetration can occur, proper maturation of the endosomal vacuoles is essential.

Not all viruses that enter the cell inside vesicles use clathrin-coated vesicles. For example, simian virus 40, a (non-enveloped) polyomavirus, enters via mechanisms mediated by **caveolae** and **lipid rafts**. Small, flask-shaped invaginations in the plasma membrane are enriched in cholesterol and sphingolipids. In some cases, these invaginations are caveolae, which contain cellular proteins called **caveolins** and **cavins**. Alternatively, the invaginations are formed by the induction of membrane curvature caused by the tight, multivalent interaction of capsid proteins on the virion with receptor gangliosides on the cell surface. Caveolae and the caveolar pathway are thought to play a role in signaling and in the uptake of cholesterol and selected ligands. After transport through early and late endosomes, simian virus 40 and other polyomaviruses are brought to the endoplasmic reticulum, where penetration to the cytosol takes place.

Another endocytic mechanism used by viruses is called **macropinocytosis**. Vaccinia virus, a large, enveloped DNA virus of the poxvirus family, triggers a complex signaling cascade that induces transient ruffling of the plasma membrane followed by the internalization of fluid and virus particles in large vacuoles called macropinosomes. Penetration into the cytosol occurs from these vacuoles.

Passage from endosomes to the cytosol is often triggered by low pH

Endosomes are a major site of virus penetration into the cytosol. The trigger for penetration is usually low pH, which induces a conformational change in the spike glycoproteins of enveloped viruses or in the capsids of non-enveloped viruses. Low pH signals to the virus that it has reached an intracellular location and that the penetration reaction should now begin. Virus particles attached to the plasma membrane can be artificially activated to penetrate prematurely simply by lowering the pH in the medium bathing the cells, showing that low pH is all that is needed as a cue for penetration. Most viruses adjust the timing of their penetration by carefully titrating the activation threshold to a pH in the 6.5–5.3 range. Depending on the threshold pH, conversion occurs 5–30 minutes after uptake into early and late endosomes, before the virus can be delivered to lysosomes.

Table 4.1 lists viruses known to enter through an endosomal, acid-triggered pathway, and viruses that use pH-independent entry. pH-independent penetration

Table 4.1 Acid-activated and pH-independent viruses

Acid-activated penetration	pH-independent penetration
Orthomyxoviruses (influenza A and B viruses)	Most retroviruses (HIV-1)
Togaviruses (Semliki Forest and Sindbis viruses)	Herpesviruses (HSV-1)
Flaviruses (tick-borne encephalitis virus)	Paramyxoviruses (Sendai, measles)
Rhabdoviruses (rabies, vesicular stomatitis virus)	Some coronaviruses
Filoviruses (Ebola virus)	Some picornaviruses (poliovirus)
Bunyaviruses (Uukuniemi virus)	Reoviruses
Adenoviruses (types 2 and 5)	
Some picornaviruses (rhinovirus)	
Some retroviruses	

is triggered by interactions between the virus and its specific receptor. Consequently, many of these viruses can penetrate directly through the plasma membrane, where they first encounter their receptors. Some viruses (e.g., adenoviruses) need both a receptor-induced and a low pH-induced conformational change to penetrate successfully.

Membrane fusion is mediated by specific viral "fusion proteins"

Penetration of enveloped viruses involves a membrane fusion event mediated by viral glycoproteins. Viral fusion proteins are generally **type I transmembrane proteins** most of whose mass lies external to the viral membrane, and they are usually organized as oligomers. Because of their unusual function and their central role in infection, many fusion proteins have been characterized in great detail.

Like other viral glycoproteins, fusion proteins are synthesized, folded, and assembled into oligomers in the endoplasmic reticulum of infected cells. To become fusion-active, they generally must undergo several maturation steps. In transit from the endoplasmic reticulum through the Golgi complex to the plasma membrane, fusion proteins may be cleaved by cellular proteases. An example is the influenza virus hemagglutinin, HA, which is cleaved into a peripheral N-terminal fragment called HA_1 and a C-terminal transmembrane fragment called HA_2 (Figure 18.4; see also Figure 2.12). This cleavage renders HA conformationally metastable, like a loaded spring; it responds to low pH with a major conformational change.

In the case of HA and some other viral spike glycoproteins, the activating cleavage occurs just to the N-terminal side of a hydrophobic sequence of about 10–12 amino acid residues. In the cleaved protein, this conserved sequence, the **fusion peptide**, often constitutes the N-terminus of the transmembrane subunit, and

plays a central role in the fusion reaction. The energy needed to induce fusion is derived from the conformational change in the fusion protein. Fusion by viral fusion proteins is therefore independent of metabolic energy in the form of ATP.

Fusion proteins undergo major conformational changes that lead to membrane fusion

When activated by low pH or by receptor binding, fusion proteins undergo a series of changes resulting in a dramatic alteration in protein architecture, properties, and membrane association. These changes have been studied in most detail for the trimeric influenza virus HA (Figure 4.4). Immediately after exposure to low pH in late endosomes (step 1, Figure 4.4), the fusion peptides in the HA_2 subunits (red filaments shown at top of the extended polypeptide chains) are exposed and insert into the outer bilayer leaflet of the target membrane (step 2). The HA_1 subunits (not shown in Figure 4.4) remain attached but no longer form a trimeric domain at the top of the oligomer. As a result, the viral HA_2 protein is now attached both to the viral membrane (via the transmembrane sequence) and to the target membrane (via the fusion peptide). This attaches the membranes to each other but does not cause them to fuse because the distance between them is still too great.

A further conformational change is needed to close the gap. Several of the HA molecules come together to form a cluster (step 3). Each HA trimer then folds back on itself like a jackknife, forcing the two membranes closer together in the middle of the cluster (step 4). When the layer of bound water molecules that normally prevents fusion is eliminated in a small local spot in the middle of the HA cluster, the two bilayers proceed to fuse (step 5). The outermost bilayer leaflets fuse first (this is called **hemifusion**), followed by the inner bilayer halves. A narrow aqueous channel is formed between

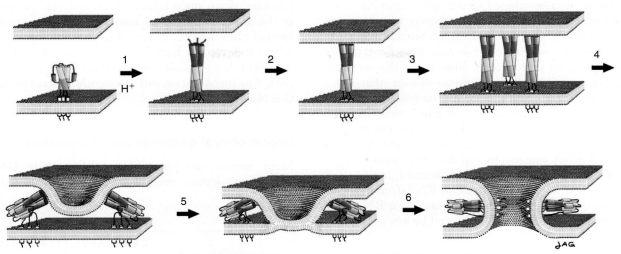

Figure 4.4 A model for influenza virus hemagglutinin-mediated membrane fusion. After binding to the cell membrane (above) via its HA₁ subunit (not shown in this figure) and being brought into an endosome, the trimeric HA molecule, shown as colored cylinders, rearranges upon exposure to low pH (step 1), exposing the fusion peptides (red filaments) at its extremity. These fusion peptides insert into the cell membrane (step 2), and a cluster of such trimers forms (step 3). These trimers then close like a jackknife, bringing the two membranes in close proximity, resulting in fusion (steps 4 to 6).

the internal cavities of the virus and the cytosol (step 6). After "flickering" open and shut briefly, the fusion channel remains permanently open and starts to expand. The nucleocapsids (not shown in the figure) are now exposed, and can move from the virion into the cytosol. Altogether, the fusion process is rapid and efficient.

Although the step-by-step process described above for fusion is best characterized for the influenza virus HA protein and the HIV-1 gp41 glycoprotein, it is still only partially understood. A similar mechanism probably occurs with several other enveloped viruses that have **class I fusion proteins**, including retroviruses, paramyxoviruses, and filoviruses. The three-dimensional structure of fragments of the glycoproteins of these viruses reveals a trimeric spike standing perpendicular to the viral envelope and dominated by long central alpha-helical **coiled coils**. These coiled coils can adopt two distinct conformations, which correspond to the form present in virions and the post-fusion jackknife form.

Other viruses have **class II fusion proteins** with a somewhat different design. These include togaviruses, flaviviruses, rhabdoviruses, and hepadnaviruses. For some of them the structure of the fusion proteins and the conformational changes that they undergo are well known. The fusion-activating changes involve major alterations in the interactions between folded domains and between protein subunits in oligomers. However, contrary to the class I fusion proteins, these proteins undergo only minor conformational changes at the level of secondary structure.

For example, in the case of dengue virus, at neutral pH the fusion proteins are homodimers that lie parallel to the surface of the viral envelope (see Figure 12.3). The fusion peptides—present as polypeptide loops—are hidden in the dimer interphase. After exposure to low pH, the dimer dissociates and homotrimers are formed. These homotrimers are extended perpendicular to the membrane, exposing the fusion peptides, which insert into the target membrane. The bridge between the membranes is thus established, and further structural changes induce bilayer contact and fusion.

Non-enveloped viruses penetrate by membrane lysis or pore formation

The proteins of non-enveloped viruses can also undergo major conformational changes in response to receptor binding or low pH in endosomes. As a result, these proteins either induce the formation of pores in the cellular target membrane through which the genome can escape, or they actually lyse the membrane barrier.

Adenoviruses are the best-studied example of viruses that penetrate by acid-activated, virus-dependent endosome lysis (see Figure 4.5c). The virion is subjected to a series of changes prior to its activation in the endosome. The fibers dissociate, the penton base proteins bind to arginine-glycine-aspartate sequences in the $\alpha_v\beta_{3/5}$ integrin (the coreceptor), and when the virion is exposed to low pH, it converts to the lytic form. As a result, the endosome membrane ruptures, and the virion together with other contents of the endosome is released into the cytosol.

Other non-enveloped viruses pass through cellular membrane barriers without apparent lysis or major membrane perturbations. Many picornaviruses, such as poliovirus, undergo pH-independent entry, but still depend on endocytosis. Binding to the receptor triggers an initial conformational change that loosens up the capsid wall and makes the particles hydrophobic. These so-called "A-particles" then interact hydrophobically with the endosomal membrane and induce the formation of a membrane channel, which the viral RNA can use to slip into the cytosol, leaving an empty capsid in the endosomal lumen (see Figure 11.3). What remains is an RNA-free particle, the "B-particle". Rhinoviruses release their RNA by a similar mechanism, but it is induced by low pH within endosomes.

Virions and capsids are transported within the cell in vesicles or on microtubules

For most viruses, reaching the cytosol is not the end of the journey. They still have to find the proper site for replication, either in the nucleus or in some defined location in the cytoplasm. In extreme cases, virus particles or capsids may still have to travel extremely long distances inside the cell. For example, upon entering a peripheral neuron, herpes simplex virus type 1 or rabies virus must move distances measured in centimeters or meters along the nerve axon to reach the cell body, where the nucleus is located. The travel distance can be a million times the capsid diameter!

Diffusion of macromolecular assemblies in the cytosol is severely limited due to molecular crowding by organelles, the cytoskeleton, and macromolecules. The cortical actin region just beneath the plasma membrane is highly structured and constitutes a difficult obstacle. It has been calculated that free diffusion of particles the size of herpesvirus capsids within the cytoplasm over a distance of 10 micrometers would take 2 hours. However, the measured rate of movement during retrograde transport of herpesvirus capsids in neurons is 400 times faster than that. Not surprisingly, a variety of viruses make use of cellular transport systems that are normally used for moving membrane vesicles and other large objects within the cell.

Virions can be transported inside endocytic vesicles (prior to penetration), or as free nucleocapsids within the cytoplasm (after penetration) (Figure 4.5). The former strategy has the advantage that the cell provides all the systems and signals needed to move the incoming vesicle in the direction of the nucleus. The cell therefore grants free passage for its lethal cargo through a maze of obstacles within the cytoplasm. The molecular motors used include the microtubule-dependent motor **dynein** and an adaptor complex called **dynactin**. Dynein moves

particles along microtubules from the cell periphery in the direction of the microtubule-organizing center, located close to the nucleus. Herpesvirus capsids and adenovirus capsids have been shown to interact with dynein in the cytoplasm, and use it for retrograde transport. The incoming capsids apparently contain proteins that bind to the motors.

Import of viral genomes into the nucleus

Many animal and plant viruses, particularly DNA-containing viruses, use the nucleus as their site of

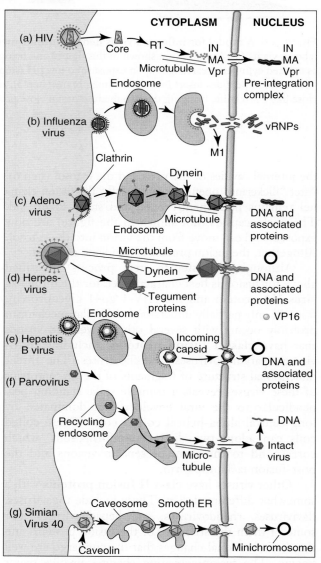

Figure 4.5 Entry and intracellular transport of viruses that replicate in the nucleus. Distinct entry and transport pathways of seven viruses are shown schematically. The extracellular space is on the left, the cytoplasm is in the center, and the nucleus is on the right. The double nuclear membrane is shown traversed by nuclear pores.

replication. There viruses profit from the presence of the cellular machinery for transcription, RNA processing, RNA export, and DNA replication. Alternatively, viruses can establish a latent state in the nucleus, or integrate their genes into host chromosomes.

While dependence on the host cell nucleus carries with it many advantages, it also poses a problem: the incoming virus must deliver its genome and necessary viral proteins to the nucleus. To accomplish this, most viruses piggyback on the machinery that cells utilize for protein and nucleic acid trafficking. The mechanisms and strategies by which incoming viruses enter the nucleus differ from one family to the next. A number of viruses interact with nuclear targeting receptors such as **importins** α and β (also called **karyopherins**), and their genomes enter the nucleus through the nuclear pore complexes.

Dissociation of the nuclear membrane during mitosis.

Most retroviruses (with the exception of lentiviruses) have no means of entering the intact, interphase nucleus. Therefore, their capsids ("preintegration complexes"), which contain proviral DNA made by reverse transcription as well as several viral proteins, must wait in the cytoplasm until the cell divides. During cell division the nuclear envelope disintegrates, allowing the preintegration complexes to gain access to the nucleus when the nuclear envelope reforms. The proviral DNA is therefore able to integrate into the cellular genome in the reformed nucleus. Consequently, infection is only possible in cells that undergo mitosis. This explains why terminally differentiated (non-dividing) cells in the body are not usually targets for retrovirus infection.

Import after partial disassembly in the cytoplasm.

Lentiviruses like HIV-1, in contrast to other retroviruses, direct their preintegration complexes to enter the nucleus via the nuclear pores. The matrix and Vpr proteins within these complexes presumably interact with importins (Figure 4.5a). Each of the eight distinct genome RNAs of influenza virus is individually packaged into ribonucleoprotein complexes small enough to pass through the nuclear pores. After release of these complexes in the cytoplasm by fusion with the endosome membrane, they interact with the importins, and are rapidly transported to and through the nuclear pore complexes (Figure 4.5b).

Import after disassembly at the nuclear pore.

Adenoviruses and herpesviruses have large isometric capsids that are too bulky to pass through the nuclear pores. These capsids bind to fibers that extend from the cytoplasmic face of the pore, and the capsids are subsequently disassembled, allowing the viral DNA and associated proteins to enter through the nuclear pore (Figure 4.5c, d). Adenovirus capsids disintegrate in the process, but intact, empty herpesvirus capsids remain bound to the pore fibers after the viral DNA has entered the nucleus.

Transport of intact virions through the nuclear pore complex.

Only a few viruses have capsids small enough to penetrate through the nuclear pore complexes without dissociation or deformation. Electron microscopy

---

How can virus entry be prevented?

Since viruses are pathogens and cause major medical and veterinary problems, every step in the replication cycle must be viewed as a potential target for inhibition by antiviral drugs. The most direct means of preventing virus entry is to intercept the virus with neutralizing antibodies before it reaches the cell. This is the basis of vaccination (Chapter 35), still by far the most successful antiviral strategy available.

An alternative method is to flood the extracellular space with truncated, soluble versions of the cellular receptor protein for a particular virus. These soluble receptors can bind to virions, thus preventing their interaction with the natural receptor on the cell surface. Conversely, proteins or antibodies that bind to the receptor on the cell surface can be used. Such agents work either by occupying the receptor sites, or by inducing receptor internalization and down-regulation.

Entry can be blocked by interfering with cellular processes needed for virus internalization or penetration. This can be achieved by treatment with agents that interfere with the formation of clathrin-coated vesicles or block transport via the nuclear pore complexes. Compounds that prevent acidification of endosomes are often used as tools to study entry of acid-activated viruses. They include **lysosomotropic agents** such as ammonium chloride, chloroquine, and other organic weak bases that allow protons to escape from acidic compartments. **Carboxylic ionophors**, such as monensin and nigericin, are also effective in breaking down proton gradients, as are inhibitors of the vacuolar proton ATPase such as **bafilomycin A1**.

To prevent membrane fusion more directly, peptides have been designed that bind to conformational intermediates of viral fusion proteins and inhibit their function. Inhibitors of influenza virus neuraminidase are being used as drugs against flu infections. Finally, there are agents that inhibit uncoating. These include **amantadine**, a small molecular weight molecule that specifically blocks the M2 proton channel of influenza A virus and thus prevents acidification of the virus interior. "WIN compounds" (named for the company that originally synthesized them) inhibit human rhinoviruses by binding to a pocket under the receptor-binding canyon. They stabilize the virus capsid and prevent uncoating.

shows that hepatitis B virus capsids (diameter of 34 nm) and simian virus 40 capsids (diameter of 48 nm) can pass intact through the nuclear pores. Parvoviruses (diameter of 20 nm) are also thought to enter the nucleus in this fashion (Figure 4.5e–g).

The many ways in which viral genomes are uncoated and released

The ultimate goal of virus entry is to release the viral genome within the host cell in a form able to be translated, transcribed, or replicated. Depending on the virus, this "uncoating" step may occur simultaneously with the first step of entry, or it may only come after a complex series of membrane fusion and transport steps as outlined above. The mechanism of uncoating of only a few viruses is understood in any detail.

The non-enveloped picornaviruses release their genome RNA directly into the cytosol, leaving the viral capsid within an endosomal vesicle. Therefore virus entry and uncoating are the same event for this virus family.

At the opposite extreme, uncoating of the DNA genomes of hepatitis B virus, parvoviruses, and polyomaviruses takes place only after the capsids have been transported into the nucleus through the nuclear pores. Uncoating of herpesvirus and adenovirus genomes takes place at the nuclear pore, as the capsids are disassembled there. In the case of adenovirus, a viral cysteine protease is activated by entry into the reducing cytosolic environment. This protease cleaves some capsid proteins, releasing the DNA from its attachment to the inside surface of the capsid shell. The DNA genomes of most of these viruses are not released as naked DNAs, but are complexed with cellular or viral proteins that compact the DNA and neutralize its charge. For example, polyomaviruses release their circular DNA as a minichromosome, covered with nucleosomes made from cellular histones.

In the case of togaviruses, uncoating of the viral RNA occurs only minutes after low pH-induced fusion of the viral envelope with an endosomal membrane. Ribosomal 60S subunits serve as cellular uncoating factors. They have affinity for viral capsid proteins, and by binding to the capsid they help to release the viral RNA, which is then ready for translation.

Influenza A viruses encode a protein (M2) that forms membrane channels allowing free passage of H^+ ions at low pH (see Figure 36.3). Small amounts of this protein are incorporated into the virion envelope. As a result, when virions are present in acidifying endosomes, the interior of the particle is also exposed to a low pH, and this releases the individual nucleoproteins from their association with each other and with the matrix protein (M1). Upon fusion of the viral envelope with the endosomal membrane, the individual genome segments are released freely into the cytosol. However, like other negative-strand RNA viruses, influenza virus genomes remain associated with nucleocapsid proteins at all times.

Another variation is seen for reoviruses, which are non-enveloped double-stranded RNA viruses. They enter the cytosol by a pH-independent mechanism through the plasma membrane or an endosomal membrane, during which an outer layer of capsid proteins is removed, but the inner capsid remains intact. Therefore uncoating is incomplete, and viral double-stranded RNA is never released to the cytosol. Synthesis and capping of viral mRNAs occurs within these capsids, and the mRNAs are extruded through channels in the "turrets" that lie at the axes of fivefold symmetry (see Chapter 19).

KEY TERMS

Amantadine	Hemifusion
Attachment factor	Immunoglobulin family
Bafilomycin A1	Importin
Budding	Integrins
Carboxylic ionophors	Karyopherin
Caveolae	Ligand
Caveolin	Lipid raft
Cavin	Lumen
Class I fusion proteins	Lysosome
Class II fusion proteins	Lysosomotropic agent
Clathrin-coated pits	Macropinocytosis
Coiled coil	Neuraminidase
Cytosol	Plasmodesmata
Dynactin	Proteoglycan
Dynein	Receptor
Early endosome	Receptor-mediated endocytosis
Endosome	
Filopodia	Sialic acid
Fission (membrane)	Syncytium (plural: synctia)
Fusion (membrane)	Type I transmembrane protein
Fusion peptide	
Glycolipid	Vesicle transport
Glycoprotein	Virological synapse

FUNDAMENTAL CONCEPTS

- Viral capsid or envelope proteins bind to specific receptors on the cell surface. These may be sugar residues on glycolipids or glycoproteins, or a variety of cellular membrane proteins.

- In some cases, virions subsequently bind to a second receptor, which mediates entry into the cell.

- Enveloped virus particles deliver their nucleocapsids into cells by fusion of the viral envelope with a cellular membrane.

- In some cases, fusion occurs directly at the cell surface, with the plasma membrane.

- In other cases, enveloped virions are first taken up in endosomes, caveosomes, or pinocytic vesicles before fusion occurs.

- Many viruses enter by fusion mediated by low pH in endosomes. A well-studied example is fusion carried out by the HA protein of influenza virus.

- Viruses that do not have envelopes can penetrate membranes by lysis of the endosomal membrane, or by insertion of viral proteins into the membrane, allowing passage of the viral genome directly into the cytosol.

- The nucleocapsids of many DNA viruses are transported to the nucleus by dyneins on microtubules, and either pass directly through the nuclear pores or are disassembled at the nuclear pore.

REVIEW QUESTIONS

1. What are the four basic steps of virus entry into a host cell?
2. What are the most important factors determining which species and cell types can be infected by a given virus?
3. How do enveloped viruses penetrate the cellular membrane?
4. How do non-enveloped viruses penetrate the cellular membrane?
5. What is the key trigger for virus penetration into endosomes? What impact does this trigger have?

VIRUSES OF BACTERIA AND ARCHAEA

Viruses that infect bacteria are called "bacteriophages", or "phages" for short. Research beginning in the 1940s by the "Phage Group", an informal grouping of scientists, established many of the basic principles of molecular biology. They recognized that phages are sets of genes wrapped in a protein coat, and therefore were ideal objects for studying how genes (RNA or DNA) are replicated and expressed.

Chapters 5 through 8 discuss a selection of well-studied bacteriophages that have served as models for understanding cell processes and the replication cycles of other viruses. The discovery of single-stranded RNA phages (Chapter 5) allowed scientists for the first time to study a purified messenger RNA. These are among the simplest of viruses, and they control the translation of their genome RNA by modulating RNA secondary structure. Microviruses, represented by bacteriophage φX174 (Chapter 6), first revealed that genes can overlap each other in a viral genome. This phage was used as a model system for studying DNA replication and virion assembly using scaffolding proteins.

Bacteriophage T7 (Chapter 7) is a double-stranded DNA virus with a stubby tail structure. It injects its DNA part way into the host cell, and uses transcription by cellular RNA polymerase to complete entry of its DNA. It also makes its own RNA polymerase, which transcribes late genes and has been used in many applications in molecular biology and genetic engineering.

Bacteriophage lambda (Chapter 8) is a temperate virus that can integrate its DNA into the host cell and remain latent until the conditions are right to make new virus particles. It controls gene expression by making a variety of repressors that bind to viral DNA, and by controlling termination of RNA polymerase at specific sites on viral DNA. It also uses scaffolding proteins to assemble virions, cleaving genome-length fragments from oligomers made during DNA replication, and stuffing them into procapsids, which are subsequently joined to a helical tail structure.

Archaea were only recently distinguished from bacteria; although they are small, unicellular organisms with circular DNA genomes and no nucleus, they are as different from bacteria as they are from eukaryotes. The discovery and description of viruses of archaea (Chapter 9) is in its infancy. Many of these viruses have lipid membranes and bizarre shapes. All except one have double-stranded DNA genomes, and most archaeal viral genes have no counterparts in the database of sequenced genes from any of the three domains of life. How they enter and exit their host cells is largely unknown.

CHAPTER 5

Single-Stranded RNA Bacteriophages

Jan van Duin

Leviviridae

From Latin *levis* (light), referring to small size of phage

VIRION

Naked icosahedral capsid made from 180 copies of coat protein (T = 3).
Diameter 26 nm.
One copy of maturation protein bound to RNA genome.

GENOME

Linear single-stranded RNA, positive sense, 4 kb.
High degree of internal base pairing.

GENES AND PROTEINS

Four virus-coded proteins:
- *Coat protein*
- *Maturation protein*
- *Lysis protein*
- *RNA replicase*

VIRUSES AND HOSTS

Two genera, found in a variety of gram-negative bacteria:
- *Levivirus (Examples: f2, MS2, R17)*
- *Allolevivirus (Example: Qβ)*

DISTINCTIVE CHARACTERISTICS

One of the smallest and simplest viruses known.
Binds to F-pili projecting from surface of F1 (male) *Escherichia coli.*
Translation of genes is controlled via RNA secondary structures at ribosome binding sites.
RNA polymerase contains virus-coded RNA replicase plus three cellular proteins.

The discovery of RNA phages stimulated research into messenger RNA function and RNA replication

The RNA coliphages were discovered by Tim Loeb and Norton Zinder in 1961 as the result of a search for phages whose replication depends on *E. coli* F-pili, used for bacterial conjugation. Loeb and Zinder plated filtered samples of raw New York City sewage on *E. coli* strains and screened for phages that would produce plaques on male (F+) but not female (F−) bacteria. The first isolate, named f1, made turbid plaques but the second isolate, named f2, made clear plaques; Loeb and Zinder decided to concentrate their work on phage f2. f1 turned out to be a filamentous phage with a single-stranded DNA genome, and f2 was the first RNA-containing phage to be discovered.

Over the next 20 years, study of the RNA phages became a growth industry. These phages came on the scene just as the first evidence for the existence of messenger RNA appeared, and the genetic code was being cracked. They represented a superb source of pure messenger RNA that could be produced in large amounts: up to 10^{13} phage particles per mL are made within a few hours after infection of bacterial cultures, and the phages can be easily purified. The RNA phages also attracted attention because scientists were intrigued about how their RNA genomes were replicated, since cells normally make RNA only from a DNA template. Many scientists who were to become dominant figures in molecular biology in the latter part of the twentieth century got their first taste of research by working with RNA phages.

RNA phages are among the simplest known organisms

RNA phages occur wherever *E. coli* lives, for example, in the intestinal tract of humans and other animals. Studies have shown that sewage samples worldwide contain from 10^2 up to 10^7 RNA phage particles per mL. However, for humans RNA phages are harmless creatures. In fact, RNA phages share the same habitat and the same inactivation characteristics in sewage treatment processes as numerous pathogenic enteroviruses (polioviruses, hepatitis A virus, etc.). However, the RNA phages are much easier to detect by a rapid plaque assay on male *E. coli* hosts. For this reason RNA phages are used as index organisms to reveal the potential presence of pathogenic enteroviruses in drinking water.

RNA phages are among the smallest autonomous viruses known; their capsid is made from a single coat protein, and they code for only three other proteins. Their task upon infecting a cell is very straightforward: make their proteins, replicate their RNA, assemble progeny phage particles, and leave the cell. However, the control of protein synthesis and RNA replication turns out to be carefully regulated, both by RNA secondary structure and by binding of coat and replicase proteins to the RNA at specific sites. Unraveling the mechanisms by which these processes are controlled has been a fascinating puzzle.

Two genera of RNA phages have subtle differences

RNA phages have been placed into two genera on the basis of genetic maps, serological cross-reactions, and RNA size. Members of the *Levivirus* genus (also called supergroup A) have RNAs approximately 3500 nt in length, and code for four proteins: maturation (or A) protein, coat protein, lysis protein, and the RNA replicase, encoded in that order proceeding from the 5' end of the genome (Figure 5.1). The open reading frame for the lysis protein overlaps with those of the coat and

replicase proteins. MS2 phage will be used to discuss translational control in this chapter. In contrast, members of the *Allolevivirus* genus (supergroup B) have a longer genome of approximately 4200 nucleotides. The difference comes mostly from a region on the genome used to code for a protein called readthrough protein or protein A1.

Readthrough in phage Qβ arises when the UGA stop codon at nt 1742 that terminates the normal coat protein gene is occasionally read as a tryptophan codon (UGG). This occurs with a probability of about 6%. A small number of copies of the readthrough protein are incorporated into the capsid; their function is unknown. These phages have no distinct lysis protein; lysis is carried out by the maturation (A2) protein. Phage Qβ is the most intensively studied allolevivirus. The RNA replicase of Qβ can be easily purified, while the replicases of leviviruses are much more difficult to work with, and so most of the work on RNA replication has been done with Qβ.

RNA phages bind to the F-pilus and use it to insert their RNA into the cell

RNA phage capsids contain 180 copies of the coat protein arranged in a T = 3 icosahedral shell that encloses the RNA (Figure 5.2). A single copy of the maturation protein is bound to the encapsidated RNA at sites near both its 5' and 3' ends. The encapsidated RNA is resistant to ribonuclease treatment. However, RNA in virions made with defective or missing maturation protein is sensitive to RNAse degradation.

RNA phages do not have the specialized tail assemblies that are used by many DNA phages to inject their genomes. These phages instead subvert the bacterium's

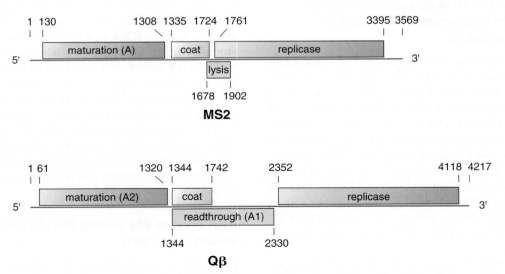

Figure 5.1 Genetic maps of phage MS2 and Qβ RNAs. First and last nucleotides are shown for all open reading frames (boxes).

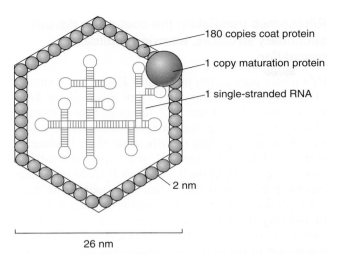

Figure 5.2 **Diagram of an MS2 virion.** Note that the RNA is highly structured. Alloleviviruses have in addition about 12 copies of the readthrough protein in their capsid.

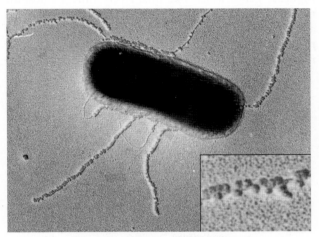

Figure 5.3 **Binding of RNA phage to F-pili.** Electron micrograph of a male *E. coli* cell showing numerous MS2 phages attached to F-pili (enlarged in inset). Normally, only one or a few phages bind and initiate infection.

F-pili (singular: pilus) to maneuver their single-stranded RNA genomes into the cell. F-pili are filamentous structures of *E. coli* and other bacteria that enable F+ bacteria to transfer a partial (single-stranded) copy of their chromosome to F− cells, which do not possess pili.

F-pili are made from a single protein polymerized into a long (1–2 μm), ribbon-like structure that protrudes from the cell. Virions attach to the side of the pilus via their maturation protein (Figure 5.3). Upon contact with the pilus, the maturation protein is cleaved into two fragments. This releases the RNA from the virion, and it becomes sensitive to RNAse degradation. How the RNA then gets inside the cell is not known. A likely possibility is that the pilus with the attached RNA retracts into the cell, dragging the RNA and the cleaved maturation protein along with it.

Phage RNA is translated and replicated in a regulated fashion

Once inside the cell, the phage RNA begins to function as a messenger RNA for the synthesis of phage proteins. This turns out to be a highly regulated process, for two reasons. First, different amounts of each protein are needed. For every 180 copies of the coat protein, the phage needs only 1 copy of the maturation protein to construct virions, and a few copies of the lysis protein to leave the cell. Also, since a single replicase protein can make multiple copies of RNA, fewer replicase proteins are needed than coat proteins.

Second, replication and translation of the same RNA molecule can lead to problems. Replication starts at the 3' end of the RNA and proceeds toward the 5' end. Translation, on the other hand, moves in the opposite direction on the RNA. If the phage had not made the proper arrangements, the replicase and the translating ribosome would meet somewhere on the RNA and sit facing each other forever (Figure 5.4)! Thus a molecule of phage RNA that begins to be translated must not be allowed to begin replication, and vice versa.

To regulate translation, the access of ribosomes to the AUG start codons of phage genes is strongly restricted. In fact, only the coat gene translational start site (nt 1335 on MS2 RNA) is able to bind directly to ribosomes. Experiments have shown that if the start codon of the coat gene is deleted, preventing ribosomes from translating this gene, there is no synthesis of either the lysis or the replicase protein.

RNA secondary structure controls translation of lysis and replicase genes

How is the access of ribosomes to the start sites of the lysis and replicase genes made dependent on translation of the coat gene? The general answer is: by changing

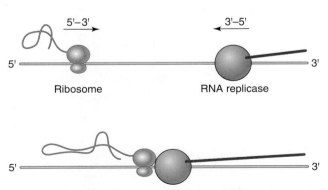

Figure 5.4 **Collision between ribosome and RNA polymerase.** Shown is an encounter between a ribosome translating an RNA in the 5' to 3' direction, and an RNA polymerase making a copy of the RNA in the 3' to 5' direction.

RNA secondary structures. To understand this it is useful to consider the forces that drive the binding of ribosomes to translational start regions on prokaryotic messenger RNAs. There are at least three contributors to this binding energy: (1) base complementarity between 16S ribosomal RNA and the **Shine-Dalgarno sequence** just upstream of the start codon on the messenger RNA, (2) interaction of the anticodon on the initiator fmet-tRNA with the AUG start codon on the messenger RNA, and (3) binding of ribosomal protein S1 to pyrimidine-rich sequences frequently found upstream of the Shine–Dalgarno sequence.

If a strong secondary structure in the messenger RNA prevents interaction with one of the above components, ribosomes will not efficiently bind and therefore protein synthesis will not begin at that site. Secondary structure in RNA consists of regions of self-complementary base sequence that form stem-loop structures. RNA phage genomes contain extensive regions of secondary structure. In fact, the beginning of the lysis and replicase genes lie within RNA secondary structures that are too stable to allow ribosome binding (Figure 5.5), and this is the reason that they are not translated on their own.

Ribosomes translating the coat gene disrupt secondary structure, allowing replicase translation

Ribosomes contact the messenger RNA over a stretch of about 20 nt upstream and 15 nt downstream from the initiator AUG. Figure 5.5 shows that three regions of secondary structure near the replicase start site could contribute to impeding the ribosome from binding to that site. These are stem MJ, the "operator hairpin" containing the AUG itself, and stem R32. Mutagenesis experiments showed that when base pairing at stem MJ is abolished by the introduction of mismatches or by deleting the sequence 1427–1434, there is a large increase in replicase synthesis, which is independent of translation of the coat gene. Apparently, the remaining operator and R32 hairpin structures together are not strong enough to block the entry of ribosomes.

The coupling between translation of the coat and replicase genes works like this. Every time a ribosome reads the coat gene, it disrupts base pairing at stem MJ as it passes through that part of the coat gene open reading frame (Figure 5.6). Once this happens, other ribosomes can bind to the replicase start site and initiate

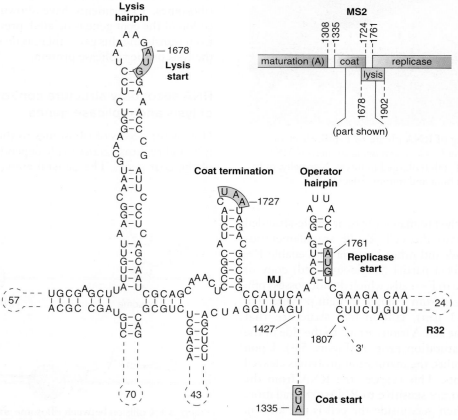

Figure 5.5 Translational control by secondary structure. Secondary structure of MS2 RNA between nts 1427 and 1807. Start and stop codons are shaded in green. Regions irrelevant for translational controls are shown as dashed lines; numbers in dashed hairpins show how many nucleotides have been omitted.

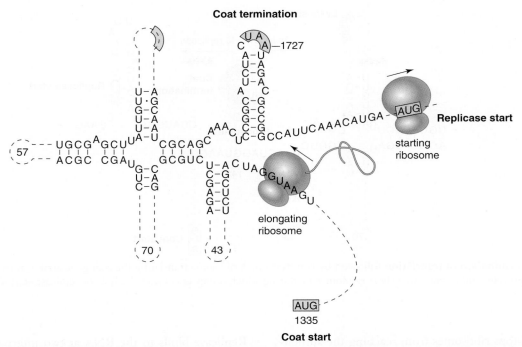

Figure 5.6 **Activation of replicase translation.** A ribosome translating the coat gene disrupts the MJ interaction, thus opening up the replicase start site for ribosome binding.

translation there, even though this site is some 340 nt downstream from the position where the ribosome is translating the coat protein gene.

Further support for this model comes from the finding that introduction of translational stop codons into the coat protein gene upstream of nt 1427 inactivates translation of the replicase gene (the ribosome never gets to stem MJ). However, stop codons placed beyond nt 1434 allow replicase synthesis to proceed at wild type levels.

There is a second level of control of the translation of replicase. When the concentration of coat protein becomes sufficiently high in the cell, dimers are formed. These dimers bind to the operator hairpin, precluding further translational starts of the replicase gene by excluding ribosome binding. This protein-RNA interaction is believed to lead to capsid formation by serving as a nucleation site for the further addition of coat dimers.

Ribosomes terminating coat translation can reinitiate at the lysis gene start site

Independent access of ribosomes to the lysis gene start site (nt 1678) is precluded by the "lysis hairpin" (Figure 5.5). In the absence of coat gene translation, the lysis gene is not expressed. Mutations that destabilize the lysis hairpin by introducing base mismatches allow lysis gene translation independent of the coat gene.

However, activation of the lysis gene is not directly coupled to opening of this hairpin by translation of the overlapping coat gene, as with the replicase gene. Instead, initiation of translation at the lysis gene depends on termination of translation at the end of the coat gene, 50 nt downstream (Figure 5.5). If this UAA stop codon and the UAG stop codon that immediately follows it are mutated so that termination does not take place until the next UGA codon (25 nt farther along the RNA), there is no expression of the lysis gene. On the other hand, if stop codons in the coat gene are introduced by mutagenesis between nts 1678 and 1725 or even somewhat upstream of the lysis protein start codon, lysis protein is made. In fact, the closer the coat gene stop codon is to the lysis start codon, the more lysis protein is made.

These results suggest that a ribosome that has reached the coat stop codon and released the coat protein does not immediately leave the mRNA, but makes a short random walk along the RNA in either direction. If a start codon is encountered, the ribosome may reinitiate translation (Figure 5.7); if not, it will eventually detach from the mRNA. From the proportion of lysis to coat protein synthesized by phage MS2, the probability for such successful re-initiation is only 5%.

This "scanning" model was further tested by introducing an additional start codon a short distance downstream from the authentic lysis start. The new start codon is used in place of the authentic start codon, as it is closest to the termination codon and thus forms a

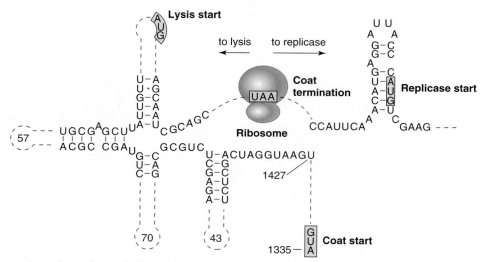

Figure 5.7 Termination of translation followed by reinitiation. A ribosome translating the coat gene arrives at its stop codon, releases coat protein, and carries out a short random walk during which it may encounter the lysis or replicase start sites.

barrier that stops ribosomes from reaching the authentic start site. This shows that ribosomes look for the lysis start codon from the 3' direction.

However, when a termination codon for coat gene translation was engineered upstream of the lysis start codon, the authentic lysis initiation site was again used, because it is closer to the termination site. This shows that ribosomes that have recently terminated can search for nearby start codons in both the 5' and 3' directions.

In vitro experiments have shown that on short messenger RNAs containing only 5 codons, the ribosome shuffles back and forth between the start and stop codons, making multiple copies of a pentapeptide without ever leaving the template. This is a rarely used mechanism of initiation of protein synthesis. Normally when ribosomes come to the end of a coding region they are released from the messenger RNA if no reinitiation sites are nearby.

Replication versus translation: competition for the same RNA template

Once the replicase is made, it can begin to generate new copies of phage RNA. The plus-strand genome is first copied into a complementary minus-strand RNA. This RNA is in turn used as a template to produce more plus-strand RNAs. Although they are fully complementary, plus- and minus-strands do not anneal to form a double-stranded RNA. Such annealing is inhibited by the high degree of internal secondary structure found in each single-stranded RNA.

How does the replicase copy the genome RNA starting at its 3' end while ribosomes are translating the same RNA in the opposite direction (Figure 5.4)? The switch from translation to replication works as follows.

Replicase binds to the RNA at two internal positions, one of which is the start of the coat gene. As a result there is competition between replicase and ribosomes for this site. If the replicase arrives first, there will be no new translation, allowing replicase to copy the RNA with no oncoming traffic. On the other hand, if the ribosome binds first to the coat protein gene, replicase will not be able to bind to its template, and therefore the ribosome is free to complete its voyage unhindered.

Genome replication requires four host cell proteins plus the replicase

How the replicase actually binds to the coat protein start site is fascinating. When scientists isolated the RNA polymerase activity from infected cells, they found to their surprise that it is a complex made of four proteins, only one of which (the replicase protein) is coded by the phage genome. The three other proteins in the complex are cellular proteins: the S1 protein of the small ribosomal subunit, and two translational elongation factors called EF-Tu and EF-Ts (Table 5.1). The active polymerase complex contains one copy of each of the four proteins (we will refer to the active enzyme as *polymerase* while reserving the term *replicase* for the phage-coded subunit).

The three host cell proteins are crucial factors in the cellular translational apparatus. The S1 protein is instrumental in binding messenger RNA sequences to the ribosome during initiation of translation in bacteria. EF-Tu positions charged tRNAs on the ribosome during elongation of growing peptide chains. This activity involves binding of GTP and its hydrolysis to GDP. EF-Ts recycles the resulting EF-Tu/GDP to EF-Tu/GTP; an intermediate in this process is a complex of EF-Tu and EF-Ts.

Table 5.1 Proteins needed for Qβ RNA replicase activity

Protein	Coded by genome of	Required for synthesis of:		Function
		Minus-strand	Plus-strand	
Replicase	Qβ	Yes	Yes	RNA-dependent RNA polymerase
EF-Tu	E. coli	Yes	Yes	Initiation of RNA synthesis by bound GTP
EF-Ts	E. coli	Yes	Yes	Unknown
S1	E. coli	Yes	No	Binding to plus-strand RNA; inhibition of ribosome binding to coat start site
Host Factor	E. coli	Yes	No	Unwinding of 3'-terminal hairpin on plus-strand RNA

A host ribosomal protein directs polymerase to the coat start site

It turns out that binding of the polymerase to the coat protein start site is facilitated by the S1 protein, which carries out the same function for the ribosome when it is binding to the coat protein start site. Thus the competition between ribosomes and replicase for the coat protein start site is in fact mediated by the same (cellular) protein!

Once the polymerase complex binds, assuring absence of ribosomes on the RNA it is about to copy, it looks for the 3' end of the RNA. This requires a fifth protein, also a host cell protein called Host Factor. Host Factor is a poly(A) binding protein that, in the uninfected cell, stimulates translation of the mRNA of a transcriptional sigma factor. This protein, active as a hexamer, does not associate with the replicase but acts directly on the phage RNA and is thought to expose the 3'-terminal residues (CCA_{OH}), which are base-paired to nearby G residues. If two more Cs are added at the 3' end of phage RNA ($CCCCA_{OH}$) the CCA_{OH} sequence will stick out. This mutant is no longer dependent on Host Factor for its replication.

Polymerase skips the first A residue but adds a terminal A to the minus-strand copy

Surprisingly, polymerase initiates RNA synthesis not at the 3'-terminal A, but at the second nucleotide, a C. Thus the 5' end of the new minus-strand starts with a G residue. Here is where EF-Tu may act. Since EF-Tu has a GTP binding site, it could bring GTP to the replicase to initiate RNA synthesis. Therefore, EF-Tu would be acting as a kind of primer for the initiation of minus-strand RNA synthesis. The role of EF-Ts remains obscure.

Once synthesis has begun, it can continue in the absence of S1, EF-Tu, and EF-Ts, since the replicase protein has RNA polymerase activity by itself (but the complex usually stays intact). When the polymerase reaches the end of the plus-strand template, it completes the minus-strand by copying the G's at the 5' end of the plus-strand into C's. Then the replicase adds on a single, untemplated A residue—so the 3' end of the minus-strand looks like the initial 3' end of the plus-strand, C's with a terminal A.

Synthesis of plus-strands is less complex and more efficient than that of minus-strands

The next step is to copy the minus-strand RNA into plus-strands. This does not require either S1 or the Host Factor. The polymerase is not competing with ribosomes for binding to the minus-strand RNA, explaining the lack of requirement for S1. The Host Factor is not needed because the 3' end of the minus-strand RNA is apparently not hidden in a stem structure and is therefore available as a template. Perhaps for all these reasons, plus-strand synthesis is more efficient than minus-strand synthesis, and therefore more plus-strands are made. This makes sense, for the phage encapsidates only plus-strand RNA, and only this RNA can be translated to make more phage proteins.

A final question is why the Qβ plus-strand conceals its 3' end even though replication needs unpaired nucleotides to initiate. The best guess is that any protruding nucleotides will be a target for host exonucleases (tRNA, which has a protruding CCA_{OH} 3' end, is continuously repaired by tRNA nucleotidyl transferase). One has also to bear in mind that the plus-strand needs extra protection because it is the infectious agent. A single molecule must survive during infection long enough to be translated and then replicated at least once. The minus-strand, which does have protruding C's, may occasionally perish. It can always be replaced by a new transcript.

The start site for synthesis of maturation protein is normally inaccessible to ribosomes

So far we have ignored translation of the maturation protein, whose coding region lies near the 5' end of the phage RNA. Figure 5.8a shows the secondary structure of the 5' untranslated region of MS2 RNA, just upstream of the maturation start codon. A hairpin at

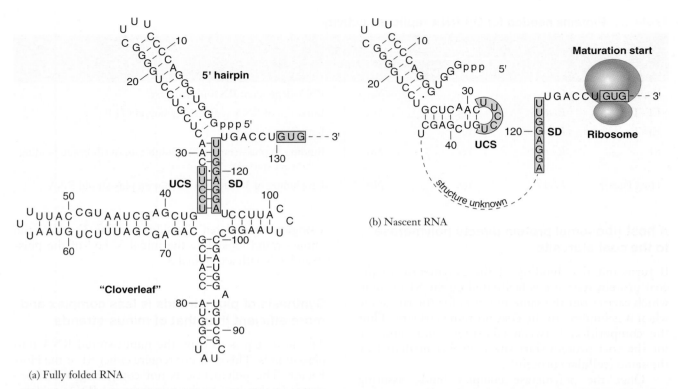

Figure 5.8 Translation of maturation gene on nascent RNA. (a) Equilibrium folding of the 5' untranslated region of MS2 RNA. The secondary structure formed between the upstream complementary sequence (UCS) and the Shine–Dalgarno sequence (SD) does not allow ribosome binding. (b) Folding intermediate of the same region on nascent RNA, allowing translation of the maturation gene.

the 5' end is followed by a "cloverleaf" (three "leaves", one stem) structure in which the Shine–Dalgarno sequence (SD) of the maturation gene is base-paired to an upstream complementary sequence (UCS). This long-distance interaction effectively prevents ribosome binding. How then is the maturation protein made? The answer is that ribosomes only bind to the maturation start site on RNA chains that are in the process of being synthesized, that is, "nascent" chains. Furthermore, these nascent RNAs must not yet have reached their equilibrium folding configuration.

Consider a growing plus-strand RNA being made by polymerase from a minus-strand RNA template. By the time nt 123 is added, all nucleotides needed to build the cloverleaf shown in Figure 5.8a have been incorporated. However, translation cannot yet start because ribosome binding needs the start codon and sequences downstream of the initiator codon, up to about nts 140–145. It is known that small RNAs such as tRNA (~75 nts) fold up correctly within milliseconds after denaturation. On the other hand, phage RNA replicases incorporate approximately 35 nucleotides per second into a growing chain. Thus one would expect the cloverleaf structure to be formed long before the ribosome binding site is made, as addition of a further 20 nucleotides beyond nt 123 could take as long as half a second.

Synthesis of maturation protein is controlled by delayed RNA folding

However, the folding of the 5' untranslated region of MS2 RNA to its final equilibrium configuration is unusually slow; *in vitro* experiments show that it takes minutes to refold the heat-denatured cloverleaf structure. Generally speaking, such a delay can be caused only by a "kinetic trap". This is an alternative folding configuration that is stable enough to temporarily prevent the RNA from reaching the equilibrium structure.

Mutational analysis has indicated that the kinetic trap involves the small hairpin between nts 25 and 43 shown in Figure 5.8b. Formation of this hairpin excludes the inhibitory long-distance base pairing between the upstream complementary sequence and the Shine–Dalgarno sequence, thus permitting ribosome binding. However, this secondary structure is not highly stable; the RNA will eventually fold up into the structure shown in Figure 5.8a, and translation will be shut down.

It is not difficult to see what the biological purpose of this control is; it ensures that the maturation gene will be translated only once or a few times per newly synthesized plus-strand phage RNA. Since one molecule of maturation protein is packaged per plus-strand RNA molecule, this mechanism assures the appropriate level of its synthesis. Furthermore, since maturation protein cannot be translated from the full-length genome, replicase needs only to compete with ribosomes that bind at the coat start site to create a ribosome-free template, which is needed for unimpeded synthesis of minus-strand RNAs.

Assembly and release of virions

The assembly of virions has not been studied in great detail. It is known that the first step in this process must be the binding of the maturation protein to the RNA. This complex can be made *in vitro* and is infectious for F+ cells; thus it can bind to pili in the absence of the capsid, and direct the phage RNA into the cytoplasm.

The second step in virion construction is the assembly of the coat protein around the RNA to form an icosahedral shell. The nucleation point for this coating reaction is the replicase operator hairpin (Figure 5.5). The coat protein will form capsids even in the absence of phage RNA. However, capsid formation occurs at a much lower concentration of coat protein in the presence of phage RNA, assuring that empty capsids will not form and all phage RNA becomes packaged.

The infectious cycle ends when the accumulation of the lysis protein in the cytoplasmic membrane has caused the collapse of the cell wall. This 75-amino acid hydrophobic protein short-circuits the membrane potential, eventually leading to degradation of peptidoglycan. Electron micrographs show that only a very small section of the sacculus (the cell wall network) has dissolved, usually at the equatorial growth zone.

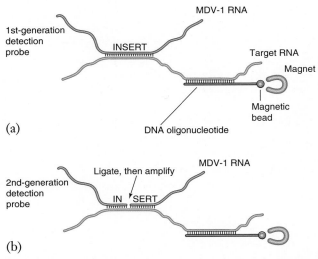

(a)

(b)

Figure 5.9 Detection of minute amounts of target RNAs using Qβ replicase. See text for details.

Detection of minute amounts of RNAs by using the power of Qβ replicase

Among all the single-stranded RNA phages, Qβ is famous for its replicase. This is the best-characterized RNA-dependent RNA polymerase and can be readily purified. In addition to Qβ RNA itself, a number of small (100–200 nt) RNAs can be efficiently replicated in vitro by this RNA polymerase. Many of these evolved spontaneously in cells infected with the phage or in amplification reactions carried out in vitro.

One of these templates is called midi variant-1 (MDV-1). This RNA can be amplified *in vitro* some 10^{11} times in about 30 minutes by Qβ replicase. Several groups are trying to exploit this amplification capacity to detect minute amounts of DNA or RNA molecules in clinical samples, for instance HIV-RNA. The concentration of these molecules is usually too small to be detected by conventional techniques. The principle is outlined below (Figure 5.9).

MDV-1 RNA is first modified to contain a short sequence complementary to the target RNA; this sequence does not interfere with the replication of the RNA by Qβ polymerase. RNAs can be produced by in vitro transcription of plasmids containing cloned copies of the modified MDV-1 RNA, using bacteriophage T7 RNA polymerase. The target RNA in the clinical sample is first hybridized to a complementary DNA oligonucleotide that carries a magnetic bead at one end, which allows binding to a solid support (the magnet). In this way the target can be separated from the bulk of the other RNAs in the sample.

The immobilized target RNA is then hybridized to the modified MDV-1 RNA (Figure 5.9a). After thorough washing to remove all unbound RNAs, the hybridized MDV-1 RNA is released by denaturation and amplified by Qβ polymerase. The amplified RNA can be detected by gel electrophoresis and ethidium bromide staining. The presence of the target RNA is therefore signaled by the appearance of a band of MDV-1 RNA in the gel.

False positive reactions can arise if there is partial homology between the MDV-1 RNA and some other cellular or viral RNA in the sample. To reduce such reactions, MDV-1 RNA can be added as two half-molecules (Figure 5.9b), which are joined when properly aligned on the target RNA by adding RNA ligase to the reaction. Since Qβ replicase needs specific sequences at both ends of the RNA to replicate RNAs, half-molecules by themselves cannot be replicated. Any partial hybridization of the insert to an RNA molecule other than the target would be very unlikely to align both ends of the MDV-1 RNA half-molecules in such a way that ligation could occur. The ligated MDV-1 RNA is then released and amplified by Qβ polymerase.

KEY TERMS

F-pili   Shine–Dalgarno sequence

FUNDAMENTAL CONCEPTS

- When first discovered, single-stranded RNA bacteriophages represented an ideal source of a purified messenger RNA for molecular biology researchers.

- Single-stranded RNA bacteriophages have four genes on a linear RNA of 3500–4200 nucleotides.

- RNA phages bind to F-pili and enter the cell by release of RNA from the virion and probably by retraction of the F-pilus bound to viral RNA into the cell.

- Only the coat protein gene directly binds to ribosomes within the cell.

- Alterations in secondary structures within phage RNA control translation of the four phage genes.

- Ribosomes translating the coat protein gene disrupt secondary structure nearby the replicase gene start site, allowing ribosomes to bind to and translate that gene.

- Ribosomes that terminate translation of the coat protein gene can scan forward or backward on viral RNA and initiate translation at the nearby lysis gene start site.

- Translation of the maturation protein only occurs on newly synthesized viral RNAs during the short time when they have not yet folded into their stable secondary structure.

- Phage RNA polymerase contains one viral protein subunit and three or four cellular protein subunits.

- Phage RNA polymerase binds to the start of the coat gene at the same site where ribosomes bind; therefore an RNA can be either replicated or translated, but not both at the same time.

REVIEW QUESTIONS

1. The explosion of research on RNA phages laid the foundations for the careers of a number of modern day molecular biologists. Why were RNA phages so interesting to researchers during the 1960s through the 1980s?

2. RNA viruses are said by some to be the simplest viruses. Why would this be true?

3. Most single-stranded RNA animal viruses form a double-stranded RNA replication intermediate when replicating their genome. In contrast, RNA phages rarely if ever produce double-stranded RNA. How do RNA phages address the requirement for single-stranded RNAs to serve both as templates for translation and templates for genome replication without producing a double-stranded RNA intermediate?

4. How does phage MS2 ensure that only a limited number of copies of the maturation protein are produced?

5. The phage MS2 RNA polymerase initiates RNA synthesis at the second nucleotide from the 3'-terminus. How is the terminal nucleotide replaced?

6. Initiation of synthesis of the MS2 lysis protein is explained by the "scanning" model. What is the scanning model?

Microviruses

Bentley Fane

Microviridae
From Greek *micro* (small), referring to size of phage

VIRION
Naked icosahedral capsid (T = 1).
Diameter 33 nm.
12 spikes project from surface.

GENOME
Circular single-stranded DNA, 4.4–6.1 kb.

GENES AND PROTEINS
φX174: 3 transcriptional promoters and 4 terminators control gene expression.
Genome codes for 11 proteins including:
- *2 DNA replication/packaging proteins*
- *2 scaffolding proteins*
- *4 virion structural proteins*
- *1 lysis protein*

VIRUSES AND HOSTS
Two subfamilies of *Microviridae*:
- *Microviruses infect free-living bacteria.*

- *Gokushoviruses, from Japanese* gokusho *(very small), infect obligate intracellular parasitic bacteria.*

DISTINCTIVE CHARACTERISTICS
Genomes contain extensive overlapping reading frames.
DNA is delivered into the cell through spikes on the virion surface.
DNA becomes double-stranded before being transcribed.
DNA replicates by a rolling circle mechanism.
Scaffolding proteins are used for virion assembly.
DNA is inserted into procapsids during DNA replication.

φX174: a tiny virus with a big impact

Although viruses were first discovered and distinguished from other microorganisms by Martinus Beijerinck and Dimitrii Ivanowski in the late 1890s (tobacco mosaic virus), the first bacteriophage was not identified until 1917 by Felix d'Herelle. Throughout the 1920s and 1930s, many different bacterial viruses were identified and classified, including the *Microviridae* (Table 6.1). The name of the best-known family member, φX174, reflects its history. Without electron microscopes, Sertic and Bulgakov were able to distinguish different groups, or "races", of bacteriophage, primarily by plaque morphology and ultrafiltration. The φ (Greek letter *phi*) in φX174 refers to "phage"; the X is the Roman numeral 10; "race X" bacteriophages were unusually small, as determined by ultrafiltration, and the first isolate was placed in a test tube labeled 174.

This small virus would have a big impact on molecular biology. Recognizing its unique properties,

Robert Sinsheimer began detailed investigations in the late 1950s. The first surprise was the discovery that its genome consisted of single-stranded DNA; other known phage genomes were made of double-stranded DNA, as are all genomes of cellular organisms. Masaki Hayashi and colleagues defined the φX174 assembly pathway, and demonstrated that viral DNA could be packaged into capsids *in vitro*. Using φX174, Arthur Kornberg and coworkers elucidated many of the fundamental mechanisms of DNA replication.

In 1978, Fred Sanger and colleagues were the first to determine the nucleotide sequence of the entire genome of an organism, and that organism was φX174. The results were astounding: many of the genes overlapped! One region of DNA could even encode part of four different genes (Figure 6.1). This finding so challenged the concept of the gene, reported *The New York Times*, that some scientists were searching for hidden messages in the DNA sequence, hypothesizing that φX174 was a

Table 6.1 Phages in the *Microviridae* family

Genus	Hosts	Examples
Microvirus	*E. coli*, Shigella, Salmonella	φX174, G4, α3
Bdellomicrovirus[a]	Bdellovibrio (obligate intracellular bacteria)	φMH2K
Chlamydiamicrovirus[a]	Chlamydia (obligate intracellular bacteria)	CHP1 and CHP2
Spiromicrovirus[a]	The Mollicute bacteria, Spiroplasma	SpV4

[a]*These genera have been grouped into a distinct subfamily, called the Gokushoviruses.*

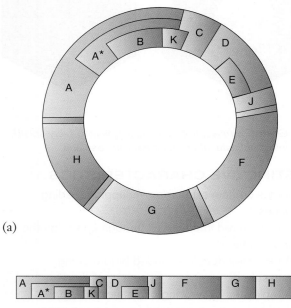

(a)

(b)

Figure 6.1 The φX174 genome. (a) Map of genes on the circular DNA genome (see Table 6.2 for functions of gene products). (b) Transcriptional map. Promoters (P) and transcription terminators (T) are indicated on a linearized map of φX174 DNA. The major transcripts are illustrated below. Line thickness indicates the relative abundance of the transcripts.

genetically engineered extraterrestrial organism! More recently, the atomic structure of the φX174 **procapsid**, an assembly intermediate containing a full complement of **scaffolding proteins**, was solved by a collaborative effort of the laboratories of Michael Rossmann, Bentley Fane, and Nino Incardona.

Overlapping reading frames allow efficient use of a small genome

Seven of the eleven φX174 genes (genes A through E and gene K) lie in a region of extensive overlapping

reading frames that covers nearly one-half the genome (Figure 6.1). These seven genes code for nonstructural proteins (Table 6.2). The remaining four genes (F, G, H, and J) code for the structural proteins of φX174, and these four genes do not overlap. The use of overlapping reading frames can be understood as a way of increasing the amount of genetic information encoded in a small viral genome. The acquisition of new genes is achieved by nesting them within the sequences of preexisting genes rather than inserting additional DNA sequences in the genome.

The A gene overlaps with genes A*, B, and K. The A protein plays a critical role in genome replication and interacts with many host cell proteins. Therefore, it must be able to coevolve with its host. Adaptive mutations in this gene can simultaneously create mutations in the overlapping genes and change their functions. However, the A* and K proteins are not essential for virus replication, and the B protein has a flexible structure that is highly tolerant of amino acid substitutions (up to 70% of its sequences can be altered with minimal effects on its functions). Similarly, the E gene, which lies within the essential D gene, leads to host cell lysis; it can be considered a "luxury" gene, whose protein product is not required for replication, but increases the efficiency of release of newly replicated phage.

φX174 binds to glucose residues in lipopolysaccharide on the cell surface

The icosahedral capsid (Figure 6.2) contains 60 copies of the viral capsid protein F. Each large protrusion found at the 12 icosahedral vertices is composed of five copies of the major spike protein G (total 60 copies) and one copy of the DNA pilot protein H. In addition, there are 60 copies of the DNA binding protein J inside the virion.

φX174 attaches to host cells via sugar residues, most likely glucose, in the **lipopolysaccharide** that is a part of the outer membrane of gram-negative bacteria. A site on the surface of the capsid, near the threefold axes of symmetry, binds reversibly to glucose. The glucose-binding site was identified by accident. Large quantities of virions were purified by sucrose gradient sedimentation for crystallization and subsequent x-ray structure determination. The structure showed that sucrose molecules (a dimer of glucose and fructose) were bound

Table 6.2 Gene products of φX174

Protein	Function
A	DNA replication. Cleaves plus-strand viral DNA at a specific site, initiating rolling circle replication. Ligates linear plus-strand DNA to form full-length, single-stranded circles.
A*	Not essential for virus growth. May play a role in the inhibition of host cell DNA replication and in exclusion of superinfecting phage.
B	Internal scaffolding protein, required for procapsid morphogenesis; 60 copies in procapsid.
C	DNA packaging. Binds to Stage II DNA replication complex, directing concomitant Stage III replication and entry of plus-strand viral DNA into the procapsid.
D	External scaffolding protein, required for procapsid morphogenesis; 240 copies in procapsid.
E	Host cell lysis.
F	Major capsid protein; 60 copies in virion.
G	Major spike protein; 60 copies in virion.
H	DNA pilot protein, required for DNA delivery; also called the minor spike protein; 12 copies in virion.
J	Small DNA-binding protein, needed for DNA packaging; 60 copies in virion.
K	Not essential for virus growth. May play a role in the optimization of burst size in various hosts.

Figure 6.2 The atomic structure of φX174. Spikes at fivefold symmetry axes are shown in blue.

via the glucose moiety to a distinct site on the major capsid protein F. Members of the Gokushovirus subfamily appear to recognize proteins rather than sugars on host cell surfaces.

φX174 delivers its genome into the cell through spikes on the capsid surface

When φX174 virions are placed in solutions of high ionic strength, the DNA genome is extruded from the capsid via the spikes, located at the axes of fivefold symmetry. This implies that during infection of *Escherichia coli*, genome DNA may be delivered into the host cell through the spike. Host range mutations change amino acids primarily in the spike proteins G and H, indicating that the spikes are also involved in attachment to host cells, and arguing that a second host cell receptor may be required for DNA penetration. The identity of this second factor is unknown. However, penetration may follow a pathway similar to that of the large, tailed bacteriophage T4.

Bacteriophage T4 interacts reversibly with lipopolysaccharide via its long tail fibers. The phage then "walks" along the surface of the cell, binding to lipopolysaccharide residues one after the other, until it finds a second receptor, which leads to delivery of the viral DNA. Instead of walking, the tailless, spherical φX174 may "roll" along the cell surface in a similar fashion. Electron micrographs show that the vast majority of adsorbed φX174 particles are imbedded at points of adhesion between the cell wall and the inner membrane. This suggests the location of the second receptor and indicates that the genome is delivered directly into the cytoplasm, where replication occurs, and not into the periplasmic space.

Penetration requires the viral H protein, also called the DNA pilot protein. This protein contains an N-terminal transmembrane helix that most likely mediates penetration. Initially, protein H and the incoming DNA are associated with the outer cell membrane at the site of cell wall adhesion regions. From there, the single-stranded DNA is delivered to the (inner) cytoplasmic membrane, the site of DNA synthesis. Although cells can be infected by transfection with naked single-stranded DNA, transfection efficiency is enhanced if H protein is added to the DNA. This suggests that the H protein may stimulate stage I DNA synthesis (see page 72) by piloting the penetrating DNA to the proper site of replication.

Stage I DNA replication generates double-stranded replicative form DNA

Microviruses depend on the host cell RNA polymerase to synthesize their mRNAs. Because cellular RNA polymerases require a double-stranded DNA template, microviruses must transform their single-stranded DNA genome into double-stranded DNA before viral genes can be transcribed. This double-stranded DNA, called replicative form (RF) DNA, is synthesized during Stage I DNA replication (Figure 6.3), using only host cell proteins.

Single-stranded DNA binding protein (SSB) first coats the intracellular phage DNA except for a specific short, highly base-paired "hairpin" region. That region serves as a signal for the assembly of a **primosome** by seven different host cell proteins. The primosome enables primase, one of its components, to generate short complementary RNA primers at different locations around the single-stranded phage DNA. These are then extended by cellular DNA polymerase III to make a complete complementary copy of the phage DNA. DNA polymerase I subsequently removes the RNA primers and fills in the gaps, and DNA ligase joins the DNA chains to form a covalently closed, circular, double-stranded RF DNA. The expression of phage genes that allow further genome replication and capsid production can now take place.

Gene expression is controlled by the strength of promoters and transcriptional terminators

Unlike most DNA viruses, microviruses do not produce regulatory proteins to ensure temporal control of gene expression. The production of different amounts of viral proteins is entirely dependent on transcriptional promoters, transcription terminators, and ribosome binding sites on mRNAs (Figure 6.1b). Three promoters (P_A, P_B, and P_D) and four terminators (T_J, T_F, T_G, and T_H) have been detected. None of the terminators is highly efficient, leading to the production of a variety of transcripts with the same 5' ends but extending for different lengths.

In general, higher levels of messenger RNAs are made for proteins that are required in greater abundance. Transcripts initiated at P_A are unstable, keeping the levels of the A and A* proteins low; the terminator for P_A transcripts is unknown. Gene D mRNAs are more abundant than any other transcript, and more of this protein is needed than any other: 240 copies of the D scaffolding protein are required to build one virion. Similarly, there are more mRNAs that encode the J, F, and G proteins than mRNAs that encode the H protein, because transcription is successively reduced at a series of terminators at the ends of each of these four genes. The numbers of these proteins per virion are 60:60:60:12, respectively.

Regulation is also achieved on the translational level. Early and abundant synthesis of the E protein would lead to premature cell lysis, before sufficient infectious progeny could be made. Gene E resides within gene D, and therefore many E transcripts are made. However, few E proteins are synthesized because the E gene has a very weak ribosome binding site, and therefore initiation of protein synthesis is strongly reduced at this site. Furthermore, the presence of many non-optimal codons in both the E and H genes most likely reduces the level of synthesis of these proteins.

Replicative form DNAs are amplified via a rolling circle mechanism

A single molecule of replicative form DNA cannot generate sufficient mRNAs and progeny single-stranded DNAs to construct the 100 phage particles made in a productive infection. Stage II DNA replication (Figure 6.4) amplifies the number of replicative form DNA molecules to about 10 per cell. This occurs in two distinct phases.

First, progeny plus-strand DNA is made by a **rolling circle** mechanism using replicative form DNA as a template. The viral A protein cleaves the plus-strand on replicative form DNA at a specific site (step 1) and is covalently linked to the resulting 5' end of the DNA via a tyrosine residue in the protein. The host cell rep protein acts together with the viral A protein as a **helicase** to unwind the plus-strand from its association with its complementary minus-strand (step 2). As the plus-strand is freed, it is coated with the single-stranded DNA binding protein. Simultaneously, host cell DNA polymerase

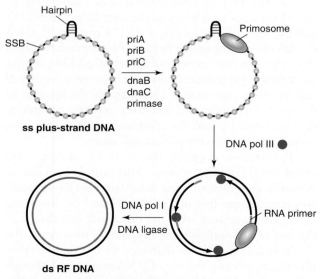

Figure 6.3 Stage I DNA replication. Formation of double-stranded replicative form DNA from single-stranded genome DNA, carried out entirely by host cell proteins. SSB: single-strand DNA binding protein; priA–C and dnaB–C: primosome proteins. RNA primers shown in orange.

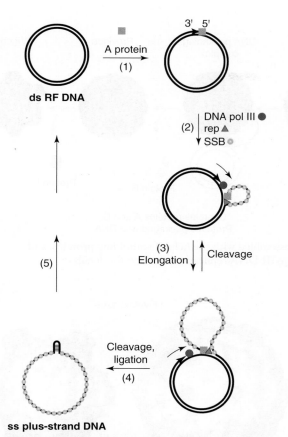

ds RF DNA

A protein
(1)

3' 5'

DNA pol III ●
rep ▲
SSB ◉
(2)

(5)

(3)
Elongation   Cleavage

Cleavage,
ligation
(4)

ss plus-strand DNA

Figure 6.4 Stage II DNA replication. Rolling circle replication forms single-stranded DNA from double-stranded replicative form DNA, using viral protein A (black circle) and three host cell proteins. Steps (1) to (5) are described in the text.

Summary of viral DNA replication mechanisms

Figures 6.3 and 6.4 illustrate the contrasting mechanisms used for plus- and minus-strand synthesis. To summarize: minus-strand DNA is made on a plus-strand template by a complex RNA-primed mechanism utilizing a host cell primosome, DNA polymerase III, DNA polymerase I, and DNA ligase (Figure 6.3). The newly synthesized minus-strand DNA remains associated with its template plus-strand in a double-stranded replicative form.

In contrast, plus-strand DNA is made from the double-stranded replicative form template by a continuous rolling circle mechanism utilizing the viral A protein, DNA polymerase III, the single-stranded DNA binding protein SSB, and rep helicase (Figure 6.4). The primer for elongation by DNA polymerase III is the 3' end of plus-strand DNA cleaved by the viral A protein. Progeny plus-strand DNA molecules are cleaved from the growing strands on the rolling circle by the A protein, and are ligated to form single-stranded circles. The A protein remains on the replicative form DNA to help synthesize further progeny plus-strand DNA.

Procapsids are assembled by the use of scaffolding proteins

The assembly pathway of φX174 (Figure 6.5) was elucidated by analyzing the intermediate capsid structures that accumulate in cells infected with virus mutants. Small particles denoted "9S" and "6S" (named for their relative sedimentation rate in sucrose gradients) accumulate in cells infected with phage mutated in the internal scaffolding protein B. The 9S particles are pentamers of the major capsid protein F, and the 6S particles are pentamers of the major spike protein G. This finding demonstrates that the 9S and 6S particles are the first intermediates in the assembly pathway and that the internal scaffolding protein B is required for capsid assembly to proceed. Much like the scaffolds used in construction, these scaffolding proteins facilitate capsid assembly but are removed when construction is complete.

When protein B is present, five copies will bind to the underside of a 9S particle, yielding the 9S* particle. This binding triggers conformational changes on the upper surface of the particle. These conformational changes allow 9S* particles to interact with 6S spike particles and external scaffolding proteins, and prevent the premature association of 9S coat protein pentamers into aggregates, a function reminiscent of molecular chaperones.

Besides stimulating 9S*–6S interactions, protein B also facilitates the incorporation of the DNA pilot protein H. The resulting 12S* particle contains one

III extends the new 3' end created by cleavage of the plus-strand, and thereby synthesizes more plus-strand DNA. The DNA polymerase and helicase make their way around the circular double-stranded DNA, eventually passing through the site where the A protein originally cleaved the plus-strand (step 3).

At this point, the A protein cleaves off a full-length plus-strand and ligates it to form a single-stranded circle (step 4). The A protein is simultaneously transferred to the newly formed 5' end resulting from this cleavage, recreating the original replication complex. The result is the synthesis of a progeny viral plus-strand DNA circle and regeneration of the double-stranded replicative form DNA each time the replicative machinery makes a full turn on the template DNA. The second phase of Stage II DNA replication is the formation of more replicative form DNA from the progeny plus-strand viral DNA (step 5). The same host cell proteins used in Stage I DNA replication (Figure 6.3) mediate this reaction.

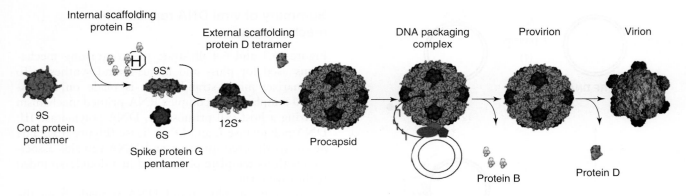

Figure 6.5 The φX174 morphogenetic pathway. Procapsids are assembled with the help of scaffolding proteins, and single-stranded genome DNA is inserted into procapsids during Stage III DNA replication. See text for details.

molecule of protein H and five each of proteins F, G, and B. Sixty tetramers of the external scaffolding protein D then organize twelve 12S* particles into "procapsids". Inside the procapsid, B proteins interact with each other across the twofold symmetry axes. On the outer surface of the procapsid, D proteins self-associate across both the twofold and threefold axes of symmetry.

Scaffolding proteins have a flexible structure

Several lines of evidence suggest that scaffolding proteins have structures that allow a high degree of adaptability. For example, the internal scaffolding proteins of the related phages φX174 and α3 have diverged 70% in amino acid sequence, yet they are cross-functional. In other words, the φX174 internal scaffolding protein can productively interact with the coat, spike and external scaffolding proteins of phage α3 to build procapsids. A similar phenomenon has been observed for the scaffolding proteins of various herpesviruses.

These findings suggest that internal scaffolding proteins may be inherently flexible and may undergo folding directed by their contacts with structural proteins. Considering the dynamics of virus assembly, some inherent flexibility is probably required. Internal scaffolding proteins must first assume a structure that directs the assembly of pentameric intermediates into a rigid capsid. Afterward, these proteins must be extruded through holes in the capsid, presumably by assuming an alternate, flexible structure.

The atomic structure of the φX174 procapsid (Figure 6.6) demonstrates that the external scaffolding protein can assume at least four different conformations. There are four D proteins for every viral capsid protein. Each D protein makes different contacts with the underlying capsid, neighboring D, and spike proteins.

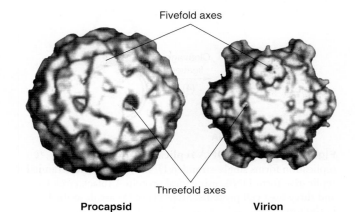

Figure 6.6 The φX174 procapsid and virion. Image reconstruction from cryoelectron microscopy. Note the pore at the 3-fold axis on the procapsid, absent in the virion.

In order for one protein to carry out such a wide variety of interactions, it must assume several different conformations. The ability to assume alternate structures challenges the widely held assumption that folded proteins are rigid and assume only one conformation.

Single-stranded genomes are packaged into procapsids as they are synthesized

Genome biosynthesis and packaging occur concurrently. In what is called Stage III replication, the viral C protein redirects plus-strand synthesis by binding to the replicative complex (shown in Figure 6.4) consisting of double-stranded DNA, the A protein, the rep protein, and DNA polymerase III. Instead of releasing free progeny plus-strands, this complex now associates with preformed procapsids (Figure 6.5) and inserts the plus-strand DNA into the procapsid as this DNA strand is displaced from the double-stranded template. The viral J protein

The evolution of a capsid assembly system using two scaffolding proteins

The microviruses are the only known non-satellite viruses requiring two separate scaffolding proteins. However, the related gokushoviruses encode only a single, internal scaffolding protein. It is possible that microviruses evolved from a gokushovirus ancestor. Once microviruses acquired an external scaffolding protein, which appears to be the more critical of the two, the internal scaffolding protein may have evolved into an "efficiency protein" that aids capsid assembly but is not strictly required for any one reaction.

If this hypothesis is correct, an inherent plasticity would allow other proteins to compensate for reduced B protein function or for its absence. To test this theory, viral mutants that allow capsid assembly in the absence of B protein were isolated. The resulting "B-free" variant of φX174 contains six mutations. Some of the mutations are located in the external scaffolding and coat protein genes, and allow these proteins to interact in the absence of protein B. Other mutations are regulatory and lead to the over-expression of the external scaffolding protein. In the absence of protein B, higher concentrations of the external scaffolding protein are required to drive capsid assembly.

(not SSB, as in normal replication) binds to the phage DNA being packaged and accompanies it into the procapsid. Packaging most likely occurs through one of the pores that are found at the threefold axes of symmetry in the procapsid but not in the virion (Figure 6.6). During packaging, B proteins are extruded from the procapsid, probably through the same pores.

After one round of replication, the viral A protein cleaves the plus-strand DNA and ligates the 3' end of the cleaved genome to its 5' end, creating a single-stranded circular molecule as in Stage II replication, but this time wholly contained within a procapsid. The replication complex, which retains proteins A and C, rep, and DNA polymerase III, can presumably bind to another procapsid to synthesize and package another phage genome. The packaged particle is called the provirion because it still contains the external scaffolding protein D. The provirion may represent the end of the intracellular assembly pathway. Upon cell lysis, the influx of divalent cations probably dissociates protein D, yielding the virion.

Role of the J protein in DNA packaging

The J protein is not only required for DNA packaging but may also play a role in organizing the DNA within the symmetrical capsid. Microvirus J proteins are small, basic polypeptides ranging from 20 to 40 amino acids. Their amino termini are rich in positively charged lysine and arginine residues, but their carboxy termini contain aromatic and hydrophobic amino acids. The positively charged amino acids in the amino terminus bind to and neutralize the negatively charged phosphate backbone of the viral DNA, while the hydrophobic carboxy terminus lodges into a cleft in the capsid protein, tethering the genome to the inner surface of the capsid. This tether prevents the single-stranded genome from forming secondary structure by base pairing with itself. Virions packaged with mutant J proteins have altered dimensions and lower infectivity, showing that the J tether is important in virion structure.

Cell lysis caused by E protein leads to release of phage

The E protein is responsible for cell lysis. The mechanism of lysis mediated by E protein is similar to that of antibiotics like penicillin, which inhibits cell wall biosynthesis. Recent studies demonstrate that the E protein most likely inhibits the host cell Mra Y protein. This protein is a translocase that catalyzes the formation of the first lipid-linked intermediate in cell wall biosynthesis. Penicillin inhibits the formation of cross-links within peptidoglycan, a major component in bacterial cell walls. In either case, the weakening of the cell wall causes an increased sensitivity to osmotic pressure, which leads to cell lysis.

Did all icosahedral ssDNA virus families evolve from a common ancestor?

There are four families of icosahedral single-stranded DNA viruses: *Microviridae* (bacteriophages), *Geminiviridae* (plant viruses), and *Circoviridae* and *Parvoviridae* (animal viruses). These families have remarkably similar features. They all have very small T = 1 capsids. Similarities between the atomic structure of parvovirus and microvirus capsid proteins are striking, as are the mechanisms employed for DNA packaging (however, parvovirus genomes are linear, not circular). Are parvoviruses and microviruses, one family replicating in animal cells, the other in bacteria, distantly related? Are they two extremes of the same superfamily?

Answers to these questions may likely be found by studying the recently discovered and poorly understood gokushoviruses (Table 6.1). The genomes of these viruses are smaller than the microvirus genomes. Their genomes encode primarily an A protein homologue, needed for DNA replication, and a capsid protein, although other genes are present. Parvoviruses (Chapter 20) similarly contain one DNA replication gene and one capsid protein gene, although alternative initiation sites and splicing produce several proteins from each gene. The prominent G protein spikes seen in the microviruses are not present in the gokushoviruses, nor is the external scaffolding protein. An added domain of the goshukovirus capsid protein performs some of the functions of the missing external scaffolding protein. This domain, located at the threefold axes of symmetry, forms protrusions called "mushrooms" on the surface of the virion, similar to those seen in some parvoviruses.

The hosts of gokushoviruses are obligate intracellular parasitic bacteria, some of which (chlamydia) multiply within eukaryotic cells. These hosts therefore provide a potential vehicle for bacteriophage genomes to gain access to a eukaryotic environment. Since most of the protein machinery used to replicate the genomes of these viruses is host-derived, substitution of the small part of the viral DNA that directs replication by bacterial host cell proteins (the replication origin) with a eukaryotic replication origin could potentially allow replication of a prokaryotic virus in a eukaryotic cell. Such a jump could have occurred during evolution.

Scaffolding proteins may increase rate of assembly and help small phages to compete against larger phages

The assembly pathways of all viruses include a rate-limiting step. Before this step can occur, capsid proteins must reach the critical concentrations needed to begin assembly. Once the critical concentrations are achieved, subsequent steps occur quickly, ensuring that the assembly of each capsid goes to completion.

For wild-type φX174, the time required to reach these critical concentrations is exceptionally short. Progeny virions can be detected as early as 5 minutes after infection. In contrast, the B-free variant (see text box on page 75) requires 20 minutes and a much higher critical concentration of the external scaffolding protein D. Thus, one of the advantages of the two-scaffolding protein system is speed.

Considering the niche in which the microviruses must compete, rapid replication may be the key to their survival. Approximately 97% of the known phages of free-living bacteria are large, double-stranded DNA viruses. Many of these phages encode proteins that participate in host cell takeover and that exclude the replication of other phages. While this may allow double-stranded DNA phages to produce high numbers of progeny particles, it delays their production.

The small coding capacity of microvirus genomes precludes the evolution of such elaborate mechanisms. Their survival strategy appears to be speed, producing progeny at a time when most double-stranded DNA phages are just beginning to make the transition to middle or late gene expression (see Chapters 7 and 8). The gokushoviruses, on the other hand, infect obligate intracellular bacteria, a niche where very few double-stranded DNA phages have been isolated. In fact, all six known phages of chlamydia, which grow only within eukaryotic cells, are gokushoviruses. It is tempting to speculate that the gokushoviruses, like the ancient coelacanth fish, have persisted by occupying a niche free of competition from larger phage species.

KEY TERMS

Helicase	Procapsid
Lipopolysaccharide	Rolling circle
Primosome	Scaffolding proteins

FUNDAMENTAL CONCEPTS

- Overlapping reading frames in genomes of microviruses maximize genetic coding capacity.
- Single-stranded DNA viruses must produce a double-stranded DNA before host cell RNA polymerase can transcribe viral genes.
- Replication of microvirus DNA occurs on a double-stranded DNA intermediate via a rolling circle mechanism.
- Synthesis of single-stranded microvirus genome DNA takes place concurrently with DNA packaging.
- Scaffolding proteins mediate capsid assembly by lowering thermodynamic barriers and ensuring the temporal nature and fidelity of the assembly pathway.
- Small, icosahedral, single-stranded circular DNA viruses may be evolutionarily related.

REVIEW QUESTIONS

1. What early discovery about bacteriophage φX174 challenged the concept of the gene?

2. What key advantage does the overlapping of reading frames impart?

3. How is minus-strand DNA synthesis carried out?

4. How is plus-strand DNA synthesis carried out?

5. What is the role of the J protein in φX174?

CHAPTER 7

Bacteriophage T7

William C. Summers

T7-like bacteriophages

T7 ("type 7"), from the collection of phages chosen
for study by the phage group

VIRION

Naked icosahedral capsid (T = 7).
Diameter 60 nm.
Short tail with short tail fibers.

GENOME

Linear double-stranded DNA, 40 kbp.
Terminal direct repeat sequence of 150 bp.

GENES AND PROTEINS

56 genes code for 59 proteins.
Three sequentially expressed classes of genes,
  organized from left to right end of genome:
  • *Class I ("early") genes: control of phage and host gene
    expression*
  • *Class II ("middle") genes: phage DNA replication*
  • *Class III ("late") genes: structural and assembly
    proteins, DNA packaging, cell lysis*

VIRUSES AND HOSTS

A genus within family *Podoviridae.* From Greek *podos*
  (foot), refering to short tail.

Similar phages infect a variety of bacterial hosts,
  including marine bacteria.
Examples: bacteriophages T3, φII, H, SP6,
  and gh-1.

DISTINCTIVE CHARACTERISTICS

Entry of DNA into cell is powered by transcription.
Novel 100 kDa phage-coded RNA polymerase
  transcribes class II and III genes.
Phage transcripts are cleaved by host RNAse III to
  small, stable mRNAs.
DNA synthesis is initiated by phage RNA polymerase.
Phage-coded DNA polymerase.

T7: a model phage for DNA replication, transcription, and RNA processing

Bacteriophage T7 is one of seven phages selected by a group of pioneer molecular biologists, headed by Max Delbrück, in the early 1940s to serve as prototypes on which the phage research effort should be focused. Some of these phages, including T7, were isolated from a commercially available mixture of phages being sold for treatment of intestinal infections. Each phage was designated by T (for "type") and arbitrarily assigned a number. T7 and its relative, T3, are the smallest of the T phages. They have a 60-nm icosahedral capsid and a short, stubby tail attached to one of the icosahedral vertices. T7 makes large plaques on susceptible host cells and has a linear genome of double-stranded DNA.

In comparison to the T-even phages (T2, T4, T6), phage T7 was ignored until the mid-1960s when its small size and easy-to-isolate DNA attracted the attention of both physical chemists and molecular geneticists. About this time, phage genetics blossomed because of the discovery of conditional-lethal nonsense and temperature-sensitive mutants. Prior to the development of these tools, phage genetics was limited to plaque-morphology and host-range mutants. Also in the 1960s, it became possible to study the genome of a phage by physicochemical techniques.

T7 DNA became one of the first well-studied phage genomes (along with that of phage φX174) and its molecular weight was established as 25×10^6 daltons, or 40,000 base pairs, both by classical physical determinations as well as by measurement of its contour length (12 μm) in electron microscope images.

Chemical analysis indicated that the ends of T7 DNA have unique sequences; this was a major achievement prior to the development of DNA sequencing methods. Thus T7 DNA does not have a circularly permuted sequence, as does the DNA of the unrelated phage T4. The separation of the complementary strands of T7 DNA was particularly easy by equilibrium sedimentation in CsCl in the presence of excess poly riboG. This technique facilitated the use of T7 DNA in many experiments requiring intact, complementary DNA strands.

F. William Studier and Rudolf Hausmann independently obtained large collections of conditional-lethal T7 mutants, which they mapped and classified in terms of phenotypes. Genes were originally given sequential numbers from 1 to 19; additional genes that were discovered subsequently were given decimal numbers such as 0.3 or 1.2 to indicate their position in the genetic map (Figure 7.1). In 1983, the complete sequence (39,937 base pairs) of T7 DNA was published along with identification of the major functional landmarks on the genome.

Searches for phages with properties similar to T7 have yielded many isolates, some of which are nearly identical to T7 and some of which are more distantly related. Phages in the T7-like group infect many different host bacteria, including marine organisms, the plague bacillus (*Yersina pestis*), *Serratia*, *Klebsiella*, and *Pseudomonas*.

T7 genes are organized into three groups based on transcription and gene function

The sequence of T7 DNA predicts that it can encode 56 proteins, defined by open reading frames with recognizable ribosome binding sites and protein synthesis initiation sequences. Fifty-nine phage proteins have been identified by genetic or biochemical evidence; some genes give rise to more than one protein by initiation of translation at internal AUG codons.

T7 has a logical arrangement of genes that could have been designed by an engineer: three groups of genes are organized in continuous blocks from left to right on the genome (Figure 7.1). These genes are clustered both by function and by the timing of their expression during the life cycle of the phage. Class

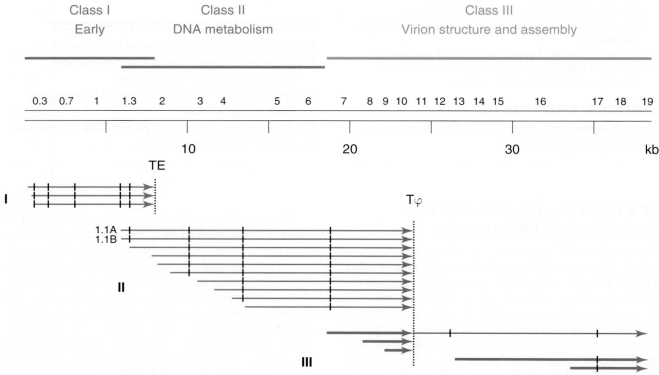

Figure 7.1 Genetic and transcriptional map of the T7 genome. The 40 kb T7 genome is shown with the positions of selected genes immediately above the map. Positions of major gene classes I, II, and III are indicated by heavy lines. Major transcripts from the three class I promoters, the ten class II promoters, and the five class III promoters are shown below in light blue, with vertical hash marks indicating ribonuclease III cleavage sites. TE and Tφ show positions of transcriptional terminators. Thickness of lines indicates approximate transcript abundance. Arrowheads denote 3' ends of RNAs. Low level of readthrough of Tφ is indicated on one class III transcript but can occur on all transcripts.

I ("early") genes, located in the left-hand one-fifth of the genome, are expressed during the first 6 to 8 minutes after infection (at 30 C). Class I gene products are needed for expression of the remainder of phage genes and to shut down host cell functions. Class II ("middle") genes occupy the next one-third of the genome. Most of the class II gene products are concerned with phage DNA replication. Class III ("late") genes, encoded in the right half of the genome, code for phage structural proteins and proteins needed for DNA maturation and packaging. All T7 genes are transcribed from left to right on the map as shown.

Entry of T7 DNA into the cytoplasm is powered by transcription

The icosahedral T7 capsid (see page 77) is constructed from a single capsid protein, gp10. The tail is attached at one vertex to a portal consisting of 12 molecules of gp8. The tail consists of two phage proteins, gp11 and 12, as well as six fibers made from trimers of gp17. Three additional proteins (gp14, gp15, and gp16) form an internal core around which the DNA is wrapped. The complex structures of the tail and the internal core provide the machinery for packaging the viral genome and for inserting it into cells during infection.

T7 phage binds to **lipopolysaccharide** molecules in the outer cell membrane via its tail fibers (Figure 7.2). The three internal core proteins are then inserted via the tail into the cell, forming a channel through the outer membrane, the peptidoglycan cell wall, and the cytoplasmic membrane. There is evidence that one of

these proteins, gp16, is able to enzymatically degrade the peptidoglycan layer as it bores through, creating a hole for the passage of the DNA.

Gp16 is also responsible for transporting the first 850 base pairs of the left end of T7 DNA through this channel into the cytoplasm, exposing three closely spaced promoters recognized by the host *Escherichia coli* RNA polymerase. At this point no more DNA enters the cell unless transcription is allowed to proceed. The host cell RNA polymerase transcribes the class I genes, simultaneously spooling the DNA out of the capsid into the cytoplasm at a rate of 40 bp per second. Thus RNA polymerase serves as a winch to transfer the DNA from the capsid into the cell while making messenger RNAs (Figure 7.2). *E. coli* RNA polymerase encounters a strong termination signal (TE) about 8 kb into the genome from the left end (Figure 7.1), where it releases the mRNA and detaches from the DNA.

Transcription of class II and III genes requires a novel T7-coded RNA polymerase

The initial steps of infection described above can take place in the presence of an inhibitor of protein synthesis, but further transcription and entry of the remainder of the genome is blocked unless phage class I proteins are synthesized. Systematic examination of infections by phage with mutations in class I genes revealed that the protein product of gene 1.0 is essential for class II and class III gene transcription. This protein was isolated and found to be a monomeric, 100 kDa RNA polymerase that is highly specific for class II and class III T7 phage promoters. Up to that time, all known RNA polymerases were multi-subunit enzymes, so the discovery of this single-subunit RNA polymerase, which does not depend on any host proteins for function, galvanized the transcription field. It has since become an important part of the toolbox of molecular biologists (see below).

T7 RNA polymerase not only takes over transcription of the phage genome, but it also continues the work begun by the host RNA polymerase of spooling the phage DNA into the cell. DNA entry at this stage progresses at about 250 bp/second, reflecting the high rate at which the phage RNA polymerase transcribes DNA.

Two other class I gene products play important roles in the infection (Table 7.1). The product of gene 0.3 inactivates host type I restriction endonucleases EcoB and EcoK, protecting the infecting DNA from cleavage and degradation. Notably, this is the first T7 gene transcribed as the DNA enters the cell, so the protection is established rapidly. Gp0.7 inactivates the host cell RNA polymerase by phosphorylation of a specific tyrosine residue, limiting the amount of class I mRNAs synthesized and interrupting further class I gene expression. This also shuts off cellular mRNA synthesis, freeing up energy and nucleotides for phage transcription.

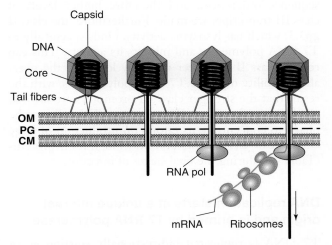

Figure 7.2 Entry of T7 genome powered by transcription. T7 virion is shown attached to outer cell membrane via tail fibers (left image). Subsequent images show channel formation and entry of the end of the DNA (black), followed by binding of RNA polymerase and transcription-mediated entry of the remainder of the DNA. OM: outer membrane; PG: peptidoglycan layer; CM: cytoplasmic membrane.

Table 7.1 Selected class I gene products

Gene	Function
0.3	Inhibits host type I restriction endonucleases
0.7	Inhibits *E. coli* RNA polymerase
1.0	T7-specific RNA polymerase

Class II genes code for enzymes involved in T7 DNA replication

Promoters recognized by the phage RNA polymerase are highly conserved, extending from 17 nt before to 6 nt after the transcription start site. Ten different class II promoters are used to initiate transcription over a 10-kb region of the T7 genome (Figure 7.1). Most of the RNAs initiated at these promoters end at a single, stong terminator (Tφ) located 24 kb from the left end of the genome. A minority of polymerases reads through this terminator to form longer RNAs.

The use of multiple promoters leads to the production of a **nested set** of overlapping primary transcripts of the class II region with a common 3' end. This allows the phage to make lower levels of gene products coded at the left end, and higher levels of gene products coded at the right end of this part of the genome. Twenty-nine class II gene products are made from the resulting mRNAs.

Most of the class II proteins participate in T7 DNA replication and recombination (Table 7.2). The products of genes 3 and 6 degrade the host DNA to provide nucleotide precursors for the synthesis of T7 DNA. Gene products 1.3, 2.5, 4, and 5 act at the replication fork to synthesize progeny DNA molecules. Gene 2, like gene 0.7, inhibits host RNA polymerase, ensuring that class I genes and host transcription are shut down. The products of genes 1.2 and 1.7 also contribute to the cessation of host transcription by complex effects on nucleotide pools.

Table 7.2 Selected class II gene products

Gene	Function
1.3	DNA ligase, joins lagging-strand DNAs
2	Inhibits *E. coli* RNA polymerase
2.5	Single-stranded DNA binding protein
3	Endonuclease, degrades host cell DNA and cleaves T7 DNA concatemers
3.5	Lysozyme, inhibits T7 RNA polymerase
4	RNA primase/helicase, initiates lagging-strand DNA synthesis
5	DNA polymerase
6	Exonuclease, degrades host cell DNA

T7 RNAs are cleaved by host cell ribonuclease III to smaller, stable mRNAs

The primary RNA transcripts of T7 phage undergo post-transcriptional processing by the host enzyme, ribonuclease III. This ribonuclease recognizes double-stranded RNA regions and cleaves them at specific sites. The long class I transcripts synthesized by the host RNA polymerase are cleaved into 2 monocistronic mRNAs and 3 polycistronic mRNAs (Figure 7.1). The class II and class III primary transcripts are also cleaved at specific sites by ribonuclease III to form smaller functional mRNAs. The stem structures that remain at the 3' ends of the mature T7 mRNAs after cleavage render them relatively resistant to degradation by other host nucleases, making the phage mRNAs substantially more stable and long-lived than most host cell mRNAs.

Class III gene expression is regulated by delayed entry and by promoter strength

As the phage DNA enters the cell via transcriptional winching, the five class III promoters become available for transcription by T7 RNA polymerase. At an entry rate of 250 bp/second, it takes about 1 minute of transcription starting at the first class II promoter (5.9 kb from the left end) until the first class III promoter (19 kb from the left end) enters the cell, and about 2 minutes until the most distal class III promoter (34 kb from the left end) enters (Figure 7.1). Thus transcription-dependent DNA entry is one factor responsible for the delayed transcription of class III genes compared with class II genes.

Class III promoters have a higher affinity for T7 RNA polymerase than class II promoters due to slight sequence differences, and therefore higher levels of class III transcripts are made. Furthermore, the class II gp3.5, which has lysozyme activity, binds specifically to T7 RNA polymerase and inhibits its activity. This can reduce class II promoter usage by lowering the overall availability of active T7 RNA polymerases, with the result that the stronger class III promoters are favored over the weaker class II promoters. This allows the phage to concentrate its efforts on production of large amounts of virion structural and maturation proteins (Table 7.3) during the final stages of infection.

DNA replication starts at a unique internal origin and is primed by T7 RNA polymerase

T7 DNA is replicated bidirectionally starting at an origin about 5.9 kb from the left end of the DNA. Alternate origins are known to exist because T7 mutants that have lost the primary origin are able to replicate well. Short DNA fragments joined to RNA primers are found in replicating T7 DNA, so it is clear that T7 DNA replication resembles that of the *E. coli*

Table 7.3 Selected class III gene products

Gene	Function
8	Portal protein, connector for tail
9	Scaffolding protein for formation of capsids
10	Capsid protein
11	Tail protein
12	Tail protein
14, 15, 16	Core proteins, form channel for DNA entry
17	Tail fiber, attachment to cell
17.5	Holin, access of lysozyme to cell wall for lysis
18, 19	Maturation proteins, DNA packaging

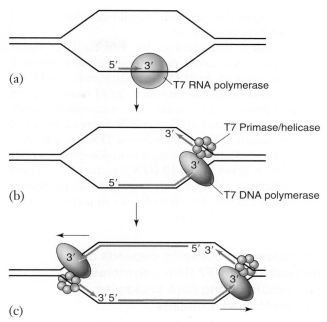

Figure 7.3 Initiation of DNA replication at the leftward T7 promoter. (a) T7 RNA polymerase (green sphere) initiates an RNA chain (light blue arrow) at promoter 1.1. (b) T7 RNA polymerase is displaced by T7 DNA polymerase (dark blue oval) in an AT-rich region shortly downstream of the promoter. T7 DNA polymerase uses the RNA as a primer for synthesis of DNA (gray arrow). T7 primase/helicase (orange hexamers) loads onto the replication fork and generates short RNA primers (light blue arrows) for DNA synthesis on the opposite strand. (c) Primers are extended by T7 DNA polymerase, and replication forks move toward opposite ends of DNA (horizontal arrows).

host cell, using continuous synthesis on the leading-strand and discontinuous synthesis with Okazaki fragments on the lagging-strand (Figure 7.3). However, T7 uses its own replication proteins.

Surprisingly, transcription by T7 RNA polymerase was found to be necessary for the initiation of T7 DNA replication. It turns out that the RNA polymerase provides the first primer for leading-strand synthesis. The two class II promoters closest to the left end of the DNA, promoters 1.1A and 1.1B, precede an AT-rich region on the DNA. When sufficient levels of class II proteins have been made, the T7 DNA polymerase can displace the RNA polymerase in the AT-rich region shortly after it has begun transcription at one of these promoters. The DNA polymerase will then use this nascent RNA chain as a primer, and begin synthesizing DNA in a rightward direction. The gene 4 primase/helicase protein subsequently binds to this replication fork, laying down a short RNA primer on the opposite ("lagging") DNA strand and simultaneously unwinding the DNA to allow the T7 DNA polymerase to continue forward on the leading-strand. Thus a replication fork is born.

T7 DNA polymerase can then extend the new primer toward the left, setting up a second replication fork and establishing bidirectional replication. Note that initiation of DNA replication at the leftmost T7 RNA polymerase promoter avoids head-on collisions between DNA polymerases and RNA polymerases, because they all travel in the same direction; transcription from class I genes, which could affect replication moving leftward, has ceased.

The T7 DNA polymerase (gp5) has structural similarity to mammalian DNA polymerase β and HIV reverse transcriptase. It forms a stable 1:1 complex with a normal host cell protein, E. coli thioredoxin. This cell protein is the only host factor required for T7 DNA replication. While T7 DNA polymerase alone can catalyze the addition of dNTPs to a growing DNA chain, the binding of thioredoxin greatly increases its rate and **processivity** (the tendency to continue on the same DNA template-primer complex). Thioredoxin interacts electrostatically

with the DNA polymerase to tether the polymerase to its template DNA strand. It also interacts with the gene 4 helicase to stabilize the replication complex. T7 DNA polymerase can also act as a "proofreading" 3' to 5' exonuclease to remove misincorporated nucleotides.

Recombination between both input and progeny T7 genomes occurs and allows standard genetic mapping by recombination frequency measurements. This recombination depends on the products of genes 3, 4, 5, and 6 as shown both *in vivo* and *in vitro*.

Large DNA concatemers are formed during replication

A classical problem with RNA primer-directed replication of linear DNAs is that the last primer laid down at the 3' end of each template DNA strand cannot be removed and replaced with DNA, because there is no sequence beyond the 3' end to which the next primer could bind (see also Chapter 20). Thus during each round of replication, a short sequence would be lost at each end of the DNA. Some viruses form circular intermediates that overcome this problem, but no circular forms of T7 DNA have been found.

T7 solves this problem by forming large linear DNA oligomers, called **concatemers**, via a terminal repeat sequence of about 150 bp at each end of its DNA. This repeated sequence may be a site of recombination between genomes to generate the concatemers, which contain only one copy of the repeat between each copy of genome-length DNA. Concatemeric DNAs accumulate in cells infected with T7 phage mutated in gene 3 (endonuclease) or genes 18 or 19 (DNA maturation). These gene functions are therefore implicated in processing concatemers to unit-length molecules in preparation for packaging into phage heads.

Concatemer processing depends on transcription by T7 RNA polymerase and occurs during DNA packaging into preformed proheads

Surprisingly, transcription by the T7 RNA polymerase is also required for cleavage of concatemers. RNA polymerases that initiate transcription at class III promoters do not encounter strong termination signals before the right-hand end of the viral DNA. When they are transcribing concatemers, RNA polymerases therefore pass through the concatemer junction. At a specific 8-nt sequence in the concatemer junction, RNA polymerases pause, forming an unusual DNA structure that recruits the DNA processing machinery, which cleaves the DNA at this 8-nt sequence. Copying of the remainder of the repeated sequence allows formation of a second repeat and release of a full-length DNA molecule containing repeats at both ends. For an idea of how this is carried out, see Figure 20.4.

Assembly of progeny virions begins by the formation of an empty **prohead** structure that includes the gene 9 product as well as the proteins of the mature capsid. Gene 9 protein functions as a **scaffolding protein** for assembly of the prohead. Cleavage of concatemers and copying of the terminal repeat is carried out during packaging of the DNA into the prohead, with the help of gene products 18 and 19. After head maturation, the scaffolding protein dissociates, leaving T7 virions.

The lysis of the infected cell to liberate approximately 100 progeny phage particles is dependent on two T7 functions. The gene 17.5 protein is a small membrane-specific protein called a **holin**. Under the appropriate conditions it assembles in the cytoplasmic membrane of the infected cell to create a hole that allows access to the bacterial cell wall of the phage endolysin, the product of gene 3.5 (T7 lysozyme). The endolysin, once in the periplasmic space, degrades the cell wall, leading to cell lysis.

Special features of the T7 family of phages

T7 and related phages have a number of properties that make them objects of special interest to biologists as well as to biotechnologists: (1) These phages have mechanisms for modulating and overcoming the classic host restriction-modification system; (2) T7 and many of its relatives are able to grow efficiently only in *E. coli* lacking the fertility plasmid F; that is, they are female-specific or "F-restricted" phages; (3) finally, because T7 is so well-studied and replicates rapidly, it can be used to study microbial evolution.

Overcoming host restriction/modification. Because of the widespread possibility of horizontal gene transfer by exchange of DNA between bacteria, many bacteria have developed mechanisms to recognize and degrade foreign DNA, which might otherwise carry genes of lesser fitness from unrelated organisms. One such mechanism employs restriction endonucleases that recognize specific DNA sequences as sites for degradation. These enyzmes are prevented from acting on the cell's own DNA by modification of the host cell DNA, frequently by methylation of one of the nucleotides in the restriction enzyme recognition sequence.

Since phage infections involve introduction of foreign DNA into the host cell, phages are often the target of this sort of restriction. Unless the immediately prior host cell has "marked" the phage DNA with the host-specific modification, allowing it to escape the restriction endonucleolytic attack, the phage DNA may be cleaved and the infection aborted. Indeed, it was the discovery of this host-controlled modification in the early 1950s by Salvador Luria and Mary Human and by Giuseppe Bertani and Jean Weigle that led to the identification and characterization of site-specific restriction endonucleases.

Bacteriophage T7 and related phages have evolved a strategy to avoid these methylation-dependent restriction systems by inactivating the restriction endonuclease. Some members of the T7 family of phages supplement this strategy by destroying the methyl group donor substrate (*S*-adenosyl methionine) present in the host cell. The product of the class I phage gene 0.3 (OCR, overcoming classical restriction) is a protein with an unusual form of molecular mimicry. Its surface displays an array of acidic residues that mimics the phosphate backbone of DNA, similar to a 24 base-pair segment of bent DNA. This protein acts as a competitive inhibitor of bacterial type I restriction enzymes by mimicking their DNA substrate. In some cases (but not in T7) this gene product also exhibits S-adenosylmethionine hydrolase activity (SAM-ase), which depletes the source for methyl groups for DNA modifications.

F-restriction. T7 (but not T3) is unable to replicate in cells containing the F-factor. The abortive phage infection progresses more or less normally in the early phase, but the class II and class III mRNAs fail to express proteins. This restriction is under the control of several genes carried by the F-factor. These genes are designated *pif*A and *pif*B (for *phage inhibition by F*). The

Genetic engineering with T7

The unique properties of the RNA polymerases of the T7 family of phages make them very useful tools for biotechnological and genetic engineering applications. T7 RNA polymerase is easily purified, is a very efficient and rapid enzyme, and is highly specific for its own late promoter sequences. Therefore, insertion of a T7 late promoter sequence upstream of any DNA sequence allows *in vitro* synthesis of substantial quantities of homogeneous RNA from the desired DNA sequence. This strategy has been used to prepare specific RNA substrates for study of RNA conformation, ribozymes, and RNA processing.

Because the RNA polymerases from the T7 group (T7, T3, and SP6 phages) differ in promoter recognition, it has been possible to make plasmids with several different promoters, so that transcription patterns can be controlled by the choice of which RNA polymerase is used. For example, cloning vectors with both T7 and T3 promoters arranged to promote convergent transcription from opposite DNA strands have been useful in applications of sequence analysis to both DNA strands of the same fragment and in preparation of antisense RNAs.

The T7 RNA polymerase/late phage gene promoter system has been exploited to generate a family of highly efficient gene expression vectors, called pET vectors (for plasmid, expression, T7 [Figure 7.4]). This expression system has the dual advantages of strict regulation of expression and high yields upon induction. The gene of interest, say gene X, is transcriptionally silent until expression is activated, at which time a massive takeover of the cellular transcription and translation capacities occurs. Gene X is placed under the control of a late T7 phage promoter and thus cannot be expressed in a normal bacterial cell. This is an important advantage if the product of gene X is toxic or even mildly detrimental to the growth of the host bacterium (in which case there is selection against plasmid retention and/or gene X functionality).

A plasmid containing the gene X sequence under the control of the T7 promoter is introduced into a bacterium in which the T7 RNA polymerase is made from an inducible lac operon promoter. When an inducer of the lac promoter is added, T7 RNA polymerase is made and gene X is transcribed. Often, a very efficient class III T7 ribosome binding site is included in the plasmid construction. Under these conditions, not only is the mRNA for gene X the major RNA in the cell, but it also is able to compete effectively for most of the protein synthetic apparatus, so that gene X product becomes the major protein in the cell. It is common to achieve expression levels so high that the desired protein product represents 25% to 50% of the total cell protein. Thus, only a few-fold purification is needed to obtain homogeneous protein X for further study.

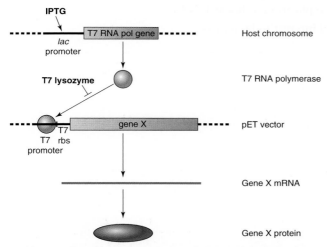

Figure 7.4 Expression of foreign genes in T7-derived expression vectors. The gene to be expressed (X) is inserted in a pET vector downstream of a T7 late promoter and a T7 gene 10 ribosome-binding site (rbs). This vector is introduced into a host cell that contains the T7 RNA polymerase gene under the control of the *lac* operon regulatory region (*lac* promoter). An inducer of the *lac* operon (IPTG) turns on T7 RNA polymerase expression and leads to expression of gene X in the pET vector. A second plasmid (not shown) expresses low levels of T7 lysozyme, which inhibits the activity of the small amount of T7 RNA polymerase made in absence of the inducer, allowing genes coding for toxic proteins to be cloned.

molecular mechanism of this apparently translational inhibition is not yet clear.

Experimental evolutionary studies. In recent years, T7 phage has emerged as a useful model system for the study of microbial evolution because of its rapid reproduction, relatively simple structure, well-understood biology, and simple regulatory strategies. Growth under various applied selections, selections starting with deliberately constructed ancestors, and tests of host/phage evolution have been employed to study predictive models, concepts of optimality, and traditional evolutionary arguments. T7 allows *in vitro* evolution in a reasonable time frame and yields results that can be analyzed at both the molecular and biological level. One general conclusion of this new field appears to be that wild type T7 is both a robust and highly optimized bacteriophage.

KEY TERMS

Concatemer	Processivity
Holin	Prohead
Lipopolysaccharide	Scaffolding protein
Nested set	

FUNDAMENTAL CONCEPTS

- The linear, double-stranded genome of bacteriophage T7 enters the cell in three phases:
  1. the first 850 bp enter spontaneously via the phage tail upon binding of the phage to the cell
  2. the next 7 kb of DNA are pulled into the cell during transcription by cellular RNA polymerase
  3. the final 32 kb are pulled into the cell during transcription by the phage T7 RNA polymerase
- Class I ("early") genes are transcribed by cellular RNA polymerase and include the gene coding for phage T7 RNA polymerase.
- A terminator of transcription, TE, releases the host cell polymerase at the end of the class I region.
- Class II ("middle") genes are transcribed by the phage T7 RNA polymerase, and include genes coding for DNA replication functions.
- A terminator of transcription, Tφ, releases the T7 polymerase at the end of the class II region.
- Class III ("late") genes are also transcribed by the T7 polymerase, and include genes coding for structural phage proteins and proteins involved in release of progeny phage.
- T7 DNA replication begins at the first transcription initiation sites for T7 RNA polymerase and is mediated by RNA primers made by T7 RNA polymerase.
- Linear concatemers of T7 DNA are formed during DNA replication, probably by recombination via a terminal direct repeat sequence.
- DNA concatemers are cleaved within the terminal repeats in a reaction dependent on transcription by T7 RNA polymerase and are simultaneously packaged in phage proheads, which mature by shedding scaffolding proteins.

REVIEW QUESTIONS

1. What attracted scientists in the 1960s to the T7 phage as a prototype for research?
2. What basic functions do bacteriophage T7 Class I–III genes perform?
3. What was discovered about T7 RNA polymerase that galvanized the field of transcription?
4. How does T7 solve the problem of RNA primer-directed replication of linear DNAs?
5. What three features make T7 and related phages of special interest to biologists?

CHAPTER 8

Bacteriophage Lambda

Michael Feiss

λ-like bacteriophages

λ is the Greek letter "lambda"

VIRION

Head is a naked icosahedral capsid (T = 7), diameter 63 nm.

Flexible tube-shaped tail, 135 nm long, with terminal fiber.

GENOME

Linear double-stranded DNA, 40–60 kb.

Ends have complementary single-stranded extensions (cohesive ends).

GENES AND PROTEINS

Codes for approximately 60 proteins.

Transcribed by cellular RNA polymerase.

Genes grouped by function.

Sequential expression of "immediate-early", "early", and "late" genes.

Transcription of DNA is controlled by:

- *CI, cro: transcription repressors*
- *N, Q: transcription antiterminators*
- *CII: transcription activator*

VIRUSES AND HOSTS

A genus within the family *Siphoviridae.* From Greek **siphon** (tube), referring to long, non-contractile tail.

Related phages: 434, φ80, 21, HK022, N15.

Some phages in the *Myoviridae* and *Podoviridae* families share segments of genome homology with λ phage. Examples: P22, L, LP-7.

DISTINCTIVE CHARACTERISTICS

DNA circularizes upon entry via cohesive ends.

A temperate phage: DNA can integrate into the host cell chromosome, creating a lysogen.

Complex lysogenic and lytic regulatory circuits.

DNA packaged by cleavage from linear concatemers.

Engineering of λ DNA has generated a variety of powerful cloning vectors for genetic and genomic studies.

Many temperate phages carry genes encoding toxins and other proteins that alter the host phenotype.

"For strange effects and extraordinary combinations we must go to life itself, which is always far more daring than any effort of the imagination." **Sherlock Holmes**

Roots . . .

Infection of *Escherichia coli* by phage λ can have two possible outcomes. The lytic growth cycle produces up to 200 progeny viruses per infected cell and causes cell lysis. Alternatively, the infecting λ DNA can enter a **prophage** state, in which the lytic genes are repressed and the viral DNA integrates into and is passively replicated with the host chromosome (Figure 8.1). Phages that can either grow lytically or exist as a repressed prophage are called **temperate phages**. The prophage makes a repressor that turns off expression of the λ lytic genes. In addition to inhibiting prophage expression, the repressor turns off the lytic genes of any incoming λ chromosome. Thus an *E. coli* cell carrying a λ prophage is immune to infection by λ. Such an *E. coli* cell is called a **lysogen**, because the expression of the prophage can

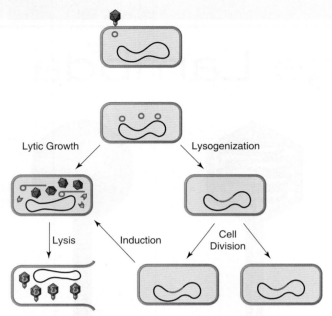

Figure 8.1 Outline of λ biology. A bacterial cell is infected by a λ phage particle. Phage DNA is in orange; phage heads are shown as hexagons, tails as bullets, and the bacterial chromosome is shown as a black circular wavy line. A decision is made during infection to progress to lytic growth (virus production and lysis) or lysogeny (integration of viral DNA and continued growth of the cells). Induction is the switch from lysogeny to lytic growth.

be induced, causing prophage excision, genome replication, and cell lysis (*lysogen* means "able to generate lysis").

The standard K-12 laboratory strain of *E. coli* was collected in 1922 from a patient with an intestinal disorder. It was not known until 1951 that K-12 carries a λ prophage. Esther Lederberg happened to spread an agar plate with a mixture of a K-12 (λ⁺) culture supernatant and cells of a strain derived from K-12 that had fortuitously lost its λ prophage, that is, was (λ⁻). The K-12 (λ⁻) cells grew to form a cell lawn, but there were plaques in the lawn caused by λ virions in the supernatant of the K-12 (λ⁺) culture.

In the late 1950s, Dale Kaiser began studying genes involved in the establishment of lysogeny. Meanwhile, Allan Campbell isolated a large set of mutants with conditional lethal mutations. Campbell proposed in 1961 that the prophage inserts into the bacterial chromosome by site-specific recombination between a circular form of the λ chromosome and the bacterial chromosome.

In 1963, Alfred Hershey and coworkers, and Hans Ris, showed that the ends of λ chromosomes could bind to each other, cyclizing the molecule. In 1971, Wu and Kaiser showed that the cohesive ends were complementary, single-stranded, 12 base-long extensions at the 5' ends of each λ DNA strand. In 1982, Fred Sanger and coworkers published the sequence of the 48,502 bp λ DNA molecule. The genetic map (Figure 8.2) shows that λ genes are arranged in functional blocks. For example, the 10 head assembly genes are followed by 11 tail assembly genes, and so on.

Phage adsorption and DNA entry depend on cellular proteins involved in sugar transport

Infection begins with binding to the cell surface and injection of the DNA into the cell. At the tip of the λ tail (Figure 8.3) is a single fiber, the J protein. J binds to an *E. coli* outer membrane protein, LamB, named because it confers susceptibility to λ infection. LamB is a maltose-inducible porin for transport of maltose and maltose oligosaccharides. Injection of λ DNA through the tail and into the cell requires the cytoplasmic membrane components of another sugar transport system, mannose permease. Injection is fascinating and more than a little mysterious. How λ DNA traverses the outer and inner lipid membranes and the periplasm of the gram-negative *E. coli* cell is by no means clear.

Following injection, λ DNA circularizes via the *cos* ("cohesive") sequences at its ends, and the circle is sealed by host DNA ligase. Transcription by host RNA polymerase then begins at two promoters, P_L and P_R (initiating transcription toward the left and right, respectively), located on either side of the repressor gene, *cI* (Figure 8.4).

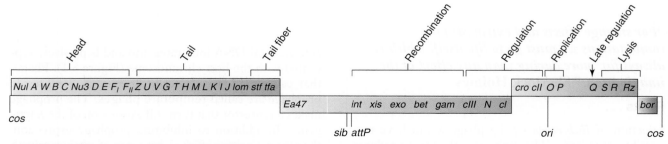

Figure 8.2 The genetic map of λ DNA. Genes are functionally grouped. Genes boxed above the central line are transcribed from left to right, those below are transcribed from right to left. *cos*: cohesive end site; *sib*: site encoding a transcription termination sequence, or if antiterminated, a site for RNAse III attack; *attP*: phage attachment site for integration; *ori*: origin of DNA replication.

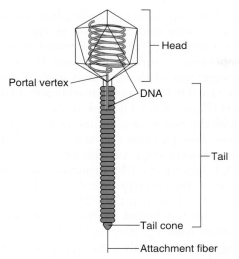

Figure 8.3 Schematic drawing of λ virion. DNA is shown as a gray coil within the capsid.

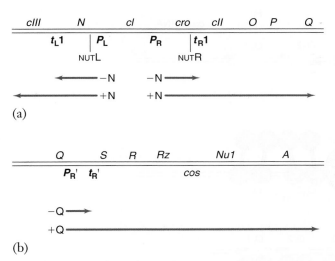

(a)

(b)

Figure 8.4 The lytic transcription program. Genes encoding proteins are drawn above the double line, and promoters and terminators are below the double line. Blue arrows represent transcripts. (a) Immediate early transcription (−N) and early, N-antiterminated transcription (+N) initiated at P_L and P_R. (b) Late transcription, sponsored by Q-dependent antitermination of the $P_R{}'$ transcript (which continues beyond gene A, through all late genes).

The λ lytic transcription program is controlled by termination and antitermination of RNA synthesis at specific sites on the genome

Transcripts from P_L and P_R express the N and cro genes; these "immediate-early" transcripts end at transcription terminators t_L1 and t_R1 (Figure 8.4a). N protein is a transcriptional antiterminator that, with a recruited

CRISPR elements

A recently discovered class of genetic elements found in roughly one-half of sequenced bacterial genomes provides resistance to phage infection. The resistance elements are arrays of <u>c</u>lustered, <u>r</u>egularly <u>i</u>nterspaced <u>s</u>hort <u>p</u>alindromic sequences, called CRISPR elements. A CRISPR element consists of several protein encoding genes and an array of direct repeats, 25–50 bp long and in some cases containing short palindromes. The direct repeats alternate with non-repetitive spacers, 25–75 bp long. The spacer segments frequently contain sequences that match sequences found in bacteriophage genomes.

Experimental observations suggest that when a population of CRISPR-containing cells is infected by a bacteriophage, short DNA sequences derived from the phage chromosome are processed and inserted as spacer sequences in the CRISPR element. Aquisition of these phage sequences confers host cell resistance to infection by the phage. It is proposed that the CRISPR element expresses RNAs containing the phage sequences, which act as antisense RNAs, preventing translation of phage mRNAs. Much remains to be understood about the details of this intriguing phage resistance mechanism.

ensemble of four host proteins, modifies RNA polymerase so that transcripts initiated at P_L and P_R are no longer terminated. For N to function, it binds to specific mRNA sequences, called NUT (*N-ut*ilization) sequences (Figure 8.4a). Because NUT sequences are found only on transcripts made from P_L and P_R, these are the only transcripts that are antiterminated by N. N acts by a mechanism analogous to that of the Tat protein of human immunodeficiency virus (see Chapter 29).

Once N is made in sufficient quantity, "early" transcription can take place. N-modified RNA polymerase simply transcribes through terminators t_L1 and t_R1 without taking any notice of them. The N-antiterminated transcript from P_L extends leftward through the general recombination genes *gam*, *bet*, and *exo*, the site-specific recombination genes *xis* and *int*, across the locus for site-specific recombination, *attP*, all the way to *Ea47* (Figures 8.2 and 8.4a). Similarly, the N-antiterminated transcript from P_R extends through *cII* and the replication genes *O* and *P* all the way to gene *Q*, which encodes the late gene activator.

Q is also a transcriptional antiterminator. The $P_R{}'$ promoter (Figure 8.4b) constitutively expresses a short transcript that ends at $t_R{}'$. Q antiterminates this short $P_R{}'$ transcript, leading to production of a very long (~25 kb) "late" gene transcript of the cell lysis, head, tail, and tail fiber assembly genes. Circularization of λ DNA joins the late structural genes, originally on the left half of the genome (Figure 8.2), to the right end of the DNA, allowing production of this long late transcript. Transcription from P_R antiterminated by N also proceeds beyond the Q gene into the late genes; this transcription gives a level of late gene expression equal to about 20% of the P_R', Q-sponsored level.

The CI repressor blocks expression of the lytic program by regulating three nearby promoters: P_L, P_R, and P_{RM}

When λ infects active cells in logarithmic growth, the phage grows lytically in most of the cells. However, when λ infects metabolically sluggish, stationary phase cells, most of the infected cells are lysogenized. The decision to go lytic or lysogenic rests on whether sufficient repressor is produced to shut down transcription of the lytic genes. The repressor is the CI protein (pronounced "see-one"), product of gene *cI*.

As a first step in understanding how the level of repressor is controlled, let's look at repressor synthesis in a prophage. Three repressor-binding sites, called **operators**, overlap the P_L and P_R promoters. The arrangement of the operators at P_R is shown in Figure 8.5a and b; P_L has a similar arrangement of three operators (Figure 8.5c). CI binds to the operators as a dimer with the following affinities: $O_R1 > O_R2 = O_R3$. CI dimers show pairwise cooperativity, so that when a CI dimer binds to O_R1, it facilitates the binding of a second CI dimer to O_R2 by protein–protein interactions between the two repressor dimers (Figure 8.5b).

O_R1 and O_R2 overlap the -10 and -35 segments of P_R, so that CI bound to these operators blocks transcription from P_R. Because the CI dimers cooperate only in pairs, binding to O_R3 is not facilitated by repressors cooperatively bound at O_R1 and O_R2. It has been postulated that CI cooperativity is pairwise

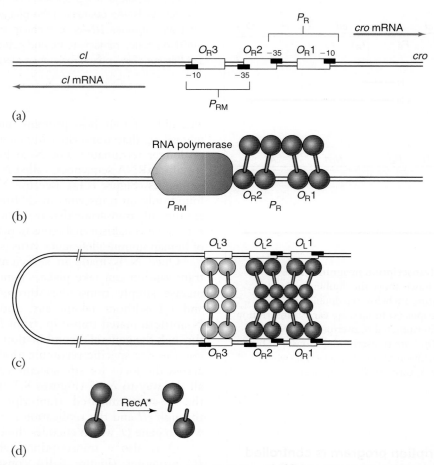

(a)

(b)

(c)

(d)

Figure 8.5 Transcription and its repression at P_{RM} and P_R. (a) Region coding for CI and Cro. Boxes indicate operator sites for binding by CI repressor. Brackets indicate promoters P_{RM} and P_R, with their -10 and -35 boxes shown. Blue arrows denote the P_{RM} (CI) and P_R (Cro) transcripts. (b) Activation of P_{RM} by CI prophage repressor. CI monomers are shown as dumbbells, and RNA polymerase is shown as a large orange asymmetric arrow. (c) The CI repression loop. The upper arm is P_L; the lower arm is the P_{RM}–P_R region (also shown in b). A CI octamer formed through long-range cooperative interactions is shown in gray. Cooperative interactions of CI dimers at O_L3 and O_R3 are required for P_{RM} repression; these dimers are shown in light green. (d) The cleavage of CI stimulated by activated RecA protein.

because the cooperating repressors must lean toward each other, so CI at O_R2 leans toward CI at O_R1 (Figure 8.5b), and cannot interact with a third repressor at O_R3.

A second promoter, called P_{RM} for repressor maintenance, is located near P_R but initiates transcription in the opposite direction (Figure 8.5a); this promoter directs transcription of cI mRNA. P_{RM} is a weak promoter, but in a prophage, P_{RM} is activated to produce sufficient CI repressor to shut down P_R and P_L, as follows. CI bound to O_R2 is physically near P_{RM} and forms favorable contacts with RNA polymerase (Figure 8.5b); these contacts make P_{RM} a much better promoter. Stated another way, CI activates its own gene to ensure that sufficient repressor is produced to maintain prophage repression. However, should the CI concentration become very high, O_R3 is also occupied and transcription from P_{RM} is turned off until the CI concentration returns to a normal level.

It turns out that effective repression of P_{RM} requires long-range interactions of the CI tetramers at O_R1 and O_R2 with CI tetramers at O_L1 and O_L2. These interactions result in formation of a CI octamer (Figure 8.5c). Formation of the octamer results in the formation of a large (2.4 kb) DNA loop. The looping that accompanies octamerization also brings O_R3 and O_L3 near to each other, enabling cooperative interactions between CI dimers at O_R3 and O_L3 (Figure 8.5c), which leads to repression of P_{RM}. Thus, effective repression of P_{RM} depends on O_R1, O_R2, O_L1, and O_L2 for CI octamerization, and on O_L3 and O_R3 for CI tetramerization. Occupancy of the O_R operators by CI is summarized in Table 8.1.

Cleavage of CI repressor in cells with damaged DNA leads to prophage induction

Prophage λ DNA within a lysogen can be induced, or derepressed, so that lytic growth ensues. Induction is a way for λ to bail out of a damaged cell and move on to more fertile hunting grounds. Induction occurs in cells that have suffered DNA damage caused by ultraviolet light or some other agent. When damaged nucleotides such as pyrimidine dimers are removed by the cell's excision repair system, a set of bacterial genes is induced whose products are involved in helping the cell to survive irradiation; these genes are called the **SOS genes**. The SOS genes of undamaged cells are repressed by the cellular LexA repressor.

Induction of the SOS genes occurs as follows. Excision repair excises single-stranded DNA oligonucleotides, which in turn bind to and activate the cell's RecA recombination protein such that it acquires a "coprotease" activity. The LexA repressor has a weak autoprotease activity that is enhanced by binding to the activated RecA protein, with the result that the LexA repressor cleaves itself into two domains and loses repressor activity, allowing induction of SOS gene transcription.

The λ CI repressor is a mimic of LexA; both consist of two globular domains connected by a linker segment. CI's weak autoprotease activity is also stimulated by activated RecA, leading to linker cleavage that inactivates CI (Figure 8.5d). Inactivation of CI leads to transcription from P_R (see Table 8.1) and P_L, so that the early genes cro and N are transcribed. N-antiterminated transcription from P_L leads to expression of both Xis (for "excision") and Int (for "integration"), which in turn sponsor excision of the prophage DNA from the bacterial chromosome, and the lytic transcription pattern ensues.

The Cro repressor suppresses CI synthesis and regulates early gene transcription

As CI is the repressor used to establish and maintain a lysogenic state, the dimeric Cro protein is the repressor that establishes and controls a lytic infection by turning off the synthesis of CI. Cro binds to the same operators as CI repressor, but with different affinities, and without cooperativity. The operator affinities for Cro are $O_R3 > O_R2 = O_R1$. Early during lytic growth, either during an infection or induction of a lysogen, Cro binds

Table 8.1 Control of P_{RM} and P_R by CI and Cro repressors

		Occupancy at operator			Activity of promoter	
Protein	Concentration	O_R3	O_R2	O_R1	P_{RM}	P_R
CI	None	−	−	−	Weak	On
	Low	−	+	+	Activated	Off
	High	+	+	+	Off	Off
Cro	None	−	−	−	Weak	On
	Low	+	−	−	Off	On
	High	+	+	+	Off	Partially off

to O_R3 and blocks P_{RM}. This prevents CI expression and channels the infected cell to lytic growth. Later during lytic growth, sufficient Cro accumulates to partially fill O_R2 and O_R1, along with O_L2 and O_L1. Partial repression of P_R and P_L by Cro reduces synthesis of early proteins that are already present in sufficient amounts to allow the lytic phase to proceed. Occupancy of the O_R operators by Cro is summarized in Table 8.1.

Making the decision: go lytic or lysogenize?

You might say, "This is all well and good, but when a λ chromosome is injected into a cell, there is no repressor in the cell! If CI activates its own gene, whence comes the repressor that initially activates P_{RM}?" The answer is that λ has a system, involving a second *cI* promoter, to provide the CI repressor needed to activate P_{RM}.

P_{RE}, the promoter for repression establishment, is used to make the initial dose of CI repressor that in turn activates P_{RM}. P_{RE} is located downstream of the *cro* gene, so that RNA polymerase initiating at P_{RE} transcribes *cro* backward, prior to transcribing *cI* (Figure 8.6). The antisense *cro* transcript may pair with the sense *cro* transcript to lower Cro expression. Whether a sufficient amount of CI is produced from P_{RE} transcripts is the crux of the choice to go lytic or lysogenic. Sufficient CI ensures that P_R and P_L are turned off and P_{RM} is activated, producing more CI and leading to stable repression of the lytic cycle. Insufficient CI results in the lytic program.

How much CI is produced from P_{RE} depends in turn on the level of the P_{RE} activator protein, CII. The *cII* gene, located just downstream from *cro* (Figure 8.6), is transcribed from the P_R promoter and sponsored by N antitermination. Like P_{RM} (activated by low levels of CI; see Table 8.1), P_{RE} is a weak promoter that, in this case, is activated by CII. Since the level of CI produced from P_{RE} depends on the level of CII, the lysis/lysogeny decision depends finally on the intracellular concentration of CII protein.

In addition to activating P_{RE}, CII activates two additional promoters (Figure 8.6). P_{Int} is a CII-dependent promoter for a gene that encodes Int, the site-specific recombinase that inserts the λ prophage into the bacterial chromosome. P_{Int} lies within the coding region for the *xis* gene, so its product, the protein that directs excision of λ prophage from the cell chromosome, is not expressed during lysogenization. The third CII-activated

promoter, P_{AQ}, is located within gene *Q*. Transcription from P_{AQ} produces an antisense transcript of *Q* that lowers Q expression, delaying late gene expression.

Since the level of CII determines the lysis/lysogeny decision, an important question becomes: What controls CII? CII is inactivated by host cell proteases, and has a half-life of only a few minutes. It is even less stable in metabolically active cells than in inactive cells. The stability of CII is enhanced by CIII, which inhibits the host proteases. The web of interactions between CII, CIII, and the proteases is not completely defined. However, it is clear that the level of CII in the cell, and hence the decision to lysogenize, depends on the metabolic state of the cell. At the cellular level, studies show that lysogenization requires that a cell be infected by more than one phage, and that the likelihood of lysogenization decreases with increasing cell size.

A quick review

The events of the lytic/lysogenic decision may be summarized as follows (see Tables 8.2 and 8.3). During an infection, immediate early transcription from P_L and P_R produces N and Cro. Antitermination sponsored by N leads to early gene expression that includes CII and CIII, along with recombination and replication proteins. At this point, two outcomes are possible, depending on the level of CII in the cell.

On the one hand, if the host cell is metabolically active, instability keeps the CII concentration low, resulting in insufficient CI production from P_{RE} to repress P_L and P_R. Continued transcription from P_L and P_R produces high levels of early proteins, including the late gene activator Q, and the lytic program ensues in more than 90% of infected cells.

On the other hand, if the host cell is metabolically inactive, resulting in fairly stable CII, sufficient CI is produced from P_{RE} to repress P_L and P_R. Repression arrests production of early proteins (including Cro and CII) and activates P_{RM} for prophage expression of CI. Int produced from CII-activated P_{int} inserts the λ chromosome into the bacterial chromosome. As CII decays, P_{RE}, P_{Int}, and P_{AQ} return to their unactivated states. CI produced from P_{RM} maintains prophage repression. More than 90% of infected stationary phase cells become lysogenized.

Breaking and entering: the insertion of λ prophage DNA into the bacterial chromosome

To be stably inherited as a prophage, the circularized λ chromosome must integrate into the bacterial chromosome; otherwise, the non-replicating circles are simply diluted out as the cells divide. So λ carries a site-specific recombination system for integration into and excision from the bacterial chromosome (Figure 8.7). Int is a phage protein that catalyzes crossing-over between the

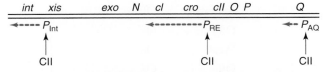

Figure 8.6 CII-activated promoters in lysogenization. See text for details.

Table 8.2 Activities of λ promoters

Promoter	Phase of growth cycle	Genes transcribed	Control mechanism
P_R	Prophage	None	Repressed by CI
	Immediate early	*cro*	Terminated at t_R1
	Early	*cro, cII, O, P, Q*	Antiterminated by N
	Late	*cro, cII, O, P, Q* (reduced levels)	Repressed by Cro
P_L	Prophage	None	Repressed by CI
	Immediate early	*N*	Terminated at t_L1
	Early	*N, cIII*, recombination genes	Antiterminated by N
	Late	*N, cIII*, recombination genes (reduced levels)	Repressed by Cro
P_R'	Prophage	None	Terminated at t_R'
	Immediate early	None	Terminated at t_R'
	Early	Lysis and structural protein genes (low)	Antiterminated by N
	Late	Lysis and structural protein genes (high)	Antiterminated by Q
P_{RE}	Prophage	*cI*, anti-*cro* (very low)	Weak constitutive activity
	Immediate early	*cI*, anti-*cro* (very low)	Weak constitutive activity
	Early	*cI*, anti-*cro* (high)	Activated by CII
	Late	*cI*, anti-*cro* (very low)	Weak constitutive activity
P_{RM}	Prophage	*cI* (moderate)	Activated by CI
	Immediate early	*cI* (very low)	Weak constitutive activity
	Early	None	Repressed by Cro
	Late	None	Repressed by Cro
P_{int}	Prophage	*int* (low)	Weak constitutive activity
	Immediate early	*int* (low)	Weak constitutive activity
	Early	*int* (high)	Activated by CII
	Late	*int* (low)	Weak constitutive activity
P_{AQ}	Prophage	Antisense *Q* (low)	Weak constitutive activity
	Immediate early	Antisense *Q* (low)	Weak constitutive activity
	Early	Antisense *Q* (high)	Activated by CII
	Late	Antisense *Q* (low)	Weak constitutive activity

phage *attP* site, near the *int* gene, and the bacterial *attB* site, located between the galactose (*gal*) and biotin (*bio*) operons. *attP* is ~250 bp long, with multiple binding sites for Int; in contrast, *attB* is about 25 bp long. *attP* is also designated POP', to symbolize its three parts. O is a 15-bp-long "core" sequence found in both *attP* and *attB*, where the DNA crossing-over reaction takes place. Flanking the *attP* core are P and P' arms.

Int binds to two different DNA sequence motifs within the arms and the core of *attP*. Int binds to the arm sequences strongly; the arms also have binding sites for the *E. coli* DNA bending protein, IHF, which bends the arms to allow Int to make bridging contacts with the core sequences. The result is a complex nucleoprotein structure called an **intasome**. In the intasome, some of the Int proteins have free binding domains; these free motifs enable the intasome to capture an *attB* site on a bacterial chromosome by binding to its core sequence. Then the crossover occurs, inserting λ into the bacterial chromosome.

Table 8.3 Roles of λ regulatory gene products

Gene product	Action	Result
CI	Represses P_R and P_L	Shuts off early transcription, excludes exogenous λ phages ("immunity")
	Activates P_{RM}	Maintains CI expression during lysogeny
	At high concentration, represses P_{RM}	Controls CI levels during lysogeny
Cro	Represses P_{RM}	Shuts off CI expression during lytic growth
	Turns down P_R and P_L	Reduces early transcription during late phase
CII	Activates P_{RE}	Turns on CI expression, directs cell to lysogeny
	Activates P_{int}	Makes Int protein, leads to prophage integration
	Activates P_{AQ}	Reduces Q expression, lowers late transcription
CIII	Blocks proteolysis of CII	Directs cell to lysogeny
N	Antiterminates at t_R1, t_L1, and downstream terminators	Activates early transcription from P_R and P_L
Q	Antiterminates at t_R'	Activates late transcription from P_R'

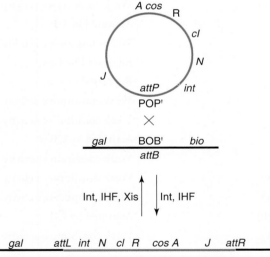

Figure 8.7 Integration and excision of λ DNA. Circular λ DNA is shown above the *E. coli* chromosome with the *attP* and *attB* sites aligned. Integration (downward arrow) and excision (upward arrow) are catalyzed by the proteins shown next to the arrows. The integrated λ DNA, flanked by *attL* and *attR*, is shown in orange at bottom.

Excision of λ DNA from the bacterial chromosome

The same two proteins, Int and IHF, plus an excision protein named Xis, are involved in the reverse reaction that generates a circular DNA molecule from the

Prophage/host cell interactions

"Lysogenic conversion" is the term for a common phenomenon in which an integrated prophage expresses one or more genes, in addition to the prophage repressor, that affect the host cell's phenotype. These additional genes express such functions as restriction systems, antibiotic resistance, or toxins. For example, many enterohemorrhagic *E. coli* (EHEC) strains carry λ-like prophages that express the gene for a shiga-like toxin, which substantially enhances the pathogenicity of the host.

Many such conversion genes appear to have hopped into the phage genome during integration/excision events as a cassette composed of a promoter, the open reading frame, and a transcription terminator. λ examples include the lom and bor genes, which encode prophage-expressed outer membrane proteins. The D gene, which encodes a head shell stabilization protein, is also thought to have been acquired by a λ ancestor phage by lateral genetic transfer. Gene D lacks a promoter but retains the transcription terminator.

integrated λ prophage. The *attP–attB* exchange during integration shuffles the *att* sequences, creating *attL* (BOP') and *attR* (POB') sites at the left and right junctions between the prophage DNA and the bacterial chromosome (Figure 8.7). When an integrated prophage is induced, N-antiterminated mRNA from P_L produces Int and Xis, the latter a protein that binds to sites only in the P arm of *attR*. Nucleoprotein structures assembled at *attL* (IHF, Int) and *attR* (Int, IHF, Xis) combine, and excision occurs.

In an intriguing variation, the prophage DNA of the λ-like phage N15 is linear, and is not integrated into the cell chromosome; such linear plasmids are rare in *E. coli*. Like λ, N15 virion DNA has cohesive ends that cyclize the molecule upon injection. However, instead of an *attP* site and Int protein, N15 has a *telRL* site and a **telomerase** enzyme. This enzyme opens the dsDNA at *telRL* and seals the resulting ends, creating covalently closed hairpins at both ends of the duplex. Upon induction, these ends must be opened and joined to recreate the *telRL* site.

Int synthesis is controlled by retroregulation

After excision of a prophage induced to lytic growth, and also during a lytic infection, high levels of Int are no longer necessary. An interesting mechanism called *retroregulation* limits Int expression during lytic growth only. A site named *sib*, located just beyond the *attP* site downstream of the *int* gene (Figure 8.2), contains a transcription termination sequence. When RNA polymerase reads through the *sib* sequence during N-antiterminated transcription from P_L, the mRNA forms a hairpin structure, SIB, that is a site for attack by double-stranded RNA-specific ribonuclease III (Figure 8.8; segments 1 and 2 pair with segments 3 and 4).

RNAse III cleavage at the SIB hairpin leads to further degradation of the mRNA into the adjacent *int* coding sequence, limiting expression of the Int protein. On the other hand, transcripts initiated at the P_{Int} promoter, which are not subject to N-antitermination, terminate upstream of segment 1 in *sib*. This RNA cannot form the complete SIB hairpin; therefore these transcripts are not cleaved by RNAse III, and are more stable. Thus, during lysogenization, CII-activated expression of Int from P_{Int} is not limited by retroregulation. Note also that in the prophage state, *sib* is separated from *int* by site-specific recombination at

attP; *sib* is only downstream of *int* after the prophage has been excised. Therefore N-antiterminated transcription from P_L on prophage DNA is not subject to retroregulation, and sufficient amounts of Int and Xis are produced for prophage excision.

λ DNA replication is directed by O and P, but carried out by host cell proteins

Expression of the early genes *O* and *P* leads to λ DNA replication. The O protein binds to multiple sites in the replication origin, *oriλ* (actually located within the *O* gene; Figure 8.2), forming a nucleoprotein structure, the "O-some". As a result, the two strands of an adjacent AT-rich segment of the DNA become unpaired. The P protein recruits the *E. coli* replication helicase DnaB to the O-some. In hijacking DnaB, P binds so tightly that *E. coli* chaperone proteins DnaJ and DnaK are needed to release it from the O-some.

Then growing points for DNA replication are set up using *E. coli* replication proteins. RNA primers are laid down on the separated DNA strands, and DNA polymerase extends these primers to begin copying the DNA. Bidirectional replication produces progeny circles at early times, prior to the lysis/lysogeny decision (Figure 8.1). During lytic growth, there is a shift to **rolling circle** replication (see Chapters 6 and 24). This mode produces multimeric double-stranded DNAs, called **concatemers**, that are the DNA packaging substrate.

Assembly of λ heads involves chaperones and scaffolding proteins

A large number of λ proteins are involved in the construction of virions; some are incorporated into the head or tail structures intact, some are cleaved during incorporation, and some are used as **scaffolding proteins** but are not incorporated into the mature virus particle. Assembly of the icosahedral capsid is initiated on a donut-shaped dodecamer of B, the "portal protein". B is modified by addition of small numbers of proteins X_1 and X_2, which are formed by a protein fusion reaction between C and E. It is thought that the X_1 and X_2 proteins form a "bushing" on the B donut, to which E can bind and form the head shell.

The host GroES and GroEL chaperone complex assists in the formation of the portal vertex. In fact, these host chaperone proteins were first discovered by isolating *E. coli* mutants that did not support growth of λ phage. The GroE gene was so named because certain mutations in the λ E gene allowed growth on these *E. coli* mutants. Chaperones have since been shown to play a major role in the proper folding of many proteins and in the assembly of multi-subunit protein complexes.

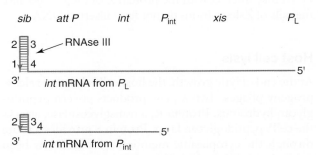

Figure 8.8 Retroregulation of int synthesis. Hairpin 1, 2/3, 4 at the 3' end of int mRNA made from P_L is long enough to be cleaved by RNAse III, resulting in instability, but hairpin 2/3 at the 3' end of int mRNA made from P_{int} escapes RNAse III cleavage and this mRNA is more stable.

The λ scaffolding protein Nu3 helps E, the major head protein, to form a closed icosahedral shell containing many (approx. 405) E subunits. B molecules at the portal vertex are simultaneously cleaved to remove 20 amino acids from their amino termini. The single B-containing portal is located at one of the 12 vertices of the icosahedral shell, called a prohead.

Recently the first crystal structure of the head shell of a tailed dsDNA bacteriophage was determined for the λ-like phage HK97. The HK97 shell is remarkable in that neighboring subunits of gp5, its major head protein, become covalently joined in pentameric and hexameric rings. The rings, in addition, loop through rings formed by neighboring pentamers and hexamers, generating protein chainmail!

DNA is inserted into preformed proheads by an ATP-dependent mechanism

Concatemeric λ DNA generated by rolling circle replication is cut to generate unit-length virion DNA molecules during packaging of the DNA into the prohead shell (Figure 8.9). Cutting requires the introduction of nicks, staggered 12 bp apart, so that the single-stranded cohesive ends are made. Terminase,

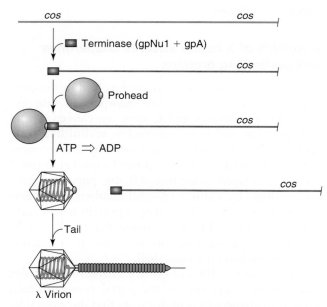

Figure 8.9 DNA packaging and virion assembly. A segment of concatemeric λ DNA is shown cleaved at a *cos* site by λ terminase, which then binds to a prohead. ATP then powers translocation of the DNA into the prohead, changing the prohead's shape. As the DNA is packaged, the downstream *cos* site is brought to, and cut by, terminase, which can then package another genome into another prohead. The filled head binds a tail to generate an infectious λ virion.

the viral cutting enzyme, contains multiple Nu1 and A subunits. Nu1 binds to an anchoring site *cosB*, and protein A cleaves the DNA at an adjacent nicking site, *cosN*. Following nicking of a concatemer, terminase separates the cohesive ends, remaining bound to the end of the DNA at *cosB*.

This DNA-protein complex docks at the portal vertex of a prohead, and DNA packaging proceeds. ATP hydrolysis by λ terminase powers translocation of the DNA into the prohead. The mechanics of the DNA packaging motor are not fully understood. Models include myosin-like contractions of terminase subunits, or rotation of the portal protein to winch or screw the DNA into the head shell.

During translocation of the DNA into the prohead, the E shell subunits undergo rearrangement, causing the shell to expand and become more angular. This rearrangement creates openings in the shell lattice for D trimers that fill the sites and greatly stabilize the shell. As the shell is filled, the next *cos* site of the concatemer approaches the terminase located at the portal. Cutting the second *cos* sequence requires recognition of *cosQ*, a site located just upstream of the nicking site *cosN*, and this completes packaging of a chromosome. Terminase undocks from the filled head, remaining bound to the end of the concatemer to form a new complex that is ready to bind to another prohead. Proteins W and FII bind to the portal vertex, stabilizing the filled head and creating a binding site for tail attachment.

Tail formation starts with J, the attachment fiber. Proteins J, K, I, H, M, and L assemble into the cone-shaped initiator, and then hexameric disks of V, the major tail subunit, stack on the initiator. The number of the hexameric V disks, 32, is dictated by the length of H, which Isao Katsura and Roger Hendrix have shown acts as a tape measure by extending up the central hole of V hexamers. When the stack of V disks reaches to the end of H, U binds, terminating tail polymerization. Finally Z is added and the tail is complete. About 200 bp of the *cosQ*-containing end of the λ DNA protrude down into the tail of a virion from within the head (Figures 8.3 and 8.9). The DNA may interact with the protein Z of the tail, because the tails of Z-less virions do not have inserted DNA.

Host cell lysis

At the end of lytic growth, the host cell is lysed to release progeny phages. Two λ gene products govern peptidoglycan hydrolysis. Protein R, a transglycosidase, attacks the cell's peptidoglycan layer. Protein S provides a route through the cytoplasmic membrane for R. S has been dubbed a *holin*, as it multimerizes in the membrane, forming a pore. S assembly is regulated so that the pore forms at the proper time.

Proteins Rz and Rz1 are required for lysis of cells in the presence of high concentrations of divalent cations.

Virus histories: mosaic chromosomes and ancient lineages

The mosaic chromosomes of bacteriophages. Bacteriophage chromosomes are mosaics of DNA segments that have been swapped due to rampant lateral exchange with other bacteriophage genomes. Comparing bacteriophage genome sequences shows that blocks of genes, often having related functions, have been swapped from phage to phage by genetic recombination. These related sequences are found not only in phages infecting the same host, but also in phages infecting distantly related bacteria, including gram-negative, gram-positive, and mycobacteria. The wide distribution of related sequences is proposed to occur by serial genetic exchanges between phages sharing common host species. That is, a phage whose normal host is bacterial species A might recombine in intermediate species B with a phage whose usual host is species C, leading to the swapping of a block of genes between the phages.

The mosaic structure of phage chromosomes challenges the classification of phages based on tail structure, because phages with related tail structures may be unrelated across much or all of their chromosomes, and conversely, phages that are genetically highly related may differ in tail structure. An example of the former: the side tail fibers of phage T4 and λ show sequence identity. An example of the latter: closely related phages λ (a member of the *Siphoviridae* that grows in *E. coli*) and P22 (a member of the *Podoviridae* that grows in Salmonella).

Ancient ancestors. There is overwhelming evidence that some eukaryotic viruses share common ancestors with bacteriophages. For example, although there is no significant amino acid sequence similarity between the major capsid proteins of bacteriophages HK97, P22 and T4, and herpes simplex virus type I, all these capsid proteins have very similar three-dimensional structures. In addition, the head shells of all tailed dsDNA viruses and herpes simplex virus contain a special vertex, the portal vertex, formed by the portal protein, a dodecamer of radially disposed subunits. Furthermore, there is extensive amino acid identity in the translocation ATPases of the tailed dsDNA phages and the herpesviruses. In a second ancient lineage, the "double barrel trimer" fold is found in the major capsid protein of viruses infecting hosts from all three kingdoms: bacteria (phage PRD1), eukaryotes (adenovirus), and archaea (*Sulfolobus* phage STIV).

The interacting Rz and Rz1 protein pair, called a *spanin*, is proposed to span the periplasm, linking and subsequently fusing the inner and outer cell membranes: Rz is an integral cytoplasmic membrane protein, and Rz1 is an outer membrane lipoprotein. The membrane fusion is expected to generate huge gaps in the cell envelopes, completing cell lysis.

KEY TERMS

Concatemer	Rolling circle
Intasome	Scaffolding proteins
Lysogen	SOS genes
Operator	Telomerase
Prophage	Temperate phage

FUNDAMENTAL CONCEPTS

- Bacteriophage lambda is a temperate phage that can integrate its DNA into the host cell chromosome and repress all phage gene expression, creating a lambda lysogen.
- Induction of lytic gene expression in a lysogen involves inactivation of the lambda CI repressor by self-cleavage stimulated by cellular RecA protein.
- In the absence of sufficient concentrations of CI repressor, transcription begins at promoters P_L and P_R on the lambda chromosome, but terminates shortly downstream at terminators t_L1 and t_R1.
- One product of the initial transcription is the lambda N-antiterminator protein, which allows transcription to proceed beyond terminators t_L1 and t_R1, making "early" mRNAs.
- Early gene products include proteins that excise and circularize lambda DNA, DNA replication proteins, and another antiterminator protein, Q.
- The Q protein allows transcription to proceed rightward on circular lambda DNA, transcribing all "late" genes in a single, long transcript.
- The level of the CI repressor is determined in part by the concentration of CI, which activates synthesis of its own transcript when at low concentrations, but represses its synthesis when at high concentrations.
- In contrast, the Cro repressor, made from the initial transcript starting at P_R, suppresses synthesis of the CI transcript when at high concentrations, allowing lytic growth.

- A complex set of interactions among repressors and operators as well as the phage CII and CIII proteins determines whether an infected cell will become a lysogen or enter the lytic cycle.
- Rolling circle replication makes long concatemers of lambda DNA that are cleaved and packaged into proheads; scaffolding proteins are shed during packaging.

REVIEW QUESTIONS

1. Lambda phage infection of *E. coli* can have two possible outcomes. What are they?

2. What is induction?

3. What affects the phage's decision to go lytic or lysogenic?

4. What is the role of CII?

5. Lambda DNA replication is directed by the early genes O and P, but what is the role of host cell proteins?

CHAPTER 9

Viruses of Archaea

David Prangishvili

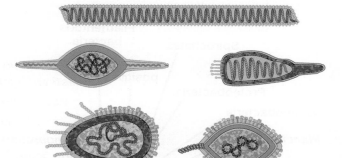

Archaea (singular: *archaeon*)

From Greek *archaios*, "ancient"

THE THREE DOMAINS OF LIFE: ARCHAEA, BACTERIA, EUKARYA

Archaea are unicellular organisms lacking a nucleus that are phylogenetically distinct from bacteria and eukaryotes

- *Cell walls made of S layer of protein*
- *Ether linkages in phospholipids*
- *Proteins initiated with methionine, not N-formyl methionine*
- *RNA polymerase and promoters resemble eukaryotic transcriptional machinery*

Many archaea thrive in extreme environments

- *Thermophiles (>45°C) and hyperthermophiles (>80°C): volcanic hot springs*
- *Acidophiles (pH 0–3): volcanic hot springs*
- *Halophiles (high salt concentrations): salt lakes*
- *Anaerobic environments (methanogens)*

Two major phyla of archaea: Crenarchaeota and Euryarchaeota.

VIRUSES OF ARCHAEA HAVE UNUSUAL MORPHOLOGIES

Lemon, droplet, rod, or bottle shapes with surface fibers.

All viruses have double-stranded DNA genomes except for one single-stranded DNA virus.

Most have internal or external lipid envelopes.

Crenarchaeota are hosts for seven distinct virus families, based on virion morphology and genome sequences.

Euryarchaeota are hosts for a reduced morphological spectrum of viruses, some which are related to head–tail bacteriophages.

LIFE CYCLES OF VIRUSES OF ARCHAEA

Many are temperate viruses that integrate their genome into host cell DNA and replicate without killing the host cell.

Some produce unique pyramidal structures that lead to exit of virions and cell death.

One virus develops long tails at both ends of virion after release from cell.

Most have a majority of genes with unknown functions and no homology to any gene in databases.

Many do not have an identifiable DNA polymerase gene.

Archaea, the third domain of life

By the mid-twentieth century, a widely accepted view of evolution postulated a single evolutionary line leading from prokaryotes to unicellular eukaryotes and then to higher, multicellular eukaryotes. Phylogenetic studies based on comparison of ribosomal RNA sequences—molecular phylogeny, pioneered by Carl Woese starting in the 1970s—radically changed this view. Woese showed that there are three different classes of cellular organisms, each with a distinct set of of ribosomal RNAs and ribosomes. This discovery led to the replacement of the procaryote/eukaryote dichotomy by a trinity of domains of life, the **Archaea**, **Bacteria**, and **Eukarya** (eukaryotes) (Figure 9.1). These three domains share a common ancestor and are as distinct from each other as any two of them are from the third.

Members of the newly recognized domain, the archaea, had previously been considered to be bacteria, as they are unicellular microorganisms that have circular chromosomes and no nucleus. However, studies of their biochemistry and cell biology revealed numerous differences between archaea and bacteria:

- Cell walls of archaea do not contain muramic acid and D-amino acid peptides as do cell walls of most

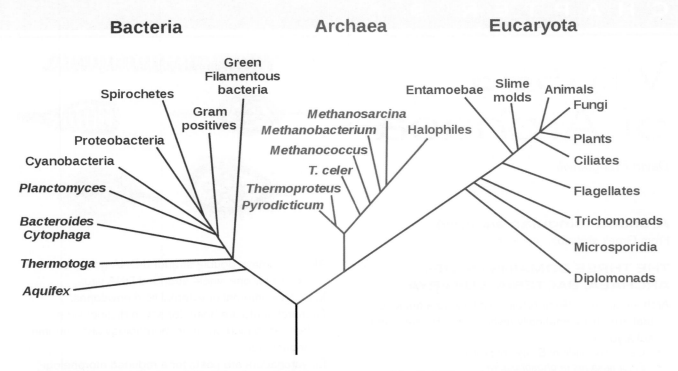

Bacteria　　　**Archaea**　　　**Eucaryota**

Figure 9.1 Phylogenetic tree of life. The three-domain system proposed by Carl Woese, based on sequence analysis of ribosomal RNAs.

bacteria; instead, they are usually made of an impermeable "S layer" of proteins that in many cases are directly associated with the cytoplasmic membrane.

- Phospholipids in membranes of archaea use ether linkages to join the glycerol phosphate polar ends and the non-polar lipid ends, which can contain cyclic moieties and branched hydrocarbon chains. In contrast, phospholipids in membranes of bacteria and eukaryotes use ester linkages, and their non-polar parts are usually made from unbranched fatty acids.
- Archaeal initiation factors for translation are more complex than in bacteria, and archaea initiate protein synthesis with methionine, as do eukaryotes, not N-formyl methionine, used by bacteria.
- Some archaea have histone-like proteins that associate with and condense genome DNA, as do eukaryotes but not bacteria.
- The RNA polymerase that transcribes genes of archaea has a number of subunits with counterparts in eukaryotic RNA polymerase II, including the TATA-binding protein TBP.
- Similarly, enzymes involved in DNA replication in archaea are more like those used by eukaryotes than bacteria.

Comparisons of ribosomal RNA sequences among archaea led to the recognition of several major divisions (**phyla**), among which best studied are the

Crenarchaeota (from Greek *crenos*, "origin") and the **Euryarchaeota** (from Greek *euryos*, "diversity"). All cultured representatives of the phylum Crenarchaeota are **hyperthermophiles**, which grow optimally above 80°C. By contrast, the phylum Euryarchaeota contains organisms of diverse phenotypes including a few hyperthermophiles, **thermophiles** (growth above 45°C), **hyperhalophiles** (growth in hypersaline waters with concentrations of sodium chloride above 1.5 molar), and **anaerobic methanogens** (organisms that produce methane by reduction of CO_2).

Nearly all cultivated archaeal species have been isolated from environments with extreme features, such as volcanic hot acid springs, as well as from undersea hydrothermal vents ("black smokers"), which can have temperatures well over 100°C at the high pressures found thousands of meters under the ocean. One such organism, *Methanopyrus kandleri*, can grow at temperatures as high as 122°C. Another organism, *Picrophilus torridus*, grows at pH 0, the equivalent of 1.2 molar sulfuric acid! However, the analysis of 16S rRNA sequence diversity in environmental samples has indicated that archaea are numerous also in "moderate" habitats, throughout the world's oceans and even in cold environments such as polar seas and buried under glaciers. Presently archaea are recognized as a major constituent of the Earth's biosphere, playing an important role in both the carbon cycle and the nitrogen cycle.

Viruses of Archaea have diverse and unusual morphologies

The first archaeal viruses that were isolated in the 1970s and early 1980s resembled bacteriophages from the family *Myoviridae* and *Siphoviridae*, with icosahedral heads and contractile helical tails. Their linear, double-stranded DNA genomes also showed strong similarity to genomes of myoviruses and siphoviruses. Consequently, it was erroneously inferred that archaeal viruses constituted a variety of the ubiquitous head-tail bacteriophages.

However, with further studies this view has changed radically. Electron microscopy of samples collected from archaea-rich extreme geothermal and hypersaline environments revealed that head-tail phage are rare among archaeal viruses. To date, slightly more than three dozen viruses infecting organisims from the archaeal phyla Crenarchaeota and Euryarchaeota have been described (Tables 9.1 and 9.2). One of these viruses has a circular single-stranded DNA genome, but all others have either linear or circular double-stranded DNA genomes.

Despite the limited number of isolates, known species of archaeal viruses encompass an exceptionally broad range of unique morphological and genomic features. Especially remarkable in their uniqueness and diversity are double-stranded DNA viruses that infect hyperthermophilic crenarchaeaota in extreme geothermal environments. These have been classified in seven recently created virus families: spindle-shaped *Fuselloviridae*, filamentous *Lipothrixviridae*, droplet-shaped *Guttaviridae*, bottle-shaped *Ampullaviridae*, spherical *Globuloviridae*, rod-shaped *Rudiviridae*, and two-tailed *Bicaudaviridae*. Several other viruses of crenarchaeaota have not yet been assigned to families.

Among viruses that infect euryarchaeota, some have typical morphological and genomic features of head-tailed bacteriophages. However, a number of other viruses of euryarchaeota are morphologically similar to viruses of crenarchaeota. Their taxonomical classification is still uncertain.

In this chapter we will describe in some detail the seven virus families that are unique to the archaea. Members of these families, all with double-stranded DNA genomes, infect exclusively hyperthermophilic crenarchaeota, the optimal growth temperature of which exceeds 80°C. These viruses have been isolated from extreme geothermal environments in regions of active volcanism and tectonics in Europe, North America, and Asia. We will mainly focus on unique morphotypes of these viruses, which are not encountered among double-stranded DNA viruses of the other two domains of life, the Bacteria and the Eukarya. Besides their unique morphologies, these viruses have exceptional genomes in which the majority of genes have no detectable homologs in other viruses or cellular life forms.

Fuselloviridae are temperate viruses that produce virions without killing the host cell

Like a number of viruses that infect crenarcheaota, the members of the *Fuselloviridae* (from the Latin *fusello*, "little spindle") integrate their genome into the host cell, and can be induced to produce virus particles by irradiation of cells with ultraviolet light or other stress factors. These viruses thus have a life cycle similar to that of temperate bacteriophages such as lambda (Chapter 8), with the important difference that they do not kill the host cell upon induction, but rather produce virions and then revert to a non-productive state.

Virions of this family are enveloped, and many have the shape of a lemon or spindle, with a bunch of thin filaments attached to one of the two pointed ends (Figure 9.2a, d). Other family members are more elongated, with three relatively thick filaments at one pointed end (Figure 9.2b). The terminal filaments are involved

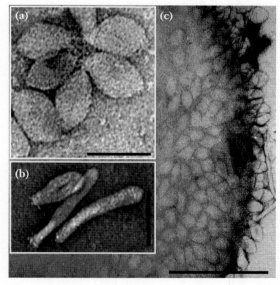

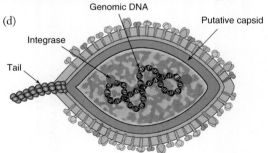

Figure 9.2 Sulfolobus spindle-shaped viruses 1 (SSV-1) and 6 (SSV-6), family *Fuselloviridae*. Negative contrast electron micrographs of (a) virions of SSV-1, (b) virions of SSV-6, (c) virions of SSV-1 being extruded from Sulfolobus shibatae. (Scale bars: a & b, 200 nm; c, 500 nm), (d) drawing of a typical fusellovirus.

Table 9.1 Viruses of Crenarchaeota

Family or genus name (No. of species)	Genome, size	Example of virus species, abbreviation	Genes	Envelope	Life cycle	Virion shape	Virion image
Fuselloviridae (9)	Circular ds DNA 15–24 kb	*Sulfolobus* spindle-shaped virus 1 (SSV-1)	34	External	Temperate, integrates genome; Productive infection does not kill cells	Lemon or elongated shape	
Lipothrixviridae (11)	Linear ds DNA 16–56 kb	*Sulfolobus islandicus* filamentous virus (SIFV)	74	External	Productive infection does not kill cells	Filament 24–40 × 400–2200 nm; terminal fibers or claws	
Unassigned (2)	Circular ds DNA 16–18 kb	*Sulfolobus* turreted icosahedral virus (STIV)	37	Internal	Lytic	Icosahedral capsid with 12 turrets	
Guttaviridae (1)	Circular ds DNA 20 kb	*Sulfolobus neozealandicus* droplet-shaped virus (SNDV)	?	External	Productive infection does not kill cells	Droplet shape, pointed end with fibers	
Ampullaviridae (1)	Linear ds DNA 24 kb	*Acidianus* bottle-shaped virus (ABV)	57	External	Productive infection does not kill cells	Bottle with fibers at blunt end	
Globuloviridae (2)	Linear ds DNA 28 kb	*Pyrobaculum* spherical virus (PSV)	48	External	Productive infection does not kill cells	Spherical	
Rudiviridae (5)	Linear ds DNA 24–35 kb	*Sulfolobus islandicus* rod-shaped virus 2 (SIRV-2)	54	None	Lytic	Rod, 23 × 600–900 nm, terminal tail fibers	
Bicaudaviridae (1)	Circular ds DNA 63 kb	*Acidianus* two-tailed virus (ATV)	72	External	Temperate, integrates genome; productive infection kills cells	Lemon shape with two tailed ends	

Table 9.2 Viruses of Euryarchaeota

Family or genus name	Genome, size	Example of virus species, abbreviation	Genes (ORFs)	Envelope	Life cycle	Virion shape	Virion image
unassigned	Circular ss DNA 7 kb	*Halorubrum* pleomorphic virus1 (HRPV-1)	9	External	Productive infection does not kill cells	Roughly spherical	
unassigned	Circular ds DNA 8 kb	*Haloarcula hispanica* pleomorphic virus 1 (HHPV-1)	8	External	Productive infection does not kill cells	Roughly spherical	
Salterprovirus	Linear ds DNA 14–16 kb	*Haloarcula hispanica* virus 1 (His-1)	35	External	Lytic	Lemon shape	
unassigned	Linear ds DNA 31 kb	SH-1	56	Internal	Lytic	Icosahedral capsid with 12 spikes	
Myoviridae	Linear ds DNA 76 kb	Halovirus HF-1	117	None	Some are temperate, some lytic	Icosahedral head, contractile tail	
Siphoviridae	Linear ds DNA 30.4 kb	Halovirus psiM-1	37	None	Lytic	Icosahedral head, flexible tail	

in adsorption to the surface of host cells. Virions have dimensions of approximately 60 × 100 nm.

The envelope of *Sulfolobus* spindle-shaped virus 1 (SSV-1) contains lipids with a composition similar to that of host cell membranes. The major capsid protein, VP1, decorates the outside of the virion, and a small basic protein, VP3, binds to the DNA genome inside the particle. Virions are assembled within the host cell and "slip" out of the cell without creating holes in the cell membrane or S layer (Figure 9.2c). It is not known how this occurs.

Genomes of fuselloviruses are positively supercoiled

Sulfolobus spindle-shaped virus 1 (SSV-1) is perhaps the best-studied archaeal virus, having been isolated by the laboratory of Wolfram Zillig in the mid-1980s and intensively studied by that laboratory. The genome of SSV-1 is a circular double-stranded DNA molecule of 15.5 kb and carries 34 open reading frames, most of them of unknown function (Figure 9.3). Other fuselloviruses have similar genomes up to 24 kb in length. These genomes have the property of being positively supercoiled when isolated from virions, and were one of

the first examples of such DNA found in nature. It turns out that all hyperthermophilic microorganisms, both archaea and bacteria, code for a "reverse gyrase" that introduces positive supercoils into DNA. This enzyme is believed to play an important role in reducing local unwinding of DNA at high temperatures.

Hosts of these viruses are members of the genera *Sulfolobus* and *Acidianus*, and experiments are usually carried out at approximately 80°C and in medium of pH 3. Although productive infection does not lyse the cells, it does slow cellular growth, and therefore plaque assays can be carried out by examining lawns of *Sulfolobus* cells for lower densities of cells, the result of slow growth of virus-infected cells. Infection of cells leads to integration of the viral genome within genes coding for cellular transfer RNAs, the specific tRNA gene depending on the cell and host combination. The virus codes for an integrase that carries out this reaction.

Transcription of SSV-1 DNA is temporally controlled

Within 1 hour after exposing cells carrying an integrated genome of SSV-1 to ultraviolet light, a short (200–300 nt) "immediate early" viral RNA named T-ind is produced

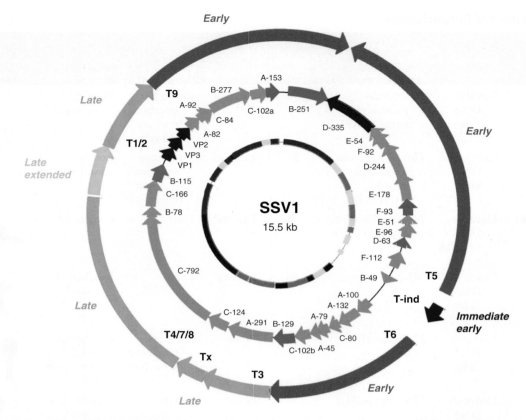

Figure 9.3 Genome of *Sulfolobus* spindle-shaped virus 1 (SSV-1). The outer circle shows groups of genes expressed in different temporal classes, and the direction of their transcription. The middle circle (green) shows location of individual genes. Dark green: known function; green: predicted function; light green: unknown function. The inner circle shows conservation of genes among the four sequenced SSV genomes: dark gray, conserved; gray, conserved in three out of four genomes; light gray, not conserved or conserved only in two of the four genomes.

(Figure 9.3). The ind gene product apparently controls the virus productive life cycle; beginning approximately 1 hour later, transcription from two "early" genes, T5 and T6, can be detected. These genes flank the ind gene and are transcribed in divergent directions, away from the ind gene. They contain unusual inverted repeat sequences in their promoter regions. However, full-length transcripts from these genes are only detected by 5–6 hours after induction. It is not known whether these early transcripts are prematurely terminated or degraded between 2 and 5 hours after induction. The distal end of the T5 transcript codes for the integrase protein, named ORF D-335.

A "delayed early" transcript from the T9 gene is detected starting 5 hours after induction, at about the time that viral DNA replication begins. Finally, transcription of the "late" genes (T1, 2, 3, 4, 7 and 8, and Tx) begins at about 6 hours after induction. One of the major late transcripts, T2, codes for the major capsid protein, VP1. Virions are assembled and released from the cell, which eventually shuts down most viral transcription and returns to the original lysogenic state and normal growth rate.

Filamentous enveloped viruses of the *Lipothrixviridae* come in many lengths

The family *Lipothrixviridae* (from the Greek *lipos*, "fat" and *thrix*, "hair") presently includes nine different species of rod-shaped or filamentous viruses of variable flexibility and lengths, depending in part on genome length. Virions have an envelope containing virus-encoded proteins and host-derived lipids. The structures of virion termini are diverse: for *Thermoproteus tenax* virus 1 (TTV-1) (Figure 9.4), the termini are stubby tails; for *Sulfolobus islandicus* filamentous virus (SIFV), and *Acidianus* filamentous virus 3 (AFV-3), the termini taper and end in mop-like structures with six or three tail fibers, respectively (not shown); while for *Acidianus* filamentous virus 1 (AFV-1) (Figure 9.5), they have characteristic claw-like structures.

The virions carry two major structural proteins in nearly equal amounts. The crystal structure of these proteins was resolved for AFV-1. The four-helix bundle structures of the two structural proteins are highly similar, and they also resemble the structure of the major capsid protein of the *Rudiviridae*.

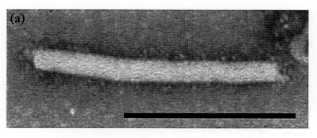

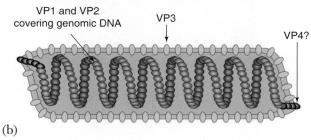

VP1 and VP2
covering genomic DNA

VP3

VP4?

(b)

Figure 9.4 *Thermoproteus tenax* **virus 1 (TTV-1),** **genus** *Alphalipothrixvirus.* (a) Negative contrast electron micrographs of virions. (Scale bar: 250 nm). (b) Drawing of TTV-1.

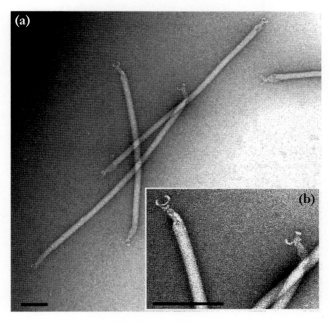

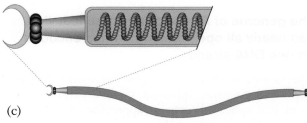

Figure 9.5 *Acidianus* **filamentous virus 1 (AFV-1), genus** *Gammalipothrixvirus.* (a) Negative contrast electron micrographs of virions (Scale bar: 50 nm), (b) terminal structures at higher magnification, (c) drawing of a typical gammalipothrixvirus.

Viral genomes are linear double-stranded DNAs, varying between 16 and 56 kb in length. They carry long inverted terminal repeats. The nature of the ends of the linear genome is presently unclear. In some cases, the termini are masked by hydrophobic ligands, precluding exonucleolytic degradation from either 3' or 5' ends. Sequences of the terminal regions of the AFV-1 genome are unusual; the short inverted terminal repeat CGGGGGGGGG is followed at either end by a 350-bp region containing numerous short direct repeats of the pentanucleotide TTGTT and close variants thereof. This structural design is reminiscent of the telomeric ends of linear chromosomes of eukaryotes.

Hosts are members of the genera *Thermoproteus*, *Sulfolobus* and *Acidianus*, isolated from terrestrial hot springs of Iceland, USA, Russia, and Italy. In the case of AFV-1, adsorption proceeds via interaction of the terminal claws of the virion with pili of the host cell, resulting in folding of the claws. How virions are released has not been investigated. During a cycle of productive infection with most of the known species, neither formation of any significant amounts of cell debris nor a decrease in cell density was observed. Integration of viral DNA into host chromosome was not detected.

A droplet-shaped virus is the only known member of the *Guttaviridae* (from the Latin *gutta*, "droplet")

A single virus species represents this family. Virions of *Sulfolobus neozealandicus* droplet-shaped virus (SNDV) resemble elongated droplets with varying dimensions, 110–185 nm in length and 95–70 nm in width (Figure 9.6). The pointed end of the virion is covered by a beard of dense filaments. The enveloped virion contains three small structural proteins. The core is covered with a beehive-like structure, the surface of which appears to be either helical or stacked. The virion adsorbs to a receptor on the host cell with its bearded end. This virus has a 20-kb circular, double-stranded DNA genome that, unlike genomes of most viruses of archaea, is extensively methylated.

Mature virions can be observed in host cells prior to their release. Virus release begins only in the early stationary growth phase of host cells and does not result in any detectable cell lysis.

Acidianus bottle-shaped virus (ABV): its name says it all!

The enveloped virion of *Acidianus* bottle-shaped virus (ABV), the only known member of the family *Ampullaviridae* (from Latin *ampulla*, "bottle"), resembles a miniature bottle. It has an overall length of 230 nm and a width varying from 75 nm at the broad end to 4 nm at the pointed end (Figure 9.7). The broad end

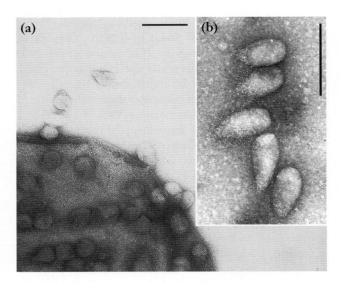

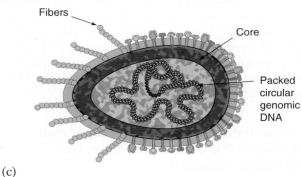

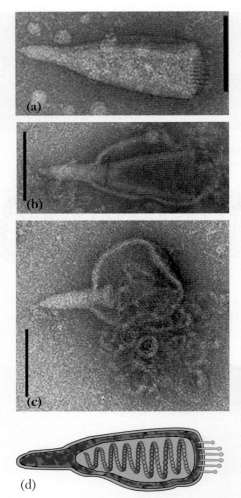

Figure 9.6 *Sulfolobus neozealandicus* **droplet-shaped virus (SNDV), family** *Guttaviridae*. Negative contrast electron micrographs of (a) virions being extruded from *S. neozealandicus;* (b) virions at higher magnification (Scale bars: 200 nm); (c) drawing of SNDV.

Figure 9.7 *Acidianus* **bottle-shaped virus (ABV), family** *Ampullaviridae*. Negative contrast electron micrographs of virions. (a) Intact virion; (b, c) partially disrupted virions revealing an inner cone-shaped core composed of a toroidally supercoiled nucleoprotein filament. (Scale bars: 100 nm); (d) drawing of ABV.

of the virion is decorated with about 20 thin rigid filaments, each 20 nm long and 3 nm thick, which appear to be inserted into a disc and are interconnected at their bases. The 9-nm-thick envelope encloses a cone-shaped nucleoprotein core formed by a toroidally supercoiled nucleoprotein filament (Figure 9.7b, c, d). It is presently unclear whether the pointed end or the filaments on the broader end are involved in adsorption to the cell surface and channeling of viral DNA into host cells.

The virion contains six major proteins in the size range between 15 to 80 kDa. The 24-kb double-stranded linear DNA genome encodes 57 proteins, including a DNA polymerase. Only two other genes have homologs in databases: a glycosyl transferase and a thymidylate kinase. The genome has inverted terminal repeat sequences of 580 bp.

Host cells survive virus infection; lysis was not observed. However, virus infection slows host cell division; the generation time of the host cells increases from about 24 hours to about 48 hours. Release of particles is observed only in the stationary growth phase.

The genome of *Pyrobaculum* spherical virus has nearly all open reading frames encoded on one DNA strand

The virions of *Pyrobaculum* spherical virus (PSV) and *Thermoproteus tenax* spherical virus 1 (TTSV-1) are enveloped, spherical particles, approximately 100 nm in diameter (Figure 9.8a, c). These viruses are the only two known members of the family *Globuloviridae* (from the Latin *globulus*, "small ball"). On the surface of the virion is a variable number of spherical protrusions about

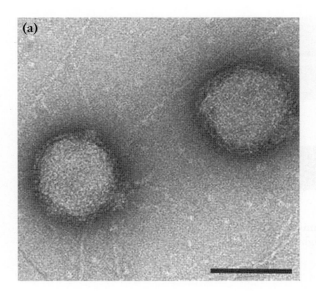

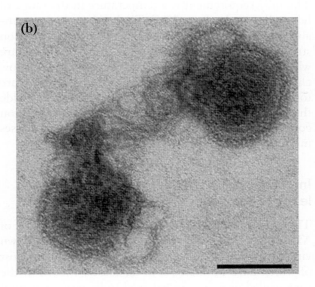

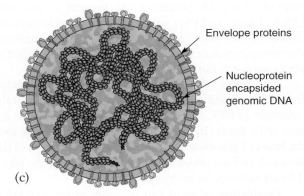

Envelope proteins

Nucleoprotein
encapsided
genomic DNA

Figure 9.8 *Pyrobaculum* **spherical virus (PSV), family** *Globuloviridae.* Negative contrast electron micrographs of (a) two intact virions, (b) two disrupted virions releasing the helical nucleoprotein core (Scale bars: 100 nm), (c) drawing of a globulovirus.

15 nm in diameter, most likely involved in adsorption to host cells. The viral envelope contains host-derived lipids and encases a tightly packed superhelical nucleoprotein (Figure 9.8b). PSV has a major virion structural protein of 33 kDa and two smaller structural proteins.

The genome of PSV is a 28-kb linear, double-standed DNA molecule. It is striking that all but 4 of the 48 open reading frames lie on the same DNA strand. The genome has 190-bp inverted terminal repeats that contain multiple copies of 5-bp direct repeats and a 16-bp direct repeat. The two DNA strands appear to be covalently linked to each other. There is evidence that one strand may be nicked 5 nt from the termini; this could be used as a primer during DNA replication, as is thought to occur with a number of linear double-stranded DNA viruses such as poxviruses.

Susceptible host cells are strains of *Pyrobaculum* and *Thermoproteus* isolated from terrestrial hot springs in the USA, Iceland, and Indonesia. Virus production does not result in cell lysis, nor does it slow growth of cells, which double every 24 hours. Moreover, host cell membranes in an infected culture remain intact, indicative of a non-cytocidal infectious cycle. Infected cells that were grown for several months continually released virus particles, showing that the virus exists in a "carrier state" in these cells.

Viruses in the family *Rudiviridae* (from the Latin *rudis*, "small rod") are non-enveloped, helical rods

These viruses replicate in a variety of strains of *Sulfolobus* and *Acidianus*, isolated from hot acidic springs on **solfataric** fields (volcanic areas emitting steam and sulfur) in Iceland, Italy, and Portugal. Virions are non-enveloped rod-shaped particles (Figure 9.9), with widths of 23 nm and different lengths in the range of 610 nm to 900 nm, proportional to the length of their linear double-stranded DNA genomes. The virions consist of a tube-like superhelix formed by the DNA and a single basic, DNA-binding protein, which for the *Sulfolobus islandicus* rod-shaped virus 1 (SIRV-1) has a mass of 14.4 kDa. At both ends of the virions are plugs about 50 nm long into which three tail fibers are anchored. These are involved in adsorption to the host cell. The second protein component of the virion, of about 124 kDa, is associated with tail fibers.

The linear DNA genomes of rudiviruses range in size from 24 to 35 kb, and carry long inverted terminal repeats, comprising up to 7% of the total genome length. The two DNA strands are covalently linked at their ends. The observation of head-to-head and tail-to-tail linked replicative intermediates suggests a

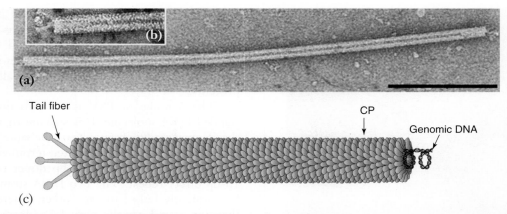

Figure 9.9 *Sulfolobus islandicus* **rod-shaped virus 1 (SIRV-1), family** *Rudiviridae.* Negative contrast electron micrographs of (a) full-length virion (Scale bar: 200 nm), (b) terminal structure at higher magnification, (c) drawing of a rudivirus.

self-priming mechanism of replication, similar to that proposed for large eukaryotic DNA viruses including poxviruses. *In vivo* studies demonstrated a rather simple transcriptional pattern for both SIRV-1 and SIRV-2, with very few cases of temporal regulation.

Rudiviruses escape from the cell by means of unique pyramidal structures

Unlike many other viruses of crenarchaeota, infection by rudiviruses kills host cells. For example, *Sulfolobus islandicus* rod-shaped virus 2 (SIRV-2), causes massive degradation of the host cell chromosome and lyses the cell. Virions are assembled in the cytoplasm and appear in aggregates within the cell. They are released from the host cell by a novel mechanism involving the formation of large pyramidal protrusions that transect the cell envelope, rupturing the S-layer. These protrusions eventually break apart the membrane, creating large apertures through which virions escape from the cell (Figure 9.10).

Acidianus two-tailed virus (ATV) has a virion with tails that spontaneously elongate

The virion of *Acidianus* two-tailed virus (ATV), the only known member of the family *Bicaudaviridae* (from Latin *bi*, "two", and *cauda*, "tail"), exists in two conformations. When released from the host cell, it is lemon-shaped, with an overall dimension of approximately 120 × 300 nm (Figures 9.11b, c). Upon further incubation at temperatures above 75°C, appendages protrude from both pointed ends of the virion, and the lemon-shaped virion body shrinks to approximately 85 × 150 nm (Figures 9.11a, d–h). The two appendages (tails) at each pointed end have variable lengths; the maximum length of the virion including tails reaches about 1000 nm.

Protrusion of tails does not require the presence of host cells, an exogenous energy source, or any cofactors.

The only requirement is a temperature in the range of host cell growth, above 75°C. Their protrusion presumably results from a conformational change in a virion structural protein. The tail has a tube-like structure with a wall about 6 nm thick. It terminates with a narrow channel, 2 nm in width, and an anchor-like structure formed by two furled filaments, each 4 nm wide. Inside the tube resides a filament 2 nm in width. The virion contains at least eleven proteins with molecular masses in the range of 12–90 kDa, and modified host lipids.

Infection with ATV at high temperatures leads to lysogeny

The circular, 62.8-kb double-stranded DNA genome of ATV (Figure 9.12) is the largest genome among known viruses of crenarchaeota. It codes for 72 proteins, most of which have no matches in databases. Eleven of the virus-coded proteins are believed to be virus structural proteins. Four **IS elements** are present in the viral DNA. These are DNA sequences that are transposable elements and code for a transposase that catalyzes the transposition of the DNA element. Their role in the viral genome is not understood.

The hosts of ATV are *Acidianus* species isolated from a hot spring on a solfataric field near Naples, Italy. Infection at 85°C, the optimal growth temperature for host cells, results in integration of the viral genome into the host cell chromosome, leading to lysogenization of the host cell and suppression of lytic functions of the virus. The virus codes for an integrase protein.

Production of virus can be induced by subjecting the host cell to stress conditions, including irradiation with ultraviolet light, treatment with mytomycin C, or freezing-thawing. Virus replication eventually leads to cell lysis. When cells growing at suboptimal temperatures (75°C) are infected with ATV, instead of lysogeny, the result is a productive infection and lysis of the cells.

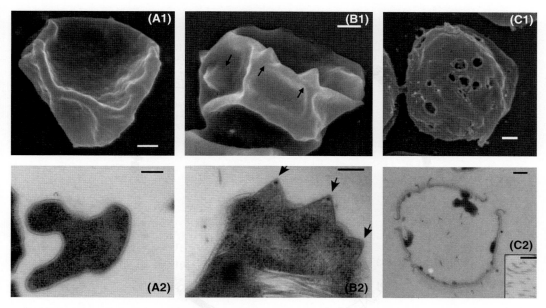

Figure 9.10 **Changes in *Sulfolobus* host cells following infection with SIRV-2.** (a1–c1) Scanning electron micrographs; (a2–c2) transmission electron micrographs (Scale bars: 200 nm). (a1, a2) Uninfected cells; (b1, b2) 10 h after infection; (c1, c2) 26 h after infection. Pyramidal structures are seen in b1 and b2; c1 and c2 show lysed cells with holes in the cell wall.

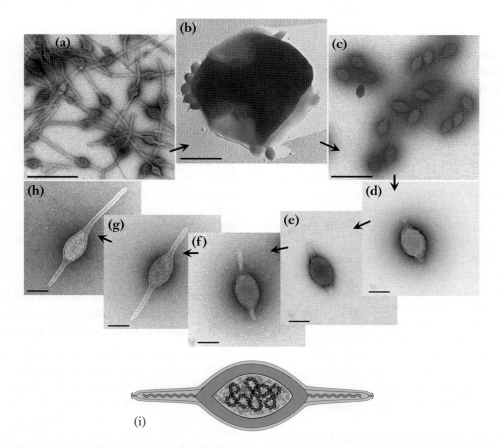

Figure 9.11 *Acidianus* **two-tailed virus (ATV), family** *Bicaudaviridae*. Electron micrographs of (a) two-tailed virions; (b) virions being extruded from a cell of *Acidianus convivator;* (c) virions in culture medium after release from infected cells, 2 days after infection; (d) as for (c), but purified by CsCl density gradient; (e–h) as for (d), but incubated at 75°C for 2, 5, 6, and 7 days, respectively. All preparations were negatively stained with 3% uranyl acetate, except for (b), which was platinum shadowed. (Scale bars: (a) to (c), 500 nm; (d) to (h), 100 nm). (i) Drawing of ATV.

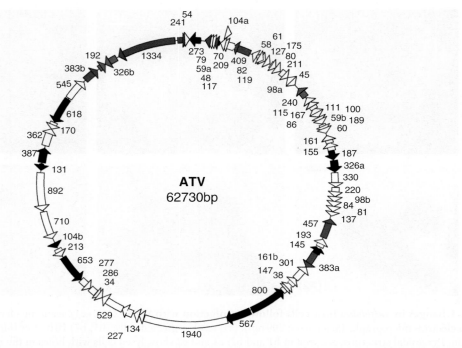

Figure 9.12 Genome of *Acidianus* two-tailed virus (ATV). Arrows show direction of transcription. White: unknown function; blue: homologues in archaeal viruses; green: homologues in archaeal plasmids; red: transposases; black: virion proteins.

Two related viruses of hyperhalophiles resemble fuselloviruses by morphology but not by genetics

The genus *Salterprovirus* has been introduced for classification of two spindle-shaped, double-stranded DNA viruses that infect hyperhalophilic euryarchaea: *Haloarcula hispanica* viruses 1 and 2 (His-1 and His-2) (Table 9.2). Morphologically these viruses are highly similar to members of the family *Fuselloviridae*; however, genome sequences reveal no relationships to viruses of that family.

Unlike the *Fuselloviridae*, neither His-1 nor His-2 are believed to integrate their DNA into host cells, and infection leads to cell lysis. Their linear, 15-kb double-stranded DNA genomes (Figure 9.13) code for a DNA polymerase of the class that uses protein-primed DNA synthesis, similar to the mechanism used by adenoviruses (Chapter 23; see Figure 23.8). There is evidence of protein covalently linked to both ends of the DNA molecule. Interestingly, all genes in the right half of the genome of His-2 are read from left to right, and all genes in the left half are read right to left, implying that transcription diverges from the center of the genome toward the genome ends.

Two unusual viruses with icosahedral capsids and prominent spikes

A number of archaeal viruses still remain unclassified. Among these are double-stranded DNA viruses with icosahedral symmetry and internal membranes, which infect

hyperthermophilic and hyperhalophilic hosts within the crenarchaeota and euryarchaeota, respectively: *Sulfolobus* turreted icosahedral virus 1 (STIV-1) (Figure 9.14a), and *Haloarcula hispanica* virus SH-1 (Figure 9.14b).

STIV-1 has elaborate turret-like structures at the fivefold vertices (Figure 9.14a). It is released from *Sulfolobus* host cells with the help of virus-induced pyramidal structures, as described for the rudivirus SIRV-2 (Figure 9.10). The unusual symmetry-mismatched spikes of SH-1 seem to play a role in adsorption to the host cell. These spikes are connected to proteins that are anchored in the underlying membrane, and that may be involved in capsid assembly and in delivery of the genome during infection.

A virus with a single-stranded DNA genome is closely related to a virus with a double-stranded DNA genome

The only known archaeal virus with a single-stranded DNA genome, *Halorubrum* pleomorphic virus 1 (HRPV-1), infects a hyperhalophile and has poorly shaped roundish virions with surface extensions. The 7 kb genome codes for 9 proteins. Virions have a lipid-containing envelope and two major structural proteins. The structure of virions suggests that they may enter cells by membrane fusion and exit by budding, but this has not yet been visualized.

Interestingly, a virus with an 8-kb double-stranded DNA genome, *Haloarcula hispanica* pleomorphic virus 1 (HHPV-1), has recently been shown to share a high

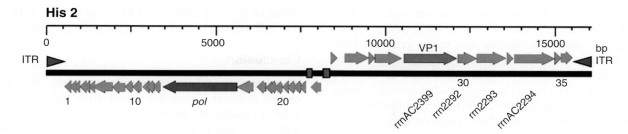

Figure 9.13 Genome of *Haloarcula hispanica* virus 2 (His-2). Top line: distance in nucleotides from left end. Purple arrowheads: inverted terminal repeated sequences. Arrows above and below the black line show genes transcribed, respectively, from left to right or from right to left. Green: unknown function; blue: DNA polymerase; red: viral capsid protein VP1.

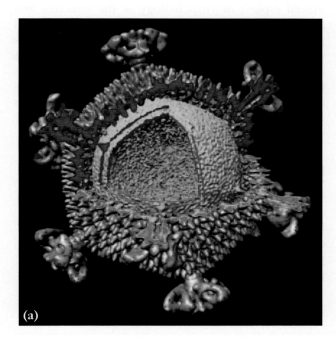

(a)

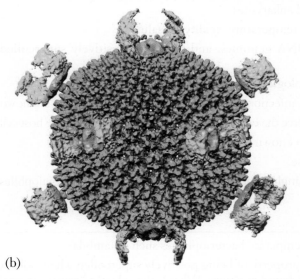

(b)

Figure 9.14 **Archaeal viruses with icosahedral symmetry.**
(a) 3D reconstruction of virion of *Sulfolobus* turreted icosahedral virus (STIV). (b) 3D reconstruction of virion of *Haloarcula hispanica* virus SH-1.

degree of sequence and organizational homology with the genome of HRPV-1. The virons of these two viruses also resemble each other, and they are clearly related. It may be that one virus (HHPV-1) packages the double-stranded replicative intermediate form of the DNA genome, and the other (HRPV-1) packages the single-stranded DNA form made during rolling circle replication (see replication of φX174 DNA in Chapter 6, Figure 6.4).

Comparative genomics of archaeal viruses

The most prominent feature of our understanding of genomes of known archaeal viruses is the extremely low number of genes with recognizable functions. A considerable majority of viral genes present in sequenced genomes of archaeal viruses have no detectable homologs other than in closely related virus species. The only exceptions are archaeal head-tail viruses, which carry genes with homologs in genomes of bacterial head-tail viruses from the families *Myoviridae* and *Siphoviridae*. The homologous genes mainly code for proteins involved in virion assembly, capsid and tail formation and DNA modification. These homologs show either that these bacterial and archaeal viruses have a common ancestor, or that there was genetic exchange between bacterial and archaeal viruses some time during evolution.

Extreme examples of the unique genetic content of archaeal viruses are provided by the globuloviruses PSV and TTSV-1, the genomes of which are almost *terra incognita*, containing no genes with homologs in extant databases. In cases where homologs are found, they are nearly exclusively from other viruses or plasmids of archaea, such as with the bicaudavirus ATV (Figure 9.12).

The genome of the ampullavirus ABV is one of those that carry the highest proportion of genes with predictable functions. ABV has the only genome besides those of the salterproviruses His-1 and His-2 that carries an identifiable DNA polymerase gene. These DNA polymerases are most similar in sequence to the type B DNA polymerases found in certain viruses of bacteria and eukaryotes, like the tectiviruses or adenoviruses, and are unusual in using proteins attached to the 5'-termini of linear dsDNA genomes to prime replication.

The failure to recognize genes involved in DNA synthesis in the genomes of most archaeal viruses is puzzling, especially considering that many genomes have exceptional features such as covalently closed ends and telomere-like terminal structures that must be involved in genome replication. This suggests that archaeal viruses may depend on the host cell replication machinery but can modify it to promote selective replication of viral DNA, as do some of the small DNA animal viruses (parvoviruses, polyomaviruses, and papillomaviruses, Chapters 20, 21, and 22). It cannot be excluded that archaeal viruses may encode unusual DNA polymerases, which cannot be identified by comparative sequence analysis.

The unique gene content of the majority of archaeal viruses may reflect exceptional features of virus–host relationships. One such example is the unique mechanism of host cell lysis and virion release using a pyramidal structure (Figure 9.10) used by viruses from at least two different families. However, this release mechanism is not universal for viruses of archaea; in many cases productive infection does not lead to cell lysis or death. Such "generosity" to the host is thought to provide a durable intracellular refuge for the virus population in harsh conditions of natural habitats (often with temperatures above 80°C and pH values below 3).

How virions are released without killing the archaeal cell remains unclear. Unclear are also the mechanisms that allow stable transmission of the viral genome during division of the host cell, especially in the case of viruses with linear DNA genomes that do not integrate into the host chromosome. Further characterization of the replication cycles of archaeal viruses and their interactions with the host cell will surely lead to interesting discoveries.

Conclusion

The discovery and description of unique and varied species of archaeal viruses has significantly influenced the field of prokaryotic virology. Future research will likely follow two major lines: (1) exploration of the diversity of archaeal viruses in different environments, and isolation and characterization of new virus–host systems; and (2) in-depth analysis of the biology of selected archaeal viruses. These studies will be focused on such fundamental aspects of virus biology as the structure and assembly of virions, their attachment to the host cell and penetration through the S layer, their egress from cells that do not die during infection, and the replication of viral genomes and regulation of their transcription. Comparison of properties of viruses of archaea, bacteria, and eukaryotes will help to reconstruct scenarios of virus evolution and will contribute to deeper understanding of the mechanisms of virus–host interactions.

KEY TERMS

Anaerobic methanogen	Euryarchaeota
Archaea	Hyperhalophile
Bacteria	Hyperthermophile
Crenarchaeota	Phyla
Domains of life	Solfataric
Eukarya (eukaryotes)	Thermophile

FUNDAMENTAL CONCEPTS

- Archaea are a third domain of life, distinct from bacteria and eukaryotes.
- Many archaea grow and survive in extreme environments of temperature, acidity, or salinity.
- Most viruses that infect archaea have double-stranded DNA genomes, and some are positively supercoiled, probably an adaptation to extreme environments.
- Many viruses of archaea have envelopes and unusual morphologies with no obvious symmetry.
- Many of these viruses do not kill the host cell during productive infection; their method of exit from the cell is unknown.
- Some, however, generate pyramidal structures that appear to pierce the cell wall, releasing virions and killing the host cell.
- Most of the genes of viruses of archaea have no homology to known genes.
- Many viruses of archaea encode no obvious DNA polymerase.
- One small, double-stranded DNA virus is closely related to a single-stranded DNA virus; both infect hyperhalophiles.

REVIEW QUESTIONS

1. How does the life cycle of *Fuselloviridae* differ from that of temperate bacteriophages such as lambda?
2. *Sulfolobus* spindle-shaped virus 1 (SSV-1) genomes have the property of being positively supercoiled when isolated from virions, and were one of the first examples of such DNA found in nature. What is the role of a "reverse gyrase" for these microorganisms?
3. Explain how rudiviruses escape from the cell.
4. What are IS elements?

POSITIVE-STRAND RNA VIRUSES OF EUKARYOTES

Viruses are the only known organisms on earth that can encode their genetic information in RNA molecules; these may be relics of the ancient RNA world. Viruses with a single-stranded RNA genome that can directly code for viral proteins are called "positive-strand RNA viruses". These viruses can be thought of as pure messenger RNA molecules packaged in a protein coat.

Upon entering a cell, the genome of a positive-strand RNA virus binds to cellular ribosomes and is translated to produce viral proteins. These proteins include an RNA-dependent RNA polymerase, which can replicate the genome first by making a complementary (negative-strand) copy, and then by making many new positive-strand copies. These new genome RNAs can either be translated, replicated, or packaged with new virion structural proteins.

A large number of plant viruses have postive-stranded RNA genomes. A well-studied plant virus, cucumber mosaic virus, is described in Chapter 10. Plant viruses must penetrate the plant cell wall, and are usually transmitted by sucking insects or mechanical means. To spread within a plant, they produce "movement proteins" that alter plasmodesmata, channels that connect the cytoplasm of neighboring cells. These movement proteins also accompany the viral genomes as they pass from cell to cell.

The remainder of this section describes four families of animal viruses, many of which cause a variety of diseases in humans or animals. Picornaviruses make non-enveloped virions, and are best known because of the paralytic disease caused by poliovirus and the annoying colds caused by rhinoviruses. Flaviviruses and togaviruses have lipid envelopes, and most are transmitted between host animals by mosquitoes or ticks. They can cause mild or severe disease, including yellow fever and fatal encephalitis. Coronaviruses have the largest known viral RNA genomes; besides causing important veterinary diseases, some cause upper respiratory infections in humans. A recently emerged coronavirus caused the unexpected SARS epidemic in the early part of this century.

CHAPTER 10

Cucumber Mosaic Virus

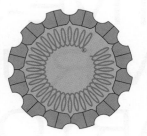

Ping Xu
Marilyn J. Roossinck

Bromoviridae

From Latin *Bromus*, a genus of grasses in the family *Gramineae*

Refers to the hosts of the type genus, *Bromovirus*

VIRION

Non-enveloped icosahedral capsid (T = 3).

Diameter 29 nm.

180 capsid protein subunits (12 pentamers and 20 hexamers).

GENOME

Linear single-stranded RNA, positive sense.

Segmented genome is composed of RNA 1 (~3.4 kb), RNA 2 (~3.1 kb), and RNA 3 (~2.2 kb).

All segments have a 5' cap structure and 3' tRNA-like structure.

Two subgenomic RNAs: RNA 4 and RNA 4A.

GENES AND PROTEINS

RNA 1 encodes the 1a protein: a capping enzyme and helicase, and a component of the replicase.

RNA 2 encodes the 2a protein: the RNA polymerase component of the replicase, and the 2b protein: a silencing suppressor (translated from subgenomic RNA 4A).

RNA 3 encodes the 3a or movement protein, and the coat protein (translated from subgenomic RNA 4).

RELATED VIRUSES AND HOSTS

The family *Bromoviridae* consists of five genera: *Alfamovirus*, *Bromovirus*, *Cucumovirus*, *Ilarvirus*, and *Oleaevirus*.

The best-characterized members of the family are brome mosaic virus and cucumber mosaic virus.

Other members of the genus *Cucumovirus* are peanut stunt virus and tomato aspermy virus.

All viruses in the family infect plants, but host ranges vary dramatically.

Cucumber mosaic virus infects monocots, dicots, herbaceous plants, shrubs, and trees, including more than 1200 species from 85 plant families.

DISEASES

Cucumber mosaic virus infections can be symptomatic or asymptomatic, depending on the host.

Most modern cultivars of cucumber are infected asymptomatically.

Cucumber mosaic virus has caused several crop disease epidemics worldwide and catastrophic crop losses in other plants, most notably tomato.

Transmission is mainly by aphids, but the virus can also be transmitted by parasitic plants, in seeds, and mechanically.

DISTINGUISHING CHARACTERISTICS

Cucumber mosaic virus has the broadest known host range of any known virus and is distributed worldwide.

Cucumber mosaic virus supports the replication of parasitic satellite RNAs that can dramatically alter the virus symptoms.

The 2b protein of cucumber mosaic virus is a multifunctional protein best known for its ability to suppress post-transcriptional gene silencing, or RNA interference (RNAi).

Like many other plant viruses with segmented genomes, individual cucumber mosaic virus genome RNAs are packaged in separate virions.

Mosaic disease in cucumber plants led to the discovery of cucumber mosaic virus (CMV)

The symptoms caused by infection with cucumber mosaic virus and the transmission of this virus among plants were first described separately by Sears Doolittle and Ivan Jagger in 1916. They studied a **mosaic disease** (discoloration and speckled appearance on leaves and fruit) occurring in cucumbers and other **cucurbit** plants (gourds, squashes, melons) in the field and in greenhouses in the United States (Figure 10.1). This disease caused a total loss of regional cucurbit crops. They found that the infectious agent could pass through a Berkefeld filter, and was therefore a "filterable virus". Healthy plants could be infected by mechanical inoculation. This virus could also be transmitted among cucurbit plants by grafting and by insects.

Ten years later, James Johnson reported that this virus also infects tobacco plants, in which it causes mosaic disease. The pathogen was first denoted cucumber mosaic virus (CMV) by Conway Price in 1940. Most plant viruses are named in a similar manner, after the host where they were first discovered and the disease symptoms caused in that host. Many virus isolates that infect other crops, and were therefore given different names, were later found to be identical to cucumber mosaic virus. Cucumber mosaic virus has the largest known host range among all viruses, infecting more than 1200 species of plants including **monocots** and **dicots**, herbaceous plants, shrubs, and trees. Different strains of cucumber mosaic virus have been found worldwide and have caused several severe epidemic crop diseases in many regions of the world.

The related virus, brome mosaic virus (BMV), has a much narrower host range, largely limited to the members of the **Gramineae** (grasses). This virus was first described in 1941, and was the first plant virus whose cloned genome was used to generate infectious RNA transcripts. Brome mosaic virus is transmitted mechanically and by beetles.

Cucumber mosaic virus has a positive-strand RNA genome enclosed in a compact capsid with icosahedral symmetry

In 1959, John Tomlinson reported that cucumber mosaic virus has an RNA genome, and he observed virus particles by electron microscopy. Hervé Lot et al. and Howard Scott et al. separately developed protocols for producing

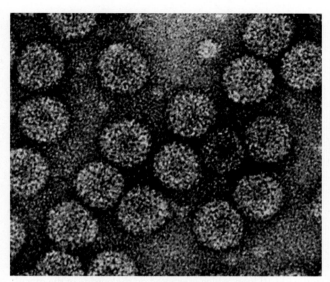

Figure 10.2 Electron micrograph of cucumber mosaic virus.

high yields of infectious cucumber mosaic virus particles, and these methods have been very useful for studies on the molecular biology of the virus. Viral RNA isolated from purified virions was found to be infectious by mechanical inoculation of plants. This showed that cucumber mosaic virus and its relatives are positive-strand RNA viruses, whose genome RNAs can act directly as messenger RNAs when introduced into a cell.

Virions of cucumber mosaic virus are compact (29 nm) spherical particles with icosahedral symmetry (Figure 10.2). The atomic structure of the virion, determined by x-ray diffraction of virus crystals, reveals a T = 3 structure with 180 identical capsid protein subunits grouped as 12 pentamers and 20 hexamers. Both pentamers and hexamers have central pores; the pentamer pores are believed to allow entry by ribonucleases, as incubation of purified virions with ribonuclease leads to degradation of the genome RNA.

A negatively charged arginine-rich domain in the N-terminal region of the coat protein is localized on the internal surface of the virion and interacts with encapsidated RNA (Figure 10.3). A highly antigenic nine-amino acid loop that is conserved among cucumoviruses is located on the outside of the virion, and mutations in this region affect efficiency of transmission by aphids. Strains of cucumber mosaic virus are divided into three subgroups (IA, IB, and II) initially based on the amino acid sequence of the coat protein and the sequence of the 5' non-translated region of RNA 3.

The genome of cucumber mosaic virus consists of three distinct RNA molecules

Virions of cucumber mosaic virus usually contain four major species of RNA (RNAs 1, 2, 3, and 4) with respective sizes of about 3360, 3050, 2215 and 1030 nt

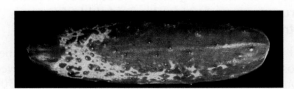

Figure 10.1 Symptoms of infection by cucumber mosaic virus on cucumber fruit.

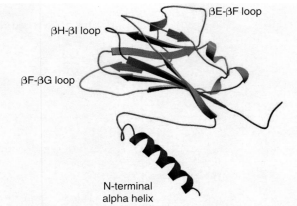

Figure 10.3 Ribbon diagram of the cucumber mosaic virus coat protein. The N-terminal α helix is involved in interactions with the genome and subgenomic RNAs packaged in the capsid. A highly conserved 9-amino acid region is located in the βH-βI loop, which is exposed on the outside of the capsid. The amino acids known to be involved in aphid transmission are found in the βE-βF, βF-βG, and βH-βI loops.

(Figure 10.4). RNAs 1, 2, and 3 are genome RNAs, and simultaneous infection with all three RNAs is required to establish a productive infection. RNA 1 has a single open reading frame (protein 1a), RNA 2 has two over-lapping reading frames (proteins 2a and 2b), and RNA 3 has two non-overlapping reading frames (protein 3a and coat protein [CP]). The 5'-termini of the genome and subgenomic RNAs are capped with 7-methylguanosine, as are cellular messenger RNAs. The 3'-terminal regions of all viral genome RNA segments can fold into a transfer RNA (tRNA)-like structure (see below).

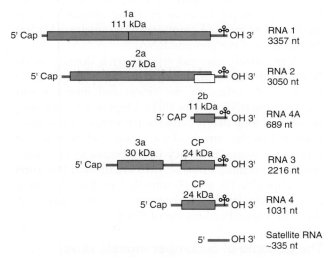

Figure 10.4 Cucumber mosaic virus genome organization. Open reading frames are shown as colored rectangles, and the names of encoded proteins and their molecular weights are shown above. The sizes of the genome RNAs are shown for the Fny strain.

RNA 4 is a subgenomic RNA equivalent to the 3' half of RNA 3; it encodes the coat protein. RNA 4A is also a subgenomic RNA, equivalent to the 3'-terminal one-quarter of RNA 2; it encodes the 2b protein, and is packaged by some strains but not by others. In some strains of cucumber mosaic virus, an additional RNA, denoted RNA 5, is detected. RNA 5 is a mixture of sequences originating from the 3'-terminal noncoding regions of RNAs 2 and 3, and these noncoding RNAs can be packaged into virions. They have no known function.

The three genome RNAs and a subgenomic RNA are encapsidated in separate but otherwise identical particles

RNAs 1 and 2 are packaged in separate particles, while RNAs 3 and 4 are packaged together in a single particle. RNA 5, when present, is packaged together with one of the three genome RNAs. Occasionally, three RNA 4 molecules are packaged in a single particle in the absence of RNA 3.

The particles containing different genome RNAs are physically and morphologically indistinguishable. Each particle contains about the same amount of RNA, ranging from 3000 to 3360 nt. Viruses within the *Bromovirus* and *Cucumovirus* genera have similar packaging strategies. However, the genome and subgenomic RNAs of viruses in the *Alfamovirus* genus are individually packaged into four virions with distinct sizes.

Packaging of individual segments of viral genomes into separate particles is commonly found in plant viruses but is rare or absent in viruses infecting other organisms. This strategy allows a relatively large genome to be packaged as distinct segments in small and simple virions. However, it has the disadvantage that productive infection by such viruses requires the delivery of at least one of each of the particles containing different genome segments into the same cell. Many plant viruses are transmitted from plant to plant on the mouthparts of insects, which can be heavily contaminated with virus particles picked up from an infected plant. This mechanism of transmission therefore often results in the delivery of many virus particles simultaneously into the same cell.

The 3'-terminal regions of cucumber mosaic virus genome segments can fold to form a transfer RNA-like structure

All genome and subgenomic RNAs of cucumber mosaic virus and its relatives contain a conserved noncoding region at their 3'-termini, which is about 300 nucleotides in length and can fold to form a tRNA-like "cloverleaf" or pseudoknot structure (Figure 10.5).

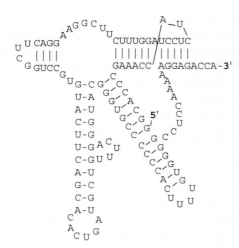

Figure 10.5 The pseudoknot configuration of the tRNA-like 3' structure of cucumber mosaic virus genome RNAs. RNA 2 of the Fny strain is shown.

Similar tRNA-like structures are found at the 3'-termini of the genomes of a number of other positive-strand RNA plant viruses, including other members of the *Bromoviridae* family, as well as turnip yellow mosaic virus and tobacco mosaic virus. However, other positive-strand RNA plant viruses have either 3'-terminal poly A tails, or complex 3'-terminal structures that do not resemble tRNAs.

Strikingly, many of these tRNA-like structures can serve as substrates for two cellular enzymes that recognize transfer RNAs: CCA-nucleotidyltransferase, which forms or regenerates the 3'-terminal CCA sequence on all tRNAs, and aminoacyl-tRNA synthetases, which covalently link specific amino acids to tRNAs, enabling their use in protein synthesis.

In the case of cucumber mosaic virus and brome mosaic virus, many encapsidated genome molecules lack the 3'-terminal A residue, and this can be added by cellular CCA-nucleotidyltransferase. Genome RNAs with a CCA 3'-terminus are substrates for tyrosine-tRNA synthetase, which covalently links a tyrosine residue to the 3'-terminal A residue. Other plant virus RNAs with a 3'-terminal tRNA-like structure are recognized by tRNA synthetases that covalently link them to histidine or valine residues.

These structures are believed to serve a number of important functions in virus replication. First, the tRNA-like structure may protect viral RNAs from degradation by cellular nucleases, as RNAs covalently bound to an amino acid are rendered resistant to ribonucleases that recognize 3' ends of RNAs. Second, the tRNA-like structure may enhance translation of the RNA by interacting with cellular proteins involved in translation elongation. Experiments with other plant virus RNAs have shown that eukaryotic elongation

factor 1A (eEF1A) binds more efficiently to aminoacylated RNAs than to RNAs lacking a 3'-terminal bound amino acid residue, and that binding of this factor is correlated with increased translation of the RNA.

Third, tRNA-like structures may initially reduce synthesis of minus-strand RNA, which begins at the subterminal C residue on genome RNA. Because ribosomes translate the (positive-strand) genome RNA in the 5'–3' direction, and RNA polymerases proceed in the opposite direction while synthesizing progeny negative strands, initiation of RNA synthesis during translation of the same RNA molecule could lead to collisions between translating ribosomes and transcribing RNA polymerases (see Figure 5.4). Reducing initiation of RNA synthesis could allow the initial translation of RNAs 1 and 2 immediately after entry into the cell, providing sufficient levels of viral RNA polymerase to subsequently carry out viral RNA replication. For at least some plant viruses with 3' tRNA-like structures, the binding of eEF1A, while stimulating translation of viral RNA, may also inhibit its use as a template for RNA synthesis.

Cucumber mosaic virus is transmitted in nature by aphids

Cucumber mosaic virus is transmitted to other plants by more than 80 species of **aphids**, small insects that feed by sucking sap from plants. Aphids are quite specific in the plants they feed on, but they test many plants to determine if a plant is appropriate, and it is this brief probing that results in virus transmission. When an aphid probes an infected plant, the virus interacts with the anterior portion of the alimentary tract, and from there it subsequently can be inoculated into an uninfected plant. Viruses can be acquired by aphids and transmitted within seconds to minutes.

Transmission efficiency varies with different hosts, aphids, and virus strains. Transmissibility and vector specificity are primarily determined by the coat protein. Several amino acids in the cucumber mosaic virus coat protein that are associated with efficient aphid transmission have been identified. These amino acids are distributed either on the external surface or the internal surface of the virion. Internal amino acids that affect transmission may be involved in virion stability; external amino acids may be recognition sites for interactions with aphids that affect uptake or release of virions from mouthparts during probing (Figure 10.3).

Cucumber mosaic virus can also be transmitted through seeds to progeny plants. Seed transmission occurs when the virus infects seed tissues after fertilization, and enters the newly developing plant during embryogenesis.

The genome of cucumber mosaic virus encodes five multifunctional proteins

The three genome segments of cucumber mosaic virus encode a total of five different proteins, denoted 1a, 2a, 2b, 3a, and CP (Figure 10.4 and Table 10.1). The two largest viral proteins, 1a and 2a, together constitute the viral RNA replicase that includes the RNA polymerase, a capping enzyme and a helicase, along with host factors. The 1a and 2a proteins normally accumulate at very low levels in infected cells.

Protein 1a (110 kDa) contains two domains: a methyltransferase domain in the N-terminal region, involved in capping the 5' ends of viral RNAs, and an RNA helicase domain in the C-terminal region. The RNA helicase domain includes two conserved motifs, one for binding of ribonucleoside triphosphates and one for binding to RNA, both of which are essential for viral replication. Intercellular movement of the virus (see next column) in some hosts also depends on protein 1a. This could be due to a requirement for replication in specific vascular-related cells for movement to occur, or could imply a more direct role for the 1a protein in virus movement. Protein 1a can also affect seed transmission, host resistance, and symptom severity.

Protein 2a (97 kDa) contains a central domain that shares sequence similarity with many viral RNA-dependent RNA polymerases, and includes four conserved motifs. The N-terminal region of protein 2a interacts with protein 1a, and phosphorylation in this region abolishes this interaction and virus replication. Protein 2a may also be involved in virus movement

or triggering specific host responses. For example, changes in two amino acids in or close to the region of the putative catalytic center of the RNA polymerase can change the host response from **hypersensitive necrosis** (localized plant tissue lesions) to **systemic mosaic** symptoms.

Protein 2b (15 kDa) is expressed in all cucumoviruses but not in members of the other four genera of the *Bromoviridae*, and it is the least conserved protein across cucumovirus species. Protein 2b has two conserved basic domains at its N- and C-terminal regions, including a nuclear localization sequence at the N-terminal region. It plays a role in the long-distance movement of cucumber mosaic virus in infected plants, and helps to counteract the host defense system of RNA silencing (see below).

Protein 3a (30 kDa), also known as the movement protein (MP), is a nonstructural protein essential for virus movement in host plants. Protein 3a targets and dilates **plasmodesmata**, the cytoplasmic channels that connect plant cells, and mediates the movement of viral RNA from one cell to another, including movement into and out of the vasculature.

Coat protein (CP; 24 kDa) is the only protein in the virus capsid. Coat protein is also required for virus movement within plants, and is involved in aphid transmission of cucumber mosaic virus between plants. The presence of coat protein increases the accumulation of plus-strand viral RNAs in infected cells, probably because it stabilizes viral RNAs by packaging them in capsids.

Table 10.1 Proteins encoded by cucumber mosaic virus

Viral protein	Translated from	M. Wt. (kDa)	Sequence features	Functions
1a	RNA 1	110	Methyltransferase domain; RNA helicase domain containing NTP-binding and RNA-binding motifs	5' capping of viral RNAs; RNA synthesis; support of satellite RNAs; virus movement and seed transmission in specific hosts
2a	RNA 2	97	RNA-dependent RNA polymerase	RNA synthesis; virus movement
2b	RNA 4A	15	Basic domains in C- and N-terminal regions; nuclear localization signal; binding of single-stranded RNA and siRNA	Long-distance movement; suppression of RNA silencing; virus accumulation
3a	RNA 3	30	Binding of single-stranded RNA	Cell-to-cell and long-distance movement; virus accumulation
Coat protein	RNA 4	24	Arginine-rich domain at N-terminal region; RNA binding	Encapsidation; aphid transmission; cell-to-cell and long-distance movement

Replication of viral RNA is associated with intracellular membranes, and requires coordinated interaction of viral RNAs, proteins, and host proteins

When viral genome RNAs are released from capsids within plant cells, they serve as messenger RNAs that direct synthesis of proteins 1a, 2a, and 3a. After initial translation of viral proteins, there is a switch from translation to the synthesis of minus-strand RNAs and then plus-strand RNAs. Replication of viral RNAs requires the assembly of a replicase complex, which involves viral RNA and proteins, host proteins, and cell membranes. In infected cucumber and tobacco plants, the cucumber mosaic virus replicase complex is associated with the **tonoplast**, the membrane surrounding the large central vacuole typically found in many plant cells.

The core promoter sequences for minus-strand RNA synthesis include a stem loop within the 3'-terminal tRNA-like structure of the three viral genome RNAs. The 5' non-coding region and the noncoding region between the two reading frames of RNA 3 contain important sequences regulating the accumulation levels of genome and subgenomic RNAs. The minimum promoter sequence for synthesis of subgenomic RNA 4 contains a stem-loop structure and lies in the noncoding region upstream of the coat protein reading frame on RNA 3.

Brome mosaic virus RNA replication has been analyzed in yeast cells

Brome mosaic virus has been extensively studied as a model system for translation and RNA synthesis by positive-strand RNA viruses. Paul Ahlquist's laboratory showed that translation, replication, and encapsidation of brome mosaic virus RNAs can be carried out in yeast transfected with viral genome RNAs. This system facilitated detailed genetic and molecular studies of the mechanism of viral RNA replication.

Synthesis of brome mosaic virus protein 2a in yeast requires the host protein DED1, an RNA helicase. DED1 is an essential gene for translation of yeast messenger RNAs in yeast. DED1 also regulates the synthesis of brome mosaic virus protein 1a by interacting with sequences near the 5' end of viral RNA 1. Regulatory sequences in genome and subgenomic RNAs that coordinate the levels of translation, and the switch from translation to replication, have been identified. For example, the 5' noncoding region of RNA 2 down-regulates the synthesis of protein 2a, which is important for replication complex assembly.

Brome mosaic virus RNA synthesis takes place on cytoplasmic membranes

The assembly of the brome mosaic virus RNA synthesis complex in yeast occurs in the membranes of the endoplasmic reticulum. Protein 1a induces the formation of spherical invaginations within the endoplasmic reticulum membrane, recruits viral RNA to these invaginations, and directs the 2a protein to the membrane (Figure 10.6). The recruitment of viral RNAs into the RNA synthesis complex involves *cis*-acting recognition elements. These recognition elements are located at the 5'-termini of RNAs 1, 2, and 3 and in the intergenic region of RNA 3, which also interacts with the 5'-terminal region of RNA 3.

Efficient recruitment of RNAs into the replication complex requires a host protein, LSM1, which functions in host mRNA turnover. The synthesis of viral proteins occurs predominantly in the replication complex, although the initial burst of viral protein synthesis must take place before replication complexes are formed. In addition to viral RNAs and proteins, the yeast chaperone protein YDJ1 (Yeast DnaJ homolog) and the levels of unsaturated fatty acids in the infected cell are also important for the initiation of minus-strand RNA synthesis. The chaperone protein may be involved in the assembly of the replicase complex, and fatty acid levels may affect the assembly of membranes within this complex.

The replicase complex includes viral proteins 1a and 2a and host proteins. Subunits of host translation initiation factor eIF3 are associated with the replicase complex purified from infected cells. The addition of eIF3 to replicase complexes isolated from transfected yeast cells stimulates RNA synthesis *in vitro*.

The 3'-terminal 135–200 nts of viral genome RNAs function as promoters for the synthesis of minus-strand RNAs, and the 3'-terminal region of minus-strand RNAs are promoters for synthesis of plus-strand RNAs. The intergenic region of the minus-strand of RNA 3 contains the promoter for synthesis of subgenomic RNA 4.

Packaging of viral genomes

Brome mosaic virus has been used as a model for the study of packaging of segmented viral genomes. Packaging of brome mosaic virus RNAs involves specific sequence- and/or structure-dependent RNA-RNA interactions and coat protein-RNA interactions. The conserved tRNA-like structures at the 3'-termini of genome and subgenomic RNAs are required for *in vitro* assembly of brome mosaic virus particles, but these structures are not required for packaging of another bromovirus, cowpea chlorotic mottle virus. Packaging of the brome mosaic virus genome is coupled to replication, a common feature of positive-strand RNA viruses. The viral replicase complex plays a role in selective packaging of viral RNAs and maintaining uniform-sized viral particles.

It remains unknown how RNA 3 and RNA 4 are packaged together into a single virion. Virions containing RNA 4 alone can be assembled efficiently *in vitro*, but 94% of virions assembled *in vivo* that contain RNA

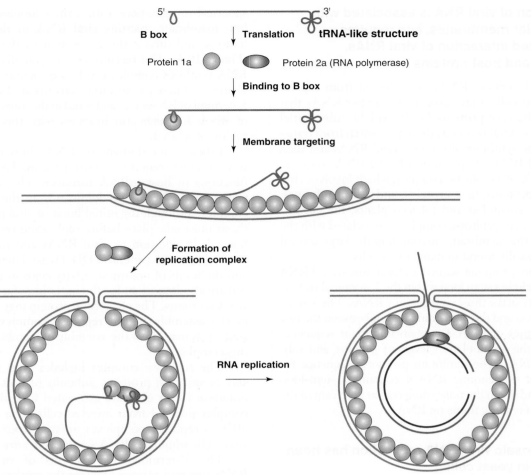

Figure 10.6 Formation of brome mosaic virus replication complexes. RNAs 1 and 2 are translated, and protein 1a subsequently binds to the B box near the 5' end of viral RNAs and targets it to cytoplasmic membranes. Protein 1a acts as a chaperone to bring protein 2a, the RNA polymerase, to the membrane where it first copies the positive-strand RNA (light blue) into a complementary negative-strand RNA (dark blue), and then synthesizes progeny positive-strand RNAs. The negative-strand RNA remains within the membrane complex.

4 also contain RNA 3. Two models for co-packaging of RNA 3 and RNA 4 have been proposed. In one model, prior packaging of RNA 3 alters the conformation of the coat protein and promotes its interaction with RNA 4, leading to packaging of both RNAs in the same capsid. In the second model, the packaging signal of RNA 3 interacts first with RNA 4, resulting in an initiation complex for co-packaging. How RNAs 4A and 5 and the satellite RNAs are packaged remains unknown. RNA 4A is not packaged in many strains of cucumber mosaic virus, and RNA 5 is only synthesized in some strains.

Cucumber mosaic virus requires protein 3a (movement protein) and coat protein for cell-to-cell movement and for long-distance spread within infected plants via the vasculature

Systemic infection of plants by cucumber mosaic virus depends on movement of viral genomes from initially infected cells to neighboring cells, vascular cells, and other distant tissues. Plant cells are surrounded by a substantial physical barrier, the cell wall, which impedes movement of macromolecules between cells. However, the cytoplasm of neighboring plant cells is connected via plasmodesmata (singular: plasmodesma). Plasmodesmata are membrane-lined channels with defined size exclusion limits that interconnect adjacent plant cells (Figure 10.7). The size exclusion limits of plasmodesmata connecting epidermal and mesophyll cells are similar. However, plasmodesmata connecting mesophyll cells and vascular-related cells (bundle sheath cells, adjacent phloem parenchyma cells, sieve elements, and companion cells; see Figure 10.8) have different morphological and transport features.

Like many other plant viruses, cucumber mosaic virus moves through plasmodesmata from the epidermis to mesophyll tissues, where the majority of replication occurs. Virus then enters minor veins in the

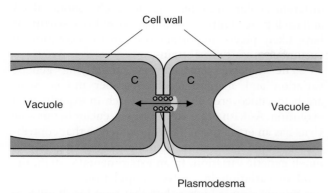

Figure 10.7 Diagram of a plasmodesma. Molecules can move between the connected cells, but the size of the molecules is restricted. Viral movement proteins (shown as small spheres) can enlarge the diameter of the plasmodesma.

infected leaf, moves along the phloem following the translocation stream, and eventually enters young plant tissues, following the same route as photoassimilates (Figure 10.8).

All proteins made by cucumber mosaic virus play some role in virus movement, sometimes in a host-specific manner, but protein 3a and the coat protein are universally required. Protein 3a targets the plasmodesmata between various cell types including sieve elements and companion cells, and alters its size exclusion limits. Plasmodesmata are found as two types, primary and secondary, based on the developmental stage. Viral movement proteins are often associated with secondary plasmodesmata, which normally originate from primary plasmodesmata.

Protein 3a binds to viral RNAs and forms a ribonucleoprotein complex that transports viral RNAs through plasmodesmata, possibly by interacting with unidentified host proteins. Coat protein plays different roles in **cell-to-cell movement** and **long-distance movement**. In cell-to-cell movement, coat protein may protect and stabilize the protein 3a-viral RNA complex or alter the conformation of protein 3a to facilitate the movement of viral RNAs, but no direct interaction between coat protein and 3a has been detected. In some experimental situations, coat protein is not necessary for cell-to-cell movement of cucumber mosaic virus. Protein 3a has been shown to bind to protein 2a in a yeast two-hybrid system. It is not clear if the protein 3a-viral RNA complex contains any replicase components or other host factors.

Very little is known about how the protein 3a-viral RNA complex moves from the sites where viral RNA and protein synthesis take place to the plasmodesmata. Very likely, host cytoskeleton components are involved. Although virions are not transported through plasmodesmata, they are found in **sieve elements**. Thus virions may be reassembled in sieve elements and move along the vascular system. This movement is long-distance movement, which transports the virus from the initially infected tissue to remote parts of the plant. Once in the vascular system, the virus moves in a passive manner along with photoassimilates. In some cases cucumber mosaic virus coat protein is associated with long-distance movement in a host-specific manner, indicating the involvement of host factors. So far, these host factors have not been identified.

Tobacco mosaic virus movement protein can direct movement of cucumber mosaic virus in infected plants

Cucumber mosaic virus 3a protein shares functional similarities with the tobacco mosaic virus (TMV)

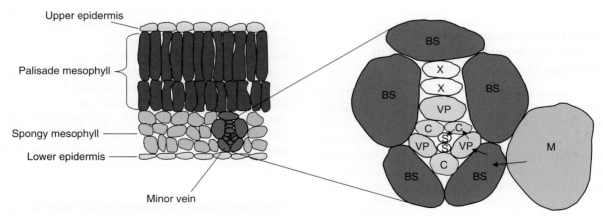

Figure 10.8 Cross-section of a leaf showing a vascular bundle. BS, bundle sheath cells; C, companion cells; M, mesophyll cells (these surround the vascular bundle, but are not part of it); S, sieve elements; VP, vascular parenchyma cells; X, xylem cells. The virus replicates predominantly in the mesophyll cells, and then moves in the direction of the arrows, from the mesophyll into the vascular bundle to the sieve elements for transport along with the photoassimilate. The vascular parenchyma, companion cells, and sieve elements make up the phloem. The virus moves out of the phloem to invade new tissue in the reverse order to that shown.

movement protein, even though they share no detectable sequence similarity. The TMV movement protein can complement the function of missing or mutated cucumber mosaic virus protein 3a. Considerably more research has been done on the cell biological and molecular mechanisms involved in movement of tobacco mosaic virus within infected plants.

The movement protein of tobacco mosaic virus binds to single-stranded RNAs, interacts with host cytoskeleton elements, and targets and dilates plasmodesmata. It contains a transmembrane domain and initially associates with the endoplasmic reticulum in infected cells. The movement protein subsequently accumulates in membrane-associated virus replication complexes (VRCs). Virus replication complexes containing movement proteins are transported rapidly within the cell and halt briefly at plasmodesmata. Then virus movement complexes containing movement protein and viral RNAs transverse the plasmodesmata and enter the adjacent cells. Some movement proteins remain in the plasmodesmata.

Actin and myosin filaments are involved in intra- and intercellular movement of tobacco mosaic virus RNA. RNA movement does not involve microtubules, although tobacco mosaic virus movement protein does bind to and colocalize with a microtubule-associating protein at a later stage of infection. This association is correlated with degradation of movement protein by the proteasome pathway.

Tobacco mosaic virus movement protein also interacts with many other host factors, including the cell wall protein pectin methylesterase, several transcriptional factors, and cytoskeleton proteins. Deletion of the domain of the tobacco mosaic virus movement protein that interacts with pectin methylesterase results in a loss of cell-to-cell movement. Pectin methylesterase can also bind to RNA. This protein may play an important role in targeting of movement protein to plasmodesmata.

Mutation, recombination, reassortment, and genetic bottlenecks are involved in the evolution of cucumber mosaic virus

Cucumber mosaic virus has been very successful in adapting to new hosts and new environments. Many different strains of this virus have been found all over the world. Based on analysis of antigenicity of the coat protein and of RNA sequences, cucumber mosaic virus strains can be divided into three subgroups: IA, IB, and II. Phylogenetic analyses using the sequences of the coat protein gene and the 5' noncoding sequences of genome RNA 3 indicate a common ancestor for each subgroup. However, while phylogenetic analyses using sequence variations in the genes coding for other viral proteins, as well as noncoding regions in viral genomes, maintain the subgroup designations, they indicate

different evolutionary histories for each genome RNA and each gene. Thus, recombination and **reassortment** have likely played important roles in the evolution of cucumber mosaic virus strains.

The molecular mechanisms underlying genetic variation and adaptation of cucumber mosaic virus are being uncovered through research in experimental evolution. As with other RNA viruses, cucumber mosaic virus has an error-prone RNA polymerase that contributes to rapid generation of mutations, giving rise to much genetic variation. In an experimental system, the level of variation in evolved populations of cucumber mosaic virus is higher than that found with tobacco mosaic virus or cowpea chlorotic mottle virus, a related bromovirus, but varies with different virus–host combinations. In addition, the generation of insertions and deletions in viral RNAs during infection in plants was shown to be dependent on both viral RNA structure and the plant host.

Cucumber mosaic virus replicase can generate recombinant RNAs *in vitro* by base pairing-independent copy-choice recombination. This involves copying of part of one template RNA, followed by dissociation of the replicase and the growing RNA chain, and association to a second template RNA upon which RNA synthesis then continues, generating a progeny RNA containing sequences from two different parental RNA molecules. Inter- and intramolecular recombination also occurs *in vivo*, both in experimental plants and in natural populations. Such recombination is particularly frequent in 3' noncoding regions, probably because they have substantial secondary structure and share common sequences. Interspecies recombination also occurs during mixed infections.

For a virus with a segmented genome, reassortment among genome segments of different strains and species is analogous to sexual recombination. Stable reassortants between subgroups of cucumber mosaic virus and between different *Cucumovirus* species are easily created experimentally; furthermore, there exist natural *Cucumovirus* isolates that originated by genome reassortment. Genetic variation can be reduced by selection or by **genetic bottlenecks**. In experimental systems, significant genetic bottlenecks are encountered by viruses during systemic spread in plants, and during transmission by aphids. The combined effects of mutations, recombination, reassortment, selection, and genetic bottlenecks all contribute to the dynamic long-term evolution of cucumber mosaic virus in nature.

Host responses to cucumovirus infections reflect both a battle and adaptation between viruses and hosts

Plants often exhibit different responses to infection with cucumber mosaic virus. These include resistant and susceptible responses, symptomatic and asymptomatic infections, disease and tolerant responses. Most modern cultivars of cucumber are either resistant to cucumber mosaic virus, meaning that they cannot be infected, or they are tolerant of cucumber mosaic virus, meaning they become infected but do not show any symptoms. Under some stress conditions, virus infection may benefit plants. For example, plants infected with cucumber mosaic virus are more tolerant to drought and to cold temperatures, compared to uninfected plants. The responses vary with specific combinations of virus strains and plant species or cultivars, and environments, indicating a constantly evolving symbiosis between virus and host.

Some plants exhibit natural resistance to infection with cucumber mosaic virus. Host resistance to cucumber mosaic virus often involves multiple genetic loci; however, in some cases resistance is conferred by a single gene. One such case is the host resistance (R) gene, whose phenotype is a hypersensitive response (HR) to infection. The hypersensitive response involves tightly controlled cell death and induced resistance of neighboring cells. These two phenomena limit viral infection to a small local lesion zone on infected plants (Figure 10.9). The hypersensitive response is normally elicited by a single viral protein of a specific strain; for example, in cowpea it is elicited by the viral polymerase gene.

Host cell elongation factors that affect translation of viral proteins can be indirectly involved in resistance. In Arabidopsis (thale cress), elongation factors that specifically reduce the synthesis of movement protein result in resistance. It is not known how these elongation factors inhibit synthesis of movement protein.

In susceptible hosts, infection with cucumber mosaic virus can be asymptomatic or can cause variable symptoms including stunting, mosaic lesions, chlorosis, necrosis, distorted morphology of various organs, and interrupted vegetative and reproductive development. In many cases, sequences in viral RNA or viral proteins are associated with specific phenotypic changes in specific hosts under certain conditions. So far little is known about how these viral proteins or RNAs or the virions reprogram cell activities in the infected plants except for the 2b protein (see below). The 2b protein interferes with the gene-silencing pathway that directly regulates plant development.

Cucumber mosaic virus infection induces comprehensive changes in the expression of a large number of host genes, and modifies cell physiology and structure. Two common defense responses are induced in infected hosts. One is a general defense and stress response to microbial infection, involving activation of salicylic acid, jasmonic acid, and ethylene-responsive pathways. These defense responses are known to be activated through analysis of plant transcriptomes, but it is not clear that they are involved in any resistance to cucumber mosaic virus. A second defense response is RNA silencing, a common and widespread antiviral defense response of plants.

Plants respond to virus infection by RNA silencing, and cucumber mosaic virus protein 2b suppresses silencing

RNA silencing (see Chapter 33) is a gene inactivation mechanism mediated by **small interfering RNAs** (siRNAs). SiRNAs are generated by the cleavage of double-stranded RNAs or imperfect hairpin RNAs by the RNase III enzyme, Dicer, or Dicer-like nuclease. One strand of an siRNA is incorporated into the *siRNA-induced silencing complex* (RISC), and this RNA guides the RISC to a complementary single-stranded RNA and directs either cleavage of the target RNA or repression of its translation (see Figure 33.11).

In plants infected with RNA viruses, double-stranded replicative intermediate viral RNAs can be generated. Hairpin structures, which result from the presence of inverted repeat sequences in viral RNAs, can also form. These double-stranded RNAs are targets of the Dicer nuclease and their digestion products become siRNAs that lead to degradation or translational repression of viral RNAs. Viral siRNAs can be replicated by a host RNA-dependent RNA polymerase (RDRP), amplifying siRNAs and leading to further degradation or translational repression.

In plants infected with cucumber mosaic virus, both virus-induced defense responses and viral double-stranded RNA-triggered RNA silencing are part of the host immune response to virus infection. There is also cross-talk between these two defense responses, such as the involvement of salicylic acid in both RNA silencing and the hypersensitive response. The 2b protein

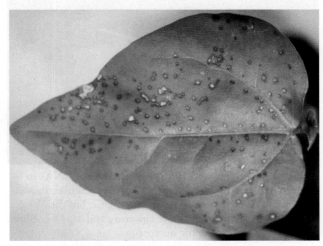

Figure 10.9 The hypersensitive response to cucumber mosaic virus on cowpea leaves. The lesions become necrotic, and virus does not spread beyond the lesions.

suppresses salicylic acid-mediated virus resistance, and the synthesis of host cell RNA-dependent RNA polymerase-1 is induced by salicylic acid.

The 2b protein of cucumber mosaic virus is also an RNA silencing suppressor. This protein counteracts host antiviral responses by blocking the systemic signal of RNA silencing, and therefore facilitates local and systemic infection. The precise nature of the systemic signal is unknown. However, this viral silencing suppressor sometimes triggers a hypersensitive response-associated resistance in specific hosts, indicating that complex molecular strategies are employed by viruses and their hosts for defense, counter defense, and perhaps eventual adaptation.

Cucumber mosaic virus supports replication of defective and satellite RNAs

Satellite RNAs are small, parasitic RNAs that are completely dependent on a helper virus for replication, encapsidation and transmission (see Chapter 31). Cucumber mosaic virus supports replication of satellite RNAs and also of defective RNAs derived from RNA 3. These RNAs replicate and are encapsidated together with genome and subgenomic RNAs in plants infected by certain viral strains.

Satellite RNAs of cucumber mosaic virus are small, linear RNA molecules first discovered as the pathogenic agent for an epidemic lethal disease in fields of tomato plants infected by cucumber mosaic virus (Figure 10.10). Since then, more than 100 **necrogenic** or non-necrogenic variant satellite RNAs have been found associated with cucumber mosaic virus worldwide. Another species of cucumovirus, peanut stunt virus, also harbors satellite RNAs. Peanut stunt virus satellite RNAs cannot be replicated by cucumber mosaic virus, and vice versa. No naturally occurring satellite RNA has been found associated with tomato aspermy virus, but this cucumovirus does support replication of cucumber mosaic virus satellite RNAs.

Cucumber mosaic virus satellite RNAs are from 332 to 405 nucleotides in length. They have capped 5'-termini and contain a 3'-terminal CCC sequence. Unlike their helper virus, the 3'-termini of these satellite RNAs do not fold into tRNA like structures, and they cannot be aminoacylated. They share no sequence similarity with cucumber mosaic virus RNAs. No translational products of these satellite RNAs have been found in infected plants, and they contain only short open reading frames, which are most likely without function as they are not conserved. Thus, the biological functions of these satellite RNAs depend on their direct interactions with viral or host cell proteins, nucleic acids, or other cellular components.

Satellite RNAs can either attenuate or increase severity of symptoms in infected plants

Cucumber mosaic virus satellite RNAs are highly structured, with more than 50% internal base pairing. They are stable and relatively resistant to ribonucleases. Sequence identity among satellite RNA variants is greater than 73%. Nevertheless, satellite RNA variants differ in the ways they modify symptoms of cucumber mosaic virus infection. Many satellite RNAs attenuate symptoms of virus infection. However, certain strains of satellite RNA cause more severe symptoms in infected plants such as systemic necrosis, severe **chlorosis**, and white leaves (Figure 10.10).

Nucleotide sequences that determine altered phenotypes have been mapped on satellite RNAs. For example, changing only three nucleotides in a stem-loop structure of a necrogenic satellite RNA transformed it into an attenuating satellite RNA in the same

Figure 10.10 Symptoms induced by satellite RNAs in tomato plants. Most satellite RNAs cause cucumber mosaic virus infection of tomato to be asymptomatic, but the WL2 satellite RNA (a) induces a white chlorosis, and the D satellite RNA (b) induces a systemic necrosis.

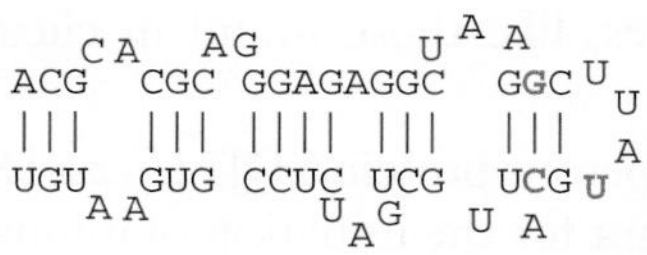

```
          CA   AG        U A
      ACG    CGC GGAGAGGC   GGC U
      |||    ||| |||||| |||    ||| U
      UGU    GUG CCUC  UCG    UCG U  A
          A A        U  G   U  A
```

Figure 10.11 The stem-loop structure of the D satellite RNA. The 3 nucleotides associated with necrosis are shown in orange.

host plant (Figure 10.11). The molecular mechanisms of symptom modification caused by satellite RNAs are not yet understood (for more discussion of this topic, see Chapter 31). The maintenance of satellite RNAs is largely determined by RNA 1, indicating that this element of the viral replicase interacts with satellite RNA.

Multimers of both positive and negative strands of satellite RNAs, probably replicative intermediates, can be detected in extracts from infected plants. The negative strands have an additional guanosine at their 3'-termini; this protrudes at the end of the replicative form double-stranded RNA, is required for efficient synthesis of plus-strand satellite RNAs, and may be required for the formation of multimers. The promoters for replication of both plus- and minus-strand satellite RNAs have not been identified and molecular mechanisms of replication of satellite RNAs remain unclear.

The presence of satellite RNAs generally reduces the accumulation of cucumber mosaic virus particles in infected plants, but surprisingly, the same satellite RNAs do not affect the accumulation of tomato aspermy virus. Satellite RNAs are encapsidated together with cucumber mosaic virus genome RNAs, and are transmitted by aphids along with the virus. Satellite RNAs spread epidemically as parasites in natural populations of cucumber mosaic virus.

KEY TERMS

Aphids	Monocots
Cell-to-cell movement	Mosaic disease
Chlorosis	Necrogenic
Cucurbit	Plasmodesmata
Dicots	Reassortment
Genetic bottleneck	Sieve elements
Gramineae	Small interfering RNAs
Hypersensitive necrosis	Systemic mosaic
Long-distance movement	Tonoplast

FUNDAMENTAL CONCEPTS

• Cucumber mosaic virus has a very wide host range, infecting many plant species. It was discovered by virtue of disease symptoms in cash crops, but it can also establish non-symptomatic infections.

• Cucumber mosaic virus is often transmitted by aphids, but it also can be transmitted mechanically, by parasitic plants and by seeds.

• Cucumber mosaic virus has a tripartite plus-stranded RNA genome. Like many plant viruses with divided genomes, its genome RNAs are packaged in three separate but otherwise identical capsids.

• Cucumber mosaic virus produces two subgenomic RNAs, one of which (RNA 4) codes for the coat protein. This RNA is packaged along with the genome RNA (RNA 3) from which it is derived.

• For the related brome mosaic virus, it has been shown that proteins 1a (capping enzyme) and 2a (RNA polymerase) act together to assemble viral replication complexes bound to cytoplasmic membranes.

• Protein 2b, made from subgenomic RNA 4A, suppresses salicylic acid-mediated virus resistance and counteracts host antiviral responses by blocking the systemic signal of host RNA silencing.

• Protein 3a binds to viral RNAs and forms a ribonucleoprotein complex that helps transport viral RNAs through plasmodesmata, facilitating movement of viral genomes and spread of the infection throughout the plant.

• Small, linear satellite RNAs that replicate and are encapsidated along with cucumber mosaic virus RNAs can either worsen or attenuate symptoms of infection by the virus.

REVIEW QUESTIONS

1. Viruses within the *Bromovirus* and *Cucumovirus* genera have similar packaging strategies, producing three different virions, each containing one or two viral genome segments. However, the genome and subgenomic RNAs of viruses in the *Alfamovirus* genus are individually packaged into four virions with distinct sizes. What are the advantages and disadvantages of the packaging strategies used by such viruses?

2. What important functions in virus replication do tRNA-like structures, like those found in cucumber mosaic virus genomes, serve?

3. In brome mosaic virus, in addition to viral RNAs and proteins, the yeast chaperone protein YDJ1 (Yeast DnaJ homolog) and the levels of unsaturated fatty acids in the infected cell are also important for the initiation of minus-strand RNA synthesis. What role might the chaperone protein and fatty acids play?

4. Brome mosaic virus has been used as a model for the study of packaging of segmented viral genomes. Why is it a good model?

5. In brome mosaic virus, it remains unknown how RNA 3 and RNA 4 are packaged together into a single virion. What two models for co-packaging have been proposed?

6. Cucumber mosaic virus infection induces comprehensive changes in the expression of a large number of host genes, and modifies cell physiology and structure. What two common defense responses are induced in infected hosts?

CHAPTER 11

Picornaviruses

Bert L. Semler

Picornaviridae

From Spanish *pico* (point; 10^{-12}), referring to small size of virions, and *RNA*

VIRION

Naked icosahedral capsid (T = 1).
Diameter 30 nm.

GENOME

Linear single-stranded RNA, positive sense.
7–8.5 kb.
Viral protein VPg is covalently bound to 5' end.
Short, genome-encoded poly(A) tail at 3' end.

GENES AND PROTEINS

All genes are translated as single polyprotein followed by proteolytic cleavage.
Three to four capsid proteins (VP1–4).
One to three proteinases.
Six to eight replication proteins.

VIRUSES AND HOSTS

Nine genera, including *Enterovirus* (poliovirus, Coxsackieviruses), *Rhinovirus* (common cold viruses), *Cardiovirus*, *Hepatovirus* (hepatitis A virus).

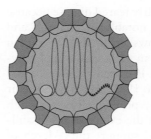 

Hundreds of known virus species.
Infect humans, mammals, birds, fish, etc.

DISEASES

Humans: poliomyelitis (paralysis), myocarditis, hepatitis, meningitis, common cold.
Cattle: foot-and-mouth disease.

DISTINCTIVE CHARACTERISTICS

Translation is initiated at internal ribosome entry sites (IRES).
Self-cleavage of polyprotein by viral proteinases.
RNA replication is associated with vesicles in cytoplasm.

Picornaviruses cause a variety of human and animal diseases including poliomyelitis and the common cold

Picornaviruses (pronounced "pic-or-na-viruses") were so named because they are small (*pico*, implying "small") RNA-containing viruses. They cause a variety of diseases, including **hepatitis** (hepatitis A virus), **myocarditis** (Coxsackievirus), the common cold (rhinoviruses), and paralytic **poliomyelitis** (poliovirus). Foot-and-mouth disease virus, caused by a picornavirus, was the first animal virus discovered (1898). This virus causes economically important outbreaks of disease in cattle and sheep; the 2001 and 2007 outbreaks in Britain required extensive quarantines and slaughter of infected animals.

Hepatitis A virus is a picornavirus that causes acute liver disease, which is usually self-limiting, but occasional fatalities do occur. It is spread by direct contact between individuals and by contamination of food or water. Contaminated food in restaurants can lead to a local outbreak of many infections in a short time. This virus is more commonly encountered in countries with poorly developed sewage systems and low hygiene. Unlike disease caused by hepatitis C and B viruses (see Chapters 12 and 30), hepatitis A infection does not usually lead to chronic disease or liver cancer. An inactivated vaccine protective against hepatitis A virus infection is available.

Rhinoviruses replicate preferentially at 34°C, the temperature of the nose; they are believed to be responsible for approximately 50% of common colds. Over

100 different types of human rhinoviruses have been described, making it unlikely that an effective vaccine against these cold viruses would ever be developed. Recently, antiviral agents active against rhinoviruses have been developed (see Chapter 36), and there is a large potential market for an effective and inexpensive agent against the common cold.

Poliomyelitis is a particularly feared disease because it causes disabling paralysis, notably in children and adolescents. Figure 11.1 shows an ancient Egyptian stele that is believed to show an early case of poliomyelitis. Franklin D. Roosevelt, President of the United States of America from 1933 to 1945, used a wheelchair because of paralysis caused by poliomyelitis, although many people were unaware of the extent of his disability. His administration helped support the "March of Dimes", an organization that funded successful efforts to develop vaccines against poliomyelitis and other childhood diseases. Table 11.1 shows the current genera in the *Picornaviridae* family and some representative members in each genus.

Poliovirus: a model picornavirus for vaccine development and studies of replication

John Enders and coworkers showed in the late 1940s that poliovirus and other human viruses could be grown in cells cultured *in vitro*. This led the way to biochemical and genetic analyses allowing detailed understanding of picornavirus gene expression. It also provided the basis for the generation of poliovirus vaccines during the 1950s.

The formalin-inactivated vaccine developed by Jonas Salk and the live, attenuated vaccine developed by Albert Sabin are two of the most successful vaccines ever produced (see Chapter 35). Poliomyelitis is now virtually absent from the developed world thanks to

Figure 11.1 Egyptian stele (Astarte Temple, Memphis, 1580–1350 B.C.). Man in center with paralysis of right leg, perhaps from poliomyelitis.

the Salk and Sabin vaccines, and the World Health Organization is using these vaccines in a far-reaching plan to eradicate poliovirus worldwide.

Table 11.1 Selected members of the picornavirus family

Genus	Members	Characteristics
Enterovirus	Poliovirus Coxsackie virus (A, B)	Replicate in gastroenteric tract; can cause meningitis, encephalitis, rashes, cardiac and muscular disease
Rhinovirus	Human rhinovirus	Over 100 serotypes; cause common cold; sensitive to low pH
Cardiovirus	Encephalomyocarditis virus Theiler's murine encephalomyelitis virus	Infect mice and other mammals; cause cardiac or neurologic disease
Aphthovirus	Foot-and-mouth disease virus	Epidemics in domestic ruminants; sensitive to low pH
Hepatovirus	Hepatitis A virus	Causes acute hepatitis in humans
Parechovirus	Human parechovirus Ljungan virus	Cause gastroenteritis and respiratory diseases

Note: There are presently three additional picornavirus genera that are not listed here.

Because poliovirus had been so intensively studied for vaccine development, it and several related picornaviruses became model systems during the 1960s for understanding how RNA viruses replicate in eukaryotic cells. David Baltimore and Richard Franklin were the first to demonstrate the presence of an RNA-dependent RNA polymerase in virus-infected cells in the early 1960s. This showed that viral RNAs could replicate directly rather than being transcribed from a DNA template.

Donald Summers and Jacob Maizel found that picornavirus proteins are synthesized as large polyproteins, which are subsequently cleaved into the various structural and enzymatic viral proteins. This foreshadowed discoveries that numerous other animal viruses also make polyproteins that are cleaved by viral proteinases, now prime targets for antiviral drugs.

In the late 1980s, independent work from the laboratories of Nahum Sonenberg and Eckard Wimmer showed that initiation of protein synthesis is cap-independent and occurs at **internal ribosome entry sites** on picornavirus RNAs. This is in contrast to most cellular mRNAs, on which ribosomes recognize the capped 5' end of the RNA and begin protein synthesis at the first AUG codon encountered.

Picornavirus virions bind to cellular receptors via depressions or loop regions on their surface

Picornaviruses have a non-enveloped, icosahedral capsid that encloses a single-stranded, positive-sense genome RNA. The three-dimensional structures of the virions of many different picornaviruses have been determined by x-ray crystallography (Figure 11.2). Mature virions contain 60 copies of each of four proteins: VP1, VP2, VP3, and VP4. Although very different in primary amino acid sequence, VP1, VP2, and VP3 all fold into a common structure that resembles a wedge shape, or a **jelly-roll**, composed of an eight-stranded **β-barrel**

20 Å

Figure 11.2 Picornavirus virion structure. Model of poliovirus virion determined by x-ray crystallography.

structure (see Figures 2.3b and 2.4). These three proteins combine to form the capsid shell of picornaviruses, and VP4 remains buried beneath the shell. Similar folds are shared by capsid proteins of other animal and plant viruses, suggesting a common functional motif required for subunit packing and assembly of icosahedral virus particles.

Table 11.2 lists some cell surface receptor molecules used by various picornaviruses to bind to target cells. Some picornaviruses have depressions (called "canyons" or "pits") on the capsid surface that are involved in binding to receptors. This is unusual; most viruses bind to cellular receptors via proteins that protrude from the surface of the virion. Other picornaviruses do not have these pronounced surface depressions but rely on surface-exposed loop regions of the capsid proteins for recognition of cellular receptors.

Table 11.2 Cell receptors and coreceptors used by some picornaviruses

Virus	Receptor or coreceptor	Characteristics
Poliovirus	Poliovirus receptor (Pvr or CD155)	Immunoglobulin superfamily
Coxsackievirus	Decay accelerating factor (CD55)	Complement cascade
	Vitronectin receptor ($\alpha_v \beta_3$)	Integrin
	Coxsackievirus–adenovirus receptor (CAR)	Immunoglobulin superfamily
	Intercellular adhesion molecule-1 (ICAM-1)	Immunoglobulin superfamily
	Integrin $\alpha_v \beta_6$	Integrin
Human rhinoviruses	ICAM-1	Immunoglobulin superfamily
	Low-density lipoprotein receptor	
Foot-and-mouth disease virus	Heparan sulfate	Glycosaminoglycan
	Vitronectin receptor ($\alpha_v \beta_3$)	Integrin
Encephalomyocarditis virus	Vascular cell adhesion molecule-1 (VCAM-1)	Immunoglobulin superfamily
	Sialylated glycophorin A	Carbohydrate

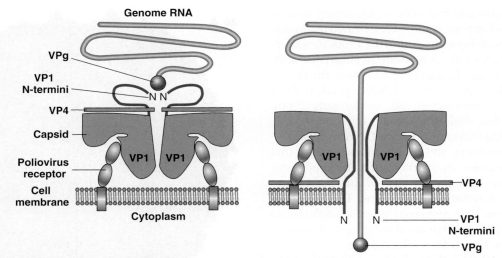

Figure 11.3 Model for entry of poliovirus RNA into the cytoplasm. Shown at left is a cross-section of one center of 5-fold symmetry on a virion docked at the cell surface by binding to the poliovirus receptor. VP4, the N-termini of VP1, and the genome RNA bound at its 5' end to VPg are all located in the interior of the capsid (above). Binding leads to a major rearrangement (shown at right) during which VP4 is externalized and the hydrophobic N-termini of the five VP1 molecules (only two shown here) insert into the membrane to form a channel through which the viral RNA can pass into the cytoplasm.

Genome RNA may pass through pores formed in cell membranes by capsid proteins

Rhinoviruses and foot-and-mouth disease virus have been shown to enter the cell by **receptor-mediated endocytosis** (see Chapter 4). Acidification within the endosomal vesicle causes a conformational change in the capsid, resulting in the dissociation of capsid subunits and release of the viral RNA. Exactly how the RNA subsequently passes from the vesicle into the cytoplasm is not known.

Other picornaviruses, including poliovirus, undergo a major conformational change directly at the cell surface upon binding to their receptor. Virions lose the small internal protein VP4, and the hydrophobic N-terminus of VP1 is extruded to the surface of the particle. This is reminiscent of the conformational changes that take place in fusion proteins of enveloped viruses like influenza and HIV-1 upon receptor binding (Chapter 4). The altered picornavirus virions may participate in the formation of a pore within the plasma membrane, allowing the genome RNA to pass directly from the virion through the membrane into the cytoplasm (Figure 11.3). Pore formation has not been directly demonstrated, but this is an attractive hypothesis to explain entry of viral RNA from altered virions at the cell surface.

Translation initiates on picornavirus RNAs by a novel internal ribosome entry mechanism

Once in the cytoplasm, the genome RNA must be translated by cellular ribosomes to produce viral proteins. Purified genome RNA can be used to infect cells, showing that none of the structural proteins present in the virion is needed to initiate the replication cycle once the RNA has entered the cell. How picornavirus genome RNAs engineer their translation and how the virus eventually takes over the cell's translation machinery is a fascinating story.

The positive-sense genome RNAs of picornaviruses range from 7 to 8.5 kb in length (Figure 11.4). A small (22 to 24 amino acids) virus-encoded protein called VPg is covalently attached to the 5' end of genome RNA, and a poly(A) tract of 30 to 100 nucleotides (depending on the virus strain) is present at the 3' end. Unlike most cellular mRNAs, the poly(A) tract of picornaviruses is part of the genome sequence and is not added by a separate poly(A) polymerase.

Picornaviruses have an unusually long (600–1200 nt) stretch of 5' noncoding sequences upstream of the AUG codon at which protein synthesis begins. These noncoding sequences contain extensive secondary and tertiary structure (see pg 130). Although multiple AUG codons are located between the 5' end and the authentic initiator AUG, these are not used by ribosomes to initiate protein synthesis as would be expected if ribosomes were "scanning" this region searching for appropriate initiation sites.

Most cellular mRNAs employ cap-dependent mechanisms to initiate translation near their 5' end. In this mode (Figure 11.5a), the eIF-4F complex, consisting of eukaryotic initiation factors eIF-4A, eIF-4E, and eIF-4G, recruits a 40S ribosomal subunit to the capped

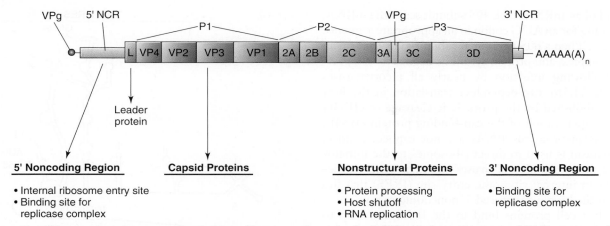

Leader
protein

5' Noncoding Region

- Internal ribosome entry site
- Binding site for
  replicase complex

Capsid Proteins

Nonstructural Proteins

- Protein processing
- Host shutoff
- RNA replication

3' Noncoding Region

- Binding site for
  replicase complex

Figure 11.4 Picornavirus genome organization. Shown are the major landmarks on the positive-strand RNA genome of picornaviruses. The single open reading frame is translated into a large polyprotein that is cleaved by viral proteinases into intermediates P1–P3 and finally into the capsid and replication proteins shown. VPg: viral protein bound to 5' end of RNA.

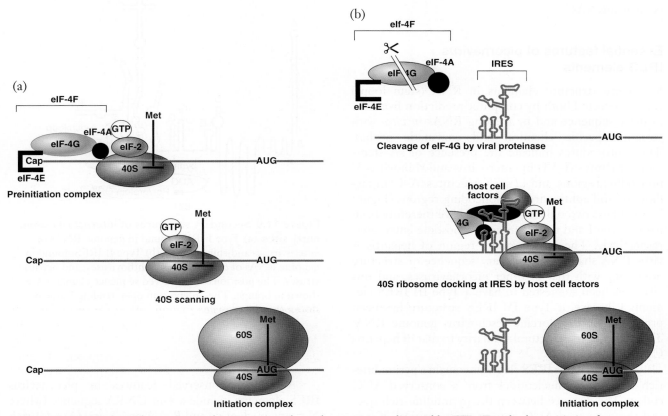

Figure 11.5 Initiation of protein synthesis. (a) Cap-dependent initiation directed by EIF-4F and other initiation factors (not all shown). (b) Cap-independent initiation at the internal ribosome entry site (IRES) of picornavirus RNAs in infected cells. 40S and 60S refer to the small and large ribosomal subunits. Met refers to methionine, used for initiation of most proteins, bound to methionine transfer RNA, which is bound to the 40S submit. GTP refers to guanosine triphosphate, a cofactor required for binding of 60S subunits to 40S subunits and the initiation of protein synthesis.

5' end of an mRNA. The 40S subunit scans the mRNA, searching for an AUG codon in an appropriate sequence context. Then a 60S ribosomal subunit joins to form an 80S complex, and protein synthesis begins.

Following infection by nearly all picornaviruses (Figure 11.5b), cap-dependent translation in the host cell is abolished by the proteolytic cleavage of eIF-4G or by sequestration of the cap-binding protein eIF-4E. Because picornavirus RNAs are not capped, a novel mechanism is used to direct ribosomes to the initiator AUG codon. All picornaviruses initiate protein synthesis via an internal ribosome entry site (IRES) that lies within the highly structured 5' noncoding region.

Host cell proteins bind to the IRES and help to dock the 40S ribosomal subunit onto the RNA. In this mode, the 40S subunit bypasses multiple upstream AUG sequences, moving by translocation or scanning from its site of initial contact with the IRES to reach the initiator AUG. The dynamics of this movement are not understood, but the ultimate outcome is the assembly of 80S ribosomes at the initiator AUG and the start of polypeptide chain synthesis. IRES sequences have also been found in other virus mRNAs and in a number of cellular mRNAs.

Essential features of picornavirus IRES elements

Secondary structure elements in RNA (stem loops) can be detected both by computer prediction from the primary sequence and by treating RNAs *in vitro* with structure-specific chemical and enzymatic probes. These procedures indicate the presence of six stem-loop structures (I–VI) in enterovirus and rhinovirus 5' noncoding regions, and 12 such structures (A–L) in cardiovirus and aphthovirus 5' noncoding regions (Figure 11.6). These two types of structure have therefore been named type I and type II internal ribosome entry sites, respectively. The 5' noncoding region of hepatitis A virus RNA shares little primary sequence or structure homology with that of other picornaviruses, and this virus has been suggested to have a Type III IRES element. Likewise, a Type IV IRES structure has been proposed for the porcine teschovirus genome RNA. This IRES bears structural similarity to that of hepatitis C virus, a flavivirus.

All picornavirus IRES elements contain a pyrimidine-rich tract 20–25 nucleotides from a conserved AUG codon. The distance between the pyrimidine-rich tract and the conserved AUG appears to be critical; if it is much smaller or larger than 20–25 nt, initiation of translation is reduced. For viruses with a type II IRES element, the conserved AUG is used for initiation of translation. For viruses with a type I element, translation is initiated at an AUG located 30–150 nucleotides downstream from the conserved AUG.

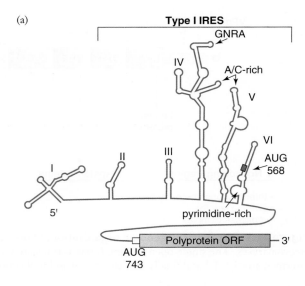

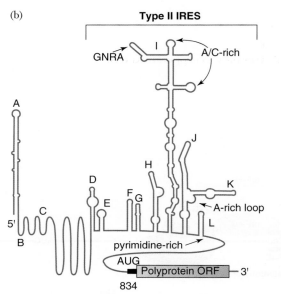

Figure 11.6 Secondary structures of internal ribosome entry sites. (a) Type I IRES, found in genome RNAs of enteroviruses and rhinoviruses. (b) Type II IRES, found in genome RNAs of cardioviruses, aphthoviruses, and hepatoviruses. The positions of conserved sequence elements are shown in orange. The polyprotein open reading frame is shown at reduced scale.

Additional conserved features in picornavirus IRES elements include a 4-nt GNRA sequence (where N = any nucleotide and R = A or G), and (A + C)-rich sequences in specific loop regions of the predicted stem-loop structures. The roles of these conserved elements have not been precisely defined, but they may provide important RNA structure and sequence information required for binding of cellular proteins and the 40S ribosomal subunit.

Interaction of picornavirus IRES elements with host cell proteins

Because no viral proteins other than VPg accompany picornavirus RNA upon entry into a cell, the factors that direct translation at an IRES must be preexisting cellular proteins. Cellular proteins known to interact with picornavirus IRES elements include the La auto-antigen, polypyrimidine tract-binding protein, poly (rC) binding protein 2, and the product of the unr gene. The precise roles of these proteins in IRES function are not known.

Cap-independent translation also utilizes some of the same eukaryotic initiation factors that are involved in cap-dependent translation. Eukaryotic initiation factors eIF-2α and eIF-4B are known to interact with picornavirus IRES elements, and eIF-2, eIF-4E, and eIF-4F all stimulate cap-independent translation. Binding of eIF-4F to an IRES is specified by the central portion of its eIF-4G subunit, replacing the function of eIF-4E, which normally recognizes the cap and directs binding of this complex to capped mRNAs.

Picornavirus proteins are made as a single precursor polyprotein that is autocatalytically cleaved by viral proteinases

Many bacterial messenger RNAs contain several different protein coding regions and can initiate protein synthesis at various internal sites. For example, the genome RNAs of bacteriophages such as Qβ and MS2 express their four viral proteins from three or four different initiation sites, and control of RNA phage gene expression is largely carried out at the level of initiation of protein synthesis (see Chapter 5). In contrast, most eukaryotic messenger RNAs code for a single protein. This poses a problem for positive-strand RNA viruses such as the picornaviruses: How can they express their numerous genes from a single RNA molecule that serves both as a genome and as a messenger RNA, in a cellular environment that favors mRNAs that code for only one protein?

The picornaviruses have solved this problem by linking all of their protein-coding regions into a single unit, and synthesizing a single, large polyprotein (Figure 11.4). This polyprotein is subsequently cleaved into distinct functional viral proteins by proteinases that are themselves part of the polyprotein; presumably, these proteinases fold into enzymatically active structures while still part of the polyprotein. The initial cleavages are rapid, so that little if any full-length polyprotein can be found in an infected cell, unless amino acid analogues that inhibit proper protein folding are incorporated during synthesis. The polyprotein

is first cleaved into three fragments, called P1, P2, and P3 (Figure 11.4). P1 is subsequently cleaved into capsid proteins VP0, VP1, and VP3; VP0 is cleaved into VP2 and VP4 during virion maturation. P2 and P3 are also cleaved into a number of smaller proteins whose functions are described below.

Picornaviruses make a variety of proteinases that cleave the polyprotein and some cellular proteins

Picornaviruses encode three distinct proteinases: the L, the 2A, and the 3C proteinases. The cleavage pathways vary among picornaviruses in different genera. Foot-and-mouth disease virus has an L proteinase, located at the very N-terminus of the polyprotein (Figure 11.4). The L proteinase cleaves itself from the VP4 protein and also cleaves cellular initiation factor eIF-4G. However, other picornaviruses do not have an L proteinase. The 2A proteinase of enteroviruses and rhinoviruses cleaves at the P1/P2 junction, and it is also responsible for cleavage of eIF-4G. However, the 2A protein in other picornaviruses has no proteinase activity.

All picornaviruses make the 3C proteinase. This proteinase cleaves at specific dipeptides of the form glutamine-X, where X can be glycine, serine, or a number of other amino acids with hydrophobic side chains. The 3C proteinase actually exists in three different forms, and each form has a different role in polyprotein cleavage. The initial cleavages of the nascent polyprotein are carried out by the uncleaved 3C polypeptide in the intact polyprotein molecule; that is, 3C can properly fold and act as a proteinase while part of the polyprotein. These cleavages may be within the same polyprotein molecule, or one 3C region can cleave within another polyprotein molecule. 3C then cleaves at its own N-terminus within P3 to generate 3AB and 3CD.

For rhinoviruses and enteroviruses, further processing of the P1 region is carried out primarily by 3CD. The molecular chaperone Hsp70, a cellular protein, has been implicated in P1 processing; it is possible that Hsp70 assists in the interaction between 3CD and P1. Alternatively, Hsp70 could help fold P1 so that its cleavage sites are recognized by 3CD. Some of 3CD is eventually cleaved to form 3C, which, along with 3CD, is able to carry out the remaining cleavages of P2 and P3 to form their constituent mature proteins. 3CD also plays a role in RNA replication (see below).

Replication of picornavirus RNAs is initiated in a multiprotein complex bound to proliferated cellular vesicles

Once viral proteins have been made, they proceed to replicate the viral RNA (Figure 11.7). Genetic and biochemical studies show that most of the proteins made from the P2 and P3 regions are involved in picornavirus RNA replication. Replication starts with the synthesis of a full-length negative-strand RNA, using the initial positive-strand RNA as a template. This negative-strand RNA is used as a template for the synthesis of additional positive-strand RNAs via multistranded replicative intermediate structures (diagrammed in Figure 11.7). These multistranded structures result from highly efficient reinitiation of synthesis of positive strands on the template RNA, before the synthesis of previously initiated positive strands is completed. The number of positive strands synthesized in infected cells exceeds the number of negative strands by a ratio of 40 to 1.

Viral RNA is synthesized in association with membrane vesicles that are induced during virus infection. These vesicles are required for picornavirus RNA synthesis and may serve as a nucleation site for the formation of replication complexes. Association of viral RNA synthesis with membranous vesicles in the

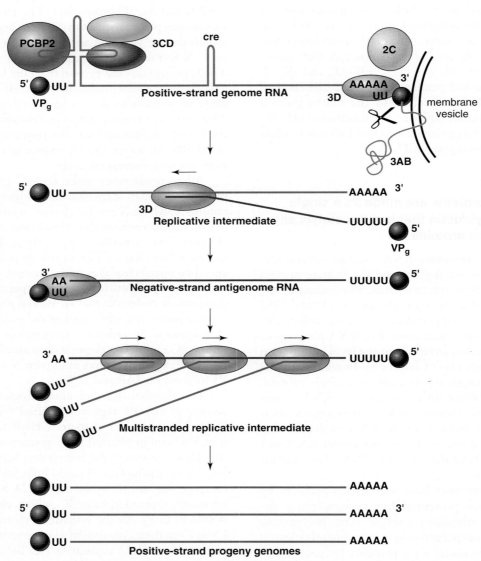

Figure 11.7 Proposed mechanism for replication of picornavirus RNA. Top: genome RNA is shown with sequence elements believed to participate in formation of replication complex (not drawn to scale). Cre = *cis*-acting replication element. Viral proteins VPg (black circle), 3AB (orange line plus black circle), 3CD, 3D, and 2C are shown bound to RNA or cytoplasmic vesicles. Poly C binding protein 2 (PCBP2) is one of several cellular proteins believed to be involved in viral RNA replication. VPg (protein 3B) linked to U residues acts as a primer for the 3D RNA polymerase, and is rapidly cleaved from protein 3A (scissors) or some other 3B-containing precursor polypeptide. Direction of transcription by 3D RNA polymerases is shown by arrows. Positive-strand RNA is shown in light blue, negative-strand RNA in dark blue. Although shown in the figure as free, single-stranded RNA, the negative-strand antigenome RNA most likely exists in a double-stranded form, hybridized to positive-strand viral RNA.

cytoplasm is a common theme encountered with a variety of positive-strand RNA viruses (see also Chapters 10, 12–14); these vesicles help to localize all the elements required for efficient viral RNA replication.

Picornavirus proteins 2B, 2C, and 3A participate in the induction of these vesicles. Proteins 2B and 3A both mediate the inhibition of cellular protein secretion, perhaps by blocking membrane trafficking from the endoplasmic reticulum through the Golgi apparatus to the plasma membrane. This could contribute to the buildup of vesicles in infected cells. In addition to inducing vesicle formation, 2C is an RNA-binding protein implicated specifically in the initiation of negative-strand RNA synthesis. It may be involved in bringing genome RNAs in contact with the vesicles.

RNA synthesis is primed by VPg covalently bound to uridine residues

The proteins that actually direct viral RNA synthesis are derived from P3. Protein 3AB is anchored to the proliferated cytoplasmic vesicles via a hydrophobic region in the 3A portion (Figure 11.7, top right). The C-terminal part of 3AB, protein 3B, is identical to VPg, a protein that is covalently attached to the 5' end of genome and antigenome viral RNAs. VPg becomes covalently attached to uridine residues via a tyrosine side chain near its N-terminus, a process known as uridylylation. These protein-bound uridine residues hybridize to the poly(A) tail at the 3' end of viral RNA, and serve as a primer to initiate negative-strand RNA synthesis.

The 3D protein is the RNA-dependent RNA polymerase that adds nucleotides to the primer, copying the template RNA strand. The 3D RNA polymerase absolutely requires a primer to copy RNA molecules, similar to most DNA polymerases but unlike many other RNA polymerases. The 3D polymerase has been crystallized and its three-dimensional structure determined. It has the familiar "hand" structure of RNA and DNA polymerases, with a finger region, a palm region, and a thumb region; the palm region includes the enzyme's active site.

The model in Figure 11.7 illustrates the steps postulated to lead to viral RNA synthesis. Protein 3CD is believed to bind to stem loop I at the 5' end of the positive-strand RNA, bringing the RNA into the replication complex. The binding of 3CD to this cloverleaf stem-loop structure is facilitated by the binding of a cellular protein, poly r(C) binding protein 2 (PCBP2). PCBP2 was previously mentioned as taking part in recognition of the IRES by 40S ribosomes, and therefore it has a role in both translation and replication of viral RNA. Since the 3D polymerase begins copying the positive-strand RNA at its 3' end, it is possible that

the 5' and 3' ends of the genome RNA are brought together at this point. Protein 2C is also believed to bind to the RNA and bring it to the vesicles where RNA synthesis takes place.

Other cellular proteins may also function in the replication complex. Sequences called *cis*-acting replication elements (cre), located in various regions of the genome RNA depending on the virus (Figure 11.7), are required to initiate picornavirus RNA synthesis by serving as templates for uridylylation of the VPg primer. The uridylylated vpg primer is then translocated to the 3' end of negative-strand templates for initiation of positive-strand RNA synthesis. Cre elements, along with stem-loop I, may also serve to distinguish viral RNA from the many molecules of polyadenylated cellular mRNA present in the cytoplasm, allowing the viral RNA polymerase to concentrate on copying only viral RNA.

Virion assembly involves cleavage of VP0 to VP2 plus VP4

Once sufficient genome RNAs are made, virions can be assembled. Following cleavage of P1, five **protomers**, each containing one molecule of VP0, VP1, and VP3, self-assemble to form a 14S pentamer structure (Figure 11.8). There are two proposed pathways leading to the production of infectious picornavirus particles.

One pathway has 12 of the pentamers coming together to form an 80S empty "procapsid" (lower part of Figure 11.8). A single molecule of newly synthesized viral RNA would then be threaded into the empty procapsid to form a "provirion". This short-lived intermediate undergoes one final proteolytic cleavage event that produces VP4 and VP2 from the VP0 precursor, thus producing the fully infectious virion. One problem with this pathway is that the threading of genome RNA into the empty capsid would require a complex mechanism that is not known to exist for other RNA viruses.

The alternative assembly pathway proposes that the 14S pentamers first associate with viral genome RNA and subsequently combine with other 14S pentamers to form the provirion directly. Whether the viral RNA is threaded into an empty capsid or is bound by 14S pentamers prior to particle assembly, only positive-strand viral RNAs with a VPg protein linked to the 5' end are packaged.

The nature of the proteinase that carries out the final (maturation) cleavage of VP0 is unknown. This cleavage is not carried out by the known viral proteinases (L, 2A, 3C, or 3CD) and, because the cleavage site is found on the inside of procapsids and provirions and would not therefore be accessible to proteolytic enzymes, there has been speculation that viral RNA or the capsid proteins themselves participate in the cleavage reaction.

Inhibition of host cell macromolecular functions

Infection by nearly all picornaviruses leads to a number of dramatic effects on host cell macromolecular functions. These include shutoff of cap-dependent translation, shutoff of host cell RNA synthesis, induction of cytoplasmic vesicles, and alteration of intracellular transport pathways between the endoplasmic reticulum and Golgi apparatus.

As noted above, some picornaviruses encode a proteinase (protein 2A for the enteroviruses and rhinoviruses or the L protein for aphthoviruses) that cleaves eIF-4G, one of the essential protein components required for recognition of the 5' end of capped cellular mRNAs. Infected cells are therefore unable to translate most cellular mRNAs, allowing the cap-independent translation of viral RNAs to take over the cellular protein synthesis machinery. Other picornaviruses inhibit cap-dependent translation by binding and sequestration of eIF-4E, the protein subunit of the eIF-4F complex that binds directly to the 5' cap structure of eukaryotic mRNAs.

Infection of cultured mammalian cells by most picornaviruses causes a shutoff of host cell RNA synthesis. Experiments performed with extracts from poliovirus-infected HeLa cells showed that RNA synthesis by all three classes of mammalian RNA polymerases (pol I, II, and III) is inhibited. These defects have been traced to the proteolytic cleavage of specific subunits of the different RNA polymerases by the 3C viral proteinase. The foot-and-mouth disease virus 3C proteinase also cleaves histone H3, a cellular protein associated with transcriptionally active chromatin in the cell nucleus.

Collectively, these effects on host cell functions ensure that the infected cell devotes its metabolic activities primarily to the expression and replication of picornavirus RNAs. It is notable that hepatitis A virus, a picornavirus with a very slow replication cycle, does not inhibit host cell protein or RNA synthesis and, for the most part, does not produce observable cytopathic effects during infection of cells in culture.

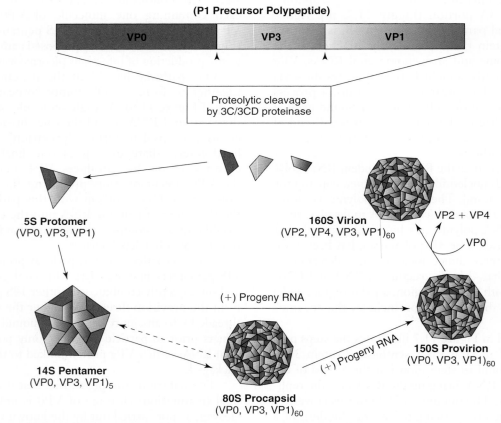

Figure 11.8 Assembly of the picornavirus virion. The P1 polypeptide is cleaved by the 3C/3CD proteinase to generate VP0, VP3, and VP1, which assemble into a 5S protomer. Five of these protomers then assemble into a 14S pentamer. Capsids are formed by the assembly of 12 such pentamers into a closed shell by one of two pathways; they may assemble directly around viral RNA (upper pathway) or they may form an empty 80S procapsid, which then incorporates viral RNA (lower pathway). VP0 subsequently is cleaved into VP2 and VP4, giving rise to the mature virion.

Picornavirus pathogenesis and disease

Picornaviruses cause a number of important and well-studied diseases in humans and animals. Most people infected with poliovirus have few or no symptoms and the infection is limited to the enteric tract, but a small percentage suffer from muscle paralysis and brain stem involvement. A major determinant of neurovirulence for all three poliovirus serotypes is found in the stem-loop V region of the IRES (Figure 11.6). Poliovirus containing a single point mutation in stem-loop V of genome RNA, a lesion present in attenuated Sabin vaccine strains, shows pronounced growth defects in neuroblastoma cells compared to the wild-type virus. These differences in growth are not observed in non-neuronal cells, indicating that a host cell factor in neurons may have a defective interaction with the mutated RNA. This translation or replication defect may therefore be a major cause of the inability of the Sabin vaccine strains to cause paralytic disease, while replicating well in intestinal cells and therefore inducing a strong immune response.

Cell-free translation of hepatitis A virus RNA is stimulated by the addition of mouse liver extract but not by other tissue extracts, indicating that specific liver cell translation factors may contribute to the preference of this virus for liver cells. Coxsackie B virus can infect heart muscle cells, causing severe and sometimes fatal damage in humans. In animal models of myocarditis, RNA sequences in the 5' noncoding regions of Coxsackie B virus genomes have been shown to be involved in virulence, again implicating specific cellular factors in translation or replication of viral RNA. Theiler's murine encephalomyelitis virus causes a neurologic disease in mice that ranges from paralytic polio-like symptoms to a persistent infection accompanied by chronic demyelination, depending on the strain of virus. Unlike the other picornaviruses discussed here, important disease determinants for this virus appear to reside in the viral capsid proteins.

Knowledge of the cellular factors involved in initiation of translation of viral RNA and of RNA synthesis, as well as the roles of viral proteins such as the RNA-dependent RNA polymerase and the 3C and 2A proteinases, may help in the development of novel antiviral agents that can target these critical steps in virus replication and reduce or eliminate the most serious disease symptoms.

KEY TERMS

Cis-acting replication
  elements (cre)
Hepatitis
Internal ribosome entry
  site (IRES)
Jelly-roll β-barrel

Myocarditis
Poliomyelitis
Protomers
Receptor-mediated
  endocytosis

FUNDAMENTAL CONCEPTS

• Picornaviruses display wide ranges of species and cell susceptibilities for viral infections.

• Picornaviruses encode their viral proteins as large precursor polyproteins that are subsequently cleaved by viral-specific proteinases.

• Synthesis of picornavirus proteins occurs via an internal ribosome entry site (IRES) found in the 5' noncoding regions of viral genomic RNAs.

• Picornaviruses carry out membrane-bound RNA replication in the cytoplasm of infected cells, yielding protein-linked progeny RNAs.

• Many picornaviruses inhibit cellular macromolecular functions during infection, including host protein synthesis, RNA synthesis, DNA synthesis, or vesicular trafficking.

REVIEW QUESTIONS

1. In the late 1980s, independent work from the laboratories of Nahum Sonenberg and Eckard Wimmer showed that initiation of protein synthesis is cap-independent and occurs at internal ribosome entry sites (IRES) on picornavirus RNAs. How does this contrast with most cellular mRNAs?

2. Some picornaviruses have depressions (called "canyons" or "pits") on the capsid surface that are involved in binding to receptors. How is this different from most other viruses?

3. All picornavirus IRES elements contain a pyrimidine-rich tract 20 to 25 nucleotides from a conserved AUG triplet. What is the role of this conserved AUG for IRES type I and type II elements?

4. Most eukaryotic messenger RNAs code for a single protein. This poses a problem for positive-strand RNA viruses such as the picornaviruses. How can they express their numerous genes from a single RNA molecule that serves both as a genome and as a messenger RNA, in a cellular environment that favors mRNAs that code for only one protein?

5. Once sufficient genome RNAs are made, picornavirus virions can be assembled. Following cleavage of P1, five protomers, each containing one molecule of VP0, VP1, and VP3, self-assemble to form a 14S pentamer structure. What are the two proposed pathways leading to the production of infectious picornavirus particles?

6. What effects does picornavirus infection have on host cells?

CHAPTER 12

Flaviviruses

Richard Kuhn

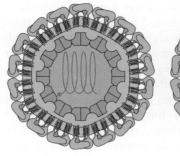
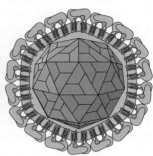

Flaviviridae

From Latin *flavus* (yellow), referring to jaundice caused by yellow fever virus

VIRION

Spherical enveloped particle.
No projections, "golf ball" appearance.
Diameter 50 nm.
Spherical nucleocapsid (25–30 nm diameter) with icosahedral symmetry (T = 3).
Envelope glycoproteins also arranged with icosahedral symmetry.

GENOME

Linear single-stranded RNA, positive sense.
10–12.3 kb.
Capped at 5' end but no poly(A) tail at 3' end.

GENES AND PROTEINS

All genes translated as single polyprotein followed by proteolytic cleavage.
One capsid protein (C).
Two envelope proteins (M and E).
Seven nonstructural proteins.

VIRUSES AND HOSTS

Three genera:
- *Flavivirus: infect humans, monkeys, bats, birds, etc.*
- *Pestivirus, from Latin pestis (plague): infect pigs, cattle, other ruminants.*
- *Hepacivirus, from Greek hepatos (liver): hepatitis C virus, in humans only.*

DISEASES

Many are associated with human disease: yellow fever, dengue hemorrhagic fever, tick-borne encephalitis, West Nile virus, hepatitis C virus.
Also affect domestic or wild animals and cause economically important diseases: swine fever, bovine viral diarrhea.

DISTINCTIVE CHARACTERISTICS

Many flaviviruses (but not hepatitis C virus) are transmitted from host to host by mosquitoes or ticks.
The major envelope protein of flaviviruses lies parallel to the lipid membrane.
Viral proteins are synthesized as a single polyprotein that is cleaved by viral and cell proteinases.
Pestiviruses and hepaciviruses have an internal ribosome entry site (IRES) for initiation of protein synthesis.
Virions are assembled by budding at endoplasmic reticulum and are released by exocytosis.

Flaviviruses cause several important human diseases

Flaviviruses have had an important impact on humans over the last several hundred years. This has been especially true in the Americas, which suffered dramatic outbreaks of yellow fever, transmitted by mosquitoes, in the 1700s and 1800s. Dengue virus and Japanese encephalitis virus, also transmitted by mosquitoes, cause serious human disease throughout tropical regions of the world. Hepatitis C virus, unknown prior to 1989, has infected 4 million people in the United States alone and is responsible not only for acute and chronic **hepatitis**, but also for **hepatocellular carcinoma**.

West Nile virus, originally discovered in Uganda and prevalent in Africa, southern Europe, and the Middle East, suddenly appeared in North America in 1999. Although not a major public health threat, infection of humans with this mosquito-borne virus can lead to neurological effects and is occasionally fatal. West Nile virus caused a mini-**epidemic** as it spread from the East coast to the West coast of North America over a five-year period via infected birds. Although certain flaviviruses have been brought under control through vaccination and vector management, a large number remain significant threats to human health.

The *Flaviviridae* family presently contains three genera. The *Flavivirus* genus includes at least 73 distinct viruses, including many human pathogens that are transmitted by bloodsucking arthropods (Table 12.1). Flaviviruses were previously grouped with the togaviruses (Chapter 13), as they are both small, enveloped viruses, and many are transmitted by arthopod vectors, within which they also replicate. However, the gene organization and replication strategies of flaviviruses and togaviruses are quite distinct.

Members of the *Pestivirus* genus cause economically important disease in cattle, pigs, and sheep. The *Hepacivirus* genus contains a single member, hepatitis C virus. Pestiviruses and hepatitis C virus are not transmitted by arthopod vectors; a major transmission pathway for hepatitis C virus was via blood transfusions due to contamination of the blood supply. Screening of the blood supply has led to dramatic drop in spread of transfusion-related hepatitis C infections. The type member of the flavivirus genus, yellow fever virus, will be used as an example throughout this chapter due to its historical importance and the wealth of information available on its biology.

Yellow fever is a devastating human disease transmitted by mosquitoes

Yellow fever was first recognized in 1648 in the Yucatan. The disease progresses in three stages: (1) an initial stage marked by high fever and accompanied by muscle pain, headaches, and severe vomiting; (2) a quiescent stage in which the fever dissipates and the patient appears to be in remission; and (3) a final stage marked by a repeat of the original symptoms in a more severe form that results in multi-organ failure with a mortality rate of between 20 and 50%. Many victims suffer from acute hepatitis, which gives the skin a yellow tinge, hence the name.

The origin of the virus was most certainly West Africa, with multiple reintroductions of the virus in the New World due to the extensive slave trade. Despite quarantine efforts, which required ships of African and Caribbean origin to anchor for several weeks prior to entry into North American ports, outbreaks of yellow fever were frequent and severe for ports as far north as Boston.

The epidemic that struck Philadelphia in 1793 is of particular note due to its impact on the fledgling government of the United States located in that city. The disease was imported from the Caribbean in July of that year and quickly established itself in the city due to the presence of an unusually high number of mosquitoes. All told, greater than 10% of the city's population of 40,000 died during the summer epidemic and a mass exodus reduced Philadelphia's population even further. President Washington and many other prominent government leaders and staff left the city for fear of infection.

The scourge of yellow fever outbreaks in the United States continued through the nineteenth century and was not brought under control until the Yellow Fever Commission headed by Walter Reed produced two significant results. First, they showed that the agent responsible for yellow fever could be passed to uninfected individuals using infected patient serum, and established that this agent was a virus by filtration methods. Second, they demonstrated that mosquitoes are capable of transmitting the virus after obtaining a blood meal from an infected host. Given the nature of the infecting agent and its transmission route, preventive steps could be taken to reduce the spread of the virus. Thus mosquito control in Panama overcame the most significant impediment to the construction of the Panama Canal.

Table 12.1 Members of the *Flaviviridae* family

Genus	Virus name	Hosts	Diseases
Flavivirus	Yellow fever	Monkeys, humans	Hemorrhagic fever
	Dengue (types 1–4)	Monkeys, humans	Fever, arthralgia, rash, hemorrhagic fever
	Japanese encephalitis	Birds, pigs, humans	Encephalitis
	West Nile	Birds, humans	Fever, arthralgia, rash
	Tick-borne encephalitis	Rodents, birds, humans	Encephalitis
Pestivirus	Bovine viral diarrhea	Ruminants	Persistent infection, mucosal disease
	Border disease	Sheep, goats	Acute and chronic hemorrhagic symptoms
	Classical swine fever	Pigs	
Hepacivirus	Hepatitis C	Humans	Acute and chronic hepatitis

A live, attenuated yellow fever virus vaccine is available and widely used

In the 1930s, a safe and effective vaccine (strain 17D) against yellow fever virus was developed by Max Theiler. Although this turned out to be one of the best viral vaccines known, yellow fever continues to inflict misery and death in unvaccinated populations in South America and Africa.

Comparison of the nucleotide sequences of the virulent Asibi strain and the 17D vaccine strain derived from it provided insight into the determinants of attenuation. Of the 10,862 nucleotides in the yellow fever genome, 68 nucleotide differences resulting in 32 amino acid substitutions were noted between the two strains. Although these mutations are spread across the genome, the envelope protein has a high number with 12 amino acid substitutions.

In 1989, a full-length cDNA clone of yellow fever virus was constructed that could be used to generate infectious viral RNA *in vitro*. This allowed scientists to introduce mutations into the viral genome and therefore to study the functions of each gene. Since then, other flavivirus cDNA clones capable of producing infectious virus have been constructed.

Hepatitis C virus: a recently discovered member of the *Flaviviridae*

Hepatitis C virus was first identified, not by the traditional method of isolation and growth of virus in culture, but rather by direct cDNA cloning of the viral genome. The use of molecular clones of the virus genome greatly facilitated the study of hepatitis C virus, because up until 2005, virus replication could not be carried out in cells cultured *in vitro* or in non-primate animals.

Although hepatitis C virus will not be discussed in detail in this chapter, it is perhaps the most interesting member of the family from a clinical perspective. It has been estimated that there are over 170 million hepatitis C seropositive individuals in the world. Acute infection with hepatitis C virus is similar to other forms of acute viral hepatitis. However, the development of chronic hepatitis, which occurs in up to 85% of infected individuals, results in an increased risk of liver cirrhosis and hepatocellular carcinoma. Therefore, it is not surprising that hepatitis C infection is one of the primary indicators for a liver transplant. Hepatitis C virus is transmitted by blood transfusion, by contaminated needles of intravenous drug users, and by sexual contact; unlike most flaviviruses, it is not known to be transmitted by an arthropod host.

The flavivirus virion contains a lipid bilayer and envelope proteins arranged with icosahedral symmetry

Flaviviruses have a small (25–30 nm) nucleocapsid enclosed within a lipid-containing envelope. Surrounding the nucleocapsid and embedded in the lipid envelope are 180 glycoproteins organized with $T = 3$ icosahedral symmetry. In this respect they are similar to togaviruses (Chapter 13), but flavivirus particles are 50 nm in diameter compared with 70 nm for togavirus particles, and togaviruses have a $T = 4$ icosahedral structure, implying 240 subunits.

Nucleocapsids are composed of multiple copies of the capsid (C) protein, and each virion also contains 180 copies of two envelope glycoproteins, M and E. Cryoelectron microscopy reveals a spherical particle with a smooth, golf ball-like appearance (Figures 12.1, 12.2c), in contrast to the fringed envelope of togaviruses and many other enveloped viruses. Immature flaviviruses at neutral pH also display distinct spikes on their surface (Figure 12.2a) and have a dramatically different appearance from the mature virus. A high-resolution map showing a cross-section of the virion reveals concentric layers of density, from the outer layer of envelope proteins, through two layers of the lipid membrane, to the surface of the nucleocapsid. A defined structure for the nucleocapsid has not been obtained.

The 101-amino acid capsid protein associates with the genome RNA and serves as a scaffold around which the envelope proteins and the lipid bilayer are organized. It is the least conserved protein of the flaviviruses, but has a high number of basic amino acid residues and a central hydrophobic region. The basic amino acids probably interact with the negatively charged phosphate groups of the genome RNA and may help to condense the RNA for encapsidation. The hydrophobic region may be involved in binding the capsid to the viral envelope.

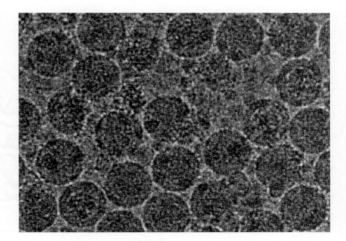

Figure 12.1 Dengue virus type 2 particles visualized by cryoelectron microscopy. The sample is slightly out of focus, as is required to obtain useful data for image reconstruction of the particles.

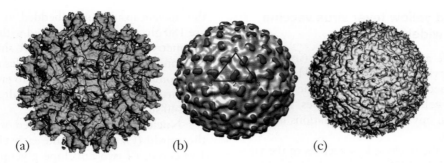

(a)   (b)   (c)

Figure 12.2 Morphology of dengue virus type 2 virions. Surface shaded views of dengue virus type 2 as determined by cryo-electron microscopy and image reconstruction. (a) The surface of an immature virion at neutral pH. The prM-E heterodimers are oriented perpendicular to the surface of the lipid bilayer. Three heterodimers converge to form the spike-like projection, with the pr peptide forming the small knob at the distal end of the spike. (b) The surface of the immature virus at the low pH found in the *trans*-Golgi network. The orientation of the prM-E heterodimer is parallel to the lipid bilayer, similar to the orientation found in mature virions. The surface of the particle is completely covered by proteins, and the underlying lipid bilayer is not visible. The pr peptide can be seen as the blue extended oval on the gray E protein surface. (c) The mature infectious virion after release of the pr peptide. The virion is 50 nm in diameter and resembles a golf ball.

Flavivirus E protein directs both binding to receptors and membrane fusion

The E protein is a **type I transmembrane protein** of approximately 500 amino acids, anchored in the membrane via two C-terminal hydrophobic regions. Its amino terminus and most of its mass reside on the outside of the envelope. This protein plays a double role in entry: it binds to cell surface receptors and directs fusion of the viral envelope with host cell membranes. The M protein is a smaller type I transmembrane protein of approximately 80 amino acids. Its precursor interacts with E during virus assembly (see below and Figure 12.2a, b).

A fragment of the E protein of tick-borne encephalitis virus (a close relative of yellow fever virus) has been crystallized and its structure was determined by x-ray diffraction analysis. This fragment is missing the C-terminal membrane anchor and an additional 50 amino acids, called the stem region. The remaining 400 amino acids are arranged as an elongated three-domain structure (Figure 12.3).

Examination of this structure revealed several unusual features for a viral envelope protein. Unlike other viral glycoproteins and many cellular membrane proteins, which protrude outward from membranes, the long axis of the E molecule lies parallel to the lipid bilayer and has a convex shape that follows the curved surface of the viral envelope. This explains the smooth, golf ball-like surface of the virion. The structures of the E proteins from multiple flaviviruses have been solved by x-ray crystallography; these proteins have a very similar architecture to the tick-borne encephalitis E protein.

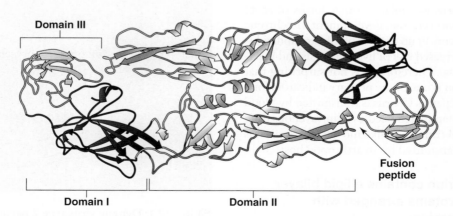

Figure 12.3 The atomic structure of the tick-borne encephaltitis virus E protein dimer. Determined by x-ray crystallography. Each monomer consists of three domains. Note the position of the fusion peptide, as it resides in a protected niche covered by the neighboring E monomer.

The E protein is found as a dimer, with domain II forming the dimer interface. At the outer ends of the dimer lies domain III, which has an immunoglobulin-like fold, and has been implicated in receptor binding. Many cell surface proteins that are used as virus receptors have immunoglobulin-like folds, and several viral structural proteins are known to have such folds. Joining these two domains is the central domain I, which contains the amino terminus of the protein.

A short hydrophobic sequence, known as the fusion peptide, is found at one end of domain II, protected from the aqueous environment by the position of the neighboring E molecule within the dimer. The fusion peptide is predicted to play a major role in the fusion of the viral and cellular membranes that leads to the release of the genome RNA into cells. Its position within a region of the dimer interface suggests that upon rearrangement of the dimer at low pH, the fusion peptide is exposed, enabling its interaction with cellular membranes.

Flaviviruses enter the cell by pH-dependent fusion

Although possible cellular receptor molecules have been suggested for several flaviviruses, none has been clearly identified. In some flaviviruses, binding of antibody to the virus particle can result in enhanced uptake into cells expressing immunoglobulin Fc receptors. This process has been observed *in vitro* and is known as **antibody-dependent enhancement**. In the case of dengue virus, such a mechanism has been proposed to explain the disease known as dengue **hemorrhagic fever** or dengue shock syndrome. This disease can occur when a patient previously infected with one dengue virus serotype is subsequently infected with a second serotype within a period of one to two years. The first infection is believed to expand the population of cells that can be infected by a second serotype via antibody-dependent enhancement of virus entry.

Following attachment of virions to cell surface receptors, entry into the cell is achieved by endocytosis within clathrin-coated vesicles. These vesicles fuse with endosomes, which subsequently undergo acidification. The reduction in pH results in a rearrangement of the E dimer into the fusion-active trimeric state. Recent studies have demonstrated that the E protein of flaviviruses and the E1 glycoprotein of togaviruses have a similar structural fold and share similar functions in both fusion and icosahedral particle formation. These proteins have been termed **class II fusion proteins**, as they differ in several physical and functional aspects from the **class I fusion proteins** represented by influenza virus HA protein (see Chapters 4 and 18).

The structures of the low pH post-fusion form of the E protein have been determined for dengue and tick-borne encephalitis viruses using x-ray crystallography.

This form of the protein is arranged as a trimer with the distal end of domain II, containing the fusion peptide, exposed and oriented toward the target membrane. Domain III has packed against domain I, suggesting that the stem region can pack against domain II, promoting the close approach of the viral and endosomal membranes and leading to their fusion.

Flavivirus genome organization resembles that of picornaviruses

The modern era of flavivirus research began with the determination of the nucleotide sequence of the 17D strain yellow fever virus RNA genome in 1985. The organization of the genome clearly established the flaviviruses as a group of enveloped, positive-strand RNA viruses different from the togaviruses, with which they had been previously grouped because of similar virion morphology, genome size, and transmission by arthropod vectors.

Unlike the togaviruses, which encode their structural proteins on a subgenomic RNA, the flaviviruses have only one messenger RNA, identical to genome RNA. This is translated into a single, long **polyprotein** that undergoes proteolytic processing to generate the individual proteins. Coding regions for virion structural proteins are in the 5'-proximal region of the viral RNA and are therefore present in the N-terminal part of this polyprotein; the viral RNA polymerase is coded at the most distal 3' region of the viral RNA (Figure 12.4). This gene arrangement closely resembles that used by picornaviruses, a family of non-enveloped positive-strand RNA viruses (Chapter 11).

The genome of yellow fever virus is an RNA molecule 10,862 nucleotides in length. Like cellular messenger RNAs, the viral RNA has a methylated cap at its 5' end, but it is not polyadenylated at its 3' end. The 118-nt 5' noncoding region is followed by a single, long open reading frame of 3411 codons. Three tandem stop codons are used to ensure termination of translation. The 511-nt 3' noncoding region contains an 87-nt sequence that is predicted to form a stem-loop structure. Similar stem-loop structures are present in other flavivirus genomes, and biochemical and genetic experiments suggest a role for this region in viral RNA replication. A complementary sequence in the 5' end of the genome is involved in the cyclization of the RNA, and is also involved in RNA replication.

Other flaviviruses, including pestiviruses and hepatitis C virus, share with picornaviruses the use of an internal ribosome entry site (IRES) to direct initiation of viral protein synthesis. This further underlines similarities between the *Flaviviridae* and the *Picornaviridae*, and suggests that these two virus families may share a common evolutionary origin: flaviviruses can be seen as picornaviruses that have acquired an envelope.

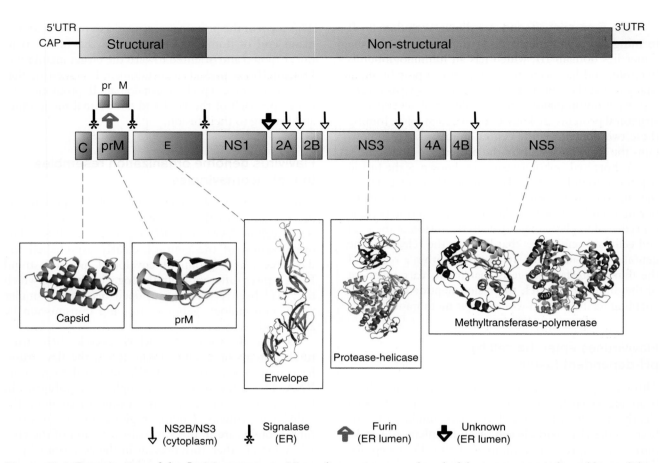

Figure 12.4 Organization of the flavivirus genome. Noncoding regions at each end of the genome are indicated by a solid line. Proteins are shown in shaded boxes. The sizes of the individual proteins are listed for yellow fever virus in Table 12.2. Cleavage by proteinases is also indicated. The genomes of the pestiviruses and hepatitis C virus are similar in layout but differ in the number and type of proteins. In addition, they have IRES elements in the 5' noncoding region that direct initiation of translation. Below the genome map are ribbon diagrams for each of the flavivirus proteins whose atomic structures have been determined. In some cases (NS3 and NS5), domain structures have been determined. Proteinases responsible for cleavage at specific sites on the polyprotein are denoted by arrows and symbols: see below for explanation of symbols.

The polyprotein is processed by both viral and cellular proteinases

After virus uptake and release of the genome RNA in the cytosol, the RNA is bound by ribosomes and translated, giving rise to the polyprotein. Proteolytic processing yields 10 mature virus proteins (Figure 12.4 and Table 12.2). The three structural proteins C, prM, and E occupy the first quarter of the polyprotein, followed by the seven nonstructural proteins (NS1–5). Both viral and cellular proteinases are required to process the viral polyprotein, which is inserted into the membrane of the endoplasmic reticulum. The viral proteinase resides within the protein NS3; however, an accessory protein, NS2B, is also required to carry out

proteolytic processing with high efficiency. The cellular proteinases responsible for processing include **signal peptidase** and **furin**.

The capsid protein is inserted into the endoplasmic reticulum via a 20-amino acid signal sequence at its carboxy terminus (Figure 12.5). The signal sequence is cleaved in the **lumen** of the endoplasmic reticulum by a cellular signal peptidase, releasing the remainder of the polyprotein from the capsid protein precursor (anchC or "anchored C protein"). The amino-terminal end of the signal sequence, on the cytoplasmic side of the endoplasmic reticulum membrane, is eventually cleaved by the viral proteinase NS2B/NS3A, releasing the mature capsid protein.

Table 12.2 Proteins of Yellow Fever Virus

Protein	Size (aa)	Function
anchC	121	"Anchored" capsid protein. Precursor to capsid protein; signal sequence inserted in membrane is cleaved by viral proteinase.
C	101	Encapsidation of genome RNA.
prM	164	Binds to E glycoprotein, inhibits membrane fusion during transit through Golgi. Cleaved to produce mature M protein, releasing pr.
M	75	Small envelope glycoprotein. Function in the mature virion is unknown.
E	493	Major envelope glycoprotein. Receptor binding, fusion of virus envelope with cell membrane.
NS1	352	RNA replicase component, possibly involved in replication complex formation.
NS2A	224	RNA replicase component, either direct role in replication or targeting replicase complex to membranes.
NS2B	130	Part of viral proteinase that cleaves viral polyprotein.
NS3	623	Viral serine proteinase involved in polyprotein cleavage; RNA replicase component; nucleoside triphosphatase and helicase activities.
NS4A	149	RNA replicase component, may recruit NS1 and other proteins into replicase complex.
NS4B	250	Unknown role. Associated with membranes in cytoplasm; may translocate to nucleus.
NS5	905	RNA-dependent RNA polymerase; methyltransferase (methylates cap structure).

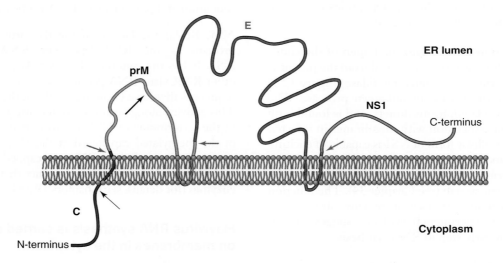

Figure 12.5 Membrane topology of the flavivirus structural proteins. Cleavages carried out by cellular signal peptidase are shown as orange arrows. The cleavage of the C-terminal signal sequence of the capsid protein by viral proteinase NS2B-NS3 is shown by the thin black arrow. Both prM and E have two C-terminal domains that span the lipid bilayer. The maturation cleavage of prM, carried out by furin (thick black arrow), releases the C-terminal fragment as mature M protein. The NS1 protein and other nonstructural proteins that follow E are not shown in their entirety.

Following the capsid protein is the precursor to the membrane protein, prM (Figure 12.5). This small protein contains a transmembrane domain at its C-terminus followed by a signal sequence that is required to insert the E protein in the correct orientation in the membrane. The prM protein associates with the E protein in the endoplasmic reticulum to form a heterodimer. This association apparently protects the E protein from a premature conformational change that would lead to exposure of the fusion peptide in the low pH environment in Golgi vesicles. At a late stage in the maturation process of the virion, a cellular furin protease cleaves prM, releasing the "pr" peptide into the extracellular medium when the virion is released from the cell, and leaving the small M protein associated with the virus particle. This cleavage results in a conversion of the prM/E heterodimer into an E homodimer as shown in Figure 12.3.

Analogous to prM, the E protein contains both a transmembrane domain and a signal sequence at its C-terminus. The signal sequence functions to insert the NS1 protein into the endoplasmic reticulum, where it functions in the early steps of RNA synthesis. Except for the NS1 protein, all other nonstructural flavivirus proteins are believed to reside within the membrane or on the cytoplasmic side of the membrane.

Nonstructural proteins organize protein processing, viral RNA replication, and capping

The nonstructural proteins (NS1 through NS5) are involved in the replication of the genome RNA. Seven different polypeptides are derived from this region. Of these seven, three are known to have specific enzymatic activities: NS5 is an RNA-dependent RNA polymerase; NS3 is a multifunctional, multidomain protein that contains both a proteinase activity and an RNA helicase acitivity; NS2B has been shown to complex with NS3 and is a required cofactor for proteinase activity. The x-ray structures of these three polypeptides have been determined and are shown in Figure 12.4.

NS1. NS1 is translocated into the lumen of the endoplasmic reticulum, where it is cleaved from the polyprotein at its N-terminus by signal peptidase (Figure 12.5) and at its C-terminus by an unknown protease. The protein is glycosylated, forms dimers, and is found both secreted and in association with membranes in infected cells. The extracellular form of NS1 is capable of eliciting a protective humoral immune response, and has been shown to help the virus evade the immune surveillance of the host complement system (see Chapter 34). The function of the NS1 protein in replication remains obscure, but several lines of experimental evidence suggest that it functions in an early step in RNA synthesis.

NS2A. The NS2A protein has also been implicated in RNA synthesis, as it has been shown to be localized in RNA replication complexes. NS2A binds to the NS3 and NS5 proteins, as well as to sequences within the 3' noncoding region of the genome RNA. Cleavage at its N-terminus occurs in the lumen of the endoplasmic reticulum, but cleavage at its C-terminus is carried out in the cytoplasm by the NS2B/NS3 proteinase. Therefore, this protein must cross the membrane of the endoplasmic reticulum at least once, and it may have additional membrane spanning sequences.

NS2B. The short NS2B protein has several membrane-spanning domains and a conserved hydrophilic domain of about 40 amino acids. NS2B is a cofactor for the proteinase activity found within NS3. The association of

NS2B with NS3 provides enhanced cleavage site specificity to the proteinase.

NS3. The proteinase activity of NS3 maps to the N-terminal one-third of the protein. The x-ray structure for the dengue virus NS3A proteinase domain confirmed sequence analyses and genetic studies that suggested that NS3 is a serine proteinase having a catalytic triad consisting of His, Asp, and Ser residues. In the NS3 C-terminal domain there are amino acid sequence motifs found in RNA helicases as well as in nucleoside triphosphatases. X-ray structure determinations of several flavivirus C-terminal domains have confirmed that the fold is consistent with known helicases and unwinding activity has been demonstrated in purified systems.

NS4A and NS4B. The functions of the small, hydrophobic NS4A and NS4B proteins are not known, although NS4A has been implicated in the formation of virus-induced membrane structures that support RNA synthesis. Like the NS2A and NS2B proteins, they contain several transmembrane domains, and NS4A also contains a signal sequence. Thus, it is likely that these proteins remain associated with the endoplasmic reticulum and play a role in RNA replication.

NS5. By far the largest of the flavivirus nonstructural proteins is the RNA-dependent RNA polymerase, NS5. It has the canonical Gly-Arg-Arg motif found in other RNA virus RNA polymerases, and it also contains sequences that have homology with **methyltransferases**. This activity, along with the **nucleoside triphosphatase** in the C-terminus of NS3, is involved in the formation of the methylated cap found at the 5' end of flavivirus genomes. Since cellular mRNAs are capped in the nucleus, cytoplasmic RNA viruses must produce their own capping enzymes (for details, see Chapters 19 and 26).

Flavivirus RNA synthesis is carried out on membranes in the cytoplasm

The synthesis of the nonstructural proteins results in the establishment of active RNA replicase complexes. These complexes include not only the RNA-dependent RNA polymerase NS5 and other viral proteins, but presumably also host proteins recruited for the purpose of genome replication. RNA replication requires the synthesis of a complementary (minus-strand) copy of the plus-strand RNA. This is followed by the synthesis of plus-strands that are then either translated, making more viral proteins; replicated, making more RNA; or packaged, making virions. The synthesis of RNA is asymmetric, generating more than 10 times as many plus-strands as minus-strands.

Significant alterations of subcellular structure have been observed in flavivirus-infected cells. New accumulations of endoplasmic reticulum membranes appear early in the infectious cycle, and numerous smooth vesicles can be found later in infection. RNA synthesis is associated with these structures (Figure 12.6), and most of the nonstructural proteins are localized there. A number of these proteins contain predicted membrane-spanning sequences. RNA synthesis directed by most positive-strand RNA viruses, including bromoviruses, picornaviruses, togaviruses, and coronaviruses (Chapters 10, 11, 13, and 14), is also associated with membranes in the cytoplasm of infected cells, and in some of these cases is understood in more detail.

Virus assembly also takes place at intracellular membranes

Unlike many other enveloped viruses, which derive their envelope from the plasma membrane of the cell, flaviviruses assemble their virions inside the cell by using intracellular membranes for envelope formation (Figure 12.6). Electron microscopy has shown that immature virions can be found in the lumen (interior) of the endoplasmic reticulum.

Assembly of the nucleocapsid takes place at the cytoplasmic face of membranes with which prM and E transmembrane proteins are associated. The carboxy-terminal signal sequence of the precursor to the C protein

anchors that protein to the membrane (see Figure 12.5); this sequence penetrates into the lumen of the endoplasmic reticulum or smooth vesicles. This enables interactions between the C protein and the envelope proteins, which are also anchored to the membrane but reside within the lumen. Multiple copies of the C protein condense on the genome RNA to form the nucleocapsid. Subsequent cleavage of the membrane anchor of the C protein by the viral proteinase (see above) releases the capsid from direct contact with the membrane.

The coupling of protein synthesis, RNA synthesis, and virion assembly on membranous structures assures that newly synthesized genome RNA can associate with C protein and initiate the assembly process. Encapsidation of the RNA initiates the budding of particles into the endoplasmic reticulum. Particles that have budded into the endoplasmic reticulum are then processed by carbohydrate addition and modification as they proceed through the Golgi membrane system. It is likely that transport to the Golgi and into the *trans*-Golgi network requires the presence of the glycosylated prM protein.

Virions follow the **exocytosis** pathway and are released to the extracellular space by fusion of vesicles containing virions with the plasma membrane (Figure 12.6). The cleavage of the prM protein by

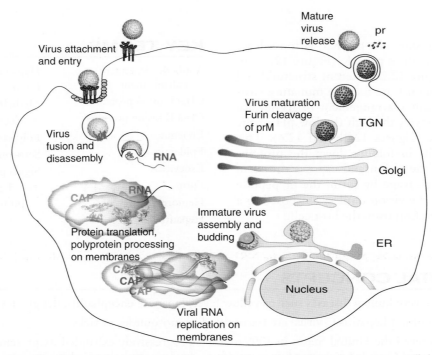

Figure 12.6 Flavivirus replication pathway. The virus life cycle initiates by binding of virions to a cellular receptor and internalization in an endosomal vesicle (upper left). The low pH within a late endosome promotes the fusion of viral and cellular membranes by the E protein and results in the release of genome RNA into the cytoplasm. Translation of the RNA leads to the establishment of membrane-bound RNA replication factories (lower left). These are spatially linked to sites of assembly and budding of immature virus on membranes of the endoplasmic reticulum (ER) (lower right). The immature virus is glycosylated and traverses the secretory network. In the *trans*-Golgi network (TGN), the low pH promotes the folding of the spike-containing immature particles into the semi-mature state and permits the cleavage of prM by host furin protease. Release of the virions into the extracellular milieu results in release of the pr peptide and a fully infectious mature virion.

Introduction of West Nile virus into North America

In the summer of 1999, residents of Queens, New York, experienced an outbreak of an encephalitic disease reminiscent of a flavivirus infection. Concurrently, the local population of crows in the greater New York City area suffered from disease, and many were found dead. First analyses of the infectious agent suggested that these diseases might be caused by a flavivirus known as St. Louis encephalitis virus. This virus appears sporadically in the eastern half of the United States, with a number of notable outbreaks in Florida near Orlando.

However, it was subsequently shown that the infectious agent affecting both the human and crow populations was West Nile virus, a virus not previously seen in North America. This prompted substantial discussion about how the virus was introduced and whether it would establish itself in the United States and become an **endemic** disease. West Nile virus was originally isolated in Uganda in 1937 and is found in Israel and throughout the Middle East, as well as in parts of Europe and Asia. The virus was likely accidentally brought into the United States via an infected bird or mosquito, perhaps from an airplane that landed at a New York airport.

The next year (2000) proved pivotal for the presence of West Nile virus in North America. The virus reappeared with the advent of the mosquito season and began its expansion outside of the New York City region, most likely carried by birds. Over the next few years, the virus expanded its presence into all of the contiguous 48 states and the adjacent Canadian provinces.

The initial human cases were widely reported in the press, leading the public to fear a widespread epidemic. However, a year or two after the virus became established in a given area, the number of human cases of encephalitis declined. In 2002, it was shown that West Nile virus could also be transmitted by blood transfusions and tissue transplantation. This quickly led to the establishment of a screening assay for the blood supply, causing a dramatic reduction in transfusion-associated West Nile virus disease. However, its presence in wild populations of mosquitoes and birds remains a threat for transmission to humans.

The ecology and epidemiology of West Nile virus is not yet understood well enough to predict its future effects on either birds or humans in North America. The level of seroconversion (an indicator of exposure to the virus) in the human population has remained relatively constant and low (~3%), but an increase in infections of the host (bird) or vector (mosquito) population could have a significant impact on the number and perhaps severity of human infections. We do know for certain that West Nile virus is now endemic in the lower 48 states and much of Canada, and is capable of causing occasional severe human disease and death.

host-encoded furin occurs just prior to virion release and converts the immature particle (Figure 12.2a) to its mature form (Figure 12.2c). Recent structural and biochemical data demonstrate that the immature virus undergoes a significant rearrangement of its surface glycoproteins as it travels through the low pH of the *trans*-Golgi network (Figures 12.2b and 12.6). This rearrangement results in folding down the proteins onto the surface of the virion and exposes the furin cleavage site. After cleavage by furin, the pr peptide is not released from the virion until it is exposed to a neutral pH upon secretion from the host cell.

KEY TERMS

Antibody-dependent
    enhancement
Class I fusion proteins
Class II fusion proteins
Endemic
Epidemic
Exocytosis
Furin
Hemorrhagic fever
Hepatitis

Hepatocellular carcinoma
Lumen
Methyltransferase
Nucleoside triphosphatase
Polyprotein
Seroconversion
Signal peptidase
Type I transmembrane
    protein

FUNDAMENTAL CONCEPTS

• Flaviviruses cause severe human diseases such as yellow fever, Japanese encephalitis, dengue fever, and hepatitis.

• Most flaviviruses (but not hepatitis C virus) are transmitted by mosquitoes or ticks.

• West Nile virus entered the United States in 1999 and has progressively expanded its presence into most of the United States and parts of Canada by virtue of its mosquito-to-bird transmission cycle.

• The flavivirus particle is covered with a flat layer of envelope proteins arranged with icosahedral symmetry, giving the virus an appearance of a golf ball.

• Flaviviruses enter the cell by uptake into endosomal vesicles; at low pH the envelope protein rearranges from a dimer into a fusion-active trimer.

• All flavivirus proteins are synthesized as a single polyprotein that is cleaved into 10 functional viral proteins by viral and cellular proteinases.

- Like other positive-strand viruses, flavivirus RNA synthesis takes place in association with cytoplasmic membranes.
- Assembly of flavivirus virions takes place on membranes of the endoplasmic reticulum, and virions mature by cleavage of the premembrane protein as they pass through the *trans*-Golgi network.

REVIEW QUESTIONS

1. Which hepatitis virus is a member of the Flaviviridae? What kind of genome does it have?

2. What were the two accomplishments of the Yellow Fever Commission that brought yellow fever under control in the United States?

3. What is structurally unusual about the E protein that contributes to the smooth surface of tick-borne encephalitis virus?

4. What type of RNA polymerase is NS5 and why does it contain a methyltransferase?

5. Where do flaviviruses acquire their envelopes?

CHAPTER 13

Togaviruses

Milton Schlesinger
Sondra Schlesinger
Revised by: Richard Kuhn

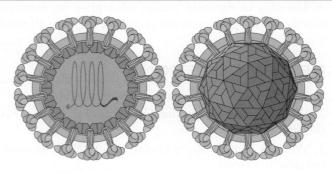

Togaviridae

From Latin *toga* (gown or cloak), referring to virus
envelope

VIRION

Spherical enveloped particle with a fringe of
projections.
Diameter 70 nm.
Nucleocapsid has icosahedral symmetry (T = 4).
Envelope glycoproteins are also arranged with icosa-
hedral symmetry, T = 4.

GENOME

Linear single-stranded RNA, positive sense.
Alphaviruses: 11 kb; rubella virus: 9 kb.
5' end is capped and methylated.
3' end has a 70-nt poly(A) tail.

GENES AND PROTEINS

Four nonstructural proteins (for viral RNA synthesis)
translated directly from genome RNA.
Five structural proteins translated from a subgenomic
mRNA:
  • *one capsid protein*
  • *three envelope proteins*
  • *a small hydrophobic protein*

VIRUSES AND HOSTS

Two genera:
  • *Alphavirus* (27 members)
  • *Rubivirus* (rubella virus)
Alphaviruses (Sindbis, Semliki Forest, Ross River)
  infect birds and mammals.
Most alphaviruses are transmitted by mosquitoes.

DISEASES

Symptoms include fever, rash, arthritis, and encephalitis;
  can be fatal.
Rubella infection (German measles) can lead to fetal
  abnormalities.

DISTINCTIVE CHARACTERISTICS

A subgenomic mRNA codes for virion structural
  proteins.
Structural proteins are synthesized as a polyprotein that
  undergoes cleavage and membrane association.
RNA replication is associated with cytoplasmic
  membranes.
Alphaviruses have the unusual ability to replicate in
  both invertebrate and vertebrate hosts.
Lipid envelope is derived from the plasma membrane
  (vertebrate cells) or internal membranes (insect cells).

Most togaviruses are arthropod borne, transmitted between vertebrate hosts by mosquitoes

Walter Reed, working in Cuba in 1900, demonstrated that
yellow fever virus (see Chapter 12) was transmitted
to humans by mosquitoes. This startling finding led to
efforts by many laboratories, particularly in tropical
regions of the world, to discover other viruses transmitted by mosquitoes. In the next half-century, scientists
isolated a large number of viruses that infect mammals
and birds and are transmitted by a variety of arthropod
vectors, including mosquitoes, ticks, and sandflies.
Such viruses were originally classified as **arthropod-borne**
viruses ("arboviruses"). Most of these viruses can replicate in both cold-blooded arthropods and warm-blooded

vertebrate hosts, and thus have an unusual ability to adapt to different growth conditions.

When studies on the molecular biology of arthropod-borne viruses were undertaken, it became apparent that not all such viruses are closely related. Group A arboviruses were reclassified into the *Alphavirus* genus of the *Togaviridae* family (Table 13.1), so named in reference to their envelope ("toga" means "cloak" in Latin). Most other arthropod-borne viruses are members of the *Flaviviridae* family (Chapter 12; named after yellow fever virus) or the *Bunyaviridae* family (Chapter 17); these viruses have distinct genome organization and replication strategies. Some members of the *Reoviridae* and *Rhabdoviridae* are also arthropod-borne.

Several alphaviruses, including eastern and western equine encephalitis viruses, Venezuelan equine encephalitis virus and Ross River virus, cause disease in animals and humans. Symptoms range from rashes and high fevers to joint pains and encephalitis, and the outcome can be fatal in both humans and domestic animals such as horses. The incidence of human disease in North America is low except for occasional local outbreaks. In most cases, the natural vertebrate hosts of these viruses are wild mammals and birds; humans (and horses) are usually dead-end hosts as far as the virus is concerned, because they do not transmit disease to other individuals.

Chikungunya virus, an alphavirus related to Ross River virus, erupted in epidemic form in the Indian Ocean region in 2005, resulting in significant disease and death. The threat of spread to developed countries was highlighted by a brief but notable outbreak in Italy in the summer of 2007. Given that one of Chikungunya's mosquito vectors, *Aedes albopictus*, is widespread, the threat of an endemic alphavirus in Europe and elsewhere is a major public health concern.

Rubella virus, the cause of German measles, is the only member of the togavirus family that is not transmitted via arthropod vectors. It causes fetal abnormalities and birth defects when women are infected in the first trimester of pregnancy.

Many togaviruses were isolated from mosquitoes or infected animals in tropical regions of the world, and their names often recall the localities where they were first found. Sindbis virus (named for an Egyptian village) and Semliki Forest virus (from Eastern Uganda) are two togaviruses that have been studied in great detail in the laboratory. They have been valuable models for investigating togavirus uptake, structure, and replication, and have also been used as important tools for studying cellular pathways such as cleavage and modification of glycoproteins and their transport to the plasma membrane.

These viruses are able to infect cultured cells from a wide variety of animal species and replicate rapidly, producing high titers of progeny (about 10,000 virions per infected cell in 12 to 16 hours). Infection of vertebrate cells quickly inhibits host cell protein synthesis and leads to cell death. Infection of mosquito cells is often not **cytopathic** but instead causes a **persistent infection**. Why infections with the same virus can lead to these two very different outcomes is not known. Sindbis and Semliki Forest viruses are not highly pathogenic in humans, allowing their safe and easy use in the laboratory.

Togavirus virions contain a nucleocapsid with icosahedral symmetry wrapped in an envelope of the same symmetry

Togavirus virions are enveloped spherical particles with a diameter of 70–80 nm. The outer surface of the virion consists of 80 flower-like structures each made up of three subunits, projecting outward in the form of petals (Figure 13.1a). The 240 petals are heterodimers composed of one molecule each of the glycoproteins E1

Table 13.1 Representative togaviruses

Genus and virus name	Vertebrate host	Arthropod host
Alphavirus		
Sindbis virus	Birds	Mosquitoes
Semliki Forest virus	Birds, rodents, primates	Mosquitoes
Ross River virus	Marsupials, rodents	Mosquitoes
Chikungunya virus	Primates	Mosquitoes
Venezuelan equine encephalitis virus	Rodents, horses, birds	Mosquitoes
Western and Eastern equine encephalitis viruses	Birds	Mosquitoes
Rubivirus		
Rubella virus	Humans	None

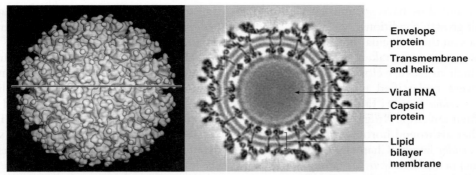

Figure 13.1 Cryoelectron microscopic image reconstruction of a togavirus virion. Left: a view of a Sindbis virion at 2.0 nm resolution. The outer surface is composed of flower-like projections of the E1/E2 heterodimers arranged as trimers. The interconnective portions form a protein shell that covers most of the lipid bilayer except for small holes. Right: a cross-section along the black line shown in the picture on the left. The lipid bilayer and transmembrane alpha helices crossing the membrane are clearly seen. The black images above and underneath the bilayer are envelope and capsid protein molecules.

and E2. They are organized with icosahedral symmetry in a T = 4 lattice. The E1 polypeptide comprises most of the protein shell or "skirt" that covers the underlying lipid membrane (Figure 13.1b); its structure is similar to that of the flavivirus envelope protein (Figure 12.3).

Cryoelectron microscopy data indicate that E2 has a shape similar to that of E1, but E2 forms the external part of the petals that project outward. X-ray crystallography has revealed that E2 contains three immunoglobulin-like domains matching the three domains of E1. Both E1 and E2 are anchored in the lipid membrane with carboxy-terminal portions of their polypeptide chains protruding into the interior of the particle, as can be seen in Figure 13.1b.

The envelope surrounds the virus nucleocapsid, which contains a single molecule of viral genome RNA. The nucleocapsid has a diameter of 40 nm and is composed of 240 copies of the capsid protein, also arranged in a T = 4 icosahedral lattice. The 240 capsid protein molecules are each in direct contact with one of the overlying 240 envelope protein heterodimers, leading to identical symmetry arrangements in both nucleocapsid and envelope, unusual for enveloped viruses. The envelope is tightly apposed to the surface of the nucleocapsid, resulting in a very uniform diameter of virions, unlike many enveloped RNA viruses with helical nucleocapsids, which can have variable size and shape. Rubella virus has not been studied in as much detail, but is likely to have similar morphology and symmetry. However, its particles exhibit greater heterogeneity than those of the alphaviruses.

Togaviruses enter cells by low pH-induced fusion inside endosome vesicles

The E2 glycoprotein binds to cellular receptors. No single cellular receptor has been unambiguously identified

for alphaviruses; different proteins have been identified as receptors for Sindbis virus in different host cells. A high-affinity **laminin receptor** was found on rodent and primate cells, and a laminin-like receptor is also believed to function as the receptor for Venezuelan equine encephalitis virus on mosquito cells. Mutations leading to amino acid changes in the viral E2 glycoprotein rapidly accumulate during passage in cultured cells; these mutations allow the virus to bind to **heparan sulfate**, which is found on many cell surface proteins. The finding that mutations in E2 can alter receptor binding confuses attempts to identify specific cellular receptors.

Uptake of virions occurs in vertebrate cells via **receptor-mediated endocytosis**. Togaviruses, particularly Semliki Forest virus, served as models to establish this means of entry, which is used by many enveloped RNA-containing viruses (see Chapter 4). After uptake of virus into endosomal vesicles, protein pumps in the vesicle membrane generate a drop in intravesicular pH, leading to conformational changes in the viral envelope glycoproteins. The flower-like envelope projections, which consist of trimers of E1-E2 heterodimers, reorganize to form trimers of E1, exposing a hydrophobic fusion domain in that protein. Fusion of the endosome membrane with the viral envelope leads to the release of the nucleocapsid into the cytoplasm. The pH-dependence of envelope-membrane fusion was shown directly by lowering the pH of medium bathing cells to which large numbers of virions were bound; virus envelopes fused directly with the plasma membrane when the pH was reduced.

The role of the 6K protein, which is found in reduced levels relative to E1, E2, and capsid in the virion, is obscure. However, it may play a role similar to the M2 protein of influenza virus, and serve as an ion channel that allows entry of hydrogen ions into the particle to promote disassembly. Thus, it could potentially serve as a target for antiviral drug intervention for the alphaviruses.

Nonstructural proteins are made as a polyprotein that is cleaved by a viral proteinase

Once in the cytoplasm, nucleocapsids release the viral genome RNA so that it can be translated by cellular ribosomes. Togavirus nucleocapsids are permeable to ribonuclease *in vitro*, suggesting that the RNA in nucleocapsids may be accessible to other proteins within the cytoplasm. It has been shown that large ribosomal subunits can bind to capsid proteins and disrupt nucleocapsids, liberating genome RNA. Interaction with ribosomes within the cytoplasm may therefore initiate nucleocapsid breakdown and release the genome RNA, allowing its translation.

The 11-kb single-stranded alphavirus RNA genome (Figure 13.2) contains two separately translated modules. The 5'-proximal two-thirds codes for the nonstructural proteins required for replication and transcription of the RNA (Table 13.2), and the 3'-proximal one-third codes for the capsid and envelope proteins (Table 13.3). These structural proteins are not translated directly from genome RNA, but from a **subgenomic mRNA** made during viral infection. Both the genome RNA and the subgenomic mRNA have methylated caps at their 5' ends and are polyadenylated at their 3' ends, resembling cellular messenger RNAs.

Translation of genome RNA begins at an initiator AUG near its 5' end. In Sindbis virus, translation terminates most of the time at a UGA codon positioned close to the end of the nsP3 coding region, generating a P123 **polyprotein** (Figure 13.3). The complete P1234 polyprotein is formed only when there is readthrough of this codon, which occurs 10 to 20% of the time, depending on the host cell.

The P1234 polyprotein is cleaved in stages to yield initially a complex of P123 and nsP4, followed by a complex consisting of nsP1, nsP23, and nsP4 and finally by a complex consisting of the individual polypeptides nsP1, nsP2, nsP3, and nsP4. The proteinase responsible for these cleavages is encoded in a domain at the C-terminal end of nsP2. The X-ray crystal structure of the nsP2 proteinase revealed two domains; the N-terminal domain is consistent with a cysteine proteinase and the C-terminal domain resembles a **methyltransferase**, but its function is not understood.

Partly cleaved nonstructural proteins catalyze synthesis of full-length antigenome RNA

The nonstructural proteins carry out synthesis of the different viral RNAs, a highly regulated event in infected cells. During the early phase of the infectious cycle, full-length negative-strand **antigenome** RNAs are synthesized by a replicase complex formed between P123 and nsP4 (Figure 13.3). Interestingly, sequences at both the 3' and 5'-termini of genome RNA are required for synthesis of negative-strand antigenome RNA, implying that the replication complex may be bound to both ends of the RNA.

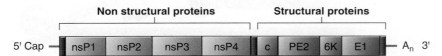

Figure 13.2 Alphavirus genome RNA. The 11 kb positive-strand RNA contains a methylated cap at its 5' end and is polyadenylated at its 3' end. Regions coding for nonstructural (ns) proteins and structural proteins are shown in boxes. Noncoding regions that include replication promoters and the internal promoter for subgenomic RNA synthesis from antigenome RNA are shown in black.

Table 13.2 Sindbis virus nonstructural proteins

Protein	Length (aa)	Properties and functions
nsP1	540	RNA capping enzyme; has methyltransferase activity. Required for initiation of transcripion of minus-strand RNA and modulation of nsP2 protease activity. Palmitoylated and membrane-associated.
nsP2	807	Cysteine proteinase, RNA helicase. Required for synthesis of subgenomic RNA. Mutations in nsP2 show reduced cytopathic effects and lower shutoff of host cell protein synthesis. A fraction of nsP2 is found in the nucleus of infected cells.
nsP3	556	Unknown function. Phosphorylated at Thr and Ser residues.
nsP4	610	RNA polymerase. Unstable, degraded by the ubiquitin pathway.

TABLE 13.3 Sindbis virus structural proteins

Protein	Length (aa)	Properties and functions
Capsid	264	Forms nucleocapsid; contains serine-like protease that autocatalytically cleaves itself from the polyprotein. Binds to genome RNA packaging signal via positively charged amino terminus.
PE2	487	Precursor to E2; palmitoylated and glycosylated. Also referred to as p63, based on its molecular weight. Cleaved in *trans*-Golgi vesicles by a furin-like cellular protease; rarely incorporated into virions.
E3	64	Amino-terminal part of PE2 containing the signal sequence. Glycosylated; retained in virions of Semliki Forest but not Sindbis virus.
E2	423	Carboxy-terminal part of PE2, a component of virus envelope projections. C-terminal cytoplasmic domain binds to capsid during assembly. Binds to cellular receptors; contains epitopes for neutralizing antibodies. Palmitoylated.
6K	55	Membrane associated; palmitoylated; C-terminus is signal sequence for E1. Small amounts in virions; enhances virus assembly and budding, possible role in hydrogen ion transport.
E1	439	Component of virion envelope projections; type 1 transmembrane glycoprotein with two positively charged C-terminal amino acids in the cytoplasm; palmitoylated and glycosylated; functions in low-pH activated membrane fusion.

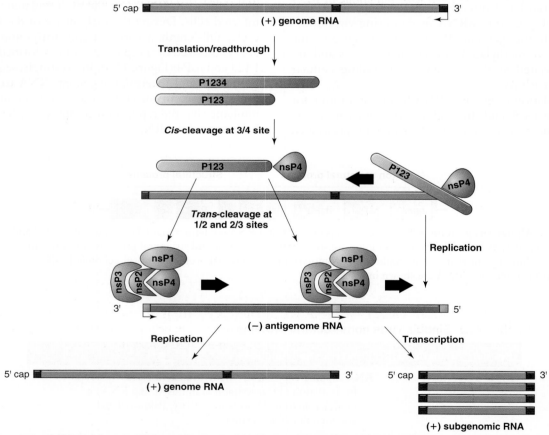

Figure 13.3 A model for the temporal regulation of minus- and plus-strand RNA synthesis. Nonstructural proteins are translated from genome RNA. A complex of proteins P123 and nsP4 initiates minus-strand RNA synthesis (replication) from a promoter (black with directional arrow) at the 3′ end of genome RNA; 5′-terminal sequences are also required. Once sufficient P123 has been made, trans-cleavage generates nsP1, P23, and finally nsP2 and nsP3. These proteins, together with nsP4, form complexes that efficiently initiate synthesis of positive-strand genome RNA (replication), and larger amounts of subgenomic mRNA (transcription), from promoters (gray squares with directional arrows) on the antigenome RNA template. Host cell proteins are probably also involved in these reactions.

At this stage, little or no positive-strand RNA is produced. In contrast, during the late phase mostly positive-strand genome and subgenomic RNAs are made. The switch from synthesis of negative to positive strands is correlated with proteolytic processing of P123 (Figure 13.3). P123 does not undergo autoproteolysis and it is not cleaved during the early phase, suggesting that the protease domain in nsP2 is only able to cleave other polyprotein molecules than the one in which it lies. During the late phase, when the concentration of P123 becomes high enough, P123 molecules can encounter each other and undergo mutual trans-cleavage.

One of the most convincing arguments for the importance of intact P123 in the synthesis of negative strands is based on a mutation in the nsP2 proteinase that leads to unusually rapid proteolytic processing of P123. RNAs containing this mutation, when transfected into cultured cells, are unable to produce new virus progeny, suggesting that intact P123 is essential for negative-strand RNA synthesis. Replicase complexes containing this nsP2 mutation regain the ability to carry out negative-strand RNA synthesis when combined with mutations in P123 that destroy the sites of cleavage between the individual nonstructural proteins nsP1, nsP2, and nsP3.

Replication and transcription: synthesis of genome and subgenomic RNAs

The antigenome RNA serves as a template for the synthesis of both full-length and subgenomic positive-strand RNAs. Genome RNA synthesis (replication) initiates at the 3' end of the antigenome RNA and proceeds to its 5' end, to make a complete complementary copy of the negative-strand RNA (Figure 13.3). Synthesis of the subgenomic RNA (transcription) initiates at a promoter that lies at the junction between the genes for nonstructural proteins and those for structural proteins. The minimal promoter sequence required for transcription is 19 nucleotides upstream and 5 nucleotides downstream from the start of the subgenomic RNA; this region is highly conserved among the various alphaviruses. The level of synthesis of subgenomic RNA is increased if flanking regions are also present.

At the peak of virus formation, about 10 times more subgenomic RNA than genome RNA has accumulated. Making higher levels of subgenomic RNA is one means by which the virus is able to produce more structural than nonstructural proteins. The structural proteins are needed in large amounts for virion assembly (240 molecules of each protein per virus particle), while only small amounts of the nonstructural proteins are required since they act catalytically.

Other activities have been associated with individual nonstructural proteins based on studies with temperature-sensitive mutants and on sequence homologies with proteins of known function (Table 13.3). The nsP1 protein contains a methyltransferase domain and functions as a capping enzyme, adding a methylated GTP cap to the 5' end of positive-strand viral RNAs (for details on enzymes involved in RNA capping, see Chapters 19 and 26). In addition to its protease activity, nsP2 contains nucleoside triphosphate-binding motifs in its amino terminal region that are homologous to those found in bacterial helicases. When nsP2 was expressed in bacteria in the absence of other viral proteins, it was shown to have an **RNA helicase** activity that may be needed for optimal RNA replication/transcription. The nsP4 protein has a Gly-Asp-Asp motif common to the active site of a number of viral RNA polymerases. It has been identified as the polymerase responsible for both genome replication and subgenomic mRNA synthesis. A specific role for nsP3 has not yet been identified.

Proteolytic processing of the rubella virus nonstructural polyprotein (p200) also appears to regulate replication and transcription of the viral RNA. The intact polyprotein can synthesize negative-strand anti-genome RNA, but cleavage of p200 by a virus-encoded proteinase to produce two proteins named p150 and p90 is required for efficient formation of new positive-strand rubella virus genome RNA and subgenomic RNA.

Structural proteins are cleaved during translation and directed to different cellular locations

In addition to being produced in large amounts, the subgenomic mRNA, which encodes viral structural proteins, is also very efficiently translated. Translation is initiated at a single site near its 5' end and proceeds uninterrupted for some 3000 nucleotides to produce a polyprotein that is co- and post-translationally cleaved. For some alphaviruses, a hairpin-like structure near the 5' end of the subgenomic RNA and within the capsid coding sequence functions to enhance its translation.

The capsid gene is translated first, and the carboxy-terminal half of this polypeptide folds to form a serine protease catalytic site that self-cleaves the capsid protein from the nascent polypeptide chain while translation continues downstream (Figure 13.4). Capsid protein molecules can then interact with each other and with the genome RNA to initiate assembly of nucleocapsids.

A domain within the amino-terminal half of the capsid protein binds to a specific region of the positive-strand genome RNA called the **packaging signal**. This is located within the coding region of the nsP1 gene (for Sindbis virus) or the nsP2 gene (for Semliki Forest virus and the related Ross River virus). The packaging signal initiates encapsidation and ensures that genome RNA is efficiently packaged, excluding subgenomic RNA, antigenome RNA, and unrelated cellular RNAs from incorporation into nucleocapsids.

Continued translation of the subgenomic mRNA beyond the capsid gene produces an N-terminal signal sequence that translocates the PE2 polypeptide to the rough endoplasmic reticulum (Figure 13.4). Additional stop-transfer and signal sequences in the PE2, 6K, and E1 regions of the polyprotein lead to insertion of the 6K protein and translocation of the E1 portion into the endoplasmic reticulum membrane. A host cell **signal peptidase** cleaves the polyprotein at two sites that define the carboxy termini of the PE2 and 6K proteins.

In the endoplasmic reticulum, PE2 spans the membrane twice at its carboxy terminus; the 6K protein spans the membrane at both its amino and carboxy termini; and E1 spans the membrane once at its carboxy terminus (Figure 13.4). As these proteins are transported to the Golgi stacks, they are modified by **palmitoylation**, the addition and trimming of oligosaccharides, and a reorientation of the carboxy terminus of PE2 from a membrane-embedded to a cytosolic location. Prior to their appearance in the plasma membrane, a host cell **furin protease** converts PE2 to E2 and E3.

Assembly of virions and egress at the plasma membrane

Two models have been proposed to explain virus assembly and release from the plasma membrane. The first model proposes that preformed nucleocapsids migrate to the plasma membrane, where they bind to sections of the membrane that are densely populated with virus envelope glycoproteins, and the enveloped particle is formed by budding from the surface of the infected cell (Figure 13.4; see also Figure 2.11a). This model is suggested by the presence in the cytoplasm of free, fully formed nucleocapsids, and by electron microscopic images showing budding virus particles that contain such nucleocapsids.

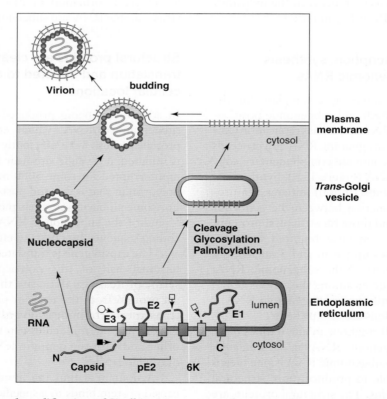

Figure 13.4 Processing and modification of Sindbis virus structural proteins. During translation of subgenomic mRNA, the capsid protein is released from the growing polypeptide chain by self-cleavage (arrow with black square). Capsid proteins oligomerize and bind to genome RNA in the cytoplasm to form nucleocapsids, which migrate to the plasma membrane, as shown here, or alternatively form during association with viral glycoproteins at the plasma membrane. The remainder of the structural polyprotein, including PE2, 6K, and E1, is translocated as a unit to the endoplasmic reticulum by means of a signal sequence at its N-terminus. Proteolytic processing by cellular signal peptidase (arrows with white squares) produces PE2, 6K, and E1. Additional protein modifications carried out during vesicle transport to the cell surface include cleavage by furin protease (arrow with white circle) in a *trans*-Golgi compartment to form E2 and E3, as well as glycosylation. Signal sequences: gray rectangles; transmembrane segments: orange rectangles; processed viral envelope proteins: orange spikes; viral RNA: blue curved line.

Initiation of the budding event is thought to occur when a nucleocapsid in the cytoplasm recognizes the cytoplasmic tails of arrays of the E1/E2 glycoprotein heterodimers localized in the plasma membrane. There is a conserved motif (Tyr-X-Leu) in the carboxy-terminal tail of the E2 glycoprotein that interacts with a hydrophobic pocket on the surface of the capsid protein.

As budding begins, each of the 240 capsid protein subunits of the core binds to an E2 protein molecule in the membrane. This forces bending of the membrane around the icosahedral nucleocapsid, culminating in fusion of the highly curved membrane and release of the virion from the cell. The precise pairing of envelope glycoproteins to capsid proteins in the core precludes incorporation of host cell surface proteins into the released virus particles.

A different assembly model proposes that budding progresses through "shell interactions" of the glycoproteins. In this model, capsid proteins bind to genome RNA in a complex that is not a structured nucleocapsid. The complex encounters the cytoplasmic tails of viral glycoproteins anchored in the plasma membrane, and the capsid proteins in the complex bind on a 1:1 basis to these tails. It is this binding that leads to the formation of the inner shell (made of capsid proteins) and outer shell (made of envelope glycoproteins), and membrane bending and budding produces virions. Evidence for this model comes from studies with capsid protein mutants in which preformed nucleocapsids are not detected in the cytoplasm of infected cells, but intact enveloped virions are still able to bud from the plasma membrane.

Effects of mutations in viral proteins on cytopathic effects and on pathogenesis

Infection of vertebrate cells with alphaviruses is usually cytopathic. Host cell protein synthesis is inhibited and in many cell types there appears to be induction of programmed cell death, or **apoptosis**. Alphavirus gene expression vectors, described in the next section, are also cytopathic. However, some mutants of these vectors are able to persist in mammalian cells without causing cytopathic effects. The mutants isolated so far all contain a single amino acid change in the C-terminal half of nsP2 and have a decreased level of viral RNA synthesis. There appears to be a correlation between the decrease in cytopathogenicity and the decrease in viral RNA synthesis, but it is not known which factors are most critical in producing cytopathic effects nor why mutations in the nsP2 gene allow cell survival.

In contrast to vertebrate cells, infection of mosquito cells by wild-type virus is often not cytopathic. Because virions are formed by budding at the plasma membrane, virus can be produced and circulate in mosquito cell cultures without disruption of infected cells. Recent studies suggest that in some viruses, the so called New World alphaviruses, the sequences controlling cytopathogenicity reside in the capsid protein and affect the survival of vertebrate cells through a different mechanism than nsP2 of the Old World alphaviruses.

Most experimental studies of pathogenesis with Sindbis and Semliki Forest viruses have been done with neonatal mice, in which disease is more apparent than with adult mice and mortality is high. Whether or not infection will produce disease is dependent both on the virus and the host. Innate host defense responses (for example, induction of **interferon**; see Chapter 33) can play an important role as does the adaptive immune response. A single amino acid change in the E2 glycoprotein of Sindbis virus has a profound effect on the neurovirulence of the virus. Differences in the 5' untranslated region of the genome have been correlated with changes in survival of infected animals. These changes are subtle, yet their effects are significant, and virologists are trying to understand what steps in infection or in the response to infection are altered by these mutations.

Alphaviruses have been modified to serve as vectors for the expression of heterologous proteins

The use of viruses as gene expression vectors is now a well-established technology (see Chapter 37). The first step in the generation of a viral vector from a positive-strand RNA virus is to convert the RNA genome to a DNA copy (cDNA) that can be cloned into a bacterial plasmid, modified in various ways, and then transcribed back into an infectious RNA.

Alphavirus cDNAs were initially placed under the control of a promoter for bacteriophage T7 or SP6 DNA-dependent RNA polymerase. The cDNA was then transcribed *in vitro* and the RNA was introduced into cultured cells by transfection, leading to virus replication. Subsequently, viral cDNAs were put under the control of a eukaryotic promoter, enabling production of viral RNA by transcription of DNA introduced into the nucleus of a eukaryotic cell by transfection.

In the most widely used "self-replicating" alphavirus vectors (Figure 13.5a), the structural protein genes are replaced by a heterologous gene—initially, **reporter genes** such as the gene producing **β-galactosidase** or the gene encoding **chloramphenicol acetyltransferase**, and subsequently any foreign gene of interest. The modified genome RNAs (referred to as **replicons**) serve as mRNAs for the translation of the nonstructural proteins, and their replication is carried out by these proteins. They produce subgenomic RNAs, which express the foreign protein.

Since replicons lack structural protein genes, production of virus particles requires cotransfection with

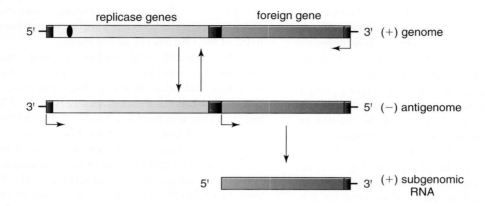

(a) **Self-replicating vector**

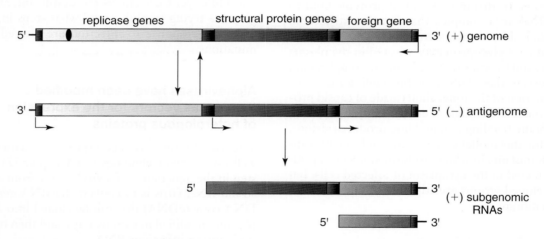

(b) **Double subgenomic RNA vector**

Figure 13.5 Alphavirus expression vectors. Black oval near left end of genome RNA is packaging signal. Small black rectangles with directional arrows represent promoters for RNA synthesis.

defective helper genomes. These are alphavirus RNAs that have undergone large deletions in the nonstructural protein genes, but retain the promoter for the subgenomic RNA and the structural protein genes. They can be replicated by the nonstructural proteins made by the replicon RNA. If the defective helper genomes are mutated so that they lack signals for RNA packaging, only replicon genomes will be packaged into virions. These viruses have been termed "suicidal", as they are able to infect cells and produce a particular protein of interest within those cells, but do not produce new particles in the absence of helper genomes, and therefore cannot spread to neighboring cells.

In the "double subgenomic" alphavirus gene expression vector (Figure 13.5b), coding regions for the structural proteins are retained and a second subgenomic RNA promoter is inserted into the genome, allowing the synthesis of both the viral structural proteins and a

foreign protein, each on different subgenomic RNAs. This type of vector produces a viable virus that can spread from cell to cell either *in vitro* or in an animal or human. These vectors tend to be less stable than replicons, particularly when the inserted foreign genes are large (greater than 2 kb). The size limitation may reflect the packaging preference of alphavirus capsids for RNAs the size of genome RNA.

Alphavirus vectors have multiple potential uses

Alphavirus vectors can be used to express a variety of important cellular proteins in different cell types. They can also be used for production and purification of selected proteins, because the levels of foreign proteins synthesized in infected cells can be quite high. These vectors are particularly useful for production of proteins that have post-translational modifications (such as phosphorylation or the addition of carbohydrates) that occur in vertebrate or insect cells but not in bacterial cells.

Alphavirus vectors have also been engineered to express proteins of unrelated viruses, with the prospect of developing them as vaccines. In initial studies, proteins derived from influenza virus, herpes virus, and human immunodeficiency virus were expressed. Injection of these vectors into animals showed them to be effective in inducing an immune response and, in the case of influenza and herpes viruses, protecting the animal from subsequent infection.

Since alphaviruses also infect insects, particularly mosquitoes, vectors are being developed for use in insects. One approach is to use a Sindbis virus strain that is not pathogenic for insects to express an interfering RNA (**RNAi**, see Chapter 33) directed against human pathogens such as yellow fever virus or dengue virus (both flaviviruses). Experiments have shown that infection of mosquitoes with such a recombinant vector will effectively inhibit replication of yellow fever virus or dengue virus in the mosquito. This opens up the possibility of disseminating DNAs that code for such RNAi in mosquito populations, perhaps by introducing them into transposons that will spread within natural populations of mosquitoes. Any such strategy would be effective only if a significant proportion of mosquitoes received the gene that expresses the antisense RNA; infection by the Sindbis vector itself would not likely achieve this result.

Another approach is to have Sindbis virus express a protein toxic to the insect, serving essentially as an insecticide. Such a vector would generate an infectious virus that contains two subgenomic RNA promoters, one for the expression of structural proteins and another for expression of the toxic protein. However, release of such a virus into nature may be difficult to justify unless extensive testing shows its safety for humans and other animals. Because these viruses circulate between vertebrate and arthropod hosts, introduction of engineered viruses into mosquito populations could involve substantial risks to the vertebrate host.

Alphavirus vectors may be useful for treatment of neurological diseases

Virologists who study alphaviruses have an interest in understanding the ability of these viruses to infect neurons. Many alphaviruses cause encephalitis; a better understanding of how these viruses enter the brain and why only some strains are able to cause symptoms may help in prevention of disease. The ability of alphaviruses to infect neurons may turn out to have some benefits.

Studies with rat **hippocampus** slice cultures show that neurons are selectively infected by Sindbis and Semliki Forest virus replicons. Live neurons expressing **green fluorescent protein** produced by engineered viral replicons can be observed for up to five days postinfection. Because the infected neurons eventually die, the current alphavirus vectors would not be appropriate for any type of treatments in the human nervous system.

Will it be possible to isolate replicon expression vectors that are not cytopathic to neurons? If so, then such vectors could be considered as a means to treat neurological diseases by replacement gene therapy. These vectors, however, can already provide valuable tools for expressing proteins and for identifying and studying neuronal functions.

KEY TERMS

Antigenome	Laminin receptor
Apoptosis	Methyltransferase
Arthropod-borne	Packaging signal
β-galactosidase	Palmitoylation
Chloramphenicol acetyltransferase	Persistent infection
	Polyprotein
Cryoelectron microscopy	Receptor-mediated endocytosis
Cytopathic	
Encephalitis	Replicon
Furin protease	Reporter gene
Green fluorescent protein	RNA helicase
Heparan sulfate	RNAi
Hippocampus	Signal peptidase
Interferon	Subgenomic RNAs

FUNDAMENTAL CONCEPTS

- Most togaviruses are transmitted among warm-blooded hosts by mosquitoes.
- Togaviruses can cause serious disease such as encephalitis in humans and horses.
- The protein subunits in both the nucleocapsid and the envelope of togaviruses are arranged with icosahedral symmetry.
- Togaviruses synthesize RNA polymerase, capping enzyme, and proteinases by translation of full-length genome RNA.
- Structural proteins (capsid and envelope proteins) are synthesized by translation of an abundant subgenomic RNA.

- Like other positive-strand RNA viruses, togavirus RNA synthesis is carried out in the cytoplasm in association with membranes.

- In vertebrates, virions are made by budding of nucleocapsids through the cytoplasmic membrane.

- A signal on the genome RNA ("packaging sequence") binds to a sequence in the capsid protein to ensure efficient packaging of the viral genome RNA into the nucleocapsid.

- An alternative assembly model proposes that a complex between genome RNA and capsid proteins forms a nucleocapsid only upon interaction with viral envelope proteins at the cytoplasmic membrane.

- Togavirus (alphavirus) expression vectors have promising potential uses as vaccines or to express useful or toxic proteins.

REVIEW QUESTIONS

1. What is the one member of the togavirus family that is not transmitted by an arthropod vector, and what diseases does it cause in humans?

2. How are the open reading frames arranged along the single-stranded alphavirus RNA?

3. What is a "packaging signal" and what is its purpose during togavirus replication?

4. Discuss the role of nsP2 in determining the fate of virus-infected cells.

Coronaviruses

Mark Denison
Michelle M. Becker

Coronaviridae

From Latin *corona* (crown), referring to spikes
projecting from envelope

VIRION

Spherical enveloped particles studded with clubbed
spikes.
Diameter 120–160 nm.
Coiled helical nucleocapsid and/or nucleocapsid shell

GENOME

Linear single-stranded RNA, positive-strand.
27–32 kb.
5'-terminal cap, 3' poly(A) tail.

GENES AND PROTEINS

Six to nine genes code for more than 20 proteins.
Replicase genes are translated directly from
genome RNA.
  • *Two large open reading frames are joined by a*
    frameshift
  • *These genes are translated as polyproteins, then*
    cleaved to 14–16 individual proteins
Other genes are translated from multiple
3'-coterminal, subgenomic mRNAs.
  • *Three to four envelope proteins (HE, S, E, M)*
  • *One nucleocapsid protein (N)*
  • *Four to six nonstructural or accessory proteins*

VIRUSES AND HOSTS

Subfamily *Coronavirinae:* three genera based on
genome homologies.
Humans: human coronaviruses 229E, OC43, NL63,
HKU-1; SARS coronavirus.
Animals including mice, dogs, cats, pigs, cattle,
turkeys, bats: murine hepatitis virus
(MHV), transmissible gastroenteritis virus
(TGEV), avian infectious bronchitis virus (IBV)

DISEASES

Coronaviruses are responsible for up to 30% of
common colds in humans.

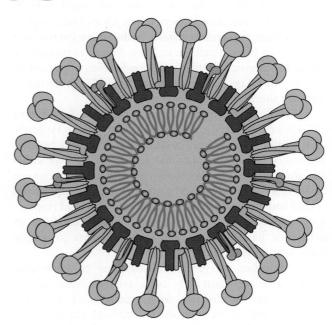

Severe acute respiratory syndrome (SARS): pneumonia
in humans with 10% fatality rate.
Important veterinary diseases: severe endemic and
epidemic respiratory, hepatic, gastrointestinal,
and neurological diseases in mammalian and
avian species.
Virus spreads between hosts by direct contact, aerosol
and fecal–oral transmission, and contact with
contaminated surfaces.

DISTINCTIVE CHARACTERISTICS

The largest known RNA virus genomes.
Large replicase polyprotein is cleaved by viral
proteinases to form numerous functional proteins.
Some unique enzymes for RNA viruses, including
a possible proofreading exonuclease.
Subgenomic mRNAs are made by an unusual
discontinuous transcription mechanism.
Bats are probable reservoir hosts.
Coronaviruses tend to move from host species into
new species.

Coronaviruses cause respiratory illnesses in humans and important veterinary diseases

The search for viruses that cause upper respiratory infections in humans (the "common cold") was intense in the mid-1960s. The laboratories of David Tyrrell, Dorothy Hamre, and Robert Chanock isolated viruses from humans that could replicate on fragments of tracheal tissue or in human cells in culture. When examined by electron microscopy, the enveloped virion appeared to be surrounded by a crown or *corona* of studded projections, and was therefore named a coronavirus.

Two distinct human coronaviruses, called 229E and OC43, were isolated and studied. These viruses typically cause mild to moderate respiratory illness in healthy individuals, being responsible for up to 30% of colds in some outbreaks. They only rarely cause more severe respiratory diseases, **gastroenteritis**, and **encephalitis**. These human coronaviruses can cause repeated respiratory infections on an alternating 2- to 3-year basis in otherwise healthy hosts, and do not appear to induce long-lasting immunity.

A number of veterinary diseases of economic importance are caused by coronaviruses. One of the first coronavirus diseases to be recognized, in 1912, was feline infectious **peritonitis**, although the virus was isolated much later. Coronaviruses cause devastating outbreaks of gastroenteritis and respiratory infections in cattle (bovine coronavirus), pigs (transmissible gastroenteritis virus), and chickens (infectious **bronchitis** virus), with mortality rates as high as 100% in newborn animals. These viruses have been targets for vaccine development and use, and have served as models for the study of coronavirus biology. Murine (mouse) **hepatitis** virus (MHV), a coronavirus, was isolated from laboratory mouse strains in the early 1960s. This virus has been very useful for experimental studies, as it grows to high titers in cultured cells and its molecular biology and pathogenesis can be readily studied.

A newly emerged coronavirus caused a worldwide epidemic of severe acute respiratory syndrome (SARS)

The field of coronavirus research took an abrupt turn when, in early 2003, an epidemic of severe and often fatal pneumonia broke out in southeast China, Hong Kong, and Vietnam, and subsequently spread to Toronto, Canada. This new disease was named severe acute respiratory syndrome (SARS), and the causative agent was quickly identified as a previously uncharacterized coronavirus. The SARS coronavirus (SARS-CoV) was isolated from people of all age groups who presented with fever and progressive pneumonia, leading to **hypoxemia** and death. Over a 6-month period this epidemic led to more than 8000 illnesses and nearly 800 deaths worldwide. Mortality was highest (up to 44%) in individuals over 65 years of age.

The spread of SARS coronavirus infections usually took place among family members of victims or in hospital care settings, and occurred principally during symptomatic illness. Virus was transmitted by direct contact with patients, but also via aerosol droplets and contaminated stool. Air travel by infected individuals quickly spread the disease to 32 countries, resulting in the first pandemic of the twenty-first century. Public health isolation measures appear to have interrupted the chain of person-to-person infection, but not without serious economic loss for the cities involved, a significant drop in travel to Southeast Asia and China, and the illness and death of hospital personnel caring for victims. Although human transmission and epidemic disease appears to have been eradicated, additional cases in 2003 and 2004 arose in laboratory settings and from animal-to-human transmission.

SARS coronavirus may have originated from related bat coronaviruses

Since SARS had not been observed in humans before the 2002–2003 pandemic, the virus causing this disease was thought to be either an avirulent human coronavirus that mutated to become more virulent, or a virus newly introduced into the human population. Based on the genome sequence of SARS coronavirus and what is known about other human coronaviruses, it seemed likely that SARS coronavirus originated as an animal virus. This hypothesis formed the basis for studies seeking similar viruses in animals.

Researchers therefore looked for possible animal carriers in open-air markets in areas where SARS was first observed. Viruses similar to SARS coronavirus were found in caged civet cats and raccoon dogs, carnivores that are captured or farmed and sold in markets for food in Guangdong province, in southern China. However, when free-ranging or farmed animals of these species were examined, no viruses similar to SARS coronavirus were found. Therefore scientists widened the net and started to sample other animal species.

Viruses genetically similar to SARS coronavirus were found in Chinese horseshoe bats. Since then, various species of bats from around the world have been examined, resulting in the discovery of a greater number of coronaviruses than had previously been anticipated. There are presently more bat coronavirus sequences in databases than coronaviruses from all other animals combined, including humans. As a result, a current hypothesis is that SARS coronavirus, and perhaps all human coronaviruses, may have originated from bat coronaviruses. Thus bats would be the main animal

reservoir for coronaviruses, as has been proposed for other "emerging" viruses such as Ebola, Hendra, and Nipah viruses.

How did a bat coronavirus mutate and enter humans to become SARS coronavirus?

It is still not clear which pathway was taken by viruses to become SARS coronavirus, but several possibilities exist. The original virus strain could have been passed from bats to an animal that served as an intermediate, or amplifying host. During this transition, the virus could have mutated to a form that was infectious in humans and could be transmitted from person to person, resulting in SARS.

Another possibility is that a bat coronavirus might have directly infected a person in a form that was not pathogenic, and was then passed to an animal that again served as an intermediate host. Finally, the virus could have been passed on to humans in a form that was virulent and transmissible, causing the pandemic. Examination of genome sequences of many coronaviruses isolated from animals and humans at different stages in the pandemic, and from bats, suggest that the second pathway, although seemingly more complicated, might be correct. This suggests that the barrier for coronaviruses to "jump" from one species to another might be much lower than previously predicted.

Coronaviruses have large, single-stranded, positive-sense RNA genomes

Coronaviruses have the largest RNA genomes (27–32 kb) of any known virus family. The single-stranded, positive-sense RNA genome contains a 5' methylguanosine cap and a 3' poly(A) tail. The RNA genome contains six to 10 genes, depending on the virus strain. Figure 14.1 shows a map of the genome of murine hepatitis virus.

Like other positive-strand RNA viruses, coronavirus genomes are infectious; introduction of purified genome RNA into a cell will initiate infection and result in production of new virus particles.

The order of genes is highly conserved among all coronaviruses. Remarkably, the first gene from the 5' end occupies two-thirds of the viral genome and is translated into **polyproteins** that are cleaved by viral proteinases into 16 or more different proteins involved in genome replication and subgenomic mRNA synthesis. The other one-third of the genome contains genes coding for virion structural proteins and group-specific nonstructural proteins. Several structural proteins are found in all coronaviruses, in the following order: hemagglutinin-esterase (HE), spike (S), small envelope protein (E), membrane protein (M), and nucleocapsid protein (N). Hemagglutinin-esterase is not present in all coronaviruses.

The conservation of gene order is notable given the high frequency of recombination observed in coronaviruses and the experimental demonstration that genes can be artificially reordered without affecting virus replication. A variable number of accessory or group-specific genes are interspersed between these structural protein genes. Some of these accessory genes have been shown to serve functions in pathogenesis, such as inhibition of interferon-regulated gene expression.

Coronaviruses fall into three groups based on genome sequences

Coronaviruses, like many viruses, were initially categorized by antigenic relationships. Viruses that could be neutralized by the same antisera were considered to be related and were classified in the same group. This approach has been replaced by comparison of nucleotide sequences of viral genome RNAs using computer

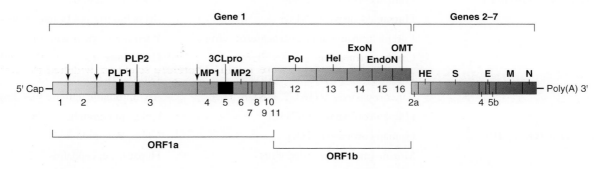

Figure 14.1 Organization of the coronavirus genome. The 31.5-kb genome of murine hepatitis virus is shown. A short 5' non-translated leader region is followed by gene 1, which consists of two overlapping reading frames (ORF1a and 1b) that are translated as polyproteins from genome RNA. Genes 2–7 are expressed from subgenomic messenger RNAs. Where known, the names or function of gene products are shown above the genome map. The nonstructural proteins nsp1–16 (shown below the map) are generated by cleavage from the 1a/1b fusion protein by viral proteinases. Cleavage by proteinases PLP1 and PLP2 (black, cleavage sites shown as vertical arrows) yields nsp1–3, and cleavage by 3CLpro (black) yields nsp4–16.

programs to determine evolutionary relatedness, or phylogeny. This process is based on the theory that the more recent the separation between two viruses, the more similar will be their genetic information.

Based on both these approaches, coronaviruses were traditionally clustered into three groups. Interestingly, sequence similarity alone does not determine other characteristics of these viruses, such as their cellular binding receptor, animal host, tissue tropism, pathogenesis, virulence, or type of disease.

More recently, a new classification system for coronaviruses (Table 14.1) has been accepted by the International Committee on Taxonomy of Viruses. In this proposal, the family *Coronaviridae* has two subfamilies: *Coronavirinae* and *Torovirinae*. The family *Coronavirinae* contains three genera: *Alpha-*, *Beta-*, and *Gammacoronaviruses*. These genera roughly correspond to the previous coronavirus groups I (alpha), II (beta), and III (gamma), but with some differences based on the criteria of relatedness of the most conserved coding sequences.

The 2002–2003 SARS epidemic sparked a flurry of investigation in the coronavirus field, including examination of archived samples of patients admitted to hospitals with respiratory symptoms of unknown etiology. In these surveys, two additional coronaviruses were identified and classified. These viruses, called NL63 (isolated from the Netherlands) and HKU1 (isolated in Hong Kong), have since been determined to circulate in human populations worldwide. In addition, coronaviruses infecting many animals such as giraffes, cheetahs, and Beluga whales have also been identified.

However, the vast majority of new coronavirus sequences that have been submitted to databases are from bat viruses; roughly half of the coronaviruses now known are bat coronaviruses. In all cases to date, the bats from which the coronaviruses were isolated were not ill, and none of the viruses have been cultured from primary isolates in the laboratory. It is likely that advances in culturing of bat cells and identification of receptors and viral determinants will result in cultivation and analysis of primary bat coronaviruses in the near future.

Coronaviruses have enveloped virions containing helical nucleocapsids

The coronavirus virion is a spherical enveloped particle, ranging in size from 75 to 160 nm, with a characteristic crown of spike proteins that gives the virus family its name (Figure 14.2). Electron micrographs of thin sections of infected cells reveal virions with a doughnut-like core structure that represents the helical nucleocapsid underlying the viral envelope; the helical nucleocapsid can also be seen in negatively stained disrupted virions.

The nucleocapsid is formed from multiple copies of the nucleocapsid protein (N) bound to the viral RNA genome. Most other positive-strand RNA viruses have

Table 14.1 Representative human and animal coronaviruses

[Order *Nidovirales*, Family *Coronaviridae*, Subfamily *Coronavirinae*]		
Genus	**Species**	**Disease**
Alphacoronavirus (Group I[1])	Human coronavirus 229E	Colds, pneumonia
	Human coronavirus NL63	Bronchiolitis, colds, pneumonia
	Porcine transmissible gastroenteritis virus	Pneumonia, gastroenteritis
	Canine coronavirus	Gastroenteritis
	Feline infectious peritonitis virus	Peritonitis, enteritis
	Bat coronaviruses (HKU2, HKU8, 512)	Asymptomatic (enteric shedding)
	Human coronavirus OC43E	Colds, pneumonia
Betacoronavirus (Group II[1])	Human coronavirus HKU1	Colds, Pneumonia
	Murine (mouse) hepatitis virus	Hepatitis, encephalitis
	Bovine coronavirus	Pneumonia, gastroenteritis
	SARS coronavirus	Pneumonia, gastroenteritis
	Bat coronaviruses (HKU4, 5, 9)	Asymptomatic
Gammacoronavirus (Group III[1])	Avian infectious bronchitis virus	Tracheitis, kidney infection
	Beluga whale coronavirus SW1	Hepatitis

[1] Previous taxonomy of virus groups within previous genus *Coronavirus* (Groups I, II, and III).

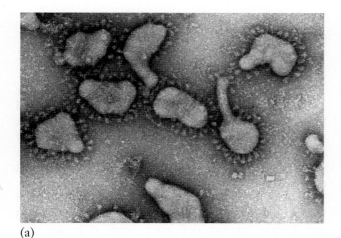

(a)

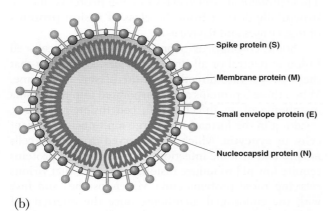

- Spike protein (S)
- Membrane protein (M)
- Small envelope protein (E)
- Nucleocapsid protein (N)

(b)

Figure 14.2 Coronavirus virion. (a) Electron micrograph of negatively stained murine hepatitis virions shows spikes protruding from the envelope, providing its crown-like appearance. (b) Schematic diagram. The three envelope proteins spike (S), membrane (M), and small envelope (E) are shown. The hemagglutinin-esterase (HE) is only present in some coronaviruses and is not shown here. The helical nucleocapsid, made from the RNA genome complexed with N protein, is shown just underneath the envelope. Some coronaviruses have a spherical core.

nucleocapsids with icosahedral symmetry; coronaviruses thus resemble negative-strand RNA viruses in having helical nucleocapsids. However, there is evidence that mature virions of some coronaviruses, including porcine transmissible gastroenteritis virus, possess a stable spherical core structure. This core is associated with the virus membrane protein (M). These coronaviruses may be unique among known viruses in having both a spherical core shell (probably with icosahedral symmetry) and an internal helical nucleocapsid.

Coronavirus virions contain multiple envelope proteins

Spike protein. The spike (S) glycoprotein is the major surface transmembrane protein protruding from the viral envelope (Figure 14.2b). S forms trimers whose stalk-and-club morphology gives the virion a crown-like fringe of proteins. The S protein is responsible for virus entry and spread, and is the primary determinant of host range. It is a target for neutralizing antibody and cytotoxic T-lymphocyte responses to virus infection. Because of this, S has been a focus of much research.

The S protein is synthesized as a single polypeptide chain that in some coronaviruses is cleaved by a cellular proteinase to yield an amino-terminal S1 domain and a carboxy-terminal S2 domain. The S1 subunit, which forms the clubbed end of the spike, recognizes specific cellular receptors and initiates attachment. The receptor-binding domain (RBD) is found within S1 in most coronaviruses. This is the smallest region that can mediate interaction with the cellular molecule that serves as receptor. The receptor-binding domain has been mapped for well-studied coronaviruses such as murine hepatitis virus and SARS coronavirus, but the exact location of this domain within the S proteins of other coronaviruses has not yet been determined.

The S2 subunit, which forms the stalk, has a short carboxy-terminal tail, a hydrophobic transmembrane domain, and an exterior domain with two long alpha-helical regions. Adjacent alpha helices within the trimer interact with each other to form a structure called a **coiled-coil**, similar to the transmembrane portion of the envelope glycoproteins of a number of other RNA viruses.

Recombination and other studies indicate that S is possibly a modular protein and that domains might be exchanged with relative ease during mixed infections with distinct coronaviruses. This could be one explanation for the potentially low barrier to species jumping by coronaviruses; if the receptor-binding domain is capable of being exchanged easily during recombination, the receptor specificity of the recombinant virus and therefore its target cell type could quickly adapt to a new species.

Hemagglutinin-esterase protein. Some coronaviruses, notably betacoronaviruses, possess a hemagglutinin-esterase (HE) protein on the surface of the virion. HE has functional similarities to two influenza virus proteins: the receptor-binding hemagglutinin protein (HA) and the receptor-destroying enzyme neuraminidase (NA) (see Chapter 18). Since not all coronaviruses express HE, it is not considered an essential structural protein. Nevertheless, HE may play a role in binding and entry of coronaviruses into target cells, and could facilitate virus release from the cell surface.

Membrane protein. The membrane (M) glycoprotein is the most abundant structural protein in the coronavirus virion. M is a highly conserved, triple membrane-spanning protein containing three hydrophobic domains separated by short hydrophilic regions. Rigid constraints on structure are likely essential for the functional requirements of M in virus assembly. The amino terminus of M is exposed on the outer surface of the envelope and is in contact with the spike protein; the carboxy terminus extends within the virion underneath the envelope, and is in contact with the viral nucleocapsid.

Small envelope protein. The small envelope (E) protein is a 9–12 kDa protein that is present only in very limited numbers in the virion. Virus-like particles can be formed by expressing only the E and M proteins together in cells, suggesting an important role for E in virion formation and budding. The E protein may serve to facilitate budding of the virion into the lumen of the endoplasmic reticulum and Golgi membranes during particle maturation (see below). Interestingly, mutant coronaviruses lacking E are viable, indicating that this protein is not required for virion formation.

Coronavirus spike proteins bind to a variety of cellular receptors

The S proteins of several alphacoronaviruses are known to bind to aminopeptidase-N, a family of zinc-binding **metalloproteinases** found in cellular plasma membranes. These are broadly distributed in epithelial and fibroblast cells in the small intestine, renal epithelium, and central nervous system, and on monocyte and granulocyte lineages. Despite the high degree of similarity among members of the aminopeptidase-N family, the utilization of these receptors by coronaviruses is for the most part species-specific.

Betacoronaviruses utilize a variety of cell receptors. Murine hepatitis virus binds to carcinoembryonic antigen-related cell adhesion molecule 1 (CEACAM1 or CD66a). CEACAM1 is a member of the immunoglobulin family and is widely distributed in all mammalian species. SARS coronavirus uses angiotensin-converting enzyme 2 (ACE2), another metalloproteinase, as a receptor. Because SARS-CoV can also infect certain cell types that lack ACE2, additional or alternate receptors are being sought. Recent studies indicate that L-SIGN, a cell surface **lectin**, may also serve as a receptor for the SARS coronavirus. NL63, a recently discovered alphacoronavirus, also binds to ACE2.

Viruses that express HE can also bind to cells by interacting with **sialic acid**, found on a variety of cell membrane glycoproteins and glycolipids. Sialic acid is a common target bound by surface proteins of many animal viruses. Coronavirus HE can cleave an acetyl group from sialic acid, resulting in the release of bound virus from the cell surface. This may help in the release and dispersal of virions from some cell types.

The virus envelope fuses with the plasma membrane or an endosomal membrane

The coronavirus S protein mediates attachment of the virion to the cell surface by interacting with a cell-surface receptor molecule, followed by fusion of the virus envelope with cellular membranes. While the external S1 subunit is responsible for attachment, the S2 subunit that forms the stalk facilitates fusion through a series of conformational changes that result in the insertion of S2 into the target cell membrane, bringing the cell membrane and the viral envelope into close contact. This mechanism appears to parallel the better-studied fusion mechanisms of influenza virus (Chapters 4 and 18) and human immunodeficiency virus (Chapter 29). These proteins are called **class I fusion proteins**, and are structurally distinct from the **class II fusion proteins** of togaviruses and flaviviruses.

Some coronavirus S proteins can induce cell-cell fusion at neutral or alkaline pH values, suggesting that these virions fuse directly with the plasma membrane. When these S proteins are incorporated into the plasma membrane of an infected cell, cell fusion can occur, resulting in the formation of large multinucleated giant cells, or **syncytia**, which are observed in infected cells *in vitro* and in some infections *in vivo*. Other S proteins require low pH to induce fusion, suggesting that virions carrying these proteins enter via endosomes and fuse with the endosomal membrane once the internal pH has dropped.

The replicase gene is translated from genome RNA into a polyprotein that is processed by viral proteinases

Because coronaviruses have positive-strand RNA genomes, the first step in their replication cycle after entry is the synthesis of viral proteins that organize and catalyze viral RNA synthesis. All known coronaviruses have a similar genome organization, in which the 20–22 kb replicase gene (gene 1) comprises the 5' two-thirds of the RNA genome. This gene is organized as two partially overlapping open reading frames, ORF1a and ORF1b (Figure 14.1). Translation by cellular ribosomes begins at an AUG initiation codon following the 60–100 nt untranslated leader sequence and progresses through ORF1a, producing a polyprotein that is subsequently cleaved into at least two intermediate precursors and subsequently into 10 or 11 mature proteins.

Some of the ribosomes translating ORF1a pause on a complex RNA structure known as a **pseudoknot** in the overlap between ORF1a and 1b, just upstream of the translation termination codon at the end of ORF1a. This leads to a shift in the reading frame, allowing continued translation through ORF1b and production of a larger ORF1ab polyprotein, which is cleaved into as many as 16 mature proteins. About four out of 10 ribosomes successfully navigate the pseudoknot structure and therefore terminate at the end of ORF1a, resulting in a lower abundance of proteins in the ORF1b region, most of which are involved in viral RNA replication and modification. ORF1b is one of the most highly conserved regions in the coronavirus genome, underscoring the importance of the proteins coded in this region.

Coronavirus polyproteins are processed by virus-encoded proteinases at cleavage sites flanking each mature protein domain. These proteinases must act autocatalytically at least initially since they are part of the polyprotein, and therefore they must be enzymatically active before they are released by self-cleavage. Polyprotein synthesis and processing is a common expression strategy among other positive-strand RNA viruses including picornaviruses, flaviviruses, and togaviruses (Chapters 11 through 13).

The viral proteinases include one or two **papain-like cysteine proteinases** (PLP1 and PLP2) and a chymotrypsin-like cysteine proteinase that resembles the **picornavirus 3C proteinases** (nsp5, also known as 3CLpro or Mpro) (Figure 14.1). The most consistent nomenclature for the replicase protein domains is "nonstructural protein" (nsp), although possible roles these proteins might play in virion structure have not been ruled out. SARS coronavirus and murine hepatitis virus have analogous replicase proteins that will be referred to as nsp1 through nsp16 (Figure 14.1 and Table 14.2).

The papain-like proteinases (PLPs) are subdomains of nsp3, a 195–213 kDa replicase protein (Figure 14.1). The PLPs catalyze cleavage between nsp1–nsp2, nsp2–nsp3, and nsp3–nsp4. Interestingly, cleavage between nsp1–nsp2 and nsp2–nsp3 is not required for viral replication in cultured cells, as these cleavage sites can be deleted without drastic effects on virus replication. In addition to its proteinase activity, nsp3 is likely involved in the formation of replication complexes, in viral RNA synthesis, and in suppression of interferon activity.

The remaining portions of the polyproteins are processed by the nsp5 proteinase, producing nsp4 through nsp16 in all coronaviruses. The activity of nsp5 is thought to be regulated by cleavage site specificity as well as by the flanking nsp4 and nsp6 proteins, both of which are hydrophobic membrane-spanning proteins that are likely membrane-associated. The crystal structure of nsp5 for several coronaviruses has recently been solved, and it is a target for development of antiviral agents specific for SARS coronavirus.

RNA polymerase, RNA helicase, and RNA-modifying enzymes are encoded by the replicase gene

The coronavirus replicase gene encodes multiple proteins that are known or predicted to direct viral RNA synthesis or to interact with or modify RNAs (Table 14.2). The predicted RNA-dependent RNA polymerase (nsp12) is a 100-kDa protein encoded in ORF1b, and a second, non-canonical RNA polymerase has been identified in nsp8. ATPase and RNA helicase activities of nsp13 may be critical in replication and packaging of genome RNA.

Proteins with RNA-modifying activity have been demonstrated *in vitro* for nsp14 (exoribonuclease and cap N7 methyltransferase), nsp15 (endoribonuclease), and nsp16 (methyltransferase, which methylates the 5' cap structure). The coordinated activities of nsps in RNA synthesis and modification remain to be shown during virus infection, but mutagenesis studies have confirmed the importance of these proteins for replication. Recently, nsp14-exonuclease mutants of murine hepatitis virus were shown to have a mutator phenotype and profoundly increased mutation rates, consistent with an RNA proofreading function in nsp14.

The functions of other nsps in replication are less well studied, but there are data suggesting roles for these proteins in virulence (nsp1), cell-cycle control (nsp1), cell membrane modification, and replication complex formation (nsp4, nsp6). When the nsp2 protein was deleted from the polyproteins of both murine hepatitis virus and SARS coronavirus, viable mutants were recovered, demonstrating that nsp2 is not required for virus replication. Thus, a large range of protein functions important for replication have been assembled and incorporated as the result of the plasticity and adaptability of the coronavirus genome.

Replication complexes are associated with cytoplasmic membranes

Membrane association of viral RNA synthesis is a common pattern among positive-strand RNA viruses of eukaryotes (see Chapters 10 through 13). All coronavirus replicase proteins tested can be found in membrane-associated complexes in the cytoplasm. These membranes are the sites of viral RNA synthesis, and are thus referred to as **replication complexes**.

Infection of cells by coronaviruses induces dramatic rearrangement of the subcellular architecture, most notably of cytoplasmic membranes. Replication complexes are often observed on double-membrane vesicles by electron microscopy (Figure 14.3). The membranes in these structures most likely originate from the endoplasmic reticulum, although other membranes have been implicated. In addition, structures not usually found in uninfected cells, such as convoluted membranes, membrane whorls and multivesicular bodies, have been observed in coronavirus-infected cells.

The virus-induced membrane changes have been described as a "reticulovesicular network" of interacting and evolving structures over the course of

Table 14.2 Coronavirus proteins and functions

Protein	Functions
nsp1	Unknown functions in replication; may mediate cell-cycle arrest during over-expression in culture; SARS nsp1 is an interferon antagonist.
nsp2	Unknown, dispensable for replication in cultured cells of murine hepatitis virus and SARS coronavirus.
nsp2–3	Intermediate product, timing of exponential phase growth in culture.
nsp3	One or two papain-like proteinase domains (PLP1, PLP2) responsible for cleavage of nsp1, nsp2, and nsp3; zinc ribbon motifs with predicted transcription factors; transmembrane sequences with membrane integration; PLP2 and homologs with ubiquitinase activity; interferon antagonist.
nsp4–10	Intermediate product, unknown function.
nsp4, nsp6	Membrane-spanning proteins, may localize replication complexes to membranes; glycosylation of nsp4 important for replication complex formation.
nsp5	Chymotrypsin-like proteinase (3CLpro or Mpro) responsible for cleavage of nsp4 through nsp16 in all coronaviruses.
nsp7, 9, 10	Unknown function, possible hetero- and homo-oligomeric structures.
nsp8	Primase, polymerase.
nsp12	RNA-dependent RNA polymerase.
nsp13	RNA helicase, nucleoside triphosphatase activity *in vitro*. Likely involved in genome unwinding, separation, and packaging; may be virulence factor.
nsp14	3' to 5' exoribonuclease; ExoN mutants of murine hepatitis virus and SARS coronavirus with mutator phenotype; possible RNA proofreading; N7 cap methyltransferase.
nsp15	Endoribonuclease.
nsp16	O-methyl transferase.
Hemagglutinin-esterase (HE)	Envelope protein expressed by some but not all coronaviruses. Virus entry, pathogenesis, virus release.
Spike (S)	Virion structural protein; receptor binding, fusion, virus entry, host range, tropism, immune response, syncytia formation.
Envelope (E)	Virion structural protein; envelope formation and stability.
Membrane (M)	Virion structural protein that contacts both spike and nucleocapsid protein. Required for virion formation.
Nucleocapsid (N)	Structural protein, forms helical nucleocapsid with genome RNA. Interacts with M protein; may form icosahedral shell.
Accessory or group-specific proteins	Three to five proteins, dispensable for replication *in vitro*, and variable between different coronaviruses. Most functions unknown, probable functions in pathogenesis. SARS ORF6: interferon antagonism.

infection. The formation of these unusual structures likely involves both stimulation of cellular processes to synthesize new membranes and modification of existing membranes by viral replicase proteins. The nsps 3, 4, and 6 are integral membrane proteins and are proposed to nucleate the formation of replication complexes and possibly induce membrane modifications. Nucleocapsid protein is also found in abundance in replication complexes, suggesting that encapsidation of newly synthesized genome RNA for packaging into virions also occurs at these sites.

Genome replication proceeds via a full-length, negative-strand intermediate

The proteins in replication complexes direct both genome replication and transcription, the production of viral mRNAs. Genome replication requires the synthesis of a full-length, negative-strand copy of the viral genome, which in turn is used to direct synthesis of full-length, positive-strand genome RNAs. Negative-strand templates account for only 1 to 2% of total viral RNA produced in an infected

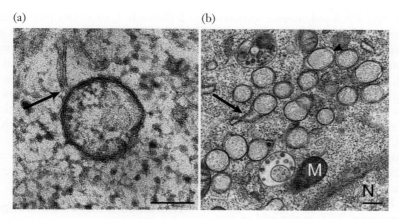

Figure 14.3 Membrane structures induced by SARS-coronavirus infection. Electron micrographs of Vero E6 cells infected with SARS-coronavirus. The cells were cryofixed and freeze substituted at 2 h (a) or 8 h (b) after infection. (a) An early double-membrane vesicle showing a connection (arrow) to a reticular membrane. (b) A cluster of double-membrane vesicles, showing a connection between the outer membranes and reticular membrane structures (arrow). M: mitochondrion; N: nucleus. (Scale bars represent 100 nm (a) or 250 nm (b)).

cell. Most negative-strand RNAs are associated with one or more growing positive-strand molecules in a **replicative intermediate**.

Transcription produces a nested set of subgenomic mRNAs

The expression of all coronavirus genes downstream of the replicase gene occurs from a series of **subgenomic mRNAs** (Figure 14.4). These mRNAs have a unique structure. The sequence of the first 60 to 100 nucleotides starting from their capped 5' ends is identical to the same region at the 5' end of genome RNA. This untranslated "leader" sequence is joined to an mRNA "body" sequence that includes one or more of the open reading frames in genes 2 through 7 (for murine hepatitis virus). All six of these subgenomic RNAs extend to the 3' end of the genome RNA and, like genome RNA, have poly(A) tails. Because they have different lengths

but share common sequences at their 3' ends, these are referred to as a **nested set** of mRNAs. Each one of the subgenomic mRNAs is translated to produce one or more proteins coded by the open reading frames closest to the 5' end of the mRNA. These include the virion envelope proteins HE, S, E, and M, and the nucleocapsid protein N, as well as several nonstructural or accessory proteins that are distinct to each virus.

Subgenomic mRNAs are transcribed from subgenomic negative-sense RNA templates made by discontinuous transcription

Experimental evidence supports a model in which each subgenomic mRNA is transcribed from a corresponding subgenomic negative-strand RNA. The negative-strand subgenomic RNA templates are themselves generated by **discontinuous transcription** of full-length positive-strand genome RNA. This model is outlined in

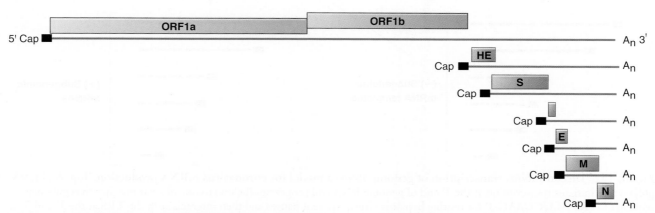

Figure 14.4 Coronavirus messenger RNAs: a nested set. The seven messenger RNAs of murine hepatitis virus are shown, along with the open reading frames they use to synthesize viral proteins. All share the same 5' leader sequence and all contain overlapping sequences at their 3' ends.

Figure 14.5. Briefly, the viral RNA polymerase initiates transcription at the 3' end of (positive-strand) genome RNA and makes a (negative-strand) RNA copy that stops at the end of a transcription-regulating sequence (TRS). These repeated, short TRS (5'-UCUAAAC-3' for murine hepatitis virus) are located at the 5' ends of each of genes 2 through 7; there is an additional TRS at the 3' end of the leader sequence in genome RNA.

At this point the RNA polymerase pauses, dissociates the nascent RNA chain from the TRS, and brings the nascent chain to the TRS located at the end of the leader sequence in the template RNA. The copy of the TRS at the 3' end of the nascent RNA chain forms an RNA-RNA hybrid with the complementary TRS at the end of the leader sequence on the genome RNA. The nascent RNA therefore serves as a primer, allowing the RNA polymerase to extend the nascent negative-strand RNA through the leader sequence up to the 5' end of the genome. This completes the synthesis of the subgenomic RNA.

As the RNA polymerase traverses the genome from right to left (Figure 14.5), it can pause and dissociate at any one of the TRSs located at the beginning of genes

2 through 7, or it can transcribe through all six TRSs and continue to the end of the genome, making a full-length, negative-strand antigenome RNA. As a result, a nested set of negative-strand RNAs with common 5' ends but different lengths is produced. Full-length negative-strand RNA (not shown in Figure 14.5) can serve as a template for production of genome RNA by the viral RNA polymerase. Each of the negative-strand subgenomic RNAs is copied by viral RNA polymerase to generate a nested set of positive-strand subgenomic mRNAs (bottom right of Figure 14.5).

The discontinuous transcription model can explain recombination between viral genomes

Discontinuous RNA synthesis also can occur by the viral RNA polymerase switching between two different positive-strand RNA genome templates within the same cell. If these two genomes are not identical, because of mutation or because two virus strains infected the same cell, the result can be homologous recombination. This

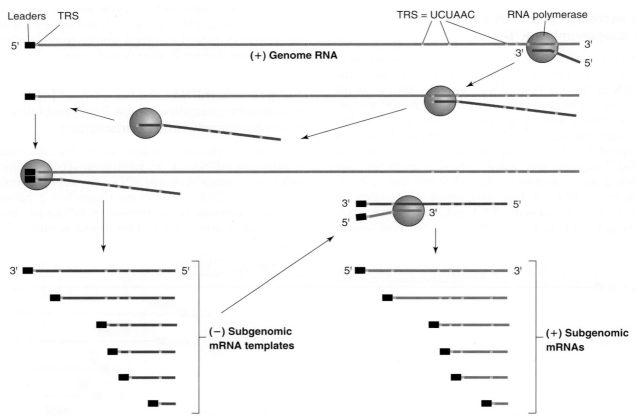

Figure 14.5 Discontinuous transcription of genome RNA: a model for coronavirus mRNA production. Top: viral RNA polymerase initiates transcription at the 3' end of genome RNA and progresses (below) to one of six transcription regulatory sequences (TRS: 5'-UCUAAC-3' for murine hepatitis virus), where it pauses and then interacts with the TRS at the 3' end of the leader sequence. The two RNAs base-pair at the TRS and the RNA polymerase extends the RNA chain to the 5' end of genome RNA, copying the leader sequence. Depending on where the RNA polymerase pauses, the product is either full-length antigenome RNA (no pause; not shown) or one of the six subgenomic negative-strand RNAs shown at bottom left. These are in turn copied by RNA polymerase to make more full-length genome RNA (not shown) or the six subgenomic mRNAs.

"template switching" may be a normal feature of coronavirus replication and may be important for genome repair and generation of new strains of coronaviruses. It is reminiscent of similar recombination events between retrovirus genomes that can occur during reverse transcription (Chapter 28).

Assembly of virions takes place at intracellular membrane structures

Once genome replication and transcription have taken place, the action shifts to a different set of intracellular membranes, the **endoplasmic reticulum–Golgi intermediate compartment (ERGIC)**. These membranes, located in the perinuclear region of the cell, lie in between the endoplasmic reticulum and the Golgi apparatus, and are part of an interconnecting system of membranes involved in modification and secretion of cellular proteins.

Virions are assembled at this membrane system (Figure 14.6). Helical nucleocapsids containing genome RNA are delivered from their site of synthesis to these membranes for packaging. While the viral and cellular determinants of this delivery system are not understood, the nucleocapsid protein and several replicase proteins of murine hepatitis virus have been detected at assembly sites in ERGIC. It remains to be determined if these are distinct compartments with a required process of delivery, or whether sites of replication and assembly are part of an interconnected and continuous replication/assembly network.

Electron micrographs of infected cells show densely staining material, probably helical nucleocapsids, lying on the cytosolic face of membranes in the intermediate compartment. Virus particles can be seen forming by budding into the lumen of these membranes. These are spherical particles that usually have a doughnut-like core morphology and are larger than mature virions (Figure 14.6, bottom).

The M and E proteins play important roles in the formation of virus envelopes by budding. Enveloped virus-like particles can be formed in the intermediate compartment when only M and E are expressed together in cells, indicating that these proteins are sufficient for assembly and budding. The carboxy-terminal cytoplasmic tail of M is thought to interact with packaging signals in N to ensure specific incorporation of only full-length viral genome RNA into virions.

HE and S are incorporated into the membrane through interactions with the M protein. These immature virions are transported within vesicles through the Golgi apparatus, where the envelope proteins are glycosylated. As they mature, virions become smaller and the doughnut-shaped core becomes more compact and dense. From the *trans*-Golgi network, mature virions are packaged into vesicles that are targeted to the plasma membrane for release of progeny virus.

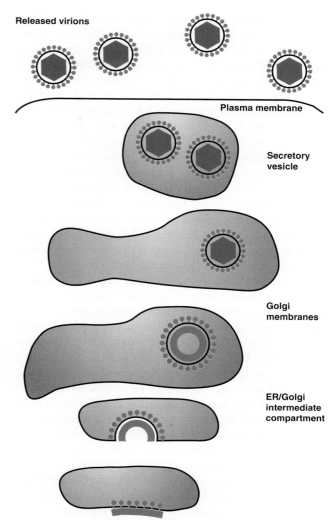

Figure 14.6 Assembly, maturation, and release of coronavirus virions. Shown schematically is the pathway of virion assembly at membranes of the endoplasmic reticulum–Golgi intermediate compartment (below): the thick blue curved lines represent helical nucleocapsids, and the orange knobbed stalks represent envelope proteins. As they bud into the lumen, virions acquire donut-shaped cores: These progress to a more uniformly dense and smaller core as virions go through the Golgi membranes, where glycosylation of envelope proteins occurs. Secretory vesicles transport virions to the cell surface, where the vesicles fuse with the plasma membrane and release virions.

Adaptability and evolution of coronaviruses

The spectrum of diseases observed in coronavirus infections reflects the capacity of these viruses to adapt to changing environments and to new hosts. The recent discovery in bats of many new coronaviruses belonging to two groups of mammalian coronaviruses (alpha and beta) has redefined the spectrum of coronaviruses and raises the intriguing possibility of bats as evolutionary pools for new coronaviruses in other animals.

The capacity of coronaviruses to undergo RNA-RNA recombination, coupled with the high theoretical error rate of the viral RNA polymerase (1 mutation in every 10,000 nucleotides), generates coronavirus genomes with changes that may be selectively advantageous in new environments or hosts. The recent discovery that coronaviruses may regulate replication fidelity suggests they may have novel mechanisms for adaptation to new hosts.

Most coronaviruses show strong species, organ, and tissue specificity, but subtle genetic changes can greatly alter their tissue **tropism**, host range, and pathogenicity. This has been experimentally demonstrated for murine hepatitis virus. Coculture of different strains of murine hepatitis virus in mixed populations of cell types generated variant viruses capable of replicating in previously non-permissive cell types including human cell lines, thus changing host range *in vitro*.

Genetic alterations in a virus can also result in changes in tissue tropism. Typically, transmissible gastroenteritis virus causes digestive tract disease in young swine. A variant form of this virus replicates preferentially in the respiratory tract. Genetic analysis has correlated this change in tropism to partial deletion of a structural protein and alterations in an accessory gene.

While some mutations may result in attenuated disease, others may increase virulence and disease. This is best illustrated by the conversion of the benign feline enteric coronavirus into lethal feline infectious peritonitis virus. These two viruses are serologically and morphologically identical. Spontaneous mutations in feline enteric coronavirus accessory proteins alter the virulence, resulting in severe immunopathology that is almost always lethal. Emergence of the SARS coronavirus provided a dramatic demonstration of the adaptability of coronaviruses, and further showed that they can cross the species barrier with sometimes devastating results.

introduce mutations in the viral genome. These include cloning a full-length cDNA in a bacterial artificial chromosome or in a recombinant vaccinia virus. Another strategy, called "targeted recombination", uses the natural recombination capacity of coronaviruses to introduce changes along with a selectable gene.

A novel reverse genetic approach that has overcome the complexity, size, and toxicity of the coronavirus RNA genome is referred to as "*in vitro* assembly" (Figure 14.7). In this system, the genome is cloned in bacterial plasmids as a set of six or seven distinct cDNA fragments. Genes within these fragments can be mutated independently, creating libraries of mutated fragments. The cDNA fragments are excised from plasmids by cleavage with restriction endonucleases, and ligated *in vitro* into a full-length cDNA. This assembled cDNA is transcribed *in vitro* by T7 RNA polymerase to produce a full-length genome RNA, which is introduced into permissive cells by electroporation. Because the genome RNA is infectious, viable mutant viruses are produced. This approach allows the testing of "mix and match" libraries of mutations in different genes to determine phenotypic effects.

Recently, a combination of reverse genetics, commercial cDNA synthesis and rational design was used to recover the first bat coronavirus to replicate in culture. The bat "SARS-like" coronavirus sequence was available from direct sequencing of bat secretions, but virus could not be grown in culture, and thus no virus or RNA was available. The sequence was used to design fragments that were commercially synthesized. DNA coding for the SARS receptor-binding domain was inserted in the bat virus spike sequence. The fragments were used to generate full-length cDNA and genome RNA for electroporation, allowing recovery of the replication competent bat/SARS virus.

The combination of these approaches will allow testing of mechanisms of coronavirus host species movement and adaptation. This approach also will allow rational design of stably attenuated viruses as possible vaccine candidates.

Making new and mutant coronaviruses: reverse genetics and synthetic biology

The ability to introduce specific mutations into a viral genome is a powerful tool to define functions of viral proteins in replication, pathogenesis, and virulence. For positive-strand RNA viruses, mutagenesis requires a reverse genetic approach, in which the RNA genome is cloned as a complementary DNA (cDNA) in a plasmid vector. Mutations are then introduced in the cDNA, the mutant RNA genome is transcribed from the mutated cDNA, and the infectious RNA is introduced into cells to generate infectious mutant viruses.

The extremely large and complex coronavirus RNA genome, and the presence of genome regions resistant to stable cloning, have been formidable obstacles to cDNA cloning and mutagenesis. In the past 10 years, however, several approaches have been successfully used to

KEY TERMS

Bronchitis	Papain-like cysteine proteinases
Class I fusion proteins	
Class II fusion proteins	Peritonitis
Coiled-coil	Picornavirus 3C proteinases
Discontinuous transcription	
Encephalitis	Polyprotein
Endoplasmic reticulum–Golgi intermediate compartment (ERGIC)	Pseudoknot
	Replication complexes
	Replicative intermediate
Gastroenteritis	Reservoir
Hepatitis	Reverse genetics
Hypoxemia	Sialic acid
Lectin	Subgenomic mRNA
Metalloproteinases	Syncytium
Nested set	Tropism

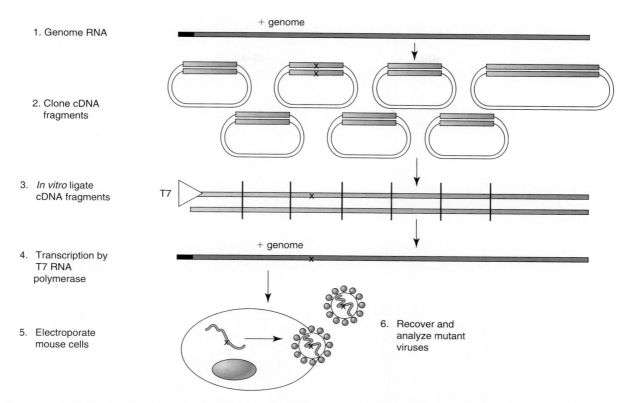

1. Genome RNA

2. Clone cDNA fragments

3. *In vitro* ligate cDNA fragments

4. Transcription by T7 RNA polymerase

5. Electroporate mouse cells

6. Recover and analyze mutant viruses

Figure 14.7 *In vitro* **assembly of genome cDNA and production of infectious genome RNA.** See text for details.

FUNDAMENTAL CONCEPTS

- Coronaviruses are responsible for common colds, a number of important veterinary diseases, and an outbreak of severe acute respiratory syndrome (SARS).
- Coronaviruses have the largest known genomes of RNA viruses (up to 32 kb).
- Coronaviruses have helical nucleocapsids, unusual for positive-strand RNA viruses.
- Some coronaviruses also appear to have an additional spherical capsid.
- The replicase genes are translated to produce two overlapping polyproteins, which are cleaved to generate up to 16 distinct viral proteins.
- All viral structural proteins are made from a 3'-coterminal nested set of subgenomic messenger RNAs.
- These subgenomic messenger RNAs are copies of subgenomic RNAs transcribed from genome RNA by an unusual discontinuous transcription mechanism.
- Like other positive-strand RNA viruses, coronaviruses induce intracellular membrane proliferation and viral RNA replication takes place in association with these membranes.
- Coronavirus virions are assembled within the endoplasmic reticulum/Golgi intermediate compartment and secreted via vesicles that fuse with the cell membrane.

REVIEW QUESTIONS

1. What is the current theory on the origin of SARS and perhaps all human coronaviruses?

2. Which coronavirus protein is associated with syncytia formation?

3. Control of transcription in coronaviruses generates a series of subgenomic mRNAs. How are these generated?

4. What aspects of coronavirus replication most likely contribute to their ability to rapidly change host range, possible even *in vitro*?

SECTION IV

NEGATIVE-STRAND AND DOUBLE-STRANDED RNA VIRUSES OF EUKARYOTES

The single-stranded RNA genomes of negative-strand RNA viruses cannot be translated, because they are complementary to the positive-strand mRNAs that encode viral proteins. Therefore, these viruses must bring along with the viral genome an RNA-dependent RNA polymerase that enables them to produce viral messenger RNAs. This enzyme is packaged within the virion.

Unlike positive-stand RNA viruses, the genomes of negative-strand RNA viruses always remain in the form of a ribonucleoprotein complex, which is usually constructed with a single major capsid protein that forms a helical nucleocapsid enclosing the RNA. This design has two advantages. First, it allows encapsidation of RNA genomes of different lengths; several families have segmented genomes that vary in size. Second, it provides a means of distinguishing mRNAs, which are not encapsidated, from genome-length RNAs of either positive or negative polarity.

Negative-strand RNA viruses have not been found in bacteria, archaea, fungi, or algae, and only a few are known to infect vascular plants. Some of the most widespread and dangerous human diseases are caused by negative-strand RNA viruses; among these are influenza, measles, rabies encephalitis, and Ebolavirus hemorrhagic fever. Highly successful vaccines help control the spread of measles, mumps, rabies, and influenza viruses.

Viruses with a double-stranded RNA genome share with negative-stranded RNA viruses two important characteristics: they package an RNA polymerase along with genome RNA, and they do not fully uncoat the genome RNA within the cell. These viruses have a double- or triple-layered capsid with icosahedral symmetry; the outer layer is removed during entry into the cell, activating transcription in the core. Messenger RNAs synthesized within the core are extruded into the cytoplasm of the infected cell, where they are translated and direct viral replication. These viruses usually have multiple genome fragments (up to 12).

Rotaviruses are a group of double-stranded RNA viruses responsible for millions of cases of infantile gastroenteritis, which each year causes the death of hundreds of thousands of young children in third-world countries where access to medical care is limited. An effective rotavirus vaccine has recently been developed.

CHAPTER 15

Paramyxoviruses and Rhabdoviruses

Nicholas H. Acheson
Daniel Kolakofsky
Christopher Richardson
Revised by: Laurent Roux

Paramyxoviridae

From Greek *para* (alternate) and *myxa* (mucus): virions bind to mucoproteins

Rhabdoviridae

From Greek *rhabdo* (rod), referring to shape of virion

VIRION

Paramyxoviruses: spherical, diameter 150–300 nm, also found as filaments.
Rhabdoviruses: bullet-shaped rods, 180 × 75 nm.
Envelope derived from plasma membrane.
Coiled helical nucleocapsid.

GENES AND PROTEINS

Five to nine genes, transcribed in series from the 3' end of the genome by viral RNA polymerase.
Most genes produce a single mRNA and a single protein.
Most proteins are packaged in the virion:
- *Nucleocapsid protein (N)*
- *RNA polymerase cofactor and accessory proteins (P/C/V)*
- *Matrix protein (M)*
- *Fusion protein (F)*
- *Hemagglutinin/neuraminidase, or envelope glycoprotein (H, HN, or G)*
- *RNA polymerase (L)*

GENOME

Single segment of linear ssRNA, negative sense.
Paramyxoviruses: 15–18 kb.
Rhabdoviruses: 11–12 kb.

VIRUSES AND HOSTS

Paramyxoviruses:
- *Humans: measles, mumps, respiratory syncytial, parainfluenza viruses*
- *Animals: Newcastle disease virus (chickens), canine distemper virus, rinderpest virus (cattle) viruses infecting birds, pigs, dogs, cats, seals, whales, etc.*

Rhabdoviruses:
- *Humans: rabies virus*
- *Cattle: vesicular stomatitis virus*
- *Numerous rhabdoviruses infecting bats, fish, insects, plants*

DISEASES

Paramyxoviruses: several important childhood diseases, including measles, mumps, respiratory diseases in humans; canine distemper, rinderpest.
Emerging paramyxoviruses Hendra and Nipah cause respiratory and neurological disease in pigs, humans, horses, and bats.
Rhabdoviruses: rabies (fatal encephalitis in humans), transmitted by bites of infected animals; vesicular stomatitis virus in cattle. Numerous insect and plant diseases.

DISTINCTIVE CHARACTERISTICS

RNA-dependent RNA polymerase in virion transcribes genome into mRNAs.
A single transcriptional promoter at the 3' end of RNA; mRNAs are made by a start–stop mechanism.
Full-length genome or antigenome RNA is always encoated in nucleocapsid proteins.
Some paramyxoviruses "edit" the P mRNA region at a specific site, generating mRNAs with distinct protein-coding capacities.
Paramyxoviruses can induce cell fusion, producing multinucleated cells (syncytia).

The mononegaviruses: a group of related negative-strand RNA viruses

Among the negative-strand RNA viruses, four families (*Paramyxoviridae, Rhabdoviridae, Filoviridae,* and *Bornaviridae*) share a number of fundamental characteristics:

- Their genome is a single, linear RNA molecule, packaged in a helical nucleocapsid.
- Nucleocapsids are packaged in an envelope derived from the plasma membrane of the cell.
- A virus-coded RNA polymerase packaged in the virion synthesizes viral mRNAs by transcribing the RNA in the nucleocapsid after it enters the cell.
- The RNA polymerase begins transcribing at the 3' end of the genome RNA and sequentially transcribes five to ten genes, terminating and releasing each mRNA before starting the next one.

Because these viruses share the same basic genome structure and replication mechanism, they have been grouped in the order *Mononegavirales* (for "single [RNA fragment], negative-strand virus"). The present chapter covers the paramyxoviruses and rhabdoviruses; filoviruses are discussed in Chapter 16.

Rabies is a fatal human encephalitis caused by a rhabdovirus

Rhabdoviruses were named because of the unusual rod- or bullet-shaped morphology of virions. Numerous different rhabdoviruses infect insects, plants, and vertebrates; this is one of the few virus families whose members infect such a wide variety of hosts. Vesicular stomatitis, an important disease of cattle, is caused by a rhabdovirus that is transmitted by insect bites. Infections of humans by this virus are mild or without symptoms. Vesicular stomatitis virus can replicate in cells cultured *in vitro* from many different species, and it has been extensively studied.

The only rhabdovirus that is of significant concern to human health is rabies virus. Humans are infected by virus-containing saliva from bites of infected animals, often domestic dogs or bats. The virus is carried in the wild by a number of small mammals, including foxes and raccoons. Infected animals suffer from neurological symptoms that can cause tameness or unusually aggressive behavior, leading to direct contact with other animals or humans and thus helping to spread virus infection. In humans, infection results in an **encephalitis** that is invariably fatal once the first symptoms have appeared, although there may be a long latent period between exposure and disease. One individual recently survived rabies encephalitis after drug treatment and induced coma.

Louis Pasteur developed an effective rabies vaccine in the mid-nineteenth century (see Chapter 35), and a modern, safer version is used to immunize veterinarians and other people at risk. Because rabies is slow to develop, people can be successfully vaccinated after exposure to the virus. Vaccination of domestic animals and control of infections in wild animals have reduced the incidence of human rabies infections in developed countries; only about one case per year now occurs in North America. However, there are upward of 30,000 human fatalities per year worldwide, mostly in less developed countries.

Measles is a serious childhood disease caused by a paramyxovirus

Measles virus was probably a newly emerging virus some 6000 years ago during the urbanization stage of human civilization. It is closely related to the virus that causes rinderpest, a disease of cattle, and it may have jumped from cattle herds to humans. The earliest known written reference to measles is by an Arab physician, Rhazes of Bagdhad, in the tenth century. Epidemics were documented during the Saracen invasions of the eighth century and continued into the Middle Ages. The virus ravaged native populations in North and South America when introduced by European explorers and colonists. The virus is efficiently transmitted in aerosol droplets derived from coughs and sneezes.

Peter Panum, a Danish physician, was the first researcher to realize (in 1846) that a measles infection conveyed lifelong immunity. In 1911, Goldberger and Anderson succeeded in transmitting the virus to macaques using filtrates prepared from the blood of infected individuals.

John Enders and his team at Harvard University made a major breakthrough in 1954, when human embryonic kidney cells cultured *in vitro* were used to propagate measles virus from a young patient named David Edmonston. The virus was subsequently adapted to monkey kidney cells and chick embryo fibroblasts, and served as the basis of the Edmonston live virus vaccine, introduced in 1961 (see Chapter 35).

Vaccination has led to a decrease from 4 million cases per year in the United States to less than 500 cases per year. However, measles remains a major problem in the developing countries of Africa and South America. There are still about 40 million cases worldwide per year resulting in nearly 1 million deaths.

A rare complication of measles virus infection is a progressive, fatal neurological disease called **subacute sclerosing panencephalitis (SSPE)**. Only one in 300,000 measles victims are affected by SSPE, which seems to result from a slow, persistent infection and strikes up to 10 years after the acute illness.

Other paramyxoviruses cause several important human diseases

Mumps virus, human parainfluenza viruses, and respiratory syncytial virus are paramyxoviruses that cause serious systemic and respiratory infections, predominantly in children. Other paramyxoviruses cause canine distemper (like rinderpest, closely related to measles virus) and diseases of mice, chickens, turkeys, seals, and dolphins.

Recent epidemics in Southeast Asia and Australia with a number of human fatalities were caused by two newly discovered paramyxoviruses named Nipah and Hendra. These viruses may have emerged as a result of clearing forests for lumbering and farming; they apparently were transmitted from infected wild fruit bats to domestic livestock and then to humans. This has raised fears of wider epidemics due to these emerging agents.

This chapter will mostly use examples from the paramyxovirus family. The best-studied paramyxoviruses are Sendai virus (a virus isolated in Sendai, Japan), Newcastle disease virus (a virus that infects chickens), simian virus 5 (in reality, a canine parainfluenza virus), and measles virus. The basic life cycle of rhabdoviruses is very similar to that of paramyxoviruses, justifying their inclusion together in this chapter.

Paramyxovirus and rhabdovirus virions have distinct morphologies

Virions of paramyxoviruses consist of a loose-fitting envelope surrounding a helical nucleocapsid. Electron microscopy reveals roughly spherical virions of size varying from 150 to 350 nm in diameter, with a visible fringe of glycoproteins projecting from the envelope (Figure 15.1a). However, filamentous particles are also made. Detergent treatment of virions releases the nucleocapsids, which can be purified and visualized by electron microscopy (Figure 15.1b).

In contrast to paramyxoviruses, rhabdoviruses have an unusual bullet shape (Figure 15.2), due to wrapping of the helical nucleocapsid in a supercoil or "skeleton" 50 nm in diameter, about 2.5 times the diameter of the nucleocapsid. This coiling is mediated by the rhabdovirus matrix (M) protein. The envelope forms around the coiled nucleocapsid during budding from the plasma membrane, giving rise to the bullet shape. The blunt end of the virion is formed by sealing of the envelope behind the coiled nuclocapsid as the virion is released from the cell membrane.

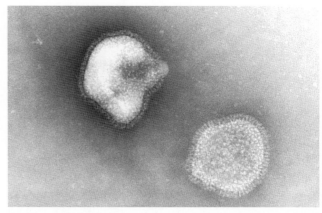

(a)

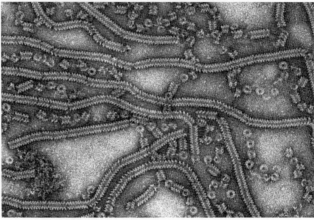

(b)

Figure 15.1 Structure of paramyxoviruses. (a) Electron micrograph of negatively stained virions of simian virus 5, showing fringe of envelope proteins projecting from the surface. (b) Electron micrograph of negatively stained nucleocapsids purified from Sendai virus.

Viral envelope proteins are responsible for receptor binding and fusion with cellular membranes

Figure 15.3 shows a schematic diagram of a typical paramyxovirus virion, emphasizing localization of proteins (not drawn to scale). The virus envelope contains two or three glycoproteins. Measles virus has a **hemagglutinin** (H) and a fusion (F) protein. The H protein is responsible for binding to receptors on the cell surface, and can also bind to receptors on red blood cells and agglutinate them (**hemagglutination**). The hemagglutinin of some other paramyxoviruses also has a **neuraminidase** activity, and is therefore called HN. The F protein directs fusion with the plasma membrane. Some paramyxoviruses make an additional small (40–60 aa) hydophobic envelope protein (SH) of unknown

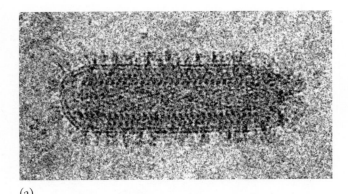

(a)

(b)

Figure 15.2 Structure of rhabdoviruses. (a) Cryoelectron micrograph of a virion of vesicular stomatitis virus. The G protein can be seen extending beyond the envelope, and the coiled nucleocapsid is visible underneath the envelope. (b) Electron micrograph of negatively stained nucleocapsids purified from vesicular stomatitis virus. These are shown in an extended, uncoiled form.

function. SH may form pentamers and create an ion channel in membranes.

The matrix (M) protein lies on the inner surface of the envelope. M is involved in virus assembly and serves as a bridge between the nucleocapsid and the transmembrane envelope glycoproteins.

A closer view of the envelope proteins (Figure 15.4) shows that H or HN is a **type II integral membrane protein**, with a transmembrane domain located near the N-terminus, a long external C-terminal projection, and a short N-terminal sequence facing inside the virion. In contrast, the F protein is a **type I integral membrane protein**, with a long N-terminal domain facing outside and a short C-terminal internal domain. The F protein is cleaved into two subunits (F1 and F2) joined by a disulfide bond on mature virions; this cleavage reveals the hydrophobic **fusion peptide** responsible for virus entry (see below).

The best-studied rhabdovirus, vesicular stomatitis virus, has a single envelope glycoprotein (G), responsible both for receptor binding and fusion activity. Unlike paramyxoviruses, rhabdoviruses enter the cell by endocytosis, and fusion with the endocytic membrane occurs at low pH, releasing the nucleocapsid into the cytoplasm. The vesicular stomatitis virus G protein has been widely used to engineer recombinant enveloped viruses, particularly retroviruses, as it allows them to infect many cell types. This process of expressing a new receptor-binding protein in a recombinant virus is called **pseudotyping**.

Genome RNA is contained within helical nucleocapsids

Genomes of paramyxoviruses contain 15 to 18 kb of single-stranded, negative-sense RNA. Rhabdovirus genomes are somewhat smaller (11–12 kb). The viral N protein forms the helical nucleocapsid along with genome RNA (Figures 15.1b, 15.2b, and 15.5). For Sendai virus and related paramyxoviruses, each N subunit binds to exactly six nucleotides in the genome; thus a 15-kb genome contains about 2500 copies of the N protein.

The nucleocapsid of Sendai virus is wound as a 20-nm-diameter, left-handed helix, with 13 subunits of N protein per turn of the helix, a pitch (distance between adjacent turns) of 5.3 nm, and a central hole 5.5 nm in diameter. In addition to the N protein, about 300 copies of P protein and 50 copies of L protein are packaged with the nucleocapsid; both are required for transcription once the virus enters the cell. Nucleocapsids protect the enclosed viral RNA from ribonuclease digestion. When serving as a template for mRNA synthesis and RNA replication (see below), the RNA genomes of paramyxoviruses and rhabdoviruses always remain associated with N protein as nucleocapsids; naked genome-length RNA is never found within the cell.

Paramyxoviruses enter the cell by fusion with the plasma membrane at neutral pH

Paramyxoviruses recognize a variety of different cellular receptors. Sendai virus and close relatives bind to terminal **sialic acid** residues on **gangliosides** (glycolipids) on the cell surface. The Sendai virus attachment protein (HN) has neuraminidase activity, which, like the influenza virus neuraminidase (Chapter 18), can release progeny virions that bind to the surface of the cell from which they are extruded. This ensures that newly made virions can leave the cell surface and proceed to infect other cells.

Unlike Sendai virus, measles virus H protein has no neuraminidase activity. Measles virus can bind via the H

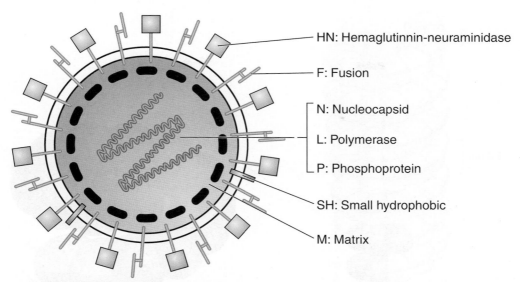

HN: Hemaglutinnin-neuraminidase

F: Fusion

N: Nucleocapsid

L: Polymerase

P: Phosphoprotein

SH: Small hydrophobic

M: Matrix

Figure 15.3 Schematic diagram of paramyxovirus virion. The HN and F proteins are shown projecting from the viral envelope. The matrix protein is shown in black just beneath the envelope, and the helical nucleocapsid is shown in green.

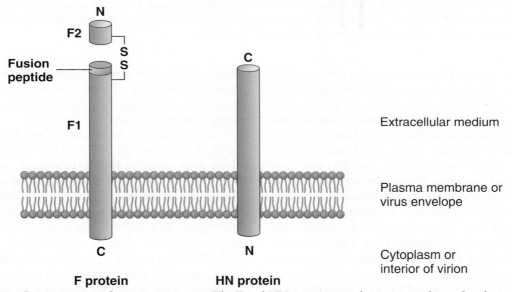

Figure 15.4 Envelope proteins of paramyxoviruses. The F and HN proteins are shown inserted into the plasma membrane of the cell or the virion envelope. C and N: carboxy and amino termini. S-S: cysteine bridge joining the N-terminal region of F1 to F2 after cleavage.

protein to at least two distinct cell surface proteins. One is CD46, a complement-binding protein found on most human cell types. The second is CD150/SLAM (signaling lymphocyte activation molecule), a member of the immunoglobulin superfamily that is expressed on activated **B lymphocytes, T lymphocytes, dendritic cells,** and some monocytes. Signal transduction mediated by CD150/SLAM may regulate T-cell proliferation and **interferon**-γ production. Measles virus infection can lead to strong immunosuppression of the host; infection of cells in the immune system by the virus may play an important role in measles pathogenesis.

Upon binding of the virion to a cell surface receptor, the F1/F2 (fusion) protein undergoes a conformational change at neutral pH that allows insertion of the fusion peptide into the plasma membrane of the cell. This brings the virus envelope close to the cell membrane and leads to fusion of the two membranes, releasing the nucleocapsid in the cytoplasm.

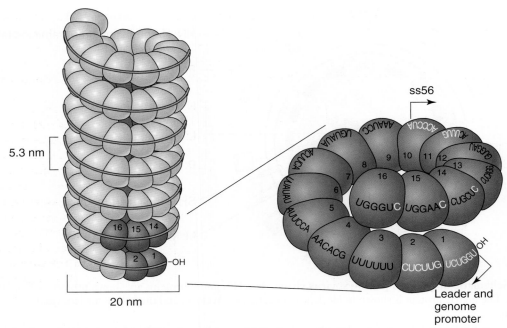

Figure 15.5 Structure of paramyxovirus helical nucleocapsid. Left: a model of Sendai virus nucleocapsid, showing N protein subunits bound to genome RNA (black) in a left-handed helix 20 nm in diameter, with 13 protein subunits per turn and a pitch of 5.3 nm. Right: expanded view of the first 16 N subunits, showing (in white) nucleotide sequences required for initiation of replication and transcription. ss56 is the start site for synthesis of N mRNA.

The exact nature of the conformational change that leads to fusion has not been determined for paramyxoviruses. It is known that parts of the F1 polypeptide chain form elongated, triple-stranded, **coiled-coil** complexes, similar to those formed by fusion proteins of Ebola virus, influenza virus, and human immunodeficiency virus (see Chapters 16, 18, and 29). This coiled-coil structure brings the transmembrane domain, which is anchored in the viral envelope, close to the fusion peptide, which is inserted into the cell membrane, and therefore brings the two membranes in close proximity (see Chapter 4 for a discussion of fusion by influenza HA protein). However, in most cases, the H or HN protein is also required for fusion, implying that F interacts with H or HN to produce the fusion-active state.

Gene order is conserved among different paramyxoviruses and rhabdoviruses

Paramyxovirus and rhabdovirus genomes contain between five and ten genes, each coding for a specific messenger RNA. A map of the Sendai virus genome is shown in Figure 15.6. A region including the first 55 nt at the 3' (left) end of the negative-strand genome RNA is denoted the "leader" sequence. The leader region contains signals for initiation of RNA synthesis by the viral RNA polymerase. Furthermore, its transcript contains signals that direct packaging of full-length plus-strand copies of the viral genome with N protein to generate nucleocapsids. The leader transcript contains no coding region, is not capped or polyadenylated, and does not serve as a viral mRNA.

Immediately downstream of the leader is the N gene, followed by the P/C/V, M, F, HN, and L genes. This gene order is highly conserved, both in paramyxoviruses and in rhabdoviruses (in which the G gene replaces the F and HN genes). Some paramyxoviruses have additional genes. At the junction between each gene is a highly conserved set of "intergenic sequences" that control transcription termination, polyadenylation, and reinitiation. Beyond the end of the L gene is a short (50–150 nt) noncoding "trailer" sequence. Like the leader sequence on genome RNA, the trailer region, located at the 3' end of the antigenome RNA, contains the promoter for genome replication, and its negative-strand transcript contains signals for genome packaging.

Viral messenger RNAs are synthesized by an RNA polymerase packaged in the virion

The RNA of negative-strand viruses cannot be directly translated upon entry into the cell; it must first be transcribed into complementary, positive-sense messenger RNAs. However, there is no cellular RNA-dependent RNA polymerase available for production of viral mRNAs. Therefore, negative-strand RNA viruses

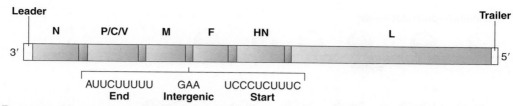

Figure 15.6 Paramyxovirus gene map. The six viral genes are shown on a map of the Sendai virus genome. Below are sequences corresponding to the end of the upstream gene, the intergenic sequences, and the start of the downstream gene.

must make their own RNA polymerase and package at least one copy of this enzyme with the nucleocapsid in the virion.

This emphasizes an important difference in replication strategy between positive-strand and negative-strand RNA viruses. Infection by positive-strand RNA viruses delivers a messenger-sense RNA directly to the cytoplasm, where it can be translated by cellular ribosomes. The translation products include a viral RNA polymerase, which can then replicate the viral genome and synthesize more viral messenger RNAs. These viruses therefore do not package an RNA polymerase in their virions. The positive-sense genome RNA alone, in the absence of any viral proteins, is infectious when introduced artificially into susceptible cells.

In contrast, the genome RNA of negative-strand viruses is not infectious by itself because it requires the viral RNA polymerase to function. This RNA polymerase, which is packaged in the virion and enters the cell along with the genome RNA, uses ribonucleoside triphosphates provided by the cell to catalyze viral mRNA synthesis. For rhabdoviruses and paramyxoviruses, the RNA polymerase is a complex between the P and L proteins.

Viral RNA polymerase initiates transcription exclusively at the 3' end of the viral genome

There are two general models by which viral RNA polymerases could initiate synthesis of viral messenger RNAs on the genomes of mononegaviruses. Each mRNA sequence could be preceded by its own promoter region, which would allow RNA polymerases to bind to each promoter region independently and begin synthesis of the different mRNAs (Figure 15.7a). These promoters could have different strengths, allowing control of the amount of each mRNA produced. Many DNA viruses organize their gene expression in this fashion. Alternatively, all RNA polymerases could begin synthesis at a single promoter at the 3' end of the genome RNA, and progress from right to left through the genome, making the different mRNAs in sequence (Figure 15.7b).

These models can be distinguished by experimental observations. Irradiation of viral genome RNA with ultraviolet (UV) light inhibits transcription by forming dimers between adjacent uridine residues on the template RNA. These dimers block progression of the RNA polymerase along the template, and therefore inhibit all RNA synthesis downstream of the lesion. Dimers can be formed at many places along a genome RNA molecule, more or less at random.

If the multiple-promoter model were correct, the level of inhibition of mRNA synthesis by UV irradiation of the viral genome should depend only on the length of the mRNA being made. This is because the chance of encountering a block to transcription should increase as a function of the distance between the promoter and the end of the mRNA.

If, on the other hand, there were a single entry site for RNA polymerase at the 3' end of the genome, synthesis of mRNAs that lie closest to the 3' end (N and P/C/V) should be least inhibited by UV irradiation. Synthesis of mRNAs that lie more distant from the promoter, toward the 5' end of the genome (HN and L), should be more strongly inhibited, because the polymerase would have to travel through most of the genome before it began to make those mRNAs, and therefore would be more likely to encounter a block.

Careful quantitation of the effect of UV irradiation on synthesis of different viral mRNAs showed that the single-entry model is correct. Messenger RNAs whose genes lie near the (proximal) 3' end are less affected by a given UV dose than mRNAs whose genes lie near the (distal) 5' end of the genome. Independent experiments examining transcription *in vitro* with disrupted virions showed that mRNAs closest to the 3' end of the genome appear first, and mRNAs closest to the 5' end appear last, a result also consistent with the single-entry model.

The promoter for plus-strand RNA synthesis consists of two sequence elements separated by one turn of the ribonucleoprotein helix

All paramyxoviruses and rhabdoviruses have conserved sequences at the 3' ends of their negative-strand genome RNAs as well as their positive-strand antigenome RNAs. These conserved sequence elements are required for initiation of transcription (mRNA

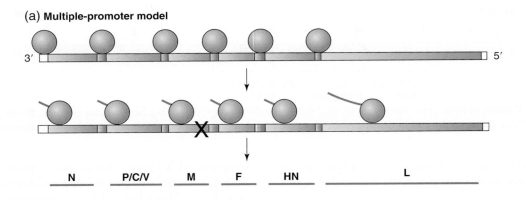

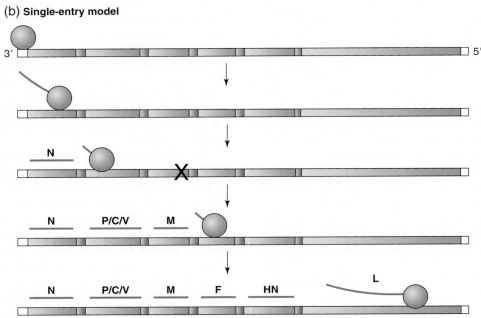

Figure 15.7 Alternative models for transcription of mononegavirus genome RNA. (a) Multiple-promoter model: RNA polymerases (orange spheres) initiate transcription independently at promoters just upstream of each viral gene. Growing RNA chains are shown as blue lines trailing the RNA polymerases, and completed mRNAs are labeled. (b) Single-entry model: RNA polymerases initiate transcription only at the 3' end of the genome, and transcribe each gene in sequence. X shows location of a possible UV-induced dimer that would block transcription only of M in model (a) but would block transcription of M, F, HN, and L in model (b).

synthesis) and replication (genome synthesis). In the case of Sendai virus, the first 12 nucleotides at the 3' end of the genome are identical to the first 12 nucleotides at the 3' end of the antigenome. Mutational analysis has shown that sequences extending up to 30 nucleotides from the 3' end are important for transcription and replication.

Surprisingly, another sequence element beginning 79 nt from the 3' end of the genome is also required for proper transcription and replication. In the case of Sendai virus, the only conserved nucleotides are C residues at positions 79, 85, and 91 (Figure 15.5). These three C residues are located at the first position

of the 14th, 15th, and 16th hexanucleotides (groups of six nucleotides) on the genome. Their precise position and spacing is vital for optimal transcription.

Since one turn of the helical nuclecapsid contains 78 nucleotides wrapped in 13 molecules of N protein (each N molecule binds to six nucleotides), these three conserved nucleotides lie exactly one turn of the helix from the 3' end of the RNA (Figure 15.5). Therefore the transcriptional promoter consists of two distinct sequence elements, brought into proximity by the structure of the helical nucleocapsid. Presumably, the RNA polymerase must interact with both sequence elements, aligned on one side of the nucleocapsid, to initiate transcription.

mRNAs are synthesized sequentially from the 3' to the 5' end of the genome RNA

The viral RNA polymerase (L protein) begins transcription at the 3' terminal nucleotide of the viral genome (Figure 15.8). In conditions where there is little or no free N protein (such as at the beginning of infection), the RNA polymerase transcribes for a short distance (40–60 nt), then terminates and releases an uncapped "leader" RNA. The RNA polymerase then scans the genome for a nearby mRNA start site, and begins mRNA synthesis. The nearest start site on Sendai virus RNA is for the N messenger RNA, 56 nt from the 3' end of the genome RNA (Figure 15.5). The L protein adds a methylated G cap at the 5' end of the growing mRNA chain, and elongates the chain until the end of the N gene.

Conserved regulatory sequences (Figure 15.6) that lie at the end of each gene signal termination of transcription of the gene, polyadenylation of the mRNA, and initiation of transcription of the next gene in line. The intergenic sequences between the transcriptional stop and start signals are always 3 nt long for Sendai virus, but range from 1 to about 60 nt for other paramyxoviruses. At the termination site there is a run of four to seven uridine residues in the template RNA, which are transcribed into adenine residues in the mRNA. The RNA polymerase repeatedly transcribes this U run by "stuttering" (see below), giving rise to a poly(A) tract some 200 nt long, and then releases the mRNA. Thus viral messenger RNAs are both capped and polyadenylated, like cellular mRNAs.

After releasing the mRNA, the RNA polymerase has two choices: it can again scan the genome RNA for a start site and begin synthesis of the next mRNA, or it can dissociate from the genome. This choice is dictated partly by the precise sequence elements in the end/start region. A fraction of RNA polymerase molecules dissociate from the genome at the end of each gene, and therefore cannot reinitiate transcription at the next downstream gene. This results in diminishing quantities of mRNAs as their genes lie farther from the 3' end of the genome. This mechanism provides for the synthesis of large amounts of mRNAs for proteins needed in high concentration, such as N and M,

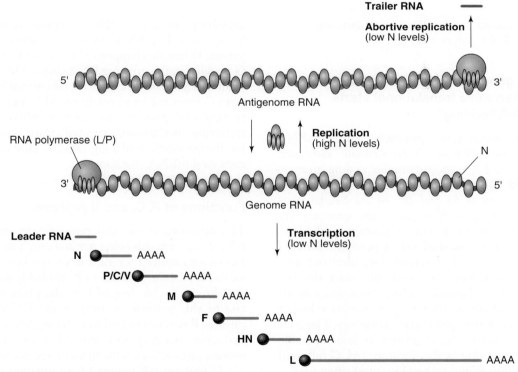

Figure 15.8 Paramyxovirus transcription and replication. The viral genome (dark blue) is shown as a helical nucleocapsid covered with N protein (green ovals) and containing RNA polymerase components L (orange spheres) and P (thin gray ovals). When N levels are low (lower part of figure), RNA polymerase terminates and reinitiates transcription at multiple sites, producing mRNAs as shown. Small gray spheres denote 5' caps, AAAA denotes poly(A) tails. When N levels are high (central part of figure), the RNA leader is encapsidated and RNA polymerase does not terminate but copies the entire genome, to form full-length antigenome nucleocapsids. Antigenome RNA is in turn replicated to form genome RNA. When N levels drop (top), trailer RNA is not encapsidated; RNA polymerase terminates and does not complete replication of the antigenome.

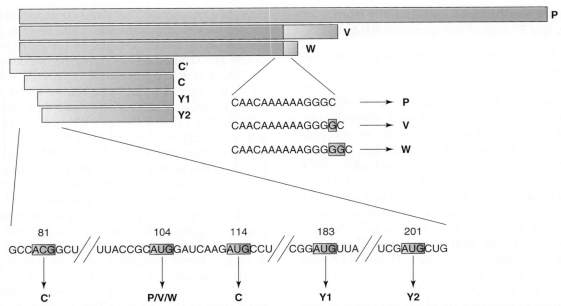

Figure 15.9 Sendai virus P/C/V gene: open reading frames, translational start sites, and "stuttering" site. Open reading frames are shown at top, with the protein produced shown at the right end. Five alternative start sites (ACG or AUG) in a 120-nt region are shown at bottom. Transcriptional stuttering occurs within the P reading frame at sequence shown in the middle, adding one or two G residues that change the reading frame, producing proteins V or W.

and smaller amounts of mRNAs for envelope proteins and the RNA polymerase (L).

The P/C/V gene codes for several proteins by using alternative translational starts and by mRNA "editing"

Most paramyxovirus genes produce a single mRNA species that codes for one viral protein. However, the P/C/V gene has a more complex coding strategy (Figure 15.9). First, there are multiple translational start sites between nt 81 and 203 of the Sendai virus P/C/V mRNA. Four of these are in the same reading frame, and ribosomes that initiate protein synthesis at these sites produce a **nested set** of proteins, denoted C', C, Y1, and Y2, with different N-termini but with a common C-terminus. An alternative start site lies in a distinct and overlapping reading frame that extends through most of the mRNA to give rise to a longer protein denoted P (the polymerase cofactor). The start site for translation of the P protein is in a favorable context for recognition by the ribosome (AUG followed by a G at position 4) and is therefore most often used.

The second complexity arises at a specific *editing* site in the middle of the P reading frame. At this sequence, which contains a stretch of 6 A's followed by 3 G's, the RNA polymerase occasionally "stutters", and adds one or more extra G residues within the stretch of 3 G's (Figure 15.9). Stuttering most likely happens when the RNA polymerase pauses, and the growing RNA chain slips backward on the RNA template by one or more nucleotides. The RNA polymerase then resumes elongation. When this happens in the run of 3 G's, an extra G is added to the growing RNA chain. This changes the reading frame; in Sendai virus, edited mRNA with one extra G gives rise to a protein called V, and mRNA with two extra G's gives rise to a protein called W. A similar stuttering mechanism, repeated many times, accounts for the production of long poly(A) tails at the end of each viral mRNA (see above).

Functions of P, C, and V proteins

The functions of the different proteins made by the P/C/V gene are still being worked out. The P protein is a cofactor required for viral RNA synthesis carried out by the L protein. P binds to L, probably as a tetramer, and directs the binding of L to the promoter at the 3' end of the genome or antigenome RNA in nucleocapsids. P remains bound to L during RNA synthesis. P also serves as a chaperone that aids in the assembly of N protein onto RNA chains to form nucleocapsids.

C proteins are required for formation of infectious paramyxovirus virions. However, their exact role in virion formation is not yet understood, and little or no C is incorporated into virions. C proteins also antagonize the antiviral effects of interferons by interfering with the function of Stat proteins (see Chapter 33). Mutations in C reduce virulence in experimental animals, probably because interferon-induced pathways restrict virus replication.

The "rule of six"

Mutational analysis of paramyxovirus genomes revealed an intriguing length restriction for genome replication. Many paramyxoviruses have a strict requirement that the length of their genomes must be an exact multiple of six nucleotides. This "rule of six" was discovered by making artificial "minigenomes" containing the two ends of the genome but lacking most internal coding sequences. When expressed in cells infected by wild-type helper virus, these minigenomes can replicate, because they contain the essential binding sequences for the viral RNA polymerase at both ends. It was found that minigenomes whose total length is an exact multiple of six nucleotides replicate much more efficiently than minigenomes of other lengths. Furthermore, most of the genomes of the *Paramyxovirinae* subgroup (measles, mumps, Sendai, etc.) sequenced to date have lengths that are exact multiples of six nucleotides.

The rule of six is explained as follows. Replicating genomes and antigenomes are assembled into nucleocapsids during elongation of RNA chains. As soon as the first six nucleotides at the 5' end of the growing viral RNA chain have been made, they bind to a molecule of N protein. This packaging reaction proceeds sequentially on each hexanucleotide sequence as the RNA is elongated. Therefore, RNAs whose length is an exact multiple of six nucleotides will have the last six nucleotides at their 3' end fully bound to an N protein subunit in a precise manner. However, RNAs that have between one and five additional nucleotides will not properly bind to the last N protein subunit, leaving some RNA sequences at the 3' end unpackaged. It is likely that when the RNA polymerase comes across a 3' end that is not properly packaged, it cannot begin transcription or genome replication efficiently, and that RNA molecule is therefore lost to the replication system.

It is interesting to note that all paramyxoviruses for which the rule of six strictly applies use RNA editing for expressing one or more viral proteins. Editing, the addition of one or more nucleotides to a growing mRNA molecule at an internal editing site, can also take place during genome replication. In mRNA transcribed from an edited genome, the reading frame of the protein that is made normally from the unedited mRNA would no longer be accessible by ribosomes. The rule of six may have evolved as a mechanism to prevent replication of these edited genomes, which would not produce viable virus if packaged.

C has also been shown to modulate the activity of the RNA polymerase. In its absence, the RNA polymerase appears less stringent in the recognition of the RNA sequences that direct its activity. The rule of six (see text box) is less strictly observed, and the difference in strength between the promoters on the genome and the antigenome is reduced.

The V protein, which shares common N-terminal sequences with P but has a distinct and highly conserved C-terminal region, is not normally required for replication of paramyxoviruses in cells cultured *in vitro*. However, elimination of V by mutating the editing site, or mutating the amino acid sequence of V in its specific C-terminal domain, reduces the virulence of Sendai virus in mice. This suggests that the V protein normally interferes with some host mechanisms that reduce virus replication or spread. Like C, V can inhibit the host interferon response, and it is known to stimulate degradation of Stat proteins.

Thus both C and V proteins are virulence factors that modulate host defense mechanisms. These functions may have been added on to the original functions of the P gene as the virus evolved, by the addition of new translation initiation sites and the RNA editing mechanism.

N protein levels control the switch from transcription to genome replication

Once sufficient N protein is made, genome replication can begin (Figure 15.8). There are two major differences between genome replication and transcription. First, the full-length RNA produced during genome replication is encapsidated by binding to N protein molecules, while mRNAs are not encapsidated. P proteins serve as chaperones that help to assemble N proteins onto the growing helical nucleocapsid.

Second, the viral RNA polymerase changes its mode of synthesis, and no longer recognizes the intergenic termination-polyadenylation-reinitiation sequences. This change may be signaled by the binding of N molecules to the growing RNA chain, or it may be a result of interaction between newly synthesized molecules of P protein and the RNA polymerase. As a result, the RNA polymerase extends the growing chain to the end of the template RNA without stopping, making a full-length antigenome nucleocapsid.

This antigenome nucleocapsid can in turn be replicated by the viral RNA polymerase, producing full-length encapsidated genomes. A delicate control mechanism senses the amount of N protein present in the cell, allowing full-length genome RNA to be made only if there is sufficient N protein to encapsidate it. This mechanism works as follows (Figure 15.8, top). Like the leader sequence at the 3' end of the genome RNA, there is a "trailer" sequence at the 3' end of the antigenome RNA. Transcripts of both the leader and the trailer sequences contain sites that initiate nucleocapsid formation by specifically binding to N protein. Once initiated, nucleocapsid formation continues by adding N protein subunits every six nucleotides without a further requirement for specific binding sites on the RNA.

If there is not sufficient N protein to initiate encapsidation within the newly formed leader or trailer RNAs, the RNA polymerase terminates some 40 to 60 nt downstream from its start site and releases a short

RNA molecule. When the template is genome RNA, the RNA polymerase can then produce a set of viral mRNAs by the transcription mechanism described above; these mRNAs do not interact with N protein. When the template is antigenome RNA, there are no further transcription initiation signals downstream of the promoter within the trailer sequence, and RNA polymerase therefore dissociates from the template. This mechanism avoids the production of genome RNAs when insufficient N protein is available to encapsidate them; these RNAs could form double-stranded hybrids with viral mRNAs, leading to induction of antiviral proteins (see Chapter 33).

As the antigenome RNA serves only as a template for viral genome synthesis, it is made in about 10-fold lower amounts than the genome. This is regulated by the strength of the respective promoters: the promoter at the 3' end of the antigenome is 10-fold stronger than the promoter at the 3' end of the genome.

Virions are assembled at the plasma membrane

Both paramyxoviruses and rhabdoviruses assemble their envelopes at the plasma membrane. Envelope proteins are inserted into membranes of the endoplasmic reticulum during synthesis, and are transported via the Golgi apparatus to the plasma membrane. The proteins are glycosylated and in some cases cleaved during transport.

Envelope glycoproteins form stable trimers or tetramers in the plasma membrane. Matrix protein associates with the cytoplasmic face of the plasma membrane, probably by interacting with the cytoplasmic tails of the viral envelope glycoproteins. Nucleocapsids, formed in the cytoplasm during genome replication, migrate to the plasma membrane where they interact with the matrix protein. When the complex between envelope proteins, matrix protein, and nucleocapsids is formed, virions are made by budding from the plasma membrane (for a diagram showing budding, see Figure 2.11). In the case of rhabdoviruses, the matrix protein is responsible for condensing the helical nucleocapsid into a coil or "skeleton" that defines the unusual bullet shape of the virion.

Infected cells often undergo fusion with neighboring cells as a result of the presence of fusion-competent envelope proteins in their membranes. Fusion can occur among numerous nearby cells, leading to the production of giant multinucleated cells, called **syncytia**. Such syncytia can form both in cells cultured *in vitro* and in tissues of infected animals. Cell fusion causes tissue destruction and may be responsible for pathogenic effects of some paramyxoviruses.

KEY TERMS

B lymphocytes

Coiled-coil

Dendritic cells

Encephalitis

Fusion peptide

Gangliosides

Hemagglutination

Hemagglutinin

Interferon

Nested set

Neuraminidase

Pseudotyping

Sialic acid

Subacute sclerosing panencephalitis (SSPE)

Synctia (singular: syncytium)

T lymphocytes

Type I integral membrane protein

Type II integral membrane protein

FUNDAMENTAL CONCEPTS

• Rhabdoviruses are enveloped, bullet-shaped viruses with a negative-strand RNA genome.

• Rabies virus is a rhabdovirus that infects bats, small mammals, dogs, and humans; the ensuing encephalitis is nearly always fatal.

• Louis Pasteur developed an effective rabies vaccine in the mid-nineteenth century; modern rabies vaccines are safe and widely used in domestic animals, in at-risk humans, and as prophylactic agents that can interrupt rabies infection before disease symptoms have appeared.

• The paramyxovirus measles causes a highly contagious disease that is often fatal in children in developing countries. An effective live vaccine is available.

• Paramyxoviruses are spherical or elongated enveloped viruses with a negative-strand RNA genome, and share many other characteristics with rhabdoviruses and filoviruses.

• Paramyxovirus envelopes fuse with the external cell membrane at neutral pH upon structural rearrangement of the viral fusion protein.

• Paramyxoviruses and rhabdoviruses have six to ten genes arranged in a highly conserved order on the viral genome, beginning with nucleocapsid protein and ending with the viral RNA polymerase.

• Both virus families transcribe their genomes in a sequential manner; the viral RNA polymerase initiates transcription exclusively at the 3' end of the genome and travels along the genome, terminating and releasing individual mRNAs and then reinitiating transcription of the subsequent mRNA, until it reaches the 5' end of the genome.

• In contrast to transcription, replication of viral RNA begins when sufficient quantities of nucleocapsid protein are available to encapsidate nascent transcripts; in this mode, the RNA polymerase makes a full-length copy of the genome RNA and does not terminate transcription before the 5' end of the genome RNA.

• Some paramyxoviruses "edit" transcripts in the P/C/V gene by pausing and "stuttering", adding one or more nucleotides to the transcript and changing the reading frame downstream of the editing site.

REVIEW QUESTIONS

1. Measles virus infection can lead to immunosuppression due to its choice of cellular receptors. What are these cellular receptors?

2. What is the "single-entry model" of transcription of viral genomes, and how does it control the levels of individual viral mRNAs?

3. Explain how the transcriptional promoter of Sendai virus can contain two elements separated by more than 60 nucleotides.

4. Where in the cell are the envelopes of paramyxoviruses and rhabdoviruses assembled?

Filoviruses

Heinz Feldmann
Hans-Dieter Klenk
Nicholas H. Acheson

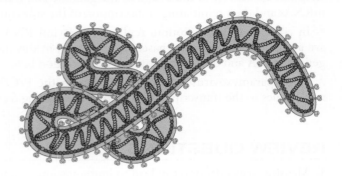

Filoviridae
From Latin *filum* (filament or thread), referring to shape of virion

VIRION
Filamentous enveloped particles.
Diameter 80 nm, length 800 nm or more.
Helical nucleocapsid.

GENOME
Linear ss RNA, negative sense.
Single segment, 19 kb.

GENES AND PROTEINS
Seven genes, transcribed in series from 3' end of genome by viral RNA polymerase
Most genes produce a single mRNA and a single protein.
Most proteins are packaged in virion:
- *Nucleocapsid protein (NP)*
- *RNA polymerase cofactor (VP35)*
- *Matrix protein (VP40)*
- *Envelope glycoproteins (GP, cleaved into GP1 and GP2)*
- *Minor nucleocapsid protein (VP30)*
- *Membrane protein (VP24)*
- *RNA polymerase (L)*

Ebola makes additional secreted glycoproteins (sGP, δ-peptide).

VIRUSES AND HOSTS
Two genera: *Marburgvirus* and *Ebolavirus*.
Ebolavirus species are named after their original site of discovery: Bundibugyo (tentative), Cote d'Ivoire, Reston, Sudan, and Zaire.
There is a single *Marburgvirus* species.
These viruses are probably transmitted to primates from fruit bats.

DISEASES
Infections with Marburg and Ebola viruses cause severe hemorrhagic fever.
Major symptoms are fever, hemorrhages, liver dysfunction, intravascular coagulation, and shock.
Lethality can be high (up to 90%).
No treatment or licensed vaccine is available.
Local outbreaks are often followed by lengthy absence of disease.

DISTINCTIVE CHARACTERISTICS
Filoviruses are prime examples of (re)emerging infectious agents.
Gene order and expression are similar to those of paramyxoviruses and rhabdoviruses.

Marburg and Ebola viruses: sporadically emerging viruses that cause severe, often fatal disease

In August 1967, three laboratory workers of the Behringwerke in Marburg, Germany, became ill with a hemorrhagic disease after processing organs from African Green monkeys (*Cercopithecus aethiops*). In the course of the epidemic, 17 more persons engaged in this kind of work were admitted to hospital, and two medical staff members became infected while attending to the patients. The last patient was admitted in November of the same year. Six more cases, including two people with secondary infections, occurred at the same time in Frankfurt, Germany. Furthermore, in Belgrade, Yugoslavia, a veterinarian performing an autopsy on dead monkeys became infected, together with his wife, who nursed him during the first days of the illness. Altogether, there were 32 cases, of whom seven died.

To Gustav-Adolf Martini of the Department of Internal Medicine at the University of Marburg it was immediately clear that he was faced with an unusual infectious disease. By the end of October of that year, Werner Slenczka and Rudolf Siegert of the Institute of Hygiene had identified the causative agent as a new virus that soon received the name of the town where it was discovered. The identification of Marburg virus (MARV) within two months of its first appearance is even by present-day standards a remarkable accomplishment of medical virology.

A related virus emerged in 1976, when two epidemics occurred simultaneously in Zaire (now Democratic Republic of Congo) and Sudan. The agent was isolated from patients in both countries and was named after the Ebola River in northwestern Zaire. Later it was discovered that they represented two distinct Ebola virus (EBOV) species, *Sudan ebolavirus* (SEBOV) and *Zaire ebolavirus* (ZEBOV). Over 500 cases were reported, with lethality rates of 88% in Zaire and 53% in Sudan. Subsequently a smaller outbreak of Ebola disease occurred in Sudan in 1979. In 1994, the third Ebola virus species, *Cote d'Ivoire ebolavirus* (CIEBOV), was isolated from an ecologist who was infected after having examined a dead chimpanzee.

Zaire ebolavirus re-emerged twice in the Democratic Republic of Congo: in Kikwit in 1995, with 315 cases (81% case fatality rate); and in Kasai Occidental Province in 2007, with 264 cases (71% case fatality rate). From 1994 to 2005, several outbreaks of *Zaire ebolavirus* disease occurred in the border region of Gabon and the Republic of Congo. A major outbreak of *Sudan ebolavirus* disease occurred in Gulu, northern Uganda, in 2000/2001 with 425 cases (53% case fatality rate). More recently, a new species, *Bundibugyo ebolavirus* (BEBOV), was described as the causative agent of a disease outbreak in Uganda with 149 cases (25% case fatality rate).

From 1989 to 1996, several Ebola virus infections occurred in cynomolgus monkeys imported from the Philippines to different facilities within the United States including Reston, Virginia, and a facility in Siena, Italy. This virus, which represents another species named *Reston ebolavirus* (REBOV), spread among monkeys housed within the facilities by different routes, potentially including aerosol transmission. This virus was subsequently found in cynomolgus macaques in the export facility in the Philippines, and more recently it was discovered in pigs in the Philippines, raising the possibility of an amplifying host with potential implications for public health, agriculture and food safety. Fortunately, *Reston ebolavirus*, the first Ebola virus species with an origin outside of Africa, does not appear to be pathogenic to humans despite several serologically documented human exposures.

Filoviruses are related to paramyxoviruses and rhabdoviruses

Marburg and Ebola viruses have enveloped virions 80 nm in diameter and of varying lengths, including filamentous particles between 800 and 1200 nm long (Figures 16.1 and 16.2). Virions contain a negative-sense, single-stranded RNA genome 19 kb long, wrapped in a helical nucleocapsid. These viruses were first classified as rhabdoviruses, but differences in virion morphology and conserved protein sequences led to the establishment of a new family, the *Filoviridae*, a name based on the elongated shape of virions. Filoviruses belong to the so-called "exotic viral pathogens" and are prime examples of (re) emerging infectious agents. *Marburgvirus* and *Ebolavirus* are two distinct genera within the filovirus family. *Ebolavirus* species are named after their original site of discovery: Cote d'Ivoire, Reston, Sudan, Zaire, and tentatively Bundibugyo. There is only a single species of Marburg virus, designated *Lake Victoria marburgvirus*.

Filoviruses cause hemorrhagic fever

Infections with Marburg and Ebola viruses cause severe viral **hemorrhagic fever**. The major symptoms are fever, internal bleeding, liver damage and dysfunction, **disseminated intravascular coagulation**, and shock. Lethality is high and varies between 20 and 90%, except for *Reston ebolavirus*, which has not been associated with human disease.

It is clear from the recorded history of filovirus outbreaks that all of them have so far been self-limited and that the total number of human infections hitherto recorded does not exceed 3000 cases. However, because of the dramatic course of the disease, the high case fatality rates, and the lack of effective vaccines and therapeutic measures, re-emergence of these viruses regularly causes high public concern. Furthermore, there is concern that filoviruses could be intentionally spread as a bioterrorist measure. For these reasons, their study is important, and development of effective vaccines and chemotherapeutic measures is a priority.

This chapter will concentrate on aspects of filoviruses that distinguish them from the related paramyxoviruses and rhabdoviruses, and should be read in conjunction with Chapter 15, where a more complete description of the replication cycle of this group of negative-strand RNA viruses is given.

Filovirus genomes contain seven genes in a conserved order

The complete nucleotide sequences of both Marburg and Ebola virus strains have been determined, allowing mapping of their coding regions. These viruses contain seven genes, arranged in an order similar to those of paramyxoviruses and rhabdoviruses. A map of the *Zaire ebolavirus* genome is shown in Figure 16.3.

As with paramyxoviruses and rhabdoviruses, the gene coding for the nucleocapsid protein (NP) is located closest to the 3' end of the genome, and the gene coding for the RNA polymerase (L) is located closest to the 5' end. All genes are flanked by highly conserved sequences that signal transcription termination, polyadenylation, and transcription reinitiation. Some genes are separated by short untranscribed intergenic sequences. However,

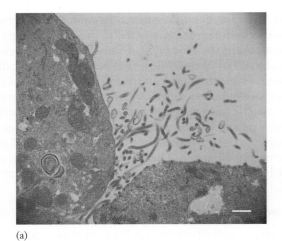

(a)

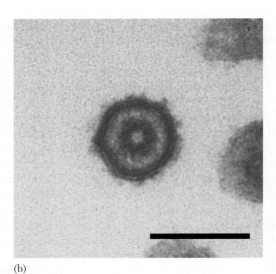

(b)

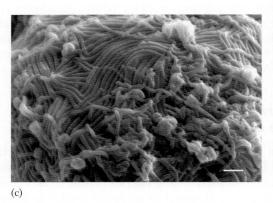

(c)

Figure 16.1 The structure of filoviruses. (a) Transmission electron micrograph of a thin section of Vero E6 cells from which Ebola virus particles are budding. Filamentous particles are seen in longitudinal, transverse, or cross-sections (bar: 500 nm). (b) Enlargement of a single particle in cross section. Particles consist of a nucleocapsid with a central channel, surrounded by a membrane; glycoprotein spikes are seen as a fringe outside membrane (bar: 100 nm). (c) Scanning electron micrograph of Ebola virus particles budding from the surface of Vero E6 cells (bar: 500 nm).

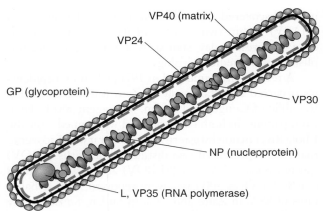

Figure 16.2 Schematic view of a filovirus virion. The RNA genome is in a helical nucleocapsid associated with 4 viral proteins: the viral polymerase (L), the nucleoprotein (NP), and proteins VP35 and VP30. VP40 is a matrix protein, the spikes are formed by trimeric GP_1/GP_2 complexes, and VP24 is associated with the envelope.

the stop and start sites (but not coding regions) of several Ebola virus genes and one Marburg virus gene actually overlap by a few nucleotides, so that the 3' end of one mRNA lies downstream from the 5' end of the next mRNA (Figure 16.3). The effect of these overlaps on mRNA synthesis is not known, but they may control the level of initiation of synthesis of mRNA of the downstream genes. Noncoding leader and trailer sequences comparable to those found in paramyxoviruses and rhabdoviruses are found at the 3' and 5' ends of filovirus genomes.

With one exception, each gene codes for a single protein, and seven proteins are incorporated into virions (Figure 16.2). The proteins are named based on their function, glycosylation status, or molecular weight, and are listed in Table 16.1 in the order of their genes from the 3' end of the genome. Four viral proteins are associated with genome RNA in nucleocapsids: NP (nucleoprotein), VP35 (RNA polymerase cofactor), VP30 (transcription activator; only shown for Ebola virus), and L (viral RNA polymerase). The other three proteins are associated with the viral envelope: GP (envelope glycoprotein), VP40 (matrix protein), and VP24 (membrane protein).

Filovirus transcription, replication, and assembly

As with paramyxoviruses and rhabdoviruses, transcription and replication of filovirus genomes take place in the cytoplasm of infected cells. The template for both mRNA synthesis and genome replication is the negative-strand genome assembled into an intact nucleocapsid. The 3' leader contains the promoter for the

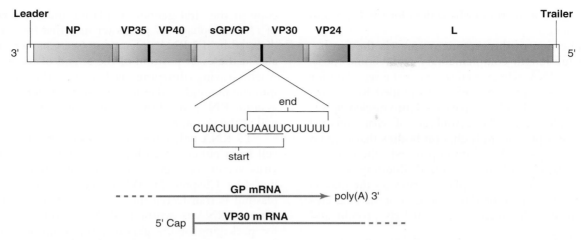

Figure 16.3 Map of Ebola virus genome. The seven viral genes are shown above the genome map along with the leader and trailer sequences. Nontranscribed sequences between genes are indicated by gray vertical bars; overlapping gene sequences are indicated by black vertical bars. The sequence overlap between the GP and VP30 genes, and the positions of the 3' and 5' ends of their mRNAs, are shown below.

Table 16.1 Ebola virus proteins

Protein	Function
NP (nucleoprotein)	Phosphorylated protein that forms the helical nucleocapsid of the virus.
VP35 (virus protein, 35 kDa)	Phosphoprotein, associated with the nucleocapsid; required cofactor for viral RNA polymerase activity; interferon antagonist.
VP40 (virus protein, 40 kDa)	Associates with membranes and resembles matrix (M) protein; directs budding of virus envelopes at the plasma membrane to form virions.
GP (glycoprotein)	Heavily N- and O-glycosylated type I transmembrane protein; single envelope glycoprotein of filoviruses; responsible for both receptor binding and fusion with cellular membranes; is cleaved into two subunits: GP_1, N-terminal portion, and GP_2, C-terminal portion with membrane anchor and fusion peptide.
sGP (secreted glycoprotein)	Synthesized as a glycosylated precursor protein that shares N-terminal 295 amino acids with GP but has a distinct C-terminus because of RNA editing; has a different oligomeric structure; lacks a membrane anchor; is cleaved into mature sGP and delta-peptide which are both secreted; not present in virions; unknown function; not found in Marburg virus.
VP30 (virus protein, 30 kDa)	Associated with nucleocapsids; has a strong activating effect on synthesis of viral messenger RNAs for Ebola but not Marburg virus.
VP24 (virus protein, 24 kDa)	Associated with membranes; interferon signaling antagonist; involvement in host adaptation.
L (large protein)	Viral RNA-dependent RNA polymerase.

viral RNA polymerase as well as a packaging signal for assembly of nucleocapsids.

Filovirus genes are believed to be transcribed in sequence starting at the 3' end of the genome to yield seven different subgenomic messenger RNA species by a start–stop mechanism described in Chapter 15 (see Figure 15.7). These mRNAs are capped at their 5' ends and contain a 3'-terminal poly(A) tail generated by stuttering of the viral RNA polymerase at a run of uridine residues located adjacent to each transcription termination signal sequence. Transcription efficiency may be influenced by gene order, formation of RNA secondary structures at the 3' ends of the genes or within the intergenic sequences, overlapping genes, and presence of duplicated termination sites. These all tend to reduce the level of transcripts as the RNA polymerase passes from the 3' end to the 5' end of the genome. This results in higher levels of mRNAs for the nucleocapsid protein (the first gene to be transcribed) than for the L protein (the last gene).

Replication of the genome occurs via the synthesis of a full-length complementary antigenome (positive sense) RNA, which then serves as the template for the synthesis of progeny negative-stranded RNA molecules. Genome RNAs are packaged into nucleocapsids during replication. The cytoplasm of virus-infected cells contains prominent **inclusion bodies** that contain viral nucleocapsids. As infection proceeds, these bodies grow and become highly structured. Budding of completed virus particles takes place at sites on the plasma membrane that are altered by insertion of the viral glycoprotein and alignment of viral matrix protein and nucleocapsids.

Cloned cDNA copies of viral mRNAs and viral genome RNA are used to study filoviruses

Because of the severity of human filovirus disease and the lack of vaccines and therapeutic agents, experimental study of live infectious Marburg and Ebola viruses is restricted to a very small number of laboratories that have high-level biological containment (Biosafety Level 4 [BSL-4]). This has obviously limited work on these viruses. However, complementary DNA copies of mRNA sequences for each of the viral proteins have been cloned in bacterial plasmids, and the proteins can be expressed either in bacterial or mammalian cells. This allows studies of the structure and activity of viral proteins in absence of infectious virus. Much work has therefore been accomplished by this route.

Furthermore, recent developments allow recovery of fully infectious virus from cloned DNA copies of the genomes of negative-strand viruses. Until recently this was difficult to accomplish because, unlike positive-strand RNA viruses, expression of the RNA genome of a negative-strand virus from a transfected plasmid will not by itself generate virions. The genome RNA of a negative-strand virus must be packaged in a nucleocapsid to be transcriptionally active, and it needs the virus-coded RNA polymerase and in some cases appropriate virus-coded cofactors to be expressed. Therefore, nucleocapsid, RNA polymerase, and cofactor proteins must be expressed in a cell independently of genome RNA in order to recover virus from transfected plasmids. In addition, copies of the virus genome with authentic 5' and 3' ends must be made, for without the authentic end sequences, transcription and replication will not occur.

Multiplasmid transfection systems allow recovery of infectious filoviruses

Both of these complications have been solved by the use of multiple-plasmid systems (Figure 16.4). A cDNA copy of the viral genome is placed under the control of a bacteriophage T7 promoter such that transcription directed by T7 RNA polymerase (see Chapter 7) initiates at the exact position of the 5' end of the genome. A self-cleaving **ribozyme** (see Chapter 31) is engineered into the plasmid so that it is transcribed along with the genome RNA, and cleaves itself to generate the exact 3' end of the genome.

T7 RNA polymerase is expressed in a mammalian cell from either a transfected plasmid, a recombinant virus vector, or a gene in a specially engineered cell line (see Chapter 7). When the genome-expression plasmid is transfected into such a cell, authentic viral genome RNA is synthesized. Viral proteins required for packaging of the genome, transcription and replication are expressed on a set of plasmids transfected along with the genome-expression plasmid. This allows production of nucleocapsids that can actively transcribe the genome, producing all viral messenger RNAs. In turn, newly synthesized viral proteins can replicate the genome, and viral envelope and matrix proteins incorporate nucleocapsids into infectious virus particles.

This system allows researchers to introduce directed mutations into viral genes and to study their effects on virus production, infectivity, and pathogenicity in infected animals. Some examples of such experiments are described below. However, even production of infectious filoviruses by plasmid transfection is a hazardous procedure because of the deadly consequences of infection, and must be done under biocontainment conditions.

Virus-like particles that contain a mini-genome, coding only for **green fluorescent protein** or **chloramphenicol acetyltransferase**, convenient markers of infected cells, have been made by modifications of the above multiplasmid procedure. These particles resemble authentic filovirus particles and can enter susceptible cells, but will replicate only if filovirus nucleocapsid and envelope proteins are expressed by plasmid transfection in the same cells. Therefore these virus-like particles are replication-deficient and not pathogenic, but can be used to study a variety of aspects of virus replication in cultured cells under lower biocontainment levels (BSL-2).

Filovirus glycoprotein mediates both receptor binding and entry by fusion

The glycoprotein (GP) (Figure 16.5) is synthesized as a 676- (*Zaire ebolavirus*) or 681- (Marburg virus) amino acid precursor protein that is inserted into the lumen of the endoplasmic reticulum. An N-terminal signal sequence is cleaved from the precursor during insertion into the membrane. A hydrophobic sequence near the C terminus serves to anchor the protein in the membrane;

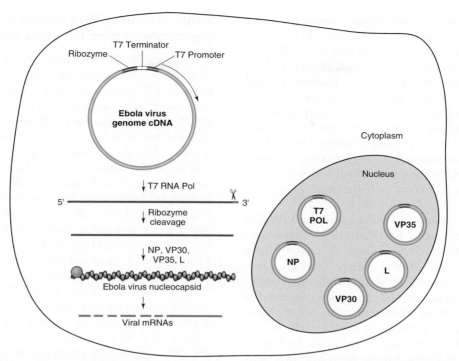

Figure 16.4 Production of infectious filoviruses by transfection of cells with multiple plasmids. Shown in the nucleus are plasmids used to make bacteriophage T7 RNA polymerase and four Ebola virus proteins that form nucleocapsids. These are driven by promoters (gray) that are recognized by cellular RNA polymerase II; mRNAs are exported to the cytoplasm where they are translated. A system fully driven by T7 RNA polymerase is also available; in that case, all plasmids are transcribed in the cytoplasm. Shown in the cytoplasm is a plasmid that expresses full-length Ebola virus RNA genome from a T7 promoter. Once all viral mRNAs are made from newly assembled nucleocapsids, virus replication and virion formation proceed. See text for details.

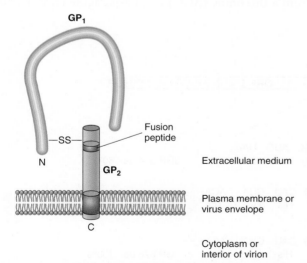

Figure 16.5 Ebola virus envelope glycoprotein. The cleaved glycoprotein fragments GP_1 and GP_2 are shown linked by a cysteine bridge (SS). GP_2 is inserted into a lipid bilayer (initially the endoplasmic reticulum, later the plasma membrane and the virion envelope) via its C-terminal transmembrane domain (shown in gray). A region thought to be responsible for membrane fusion is shown. GP_1 interacts with cellular receptors and is exposed to the extracellular medium.

only four to eight C-terminal amino acids project into the cytosol.

GP subsequently undergoes extensive N- and O-glycosylation in the endoplasmic reticulum and the Golgi compartments as it is transported to the plasma membrane; more than 50% of its mass consists of oligosaccharide residues. It is also cleaved by a cellular **furin** protease into two subunits: a large amino-terminal **ectodomain**, GP_1, and a shorter carboxy-terminal transmembrane protein, GP_2. The two subunits are held together by a cysteine bridge. GP forms trimers in the plasma membrane and in the virion envelope. In many of these respects, the filovirus GP resembles surface glycoproteins of paramyxoviruses, rhabdoviruses, and orthomyxoviruses.

Being the only virion surface protein, GP mediates binding to cellular receptors via sites in GP_1. A number of different cellular proteins that mediate filovirus entry have been detected: **asialoglycoprotein receptor**, **folate receptor-α**, **integrins**, and **DC-SIGN** (*d*endritic *c*ell-*s*pecific *i*ntercellular adhesion molecule-grabbing *n*onintegrin) are among these. Most experiments on virus binding and entry have been done with replication-defective **pseudotypes**, using filovirus GP incorporated into the envelopes of unrelated recombinant viruses; these systems may not reflect the authentic entry pathways of filoviruses. However, there are indications that the virus enters **clathrin-coated pits** and undergoes low pH-dependent fusion within vesicles in the cytoplasm. A region corresponding to the **fusion peptide** of other enveloped viruses lies near the N-terminus of GP$_2$ and is believed to mediate fusion by insertion into cellular membranes, followed by conformational changes in GP that bring the virus envelope into close contact with the membranes (see Chapter 4).

Surprisingly, cleavage of GP to produce GP$_1$ and GP$_2$ is not required either for production of infectious virions or for pathogenicity in animals. The highly basic furin cleavage site of GP was mutated and an infectious virus clone was made by the multiplasmid system. Experiments confirmed that GP remained uncleaved at the furin cleavage site in this mutant virus, but the virus was able to infect and replicate in monkey cells in culture and in infected cynomolgus macaques, leading to a lethal outcome indistinguishable from infection with wild-type Ebola virus. This is in contrast to a number of paramyxoviruses and orthomyxoviruses (Chapters 15 and 18), in which infectivity, and therefore pathogenicity in animals, depends on cleavage of the envelope glycoprotein. In those viruses, cleavage reveals the fusion peptide and allows it to insert into cellular membranes and direct virus fusion. More recently the activity of endosomal cathepsins B and L were identified to be key players in filovirus entry by cleavage of GP within **endosomes** during virus entry.

Ebola virus uses RNA editing to make two glycoproteins from the same gene

In addition to the envelope glycoprotein, all Ebola viruses (but not Marburg virus) produce a second protein from the GP gene. This protein contains the same 295 N-terminal amino acids as GP, including the N-terminal signal sequence, but has a different C-terminal region that does not include the transmembrane domain of GP (Figure 16.6). As a result, this glycoprotein is secreted into the endoplasmic reticulum but is not anchored in the membrane. It undergoes N- and O-glycosylation and proteolytic processing by a furin-like endoprotease, and is released into the medium of infected cells as two distinct molecules: mature sGP (N-glycosylated) and delta-peptide (O-glycosylated), the approximately 40 amino acid C-terminal portion of the protein. sGP is found in the serum of infected patients.

The GP gene makes both GP and sGP by producing two messenger RNAs that differ by a single A residue, added at an editing site within the gene (Figure 16.6) in a manner similar to RNA editing of the P/C/V gene of paramyxoviruses (Chapter 15). This site contains a stretch of seven U residues, which are transcribed into A residues in positive-sense messenger RNA. The unedited mRNA has an open reading frame (frame 0) that stops at a UAA codon and produces the 364-amino acid precursor of sGP. About 20% of the time, the viral RNA polymerase pauses and "stutters" over the U stretch, usually adding a single additional A residue in the messenger RNA transcript. This changes the amino acid at position 295 from a threonine (ACC) to an asparagine (AAC), and

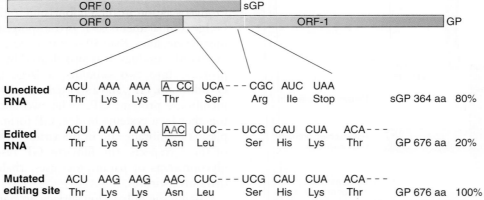

Figure 16.6 Editing of Ebola virus GP mRNAs. Shown above (in green) is the open reading frame (ORF 0) translated to produce precursor sGP and the N-terminal portion of GP, and below (in orange) is the open reading frame (ORF –1) for the C-terminal portion of GP. The mRNA sequences at the editing site, and their products of translation, are shown below. Editing introduces an additional A residue (shown in orange) that shifts the reading frame to that of GP. Only 20% of mRNAs are edited. Parts of the overlapping reading frames at the stop site for translation of sGP are shown at right. Below: an engineered virus mutant that does not undergo editing because the run of U residues in the editing site (A's in the complementary mRNA) is interrupted by two C residues (underlined G's in the mRNA). The reading frame is shifted to –1 because of an extra U added at the end of the editing site (underlined A in mRNA). This mutant over produces GP and makes no sGP.

changes the reading frame downstream of that site by one nucleotide (frame –1). As a result, ribosomes translating the edited mRNA make GP, a protein of 676 amino acids. The other 80% of the GP mRNA codes for the precursor of sGP, which is therefore made in larger amounts than GP.

Do the secreted glycoproteins play a role in virus pathogenesis?

The role of sGP in the life cycle of Ebola viruses remains obscure. All Ebola virus species tested make sGP as well as GP, but Marburg virus makes only GP. Because sGP it is released into the medium of infected cells and is found in the serum of patients, it has been suggested that sGP may act as a soluble factor that targets elements of host defense systems. sGP could bind to **antibodies** and contribute to the immunosuppression found in patients and animals infected with Ebola viruses. So far, there is no evidence that sGP is involved in activation of **macrophages** or endothelial cells, which could lead to capillary leakage and coagulation disorders.

To help understand the role of sGP, an infectious virus with a mutation in the editing site of the GP gene was made by the multiplasmid system. In this mutant genome, two U residues in the stretch of 7 were exchanged for C residues, and an additional U residue was added (Figure 16.6). As a result, editing during mRNA synthesis did not take place, and only mRNA coding for GP was made, because the extra A residue in the mRNA transcript changes the reading frame.

This virus replicated and produced infectious virions, but induced cytopathic effects in cultured cells more rapidly than wild-type virus, and gave smaller plaques and lower titers of infectious virus. The increased cytopathic effects caused by the mutant virus may be due to the increased rate of synthesis of GP, which affects the protein processing and transport system of the cell. Thus RNA editing in Ebola viruses limits the amount of GP made, allowing the infected cell to produce higher levels of virus. The effects of the mutant virus on pathogenesis in infected animals remain to be tested. There may be other roles that sGP plays, particularly in the (unknown) natural host of this virus. There are no data on the possible role of the delta-peptide.

Minor nucleocapsid protein VP30 activates viral mRNA synthesis in Ebola virus

Filovirus nucleocapsids contain four proteins, three of which (NP, VP35, and L) correspond to their counterparts (N, P, and L) in nucleocapsids of paramyxoviruses and rhabdoviruses. The fourth nucleocapsid protein, VP30, is required for optimal levels of viral messenger RNA synthesis from Ebola virus genomes. Thus, one role of VP30 is to activate Ebola virus transcription.

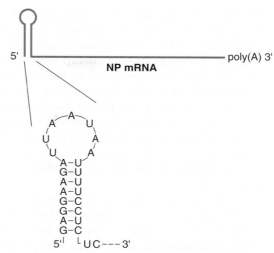

Figure 16.7 Stem-loop at 5' end of NP mRNA that controls transcription of Ebola virus genome. VP30 expression activates Ebola virus transcription by counteracting the inhibitory effect of this stem-loop structure on the initiation of transcription of the NP gene by viral RNA polymerase.

Transcription of filovirus genomes begins at the NP gene start site, shortly downstream from the 3' end of the genome (Figure 16.3). The nucleotide sequence at this start site can form a stem-loop structure in either the genome RNA or the complementary NP mRNA (Figure 16.7). When this stem-loop structure was disrupted by mutation, transcription of Ebola virus minigenomes was no longer dependent on VP30 expression. Furthermore, when the NP stem-loop was placed at the beginning of the next downstream gene, VP35, transcription of that gene also became dependent on expression of VP30. Thus, the stem-loop structure at the beginning of the NP gene inhibits viral RNA polymerase from initiating mRNA synthesis, and VP30 activates transcription by reversing this inhibition.

Although VP30 is present in nucleocapsids, it is not known if VP30 directly binds to this stem-loop, either in the genome RNA or in the growing mRNA chain. VP30 is a zinc-binding protein that forms oligomers, and both zinc-binding and oligomerization are required for its transcription-stimulating activity.

Matrix protein VP40 directs budding and formation of filamentous particles

VP40, the most abundant protein in filovirus particles, is associated with the viral envelope. It does not appear to be a transmembrane protein, but is located on the cytoplasmic face of the plasma membrane, or the inside of the virus envelope. Thus it resembles the matrix proteins of other enveloped viruses. Matrix proteins often serve as a bridge between the envelope glycoproteins incorporated into the lipid membrane and the nucleocapsids enclosed within the envelope.

VP40 of both Marburg and Ebola viruses binds strongly to membranes via its C-terminal region and forms oligomers when membrane-bound. These oligomers can be dimers or ring-shaped hexamers or octamers. Strikingly, when expressed by itself in mammalian cells, VP40 can direct the formation of virus-like particles that bud from the plasma membrane and are released into the medium. Co-expression of VP40 with GP and NP increases production of such particles, which resemble mature filovirus virions. Thus VP40 appears to be the major viral protein that directs budding and virion formation.

Matrix proteins of a number of other enveloped viruses contain short, conserved amino acid sequences, called "late domains", which are required for production of virus-like particles. Both Ebola and Marburg viruses have the late domain motif Pro-Pro-X-Tyr (where "X" means any amino acid) in the N-terminal region of their NP40. Mutations or deletions in this late domain motif impair or block production of virus-like particles by VP40.

A number of cellular proteins known to be involved in trafficking and sorting of intracellular vesicles have been shown to bind to proteins that contain such late domains. These include Nedd4, a **ubiquitin ligase**, and Tsg101, which is part of a multiprotein complex that directs vesicles to endosomes, where they bud into the lumen and form **multivesicular bodies**. There is evidence that these cellular proteins are required to help VP40 and other viral matrix proteins form virions by budding at the plasma membrane. Although the exact mechanism by which these proteins interact with VP40 to form budding virions is not clear, it may involve **ubiquitination** of VP40 or associated proteins, and association with endosomes before transport to the plasma membrane.

Most filovirus outbreaks have occurred in equatorial Africa

Marburg virus and the Ebola virus species *Bundibugyo*, *Cote d'Ivoire*, *Sudan*, and *Zaire ebolavirus* are indigenous to the African continent (Figure 16.8). Marburg virus was first isolated from patients in Europe, but the origin of the European cases could be traced back to foci in Uganda from where vervet monkeys held in Entebbe were imported into Germany and former Yugoslavia.

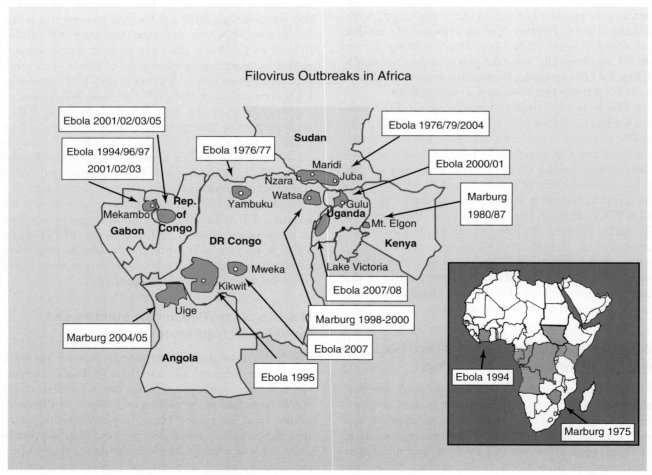

Figure 16.8 Locations of filovirus outbreaks in Africa. The sites and years of filovirus episodes and outbreaks in African countries are shown.

Both index case patients of the episodes of Marburg virus hemorrhagic fever in Kenya (1980/87) traveled in the Mt. Elgon region. This region is near the shores of Lake Victoria and is close to the trapping place and holding station of the monkeys that initiated the 1967 outbreak in Marburg. The first known community-based outbreak of Marburg virus in Africa occurred in 1998–2000 in the eastern part of the Democratic Republic of Congo, close to the border with Uganda. Another outbreak of Marburg virus disease began in Angola in 2004–2005, and is the biggest outbreak so far encountered.

Outbreaks of Ebola hemorrhagic fever caused by *Bundibugyo, Sudan,* and *Zaire ebolaviruses* have taken place in Gabon, the Republic of Congo, the Democratic Republic of Congo, Sudan, and Uganda. In 1994, the first West African species of Ebola virus (*Cote d'Ivoire ebolavirus*) was isolated during an outbreak of hemorrhagic fever among wild chimpanzees in a rain forest of Ivory Coast.

These data strongly suggest **endemic** foci for Marburg and Ebola viruses in an area lying between the 10th parallels north and south of the equator in Central Africa. Additional endemic areas in Africa or on other continents may exist. The identification of *Reston ebolavirus* in 1989 suggested for the first time the presence of an Asian filovirus that may be associated with wild nonhuman primates.

Filovirus infections are transmitted to humans from an unknown animal origin

It is believed that filoviruses normally infect and circulate in one or more animal species in Central Africa and perhaps in South Asia. These species may or may not show overt signs of infection. They would serve as a **reservoir** for the virus, from which it occasionally emerges to infect primates, probably via direct contact or contact of secretions/excretions.

Many species have been discussed as possible natural hosts; however, no nonhuman vertebrate hosts or arthropod vectors have yet been identified. *Reston ebolavirus* infection within a colony of Asian monkeys raised the question whether nonhuman primates may serve as a reservoir. The fact that Marburg and Ebola virus species are pathogenic for monkeys and apes argues against the idea that these animals may be a reservoir for the virus; they are more likely secondary hosts, as are humans.

Studies on experimentally infected wild animals showed that fruit-eating and insect-eating bats support replication and circulation of high titers of Ebola virus without necessarily becoming ill. Bats have been discussed before as potential reservoirs for the Ebola virus outbreaks in Sudan and the Democratic Republic of Congo, the Kenyan Marburg virus disease cases, and the more recent Marburg virus disease outbreak in eastern Democratic Republic of Congo. More recently,

viral RNA and antiviral antibodies were detected in fruit bat species from known outbreak sites, and Marburg virus has been isolated from wild-caught *Rousettus aegypticus*, a cave-dwelling fruit bat species. Since a persistently infected host is the most likely reservoir for a **zoonotic disease**, chronic infection in bats or other small mammalian species may be considered as a possible mechanism that regulates survival of filoviruses in nature.

Spread of filovirus infections among humans is limited to close contacts

The usual pattern seen in large outbreaks of filovirus hemorrhagic fever begins with a single case, who disseminates infection to several other people. Secondary and subsequent infections occur in close family members or among medical staff. The epidemic eventually ends because transmission of the infection is transient and the spread of the virus is inefficient. Person-to-person transmission by physical contact with case patients or their secretions/excretions is the main route of infection in human outbreaks. Activities such as nursing and preparing bodies for burial are associated with an increased risk. During the Ebola virus disease outbreaks in 1976 and 1995, transmission within hospital settings via contaminated syringes and needles was a major problem.

Transmission is inefficient, as documented by secondary infection rates that rarely exceed 12%. Nevertheless, extreme care should be taken with blood, secretions, and excretions of infected patients. Sexual transmission has been suspected for Marburg and Ebola viruses, and neonatal transmission was reported for the 1976 outbreak in the Democratic Republic of Congo. Semen or other body fluids may be infectious for some time during convalescence. Transmission by droplets and small-particle aerosols has been observed among experimentally infected and quarantined imported monkeys. This is confirmed by identification of filovirus particles in lungs of naturally and experimentally infected monkeys and human postmortem cases. However, the contribution of aerosol transmission to the course of human outbreaks seems to be low.

Pathogenesis of filovirus infections

Marburg and Ebola viruses cause similar pathological changes in humans. The most striking lesions are found in the liver, spleen, lymph nodes, adrenal glands, and kidney. In late stages of the disease, hemorrhage often occurs in the gastrointestinal tract, the spaces surrounding the lungs and heart, the peritoneal spaces, and the renal tubules. Disseminated intravascular coagulation is a terminal event. There is usually also a profound drop in white blood cell counts, and secondary bacterial infections can be found.

In animals infected with Ebola virus, replication was extensive in macrophages, monocytes, dendritic cells, and fibroblasts of many organs. There was less replication in vascular endothelial cells, hepatocytes, and tubular epithelium of the kidney. Macrophages/monoctes and dendritic cells seem to be the first and preferred site of filovirus replication.

Fluid distribution problems and platelet abnormalities in disease victims indicate dysfunction of capillary endothelial cells and platelets. Virus-induced release of **cytokines** may increase endothelial permeability and may be a major factor in the **shock** syndrome regularly observed in severe and fatal cases. Fatal filovirus infections usually end with high levels of virus in the circulation and little evidence of an effective immune response. In fact, filoviruses induce immunosuppression in the infected host, which appears to be a major factor for the rapid generalization and the severity of the disease.

Clinical features of infection

The incubation period of Marburg and Ebola virus infections varies between 4 and 21 days. The onset of the disease is sudden, with fever, chills, headache, muscle pain, and loss of appetite. This may be followed by symptoms such as abdominal pain, sore throat, nausea, vomiting, cough, joint pain, diarrhea, and a rash. Prostration, lethargy, wasting, and diarrhea seem to be more severe in filovirus hemorrhagic fever than in patients suffering from hemorrhagic fevers caused by other viruses. Severe diarrhea is often misleading, since it is also found in many bacterial enteric infections.

Patients are dehydrated, apathetic, and disoriented, and they may develop a characteristic rash associated with peeling skin by day 5 or 7 of the illness. The rash is extremely useful in differential diagnosis. Hemorrhagic manifestations develop during the peak of the illness, and can predict the likely progress of the disease. Bleeding into the gastrointestinal tract is most prominent, and localized skin hemorrhages and hemorrhages from puncture wounds and mucous membranes are common. Nonfatal cases show fever for several days. Fatal cases develop clinical signs early during infection, and death commonly occurs between days 6 and 16 after the development of hemorrhage and shock due to low blood volume.

Patients infected with filoviruses must be isolated, and protection of medical and nursing staff is required. This can be achieved by strict barrier nursing techniques and addition of HEPA-filtered respirators for aerosol protection when feasible. Filoviruses seem to be resistant to the antiviral effects of **interferon**. Ribavirin, a drug that is effective against other agents that cause viral hemorrhagic fever, has no effect on filoviruses. A protective vaccine (see text box) would be extremely valuable for at-risk medical personnel in Africa and researchers working with infectious filoviruses.

Filovirus diagnosis, treatment options, and vaccine candidates

Diagnosis of Marburg and Ebola virus infections is generally achieved in two ways: measurement of host-specific immune responses and detection of viral particles or particle components. Reverse transcriptase-polymerase chain reaction (RT-PCR) and antigen detection enzyme-linked immunosorbent assays (ELISA) are most frequently used for acute laboratory diagnosis. The diagnostic window for blood using these assays lies within 3 to 16 days after onset of disease. For antibody detection direct IgG and IgM ELISAs and IgM capture ELISA are used. The diagnostic window for IgM antibodies lies within 2 days and about 5 months after onset of disease, and IgG-specific antibodies can be detected for many years starting as early as 6 days after onset of disease.

Currently, treatment of Marburg and Ebola hemorrhagic fever is symptomatic and supportive. Patient isolation and contact tracing are key elements for the public health response. There is no proof of any successful strategy in specific pre- and post-exposure treatment of human filovirus infections. Surprising results, however, have been achieved in nonhuman primates with post-exposure vaccination using live attenuated recombinant strains of vesicular stomatitis virus expressing Marburg or Ebola virus GP (50–100% survival if given early after infection). Furthermore, treatment with the nematode-derived anticoagulation protein rNAPc2 resulted in partial survival (33%) of Ebola virus-infected nonhuman primates but failed against Marburg virus infection. Promising strategies in rodent models are the use of antisense oligonucleotides and RNA interference (RNAi), which interfere with virus transcription and RNA replication. Finally, immune therapy, although not being convincingly successful, should not be dismissed at this time.

The development of prophylactic vaccines has also been an important issue in research on filoviruses in recent years. The first Ebola virus vaccine to protect nonhuman primates was based on injection of DNA plasmids coding for both GP and NP, followed by boosting with an adenovirus recombinant vector expressing GP. This approach required several months for immunity to develop. Subsequent modification of the immunization protocol accelerated protection. Despite the intriguing success of these vaccines, preexisting immunity to adenovirus in the human population may eventually limit the utility of this approach.

Togavirus replicons expressing Marburg virus proteins protected cynomolgus monkeys from infection with the virus. However, subsequent studies evaluating this strategy for an Ebola virus vaccine were less encouraging. More recently, strains of live attenuated recombinant vesicular stomatitis virus and human parainfluenza virus that express either Marburg or Ebola virus GPs were constructed. Inoculation of cynomolgus monkeys with these vectors resulted in complete protection against disease from an otherwise lethal challenge dose. In addition, virus-like particles (VLPs) generated by transient expression of the viral proteins VP40, NP, and GP have shown promise as vaccine candidates. The VLPs have a safety advantage over live attenuated vaccines but still need booster immunization to protect nonhuman primates against lethal challenge.

KEY TERMS

Antibody

Asialoglycoprotein
   receptor

Chloramphenicol
   acetyltransferase

Clathrin-coated pits

Cytokines

DC-SIGN

Disseminated intravascular
   coagulation

Ectodomain

Endemic

Endosome

Folate receptor-α

Furin

Fusion peptide

Green fluorescent protein

Hemorrhagic

Inclusion bodies

Integrins

Interferon

Macrophage

Multivesicular bodies

Pseudotype

Reservoir

Ribozyme

Shock

Ubiquitin ligase

Ubiquitination

Zoonotic disease

FUNDAMENTAL CONCEPTS

• Marburg and Ebola viruses are members of the family *Filoviridae*, and are related to paramyxoviruses and rhabdoviruses.

• Filoviruses cause sporadic outbreaks of severe hemorrhagic fever with high case fatality rates in Central Africa.

• Disease outbreaks are self-limiting, because human-to-human transmission requires intimate contact with patients and is inefficient.

• Although not yet proven, it is likely that fruit bats serve as a reservoir for filoviruses in nature.

• Filovirus genomes have seven genes, and seven viral proteins are incorporated into virions.

• One Ebola virus gene makes two distinct proteins by RNA editing; one of these proteins, sGP, is secreted by infected cells and is not present in virions, but has unknown functions.

• The viral matrix protein, VP40, can stimulate budding at membranes and produces virus-like particles when expressed in mammalian cells.

• Because they are dangerous pathogens, research on infectious filoviruses is restricted to laboratories with high biological containment. However, much work has been done with cloned filovirus genes or with nonpathogenic virus-like particles made with filovirus proteins.

REVIEW QUESTIONS

1. How do filoviruses ensure the production of high levels of their nucleocapsid protein?

2. In order to create an *in vitro* plasmid-driven system capable of producing infectious filovirus, the plasmids were engineered to ensure they would generate an authentic 3' end on viral RNAs. How was this accomplished?

3. How does the filovirus GP gene make both a secreted and membrane-bound protein from the same gene?

4. Where were the known index cases for Marburg virus outbreaks located, and how does the location of additional outbreaks differ from those of Ebola virus?

5. Are there any treatments for Marburg and/or Ebola infections in humans?

Bunyaviruses

Richard M. Elliott

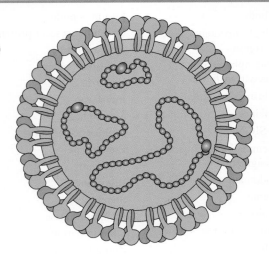

Bunyaviridae

From *Bunyamwera,* locality in Uganda where type virus was isolated

VIRION

Spherical enveloped particles with a fringe of projections.
Diameter 90–100 nm.
Helical nucleocapsids.

GENOME

Linear single-stranded RNA, negative or ambisense.
Three genome segments: S, 1–3 kb; M, 3–5 kb; L, 6–12 kb.
Total genome length 11–19 kb.

GENES AND PROTEINS

Each genome segment codes for 1–3 proteins:
- *L segment: RNA polymerase (L)*
- *M segment: Two viral envelope glycoproteins (Gn, Gc), cleaved from a precursor during translation*
- *S segment: Nucleocapsid protein (N)*

Zero to three nonstructural proteins are also made, depending on virus strain.

VIRUSES AND HOSTS

Viruses in four genera (*Orthobunyavirus, Phlebovirus, Nairovirus,* and *Hantavirus*) infect vertebrates.

Many (La Crosse, Rift Valley fever, Nairobi sheep disease viruses) are transmitted between animals by arthropod vectors.
Hantaviruses (Hantaan, Sin Nombre viruses) are transmitted by rodents and are not arthropod-borne.
Viruses in *Tospovirus* genus infect plants, are transmitted by thrips.

DISEASES

Can cause severe fever, encephalitis, hemorrhagic fever, and fatal respiratory syndrome.

DISTINCTIVE CHARACTERISTICS

Can replicate in both vertebrate and invertebrate hosts.
Segmented genome.
Ambisense coding strategy in some viruses.
Viral mRNA synthesis is primed by stealing capped 5' ends of cellular mRNAs.

Most bunyaviruses are transmitted by arthropod vectors, including mosquitoes and ticks

The *Bunyaviridae* family includes over 300 named viruses in five genera (Table 17.1). All the viruses in this family share the following common characteristics: (1) a single-stranded RNA genome composed of three unique segments, each wrapped in a helical ribonucleoprotein complex; (2) cytoplasmic site of replication; (3) viral mRNA synthesis primed by capped oligonucleotides derived from host cellular mRNAs; and (4) virion assembly at Golgi membranes.

Bunyamwera virus, isolated in 1943 in Uganda, is the prototype of both the *Orthobunyavirus* genus and the *Bunyaviridae* family. It was the first bunyavirus whose complete genome sequence was determined, and it was the first segmented negative-strand virus from which infectious virus was recovered by transfecting cells with cloned cDNAs. Rift Valley fever virus, a member of the *Phlebovirus* genus, uses an **ambisense** coding strategy on its S genome segment (see below). Ambisense coding strategies are also employed by tospovirus M and S segments, and by both the L and S segments of viruses in the *Arenaviridae* family.

Most bunyaviruses are transmitted from animal to animal by arthropod vectors, and these viruses are able to replicate in both their invertebrate and vertebrate hosts. Thus their lifestyle is very similar to those of the positive-strand RNA flaviviruses and togaviruses

Table 17.1 **Some pathogens in the family *Bunyaviridae***

Genus/Virus	Host: disease	Vector	Distribution
Orthobunyavirus			
Oropouche virus	Human: fever	Midge	S. America
La Crosse virus	Human: encephalitis	Mosquito	N. America
Akabane virus	Cattle: abortion and congenital defects	Midge	Africa, Asia, Australia
Hantavirus			
Hantaan virus	Human: severe hemorrhagic fever with renal syndrome (HFRS); 5–15% fatality	Field mouse	Eastern Europe, Asia
Puumala virus	Human: mild HFRS, nephropathica epidemica; 0.1% fatality	Bank vole	Western Europe
Sin Nombre virus	Human: hantavirus pulmonary syndrome; 40% fatality	Deer mouse	N. and S. America
Nairovirus			
Crimean-Congo hemorrhagic fever virus	Human: hemorrhagic fever; 20–80% fatality	Tick	Eastern Europe, Africa, Asia
Nairobi sheep disease virus	Sheep, goats: fever, hemorrhagic gastroenteritis, abortion	Tick	E. Africa
Phlebovirus			
Rift Valley fever virus	Human: encephalitis, hemorrhagic fever, retinitis; 1–10% fatality	Mosquito	Africa
	Domestic ruminants: necrotic hepatitis, hemorrhage, abortion		
Sandfly fever virus	Human: fever	Sandfly	Europe, Africa, Asia
Tospovirus			
Tomato spotted wilt virus	More than 650 plant species; various symptoms	Thrips	Worldwide

(Chapters 12 and 13). Virus infection has little effect on the invertebrate host even though the virus causes a systemic infection. The vector transmits the virus to a vertebrate when taking a blood meal. Virus can also be transmitted to other individuals of the same invertebrate species by mating or to offspring via the egg. **Transovarial transmission** is an important mechanism for the maintenance of these viruses in nature, as the virus can survive in the egg until conditions are favorable for the arthropod to hatch.

Viruses of a particular genus tend to be transmitted by specific types of invertebrate vectors. Mosquitoes and midges transmit orthobunyaviruses, ticks transmit nairoviruses, and sandflies or ticks transmit phleboviruses. Tospoviruses are plant viruses that are transmitted by **thrips** and are able to replicate both in vector and in plant cells. Hantaviruses are exceptions; they are not transmitted by arthropod vectors but are maintained in nature as persistent infections of rodents. Although hantaviruses can be present in large amounts in lungs and kidneys of infected rodents, they do not cause obvious disease in the animal host. Humans become infected by inhaling aerosolized excretions from infected rodents.

Some bunyaviruses cause severe hemorrhagic fever, respiratory disease, or encephalitis

Although most bunyaviruses have little impact on humans, some are important pathogens, either directly by causing disease in humans, or indirectly by affecting domestic animals or food crops (Table 17.1). Bunyaviruses were among the earliest animal viruses discovered: Nairobi sheep disease (in 1910) and Rift Valley fever virus (in 1930).

Four types of illness are caused by bunyaviruses in humans: fever, **encephalitis**, **hemorrhagic fever**, and a fatal respiratory syndrome. Some viruses cause a self-limiting febrile illness, often with rash and rarely fatal. In South America, Oropouche virus can cause **epidemics** involving tens or hundreds of thousands of people. Rift Valley fever infection results in a range of disease syndromes, from severe febrile illness to retinal disease to a fatal hemorrhagic disease. La Crosse virus is a significant cause of childhood encephalitis in the Midwestern United States, averaging about 100 cases per year. Although rarely fatal, the acute illness is severe, frequently involving seizures, and can lead to unpleasant aftereffects such as chronic seizures or epilepsy.

Hemorrhagic fevers are caused by some nairoviruses and hantaviruses. Crimean–Congo hemorrhagic fever, a tick-borne disease found in Eastern Europe, Asia, and Africa, results in severe illness and high case fatality. Its symptoms resemble those of Lassa fever and Ebola (see Chapter 16) viruses. Hantaviruses are responsible for three forms (mild, moderate, or severe) of **hemorrhagic fever with renal syndrome**. Hundreds of thousands of cases of hemorrhagic fever with renal syndrome are reported throughout Europe and Asia every year, with case fatality rates ranging from less than 0.1% for Puumala-like viruses in Western Europe, to 15% for Hantaan-like viruses in China.

A new type of hantavirus disease called **hantavirus pulmonary syndrome** was described in the southwestern United States in 1993, and was linked to a virus called Sin Nombre ("no name") virus. Several hundred cases of hantavirus pulmonary syndrome have since occurred in North and South America. This severe respiratory disease, which is a classic example of an emerging viral disease, has an overall fatality rate of 40%.

Bunyaviruses encapsidate a segmented RNA genome in a simple enveloped particle

Like the orthomyxoviruses and the reoviruses (Chapters 18 and 19), the genome of bunyaviruses consists of multiple distinct RNA molecules. All bunyaviruses contain three genome segments, named L, M, and S for large, medium, and small. These genome segments are encapsidated by the nucleocapsid (N) protein to form ribonucleoprotein complexes with helical symmetry. Associated with the nucleocapsids is the RNA polymerase (L) protein.

Virions enclose the nucleocapsids within a lipid envelope derived from the membranes of the Golgi complex. The envelope contains two viral glycoproteins, named Gn and Gc. Thus bunyavirus particles contain only four viral proteins: N, L, Gn, and Gc. Virions examined by cryoelectron microscopy are spherical, with a diameter of about 100 nm (Figure 17.1a). Smaller particles are also seen; these may contain fewer than three nucleocapsids, and therefore would be non-infectious. The mechanisms controlling genome packaging are unknown; as with other viruses that have segmented genomes, the question of how virions package at least one of each genome segment into the virion is not resolved.

The total genome size ranges from just over 12 kb for orthobunyaviruses to nearly 19 kb for nairoviruses;

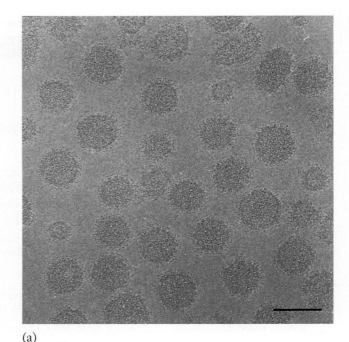

(a)

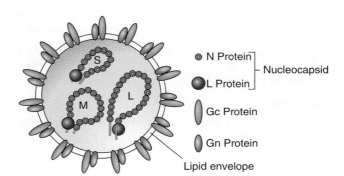

(b)

Figure 17.1 Bunyavirus virions. (a) Cryoelectron microscopic image of purified virions of La Crosse virus (Scale bar: 100 nm). (b) Schematic diagram showing the envelope and three nucleocapsids (L, M, and S) of a bunyavirus virion, with the four structural proteins labeled. Nucleocapsids are shown as circular; they are joined by complementary RNA sequences at the 3' and 5' ends.

much of this difference is accounted for by the very large nairovirus L segment. The terminal 8 to 11 nucleotides of genome RNAs are highly conserved among all three segments for a given virus, and among all viruses within a genus (Table 17.2). The conserved sequences at the 5' termini are complementary to the sequences at the 3' termini, which permits the ends of the RNA to base pair and form panhandle structures. Circular forms of both the naked RNAs and the helical ribonucleoprotein nucleocapsids (Figure 17.1b) have been observed by electron microscopy.

Table 17.2 Consensus 3' and 5' terminal nucleotide sequences of *Bunyaviridae* genome RNAs

Orthobunyavirus	3' UCAUCACAUGA..................UCGUGUGAUGA 5'
Hantavirus	3' AUCAUCAUCUG..........................AUGAUGAU 5'
Nairovirus	3' AGAGUUUCU.............................AGAAACUCU 5'
Phlebovirus	3' UGUGUUUC................................GAAACACA 5'
Tospovirus	3' UCUCGUUAG...........................CUAACGAGA 5'

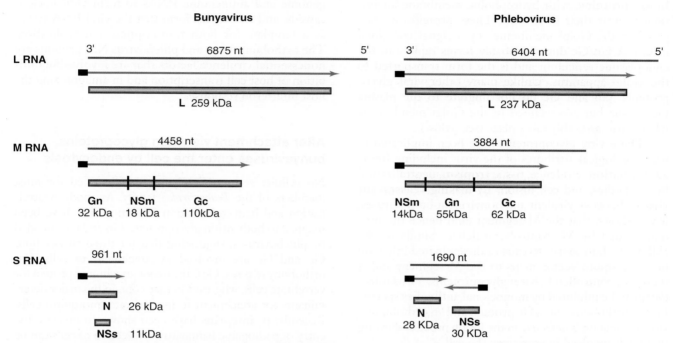

Figure 17.2 Coding strategies of genome segments of orthobunyaviruses and phleboviruses. Genomic RNAs are represented by dark blue lines (length in nucleotides above each segment) and messenger RNAs as light blue arrows; host-derived sequences at the 5' end of mRNAs are shown as black rectangles. Coding regions for viral proteins, with their molecular weights and cleavage sites, are shown below mRNAs.

Bunyavirus protein coding strategies: negative-strand and ambisense RNAs

The genome coding strategies of two representative bunyaviruses are shown in Figure 17.2. Most bunyavirus proteins are coded by messenger RNAs copied from the genome, and therefore the genomes are considered negative-strand RNAs. However, some bunyaviruses have "sense" and "antisense" protein coding regions on the same RNA genome; for an example, see the phlebovirus S RNA segment in Figure 17.2. In this case, the messenger RNA for the N protein is copied from the genome RNA, but the messenger RNA for the NSs protein is of opposite polarity, and must be copied from the antigenome RNA (see below). Such ambisense genomes are relatively rare in the RNA virus world, but arenaviruses (not covered in this book) also use this strategy to express their proteins. On the other hand, it is commonplace for many double-stranded DNA viruses to have coding regions for mRNAs and

proteins on both DNA strands, in different regions of their genomes.

L RNA codes for viral RNA polymerase

The L genome segments of all bunyaviruses encode a single large protein, the L protein, which is translated from an mRNA complementary to the genome RNA. The L proteins contain motifs characteristic of all RNA polymerases, and can both transcribe viral genome RNA, making viral messenger RNAs, and replicate the genome, making full-length genome copies. L proteins vary in size from 237 to 459 kDa, depending on the genus. Thus, like RNA polymerases of many positive- and negative-stranded RNA viruses, the L proteins are by far the largest viral proteins made by bunyaviruses.

M RNA codes for virion envelope glycoproteins

The M segment encodes the two virion glycoproteins, Gn and Gc, in the form of a precursor polyprotein. These glycoproteins are named for their relative positions in the precursor polyprotein; Gn lies toward the N-terminus and Gc lies toward the C-terminus. This replaces a confusing terminology based on relative sizes of these two proteins.

The polyprotein is cleaved during translation and transport into the **lumen** of the endoplasmic reticulum to yield the two proteins, which are notably rich in cysteine residues. Both Gn and Gc are **type I transmembrane proteins**, with hydrophobic membrane anchor regions near their C-termini. These proteins are targeted to the Golgi membranes by a signal contained in Gn. A Gn/Gc dimer probably forms rapidly in the endoplasmic reticulum and is the form transported to the Golgi apparatus. Unlike many other viral glycoproteins, Gn and Gc do not migrate to the plasma membrane but are retained in the Golgi membranes, where virus assembly takes place (see below).

These viral glycoproteins have been implicated in many biological attributes of the virus including **hemagglutination**, virulence, tissue **tropism**, neutralization by antibodies, and cell fusion. By making **reassortant viruses** between virulent and nonvirulent bunyaviruses, it was shown that the M segment (and hence the proteins coded by M) controls virulence. Similarly, the ability of a bunyavirus to cause a disseminated infection in its mosquito vector maps to the M segment and is therefore controlled by viral glycoproteins. Virus infectivity can be inhibited by monoclonal antibodies specific for the orthobunyavirus Gc protein, and an orthobunyavirus Gc mutant is defective in the fusion process, showing that Gc is involved in entry into the cell.

For orthobunyaviruses and some phleboviruses a small nonstructural protein (NSm) is encoded as part of the polyprotein and is cleaved during translation (Figure 17.2). Tospoviruses also encode an NSm protein, but they use an ambisense strategy. The NSm coding region is contained in the 5'-terminal part of the tospovirus genomic RNA, but is translated from a subgenomic mRNA. The phlebovirus NSm protein is not essential for virus viability but plays some role in virulence, while the orthobunyavirus NSm protein has a function in virion assembly and morphogenesis.

S RNA codes for nucleocapsid protein and a nonstructural protein

The S RNA segments also show diversity in their coding strategies. The S RNA of most orthobunyaviruses (Figure 17.2) encodes two proteins, the nucleocapsid protein (N) and a nonstructural protein (NSs). These proteins are both synthesized from the same complementary-sense mRNA but from different, overlapping reading frames, as the result of alternative initiation at different AUG codons. In contrast, the S RNAs of most hantaviruses and nairoviruses contain only a single open reading frame, encoding the nucleocapsid protein.

The S RNAs of phleboviruses (Figure 17.2) and tospoviruses employ an ambisense coding strategy. The N protein is encoded in an mRNA corresponding to the 3' half of the S RNA; this mRNA is copied from genome RNA. The NSs protein is encoded in an mRNA corresponding to the 5' half of the S RNA; this mRNA is copied from antigenome RNA (see below). The N protein binds specifically to full-length viral genome and antigenome RNAs to form viral nucleocapsids, and it is in this form that the viral RNA serves as a template for both transcription and replication. The orthobunyavirus and phlebovirus NSs proteins are nonessential virulence factors that are involved in inhibition of host cell transcription and in antagonizing the host interferon response.

After attachment via virion glycoproteins, bunyaviruses enter the cell by endocytosis

No cellular receptor has yet been identified for most members of the *Bunyaviridae* family. Antibody neutralization and hemagglutination inhibition sites have been mapped to both virion glycoproteins Gn and Gc encoded by phleboviruses, suggesting that for these viruses both Gn and Gc are involved in attachment to cells. For orthobunyaviruses, Gc is the major attachment protein for vertebrate cells, whereas Gn may contain the major determinants for attachment to invertebrate (mosquito) cells. Recently, β_3 **integrins** have been shown to mediate the entry of pathogenic hantaviruses; infection of cells can be blocked by anti-integrin antibodies or the integrin **ligand**, vitronectin. Which of the two hantavirus glycoproteins is involved in the interaction so far remains unknown.

Based on electron microscopic studies of phleboviruses, entry into cells is by endocytosis. It is probable that uncoating occurs when endosomes become acidified, initiating fusion of the viral membrane and endosomal membrane, followed by release of the nucleocapsids into the cytoplasm. In common with many other enveloped viruses, bunyaviruses can fuse with cellular membranes at acidic pH, and this is accompanied by a conformational change in Gn.

Bunyavirus mRNA synthesis is primed by the capped 5' ends of cellular mRNAs

Like other negative-strand RNA viruses, bunyaviruses begin their replication cycle by transcription of the infecting genome segments into messenger RNAs, using a virion-associated RNA polymerase, the L protein (Figure 17.3). This process, termed primary transcription, can begin as soon as the viral nucleocapsids are released into the cytoplasm. Transcriptase activity can be detected in detergent-disrupted bunyavirus preparations.

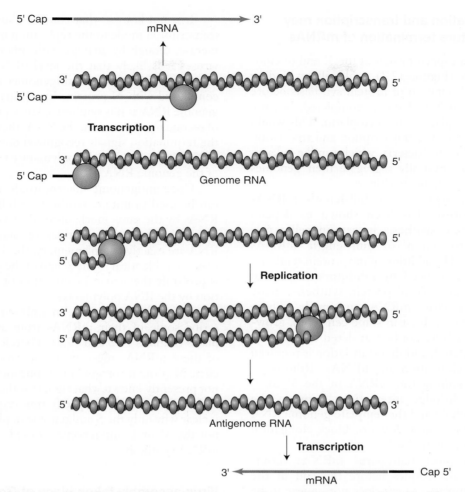

Figure 17.3 Transcription and replication of genome segments. Genome RNA is encapsidated by N protein (green ovals). Above, transcription: viral L protein (orange sphere) cleaves capped cellular mRNAs and uses the 5' end fragment (black line) as a primer for transcription of genome RNA. A termination signal releases the mRNA. Below, replication: when sufficient N protein is made, the L protein initiates RNA synthesis directly at the 3' end of genome RNA and replicates the genome via an antigenome copy, which is encapsidated. Antigenome RNA serves as a template for synthesis of progeny genomes (not shown) and in some cases ambisense mRNAs (bottom).

Surprisingly, mRNA synthesis is stimulated by addition of methylated cap analogues such as mGpppAm, or naturally capped mRNAs such as alfalfa mosaic virus RNA 4. Both cap analogues and capped mRNAs appear to act as primers for transcription. Further support for this notion was provided by sequencing the 5' ends of bunyavirus mRNAs. They were found to contain an additional 10 to 18 nucleotides at their 5' ends whose sequence is unrelated to the viral genome sequence. An endonuclease activity that specifically cleaves capped mRNAs was detected *in vitro*, and this activity has been mapped to the L protein by expression studies using a recombinant vaccinia virus.

Taken together, these data indicate that bunyaviruses, like influenza viruses (see Chapter 18), use a "cap-snatch" mechanism to prime messenger RNA synthesis. This involves binding of the capped 5' end of a cellular mRNA by the L protein, cleavage at a site 10 to 18 nt from the 5' end, and alignment of the 3' end of the cleaved RNA fragment with the 3' end of the genome RNA. The L protein then adds nucleotides to this primer RNA, using the genome RNA as a template.

Influenza virus transcription takes place in the cell nucleus, and uses newly synthesized cellular transcripts as a source of its capped primers; therefore, it is inhibited by agents that interrupt cellular RNA polymerase II transcription, such as **actinomycin D** or **α-amanitin**. In contrast, bunyavirus transcription takes place in the cytoplasm, and uses preformed cellular mRNAs as primers; thus it is not sensitive to these drugs, because cellular messenger RNAs are abundant and stable. Using capped cellular RNAs as primers for transcription allows these viruses to make their own capped mRNAs even though they do not code for a capping enzyme.

Coupled translation and transcription may prevent premature termination of mRNAs

The presence of a capped primer at the 5' end of elongating transcripts (Figure 17.3) could allow binding by ribosomes and initation of protein synthesis before synthesis of the mRNA chains is completed, because transcription takes place in the cytoplasm. This would resemble the coordinated transcription and translation known to take place in bacteria, which (unlike eukaryotic cells) do not physically segregate their genomes from the protein synthetic machinery.

Interestingly, synthesis of full-length mRNAs of some bunyaviruses has been shown to depend on ongoing protein synthesis in certain cell types. Furthermore, production of full-length mRNAs *in vitro* is stimulated by addition of the protein synthetic machinery and is reduced by inhibitors of protein synthesis. In the absence of protein synthesis, many prematurely terminated mRNAs are made. These results suggest that, at least in some cell types, growing bunyavirus mRNAs can be translated before their synthesis is completed, and that translation is required for the proper elongation of mRNAs. Ribosomes translating or scanning the mRNA in the 5' to 3' direction could dislodge interactions between the growing mRNA chain and the template genome RNA that might otherwise slow down or block elongation by the viral RNA polymerase.

Synthesis of most bunyavirus mRNAs terminates 50 to 100 nucleotides before the end of the genome RNA template. There does not appear to be a universal termination signal for bunyavirus transcription; however, two sequences that have been suggested as termination sites are 3'-G/CUUUUU-5' and 3'-GGUGGGGGGUGGGG-5'.

Bunyavirus mRNAs are not polyadenylated at their 3' ends. It was shown recently that orthobunyavirus mRNAs, unlike cellular mRNAs, are efficiently translated in a process independent of polyA-binding protein (PABP). Since viral infection corrupts PABP function, this suggests a mechanism by which the virus can promote translation of its own proteins while host protein synthesis is impaired.

Genome replication begins once sufficient N protein is made

Viral genome replication requires synthesis of a full-length complementary positive-sense RNA, called the antigenome or replicative intermediate RNA. This molecule is an exact complementary copy of the genome, and it does not contain the 5' primer sequences used to make mRNAs. Therefore the L protein initiates genome replication at the first nucleotide at the 3' end of the genome.

It is not known what controls the switch from the transcription mode to the replication mode of the polymerase, though by analogy with other negative-strand viruses it is likely that the level of N protein plays a role. Both genomes and antigenomes are always found complexed with N protein. Encapsidation of the antigenome RNA as it is synthesized may prevent formation of secondary structure in the RNA, thereby overcoming the termination signals recognized during formation of mRNA, and allowing the polymerase to copy to the end of the genome RNA.

Once antigenome nucleocapsids are formed, they can be used to initiate synthesis of full-length genome RNAs by the same mechanism. Because the 5'- and 3'-terminal 8 to 11 nucleotides of bunyavirus genome RNA are exactly complementary, the 3' end of the antigenome is identical to the 3' end of the genome, and this is probably the signal for initiation of genome replication by the RNA polymerase.

Bunyaviruses that use an ambisense strategy transcribe some of their mRNAs from antigenome templates (Figures 17.2 and 17.3). Therefore, the synthesis of these mRNAs must await the production of sufficient N protein for genome replication to begin. It is not presently known what signals at the ends of genome and antigenome RNA specify transcription. Neither is it clear why only the S antigenome of phleboviruses, but not the M or L antigenome, is used as a template for mRNA synthesis.

Virus assembly takes place at Golgi membranes

Maturation of bunyaviruses occurs at the smooth membranes in the Golgi apparatus (Figure 17.4), and is inhibited by the drug **monensin**, a monovalent **ionophore** that disrupts protein transport through Golgi membranes. The viral glycoproteins accumulate in the Golgi complex and cause a progressive vacuolization. However, the morphologically altered Golgi complex retains its ability to glycosylate and transport glycoproteins destined for the plasma membrane. Using vaccinia virus recombinants it has been shown that the targeting of bunyavirus glycoproteins to the Golgi complex is a property of the glycoproteins alone, and does not require other viral proteins or virus assembly.

Electron microscopic studies revealed that viral nucleocapsids condense on the cytoplasmic side of areas of the Golgi vesicles at sites where viral glycoproteins are present. Bunyaviruses lack a matrix protein, which for many other negative-strand viruses serves as a bridge between the nucleocapsid and the envelope glycoproteins. Therefore, direct transmembrane interactions between the nucleocapsid and the glycoproteins may be involved in budding of bunyaviruses.

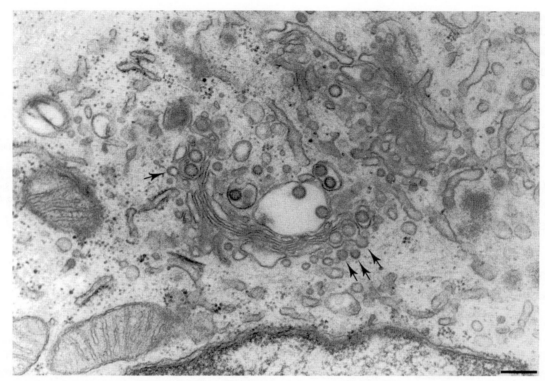

Figure 17.4 Electron micrograph of cells infected with Uukuniemi virus (a phlebovirus) showing virus particles budding into Golgi vacuoles (center) and small vesicles (arrows) budding from Golgi cisternae (Scale bar: 200 nm).

Virions form by budding into the Golgi cisternae. Vesicles containing virions are then transported to the cell surface via **exocytosis**, eventually releasing their contents to the exterior. Although most bunyaviruses mature by this pathway, Rift Valley fever phlebovirus has been observed to bud directly at the surface of infected rat hepatocytes.

Evolutionary potential of bunyaviruses via genome reassortment

Like other viruses with segmented genomes, bunyaviruses evolve by two major pathways: **genetic drift**, or the slow accumulation of point mutations, insertions, or deletions; and **genetic shift**, caused by the exchange of genome segments via reassortment. Viral RNA-dependent RNA polymerases are known to be error-prone because, unlike DNA polymerases, they lack a proofreading ability. This is a significant factor in the accumulation of mutations leading to genetic heterogeneity ("drift") in viral RNA populations. RNA oligonucleotide fingerprinting studies on field isolates of La Crosse virus revealed that no two isolates were identical. However, virus isolates obtained from the brains of children who died from infection by La Crosse virus in 1960, 1978, and 1993 revealed remarkable genetic stability: only 51 nucleotide changes were detected in the 4526 nt M genome segment over this 33-year period.

This suggests that only some genetic variants of the virus are virulent for humans.

Genome segment reassortment results in more dramatic antigenic changes ("shift"), as exemplified by influenza viruses (Chapter 18). Following a mixed infection of a cell with two parental viruses, progeny containing segments derived from the different parents may be produced (Figure 17.5).

Bunyavirus reassortment has been studied experimentally both in cells cultured *in vitro* and in mosquitoes. Reassortment is restricted to closely related bunyaviruses, and even then certain combinations of viruses appear genetically incompatible. In mosquitoes infected with two virus strains, reassortment has been documented in salivary glands and ovaries, and reassortant viruses can be transmitted both transovarially and by feeding.

Genomic analysis of bunyaviruses isolated from mosquitoes caught in the field shows evidence for naturally occurring reassortment. The genetic potential for reassortment of the vast number of known bunyaviruses should not be underestimated when considering the possibility of new and emerging pathogenic viruses. For instance, Ngari bunyavirus, isolated from humans suffering hemorrhagic disease, was shown to be a reassortant between Bunyamwera and Batai viruses, each of which causes relatively mild febrile illness.

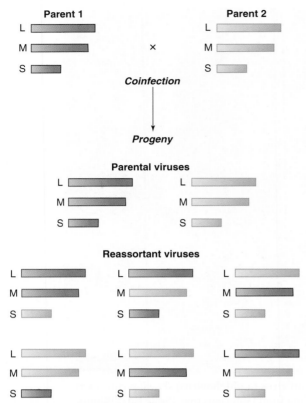

Figure 17.5 Reassortment of bunyavirus genome segments. A cell is coinfected with two closely related bunyavirus strains (parent 1 and parent 2). Eight different genotypes can be represented in progeny virus: the two parental genotypes and six possible reassortant genotypes, each with two segments from one parent and one segment from the other parent.

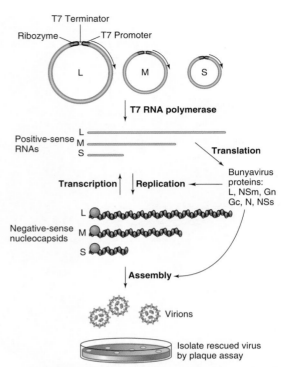

Figure 17.6 Rescue of infectious bunyavirus from cloned cDNAs. Mammalian cells constitutively expressing bacteriophage T7 RNA polymerase are transfected with a set of three plasmids containing cDNA copies of each genome segment. The transcribed antigenome (positive-strand) RNAs are directly translated to give bunyaviral replication and structural proteins. The antigenome RNAs are then replicated and assembled into virions.

Reverse genetics of bunyaviruses

For many years detailed molecular analysis of the replication of negative-strand RNA viruses was hampered because of our inability to manipulate their genome RNAs directly. Analysis of gene function by site-directed mutagenesis was not possible, because infectious RNA cannot be expressed from a cloned DNA copy. Negative-strand genome RNA must be encapsidated to be transcribed and replicated, and therefore synthesis of the genome RNA alone does not lead to virus production. However, recent dramatic progress has been made in the "reverse genetics" of negative-strand RNA viruses—the mutation of complementary DNA copies of the viral genome followed by the rescue of the mutated cDNA into an RNA genome.

Recently, a simple system for recovery of infectious bunyaviruses from cloned cDNAs was established. This system, outlined in Figure 17.6, involves the transfection of cells with a set of plasmids containing cDNAs of the L, M, and S genome segments flanked by bacteriophage T7 promoters and hepatitis delta virus ribozyme sequences (see Chapter 31). A mammalian cell line engineered to express T7 RNA polymerase enables transcription of these three cDNAs into positive-strand antigenome RNAs with the authentic 5'- and 3'-terminal sequences. Surprisingly, these antigenome RNAs can act as messenger RNAs even though they are neither capped nor polyadenylated, and small amounts of viral proteins are made in the transfected cells. Further transcription of the cDNAs provides antigenome RNAs that are encapsidated during transcription by newly synthesized N protein, and these RNAs can then be replicated by the bunyavirus L polymerase. These nucleocapsids are then packaged into infectious particles containing the Gn and Gc envelope proteins. Individual virus clones can be isolated from the supernatant medium by plaque formation in mammalian cells.

An early use of this technology was to create a virus lacking the NSs gene, leading to the discovery that the NSs protein is an antagonist of the interferon response mechanism. In the future, the system will be exploited for vaccine design by making attenuating mutations in the genome, either by mutating or deleting virulence genes or by mutating noncoding control sequences. Specific marker sequences can be incorporated into the viral RNAs and/or proteins that will allow distinction of vaccinated subjects. In addition, bunyaviruses that express autofluorescent proteins have been created, permitting real-time microscopical analyses of the infection cycle.

KEY TERMS

α-amanitin
Actinomycin D
Ambisense
Encephalitis
Epidemic
Exocytosis
Genetic drift
Genetic shift
Hantavirus pulmonary
   syndrome
Hemagglutination
Hemorrhagic fever
Hemorrhagic fever with
   renal syndrome

Integrins
Ionophore
Ligand
Lumen
Monensin
Reassortant viruses
Ribozyme
Thrips
Transovarial transmission
Tropism
Type I transmembrane
   protein

FUNDAMENTAL CONCEPTS

- Most bunyaviruses are transmitted by arthropod vectors.
- Some bunyaviruses can be maintained in nature by transovarial transmission within the arthropod host.
- Bunyaviruses can cause encephalitis, hemorrhagic fevers, or severe respiratory syndrome in humans.
- Like other negative-strand RNA viruses, the template for both transcription and replication of bunyaviruses is the helical nucleocapsid, not naked RNA.
- Bunyavirus transcription is coupled to translation and is primed by a cap-snatching process.
- Bunyaviruses assemble and mature intracellularly at the Golgi apparatus.
- The bunyavirus NSs protein is a nonessential virulence factor that antagonizes the host interferon response.

REVIEW QUESTIONS

1. What hantavirus disease appeared in the Four Corners region of the United States in 1993?

2. RNA virus genomes are generally either plus or minus sense. What is an ambisense genome, present in some bunyaviruses?

3. Bunyavirus transcription is not sensitive to inhibition by actinomycin D but shares the same RNA capping strategy as influenza virus, which is sensitive to actinomycin D. What transcription process do these viruses share and what is different?

4. Why is maturation of bunyaviruses inhibited by monensin?

CHAPTER 18

Influenza Viruses

Dalius J. Briedis

Orthomyxoviridae

From Greek *ortho* (correct or normal) and *myxa*
    (mucus); virions bind to mucoproteins

VIRION

Enveloped particles, quasi-spherical or filamentous.
Diameter 80–120 nm.
Envelope is derived from host cell plasma membrane
    by budding.
Compact helical nucleocapsids.

GENOME

Linear single-stranded RNA, negative sense.
Six to eight different segments.
Total genome length 10–15 kb.

GENES AND PROTEINS

Each genome segment codes for one or
    two proteins
- *Envelope glycoproteins: hemagglutinin (HA) and
  neuraminidase (NA)*
- *Integral membrane protein (M2) with ion channel
  activity*
- *Matrix protein (M1)*
- *Nucleocapsid protein (NP)*
- *Three RNA polymerase proteins (PA, PB1, and PB2)*
- *Nonstructural protein (NS1)*
- *Minor structural protein (NS2)*
- *Additional nonstructural protein (PB1-F2)*

VIRUSES AND HOSTS

Influenza types A, B, and C: hosts include birds,
    various mammals and humans.
Two additional genera: *Thogotovirus* (transmitted
    by ticks) and *Isavirus* (infects fish, particularly
    salmon).

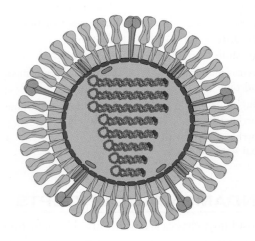

DISEASES

Symptoms include high fever, sore throat, cough,
    headache, muscular pain.
Can be fatal in elderly, infants, and chronically ill, often
    by secondary bacterial infections.
In 1918, an influenza pandemic killed 20 million people
    worldwide.
A new pandemic strain (H1N1 "swine flu") spread
    worldwide in 2009.
Emerging avian influenza virus strains threaten domestic
    fowl and another possible human pandemic.

DISTINCTIVE CHARACTERISTICS

Orthomyxoviruses replicate in the nucleus, unlike most
    RNA viruses.
Viral mRNA synthesis is primed by stealing capped 5'
    ends of cellular pre-mRNAs in the nucleus.
Viruses can undergo reassortment ("antigenic shift")
    by exchanging genome segments between related
    strains.
Reassortment generates new viruses that can cause
    pandemics because of changed surface antigens
    (HA and NA).

Influenza viruses cause serious acute disease in humans, and occasional pandemics

Influenza virus was responsible for one of the worst
human **epidemics** since those caused by the plague
bacterium in the Middle Ages. The so-called "Spanish
flu" **pandemic** (worldwide epidemic) of 1918–1920

caused the deaths of 20 million people—more than
all the battles of World War I. Influenza received its
name in eighteenth-century Italy as a reference to the
imagined cause of epidemics—the *influence* of the stars.
Despite the present availability of vaccines and antiviral
drugs, influenza viruses continue to kill hundreds of
thousands of humans worldwide every year. A major

reason is the ability of influenza viruses to escape immune surveillance by changing the amino acid sequences of their surface glycoproteins.

What the general population calls "flu" is usually a mild upper respiratory illness (the "common cold") caused not by influenza virus but by a number of upper respiratory viruses including rhinoviruses and coronaviruses. Real influenza is a severe acute disease in all age groups with symptoms of sore throat, cough, high fever, headache, and muscular aches and pains. A rule of thumb for distinguishing influenza from other respiratory virus infections is: if you really can't get out of bed, it's probably influenza! The acute stage lasts about three days, but cough and malaise may last for several weeks.

Epidemics of influenza occur every winter, usually between December and March in the Northern Hemisphere. More than 20,000 deaths per year in North America continue to be caused by influenza virus, most among elderly or otherwise debilitated or immunocompromised individuals. In May 1790, a severe influenza epidemic struck Philadelphia, then the capital of the fledgling United States of America, and George Washington, the newly elected President, was close to death but eventually recovered. In addition to these regular epidemics, more severe pandemics occur at irregular 10- to 50-year intervals. These have included not only the "Spanish flu" pandemic in 1918–1920, but also the "Asian flu" in 1956 (60,000 deaths in North America) and the "Hong Kong flu" in 1967. The "swine flu" pandemic of 2009–2010, which caused upward of 16,000 deaths worldwide, did not lead to widespread serious illness as was feared.

Influenza virus infections of the respiratory tract can lead to secondary bacterial infections

The virus initially infects epithelial cells in the upper respiratory tract (Figure 18.1). Virus infection causes loss of the **ciliated epithelium**, leading to loss of the ability of the respiratory tract to clear viruses or bacteria by **mucociliary flow** that normally traps these agents in mucus and disposes of them. Virus replication furthermore induces production of **interferons, cytokines**, and other soluble response factors, leading to local and systemic inflammatory responses. This results in the symptoms that define the "flu" syndrome: fever, headache, chills, malaise, and muscle aches. The loss of ciliated epithelium allows relatively easy entry of bacteria into the lower respiratory tract. Resulting secondary bacterial pneumonias cause many of the deaths attributed to influenza virus infection.

Orthomyxoviruses are negative-strand RNA viruses with segmented genomes

Myxoviruses were originally named because virions of influenza and several other enveloped, negative-strand RNA viruses bind to **sialic acid** residues in

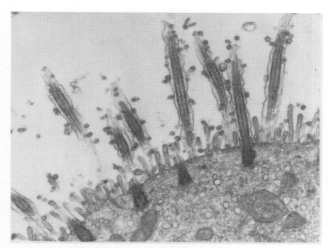

Figure 18.1 Epithelial cells of the upper respiratory tract infected with influenza virus. Cilia cut tangentially in this electron micrograph of a thin section are shown projecting from an epithelial cell. A number of influenza type A virions (small spherical particles) are shown attached to the cilia.

mucoproteins (*myxo* = mucus). When several of these viruses were understood to have a very different lifestyle from the influenza viruses, the myxoviruses were split into two families, orthomyxoviruses and paramyxoviruses (*ortho* = correct; *para* = alternative). Paramyxoviruses are discussed in Chapter 15.

The *Orthomyxoviridae* family includes influenza virus types A, B, and C, as well as two distinct tickborne mammalian viruses in the *Thogotovirus* genus and a virus that infects Atlantic salmon in the *Isarvirus* genus. Influenza type A viruses are the best-studied orthomyxoviruses; they are responsible for most flu epidemics, and are widespread in many avian and mammalian species. Influenza type B is limited to humans, and type C can infect both pigs and humans. Multiple subtypes are distinguished within each influenza virus type by means of variation in the structure of their two surface glycoproteins. Most of the information in this chapter comes from studies with influenza type A virus.

Orthomyxoviruses share a common structure and mode of replication. Their negative-strand RNA genome is composed of between six and eight distinct gene segments. These are individually wrapped within helical nucleocapsids, which are packaged in a lipid envelope derived from the plasma membrane to form the virion. Virions can be either roughly spherical or filamentous (Figure 18.2).

Fresh isolates tend to make predominantly filamentous particles, while virus repeatedly passaged in eggs or cell cultures tend to make smaller, elongated, or quasi-spherical particles. It has been speculated that filamentous virions, which can grow as long as the diameter of an epithelial cell, are responsible for cell-to-cell transmission of virus; such virions could infect a neighboring cell while still being extruded. Quasi-spherical virions are much smaller and may be more readily incorporated

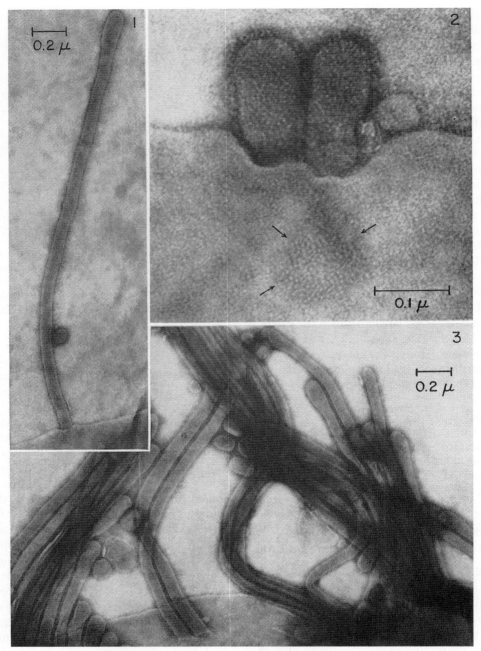

Figure 18.2 Electron micrographs of spherical and filamentous influenza virus A particles. Top right: quasi-spherical particles budding from the cell surface. The glycoprotein projections can be readily seen. Top left and bottom: filamentous particles being extruded from the cell surface. These particles can extend for several micrometers and may be an efficient means for cell-to-cell transmission. Negatively stained preparations of infected cells.

into tiny droplets in respiratory aerosols; therefore these may be responsible for person-to-person transmission.

Eight influenza virus genome segments code for a total of 11 different viral proteins

Of the eleven viral proteins produced by influenza type A viruses, nine are packaged in virions (Figure 18.3 and Table 18.1). Genome segments 1 to 6 each code for a single protein: three RNA polymerase subunits

(PB2, PB1, and PA), the envelope glycoprotein **hemagglutinin** (HA), the nucleocapsid protein (NP), and a second envelope glycoprotein, **neuraminidase** (NA). In some strains of influenza A, a second reading frame on RNA 2 codes for an additional short protein named PB1-F2.

Messenger RNAs transcribed from both genome segments 7 and 8 can be spliced, each giving rise to two distinct mRNA species that code for different viral proteins. Genome segment 7 codes for the

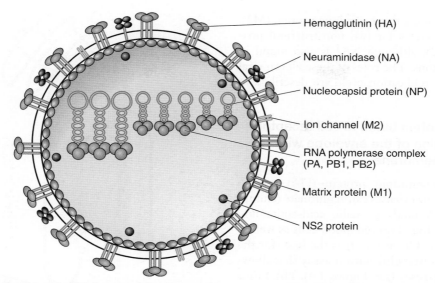

Figure 18.3 Schematic diagram of influenza virus virion. Nine virus-coded proteins are shown. Hemagglutinin, neuraminidase, M2 ion channel, and matrix protein are associated with lipid bilayer membrane. Eight different helical nucleocapsids containing NP and the RNA polymerase trimer of PA, PB1, and PB2 are enclosed within the envelope. There is evidence that one copy of each of the eight genome segments is incorporated in a bundled form into each virion. A small amount of NS2 protein is also present in virions.

Table 18.1 Influenza A virus strain A/PR/8/34 genome RNAs and proteins

RNA segment	Length (nucleotides)	Proteins encoded	Protein size (aa)	Protein function
1	2341	PB2	759	Binds to cap structures on cellular pre-mRNAs; part of transcriptase complex
2	2341	PB1	757	Cleaves cellular pre-mRNAs to create primer; RNA polymerase activity for both transcription and replication
		PB1-F2	87-90	A protein that is localized in mitochondria and enhances apoptosis in immune cells
3	2233	PA	716	Part of transcription and replication complexes, exact role unknown
4	1778	HA	566	Hemagglutinin; major surface glycoprotein; receptor-binding; mediates membrane fusion at low pH; antigenic determinant
5	1565	NP	498	Nucleocapsid protein; binds to and encapsidates viral RNA; control functions in RNA synthesis
6	1413	NA	454	Neuraminidase; major surface glycoprotein; receptor destruction; dissociation of virus aggregates; antigenic determinant
7	1027	M1	252	Matrix protein; interacts with envelope, nucleocapsids, and NS2
		M2	97	Integral membrane protein; ion channel activity essential for virus uncoating and maturation
8	890	NS1	230	Nonstructural protein that down-regulates host cell mRNA processing; suppresses RIG-1 signaling and interferon induction; sequesters dsRNA and reduces interferon response; contains a PDZ domain for protein–protein recognition that can interact with diverse cell-signaling assemblies and reduce apoptosis of infected cells.
		NS2	121	Nonstructural protein that directs nuclear export of viral nucleocapsids; interacts with M1

matrix protein (M1) and a third envelope protein (M2). Genome segment 8 codes for two nonstructural proteins (NS1 and NS2), although NS2 is also found in small amounts in virions. The roles of these eleven viral proteins in the viral replication cycle are described in detail below.

Hemagglutinin protein binds to cell receptors and mediates fusion of the envelope with the endosomal membrane

The influenza hemagglutinin protein (HA) was so named because influenza viruses can agglutinate (clump) red blood cells. HA binds to sialic acid-containing receptors, which are found on red blood cells as well as most other cell types. This property is the basis for the rapid and sensitive **hemagglutination assay** that allows detection of many viruses (see Figure 1.4). HA forms a trimer in the virus envelope, and is responsible not only for binding of virions to the cell surface but also for entry of the virus genome into the cell via the fusion pathway.

HA is a 549-amino acid **type I transmembrane protein**, with its N-terminus exposed to the outside of the virion, a hydrophobic transmembrane domain near the C-terminus, and a short C-terminal tail projecting inward (Figure 18.4). HA is first synthesized in the cell as a fusion-incompetent protein that is inserted into the membrane of the endoplasmic reticulum and transported via the Golgi apparatus to the plasma membrane, where it is incorporated into virions (see below).

Cleavage of HA by cellular proteases into two subunits activates its ability to carry out membrane fusion. HA_1, the surface subunit, contains the sialic acid-binding domain. HA_2 is anchored in the virus envelope and contains a hydrophobic **fusion peptide** near the amino terminus created by protease cleavage. This fusion peptide is hidden within a hydrophobic region of HA, near the lipid membrane of the virus envelope.

After binding to cell receptors, virions enter the cell within an endosomal vesicle. As the pH of the endosome drops, HA undergoes a major conformational change that is similar to the opening of a coiled spring. The N-terminal part of HA_2 forms a rigid alpha-helical structure, moving the fusion peptide out of the hydrophobic region of HA and toward the endosomal membrane, where it inserts itself into the membrane (see Figure 4.4 in Chapter 4 for details). At the same time, the part of HA_2 that is inserted into the virus envelope is brought closer to the extended N-terminal region. One end of HA_1 is now bound to the endosomal membrane, and the other end is bound to the virus envelope; these two membranes are brought into close proximity by this conformational change, resulting in membrane fusion.

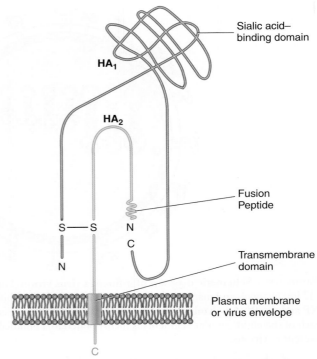

Figure 18.4 Influenza A hemagglutinin protein (HA). A schematic diagram showing the HA_1 (blue) and HA_2 (green) fragments created by proteolytic cleavage of the full-length polypeptide (HA_0) shown inserted in the plasma membrane or the virus envelope at neutral pH. N- and C-termini are shown. The sialic acid–binding domain is shown folded at the top. The two polypeptide chains are linked by a S–S bond between cysteines. This diagram does not attempt to describe the exact three-dimensional structure of the protein (see Figure 2.12).

M2 is an ion channel that facilitates release of nucleocapsids from the virion

The M2 protein is crucial in the next phase. It is a small 97 aa protein with an N-terminal external domain (residues 1–23), a single transmembrane domain (residues 24–44), and a larger internal domain (residues 45–97). M2 is an essential component of the virus envelope because of its ability to form a highly selective transmembrane ion channel that allows H^+ ions to penetrate the membrane. M2 forms tetramers by lining up four parallel transmembrane α-helices, thus creating a small pore in the viral envelope.

Release of the viral nucleocapsids from the virion is facilitated by the M2 H^+ ion channel. This channel allows protons to enter the interior of the virion during acidification within the endosome (see Figure 36.3 in Chapter 36 on antiviral chemotherapy). The low pH within virions weakens the interaction of the matrix protein M1 with viral nucleocapsids, facilitating their release into the cytoplasm upon membrane fusion.

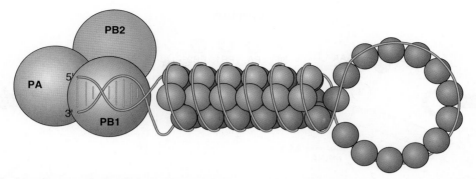

Figure 18.5 Helical nucleocapsid of influenza virus. The nucleocapsid protein (NP, small green spheres) binds to negative-strand viral genome RNA (blue line) to form a nucleocapsid with a twin helical conformation. A trimer consisting of RNA polymerase proteins PA, PB1, and PB2 is bound via PB1 to the 5' and 3' ends of the RNA. Influenza type A has eight distinct genome segments each assembled in a helical nucleocapsid.

The influenza A virus M2 protein contains a His-XXX-Trp motif (where X can be any amino acid) that appears to constitute the main functional element of the M2 channels. This sequence motif (positions 17–21) is the only conserved region shared with the similarly functioning M2 protein of influenza B virus; it is also found in the p7 proteins of some genotypes of hepatitis C virus (a flavivirus). The anti-influenza virus drug **amantadine** is a specific blocker of the M2 H$^+$ channel. In the presence of amantadine, release of viral nucleocapsids is incomplete, blocking infection (see Chapter 36).

Nucleocapsids enter the nucleus, where mRNA synthesis and RNA replication occur

Influenza virus nucleocapsids contain multiple copies of the NP protein that wrap the single-stranded genome RNAs in an unusual twin helical conformation with a central loop (Figure 18.5). Each NP subunit binds to about 20 nucleotides of viral RNA. Nucleocapsids also contain a trimer of RNA polymerase proteins PA, PB1, and PB2. Sequences of the terminal 12 to 13 nucleotides at both the 5' and 3' ends of viral RNAs are highly conserved among all genome segments of a given virus type, and are highly self-complementary. Distinct sites within the PB1 protein bind to the conserved sequences at the 5' and 3' ends of the genome RNA (Figure 18.5). The two ends of genome RNAs are therefore maintained in close proximity. In spite of a high degree of self-complementarity, however, they are not believed to be hydrogen bonded to each other.

Unlike most other RNA-containing viruses, orthomyxoviruses replicate in the nucleus. This complicates the machinery required for virus replication (Figure 18.6). Genomes of infecting virions are first transported into the nucleus, where they are transcribed; viral mRNAs are then sent to the cytoplasm to be translated; viral proteins are subsequently sent to the nucleus where they direct genome replication and form nucleocapsids

with newly replicated genomes; and these nucleocapsids are finally transported back to the cytoplasm for assembly of virions at the plasma membrane.

Nucleocapsids are imported from the cytoplasm to the nucleus via the nuclear pore complexes (step 1, Figure 18.6). The NP protein and polymerase proteins contain nuclear localization signals that mediate interactions with cellular **importin**-α. This complex interacts with importin-β, which docks with nuclear pore complexes and promotes entry of its cargo into the nucleus. At this point, the virus must begin making messenger RNAs ("transcription") by copying its negative-sense RNA genome segments (step 2). Surprisingly, the viral transcriptional machinery cannot make mRNAs on its own, but uses cellular RNAs as primers for initiating viral mRNA synthesis (Figure 18.7).

Capped 5' ends of cellular premessenger RNAs are used as primers for synthesis of viral mRNAs

The PB2 protein in nucleocapsids contains a domain that recognizes the 5' cap structure found on all eukaryotic mRNAs. Capped cellular premessenger RNAs are synthesized in the nucleus by cellular RNA polymerase II, and undergo polyadenylation and splicing before being transported as mature mRNAs to the cytoplasm. PB2 binds to the capped end of a cellular premessenger RNA, and PB1 then acts as a nuclease, cleaving the bound premessenger RNA at an A or G residue 10 to 13 nt from its 5' cap (Figure 18.7, top left).

This capped RNA fragment, still bound to PB2, is then used as a primer to begin copying the genome RNA. The terminal A or G residue is hydrogen-bonded to the 3'-terminal U residue on genome RNA, and PB1 acts as an RNA polymerase by adding ribonucleotides complementary to the genome RNA sequence, proceeding toward its 5' end. The role of PA in transcription is not well understood, and it is not shown in Figure 18.7.

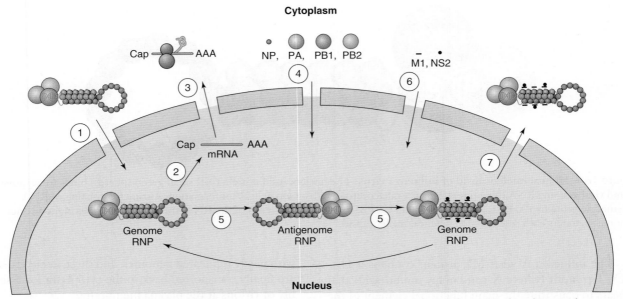

Figure 18.6 **Transport of influenza virus RNAs and proteins between nucleus and cytoplasm.** Viral nucleocapsids are transported through the nuclear pores into the nucleus (1), where viral mRNAs are synthesized (2). These mRNAs are transported back to the cytoplasm (3) where they are translated. Viral proteins are transported back into the nucleus (4 and 6), where they mediate genome replication (5) and transport of progeny nucleocapsids back to the cytoplasm (7).

The resulting capped viral mRNAs have heterogeneous sequences for the first 10 to 13 nt at their 5′ ends, derived from whatever cellular pre-mRNAs were used to prime transcription. Orthomyxoviruses therefore can make capped messenger RNAs even though they do not code for a capping enzyme. Bunyaviruses use a similar cap-stealing mechanism to prime viral mRNA synthesis, but they replicate in the cytoplasm and use mature cellular mRNAs as a source of caps (Chapter 17). In contrast, a number of cytoplasmic RNA viruses such as togaviruses, paramyxoviruses, and reoviruses cap their transcripts with virus-coded enzymes.

This intricate mechanism explains why the replication of influenza virus is blocked by treatment of infected cells with inhibitors of cellular RNA synthesis such as **actinomycin D** or **α-amanitin**. Cellular pre-messenger RNAs are rapidly spliced and exported to the cytoplasm, and therefore the pool of pre-messenger RNAs is quickly depleted when their synthesis is inhibited. In the absence of sufficient pre-messenger RNAs to serve as primers, influenza virus cannot make its own mRNAs. Thus virus replication is indirectly dependent on a healthy cellular RNA synthesizing system.

Viral mRNAs terminate in poly(A) tails generated by "stuttering" transcription

Each genome RNA segment contains a short poly(U) stretch 15 to 22 nt from the 5′ end of the RNA. When the viral RNA polymerase reaches this poly(U) sequence, it pauses and stutters, reading through the

poly(U) sequence a number of times and repeatedly adding complementary A residues to the transcript to generate a poly(A) tail (Figure 18.7). The RNA polymerase eventually terminates transcription at that position. This stuttering mechanism recalls a similar mechanism used by paramyxoviruses and rhabdoviruses (Chapter 15) for polyadenylation of their mRNAs.

Stuttering and termination of transcription on influenza genome RNAs is believed to result from blockage of further movement of the RNA polymerase by PB1 itself, because it remains bound to the terminal 12 to 13 nt at the 5′ end of the same genome RNA it is copying. This implies that the template RNA is actually threaded through the RNA polymerase complex as elongation proceeds, while its 5′ end remains bound to the complex. Figure 18.7 does not attempt to show this spatial relationship between the polymerase complex and the ends of the genome RNA.

Two influenza A mRNAs undergo alternative splicing in the nucleus

Transcription generates a complete set of eight viral mRNAs, one from each genome segment. Six of these mRNAs are exported directly to the cytoplasm where they are translated by ribosomes (step 3, Figure 18.6). However, RNAs generated from genome segments 7 and 8 (those encoding the M1 and NS1 proteins) contain splicing consensus sequences that are recognized by the cellular splicing machinery in the nucleus. Unlike many host cell mRNAs, which are usually spliced to

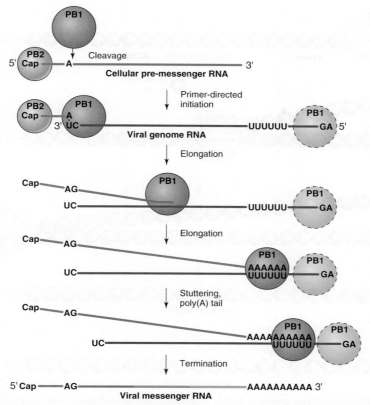

Figure 18.7 Production of influenza virus messenger RNAs by cap-stealing. In viral nucleocapsids, PB2 (small green sphere) binds to the capped 5' end of a cellular premessenger RNA (orange line) and PB1 (orange sphere) cleaves the RNA 10–13 nt from its 5' end, preferentially after an A or G residue. PB1 then uses this capped fragment as a primer for copying genome RNA (dark blue line) into viral mRNA (light blue line). Transcription ends by stuttering and termination at an oligo U region near the 5' end of genome RNA. PB1 (light orange sphere with dotted outline) is probably also bound to the 5' end of genome RNA as it copies from the 3' end, and may block transcription at the oligo U region. See text for details.

completion, only a fraction of these two viral mRNAs are spliced, and both unspliced (M1 and NS1) mRNAs and spliced (M2 and NS2) mRNAs are exported from the nucleus. The ratio of unspliced to spliced segment 7 and 8 viral mRNAs in the cytoplasm is approximately 9:1. This ensures that M2 and NS2 are made in lesser amounts than the highly abundant M1 and NS1 proteins.

Genome replication begins when newly synthesized NP protein enters the nucleus

Viral proteins that contain nuclear localization signals, including NP and the RNA polymerase subunits, are transported from their site of synthesis in the cytoplasm back into the nucleus (step 4, Figure 18.6). Their presence enables the replication machinery to begin to produce full-length, plus-strand copies of the eight viral genome segments (step 5, Figure 18.6; Figure 18.8). In contrast to capped viral mRNAs, these full-length transcripts contain 5' triphosphate groups and result from unprimed transcription initiation.

The mechanism of switching from primed to unprimed RNA synthesis depends on the presence of

free NP protein, but is not fully understood. NP interacts either with the RNA polymerase complex or with the 5' and 3' ends of the genome RNAs in nucleocapsids, and the RNA polymerase begins copying the genome RNA directly from its 3' terminus, with no intervening primer. As this plus-strand copy is synthesized it is complexed with NP protein, eventually forming nucleocapsids containing plus-strand antigenome RNAs. In this mode, the RNA polymerase does not stutter and terminate at the poly(U) stretch, but continues to the 5' end of the genome RNA, generating a full-length plus-strand copy. This presumably occurs because the PB1 subunit has been displaced and is no longer bound to the 5' end consensus sequence on the RNA genome.

These plus-strand antigenome RNAs are then themselves copied in a similar fashion by the RNA polymerase complex, in the presence of NP, to generate nucleocapsids containing full-length, minus-strand genomes (Figure 18.8). Newly synthesized nucleocapsids containing minus-strand genomes can be transcribed to produce more viral mRNAs or can generate more full-length, plus-strand RNAs for genome replication (Figure 18.6); the balance between these activities depends on the amount of NP

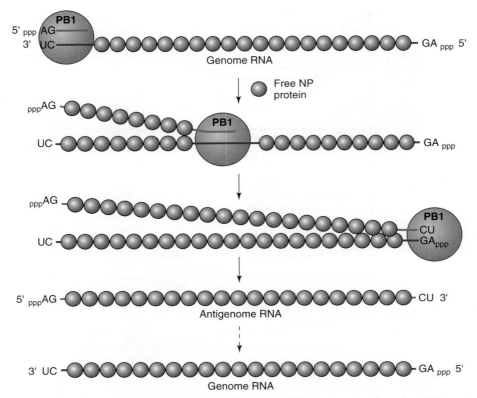

Figure 18.8 Replication of influenza virus genome RNA. Genome replication begins when free NP protein enters the nucleus. Unlike transcription, genome replication does not use a primer but begins at the 3' end of genome RNA and continues to the 5' end, making a complete copy. PB1 (orange sphere) copies genome RNA into a plus-strand complementary RNA that is coated with NP protein as a viral nucleocapsid. This RNA is subsequently copied into progeny genome RNA by the same mechanism.

protein present in the nucleus. Because NP is bound to growing RNA chains and incorporated into nucleocapsids, its concentration in the nucleus will drop during genome replication, favoring mRNA synthesis until more NP is imported into the nucleus. NP does not bind to influenza virus mRNAs, presumably because they are capped and contain cellular mRNA sequences, not viral sequences, at their 5' ends.

Nucleocapsids are exported from the nucleus in a complex with matrix protein and NS2

Remember that during the initial stages of infection, nucleocapsids are dissociated from the matrix protein as a result of low pH within the virion, brought about by the activity of the M2 ion channel. The nucleocapsids are subsequently transported into the nucleus via nuclear localization signals on the NP and polymerase proteins. Now, nucleocapsids must be transported back out of the nucleus to the cytoplasm, so that they can be incorporated into progeny virions. How does this reverse transport occur?

Current evidence indicates that the matrix protein (M1) is imported into the nucleus and binds to

nucleocapsids, forming a complex (step 6, Figure 18.6). At this point the NS2 protein comes into play. NS2 contains a nuclear export signal, and it can also bind to the matrix protein. It is believed that NS2 therefore piggybacks onto nucleocapsid/matrix protein complexes and signals the cellular protein export system to transport the complexes through the nuclear pores to the cytoplasm (step 7, Figure 18.6).

The NS1 protein interferes with polyadenylation of cellular mRNAs

Although the NS1 protein is the most abundant viral protein in virus-infected cells, its roles have only recently begun to be appreciated. NS1 can bind to both single- and double-stranded RNA and to two host cell proteins: (1) cleavage and polyadenylation specificity factor (CPSF) and (2) poly(A)-binding protein II (PABII). These two proteins are crucial players in the formation of polyadenylated cellular messenger RNAs in the cell nucleus.

CPSF, as its name implies, is responsible for the cleavage of cellular pre-mRNAs shortly downstream of the consensus AAUAAA polyadenylation signal; PABII facilitates production of full-length poly(A) chains on the 3' terminus created by cleavage. Binding of NS1 to these proteins blocks their functions, and therefore inhibits the normal 3' end cleavage and polyadenylation of cellular mRNAs. Because the polyadenylation of viral mRNAs occurs by a distinct mechanism and therefore does not depend on these cellular enzymes, the synthesis of viral mRNAs is not affected.

The cap-binding and endonuclease activities of the influenza PB2 and PB1 proteins lead to the degradation of many cellular pre-mRNAs in virus-infected cells. As a combined result of interference with proper 3' end processing and cleavage near their 5' cap structures, most cellular pre-mRNAs are degraded in the nucleus, and cellular mRNA production is therefore suppressed.

The NS1 protein also suppresses a variety of host cell antiviral response pathways

Perhaps more important to the virus is inhibition of the innate cellular antiviral responses (see Chapter 33) by NS1. NS1 binds to and inhibits signaling by RIG-1, a cytoplasmic sensor that responds to RNA virus infections by the induction of interferon synthesis via interferon response factor-3. NS1 also activates the phosphatidylinositol 3-kinase (PI3K)/ Akt pathway, leading to inhibition of caspase-9 activity and suppression of apoptosis. Furthermore, NS1 binds to and sequesters double-stranded RNA, thus suppressing activity of two major interferon-stimulated antiviral mechanisms, **double-stranded, RNA-dependent protein kinase (PKR)** and 2', 5'-oligo(A) synthetase/Ribonuclease L. Both of these pathways are activated by double-stranded RNAs in virus-infected cells.

The four C-terminal residues of the NS1 protein constitute a PDZ ligand domain of the consensus sequence X-Ser/Thr-X-Val (where X is any amino acid). This domain is a virulence determinant that can interact with PDZ-binding protein(s) and modulate pathogenicity through additional mechanisms; mutations in this small domain have strong effects on virus pathogenicity. PDZ domains are protein–protein recognition modules that organize diverse cell-signaling assemblies. NS1 has also been shown to prevent the maturation of human primary dendritic cells, thereby limiting host T-cell activation.

The importance of NS1 in facilitating virus replication was shown by poor replication of influenza viruses that have a mutated NS1 gene. These mutant viruses nevertheless replicate well in cells that have a defective interferon response. The NS1 protein therefore can be considered as a potential target for chemotherapeutic agents to suppress influenza virus infection.

PB1-F2 may contribute in suppression of the host immune response

A novel alternate reading frame that encodes the 87-90 amino acid PB1-F2 protein has recently been discovered in the PB1 polymerase gene segments of certain influenza A virus strains. PB1-F2 is localized in mitochondria. It has been shown to overcome host defense mechanisms and to enhance pathogenicity, apparently by inducing increased apoptosis in host immune cells responding to influenza virus infection.

Viral envelope proteins assemble in the plasma membrane and direct budding of virions

Viral hemagglutinin, neuraminidase, and M2 ion channel proteins are synthesized and inserted into membranes of the endoplasmic reticulum. The cellular protein-trafficking machinery translocates these proteins via the Golgi network to the cell surface, where virus assembly takes place. These three proteins accumulate in regions of the membrane enriched in cholesterol, named **lipid rafts**, which are believed to be the site of virion formation by budding.

The cytoplasmic tails of HA (10 to 11 amino acids, Figure 18.4), NA (6 amino acids), and M2 (54 amino acids) are crucial to virion formation, probably because they interact with matrix protein (M1) bound to virus nucleocapsids and transported from the nucleus to the cytoplasm. Artificial virus-like particles (VLPs) containing influenza surface proteins can be assembled in the absence of M1. However, M1 is required for the production of infectious VLPs. The function of M1 may predominantly be to package the genome in virions. M1 seems to interact directly with the cytoplasmic tail of M2 to promote recruitment of internal viral proteins and viral nucleocapsids to the plasma membrane, thus facilitating efficient assembly and budding of infectious virions.

Orthomyxovirus virions must contain at least one copy of each genome segment in order to be able to initiate an infection. Recent evidence suggests that most influenza A virus particles contain exactly one copy of each genome segment in an ordered bundle that can be seen in high-resolution electron micrographs (Figure 18.9). Seven nucleocapsids are arranged parallel to each other as a cylindrical barrel and an eighth nucleocapsid lies in the center of the bundle. The slightly enlarged "head" ends of filamentous particles (see Figure 18.2) also contain a single bundle of eight nucleocapsids, and their thinner tails appear to be empty extensions of the envelope.

This structural information implies that one copy of each of the eight different nucleocapsids is incorporated into a bundle either before or during virus budding. Experiments using deletion mutants have shown

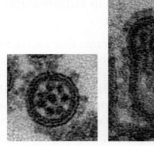

Figure 18.9 Selective packaging of influenza virus nucleocapsids. Electron micrographs of elongated influenza virions budding from infected cells cultured *in vitro*, and sectioned in a horizontal plane (left) or a vertical plane (right), show a specific arrangement of eight internal rod-like structures of different lengths (seen in serial sections through particles, not shown). These correspond to the eight viral genome RNAs assembled in nucleocapsids.

that RNA sequences in the 5' and 3' coding and noncoding regions of each genome segment are required to incorporate the segment into virions. This implies that viral proteins recognize and interact with specific RNA sequences in the nucleocapsids to incorporate them, one by one, into the bundles that are packaged into virions during budding.

Neuraminidase cleaves sialic acid, the cellular receptor that binds to HA

The roles of HA and M2 in virus entry were discussed above. Influenza neuraminidase (NA), in contrast to HA, is a **type II transmembrane protein** with a short cytoplasmic N-terminus, a membrane-spanning domain, and a long C-terminal domain that extends outward from the virus envelope.

Neuraminidase was named for its ability to cleave terminal N-acetyl neuraminic acid (sialic acid) from oligosaccharides located on mucoproteins, cell-surface glycoproteins, and glycolipids. When HA binds to sialic acid residues on these macromolecules, NA can reverse this binding by cleaving the bound sialic acid residue from the remainder of the oligosaccharide. The ying-yang relationship between the activities of these two influenza A surface glycoproteins may help virions reach their target cells by releasing them from the surface of virus-producing cells or from mucoproteins that are abundant in the respiratory tract.

Influenza virus strains vary in both transmissibility and pathogenicity

Most influenza A viruses that circulate in wild or domestic animals (particularly in ducks, chickens, and pigs) do not readily infect humans. This host specificity resides in

certain viral proteins that have adapted to function optimally in a particular species. These include HA, NA, and the RNA polymerase proteins PA, PB1, and PB2.

For example, HA proteins of most avian influenza A viruses bind preferentially to α-2,3 sialic acid, the major form of sialic acid found in the avian enteric tract. In contrast, human influenza A viruses code for HA proteins that bind preferentially to α-2,6 sialic acid, the major form found in the human respiratory tract. Furthermore, some HA proteins are readily cleaved by intracellular **furin** proteases, allowing them to infect many cell types because their fusion mechanism is activated upon cleavage. Other HA proteins are not cleaved intracellularly during virus formation, but require activation by trypsin cleavage after release of virions from an infected cell. Trypsin expression is limited to the upper respiratory tract in humans, and therefore these viruses are less likely to infect lung and other tissues in the body, leading to milder and more self-limiting infections.

Genetic variability generates new virus strains that can cause pandemics

Unlike what is seen with a number of other respiratory viruses, protective immunity after infection with influenza A virus does not last for a lifetime. This results from the ease with which antigenic domains in the virus envelope glycoproteins, HA and NA, can change. Antigenic change occurs in two ways: (1) slowly but continuously ("antigenic drift"), and (2) suddenly but episodically ("antigenic shift").

Antigenic drift results from the accumulation of point mutations within regions of the envelope glycoprotein genes that code for the antigenic domains of these proteins. RNA polymerases can make errors in incorporation of nucleotides into genome RNAs during replication, and do not have the proofreading mechanisms possessed by many DNA polymerases. Therefore, as RNA viruses replicate and spread from cell to cell or animal to animal, variants in antigenic domains of the envelope glycoproteins appear. Because these protein variants are not as well recognized by the immune system as those of the original infecting virus strain, these viruses can selectively replicate and be propagated.

This drift mechanism is common to a number of RNA viruses, including human immunodeficiency virus, and is a major complicating factor in production of vaccines. Antigenic drift has resulted in the generation of numerous influenza A subtypes with antigenically distinct HA and NA surface glycoproteins. So far, 16 different HA subtypes (H1 to H16) and 9 NA subtypes (N1 to N9) have been identified in naturally occurring influenza type A viruses in birds, particularly in waterfowl.

Antigenic shift results from **reassortment** of influenza virus genes during mixed infections with two or more virus subtypes. Cells infected simultaneously with two virus strains can produce reassortant virions that contain some genome segments from one strain and some genome segments from the other strain (see Figure 17.5). If such a reassortant virus can replicate well in the animal host, it is propagated. By this mechanism, viruses with many combinations of the different HA and NA subtypes, as well as other viral genes, are generated in multiply infected animals in the wild.

Reassortment between animal and human strains can create a virus that can replicate in humans but has acquired an HA or NA subtype derived from the animal virus. Such a virus can rapidly spread throughout the human population if antibodies against that subtype are not present, and can cause a pandemic. This seems to be a rare event, as it has occurred only a few times in the past century, but the results can be disastrous.

The 1918 pandemic influenza A virus was probably not a reassortant virus

In an experimental *tour de force*, scientists recently recovered genome fragments of the influenza A virus implicated in the 1918 pandemic from preserved human tissue, and determined the complete genome sequence of this virus. The virus was demonstrated to be subtype H1N1. Surprisingly, analysis of its genome sequence suggests that this pandemic virus strain arose from an avian virus by mutation ("drift"), rather than by reassortment.

The polymerase proteins PA, PB1, and PB2 of the 1918 human virus differ by a total of only 10 amino acids from the consensus sequences of the same proteins made by most avain strains. The HA protein of the 1918 virus acquired the ability to bind to the α-2,6 sialic acid present in the human respiratory tract by virtue of a single amino acid change from an HA that binds to the avian α-2,3 sialic acid. The 1918 strain HA lacks a furin cleavage site, but the accompanying NA protein enhances the cleavage of HA in a trypsin-independent fashion that is not yet fully understood, increasing the infectivity and spread of this virus.

An infectious influenza virus with the sequence of the 1918 strain was generated in a high-level biosecurity laboratory by reverse genetic engineering and its properties were tested. This virus showed extremely high pathogenicity for mice and for embryonated chicken eggs, and these properties required the 1918 versions of HA, NA, and the three polymerase proteins.

Genome sequences from some previous influenza A virus strains confirm the antigenic shift hypothesis

The H1N1 subtype that emerged in 1918 was still present in 1933, when the first human influenza virus was isolated in the laboratory, and continued to circulate among humans into the 1950s. In 1957, a new influenza A virus suddenly appeared in Southeast China. This subtype H2N2 virus caused the "Asian flu" pandemic. At some point during this pandemic, H1N1 strains disappeared completely from the human population.

H2N2 strains persisted until 1968, when another new virus suddenly appeared in China. This virus, subtype H3N2, had a new hemagglutinin gene and was the cause of the "Hong Kong" influenza pandemic, during which H2N2 viruses disappeared completely from people. In 1977, the H1N1 virus reappeared in humans in China and then spread around the world. Presently, variants of H3N2 and H1N1 viruses continue to circulate in human populations along with influenza type B strains.

The Asian (H2N2) and the Hong Kong (H3N2) viruses are reassortant viruses. In 1957, the new Asain human virus contained an HA gene, an NA gene, and a PB1 gene that were all apparently derived from an avian influenza virus, while the other genes were the same as those in the original H1N1 human virus. In similar fashion, the 1968 Hong Kong virus contained an HA gene and a PB1 gene derived from an avian virus but other genes derived from the previously circulating human H2N2 virus. Up to the present, influenza viruses that cause widespread human disease have been limited to strains that contain HA types 1, 2, or 3.

Highly pathogenic avian influenza A H5N1 strains in poultry farms are a potential threat but are poorly transmitted among humans

Starting in 1997, highly pathogenic avian H5N1 strains spread through commercial poultry farms in Hong Kong, and a small number of humans were infected with this virus, causing six deaths. These strains reappeared in multiple sites in Southeast Asia during subsequent years, requiring destruction of millions of chickens and resulting in serious and often fatal disease in a number of people directly exposed to infected birds. Further spread throughout the world is occurring, probably via infected migrating waterfowl.

Fortunately, these strains have not yet shown signs of spreading efficiently among humans. However, the danger of these highly pathogenic strains adapting to humans and causing a serious pandemic is present. The high density of domestic animals living in close proximity to humans in Southeast Asia, as well as the spread of these viruses worldwide by migratory wildfowl, increase the likelihood of generation of mutants or reassortant viruses that will acquire the ability to be transmitted among humans. However, this has not yet happened, and it is very difficult to predict whether such a virus will appear, particularly since determinants of pathogenicity and transmission are still incompletely understood.

Because of increased travel and rapid transportation, the next pandemic could be particularly hard to deal with. These developments have led to constant surveillance of avian viruses circulating in poultry farms worldwide, coupled with attempts to produce a human vaccine that could protect against H5N1 and other highly pathogenic influenza virus strains. National public health officials are also stockpiling chemotherapeutic agents active against influenza A, such as the neuraminidase inhibitors **zanamivir** and **oseltamivir** (see Figure 36.10a). These could slow down the spread of influenza A virus during a pandemic, or reduce symptoms in infected people.

A new pandemic strain of influenza A virus arose by genetic shift and spread worldwide in 2009

In April 2009, a new strain of human H1N1 influenza A virus was identified in the United States and Mexico, where it was initially associated with severe respiratory disease. Genomic analysis of this human virus showed that it was related to common reassortant swine influenza A viruses isolated in North America, Europe, and Asia, with the contribution of a single genome RNA segment from an avian strain (Figure 18.10). Because the versions of HA and NA in this strain had not previously been circulating among humans, a large part of the population was susceptible to virus infection.

Curiously, the HA protein of this virus strongly resembles the HA protein of the original 1918 pandemic virus. This protein "drifted" during the 40 or so years when it dominated human influenza strains, so that current "seasonal" H1N1 viruses do not induce immunity to the 2009 H1N1 virus. But the 1918 HA protein was present in viruses that circulated in swine, and it had varied little over the past 90 years, perhaps because commercial swine live short lives and therefore herds develop little immunity to the virus.

On June 11, 2009, the World Health Organization raised the warning level for this virus to the highest available—Phase Six—indicating that influenza had entered a pandemic stage of worldwide spread. As of March 7, 2010, WHO had counted more than 213 countries and overseas territories or communities that had reported laboratory-confirmed cases of pandemic influenza H1N1 2009, including at least 16,713 deaths. Surprisingly, many of the most serious cases of illness, and many deaths, occurred in young, healthy people, unlike the "seasonal" flu, which mostly affects older people. Most patients, however, had only mild symptoms and a short course of disease without severe complications or mortality.

All virus isolates were initially susceptible to the neuraminidase inhibitor oseltamivir (Tamiflu), which helped to shorten the fever period and quickly improved disease symptoms. However, strains resistant to the drug were reported within months, leading to controversy about its use in patients with mild disease.

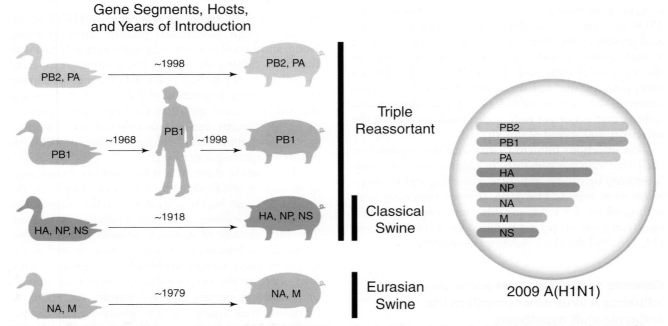

Gene Segments, Hosts, and Years of Introduction

Figure 18.10 Origin of genome segments of the 2009 H1N1 pandemic influenza A strain. The eight segments shown within the circle at the right are labeled to denote major proteins coded by each segment. Segments are color-coded to denote their origins, shown at the left, in birds, humans or swine, and the approximate year of transmission from one animal host to another is also shown. The 2009 H1N1 virus presumably arose as a recombinant between viruses of Eurasian and North American swine.

A massive vaccine development campaign allowed the first human vaccination campaigns against pandemic H1N1 virus to begin in October 2009. In early 2010, seasonal strains of influenza A H3N1 and H1N1 as well as influenza B virus began to cocirculate with pandemic H1N1 virus.

It is not yet clear why this pandemic virus was so mild in its impact. One answer may be that many people, particularly older individuals, had partial immunity against this virus because of previous exposure to strains of H1N1 virus that have circulated over the last 50 years in the human population. The pathogenicity of influenza virus strains is not yet completely understood, but appears to be caused by a constellation of separately assorting virulence determinants in multiple virus genes.

Vaccination against influenza is an effective prevention method

Inactivated influenza vaccines are manufactured every year; each year's vaccine reflects the strains of influenza A and B viruses circulating during the end of the preceding season. It is recommended that influenza vaccine be administered yearly to all individuals with significant chronic respiratory disease as well as all individuals over the age of 65. Vaccination in healthy adults is 60 to 90% effective in preventing clinical illness caused by virus strains similar to those used to manufacture the vaccine. Although this effectiveness is diminished in the elderly and chronically ill, vaccination still reduces illness and death in these populations. Current vaccines are highly purified, and the only common side effect is soreness at the site of vaccination. Because these vaccines are inactivated, they cannot cause influenza.

A new live influenza vaccine has recently been approved for use in humans between 6 and 55 years old. This cold-sensitive strain was developed and tested over a 30-year period. It can replicate in the (cooler) nasal passages but not in the bronchi or lungs, and therefore causes few or no disease symptoms. It is administered by a nose spray, in contrast to inactivated vaccines, which require injection. Furthermore, it protects 90% of recipients against infection by viruses of the same subtype. Further clinical studies are needed before this live vaccine can be used in people who are most susceptible to serious influenza disease, the very young and the very old.

KEY TERMS

α-amanitin
Actinomycin D
Amantadine
Antigenic drift
Antigenic shift
Ciliated epithelium
Cytokine
Double-stranded RNA-
    dependent protein
    kinase (PKR)
Epidemic
Furin
Fusion peptide
Hemagglutination assay
Hemagglutinin

Importin
Interferon
Lipid rafts
Mucociliary flow
Neuraminidase
Oseltamivir
Pandemic
Reassortment
Sialic acid
Type I transmembrane
    protein
Type II transmembrane
    protein
Zanamivir

FUNDAMENTAL CONCEPTS

• Influenza viruses cause serious acute disease in humans, and occasional pandemics.

• Influenza viruses are negative-strand RNA viruses whose eight genome segments code for a total of 11 different viral proteins.

• The influenza virus hemagglutinin (HA) protein binds to cell receptors and mediates fusion of the viral envelope with the endosomal membrane after endocytic entry into cells.

• The influenza virus NS1 and PB1-F2 proteins suppress a variety of host cell antiviral responses.

• The influenza virus M2 protein is an ion channel that facilitates release of nucleocapsids from the virion.

• Influenza virus nucleocapsids travel to the nucleus, where three viral RNA polymerase proteins (PA, PB1, and PB2) direct mRNA synthesis, using capped RNAs from the cell as primers, and primer-independent RNA replication.

• Influenza virus neuraminidase (NA) protein cleaves sialic acid, the cellular receptor that binds to HA, facilitating efficient virus release. Neuraminidase is the target for currently available anti-influenza chemotherapy.

• Influenza viruses exhibit unusual genetic variability by both rapid point mutations ("drift") and genome reassortment ("shift"), generating new virus strains that can cause pandemics.

REVIEW QUESTIONS

1. What are the host ranges of influenza virus types A, B and C?

2. Discuss the process of membrane fusion carried out by the influenza hemagglutinin.

3. How do influenza viruses manage to cap their own mRNAs and greatly reduce cellular protein synthesis in one fell swoop?

4. Which elements of the host antiviral responses does the influenza NS1 protein inhibit?

5. Are pandemic strains of flu always the result of reassortment, or have some been the result of antigenic drift? Give specific examples to support your answer.

Reoviruses

James D. Chappell
Terence S. Dermody

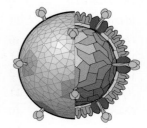

Reoviridae

From *r*espiratory *e*nteric *o*rphan viruses

VIRION

Naked icosahedral capsid (T = 13), Diameter 60–85 nm.
Capsid consists of two or three concentric protein shells.
Inner capsid (T = 1), or *core,* contains RNA polymerase and capping enzymes.

GENOME

Linear double-stranded RNA, 10–12 segments.
Total genome length 18–24 kb.

GENES AND PROTEINS

mRNAs are full-length copies of each genome segment.
Typically one protein is encoded per genome segment.
6–8 capsid proteins.
3–6 nonstructural proteins.

VIRUSES AND HOSTS

12 genera, including *Orthoreovirus, Rotavirus, Orbivirus.*
Infect humans (reoviruses types 1–3, rotaviruses, and Colorado tick fever virus), mammals, birds, fish, mollusks, plants, insects, and fungi.

DISEASES

Members of genus *Orthoreovirus* cause little or no disease in humans.
Rotavirus is an important cause of gastroenteritis worldwide and infant mortality in the developing world.
Viruses spread between hosts by direct transmission, contaminated food or water, or arthropod vectors.

DISTINCTIVE CHARACTERISTICS

Reovirus family has members that infect a broad range of hosts from fungi to humans.
mRNAs are synthesized and capped inside intact cores and extruded through channels into the cytosol.
Synthesis of double-stranded genome RNAs occurs within core-like subvirion particles.
A single copy of each gene segment is packaged into each virion by an unknown sorting mechanism.
Gene segments can be reassorted during coinfection of cells by different strains.

Reoviruses were the first double-stranded RNA viruses discovered

Reoviruses were first isolated from stool specimens of children during the 1950s by Albert Sabin, Leon Rosen, and their colleagues. *Reovirus* is an acronym for *r*espiratory *e*nteric *o*rphan virus; infections of humans by reoviruses types 1 to 3 usually involve the respiratory and intestinal tracts with minimal or no associated disease symptoms (therefore the virus is an "orphan").

In the early 1960s, Peter Gomatos and Igor Tamm noticed that **inclusion bodies** in reovirus-infected cells fluoresce greenish-yellow when stained with acridine orange, whereas single-stranded RNA fluoresces orange. They then demonstrated by chemical and physical studies that the viral genome consists of double-stranded RNA. These were the first double-stranded RNA viruses to be described.

The *Reoviridae* family presently includes 12 genera of double-stranded RNA viruses that infect a wide variety of animals, from mammals to insects, as well as plants and fungi (Table 19.1). The mammalian reoviruses are the type species of the *Orthoreovirus* genus, which also contains viruses that infect birds and reptiles. The term *reovirus* in its common usage refers to the mammalian reoviruses and will be used as such in this

Table 19.1 Genera within the *Reoviridae*

Genus	Number of gene segments	Hosts
Orthoreovirus	10	Mammals, birds, reptiles
Rotavirus	11	Mammals, birds
Orbivirus	10	Mammals, birds, arthropods
Coltivirus	12	Mammals, arthropods
Aquareovirus	11	Fish, mollusks
Cypovirus	10	Insects
Idnoreovirus	10	Insects
Fijivirus	10	Plants, insects
Oryzavirus	10	Plants, insects
Phytoreovirus	12	Plants, insects
Seadornavirus	12	Insects, humans
Mycoreovirus	11–12	Fungi

chapter. To date, only three mammalian reovirus serotypes have been identified despite isolation from a broad range of hosts. Three virus strains—type 1 Lang (T1L), type 2 Jones (T2J), and type 3 Dearing (T3D)—are the prototype members of the three reovirus serotypes. A distinct reovirus strain was isolated from a mouse in the Cameroon; this virus may represent a fourth mammalian reovirus serotype.

Some members of the *Reoviridae* are important pathogens

Other recognized genera of *Reoviridae* that infect animal hosts are *Rotavirus*, *Orbivirus*, *Coltivirus*, and *Aquareovirus*. Rotaviruses are responsible for gastroenteritis in animals and humans; they cause a large proportion of childhood gastroenteritis and are a major cause of infant illness and death in the developing world. New rotavirus vaccines show promise in reducing the morbidity and mortality associated with rotavirus infection.

Orbiviruses are transmitted by arthropod vectors, as are viruses of the very different bunyavirus, flavivirus, and togavirus families. These virus groups can replicate in both warm-blooded mammals and cold-blooded arthropods. The best-studied orbivirus is bluetongue virus, an economically significant pathogen of sheep and cows named for a symptom affecting sick animals. Coltiviruses are also transmitted by arthropod vectors, and the prototype member, *Colo*rado *ti*ck fever virus, can cause serious neurologic disease in humans. Aquareoviruses infect fish and mollusks.

Reoviruses are excellent models for studies of viral replication and pathogenesis. Bernard Fields and his coworkers used genetic approaches to study the pathogenesis of reoviruses in rodent models of infection and disease, especially viral neurologic illness.

Reoviruses also have been used to study RNA synthesis. Early work by the groups of Wolfgang Joklik and Angus Graham showed that the isolated viral core contains all of the enzymes required to synthesize messenger RNA molecules from the double-stranded genome templates. Aaron Shatkin and colleagues were the first to describe the methylated cap structure that adorns most eukaryotic messenger RNA molecules and is required for normal translation; this work was done with RNAs synthesized *in vitro* from reovirus cores.

Reoviridae have segmented genomes made of double-stranded RNA

Reoviridae that infect animals have genomes consisting of 10 to 12 discrete, linear segments of double-stranded RNA (Figure 19.1). Reovirus gene segments range from approximately 1200 to 3900 bp in length. They are classified as large (L), medium (M), or small (S) based on their relative mobility in polyacrylamide gels; there are three L genes, three M genes, and four S genes.

Mixed infection with different virus strains produces a collection of progeny **reassortant viruses** in which the inheritance of specific parental gene segments can be deduced (Figure 19.2). This feature allows the determination of which viral proteins are involved in particular biologic properties of the virus. Analysis of reassortant viruses to delineate the genetic basis of specific viral phenotypes is termed *reassortant genetics*.

Like most viruses, reoviruses demonstrate an extraordinary degree of genetic economy; nontranslated regions on different gene segments range from 12 to 83

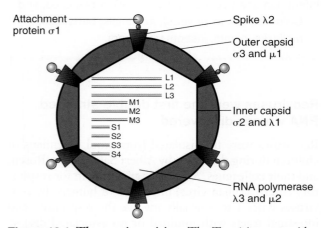

Figure 19.1 The reovirus virion. The T = 1 inner capsid encloses the 10 double-stranded RNA gene segments and is covered by a T = 13 outer capsid. Twelve spikes project from the core through the outer capsid. Gene segments are classified as large (L), medium (M), or small (S) based on relative electrophoretic mobilities. Viral capsid proteins are labeled.

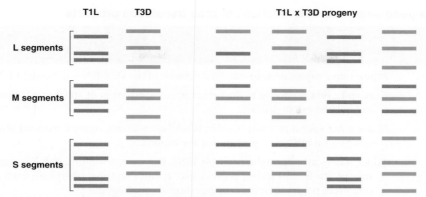

Figure 19.2 Generation of reassortant viruses. Reovirus strains are distinguishable by unique electrophoretic profiles of their double-stranded RNA gene segments. Coinfection of cells with different viral strains produces a collection of progeny reassortant viruses containing various combinations of gene segments from each parent. Shown here are schematized diagrams of electrophoretic profiles of prototype strains type 1 Lang (T1L) and type 3 Dearing (T3D) along with some reassortant progeny.

nucleotides in length (Figure 19.3). The extreme 5'-terminal four nucleotides (5'-GCUA-3') and 3'-terminal five nucleotides (5'-UCAUC-3') are conserved among all 10 reovirus gene segments. Furthermore, sizable regions of subterminal sequence are conserved among homologous genes of different virus strains. Nucleotide sequence conservation extends inward from both nontranslated regions into the protein-coding region. This pattern indicates that there are selection pressures on reovirus gene segments to maintain certain nucleotide sequences independent of their protein-coding function. Suggested functions of these conserved sequences are recognition by viral RNA polymerase, RNA packaging into capsids, and translational control, but the role of these sequence elements has not been precisely defined.

Nine of the 10 reovirus gene segments code for a single protein; the exception is the S1 gene, which codes for both the σ1 attachment protein and the σ1s nonstructural protein in overlapping reading frames. Viral proteins are named with the Greek letters λ (lambda), μ (mu), and σ (sigma), corresponding to the size

classes (L, M, and S) of their respective gene segments. Unfortunately, the numbering schemes used for genes and proteins, based on their relative migration rates in polyacrylamide gels, do not perfectly correspond. Salient features of the eight structural and three replication proteins are provided in Table 19.2.

Reovirus virions contain concentric layers of capsid proteins

Virions of mammalian reoviruses are non-enveloped particles approximately 85 nm in diameter (Figure 19.1). The virion is formed from eight proteins arranged in an unusual two-layered structure of concentric protein shells, each with icosahedral symmetry. The T = 1 inner capsid surrounds the 10 tightly packed, double-stranded RNA gene segments, forming the viral core. It is assembled from two major proteins, λ1 and σ2 (120 and 150 copies each, respectively), and minor core proteins μ2 (probably 24 copies) and λ3 (probably 12 copies), which constitute the viral RNA polymerase.

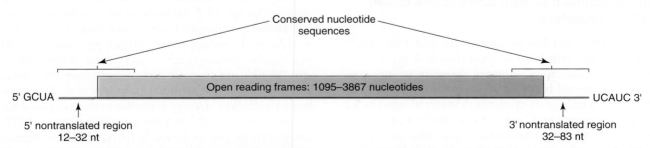

5' GCUA

Conserved nucleotide sequences

Open reading frames: 1095–3867 nucleotides

UCAUC 3'

5' nontranslated region
12–32 nt

3' nontranslated region
32–83 nt

Figure 19.3 General characteristics of a mammalian reovirus gene segment. Only the plus strand is shown here. Four nucleotides at the 5' end and five at the 3' end are invariant among all 10 viral gene segments. Sequence ranges are based on strain T3D.

Table 19.2 Reovirus gene segments and functions of their translation products

Gene segment	Protein	Functions
L1	λ3	Minor inner-capsid protein; catalytic subunit of the viral RNA-dependent RNA polymerase
L2	λ2	Core spike protein; forms pentameric turret for insertion of attachment protein σ1; adds a 5' methylated cap to mRNAs
L3	λ1	Major inner-capsid protein; possible RNA helicase; may remove terminal phosphate from newly synthesized mRNA in preparation for capping
M1	μ2	Minor inner-capsid protein; probable RNA polymerase subunit; stabilizes microtubules; nonspecific RNA-binding protein; may remove terminal phosphate from newly synthesized mRNA in preparation for capping; interferon antagonist
M2	μ1	Major outer-capsid protein; important for penetration of endosomal membrane by the viral core; probably regulates late mRNA synthesis from secondary transcriptase particles; affects efficiency of reovirus-induced apoptosis
M3	μNS	Nonstructural replication protein; forms cytoplasmic inclusion structures; associates with nonstructural protein σNS and minor core protein μ2; first viral protein found in association with viral mRNA
S1	σ1	Attachment protein that binds to carbohydrate and junctional adhesion molecule A (JAM-A); determines virus serotype; determines cell and tissue tropism and route of virus spread
	σ1s	Nonstructural replication protein; required for reovirus-induced cell-cycle arrest
S2	σ2	Major inner-capsid protein; weakly binds to dsRNA
S3	σNS	Nonstructural replication protein found early in viral inclusions; associates with nonstructural protein μNS; nonspecific RNA-binding protein with high affinity for single-stranded RNA
S4	σ3	Major outer capsid protein; forms protective cap over μ1; nonspecific dsRNA-binding protein; influences efficiency of translation; associated with viral mRNA early in replication cycle; inhibits activation of PKR

The T = 13 outer capsid is constructed from complexes of μ1 and σ3 proteins, present in 600 copies each. A spike, formed by pentamers of the λ2 protein (60 copies), projects from the inner capsid through the outer capsid at the 12 axes of fivefold symmetry. The RNA polymerase complexes are located at the base of each spike in the core. Inserted within a turret-like opening of the spike pentamer is the trimeric attachment protein, σ1 (36 copies).

The attachment protein binds to one or two cellular receptors

The reovirus attachment protein σ1 is a long, fiber-like molecule with head-and-tail morphology. The σ1 protein has the most sequence diversity of any reovirus protein and is the serotype-specific antigen. The N-terminal tails of σ1 trimers are inserted into the λ2 spikes, whereas the C-terminal globular heads project away from the virion surface. The structure of this protein closely resembles that of the adenovirus fiber protein, also a trimer (see Chapter 23).

The attachment protein of type 3 reovirus contains two receptor-binding domains: one in the tail, which binds to **sialic acid**, and another in the head, which binds to **junctional adhesion molecule A (JAM-A)** (Figure 19.4). JAM-A is a transmembrane protein with two immunoglobulin-type domains in the extracellular region. It is located at **tight junctions** of epithelial and endothelial cells and mediates cell-to-cell interactions. Attachment proteins of the two other reovirus serotypes also bind to JAM-A but not to sialic acid. Binding of reoviruses to JAM-A on endothelial cells that line blood vessels is required for spread of the virus via the circulatory system.

Reovirus serotype 3 virions are likely to bind first to sialic acid residues on cell-surface glycoproteins and glycolipids. Subsequently, virions can diffuse laterally ("walk") on the cell surface until they encounter and bind to JAM-A within a tight junction. This two-stage attachment strategy is similar to that proposed for several other viruses, including bacteriophage φX174, enteroviruses, adenoviruses, herpesviruses, and human immunodeficiency virus (Chapters 6, 11, 23, 24, and 29). Reovirus strains that do not recognize sialic acid may bind directly to JAM-A, although these strains may also use carbohydrate receptors other than sialic acid.

During entry, the outer capsid is stripped from virions and the core is released into the cytoplasm

Following binding to receptors on the cell surface, reovirus virions enter cells by clathrin-mediated endocytosis using a mechanism dependent on β1 integrin (Figure 19.4). It is not known how β1 integrin promotes endocytosis, but the simplest explanation is that following JAM-A engagement, the virus binds to the integrin via surface-exposed integrin-binding motifs Arg-Gly-Asp or Lys-Gly-Glu in the λ2 protein, and this promotes clathrin-dependent uptake. Adenovirus uses a similar mechanism to enter cells by first binding to the coxsackievirus and adenovirus receptor (CAR) and subsequently binding to αVβ3 and αVβ5 integrins.

In endocytic vesicles, virions undergo partial degradation by cellular proteases, resulting in the generation of **infectious subvirion particles (ISVPs). Cathepsins** B, L, or S, which are proteases located in endocytic vesicles, can carry out this reaction. Cathepsins are also required for entry of Ebola virus and SARS-coronavirus. Infectious subvirion particles can be formed extracellularly in the intestine or in the laboratory by *in vitro* treatment of reovirus virions with proteases. ISVPs may in fact be the form of virus particle that infects intestinal cells.

During conversion of virions to ISVPs, the attachment protein σ1 undergoes a conformational change and becomes elongated. The outer capsid protein σ3 is degraded and lost from virions, and the exposed μ1 protein is cleaved to form two fragments, δ and φ, that remain bound to the ISVP. These changes facilitate interactions of the ISVP with endosomal membranes, leading to delivery of the core particle into the cytoplasm and activation of the viral transcriptase. Although the precise mechanism of membrane penetration is not well understood, ISVPs are thought to breach the endosomal membrane by a process that requires μ1-mediated formation of small pores.

The reovirus disassembly intermediates can be visualized using the technique of **cryoelectron microscopy** and three-dimensional image reconstruction (Figure 19.5). Virions are spherical particles with a star-shaped density corresponding to pentamers of the λ2 protein at the fivefold axes and flower-like hexameric rings corresponding to the σ3 protein covering the remainder of the surface. Conversion to ISVPs involves removal of σ3 and extension of the attachment protein σ1 outward from the virion. Comparison of image reconstructions of the virion (85 nm in diameter) and ISVP (80 nm in diameter) reveals the loss of 600 finger-like subunits, which correspond to the 600 copies of σ3 in the virion.

During the transition from ISVP to core (70 nm in diameter), the attachment protein and the μ1 protein fragments are lost, leaving the λ2 spike protein and the viral core proteins. Concurrently, the spike protein in the core undergoes a major conformational change in which the turrets formed by λ2 pentamers open to allow egress of mRNA during transcription.

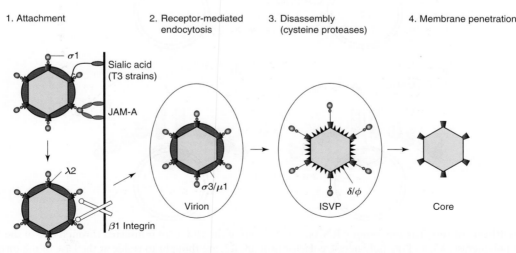

Figure 19.4 Stepwise disassembly of reovirus. Following attachment to cell-surface carbohydrate (α-linked sialic acid for type 3 reoviruses) and junctional adhesion molecule A (JAM-A), reovirus virions enter cells by clathrin-mediated endocytosis using a mechanism involving β1 integrin. Within an endocytic vesicle, the viral outer capsid undergoes cathepsin-mediated proteolysis, generating infectious subvirion particles (ISVPs). ISVPs interact with endosomal membranes, leading to delivery of transcriptionally active core particles into the cytoplasm.

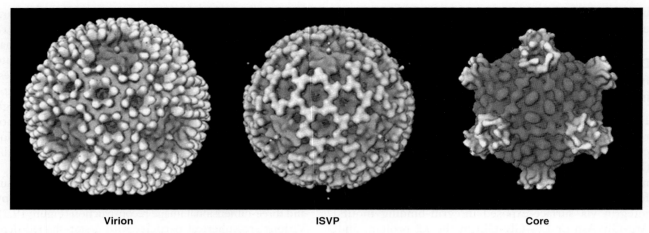

Virion	ISVP	Core

Figure 19.5 Structure of reovirus virions and subvirion particles. Surface-shaded views of the reovirus strain T1L virion, infectious subvirion particle (ISVP), and core. The images were obtained using cryoelectron microscopy and three-dimensional image reconstruction. Each particle is viewed along a threefold axis.

Enzymes in the viral core synthesize and cap messenger RNAs

The core particles released into the cytoplasm contain all of the machinery needed by the virus to make and cap messenger RNAs. These enzymes are activated on removal of the outer capsid, and ribonucleoside triphosphates required for RNA synthesis diffuse into the core from the cytoplasm.

Messenger RNA synthesis. The reovirus core is a highly structured "mRNA factory" (Figure 19.6). An RNA polymerase complex, composed of one molecule of λ3 and one or two molecules of μ2, is located at the base of each of the λ2 spikes. It is notable that different members of the *Reoviridae* family contain between 10 and 12 gene segments, which allows each segment to be transcribed at one of the 12 spike bases. All 10 reovirus double-stranded RNAs are simultaneously transcribed,

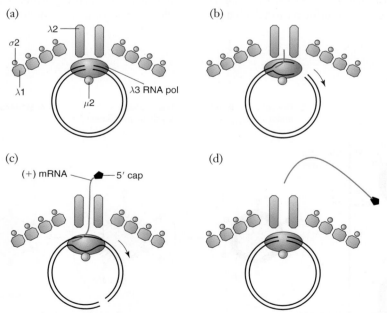

Figure 19.6 Synthesis of reovirus messenger RNAs. (a) Diagram of the viral core transected through one of the 12 spikes. The viral RNA polymerase, λ3, and the polymerase cofactor protein, μ2, are thought to reside at the base of the turret formed by pentamers of the λ2 protein. The viral gene segments may be maintained in circular form as shown here. (b–d) The double-stranded template RNA is locally unwound, possibly by a helicase function of the λ1 protein, and serves as a template for synthesis of full-length plus-strand mRNAs (light blue line). These nascent mRNAs are extruded through the λ2 turret, receiving a 5′ terminal methylated cap (black pentagon) in the process. The template RNA is pulled through the RNA polymerase in the direction shown by the arrows. The circular form of the genome template would allow for multiple rounds of transcription.

probably by translocation through fixed RNA polymerase complexes, generating a full-length, plus-strand RNA copy. This strategy has been termed the "moving template" model of transcription, because it postulates that the template RNA moves through the fixed RNA polymerase. It is possible that the double-stranded gene segments are maintained in circular form during transcription, facilitating reinitiation and multiple rounds of transcription at the same locus.

X-ray crystallographic studies of the λ3 protein complexed with oligoribonucleotides suggest that the growing messenger RNA chain passes up into the spike turret and the minus-strand template of the genome RNA passes down into the core interior via independent channels within the RNA polymerase. Core shell protein λ1 appears to possess RNA helicase activity and may locally unwind the double-stranded genome RNA to provide the polymerase access to the minus-strand template. Both λ1 and μ2 possess an enzymatic activity that can remove the γ phosphate from the 5'-terminal nucleotide to yield a 5'-diphosphate end; this prepares the RNA for cap addition.

Messenger RNA capping. During the initial stages of synthesis of each mRNA molecule, the λ2 spike protein catalyzes cap addition to the RNA 5' end. The spike protein has three distinct active sites for the three enzymatic activities required for capping: (1) formation of the cap itself by adding a GMP residue (derived from GTP by pyrophosphate release) to the 5'-diphosphate end of the RNA; (2) methylation of the cap guanine at the N7 position; and (3) methylation of the ribose sugar of the first template-coded G residue. These reactions probably occur in sequence as the RNA progresses through the λ2 spike during transcription.

Messenger RNA export. The capped RNA transcripts are extruded through channels within the spikes to the exterior, where they are released into the cytoplasm (Figure 19.6). This process can be visualized by electron microscopy of actively transcribing cores, which reveals multiple RNA transcripts emerging from core particles. The double-stranded RNA templates remain within the core and can be repeatedly transcribed, giving rise to multiple copies of all mRNAs.

Translation of reovirus mRNAs is regulated

Newly synthesized viral proteins can be detected in infected cells as early as two hours after infection, and as their synthesis increases, there is a concomitant decrease in synthesis of cellular proteins. Not all reovirus proteins are produced in the same amount. One factor affecting protein abundance is the length of the mRNA; the longer the transcript, the longer the time required for its synthesis, and therefore long mRNAs

are less abundant than shorter transcripts. Sequences surrounding the initiator AUG codons and differences in the length and secondary structure of 5' nontranslated regions also likely account for different levels of synthesis of viral proteins.

Unlike most cellular mRNAs, reovirus mRNAs do not contain 3' poly(A) tails. Translation efficiency in eukaryotic cells is usually strongly dependent on the poly(A) tail, which is bound by cellular proteins and interacts with the translational machinery. Studies of translational control have identified sequences in the 3' nontranslated region of the S4 mRNA that interact with cellular proteins and affect translation efficiency. These sequence elements are approximately 20 nucleotides in length and do not appear to affect transcript stability. It is possible that signals embedded in reovirus mRNA functionally substitute for the poly(A) tail by binding to cellular proteins.

Interferon and PKR: effects on viral and cellular protein synthesis

Reovirus double-stranded RNA is a potent inducer of type I (α and β) **interferons**. This cellular response to virus infection can block translation of mRNA transcripts by induction of **double-stranded RNA-dependent protein kinase R (PKR)** and **ribonuclease L** (see Chapter 33). PKR, a latent protein kinase activated by double-stranded RNA, interferes with translation of mRNAs by phosphorylation (and consequent inactivation) of eukaryotic initiation factor 2α (eIF-2α). Ribonuclease L cleaves mRNAs and is activated by oligonucleotides generated by another protein induced by interferon, **2', 5'-oligo(A) synthetase**, which is also activated by double-stranded RNA.

The σ3 protein modulates PKR activation. The translation of cellular mRNAs is strongly inhibited in reovirus-infected cells; by approximately 10 hours after reovirus infection, most proteins synthesized in the cell are viral. The reovirus outer capsid protein σ3 appears to influence this selective inhibition of cellular protein synthesis via its effects on PKR activity. σ3 binds to double-stranded RNA, competing with PKR for binding. In the vicinity of virus factories, the concentration of σ3 may be high, so that any double-stranded RNA would be bound to σ3 and PKR activation would be correspondingly low, allowing viral protein synthesis to continue. In other areas of the cytoplasm, lower concentrations of σ3 would allow PKR activation, leading to selective inhibition of cellular protein synthesis.

PKR regulation and cancer. Reoviruses replicate more efficiently in transformed cells cultured *in vitro* than in their normal counterparts. Activated signaling pathways such as the **Ras** pathway interfere with PKR function in transformed cells; the resulting low PKR activity might allow higher levels of viral protein synthesis and replication. Preferential replication of reovirus in transformed cells also might result from greater efficiency of disassembly of reovirus virions. These observations suggest a potential use of reoviruses as a treatment for cancer. Reovirus infection of mice with experimentally induced tumors leads to tumor regression. Because reovirus replicates in humans but is largely nonpathogenic, this virus is now being tested in clinical trials as therapy for a wide array of human cancers.

Synthesis of progeny double-stranded genomes occurs within subviral particles

In addition to being used for protein synthesis, reovirus mRNAs have a second function: they serve as templates for the synthesis of progeny double-stranded genome RNAs. Production of genome RNAs (see below) is a two-step process. First, one copy of each of the 10 reovirus mRNA molecules is assembled into a subviral particle; second, each plus-strand mRNA within this particle is copied by viral RNA polymerase to form the 10 different double-stranded genome RNAs. This mechanism ensures that one molecule of each gene segment is incorporated into every virion. Furthermore, it keeps most viral double-stranded RNA segregated

within subviral particles and away from the cytosol, reducing activation of cellular signaling pathways that suppress virus replication.

Virus factories. Reovirus RNA replication and virion assembly take place in intracytoplasmic inclusions (Figure 19.7), detectable as early as 8 hours post-infection. These "virus factories" grow larger and coalesce as the virus replication cycle progresses. They contain viral proteins and double-stranded RNA, virions at various stages of maturation, and at late times crystalline arrays of mature virions. Viral inclusions do not contain membranes but are intimately associated with the cytoskeleton. Core protein μ2 stabilizes microtubules within viral inclusions and associates with nonstructural protein μNS, forming an organizing center for formation of inclusions.

Replicase particles. Complexes that contain reovirus mRNAs bound to nonstructural viral proteins μNS and σNS and outer capsid protein σ3 are postulated to represent the first step in reovirus assembly (Figure 19.8). The next step probably involves formation of "replicase particles" engaged in minus-strand synthesis. These particles contain reduced levels of both core proteins and outer capsid proteins compared with mature virions. The mechanism ensuring that only one copy of

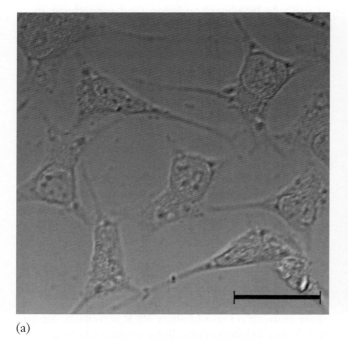

(a)

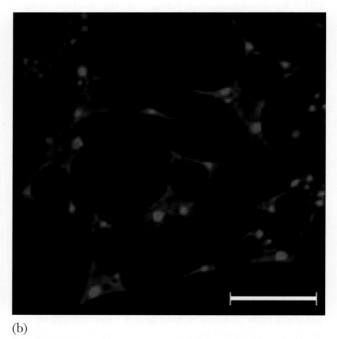

(b)

Figure 19.7 Formation of inclusions in reovirus-infected cells. Mouse L fibroblast cells were infected with reovirus strain T3D and examined 24 hours after infection using a confocal microscope. (a) Differential interference contrast image. (b) Cells were stained with fluorescent antibodies that detect reovirus proteins, which are revealed as bright patches within cells. Note the large inclusion structures in the cytoplasm that contain reovirus proteins and the absence of reovirus proteins in the nucleus (Scale bar: 25 μm).

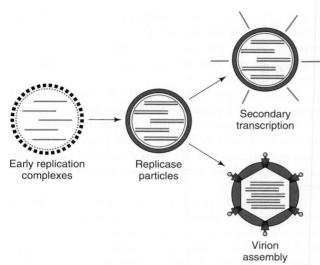

Figure 19.8 Assortment of gene segments and virion assembly. Reovirus plus-strand RNAs are assembled into particles called early replication complexes, which contain only one copy of each of the 10 different transcripts. Viral RNA polymerase then synthesizes double-stranded genome RNAs by copying the plus-strand mRNAs. These replicase particles can proceed to synthesize new mRNAs (upper arrow) or they can be assembled into virions by the addition of outer capsid proteins (lower arrow).

each of the 10 unique RNA segments is packaged into each particle is not understood. However, this process most likely involves specific recognition of nucleotide sequences and secondary structures at the termini of each mRNA (Figure 19.3) by viral proteins (for further discussion of this topic, see Chapter 18).

Replicase particles can synthesize full-length, double-stranded RNAs by copying the packaged plus-strand mRNAs using the viral RNA-dependent RNA polymerase. The same RNA polymerase therefore participates in two very different reactions: (1) synthesis of single-stranded mRNAs in viral cores, using a double-stranded RNA template that remains intact and can be used multiple times, and (2) synthesis of double-stranded genome RNAs within replicase particles, using a single-stranded mRNA template to which the negative-strand copy remains hydrogen-bonded after synthesis.

Secondary transcription and encapsidation. Once double-stranded RNA is made, the replicase particles are able to synthesize more plus-strand viral mRNAs. Transcription by these "secondary transcriptase particles" accounts for the bulk of the mRNA made in reovirus-infected cells. Secondary transcriptase particles, like replicase particles, lack the full complement of virion proteins, but the two particle forms are physicochemically distinct. When

sufficient capsid proteins have been synthesized, assembly is completed by the addition of a layer of outer capsid proteins to replicase particles, resulting in the formation of infectious double-shelled virions (Figure 19.8).

Reoviruses induce apoptosis via activation of innate immune response transcription factors NF-κB and IRF-3

Reoviruses can induce **apoptosis** (programmed cell death) in cultured cells and in host animals. Reovirus infection leads to activation of **nuclear factor kappa B (NF-κB)**, a family of transcription factors known to play important roles in regulating cellular stress responses, including apoptosis, and interferon regulatory factor-3 (IRF-3), which is required for expression of type I (or antiviral) interferons in many types of cells. NF-κB complexes activated by reovirus are comprised of NF-κB subunits p50 and p65.

Apoptosis induced by reovirus is significantly reduced in cells expressing an inhibitor of NF-κB. Furthermore, apoptosis is blocked in reovirus-infected cells deficient in the expression of the p50 or p65 NF-κB subunits. These findings demonstrate that reovirus-induced apoptosis requires functional NF-κB, which presumably enhances the expression of genes that induce this cellular response to reovirus infection.

Some NF-κB-responsive genes function in host innate immune defense mechanisms; these include interferon-stimulated genes and the gene encoding interferon-β (see Chapter 33). Like NF-κB, IRF-3 also is required for reovirus to induce maximal levels of apoptosis. Collectively, these findings point to a cellular response axis linking NF-κB, IRF-3, and reovirus-induced apoptosis. Reovirus infection also activates **caspases**, which are cellular proteases that mediate apoptosis. Caspases linked to both the intrinsic (mitochondrial) and extrinsic (death receptor) pathways of apoptosis induction are activated by reovirus.

Membrane penetration and apoptosis. In cultured cells, reovirus-induced apoptosis does not require *de novo* synthesis of viral RNA and protein, indicating that an inducer of apoptosis is contained within infecting virions. Consistent with these findings, strain-specific differences in the capacity of reovirus to induce apoptosis segregate genetically with the viral S1 and M2 gene segments, which encode components of the outer capsid: attachment protein σ1 and membrane-penetration protein μ1, respectively. Entry of reovirus into cells using a pathway that bypasses σ1 attachment to its receptors, JAM-A and sialic acid, leads to apoptosis, indicating that signaling pathways triggered by interactions between σ1 and these receptors are dispensable for this process.

Mutants of reovirus with alterations in the μ1 protein are diminished in the capacity to activate NF-κB and induce apoptosis, pointing to a key role for μ1 in the reovirus cell-death mechanism.

Cell-cycle arrest. In addition to apoptosis, reovirus also causes cell-cycle arrest in infected cells. Arrest occurs at the transition between the G2 stage of the cell cycle and mitosis. It is mediated by nonstructural protein σ1s and occurs independently of the apoptotic response to infection. Cell-cycle arrest may be a result of hyperphosphorylation and consequent inactivation of the **cyclin-dependent kinase (CDK)**, p34^{cdc2}/cdk1, by σ1s.

Studies of reovirus pathogenesis in mice

Central nervous system. Although reoviruses are mostly nonpathogenic in humans, these viruses have served as very productive experimental models for studies of viral pathogenesis. In newborn mice, reoviruses cause patterns of disease that vary with the virus serotype. The best-characterized model is reovirus pathogenesis in the mouse central nervous system. After primary replication in the intestine, type 1 reovirus strains spread through the bloodstream to the brain where they infect **ependymal cells**, leading to nonlethal **hydrocephalus**. In contrast, type 3 reovirus strains spread primarily by neural routes to the central nervous system and infect neurons, causing fatal **encephalitis**.

The use of reassortant viruses showed that these distinct patterns of central nervous system disease depend on the source of the σ1-encoding S1 gene. This analysis suggests that type 1 and type 3 reoviruses interact with different receptors expressed on ependymal cells and neurons. However, both type 1 and type 3 reovirus strains use JAM-A as a receptor, which is not required for reovirus infection in the brain. Because type 3 reovirus also binds to sialic acid, the distribution or density of cell-surface carbohydrates may determine the types of cells infected by a given reovirus strain. It is also possible that reovirus infects cells in the central nervous system using alternative receptors that have not yet been discovered.

Liver and heart. Reovirus infection of mice causes pathology in a wide range of other organs and tissues, including the liver and bile ducts, heart muscle, lungs, exocrine pancreas, and endocrine tissues. Reovirus liver and bile duct disease and **myocarditis** have become particularly well-established experimental models of viral injury at these sites.

Some type 3 reovirus strains infect the epithelial lining of bile ducts within the liver, leading to ductal obliteration and obstructive **jaundice**. The capacity of attachment protein σ1 to bind to cell-surface sialic acid

is a determinant of viral **tropism** for bile duct epithelium and inflammatory damage to bile ducts. There are indications that injury to bile ducts leading to jaundice in newborn humans may also be caused by reovirus infection.

Unlike myocarditis caused by several other viruses, reovirus-induced myocarditis is not mediated by immune responses. Rather, reovirus-induced cytopathic effects directly cause injury of heart muscle cells. Reovirus isolates that are most capable of myocardial injury induce little interferon and are resistant to its antiviral effects.

Reovirus infection of mice with a defect in NF-κB activation due to absence of the p50 subunit results in massively increased levels of myocardial apoptosis and marked cardiac dysfunction compared to wild-type animals. This cell damage is due to decreased production of type I interferon in response to viral infection. Treatment of p50-negative mice with interferon protects heart cells from apoptosis following reovirus infection. Interestingly, p50-negative mice exhibit greatly reduced levels of apoptosis in the central nervous system in comparison to normal animals, in keeping with the proapoptotic effects of NF-κB in cell culture. Thus, NF-κB appears to play organ-specific roles in reovirus pathogenesis.

Plasmid-based Reverse Genetics for Reovirus

"Reverse genetics" describes the process by which plasmid copies of viral genomes are manipulated to introduce specific mutations to test hypotheses about functional domains in viral proteins. Prior to the development of reverse genetics, selection of mutants with particular phenotypic traits was used to study the relationship between particular genes and their functions—so-called "forward genetics". Reverse genetics allows direct engineering of the viral genome without a need to devise complicated selection strategies for isolation of viral mutants. Furthermore, reverse genetics has allowed rapid generation of viral vaccines and propelled the use of viruses as gene delivery vehicles.

A long-standing problem in studies of viruses in the *Reoviridae* family was recently solved with the development of a fully plasmid-based reverse genetics technology for mammalian reoviruses (Figure 19.9). This system permits selective introduction of desired mutations into cloned cDNAs encoding each of the 10 viral gene segments, followed by isolation of mutant viruses from cells transfected with the plasmid constructs. Recombinant viruses are generated without a requirement for helper virus and free of any selection. Thus, this new technique provides a means to directly and precisely engineer the viral genome in the context of infectious virus.

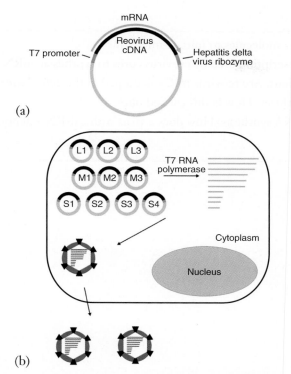

(a)

(b)

Figure 19.9 Strategy to generate reovirus from cloned cDNA. (a) Prototype reovirus gene segment cDNA in plasmid. Cloned cDNAs representing the 10 reovirus dsRNA gene segments are flanked by the bacteriophage T7 RNA polymerase promoter and hepatitis delta virus ribozyme sequences. (b) The 10 plasmid constructs are transfected into mouse L cells expressing T7 RNA polymerase from a replication-defective vaccinia virus strain. Transcription within the cells generates the ten different viral mRNAs, each with its authentic 5' end. Self cleavage by the ribozyme generates authentic viral 3' ends. These mRNAs are translated and produce all eleven reovirus proteins, which can then direct a productive infectious cycle. Following several days of incubation, transfected cells are lysed by repeated freezing and thawing, and viable virus is isolated by a plaque assay using L cells.

KEY TERMS

2', 5'-oligo(A) synthetase	Infectious subvirion particles (ISVPs)
Apoptosis	Interferons
Caspases	Jaundice
Cathepsins	Junctional adhesion molecule A (JAM-A)
Cryoelectron microscopy	Myocarditis
Cyclin-dependent kinase (CDK)	NF-κB (nuclear factor-κB)
Double-stranded, RNA-dependent protein kinase (PKR)	Ras
	Reassortant viruses
Encephalitis	Ribonuclease L
Ependymal cells	Sialic acid
Hydrocephalus	Tight junction
Inclusion bodies	Tropism

FUNDAMENTAL CONCEPTS

• *Reoviridae* have double- or triple-layered capsids with icosahedral symmetry, and from 10 to 12 different double-stranded RNA genome segments.

• Rotaviruses, members of a genus in the *Reoviridae* family, cause severe gastroenteritis and are a major cause of disease and death in children in the developing world.

• During entry in endosomes, reoviruses lose their outer capsid shell, and the cores penetrate the endosomal membrane and enter into the cytoplasm.

• Reovirus cores contain viral RNA polymerases, which make viral messenger RNAs by transcribing each of the 10 to 12 viral genome segments.

• Viral mRNAs are capped by viral capping enzymes during transcription, and are extruded from the core into the cytoplasm.

• The reovirus outer capsid protein σ3 binds to double-stranded RNA, reducing activation of PKR and therefore allowing viral protein synthesis to continue.

• Virus factories are localized areas of the cytoplasm where viral RNA synthesis and virion assembly take place.

• Viral mRNAs are selectively incorporated into replicase particles, where the viral RNA polymerase copies them, making full-length, double-stranded RNA genomes.

• The resulting virus cores can either continue to synthesize viral mRNAs, or can be further encapsidated to make mature reovirus virions.

• Reoviruses have been used to study aspects of viral pathogenesis in mice.

REVIEW QUESTIONS

1. Reovirus serotype 3 particles interact with which two cellular molecules to initiate infection?

2. What is meant by the term "moving template" model of transcription used by reovirus cores to synthesize mRNA?

3. Capping of cellular mRNAs normally takes place in the nucleus. Are reovirus mRNAs capped in the cytoplasm?

4. Reovirus infection leads to inhibition of cellular protein synthesis. How is this carried out?

5. Reovirus infection results in a rapid inhibition of cellular DNA synthesis. How does a virus with a dsRNA genome inhibit cellular DNA synthesis?

SECTION V

SMALL DNA VIRUSES OF EUKARYOTES

This section considers viruses that have DNA genomes of less than 10 kb, limiting the number of viral proteins they encode to between 5 and 10. These viruses depend on cellular enzymes to transcribe and to replicate their genes. This requires that they deliver their DNA genomes to the nucleus, where these enzyme systems are located.

Parvoviruses are one of a small number of virus families with single-stranded DNA genomes. Upon entry into the cell nucleus, parvoviruses must first make a double-stranded version of their genome, using cellular DNA polymerases, because only then can cellular RNA polymerase II recognize and transcribe their genome. Replication of linear parvovirus genomes involves self-priming and specific cleavage of hairpin sequences by viral nucleases; this is a model that may apply to linear DNA genomes of much larger viruses. Some parvoviruses can only replicate if they infect a cell that is also infected with a larger "helper" virus.

Polyomaviruses and papillomaviruses have circular, double-stranded DNA genomes. These genomes can be transcribed by cellular RNA polymerase II immediately upon arrival in the cell nucleus. Most cells in multicellular organisms are in the resting G_0 phase of the cell cycle, and do not have sufficient levels of the enzymatic machinery needed to replicate DNA. These small DNA viruses encode proteins that interact with cell signaling pathways and trick the cell into entering the S, or DNA synthesis, phase. They also make proteins that inhibit apoptosis, which is induced when the cell undergoes unscheduled DNA replication; otherwise, the infected cell would die before producing progeny virus.

Studying the interactions of viral and cellular proteins has helped scientists understand a lot about the cell cycle and signaling pathways. One result of these interactions is that infected cells can grow and divide indefinitely, and if the cell is not killed by the virus it can eventually evolve into a tumor cell. The ability of these viruses to transform normal cells into cancerous cells has led to intensive research efforts and significant understanding of the nature of tumorigenesis. Several of these viruses can cause malignant tumors in humans.

CHAPTER 20

Parvoviruses

Peter Beard

Parvoviridae

From Latin *parvus* (small), referring to size of virion

VIRION

Naked icosahedral capsid (T = 1).
Diameter 28 nm.

GENOME

Linear single-strnded DNA, 5 kb.
Self-complementary hairpin structures at genome
  termini.

GENES AND PROTEINS

One to three transcriptional promoters on same DNA
  strand. Transcribed by cell RNA polymerase II.
Splicing generates up to six different mRNAs.
Two to four replication (Rep) or nonstructural (NS)
  proteins: DNA replication and regulation.
One to three capsid (Cap) proteins: formation of virion.

VIRUSES AND HOSTS

Autonomous parvoviruses: minute virus of mice, canine
  and feline parvoviruses.
Dependoviruses: adeno-associated viruses (human,
  bovine, canine).
Erythroviruses: B19 (humans).
Densoviruses: infect insects.

DISEASES

Humans: B19 causes erythema infectiosum (rash/
  fever), aplastic crisis in anemia.
Leukopenia and enteritis in dogs, cats, mink.
Dependovirus infection has not been associated with
  disease.

DISTINCTIVE CHARACTERISTICS

Strong dependency on cell cycle.
Autonomous parvoviruses replicate only in S-phase
  cells.
Dependoviruses require a helper virus to replicate.
DNA must become double-stranded before being
  transcribed.
Site-specific integration into host genome
  (adeno-associated virus).
Selectively replicate in and kill tumor cells.
Used as vectors for gene transfer into human cells.

Parvoviruses have very small virions and a linear, single-stranded DNA genome

Parvoviruses, as their name implies, are among the smallest of all viruses. Although their genome (5 kb) is actually larger than that of hepatitis B virus (3 kb) and is approximately the same size as the genomes of the polyomaviruses, parvoviruses have small virions because their DNA is single-stranded, and can therefore be packaged very compactly. Their 28-nm diameter, non-enveloped T = 1 icosahedral capsid (Figure 20.1) resembles the capsid of the single-stranded DNA bacteriophage $\phi \times 174$ (Chapter 6).

In spite of their small size, parvoviruses have a number of unique properties that set them apart from other viruses. Parvovirus genomes are linear, not circular

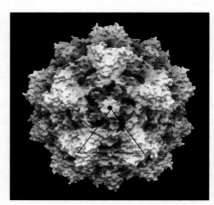

Figure 20.1 Virion of minute virus of mice. Structure determined by x-ray crystallography. The virion is oriented so that one of the 12 fivefold axes of symmetry is visible.

like those of the single-stranded DNA bacteriophages. The ends of their linear DNA molecules have self-complementary nucleotide sequences that can anneal to form hairpin structures. These unusual stuctures enable DNA replication by serving as primers for DNA polymerase. Certain parvoviruses can integrate their DNA into a specific region of a human chromosome, in a manner reminiscent of site-specific integration of bacteriophage lambda (Chapter 8). This ability to integrate has stimulated interest in using these viruses as vectors for the transfer of genes into human cells.

Parvoviruses replicate in cells that are going through the cell cycle

The small size and limited genetic coding capacity of parvoviruses make them exceptionally dependent on host cell functions. In contrast to the polyomaviruses, papillomaviruses, and adenoviruses (Chapters 21–23), parvoviruses are unable to induce cells to enter the DNA synthesis (S) phase of the cell cycle. Since parvoviruses nevertheless need the cellular DNA replication machinery for their own replication, they are able to productively infect only cells that are going through the cell cycle. This and their cell-type dependence largely account for the biology and pathology of parvovirus infections.

Parvoviruses (Table 20.1) have been isolated from numerous mammals and birds, and members of a distinct subfamily, the densoviruses, infect a variety of insects. Parvoviruses that infect animals can be divided into two major groups: the autonomous parvoviruses and the dependoviruses. The autonomous parvoviruses can replicate in cells that spontaneously enter the DNA replication phase of the cell cycle. However, the dependoviruses require coinfection with an adenovirus or a herpesvirus to undergo productive replication; these viruses are therefore named adeno-associated viruses (AAV). Several autonomous parvoviruses can cause disease in animals; canine parvovirus, feline panleukopenia virus, and Aleutian mink disease virus are well-known examples. No diseases have been associated with infection by dependoviruses. The erythrovirus B19, named for its specificity for red blood cell precursors, is noteworthy as a parvovirus that is pathogenic for humans (see below).

Table 20.1 Some representative parvoviruses

Subfamily	Example	Host
Autonomous parvoviruses	Minute virus of mice (MVM)	Mouse
	Canine parvovirus	Dog
Dependoviruses	Adeno-associated virus (AAV)	Human
Erythroviruses	B19 virus	Human
Densoviruses	Culex pipiens densovirus	Mosquito

Discovery of mammalian parvoviruses

Several autonomous parvoviruses were discovered during the hunt for tumor viruses. Material from tumors was used to infect cell cultures or laboratory rodents and the viruses that emerged were considered as possible candidate oncogenic viruses. However, it soon became clear that these parvoviruses do not cause cancer. Rather, they usually replicate more efficiently in tumor cells. This so-called **oncotropism** of parvoviruses may well turn out to be useful because it could be a basis for targeting a lytic virus to cancer cells.

Other parvoviruses were discovered by chance. For example, the B19 virus was found during screening for an unrelated virus, hepatitis B virus, and only later was it shown to cause human disease. Adeno-associated viruses were originally found contaminating preparations of adenovirus; they are fairly widespread since the majority of people have antibodies against them. Minute virus of mice (MVM), an autonomous parvovirus, was first discovered contaminating a preparation of a mouse adenovirus, suggesting that these viruses, though competent for replication, are not above using some help from a larger virus if it is available. In much of what follows, we will use minute virus of mice and adeno-associated virus type 2 (AAV2) as examples, these being the two parvoviruses whose replication is best understood at the molecular level.

Parvoviruses have one of the simplest-known virion structures

The structures of the spherical capsids of several parvoviruses have been solved at atomic resolution by x-ray crystallography (Figure 20.1). The capsid is constructed from 60 protein subunits arranged with $T = 1$ icosahedral symmetry. Each subunit contains the same 8-stranded antiparallel β-barrel shape that is found in capsid proteins of many other icosahedral viruses (see Chapter 2). The single-stranded DNA molecule is tightly compacted within the capsid. The x-ray diffraction structure shows that some of the viral DNA is distributed with the same icosahedral symmetry as the capsid proteins. This implies that there are DNA recognition sites on the inside of the parvovirus capsid, and these are probably involved in packaging the viral DNA.

Parvoviruses have very few genes

The coding region of parvovirus DNA (Figure 20.2) can be divided into two approximately equal sections. The left half, from 5 to 50 map units, codes for the nonstructural (NS) or replication (Rep) proteins (a map unit is 1% of the length of the viral genome, or approximately 50 nucleotides). The right half, from about 50 to 90 map units, codes for the viral capsid (Cap) proteins.

Despite the overall similarity of coding regions among genomes of different parvoviruses, they use

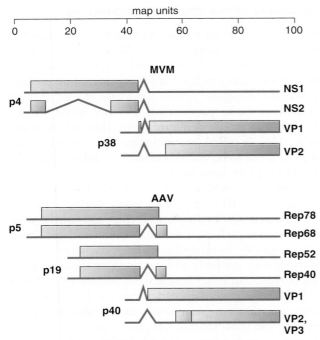

Figure 20.2 Genes and transcripts: Minute virus of mice and adeno-associated virus. Messenger RNAs are shown with open reading frames as rectangular boxes (green for early proteins, orange for late proteins) and introns as inverted V's. Transcriptional promoters are noted at the left; proteins coded by each mRNA are shown at the right. Map units divide the viral genome into 100 equal divisions of approximately 50 nt.

different strategies for controlling the expression of viral genes. Minute virus of mice uses one promoter (p4, located 4 map units from the left end) to synthesize the mRNA encoding the nonstructural proteins and a second promoter (p38) for the mRNA encoding the capsid proteins. Adeno-associated virus makes use of an additional promoter (p19) within the left open reading frame to generate mRNAs encoding shorter versions of its replication proteins. Finally, B19 virus seems to use predominantly a single promoter for transcription of the entire genome (not shown). In each case, differential splicing then generates the specific mRNAs that the virus needs.

The three capsid proteins, VP1, 2, and 3, have overlapping amino acid sequences, because they are translated in the same reading frame (Figure 20.2). For minute virus of mice, VP2 is the major capsid protein; only a few copies of VP1 are incorporated into the capsid. VP1 has additional N-terminal sequences and is translated from an alternatively spliced mRNA. VP3 is formed from VP2 by proteolytic cleavage of about 20 N-terminal amino acids, after its incorporation into the capsid. By contrast, the major capsid protein of adeno-associated virus is VP3; the larger VP2 is made by occasional initiation of protein synthesis at an ACG initiation codon upstream of the VP3 AUG start codon but in the same reading frame. The distinct roles of the

minor capsid proteins are not well understood, although their absence can reduce infectivity of virions.

Single-stranded parvovirus DNAs have unusual terminal structures

One of the most distinctive features of parvoviruses is the structure of the linear, single-stranded viral DNA. Most autonomous parvoviruses (e.g., minute virus of mice) encapsidate only the DNA strand complementary to the viral mRNA. On the other hand, dependoviruses (e.g., adeno-associated virus) encapsidate equally well strands of both polarities, but each virion contains only one of the two DNA strands. As a result, DNA extracted from virions of adeno-associated virus can self-anneal *in vitro*, producing a fully double-stranded DNA molecule.

Researchers who first extracted DNA of adeno-associated virus were puzzled by this, because the capsids of adeno-associated virus and minute virus of mice are the same size and density; however, their experiments suggested that the DNA of adeno-associated virus was double-stranded and therefore capsids of adeno-associated virus should contain twice as much DNA as capsids of minute virus of mice. When DNA of adeno-associated virus is extracted from virions under conditions that do not allow self-annealing, it remains single-stranded, resolving this paradox.

The nucleotide sequence at each end of parvovirus DNAs contains self-complementary regions. This allows one part of the terminal sequence to base pair with the other part, enabling the molecule to fold back on itself to form hairpin-shaped or T-shaped structures (Figure 20.3). The length of the self-complementary sequence element ranges from 115 to 300 nucleotides. The sequences at the 3' and 5' ends of minute virus of mice are different from each other. However, the 125-nt, self-complementary sequences at the two ends of adeno-associated virus DNA are identical, except for the orientation of the internal 42 nucleotides labeled B and C in Figure 20.3 (see below). As is described below, 3'-terminal hairpins function as primers for the initiation of viral DNA replication.

Uncoating of parvovirus virions takes place in the nucleus and is cell-specific

Parvoviruses are particularly dependent on host cell functions, a result of their small genomes and limited coding capacity. Successful infection also depends on the differentiated state of the host cell. Parvoviruses bind to cells via receptors that have in some cases been identified: globoside (blood group P antigen) for B19 virus; transferrin receptor for canine parvovirus; and heparin sulfate proteoglycan, fibroblast growth factor receptor 1, and integrin $\alpha V\beta 5$ for adeno-associated virus type 2.

Virions subsequently enter the cell by endocytosis. They migrate to the nucleus, where they enter via the

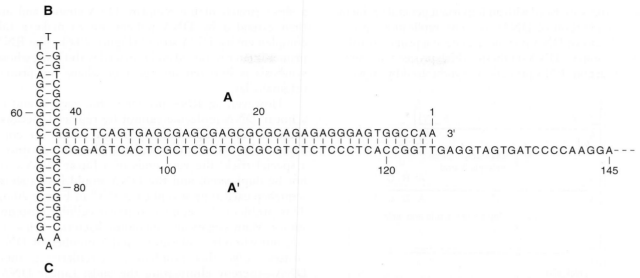

Figure 20.3 **DNA hairpin at the 3′ end of the adeno-associated virus genome.** Nucleotides 1-41 (A) and 85-125 (A′) are base-paired, as are nucleotides 42-50/54-62 (B) and nucleotides 64-72/76-84 (C).

nuclear pores and are uncoated, releasing the single-stranded viral DNA. At this point, autonomous parvoviruses like minute virus of mice must wait until the cell enters the S phase to begin their replication cycle. Dependoviruses, like adeno-associated virus rely on coinfection by a helper virus to push the cell into the S phase and for other functions.

Unexpectedly, the ability of parvoviruses to infect certain cell types or species seems to be determined at the stage of virus **uncoating**. The specificity of canine parvovirus for dog cells rather than cat cells maps to the capsid protein gene. Likewise, the **tropism** of different MVM strains for fibroblasts or lymphocytes is determined by the capsid protein. In both cases, the amino acids important for this function are located at the surface of the virion. It was therefore expected that virus-receptor interactions would be the factor that determines the specificity of infection. It was a surprise to discover that these viruses bind to and enter **permissive** and **nonpermissive cells** equally well. But when cells are transfected with naked viral DNAs instead of being infected with virions, the cell-specific block is overcome and normally nonpermissive cells produce virus. These findings point to uncoating, rather than entry, as the level of the cell-specific block to replication of certain strains of minute virus of mice. This may involve an intracellular receptor involved in uncoating.

DNA replication begins by extension of the 3′ end of the terminal hairpin

Incoming parvovirus single-stranded DNA must first be converted to a double-stranded form to be able to function as a template for transcription. This is achieved by using the terminal hydroxyl group of the 3′ hairpin as a primer

for host cell DNA polymerase. Because of the hairpin, this 3′ end is hydrogen-bonded to a complementary sequence further along on the same DNA strand (Figure 20.3), allowing the same DNA molecule to serve both as primer and template for DNA replication. Beginning at the 3′ end, the polymerase copies the single-stranded genome DNA up to its 5′ end, generating a double-stranded DNA molecule with an intact hairpin at one end (step 2 in Figure 20.4). This reaction is similar to that used to repair gaps in double-stranded cellular DNA.

Cellular RNA polymerase II then transcribes this template starting from the leftmost promoter (p4 in minute virus of mice, or p5 in adeno-associated virus, see Figure 20.2). The strong p4 promoter contains binding sites for cellular transcription factors E2F, Ets, Sp1, and TATA-binding protein TBP (Figure 20.5). Transcription from this promoter generates mRNAs encoding the NS1 and NS2 proteins. NS1 is the main regulatory protein of minute virus of mice and has several functions to ensure efficient viral replication. NS1 binds to viral DNA and stimulates transcription from the p38 promoter, which controls synthesis of the mRNAs for the coat proteins. NS1 also further stimulates transcription from the p4 promoter, leading to increased production of its own mRNA. However, it is not clear exactly how NS1 activates transcription. The p38 promoter includes a binding site for the NS1 protein (TTGGTTGGT), in addition to Sp1 and TATA sites, but apparently activation can also occur without this sequence.

The DNA "end replication" problem

Besides acting as transcriptional activators, the NS proteins also directly control the replication of viral DNA.

Parvoviruses are faced with an important general problem in the replication of DNA: the "end replication" problem. All known DNA polymerases need a primer, usually RNA, to initiate DNA synthesis, which proceeds in the 5' to 3' direction. RNA primers are synthesized by copying a short stretch of the template DNA strand, and are then extended by DNA polymerase to make a full complementary DNA strand (Figure 20.6). The RNA primer is later removed and replaced with DNA, whose synthesis is initiated upstream of where the primer originally lay.

However, an RNA primer at the very 5' end of a linear DNA molecule cannot be replaced, because there is no complementary DNA strand to copy beyond the end of the template. Therefore, without a special trick, the very ends of a linear DNA cannot be duplicated, and the DNA would lose a short sequence each time it replicates ("X" in Figure 20.6). This problem also applies to linear cellular chromosomes. Many organisms, including bacteria such as *E. coli*, mitochondria, plasmids, and a number of DNA viruses, solve this problem by circularizing their DNA—thereby eliminating the ends. Linear DNAs need another way to solve this problem.

A British cell biologist, Thomas Cavalier-Smith, suggested how cells could solve the end replication problem for their linear chromosomes. If linear DNAs are provided with hairpin ends, they effectively contain two tandem inverted copies of the sequences at the ends of their DNA. Because the hairpin end can fold back on itself, it can serve as a primer for DNA polymerase, eliminating the need for an RNA primer. Some DNA gymnastics will then reproduce the original ends.

It turns out that Cavalier-Smith's proposal is not actually used by eukaryotic cells; instead cells use the enzyme **telomerase** to extend and maintain the ends of their linear chromosomes. But his proposal does explain exactly the principle by which parvoviruses duplicate the ends of their linear DNA. Besides hairpin sequences at the ends of the DNA, the mechanism requires a site-specific endonuclease that cuts DNA within the hairpin—and both are present in parvoviruses. The steps are shown in Figure 20.4, using adeno-associated virus as the example. Although all parvoviruses employ the same principle, the scheme is simpler with adeno-associated virus since the two ends of the viral DNA are identical.

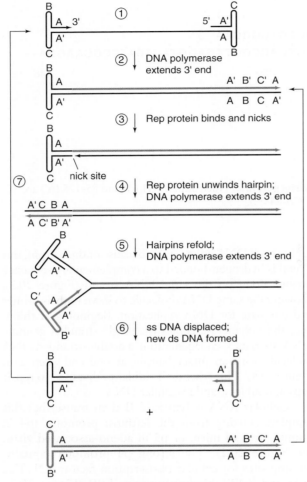

Figure 20.4 Replication strategy of adeno-associated virus DNA. See text for details. Orange lines: newly synthesized DNA strands. Arrowheads show 3' ends. Not drawn to scale: the hairpin ends are 125 nt in length, and the entire genome is 5 kb long. A' represents sequences complementary to A, B' to B, and C' to C.

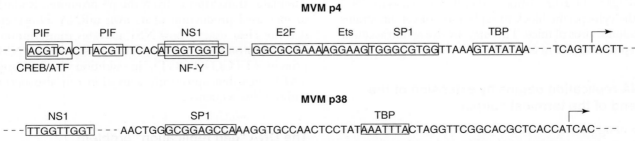

Figure 20.5 Binding sites for transcription factors in minute virus of mice promoters p4 and p38. Binding sites are boxed and the factor is noted above or below the box. Start sites for transcription are shown by arrows.

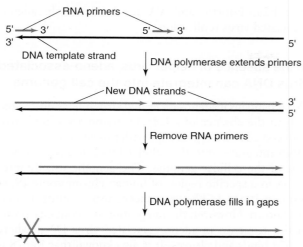

Figure 20.6 RNA primers and the DNA "end replication" problem. RNA primers (blue) initiate DNA replication by copying a short sequence on the DNA template strand, and primers are then extended (orange) by DNA polymerase. The RNA primers are subsequently removed and DNA polymerase extends upstream DNAs, filling in the gaps. Adjacent DNA chains are then sealed with DNA ligase (not shown). However, the RNA primer at the very 5' end of a linear DNA (orange "X") cannot be replaced because there is no DNA upstream.

Steps in DNA replication

1. The palindromic sequences fold to form hairpins at the ends of the DNA.

2. The 3'-hydroxyl group (arrow) of the left hairpin is used as primer for DNA chain elongation, generating a fully double-stranded DNA with one hairpin end. This can be used for transcription by RNA polymerase II.

3. The fully double-stranded DNA is nicked at a specific site, just opposite the original 3' end, by Rep78/68 of adeno-associated virus (or its equivalent NS1 in the case of minute virus of mice). This protein remains covalently attached to the 5' end of the DNA it has cut and is incorporated into new virions along with the progeny DNA (and therefore it is not strictly a nonstructural protein).

4. Starting from the 3' end at this nick, the cellular DNA polymerase then elongates the DNA chain in the other direction, to fully copy the left end hairpin. The NS1 or Rep78/68 protein helps by unwinding the hairpin.

5. The palindromic sequences at one end then reform hairpins to prime further DNA synthesis.

6. DNA synthesis using the newly formed hairpin generates a new double-stranded DNA molecule with a hairpin end, and also results in displacement of progeny single-stranded DNA. Thus the DNA replication cycle is complete. Note that refolding and priming can happen at either end, or at both ends simultaneously, in which case two double-stranded molecules are formed.

7. The displaced single-stranded DNA can begin a new replication cycle, or can be packaged if capsid proteins are present. The newly synthesized, double-stranded DNA can be transcribed to provide more messenger RNA, or can again be nicked and extended, to generate further progeny DNAs.

The attentive reader will notice that the sequence at the 3' (left) end of the original adeno-associated virus DNA molecule shown in Figure 20.4 reads ABCA' in the 3' to 5' direction, and the sequence at the 3' (right) end of the progeny single-stranded DNA shown at the bottom of Figure 20.4 reads AC'B'A', so that the order and orientation of B and C are inversed. This "flip-flop" sequence inversion, resulting from the nicking, unfolding, and copying of the hairpin, can be detected in adeno-associated virus DNA, and was instrumental in testing the validity of the replication model.

The 3' and 5' ends of the DNAs of some parvoviruses such as minute virus of mice have different sequences, which leads to a modified and more complex replication scheme, notably involving a dimer-length replicative intermediate.

Nonstructural proteins are multifunctional

NS1 of the autonomous parvoviruses and its equivalent in adeno-associated virus, Rep78/68, are the principal viral regulatory proteins and play several roles in parvovirus replication. Each protein has a number of functions and biochemical activities (Figure 20.7). They:

- bind in a sequence-specific manner to viral DNA origins and promoters
- have ATP-binding, ATPase, and helicase activities
- are sequence-specific endonucleases
- are transcriptional activators
- prevent infected cells from progressing through the cell cycle

The biological importance of the last activity may be to keep infected cells in S phase and so increase the time available for viral replication.

Much less is known about the functions of the minor nonstructural proteins. Minute virus of mice NS2 protein is needed for efficient viral replication only in certain cell types, but the reasons and mechanism are not clear. Likewise, adeno-associated virus Rep52 may be involved in encapsidation and interactions with cellular regulatory pathways, but again little is known.

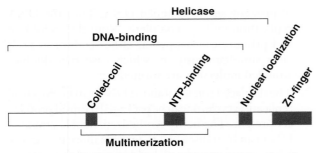

Figure 20.7 Functional domains in adeno-associated virus Rep78 protein. The protein is shown with the N-terminus at the left and the C-terminus at the right. NTP = nucleoside triphosphate.

Adenovirus functions that help replication of adeno-associated virus

For efficient replication, adeno-associated virus requires functions provided by a helper virus. If the helper is adenovirus (see Chapter 23), the adenovirus genes involved are E1A, E1B, E2, E4, and the VA RNA genes. The combined action of these gene products ensures that the conditions in the cell are right for adeno-associated virus replication. Herpesviruses can also serve as helper viruses, and exposure of the cell to certain agents that affect cell gene expression (ultraviolet light, chemical carcinogens, etc.) can allow virus replication to proceed in the absence of a helper virus. Examples of how some adenovirus gene products help adeno-associated virus replication are:

1. Adenovirus E1A protein acts as a transcriptional activator both of adeno-associated virus genes and of cellular genes involved in DNA replication, by binding to the retinoblastoma (Rb) protein and releasing it from the cellular transcription factor E2F (see Figure 23.6). As we saw above, E2F binds to the strong p4 promoter and helps activate gene expression.

2. Adenovirus E1B 55K protein and E4orf6 protein combine to form a complex that transports adeno-associated virus mRNAs from the nucleus, where they are made, to the cytoplasm.

3. VA RNA is a short, double-stranded RNA that binds to and inhibits RNA-dependent protein kinase (**PKR**). VA RNA therefore suppresses inhibition of viral protein synthesis by PKR that occurs when it is activated by longer, viral double-stranded RNA (see Chapter 33).

Interestingly, the distinction between helper-dependent and autonomous parvoviruses may not be as rigid as first thought. Autonomous replication of the human parvovirus B19, discussed below, is mostly limited to human erythroid progenitor cells. However, in some otherwise nonpermissive cells, B19 virus can replicate with the help of the same adenovirus genes

(E1, E2a, E4orf6, and VA RNA) that help adeno-associated virus replication.

In the absence of helper virus, adeno-associated virus DNA can integrate into the cell genome

What happens if adeno-associated virus infects unstressed cells in the absence of a helper? Adeno-associated virus has evolved a strategy to cope with this situation: the viral DNA integrates into the cellular DNA and replicates as part of the cellular genome. The genome usually integrates in a specific region of human chromosome 19, at position 19q13.3. The exact mechanism of integration is not known. However, the integration site contains a copy of the sequence in adeno-associated virus DNA where Rep78 binds and cleaves. It is also known that Rep78 is required to promote integration.

It would appear that infecting viral DNA undergoes the first step in DNA replication, forming a double-stranded molecule that can be transcribed. Low amounts of Rep78 are then made, even in the absence of helper virus. Because Rep78 both binds to and cleaves DNA at a specific site, it can be imagined that adeno-associated virus DNA is brought to its target sequence in the host cell chromosome via interaction between Rep78 molecules bound to the viral genome and cellular DNA. Subsequent cleavage of both DNA molecules could lead to ligation of the viral DNA to cellular sequences.

Adeno-associated virus DNA may remain integrated in the cellular genome through repeated cell divisions, thus propagating in daughter cells, until the cell is infected by a helper virus. The helper virus activates transcription of the integrated adeno-associated virus genome, which leads to synthesis of large amounts of the Rep proteins, excision of adeno-associated virus DNA from the cellular DNA, and finally production of progeny adeno-associated virus particles. The survival strategy used by adeno-associated virus, and its ability to integrate at a specific site into the human genome, are unique among animal viruses and resemble the strategy used by bacteriophage lambda (Chapter 8). It may be possible in the future to exploit site-specific integration for gene transfer applications using adeno-associated, virus-based vectors (see Chapter 37).

Parvovirus pathogenesis: the example of B19 virus

In a remarkable way the biological properties of parvoviruses help to determine the diseases they cause. Parvoviruses replicate in rapidly growing cells that are in specific states of differentiation, but not in resting cells that make up most of the body tissues. Typically parvoviruses will infect cells of the intestine, the hematopoietic

Antitumor effects of parvoviruses

Parvoviruses have a long-standing relation with cancer. Although some parvoviruses were originally isolated from tumor-bearing animals or tumor cells, they are not oncogenic. Rather, they grow productively in tumor cells, killing these cells. Parvoviruses have repeatedly been shown to reduce the formation of tumors that arise spontaneously in animals or after treatment with chemicals or other viruses. But how do they do this?

The antitumor effects of parvoviruses could be exerted in several ways. Autonomous parvoviruses grow better in transformed cells than in normal cells, partly because transformed cells are continually passing through the cell-cycle. More NS1, which is toxic for cells, is produced in transformed cells. Transformed cells are, in addition, more sensitive to the toxic effects of NS1. Moreover, parvovirus infection triggers an interferon-induced innate antiviral response (see Chapter 33) in normal cells but not in cancer cells. Parvoviruses may therefore replicate better in cancer cells in part because these cells lack an effective antiviral defense. The Rep78/68 proteins of adeno-associated virus also have antiproliferative effects on cells.

Recently, a new mechanism came to light. The unusual structure of adeno-associated virus DNA, which is mostly single-stranded with terminal hairpins, makes it appear like damaged DNA to the infected cell. This elicits a DNA damage response, even though the cellular DNA is not actually damaged. In normal cells, DNA damage leads to arrest of the cell cycle by a mechanism dependent on the cellular p53 protein, and cell-cycle arrest stops virus replication. However, most tumor cells lack the cellular p53 protein, or other proteins that it activates. Such tumor cells continue in the cell cycle in spite of the DNA damage response caused by adeno-associated virus infection, leading to cell death in mitosis. If these mechanisms used by parvoviruses can be exploited in the future for treatment of cancer, then infection with certain parvoviruses could turn out to be of benefit for the host.

---

system, and the fetus. Several parvoviruses of animals are responsible for serious diseases. Canine parvovirus infects puppies and causes disease and death. Feline **panleukopenia** virus, as the name suggests, affects blood formation in cats. Other parvoviruses that are not covered here, the densoviruses, infect invertebrate species. But there is just one parvovirus known to cause disease in humans, and that is the B19 virus.

B19 was discovered by accident during screening of blood donations. It was then realized that this virus was associated with a serious illness—**aplastic crisis** in anemia patients. B19 is very specific for a particular cell type, infecting the rapidly dividing **erythroid precursors** of red blood cells. The infection causes an acute but relatively short-lived reduction in red cell production in the bone marrow. In a healthy person this is not serious. But in patients with a **hemolytic anemia**, for example **sickle cell disease**, the red cells have a much shorter lifespan in the blood. The loss of red cell production in this case causes such a reduction in hemoglobin concentration that B19 infection can be fatal if not quickly treated by transfusion.

In addition to its involvement in aplastic crisis, B19 is also known to cause a common childhood disease called **erythema infectiosum** ("infectious rash"). Although usually not severe, this disease may lead to serious birth defects if a fetus is infected during pregnancy. The disease spectrum of B19 therefore well illustrates the basic principle of parvovirus pathogenesis resulting from infection of and damage to dividing cells.

A potentially important new human parvovirus was recently discovered, the human bocavirus (HBoV), which derives its name from the fact that it is closely related to *bo*vine parvovirus and to the minute virus of *ca*nines. Human bocavirus is frequently found in children suffering from acute respiratory and intestinal infections. Human bocavirus may be a cause of childhood pneumonia; however, more work is needed to establish the role of human bocavirus in respiratory illnesses.

KEY TERMS

Aplastic crisis	Permissive cells
Erythema infectiosum	PKR
Erythroid precursors	Sickle cell disease
Hemolytic anemia	Telomerase
Nonpermissive cells	Tropism
Oncotropism	Uncoating
Panleukopenia	

FUNDAMENTAL CONCEPTS

- Parvoviruses are among the smallest known viruses because their 5 kb DNA is single-stranded, and can therefore be packaged very compactly.
- Parvovirus replication depends on the cell entering S phase.
- Autonomous parvoviruses replicate in cells that normally cycle and therefore frequently enter S phase.

● Dependent parvoviruses replicate in cells infected with a helper virus, which induces entry of the cell into S phase.

● Linear parvovirus DNA circumvents the "end replication problem" by having terminal hairpin sequences that can prime DNA synthesis.

● Site-specific integration of adeno-associated viral DNA into the host genome depends on the viral Rep protein recognizing the same sequence in viral and cellular DNA.

● Selective replication of certain parvoviruses in tumor cells results from the dependence of those viruses on cellular factors expressed during S phase or altered by cell transformation.

REVIEW QUESTIONS

1. Is there a common theme to transcriptional control within the parvoviruses?

2. Describe two properties of the parvovirus genome that are not common among DNA viruses.

3. How do parvoviruses replicate the ends of their linear DNA genomes?

4. What is known about the site-specific integration of adeno-associated virus DNA in the host cell DNA?

5. Why are parvoviruses only able to productively infect cells going through the cell cycle?

6. How might the terminal hairpin structure of AAV single-stranded DNA lead to death of tumor cells infected by the virus?

Polyomaviruses

Nicholas H. Acheson

Polyomaviridae

From Greek *poly* (many) and *-oma* (tumor)

VIRION

Naked icosahedral capsid (T = 7).
Diameter 45 nm.
Formed from 72 capsomers, pentamers of VP1.

GENOME

Circular double-stranded DNA, 5.3 kb.
DNA is packaged as "minichromosome" with
   nucleosomes formed from cellular histones.

GENES AND PROTEINS

Two transcription units: early and late.
Divergent transcription from a central control region
   containing origin of DNA replication.
Differential splicing produces 3 to 4 mRNAs from each
   transcription unit.
Early proteins (T antigens) regulate cell cycle and
   direct DNA replication.
Late proteins (VP1, 2, 3) make virus capsid.

VIRUSES AND HOSTS

Mouse polyomavirus; simian virus 40; polyomaviruses
   infecting birds, rodents, cattle; and five human
   viruses: BK, JC, MC, WI, and MC viruses.

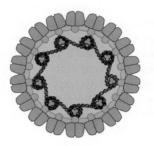

DISEASES

Usually persistent, nonsymptomatic infections.
Progressive multifocal leukoencephalopathy (PML)
   is a rare demyelinating disease caused by human
   polyomavirus JC.

DISTINCTIVE CHARACTERISTICS

Polyomaviruses are dependent on cellular enzymes
   for RNA synthesis and processing, and DNA
   replication.
T antigens interact with multiple signaling pathways
   and activate the cell cycle to facilitate viral DNA
   replication.
T antigens are oncogenes that can transform nonpermissive cells *in vitro*.
Polyomaviruses are tumorigenic in animals when
   administered at high concentrations, but rarely
   cause tumors in nature.

Mouse polyomavirus was discovered as a tumor-producing infectious agent

While studying the transmission of leukemia in mice in the early 1950s, Ludwig Gross unexpectedly discovered that extracts of mouse tissues caused salivary gland tumors when injected into baby mice. His group and that of S. Stewart and Bernice Eddy soon showed that these tumors were caused by a filterable agent. This virus was distinguished from retroviruses that cause mouse leukemia by its smaller size and greater stability. It was subsequently named polyomavirus (Greek *poly* for many and *-oma* for tumor), because it can cause tumors at numerous sites in experimental animals.

   Mouse polyomavirus is commonly found in wild mice and their inbred laboratory cousins but causes little or no overt disease. However, when injected into

susceptible rodents and other mammals at high concentrations, mouse polyomavirus can cause multiple tumors in parotid glands, skin, kidneys, bone, lungs, liver, and many other tissues. These tumors grow rapidly and can eventually kill the host animal.

Simian virus 40 was found as a contaminant of Salk poliovirus vaccine

Simian virus 40 (SV40) was discovered some 10 years later as a contaminant of rhesus monkey kidney cell cultures that were being used for production of human poliovirus vaccines. Injection of simian virus 40 into newborn hamsters also produced tumors. It was soon recognized that mouse polyomavirus and simian virus 40 are closely related, and they were initially grouped together with the papillomaviruses (Chapter 22), in a family called the *Papovaviridae*. However, differences in genome size and organization and in virus replication strategies led taxonomists to reclassify these viruses into two distinct families, the *Polyomaviridae* and *Papillomaviridae*. A number of related polyomaviruses have been discovered in rodents, birds, cattle, monkeys, and humans.

Before simian virus 40 was identified, several batches of the Salk inactivated poliovirus vaccine unknowingly contaminated with live simian virus 40 were administered by injection to millions of people. It has been estimated that virtually the entire population of the United States aged 20 and below by 1961 received such contaminated vaccine at some point. There has been controversy about the possibility that some human tumors may have been caused by simian virus 40 injected during poliovirus vaccination. Causality is hard to prove; some human tumors were shown to contain simian virus 40 DNA and express viral early proteins. However, this could be a coincidence, as simian virus 40 appears to circulate among the human population, causing no obvious symptoms. No clear excess levels of cancer or of specific tumors have been detected in people during the 45 years since they received contaminated poliovirus vaccine.

Neither mice nor monkeys naturally infected with mouse polyomavirus or simian virus 40 develop tumors. These viruses usually cause tumors only under very specific conditions: injection of high titers of virus into particularly susceptible animals shortly after birth. Polyomaviruses are therefore not "professional" tumor viruses. However, their complex interactions with cellular signaling pathways that facilitate high levels of viral DNA replication and progeny virus production can occasionally transform a normal cell into a tumor cell.

Human polyomaviruses are widespread but cause disease only rarely

Five distinct human polyomaviruses are presently known. All five of these viruses appear to be widespread in the human population and are aquired during early childhood, but usually cause no apparent disease. BK and JC viruses were first isolated in the early 1970s. A rare but fatal demyelinating disease, **progressive multifocal leukoencephalopathy (PML)**, is caused by JC virus, particularly in immunosuppressed individuals, including patients suffering from acquired immune deficiency syndrome (AIDS). Two other human polyomaviruses, named WI and KU, were recently discovered in the respiratory tracts of children, but their association with human disease is not clear.

A fifth human polyomavirus, MC virus, was recently isolated from Merkel cell carcinomas. MC DNA has been found in a large proportion of these rare but aggressive human skin tumors; the viral DNA is integrated in the cellular genome, and contains mutations that inactivate its ability to replicate. This tumor appears mainly in immunosuppressed individuals or in people over 65.

Polyomaviruses are models for studying DNA virus replication and tumorigenesis

Because they can transform cells *in vitro* and cause tumors in experimental animals, both mouse polyomavirus and simian virus 40 have been studied intensively as models for understanding tumorigenesis. Their small number of genes and their strong dependence on cellular macromolecular synthesis also make them useful model systems for studying mammalian RNA synthesis and processing, DNA replication, and signal transduction. In this chapter, emphasis in different sections is placed on results found with either mouse polyomavirus or simian virus 40, but these two viruses have a common overall replication strategy and they share many details of structure, genome organization, and gene function.

Polyomavirus capsids are constructed from pentamers of the major capsid protein

The virions of polyomaviruses are naked capsids approximately 45 nm in diameter. Their surface is constructed from 72 capsomers visible by electron microscopy, each made of five molecules of the major capsid protein VP1 (Figures 21.1 and 21.2). Twelve of

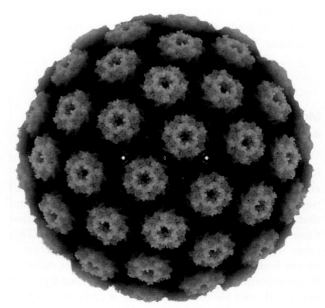

Figure 21.1 Computer-generated image of mouse polyomavirus capsid. The 72 capsomers, each composed of pentamers of VP1 and a single molecule of VP2 or VP3, are organized with T = 7 icosahedral symmetry.

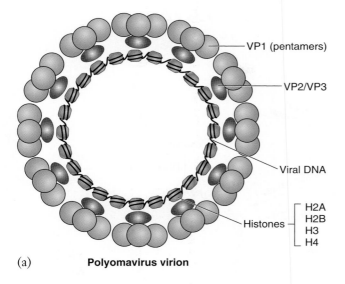

(a)　　　**Polyomavirus virion**

VP1 (pentamers)

VP2/VP3

Viral DNA

Histones — { H2A / H2B / H3 / H4 }

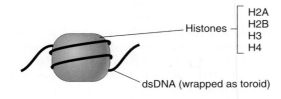

Histones — { H2A / H2B / H3 / H4 }

dsDNA (wrapped as toroid)

(b)　　　**Single nucleosome**

Figure 21.2 Components of polyomavirus virion. (a) A schematic diagram of a virion in cross-section, showing pentamers of VP1 (green spheres), with single molecules of VP2 or VP3 (gray ovals) underneath the capsid surface. Circular viral DNA is bound to nucleosomes (orange rounded squares) formed from cellular histones. (b) A larger view of a nucleosome, around which DNA is wrapped as a toroidal coil.

these capsomers are surrounded by five nearest neighbors, and define the fivefold symmetry axes typical of capsids with icosahedral symmetry. The rules of symmetry would suggest that the remaining 60 capsomers should be made from three or six molecules of VP1, because each such capsomer has six nearest neighbors. However, VP1 forms stable pentamers, not hexamers, and these pentameric capsomers fit into "sixfold holes" on the surface of the capsid.

Capsomers are linked by the flexible carboxy-terminal arms of VP1, which extend across the gap and embrace VP1 molecules in neighboring capsomers, stabilizing the capsid structure (see Figure 2.6b). Every capsomer also contains a single molecule of one of the minor capsid proteins VP2 or VP3, located below the capsid surface. The N-terminal amino acid of VP2 is linked to **myristate**, a C_{14} saturated fatty acid. This common protein modification enables interaction of proteins with lipid membranes (see below).

The circular DNA genome is packaged with cellular histones

Packaged within the capsid is the viral genome, a 5.3-kb circular, double-stranded DNA molecule (Figure 21.2a). The viral genome is associated with **nucleosomes** formed from two molecules each of the four cellular histones H2A, H2B, H3, and H4. These are the same histones that package and condense cellular chromosomal DNA. A length of approximately 200 bp of DNA is wrapped as a **toroidal coil** around the outside of each cylindrical nucleosome (Figure 21.2b). This condenses the viral DNA into a "minichromosome" containing about 25 such nucleosomes (Figure 21.3), giving a total of 200 histone molecules per viral DNA molecule.

The use of cellular histones by polyomaviruses (as well as papillomaviruses) for DNA condensation and packaging is a striking example of genetic economy. The small genomes of these viruses have limited coding capacity, and instead of coding for their internal structural proteins, they simply pick them up in the cell nucleus, where viral DNA replicates.

Circular DNA becomes supercoiled upon removal of histones

When viral genomes are gently released from the capsid without denaturing the bound nucleosomes, they can be visualized as circular molecules by electron microscopy, and the minichromosome resembles a necklace of "beads on a string" (Figure 21.3). However, when the histones are removed by treating with the strong protein denaturant **sodium dodecyl sulfate** and extraction of the proteins with the organic solvent phenol, the circular DNA now becomes coiled about itself like a rubber band held between the thumbs and forefingers of both hands and twisted. This **supercoiled** form of DNA results from the toroidal coiling of the double-helical DNA around the nucleosomes in the intact minichromosome. These toroidal coils distribute themselves around the DNA molecule after removal of histones (Figure 21.3).

If the phosphodiester bond between two nucleotides of one of the DNA strands is cleaved ("nicked"), this supercoiled molecule will unwind and become a relaxed circular DNA (Figure 21.3, top right). This relaxation occurs because one of the DNA strands can now rotate freely about the other strand, allowing the DNA to come to the lowest-energy topological state.

Alternatively, the enzyme **topoisomerase I** can bind to the DNA and unwind the supercoils by breaking a phosphodiester bond, allowing one of the broken ends to rotate around the opposite DNA strand, and then resealing the phosphodiester bond. However, both strands of this relaxed molecule are covalently closed

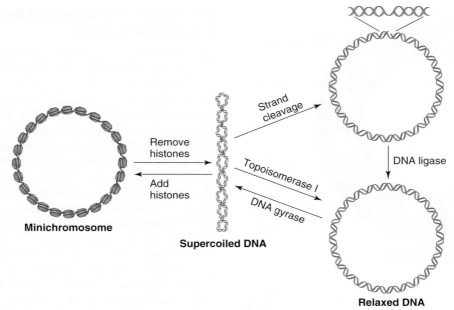

Figure 21.3 Gymnastics of circular DNA molecules. A circular, double-stranded DNA molecule is shown as a minichromosome, bound to nucleosomes (left). Removal of histones generates a supercoiled molecule (center) due to redistribution of the toroidal coils. Nicking of one strand (upper right) allows the supercoiled form to unwind to its lowest-energy relaxed circular state; a portion of the double-helix containing a nick is shown enlarged above. If the nick is sealed with DNA ligase (lower right), this becomes a relaxed, closed circular DNA molecule. Topoisomerase I can produce the same relaxed molecule by cleaving a phosphodiester bond on one strand of supercoiled DNA, allowing the strands to rotate around each other, and resealing the cleaved bond. DNA gyrase uses energy to catalyze the reverse reaction, resulting in a supercoiled DNA molecule. Addition of nucleosomal histones to a supercoiled DNA molecule can regenerate a circular minichromosome (left).

circles, as in the original supercoiled molecule but unlike "nicked" relaxed circular DNA. Supercoiling of relaxed DNA molecules can be catalyzed by specialized enzymes found in bacteria, called **DNA gyrases**, which carry out the reverse reaction of topoisomerase I, using ATP as an energy source.

Supercoiled DNA can be separated from relaxed or linear DNA molecules

The properties of circular DNA molecules were first understood by studying the genomes of polyomaviruses. Supercoiled DNA has different physical properties than relaxed circular DNA and linear DNA of the same nucleotide sequence and molecular weight. For example, supercoiled DNA migrates more rapidly during electrophoresis in agarose gels than does the same DNA when it is a relaxed circle or a linear molecule, because the supercoiled form is more compact.

Nicked circular DNA or linear DNA will also bind more of the fluorescent dye **ethidium bromide** than covalently closed circular DNA. Ethidium bromide molecules insert themselves between adjacent nucleotide base pairs, unwinding the double helix by a small amount. Nicked circular or linear DNA can bind more ethidium bromide than covalently closed circular DNA, because the strands can freely rotate about each other

and therefore unwinding is less constrained. The additional ethidium bromide in these complexes renders them less dense than complexes with supercoiled DNA, and the two forms can be separated when centrifuged in cesium chloride density gradients. Both of these methods allow scientists to purify supercoiled DNA away from other forms.

Polyomavirus genes are organized in two divergent transcription units

As a result of their small genome size, most polyomaviruses code for only 6 to 8 proteins. Viral genes are organized as two families, called early and late genes, and each gene family is expressed from a single transcription unit. Three or four different messenger RNAs, each of which codes for one viral protein, are made from each gene transcript by alternative RNA splicing. A map of the mouse polyomavirus genome is shown in Figure 21.4; the genome of simian virus 40 is organized in a very similar fashion.

Transcription of the early genes proceeds on one DNA strand from near the "12 o'clock" position in a counterclockwise direction, and transcription of the late genes proceeds on the complementary DNA strand in a clockwise direction; this pattern is said to be **divergent transcription**. Each transcription unit has one

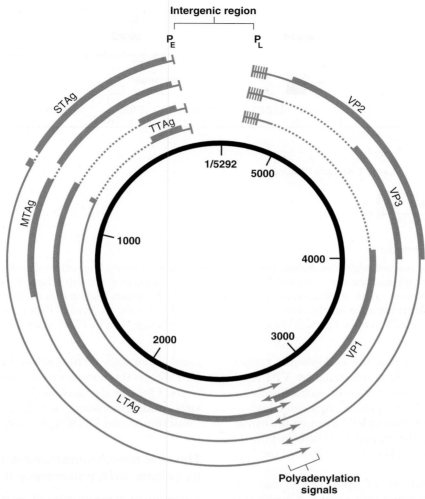

Figure 21.4 Map of mouse polyomavirus genome. Nucleotide numbers for the A2 strain of mouse polyomavirus are shown inside the circular map. Early mRNAs are shown on left side, late mRNAs on right side (light blue lines). The 5′ ends (transcription initiation sites) are shown as vertical lines and 3′ ends (polyadenylation sites) are shown as arrowheads on mRNAs. P_E and P_L: early and late promoters. Introns removed during alternative splicing are shown as dotted lines. Coding regions are shown in orange. Sequence elements in the intergenic region and the polyadenylation region are shown in Figures 21.5 and 21.6.

or several nearby sites that serve as promoters (P_E and P_L) where transcription begins. In between the early and late promoters is a noncoding region of approximately 400 nucleotides that contains several important sequence elements involved in both transcription and genome replication; this is called the **intergenic region** or the **control region** (Figure 21.5).

A **polyadenylation signal**, 5′-AAUAAA-3′, specifies the 3′ ends of each transcript (Figure 21.6). For mouse polyomavirus, the 3′ ends of the mRNAs overlap by some 40 nt, and the AAUAAA signals encoded by the two DNA strands are separated by a single nucleotide (shown by dashed lines). Furthermore, the translation termination signals for large T antigen on the early mRNA (UGA) and for VP1 on the late mRNA (UAA) are separated by only 8 nucleotides. The concentration of important sequence elements

that serve as signals for transcription, DNA replication, RNA processing, and translation in two small regions of the genome reveals the efficiency with which these genomes are organized.

Virions enter cells in caveolae and are transported to the nucleus

Both mouse polyomavirus and simian virus 40 bind to **sialic acid**-containing glycoproteins and **gangliosides** on the cell surface and are subsequently internalized in **caveolae**, which are specialized vesicles formed at the plasma membrane. Virions are then transported inside these vesicles through the cytosol to the lumen of the endoplasmic reticulum, near the nuclear membrane (see Figure 4.5). From there, viral DNA must be transported to the nucleus.

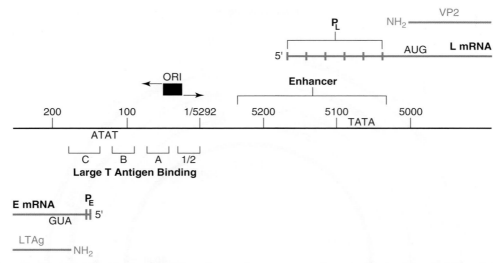

Figure 21.5 Intergenic control region on mouse polyomavirus genome. A noncoding region of ~400 nt, bounded by the divergent promoters for early and late transcription (P_E and P_L), contains sequences that control transcription and DNA replication. Transcriptional enhancer, origin of DNA replication (ORI), and large T antigen binding sites are shown above and below map. TATA boxes that direct initiation of transcription at selected promoters are shown. Vertical lines on mRNAs (light blue) show transcription start sites. N-terminal parts of the coding regions for T antigens and VP2 are shown in orange.

Passage of virions across the endoplasmic reticulum membrane appears to be mediated by VP2 molecules, which lie below the surface of each capsomere. Conditions within the endoplasmic reticulum lead to extrusion of the N-terminus of VP2 through the capsid surface, allowing the N-terminal myristate to interact with lipid membranes and leading to release of the virions into the cytosol.

Transport and release of viral DNA into the nucleus appears to be a two-step process. Capsids can be loosened or dissociated *in vitro* by treatment with reducing agents, which break disulfide bonds between cysteines, and chelating agents, which remove divalent cations bound to capsid proteins. Similar processes probably lead to uncoating and release of the viral genome. Upon entry into the cytosol, the capsid dissociates but VP1 remains associated with the viral minichromosome. The nuclear localization signal on VP1 directs this complex to the nucleus via the nuclear pores. Once in the nucleus, VP1 dissociates from the viral minichromosome, which remains intact during transcription and DNA replication.

The viral minichromosome is transcribed by cellular RNA polymerase II

Initiation of transcription of viral minichromosomes is a complex process that involves numerous cellular proteins. A **transcriptional enhancer** within the intergenic region (Figure 21.5) contains a number of consensus DNA sequences that are recognized by a variety of cellular **transcription factors**. Multicellular organisms encode hundreds of different transcription factors, which control gene expression in many circumstances,

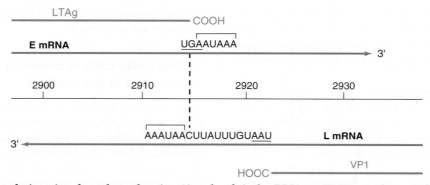

Figure 21.6 Polyadenylation signals and overlapping 3′ ends of viral mRNAs. mRNAs are shown in light blue; C-terminal parts of large T antigen and VP1 are shown in orange. Polyadenylation signals (AAUAAA) are bracketed; they are separated by only one nucleotide pair, GC, shown by a vertical dashed line. Translation termination signals for large T antigen (UGA) and VP1 (UAA) are underlined.

including embryogenesis and cellular differentiation. Each cell type in an organism may harbor a different combination of transcription factors, each of which recognizes specific binding sites on DNA.

Certain of these transcription factors recognize sequences in the viral enhancer and bind, displacing a small number of nucleosomes from that region. Bound transcription factors facilitate the association with viral DNA of the multiple cellular proteins required for transcription. Among these proteins are the TATA-binding protein (TBP), present in a complex called TFIID, which binds to a consensus **TATA box** sequence element on the viral DNA (Figure 21.5). TFIID recruits RNA polymerase II and a number of general transcription factors, named TFII A to H, to a site approximately 30 nt beyond the TATA box. Alternative transcription start sequences called initiators are present at the beginning of many viral and cellular genes that lack TATA boxes, but play a similar role. When all of these proteins are in place, transcription of early genes begins.

Four early mRNAs are made by differential splicing of a common transcript

The 2.7-kb early transcripts of mouse polyomavirus extend about half way around the circular viral DNA (Figure 21.4). They are capped at their 5' ends, and cleaved and polyadenylated at their 3' ends, by protein machinery used by the cell for generating its own messenger RNAs. Early transcripts then undergo alternative splicing within the nucleus to generate four different messenger RNA species. These mRNAs code for early proteins that were first detected in virus-induced tumors, and therefore are called tumor antigens, or T antigens. Alternative splicing occurs between two possible 5' splice sites, at nt 411 and 748, and two possible 3' splice sites, at nt 797 and 811 (Figures 21.4 and 21.7). Approximately equal amounts of the four mRNAs are made.

Two of the four possible introns (62 nt and 48 nt in length) removed during splicing of mouse polyomavirus early transcripts are unusually small. Introns smaller than approximately 70 nt are not normally found in mammalian cells. Ribonucleoproteins that make up the **spliceosome** bind to both ends of the intron as well as to an internal branch site near the 3' splice site. These factors cannot bind efficiently if the ends of the intron are too close to each other, as they mutually interfere with each other's binding ("steric hindrance"). The unusually small polyomavirus introns have apparently evolved to reduce splicing efficiency between the 5' splice site at nt 748 and either of the two 3' splice sites. If these small introns are artificially lengthened by insertion of as little as 40 nt into the viral genome, almost all splicing events target the nearest 5' splice site at nt 748, and no large T antigen mRNA is made. Besides being economical in the use of space in a small genome, these small introns therefore ensure a balanced production of the different early mRNAs.

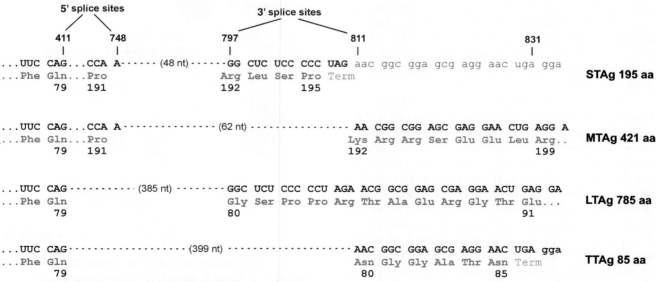

Figure 21.7 Alternative splicing of early mouse polyomavirus transcripts generates four different T antigens. Shown are the nucleotide sequences of parts of the four mouse polyomavirus early mRNAs (in black) and their translation products (in green). Translated codon triplets are shown in capitals in bold type; nucleotides that lie beyond a translation termination signal are shown in small, unbolded type. The amino acid coded by each triplet is shown directly below. Excised introns are shown as dashed lines, with the number of nucleotides excised shown at the center. Positions of 5' and 3' splice sites are noted. Translation of all four early mRNAs starts at an AUG at nt 175 (to the left). Note that all three reading frames are used between nt 811 and 831.

T antigens share common N-terminal sequences but have different C-terminal sequences

The four mouse polyomavirus early mRNAs code for four different proteins. Ribosomes initiate translation on all four early mRNAs at the first AUG codon they encounter, at nt 175 on the polyomavirus genome. This defines reading frame 0. As shown in Figure 21.7, excision of the smallest intron (48 nt) removes exactly 16 triplet codons. Ribosomes translate the first 191 codons and continue translation of this mRNA in the same reading frame downstream of the splice junction. They encounter a UAG termination codon only 12 nt after the splice junction, leading to production of a 195-amino acid protein called small T antigen (STAg).

In contrast, excision of the 62-nt intron removes 20 triplet codons plus 2 nt. Ribosomes translating this mRNA make a protein that contains the same first 191 amino acids as small T antigen, but they continue in a different reading frame (–1) downstream of the splice junction. In reading frame –1, the protein is extended to a size of 421 amino acids before a termination codon is encountered. This protein is called middle T antigen (MTAg).

Excision of the 385-nt intron removes 128 codons plus 1 nt. Ribosomes translating this mRNA make a protein that shares the same first 79 amino acids with small and middle T antigens, because the 5′ splice site lies exactly 79 triplet codons beyond the initiator AUG. However, the reading frame downstream of the splice junction is now +1; although this sequence overlaps with parts of the small and middle T antigen reading frames, it codes for a totally different amino acid sequence because different triplet codons are recognized by the ribosome. This reading frame does not terminate until very near the 3′ end of the mRNA (Figure 21.6). The result is large T antigen (LTAg), a protein 785 amino acids in length (708 amino acids for simian virus 40).

The fourth mouse polyomavirus early mRNA makes a protein with the same 79 N-terminal amino acids as the other T antigens. A 399-nt intron (133 triplet codons) keeps the ribosome in reading frame 0, as with small T antigen mRNA. However, the 3′ splice site at nt 811 lies just downstream of the UAG that terminates small T antigen translation. Another termination codon, UGA, is present just 7 codons beyond the splice junction, making an 85-amino acid protein called tiny T antigen (TTAg). This protein has not been carefully studied. Note that between nt 797 and 811, two different reading frames are used to make parts of STAg and LTAg, and between nt 811 and 831, all three reading frames are used to make parts of MTAg, LTAg, and TTAg.

Simian virus 40 has two 5′ splice sites analogous to those of mouse polyomavirus, but only one 3′ splice site. As a result, it makes a small and large T antigen, but no middle T antigen. However, a rare doubly spliced simian virus 40 mRNA codes for a protein similar to mouse polyomavirus tiny T antigen.

T antigens bring resting cells into the DNA synthesis (S) phase of the cell cycle

The principal function of the early proteins is to enable viral DNA replication. Most cells in an organism are not dividing, but are differentiated and in a resting (G$_0$) state. Under these conditions, the enzymes and precursors needed to support cellular or viral DNA synthesis are not normally present in the cell nucleus. The early proteins of DNA viruses induce entry of the cell into the DNA synthesis (S) phase of the cell cycle. The T antigens of both mouse polyomavirus and simian virus 40 have been intensively studied and have been shown to interact with numerous cellular signaling pathways, leading to stimulation of the cell cycle. Only a selection of these interactions will be discussed here.

Small T antigen inhibits protein phosphatase 2A and induces cell cycling

Small T antigen forms a complex with a subunit of the trimeric cellular regulatory protein, protein phosphatase 2A (PP2A), inhibiting the phosphatase activity of this protein (Figure 21.8). PP2A plays an important

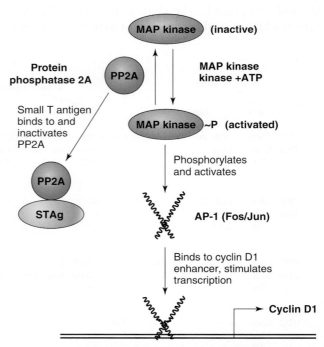

Figure 21.8 Small T antigen stimulates MAP kinase pathway. MAP kinase kinase activates MAP kinase by phosphorylation; protein phosphatase 2A inactivates MAP kinase by removing phosphates. Binding of small T antigen to protein phosphatase 2A inhibits its activity, allowing phosphorylated MAP kinase to accumulate and to activate transcription factor AP-1, which enhances cyclin D1 transcription.

regulatory role in several cellular signaling pathways by removing phosphate groups that stimulate the activities of proteins in those pathways. One set of cellular targets of PP2A are mitogen-activated protein kinases (MAP kinases). When a MAP kinase is phosphorylated at specific serine and threonine residues (by MAP kinase kinase), it can in turn activate the nuclear transcription factor AP-1, which activates transcription of the **cyclin** D1 gene among others (Figure 21.8). Cyclin D1 is an important activator of the cell cycle. Inhibition of PP2A activity by small T antigen allows MAP kinases to remain phosphorylated and therefore enzymatically active. Thus small T antigen indirectly stimulates the MAP kinase pathway and helps bring the cell into the S phase.

Middle T antigen stimulates protein tyrosine kinases that signal cell proliferation and division

Middle T antigen is localized at the plasma membrane of infected cells, where it is inserted via a hydrophobic region near its C-terminus (Figure 21.9). Because it contains most of the amino acid sequences of small

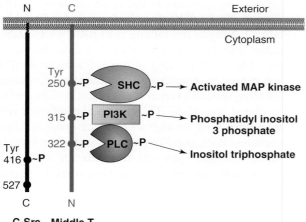

Figure 21.9 Middle T antigen binds to c-Src and activates pathways that stimulate cell metabolism and the cell cycle. Shown at left is c-Src, a protein tyrosine kinase inserted into the plasma membrane via a N-terminal myristate residue. When phosphorylated at tyrosine 527, c-Src is inactive (not shown). Middle T antigen, bound to plasma membrane via hydrophobic residues near its C-terminus, associates with c-Src. This leads to removal of the phosphate residue at tyrosine 527, and autophosphorylation of c-Src at tyrosine 416 (~P), activating its protein tyrosine kinase activity. C-Src phosphorylates middle T antigen tyrosines 250, 315, and 322, which serve as binding sites for cellular proteins involved in cell-cycle signaling: Shc, phosphatidyl inositol 3 kinase (PI3K), and phospholipase C (PLC). These are in turn phosphorylated by c-Src and activated, leading to increase in cell metabolism and cell cycling.

T antigen, middle T antigen can also bind to and inactivate PP2A. However, its major functions are carried out by association with several cellular **protein tyrosine kinases**, of which the best known is c-Src. C-Src can modify a number of other cellular proteins by phosphorylating specific tyrosine residues on those proteins. A mutated and highly active form of c-Src is encoded by the acutely transforming retrovirus Rous sarcoma virus, which produces cancer in chickens (see Chapter 28).

When associated with middle T antigen, c-Src loses a phosphate group at tyrosine 527 and phosporylates itself at tyrosine 416 (Figure 21.9). These changes greatly stimulate its protein kinase activity. C-Src then phosphorylates several tyrosine residues on middle T antigen itself. These phosphotyrosines act as part of binding sites for a set of cellular proteins including Src-homology 2 domain protein (Shc), phosphatidylinositol 3-kinase (PI3K), and phospholipase C (PLC). When bound to middle T antigen in the presence of activated c-Src, these proteins are in turn phosphorylated by c-Src and therefore activated. This leads to a cascade of signaling events too complex to detail here. Phosphorylated Shc leads, through an activated Ras protein, to the activation of the MAP kinase pathway referred to above. PI 3-kinase synthesizes phosphatidylinositol 3-phosphate, a phospholipid that controls such cellular functions as proliferation, vesicle trafficking, and organization of the cytoskeleton. Phospholipase C cleaves phosphatidylinositol, generating inositol triphosphate, which releases calcium ions from the endoplasmic reticulum. Many of these pathways activate cell metabolism and the cell cycle.

Large T antigen activates or suppresses transcription of cellular genes by binding to a number of important cellular regulatory proteins

Large T antigen is a complex protein with a surprisingly large number of different functions. Most large T antigen is localized in the nucleus of infected cells, although some is found in the cytoplasm and at the plasma membrane. The nuclear localization signal of simian virus 40 large T antigen was one of the first such signals identified. A schematic diagram showing functional domains of simian virus 40 large T antigen is shown in Figure 21.10. Like small and middle T antigens, large T antigen interacts with a number of cellular proteins that control cell metabolism and the cell cycle.

1. ***Retinoblastoma protein (pRb) controls cell cycle and S-phase gene expression.*** The retinoblastoma protein (pRb), so-named because mutations in the Rb gene can lead to retinal tumors in humans, is one of the major cell-cycle control proteins. pRb and related proteins p107 and p130 repress transcription

from a large number of cellular genes when bound to members of the E2F family of transcriptional activator proteins. When phosphorylated, pRb dissociates from its E2F partner, allowing transcription of these cellular genes, entry into the cell cycle, and expression of proteins required for both cellular and viral DNA replication (see Figure 23.6).

Large T antigen has a conserved binding site for the Rb protein (Figure 21.10). Binding to pRb begins a process that mimics pRb phosphorylation, releasing E2F and allowing transcription of cellular genes targeted by E2F. These genes include cyclins, DNA polymerase alpha, thymidine kinase, DNA ligase, histones, and others. A number of DNA viruses make early proteins that interact with pRb and stimulate the cell cycle. For a more complete description of pRb interactions, see chapters on papillomaviruses and adenoviruses (Figures 22.4, 22.5, and 23.6).

2. **The DNA J domain acts as a cochaperone to dissociate pRb from E2F.** The N-terminal region of large T antigen, called the DNA J domain, is also required for gene activation via the Rb pathway. DNA J is a cochaperone protein, first described in *E. coli* as a bacterial protein required for replication of bacteriophage lambda DNA (Chapter 8). DNA J and its eukaryotic analogues interact with a cellular chaperone called Hsp70 (known as DNA K in *E. coli*) and take part in ATP-dependent protein folding and unfolding reactions that are vital in assembling or disassembling multiprotein complexes.

It was surprising to discover that the N-terminal region of large T antigen can, by itself, complement DNA J-minus mutants of *E. coli* and enable bacteriophage lambda DNA replication! This shows the extraordinary conservation of protein functions in widely separated organisms, and reveals a connection across perhaps billions of years of evolution between a bacteriophage and a mammalian DNA tumor virus. The DNA J domain of large T antigen is believed to act by helping to separate the Rb protein from a complex containing E2F and other proteins, thus stimulating transcription of cell-cycle and S-phase genes.

3. **p53 blocks the cell cycle and induces apoptosis in response to virus infection.** The cellular control protein p53 was first discovered as a 53-KDa protein present in immunoprecipitates of simian virus 40 large T antigen. p53 responds to a variety of signals by binding to and activating transcription of genes that can lead either to blockage of the cell cycle or **apoptosis** (programmed cell death). Its central role in controlling the cell cycle is emphasized by the fact that a majority of human tumors have undergone mutations in the p53 gene. Therefore, for a tumor cell to thrive, it must rid itself of the control exercised by the p53 protein.

Normally p53 is rapidly degraded, is present in very low concentrations in the cell, and is inactive as a transcription factor. Various insults to the cell, including infection by polyomaviruses, can lead to increases in the concentration and the activity of p53. This is at least partly mediated by large T antigen interacting with pRb, which stimulates transcription of genes that raise levels of p53. However, binding of large T antigen to p53 suppresses its transcriptional activation function, allowing the cell to enter S phase and avoiding p53-mediated apoptosis. This allows the infected cell to survive long enough to produce progeny virus particles. Both papillomaviruses and adenoviruses make analogous proteins that interact with p53. For a more complete description of p53 control

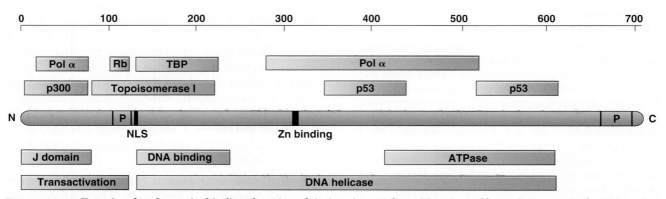

Figure 21.10 Functional and protein-binding domains of simian virus 40 large T antigen. Shown in orange is the 708-amino acid large T polypeptide. P designates regions that are phosphorylated, **NLS** designates the nuclear localization signal, and **Zn** designates a zinc-binding site. Above are shown regions that bind to cellular proteins involved in cell signaling or DNA replication: **Pol α** = DNA polymerase α/RNA primase; **Rb** = retinoblastoma protein; **TBP** = TATA-binding protein. Below are shown functional domains of large T antigen; see text for details.

and function, see the chapters on papillomaviruses and adenoviruses (Figures 22.6 and 23.7).

4. *p300 and general transcription factors control levels of transcription of viral and cellular genes.* Large T antigen interacts directly with a number of cellular proteins involved in transcription. These include the TATA-binding protein (TBP), the general transcription factor TFIIB, and the transcriptional co-activator p300, which has histone acetylase activity (Figure 21.10). Binding of large T antigen to these proteins can lead to both stimulation and repression of gene expression.

For example, p300 is associated with a number of cellular transcription factors, including p53 and AP-1. p300 is believed to cooperate with these factors in stimulating transcription by acetylating nucleosomal histones, therefore reducing their affinity for DNA and making DNA more accessible to components of the transcription machinery. Binding of large T antigen to p300 can inhibit its transcriptional coactivation function and therefore reduce transcription of genes targeted by these factors. However, large T antigen is known to increase transcription of many other genes, probably via its interactions with the machinery that assembles RNA polymerase II at promoters.

Large T antigen hexamers bind to the origin of DNA replication and locally unwind the two DNA strands

Once the cell has entered the S phase, viral DNA replication can begin. Large T antigen now plays an entirely different role: instead of interacting with cellular regulatory proteins, it binds to viral DNA and directs the assembly of a number of cellular proteins that carry out DNA replication. The DNA-binding domain of large T antigen (Figure 21.11) recognizes G(A/G)GGC pentanucleotide sequences on double-stranded DNA. Both simian virus 40 DNA and mouse polyomavirus DNA have 12 such sequences that lie within a 160-nt span in the intergenic region. In mouse polyomavirus, these 12 binding sequences are grouped in four sites denoted 1/2, A, B, and C (Figure 21.5). Large T antigen molecules bind cooperatively to these sequences and assemble at a specific site denoted the DNA replication origin (Ori) to form two circular hexamers that enclose the DNA like a wheel on an axle.

Electron microscopy and x-ray crystallographic studies have revealed the detailed structure of the hexamer form of large T antigen (Figure 21.11). Each large T antigen monomer in a hexamer binds a molecule of ATP. Hydrolysis to ADP and release of the ADP molecule lead to a conformational change that rotates adjacent protein domains with respect to each other and

changes the size of the central opening of the hexamer that is in contact with the DNA. This has been likened to an iris that can change its diameter by contracting or expanding.

As a result of this change, 5 to 6 bp of dsDNA are pulled through the hexamer by the motion of a β-hairpin that is in contact with the DNA. The two DNA strands are locally separated, and the separated strands are extruded through channels within the hexamer. Energy derived from hydrolysis of ATP to ADP powers the movement and denaturation of the DNA strands. Once the bound ADP residues dissociate from the large T antigen molecules, another set of ATP molecules binds, and the cycle continues. Thus, large T antigen acts as a DNA **helicase**.

Large T antigen assembles the cellular DNA synthesis machinery to initiate viral DNA replication

This process generates a localized region of single-stranded DNA that can now serve as a template for DNA replication. Here again large T antigen has a vital function. DNA-bound large T antigen hexamers bind to three cellular proteins required for initiation of DNA replication: replication protein A, DNA polymerase α/primase, and topoisomerase I (Figure 21.12a). Replication protein A binds to the unwound DNA strands and keeps them from renaturing into a double-helical form. DNA polymerase α/primase binds to the single-stranded DNA and lays down a short (6–9 nt) RNA primer that remains hydrogen-bonded to the DNA template. The same enzyme then extends the primer by adding a short (20–30 nt) DNA molecule. Primers are formed and elongated on both DNA strands, and as a result, bidirectional DNA synthesis is initiated. Topoisomerase I is required to help release the tension created in the circular, double-stranded DNA molecule as the DNA polymerase and helicase proceed along the DNA molecule, twisting the double helix in front as it is unwound behind.

Finally, three other cellular proteins, replication protein C (RPC), proliferating cell nuclear antigen (PCNA), and DNA polymerase δ (delta), are recruited to the complex (Figure 21.12b). Replication protein C helps to dissociate DNA polymerase α/primase and load DNA polymerase δ onto the 3' ends of the growing DNA chains. PCNA binds to DNA polymerase δ and ensures that it does not dissociate from the DNA template. DNA polymerase δ then extends the DNA chains.

As the leading strands are extended toward each replication fork, large T antigen continues to act as a helicase by unwinding the template DNA. As a result, more single-stranded DNA becomes exposed on the complementary strand at each replication fork. DNA polymerase α/primase lays down new RNA primers on

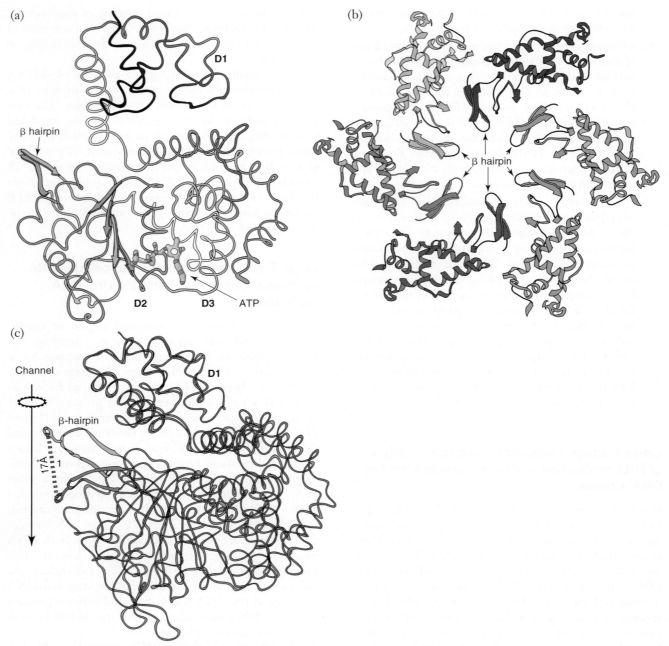

Figure 21.11 Structure of simian virus 40 large T antigen. (a) Ribbon diagram of the polypeptide chain of amino acids 251–627 of simian virus 40 large T antigen as determined by x-ray crystallography. Domains D1, D2, and D3 are shown, and bound ATP is shown at the junction between D2 and D3. (b) View of a hexamer of large T antigen molecules showing the central channel through which DNA passes and the β-hairpins that contact the DNA and pull it through the channel. (c) Side views of one monomer of large T antigen in a hexamer, showing the channel and the 17 Å (1.7 nm) movement of the β-hairpin (1) as ATP is hydrolyzed to ADP and ADP is released.

these single-stranded regions and reinitiates lagging strand DNA synthesis. These short DNAs are extended by DNA polymerase δ up to the RNA primers of the adjacent DNA chains, and the primers are removed by ribonuclease H. The short gaps that remain are filled in by DNA polymerase δ, and adjacent DNA molecules

are then ligated, forming a continuous lagging strand (Figure 21.12c).

As DNA synthesis progresses, cellular histones present in the S-phase nucleus assemble on the growing DNA chains to form nucleosomes. The final result is the synthesis of two circular, double-stranded

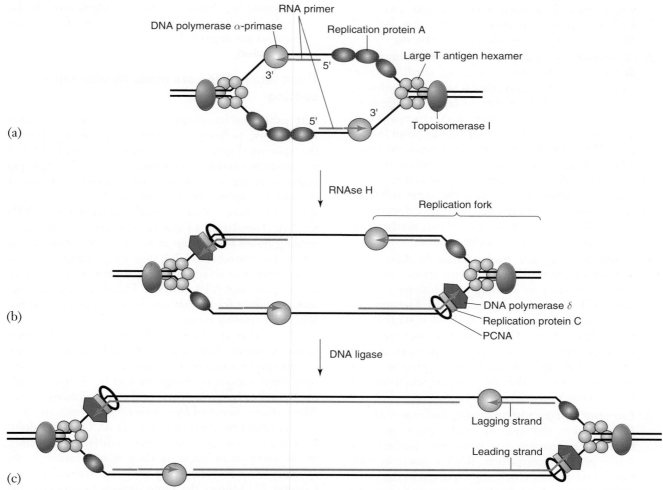

Figure 21.12 Mechanism of bidirectional replication of polyomavirus DNA. (a) Initiation of DNA synthesis at viral replication origin. DNA polymerase α/RNA primase binds to a region of single-stranded DNA unwound by the DNA helicase activity of large T antigen hexamers and bound by replication protein A. (b) Extension of DNA chains beyond the initiation region by DNA polymerase δ associated with replication protein C and proliferating cell nuclear antigen (PCNA). Ribonuclease H (RNAse H) has removed initial RNA primers. DNA polymerase α/RNA primase reinitiates synthesis on single-stranded DNA generated by continued helicase action of large T antigen as the two replication forks travel in opposite directions on the DNA template. (c) Joining of growing DNA strands. Lagging-strand DNAs are extended up to the 5′ end of the leading-strand DNA and the strands are linked by DNA ligase to produce a single, covalently joined daughter strand on each template strand. RNA is shown in light blue, newly synthesized DNA in gray.

viral minichromosomes. The progeny DNAs can be used as a template for further transcription, or can be replicated.

High levels of late transcripts are made after DNA replication begins

Initiation of late transcription on both mouse polyomavirus and simian virus 40 DNA occurs at multiple promoter sites within a 100-nt region that overlaps with the transcriptional enhancer. Most of these late promoter sequences are not associated with a consensus TATA box. Before DNA replication begins, low but detectable

levels of late transcripts are made. After DNA replication begins, much higher levels of late transcripts are made, and synthesis of early transcripts is reduced. Five different reasons for the increase in the ratio of late to early transcription have been suggested.

1. ***Repression of early transcription by large T antigen.*** The early promoters of both simian virus 40 and mouse polyomavirus overlap with one or more of the G(A/G)GGC binding sites for large T antigen. Binding of large T antigen to these sites can interfere with the binding of RNA polymerase and general transcription factors such as TFIID, reducing the amount of early RNA synthesized. Therefore,

when large T antigen concentration increases, early transcription decreases. In this way, large T antigen controls its own levels of synthesis.

2. *Activation or induction of new transcription factors.* The combined effects of small, middle, and large T antigen on cellular signaling pathways (see above), and the ability of large T antigen to bind to a variety of transcriptional repressors and activators, changes the mix of transcription factors available to bind to the transcriptional enhancer. These changed factors specifically activate the late promoters. Because numerous transcription factors are involved, and because the same enhancer region powers both early and late transcription, the exact mix of transcription factors that activate late transcription is not fully understood.

3. *Derepression of late transcription by dilution of a repressor.* Simian virus 40 late transcription is initially blocked by binding of a cellular protein to viral DNA near the late promoter. The concentration of this repressor in the cell is limited. As the number of viral DNA molecules increases due to DNA replication, a smaller proportion of viral DNA molecules contain bound repressor. This allows activation of the late promoter and increased levels of late transcription.

4. *Increased efficiency of processing and export of late RNAs.* Before the onset of DNA replication, most mouse polyomavirus late transcripts initiate at a single promoter sequence. RNA transcripts with this 5' end are not efficiently exported to the cytoplasm, perhaps because they are missing a critical RNA export sequence. Activation of late transcription leads to initiation at multiple sites upstream of this promoter sequence, and RNAs with these 5' ends are more efficiently processed into mRNAs and exported to the cytoplasm (Figure 21.5).

5. *Readthrough RNAs hybridize with early transcripts and lead to their degradation.* Most RNA polymerases that transcribe the late genes of mouse polyomavirus do not terminate transcription after passing through the late polyadenylation signal, but continue transcription along the DNA template. Because the DNA is circular, this leads to the production of very long late RNAs that can contain multiple, tandem repeats of the DNA sequence, similar to the production of long DNAs by rolling circle DNA replication. Some RNA polymerases can pass 10 or more times around the circular DNA before they terminate! These giant viral RNAs contain sequences that are complementary to early transcripts, and they are made in large amounts for reasons explained above. The presence of complementary early and late RNAs in the nucleus leads to the formation of double-stranded RNAs, which are degraded by cellular enzymes.

Because late transcripts are in excess, most of the early transcripts but only a fraction of late transcripts are degraded.

Three late mRNAs are made by alternative splicing

Three different late messenger RNA species are formed from mouse polyomavirus late transcripts by alternative splicing (Figure 21.4). An unspliced late mRNA is translated to produce viral capsid protein VP2. Splicing removes either a small or a large intron, starting at the same 5' splice site but cutting at two possible 3' splice sites. Removal of the small intron generates VP3 mRNA. Translation of this protein starts in the same reading frame as VP2 but at an AUG that codes for an internal methionine in VP2; thus VP3 is identical to the C-terminal two-thirds of VP2. Removal of the large intron generates VP1 mRNA. This is coded in a different reading frame that overlaps with the end of the VP2/VP3 reading frame.

Note that alternative splicing of *early* transcripts removes introns downstream of the AUG start codon, within open reading frames, and generates distinct proteins from a single transcription unit by changing reading frames or removing parts of the same reading frame. Alternative splicing of *late* transcripts, in contrast, generates distinct proteins by removing AUG start codons, leading to initiation of translation at different start sites, in the same or different reading frame. Interestingly, splicing of adenovirus early and late transcripts (see Chapter 23) follows the same pattern to generate distinct proteins.

Capsid proteins are imported into the nucleus, where they form capsids that package viral minichromosomes. Empty capsids can also form. Under certain conditions, fragmented cellular DNA, resulting from cleavage during cell death, can also be packaged into capsids to form "pseudovirions". Because these particles can enter into other cells or be transmitted to other animals, pseudovirions may enable passage of certain cellular DNA sequences from one host to another.

How do polyomaviruses transform cells *in vitro* and cause tumors *in vivo*?

As we have seen, productive infection of permissive cells by polyomaviruses leads to cell death and release of progeny virions. However, polyomaviruses were first discovered as agents able to induce formation of tumors in mice and other rodents. Furthermore, these viruses can transform normal rat or hamster fibroblasts grown *in vitro* into cell lines that are immortal, form dense and rapidly growing colonies, have reduced requirements for growth factors, and in some cases can cause tumors when injected into animals.

Paradoxically, polyomaviruses cause few if any tumors in their natural hosts.

The ability to study cell transformation *in vitro* by these viruses allowed scientists to study the biochemical and signaling pathways that are altered upon infection by polyomaviruses. It was quickly discovered that polyomavirus early genes act as **oncogenes** both in cells cultured *in vitro* and in intact animals. Study of the structure and activity of the early gene products (T antigens) led to the discovery of a number of important cellular regulatory pathways that are disrupted or modified during transformation and tumor formation. Some of these pathways have been outlined above. Their discovery has helped provide a detailed understanding of the process of tumorigenesis.

Only non-permissive cells can be transformed

Transformation of cells by polyomaviruses can take place only if the cells are **nonpermissive** for viral replication. This means that the virus can enter the cell and express its early genes, but can neither replicate its DNA nor express late proteins. Therefore, the infectious cycle is **abortive** and does not result in the production of progeny virus or kill the infected cell. The inability of rat and hamster cells to replicate mouse polyomavirus or simian virus 40 DNA has been traced to a lack of interaction between some elements of the cellular DNA replication machinery and the viral large T antigen.

Transformed cells integrate viral DNA into the cell chromosome

Only 1 to 10 out of every 100,000 infected nonpermissive cells end up being transformed; these can be detected as dense, rapidly growing colonies on a monolayer of normal fibroblasts growing on the surface of a Petri dish. Such transformed cells invariably contain most or all of the early region of the viral genome, integrated into the cellular genome at apparently random positions.

The low frequency of cell transformation *in vitro* (and tumor formation *in vivo*) is probably due to the low probability of integration of the early region of viral DNA into the cellular genome. Unlike retroviruses (Chapter 28), polyomaviruses do not normally integrate their DNA into the host cell chromosome during replication; this rare event is carried out by cellular enzymes that pick up degraded fragments of viral DNA and insert them into the host genome.

Cells that have integrated a functional early region can continue to express T antigens, and all daughter cells inherit this ability. Therefore the various signaling pathways that turn on cell cycling, reduce dependence on growth factors, and increase cell growth rate are permanently altered. When this occurs *in vivo*, such cells can eventually evolve to form tumors. Transgenic animals that express one or more of the early genes of polyomaviruses in specific tissues are being studied as model systems for tumor formation.

KEY TERMS

Abortive	Polyadenylation signal
Apoptosis	Progressive multifocal
Caveolae	leukoencephalopathy
Cochaperone	(PML)
Cyclins	Protein tyrosine kinase
Divergent transcription	Sialic acid
DNA gyrase	Sodium dodecyl sulfate
Ethidium bromide	Spliceosome
Gangliosides	Supercoiled
Helicase	TATA box
Intergenic region	Topoisomerase I
Myristate	Toroidal coil
Nonpermissive	Transcription factors
Nucleosomes	Transcriptional enhancer
Oncogene	

FUNDAMENTAL CONCEPTS

* Polyomaviruses are among the smallest double-stranded DNA viruses, with a circular genome of 5.3 kb.

* The polyomavirus genome is packaged with cellular histones to form a "minichromosome".

* Both mouse polyomavirus and simian virus 40 can cause tumors in experimental animals and transform cells *in vitro*, but they rarely if ever cause cancer in nature.

* Cell transformation and tumor formation results from integration of viral DNA into the genome of nonpermissive cells and expression of viral early genes, turning on cell cycling and increasing cellular growth and multiplication.

* Polyomavirus DNA becomes supercoiled upon removal of the histones, but remains in an open circular form within the cell as it is always associated with histones.

* Polyomaviruses have two divergent transcription units that lie on either side of a central control region, which includes the origin of DNA replication and transcriptional enhancer elements.

- Two to four early messenger RNAs and two to four late messenger RNAs are made by splicing primary transcripts of the early and late regions.
- Introns removed by splicing of some polyomavirus early RNAs are unusually small, and their small size reduces the efficiency of splicing.
- Polyomavirus early proteins, called T antigens, have multiple interactions with cellular signaling proteins.
- Small T antigen interacts with protein phosphatase 2A, stimulating the MAP kinase pathway and bringing the cell into the S (DNA synthesis) phase of the cell cycle.
- Middle T antigen binds to protein kinases such as c-Src and activates them, phosphorylating middle T antigen itself, and leading to activation of cellular signaling proteins such as SHC, PLC, and PI3K.
- Large T antigen binds to Rb protein, releasing it from the transcription factor E2F and allowing transcription of cellular genes that direct DNA synthesis and move the cell into the S phase.
- Large T antigen also binds to the replication origin on viral DNA, recruiting a set of DNA replication proteins that result in replication of the viral genome.
- Transcription of late genes is activated by a number of transcription factors that are induced by the action of T antigens.
- Viral capsid proteins are imported into the nucleus where they condense around progeny viral minichromosomes and assemble into mature virions.

REVIEW QUESTIONS

1. How do polyomaviruses condense their genome to package it in the virion?

2. Describe the transcription pattern used by polyomaviruses to express their genome. What is it called?

3. How do polyomavirus proteins interact with Rb to induce the cell to enter the S (DNA synthesis) phase of the cell cycle?

4. Where in the cell do polyomaviruses assemble progeny virions?

5. What role does small T antigen play in bringing resting cells into the DNA synthesis phase of the cell cycle?

6. How does middle T antigen stimulate cell proliferation and division?

7. Large T antigen hexamers bind to the origin of DNA replication and locally unwind the two DNA strands. How does large T antigen subsequently help to initiate viral DNA replication?

CHAPTER 22

Papillomaviruses

Greg Matlashewski
Revised by Lawrence Banks

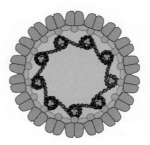 

Papillomaviridae

From Latin *papilla* (nipple) and Greek *–oma* (tumor) referring to warts (papillomas)

VIRION

Naked icosahedral capsid (T = 7).
Diameter 55 nm.
Formed from 72 capsomers, pentamers of L1.

GENOME

Circular double-stranded DNA, 8 kb.
DNA is packaged as a "minichromosome" with cellular nucleosomal histones.

GENES AND PROTEINS

Two transcriptional promoters and two polyadenylation signals (early and late) on the same DNA strand.
Splicing generates 10 or more different mRNAs.
Early proteins E1 to E7 stimulate cell proliferation and enable viral DNA replication.
Late proteins L1 and L2 form capsid.

VIRUSES AND HOSTS

Over 100 known human papillomaviruses.
Other hosts: cattle, dogs, deer, rabbits, monkeys, etc.

DISEASES

Benign warts at specific sites (skin, mucosa, larynx) depending on virus strain.
Transmitted by direct contact.
Some types cause cervical carcinoma, a sexually transmitted disease and major cause of cancer in women.
May cause other ano-genital cancers in both men and women, and cancers of the oral cavity, oropharynx and tonsil.

DISTINCTIVE CHARACTERISTICS

Generally difficult to grow *in vitro*: require specialized raft cultures of epithelial cells.
Replication pattern follows differentiation of epithelial cells in skin or mucosa.
Early proteins E6 and E7 can transform cells *in vitro*, are expressed in cervical carcinomas.

Papillomaviruses cause warts and other skin and mucosal lesions

The first papillomavirus was discovered in 1933 by Richard Shope. He showed that warts on wild cottontail rabbits could be transmitted to domestic rabbits by inoculation with extracts of the warts. It is now recognized that papillomaviruses are widespread in animals and humans. Bovine papillomavirus, which causes warts in cattle, originally served as a prototype for the study of the molecular biology and transforming genes of papillomaviruses because bovine warts are easily available, and because this virus can transform cultured rodent cells *in vitro*. However, it is now recognized that there are many important differences in pathogenesis between bovine papillomavirus and human papillomaviruses (HPV).

Over 100 types of human papillomaviruses have been identified (Table 22.1). Because they are difficult to grow *in vitro*, most of these viruses were identified by direct cloning of their DNA in bacterial plasmids, followed by nucleotide sequence analysis. Papillomavirus genotypes are considered distinct if the nucleotide sequences of specific regions of the viral genome (L1, E6, and E7 genes) differ by more than 10%. Different human papillomavirus types tend to cause lesions in different anatomical sites. For example, plantar warts of the feet are associated with human papillomavirus type 1, and common hand warts are associated with human papillomavirus types 2, 4, and 7.

Human papillomaviruses tend to infect either mucosa (genital tract, oral and nasal cavity) or skin. Types that infect skin are likely transmitted through skin

Table 22.1 Sites of infection of selected human papillomaviruses

Virus type	Site of infection	Clinical manifestation
1	Soles of feet	Deep plantar warts
2, 4, 7	Hands	Common warts
13, 32	Oral cavity	Focal epithelial hyperplasia
3, 8, 9, 10, 12, 14, 15, 17, 19–25, 36, 38	Arms, forehead, trunk	Flat warts, skin cancer
6, 11	Genital tract, larynx, nasal cavity	Cervical warts, condyloma
16, 18, 31, 33, 35, 39, 41–45, 52, 56, 58, 59	Genital tract	Cervical cancer, atypical condyloma

contact with an infected individual or through contact with a virus-containing surface. Transmission of viruses that infect genital tract mucosa is largely through sexual contact with an infected individual. The epithelia of the male and female genital tracts likely serve as virus reservoir.

Oncogenic human papillomaviruses are a major cause of genital tract cancers

Since 1976, it has been recognized that human papillomavirus infections of the genital tract frequently lead to benign genital warts, neoplasias, and invasive **squamous cell carcinomas**. Cervical cancer in women is associated with infection primarily with the human papillomavirus types 16, 18, and 31, but a number of other types are also implicated (see Table 22.1). In contrast, infection of the genital tract with non-oncogenic human papillomavirus types 6 and 11 can produce cervical warts (also known as **condylomas**), which are benign and only rarely progress to cancer. Because of the link with cancer, the focus of research on human papillomaviruses is now largely centered on the oncogenic human papillomavirus types 16, 18, and 31.

Many genital infections with human papillomaviruses are transient and result in no disease. These unrecognized infections can result in spread of virus to uninfected individuals. Only a minority of infections with oncogenic human papillomaviruses result in cervical cancer, and many of the lesions regress spontaneously without treatment. Nevertheless, infection with these viruses is the most significant and important risk factor for the development of cervical cancer.

Although these viruses infect both sexes, papillomavirus-associated genital cancer is much less prevalent in men than in women. This may be related to differences in the target cells of the virus, since both men and women appear equally susceptible to the development of HPV-induced anal cancer. In the cervix, the cells that become infected are non-differentiated basal cells of the epidermis within a region known as the "transformation zone". These cells are particularly susceptible to oncogenic transformation by the high-risk human papillomavirus types such as 16 and 18.

Papillomaviruses are not easily grown in cell culture

A long-standing problem in studying papillomaviruses has been the difficulty in propagating them in cultured cells *in vitro*. Much of what is known about papillomavirus replication is derived from studies on warts, or on cells transformed by papillomaviruses and propagated *in vitro*. However, transformed cells do not generally produce virus, and express a limited range of viral proteins. There has been recent success in propagating several human papillomaviruses in specialized **raft culture** systems, which are capable of maintaining fully differentiated epithelial cells (**keratinocytes**) over long periods of time. Keratinocytes in these cultures can be transfected with cloned viral DNA and are able to produce infectious viral particles. Use of these systems has provided valuable insights into the replication of these viruses and allows detailed genetic analyses to be performed on the roles of specific viral proteins in the virus life cycle.

Papillomavirus genomes are circular, double-stranded DNA

Papillomavirus virions are small, non-enveloped particles with T = 7 icosahedral symmetry (Figure 22.1). Virions are constructed from 72 capsomeres, each containing 5 molecules of the major capsid protein L1 as well as one molecule of the minor capsid protein L2. The viral genome, an 8-kbp molecule of double-stranded circular DNA, is associated with cellular nucleosomal histones. Papillomaviruses share many of these properties with the polyomavirus family (Chapter 21), with which they were previously grouped. However, papillomaviruses have a significantly larger genome, their gene map is organized in a different fashion, and they share a distinct lifestyle, all justifying their classification in their own family.

Most human and animal papillomavirus genomes contain 8 to 10 identifiable open reading frames. The

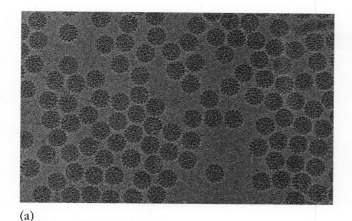

(a)

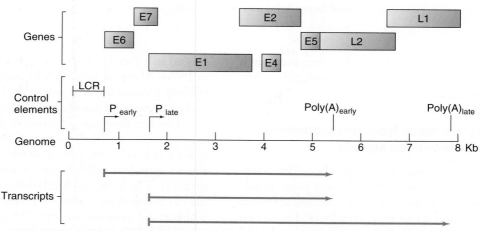

(b)

Figure 22.1 Papillomavirus virions. (a) Transmission electron micrograph of frozen-hydrated bovine papillomavirus type 1. (b) Three-dimensional image reconstruction of virions derived from cryoelectron microscopy analysis.

human papillomavirus type 16 genome contains six early genes (E1, E2, E4, E5, E6, E7) and two late genes (L1, L2). These open reading frames are shown on a linearized genome map in Figure 22.2. Unlike

the polyomaviruses, all papillomavirus mRNAs are transcribed from only one of the DNA strands. Note that several protein coding regions overlap, allowing the efficient use of a small genome. A brief description of the function of each viral gene is presented in Table 22.2.

The infectious cycle follows differentiation of epithelial cells

During infection of the **mucosal epithelium** (Figure 22.3), virions must first reach the non-differentiated **basal cells** of the epithelium where they bind to components of the extracellular matrix and then to the major viral receptors: **heparan sulfate proteoglycans** and the cellular surface protein, α_6-**integrin**. Virions are then taken up by endocytosis and viral DNA enters the nucleus, where it is maintained as a free, circular minichromosome in the basal cells. Early genes are transcribed, permitting an initial replication phase that results in 50 to 100 copies of DNA per cell. Genomes then replicate on average about once per cell cycle as the basal cells divide and the viral genomes are distributed equally among the daughter cells. This type of viral DNA replication is referred to as "plasmid replication".

When the basal cells become committed to the pathway toward terminally differentiated epithelial cells (keratinocytes), there is a burst of viral DNA synthesis, known as "vegetative replication". At the same time, late genes L1 and L2 are expressed, producing the viral capsid proteins. Progeny virus particles are assembled in the nucleus, and are released upon cell death and shedding at the surface of lesions. Little is presently known

Figure 22.2 Genetic and transcriptional map of human papillomaviruses. A linear representation of the circular genome of papillomaviruses, based on information from human papillomavirus types 16 and 31. Open reading frames of the six early (E) genes and two late (L) genes are shown in boxes. LCR: long control region. Two promoters (P_{early} and P_{late}) and two polyadenylation signals (Poly A_{early} and Poly A_{late}) control production of transcripts, shown at bottom. These transcripts are spliced in multiple alternative ways to produce a large number of mRNAs (not shown), many of which contain multiple open reading frames.

Table 22.2 Functions of human papillomavirus genes

Gene	Function
E1	DNA helicase, binds to replication origin, initiates DNA replication
E2	Binds to specific sites on viral DNA, enhances binding of E1 to replication origin, activates or represses transcription of viral genes, binds to mitotic chromosomes and aids in segregation of viral genomes between two daughter cells
E4	Function largely unknown; most abundantly expressed viral protein, binds keratins and may help virus maturation and release
E5	Small, membrane-associated protein, enhances growth factor responses in infected cell, stimulates cell proliferation
E6	Directs ubiquitin-mediated degradation of cellular p53 and other target proteins, inhibits cell-cycle block and apoptosis (oncogenic papillomaviruses only)
E7	Binds to and degrades Rb protein, releases E2F, activates cell cycling and DNA replication genes, stimulates cell and viral DNA replication
L1	Major capsid protein
L2	Minor capsid protein, helps encapsidate viral DNA

Note: Other human and animal papillomaviruses have additional genes designated E3 and E8.

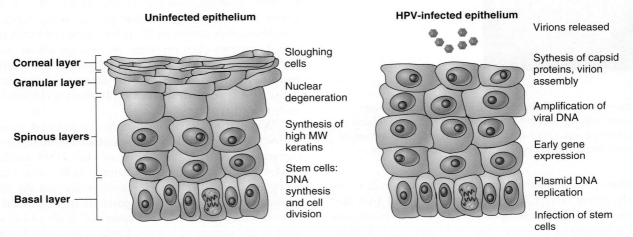

Figure 22.3 Papillomavirus infection of differentiating epithelial cells. Cells differentiate as they pass from the basal layer (bottom) to the surface (top). Activities in different cell layers are shown in the middle. HPV infection leads to increased cell division and retention of cell nuclei in outer layers. Viral functions that take place in different layers are shown at the right.

about the intracellular signals that regulate the switch from plasmid to vegetative viral DNA replication during differentiation in keratinocytes.

Viral mRNAs are made from two promoters and two polyadenylation signals

Numerous viral messenger RNAs have been described. Two major promoters are used by human papillomavirus types 16 and 31: an early promoter located upstream of the E6 gene, and a late promoter located upstream of the E1 gene (Figure 22.2). In undifferentiated basal epithelial cells, the early promoter directs transcription of RNAs that code for all six early gene products. The early primary transcript is polyadenylated at a site downstream of the E5 gene, and these transcripts are

spliced in a variety of ways to generate numerous early mRNAs. Many of these RNAs contain multiple open reading frames, but it is not clear how these multiple coding regions are translated by ribosomes.

The late promoter is active in differentiated keratinocytes (Figure 22.3). Some of these transcripts are polyadenylated at the early polyadenylation site, giving rise to abundant mRNAs that express the E1 and E2 proteins. The increased level of these proteins contributes to the shift to vegetative DNA replication, during which many thousands of viral DNA molecules per cell are made. Other transcripts from the late promoter extend through most of the viral genome to a second major polyadenylation site. Again, the alternative use of several splice sites leads to the production of a number of late mRNAs. Only mRNAs that start at the late promoter and end at the late polyadenylation site code for L1 and L2 capsid proteins. These mRNAs undergo splicing that removes most of the 3- to 5-kb region that lies between the late promoter and the beginning of the L1 and L2 open reading frames.

Viral E1 and E2 proteins bind to the replication origin and direct initiation of DNA replication

Regulatory DNA sequences known as the long control region (LCR) of human papillomaviruses extend from the end of the L1 gene to the beginning of the E6 gene (Figure 22.2). The long control region contains the viral origin of DNA replication as well as enhancer sequences that control viral gene expression.

The viral E1 protein is a DNA helicase that binds to the origin of replication and locally unwinds the viral DNA, allowing cellular DNA polymerases and other cellular proteins to carry out viral DNA replication. E1 is therefore the papillomavirus homologue of the polyomavirus large T antigens (Chapter 21). The viral E2 protein is also involved in viral DNA replication. It binds to distinct sites near the replication origin, and forms a complex with the E1 protein. Binding of E2 to E1 increases the affinity of E1 binding to the replication origin. E2 also binds to mitotic chromosomes and is believed to aid in the correct segregation of the viral genomes between the two daughter cells

Numerous cellular transcription factors including NF1, SP1, AP1, Oct1, YY1, and the DNA-binding glucocorticoid receptor protein have been shown to interact with the regulatory regions of papillomavirus DNA and to play important roles in activating viral transcription. The viral E2 protein also acts as a transcriptional regulator of viral promoters through its ability to bind to multiple specific sequences within the long control region. Depending upon the amount of E2 protein expressed, it can act as either a transcriptional activator or repressor of viral gene expression. Alternative splicing can produce a truncated form of E2 that lacks the transactivating function, and this further contributes to the regulation of viral gene expression.

Viral E7 protein interacts with cell-cycle regulatory proteins, particularly Rb

Papillomaviruses do not code for their own DNA replication enzymes. Therefore, in order to replicate their double-stranded DNA genomes, they must use the machinery of the host cell. Viral DNA replication occurs in fully differentiated keratinocytes, which do not normally divide, and therefore DNA synthesis does not normally occur in these cells. It is the role of the viral E7 protein to induce cellular DNA synthesis.

The E7 proteins of all papillomaviruses are small, multifunctional proteins approximately 100 amino acids

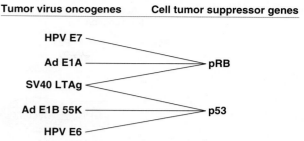

Figure 22.4 Common pathways used by viral oncogenes. Human papillomaviruses (HPV), adenoviruses (Ad), and polyomaviruses (SV40) express proteins that bind to and inhibit the activity of cellular tumor suppressor proteins involved in controlling cell-cycle progression (pRb) and apoptosis (p53).

in length. E7 proteins share some structural homology with proteins from other viruses, including the adenovirus E1A protein and polyomavirus large T antigens. E7, E1A, and large T antigens have been shown to interact with many of the same cellular targets, and these interactions result in deregulation of the normal cell cycle, resulting in the induction of cellular DNA synthesis (Figure 22.4).

The most significant cellular targets of E7 are the **retinoblastoma (Rb)** protein and its functionally related proteins, p107 and p130. Rb is considered a **tumor suppressor protein** since loss of the Rb genes is associated with the development of different forms of cancers, most notably retinoblastoma, a childhood tumor of the eye.

The Rb protein is a major regulator of the cell-cycle, and functions through its ability to bind to and repress the activity of a family of cellular transcription factors known as the E2F family (Figure 22.5). When E2F is freed from bound Rb, it can induce the expression of a number of genes whose products initiate and carry out cellular DNA synthesis. Consequently, activation of the E2F transcription factors results in progression of the cell cycle from the G1 phase to the S phase, during which DNA synthesis takes place.

It is noteworthy that the transition from G1 to S phase is a crucial point in the cell cycle; once cellular DNA synthesis begins, the cell is committed to undergo mitosis. The expression of E7 alone in resting cells can stimulate cellular DNA replication, and this activity correlates with the ability of E7 to bind to Rb, resulting in the release of E2F. E7 proteins of the highly oncogenic human papillomavirus types 16 and 18 bind more strongly to Rb than E7 proteins of non-oncogenic virus types 6 and 11. Oncogenic E7 proteins can also induce the **ubiquitin**-mediated proteolytic degradation of Rb. These activities may account in part for the differences in cell transforming capacity between oncogenic and non-ocongenic papillomaviruses.

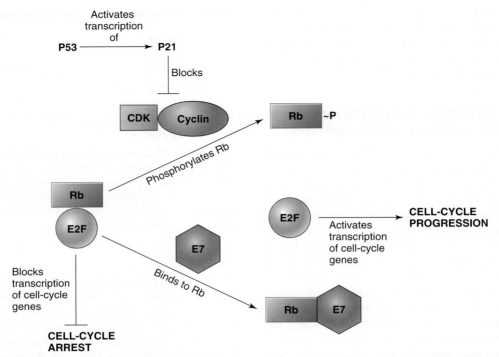

Figure 22.5 Papillomavirus E7 protein activates cell-cycle progression. Rb protein binds to E2F, blocking activation of cell-cycle genes. In normal cell cycle (top half), CDK/cyclin complex phosphorylates Rb, releasing it from E2F and leading to expression of cell-cycle genes. This can be blocked by activated p53 via p21. In HPV-infected cells (bottom half), E7 binds to Rb, activating E2F and leading to cell DNA replication and cell-cycle progression.

In summary, the expression of E7 leads to induction of unscheduled DNA synthesis in fully differentiated keratinocytes. This allows the viral genome to be replicated to a high copy number by the cellular DNA replication machinery. However, unscheduled DNA synthesis in fully differentiated cells is clearly not a normal cellular event; mechanisms have developed to eliminate such runaway cells from the body. Unscheduled DNA synthesis results in activation of **apoptosis,** leading to cell death, and this process is dependent on the activity of the cellular p53 protein. In fact, not only does E7 induce cell-cycle progression, but this also results in an increase in p53 levels. In order for the virus to keep the cell alive long enough to complete its replication, it has developed a means to counteract p53-mediated apoptosis. The virus expresses another protein, E6, which mediates the degradation of cellular p53 protein in the infected cell (Figures 22.4 and 22.6).

Viral E6 protein controls the level of cellular p53 protein

Human papillomavirus E6 proteins are approximately 150 amino acids in length. The most notable feature of E6 proteins made by oncogenic papillomaviruses such

as types 16 and 18 is their ability to mediate the destruction of p53 through the ubiquitin proteolysis pathway (Figure 22.6). E6 first associates with a cellular protein termed E6AP. The E6-E6AP complex then binds to p53, stimulating the **ubiquitination** of p53 and its subsequent degradation in **proteasomes**. In comparison, E6 proteins made by non-oncogenic human papillomaviruses such as types 6 and 11 are unable to mediate the degradation of p53.

Cellular p53, like Rb, is a tumor suppressor protein. This is best illustrated by experiments showing that p53 knockout mice, genetically engineered to contain no p53 genes, develop normally in every respect except that they succumb to a variety of tumors at a young age (a few months old).

Clinical studies have shown that mutations in the human p53 gene are present in over 50% of all human tumors. The majority of cancers that do retain wild-type p53 genes suffer mutations in other genes whose products are required to activate p53 function. Consequently, the majority of all cancers have undergone either a mutation in the p53 gene or alterations in genes encoding p53 regulatory molecules. Oncogenic papillomaviruses, simian virus 40, and adenoviruses have all evolved to express proteins that target and inhibit the activity of p53, albeit through different mechanisms (Figure 22.4).

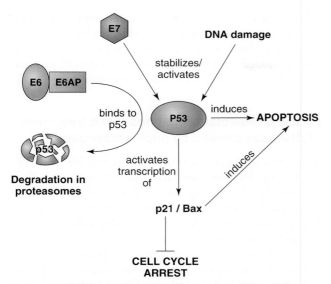

Figure 22.6 Interaction of papillomavirus E7 and E6 proteins with p53. Following DNA damage or virus infection, p53 becomes stabilized and activated, and its concentration increases. p53 activates transcription of genes involved in induction of apoptosis (bax) and arrest of cell cycle (p21). E6 reverses p53 increase by binding to E6AP and p53, leading to degradation of p53.

The p53 protein is normally present at low levels in the cell in a latent, inactive form. When the cell is exposed to stressful conditions such as virus infection or DNA damage, p53 becomes stabilized by phosphorylation, and its level increases. The accumulated p53 either induces cell-cycle arrest in G1, allowing sufficient time to repair damaged DNA, or induces apoptosis, which eliminates these abnormal cells. In a multicellular organism, the loss of abnormal cells through apoptosis can be considered as a defense mechanism against disease, including virus infection and cancer.

The biological effects of p53 are largely due to sequence-specific transcriptional activation of target genes whose products mediate cell-cycle arrest or apoptosis (Figure 22.6). Bax and p21 are two examples of such genes. However, p53-mediated apoptosis of virus-infected cells could limit the ability of viruses to complete their life cycle and thus restrict the spread of the virus. The major role of papillomavirus E6 protein is to direct the degradation of p53, thereby inhibiting apoptosis in the infected cell and allowing cell survival until sufficient new progeny virus is made.

It is noteworthy that, unlike most human cancers, cervical cancers caused by papillomaviruses do not have mutated p53 genes. This makes sense: there should be no selective pressure to mutate the p53 gene in these cancers, since the p53 protein is eliminated by the activity of the viral E6 protein.

E6 proteins from high-risk human papillomaviruses have also been shown to interact with several other cellular proteins involved in signal transduction, apoptosis, DNA replication, and calcium metabolism. The full significance of these interactions and how they affect viral replication and tumourigenesis are unknown. However, a class of intriguing cellular targets of E6 is proteins that possess PDZ domains, named after the prototype proteins of this class (PSD95, Dlg, and ZO-1). Interaction is mediated by a carboxy-terminal domain of four amino acids (X-Ser/Thr-X-Leu/Val, where X = any amino acid) on the E6 protein, and, most importantly, this is only found on those E6 proteins associated with the development of cervical cancer. This domain can therefore be considered as a candidate marker of oncogenic potential. Through this sequence E6 can target several PDZ domain-containing cellular substrates, including Dlg and Scribble. These proteins are involved in regulating cell polarity and cell proliferation control, targeting of which may be involved in the later stages of malignant progression.

Synergism between E6 and E7 and the predisposition to cancer

Although p53 and Rb are multifunctional tumor suppressor proteins, they intersect in a common regulatory pathway. Stabilization of p53 in response to DNA damage or viral oncogenes such as E7 results in transcriptional activation of the p21 gene. The resulting p21 protein is a potent inhibitor of Rb phosphorylation, and this prevents the expression of E2F-responsive genes. Therefore, p53-activated p21 expression inhibits progression of the cell cycle from G1 to S phase, inhibiting cell proliferation. When this occurs in response to the presence of damaged DNA, cells will not replicate until the damaged DNA has been repaired.

In cells infected with oncogenic papillomaviruses, this pathway is not under normal control since p53 levels are suppressed by E6, and the pRb protein is bound by E7, resulting in activation of E2F-mediated genes (Figures 22.5 and 22.6). These papillomavirus oncoproteins are therefore able to act at two distinct points in the same pathway by abrogating both p53 and Rb activity. This effectively results in the uncontrolled proliferation of the virus-infected cell, which together with the loss of the ability of p53 to induce apoptosis, makes a potent oncogenic combination. As a result of these host–virus protein interactions, cervical cancer is among the most common forms of malignancy in the world.

The ultimate "objective" of the virus is not to induce malignant transformation, but to replicate by completing a productive growth cycle. To replicate in resting cells, the virus must target the growth regulatory pathways described above. Cells transformed by papillomaviruses are in fact unable to produce new virus particles, because only a part of their genome is present in the transformed cell, and furthermore it is frequently integrated into the cellular genome and can no longer replicate independently. Cancers mediated by oncogenic papillomaviruses occur in only a small percentage of people who become infected with papillomaviruses. Nevertheless, these viruses are widespread and, in those individuals where there is long-term persistent viral infection and lack of immune clearance, oncogenic papillomaviruses are potent human carcinogens.

Cells transformed by papillomaviruses express E6 and E7 gene products from integrated viral DNA

The most important and interesting disease aspect of human papillomaviruses such as types 16 and 18 is their involvement in cervical cancer. Transformed cells derived from cervical carcinomas can be propagated *in vitro*. In addition, some cell types cultivated *in vitro* can be transformed by transfection with viral DNA. Observations made with both kinds of transformed cells have led to a detailed understanding of the molecular mechanisms by which viral gene products induce cell transformation.

Cell lines derived from cervical cancers contain copies of a part of the human papillomavirus genome integrated into the cellular genome at sites that differ in each cell line. Integration of viral DNA into the cellular genome only occurs during the development of the cancer; viral DNA in productively infected cells is present as a free plasmid. Integration in cancer cells often results in the loss of all viral genes except the E6 and E7 genes. These two viral genes are expressed in most cervical cancers. Indeed, one of the most widely used human cell lines, HeLa, was derived from a cervical cancer many years ago and has been shown to contain papillomavirus DNA and continues to express the E6 and E7 proteins. Inhibiting the expression of these proteins using siRNA technology inhibits the growth of HeLa cells and results in apoptosis, demonstrating that E6 and E7 continue to drive cell proliferation many years after the initial transforming events. This feature makes these two proteins ideal targets for potential therapeutic intervention in malignancies induced by human papillomaviruses.

No virus replication takes place in transformed cells. The process of integration of viral DNA into the host cell therefore has no obvious benefit for the virus, and can be considered as an "unplanned" byproduct of the infection process that has unfortunate consequences for the host.

A number of elegant studies using transgenic mice have been done to assess the relative contribution of E6 and E7 to cancers caused by human papillomaviruses. Expression of these viral oncoproteins in mice induces skin and cervical cancers. These studies have shown that E7 is mainly responsible for inducing cell proliferation and initiating tumor formation, while E6 has a cooperative role in increasing tumor size and inducing malignant progression. Roles for pRb, p53, and PDZ proteins have all been implicated in these processes.

Future prospects for diagnosis and treatment of diseases caused by papillomaviruses

The cervical smear test (**Pap test**, named for its inventor, George Papanicolaou) has proven to be a very effective means of reducing the incidence of death caused by cervical cancer in developed countries. This test is based on the recognition of morphologically abnormal cervical cells indicating the presence of premalignant or malignant lesions. Cervical warts and precancerous tumors are routinely removed by surgery or physically destroyed by freezing with liquid nitrogen or laser vaporization. Therefore, with appropriate screening and effective treatment options, the development of malignant cervical cancer is relatively rare in developed countries.

However, human papillomavirus-associated cervical cancer is one of the most common forms of cancer in developing countries (several hundred thousand new cases per year). The incidence of this cancer could be dramatically reduced with appropriate healthcare intervention such as a routine Pap smear and follow-up treatment, although this is rendered difficult in many countries because of the lack of appropriately trained pathologists and because of cultural inhibitions. Development of self-sampling tests based on DNA detection (see below) may offer some solutions to these problems. Without doubt such universal screening has the potential for decreasing the worldwide incidence of cervical cancer.

Treatment of papillomavirus-induced precancerous lesions involves the removal of the infected lesions. To a large extent this is made possible because these are localized surface lesions, and papillomaviruses do not give rise to systemic infections. Several topical creams are also available; these have poorly understood antiviral activity or work by stimulating a local cellular immune response against the virus.

The Pap smear is still the most widely used and effective test to detect cellular changes indicative of disease such as cervical neoplasia. However, it is a screening method, not a viral diagnostic test. Diagnostic tests are now available based on the detection of viral DNA and are often used in conjunction with the Pap test. Detection of an oncogenic papillomavirus type in patients with an abnormal Pap smear is valuable in patient management, especially when an infection appears to be persistent. This helps with decisions on directing the patient toward further tests and treatments, and more stringent subsequent monitoring.

Human papillomavirus vaccines

Two different human papillomavirus vaccines have been produced by using viral capsid protein L1 assembled into virus-like particles that do not contain the viral genome. These vaccines carry no risk of infection but can induce antibody responses to the virus. Both of these vaccines contain L1 of papillomavirus types 16 and 18, responsible for most cervical carcinomas; one of the vaccines also contains L1 of types 6 and 11, responsible for most genital warts or condylomas.

Large-scale trials have shown strong antibody production in vaccinated individuals and nearly 100% protection against infection by targeted viruses. Furthermore, vaccinated individuals were protected against development of pre-malignant cervical lesions. Based on these results, both vaccines have been licensed for use in the general population in many different countries. The consensus is that girls 11 to 12 years old should

be the principal target for vaccination since it is important to vaccinate people before most sexual activity begins and before the infection occurs.

Another approach involves the use of the human papillomavirus L2 protein, which may provide protection against a wider range of virus types. A vaccine based on the L2 protein will soon be tested. However, vaccines based on virus-like particles are expensive to manufacture and require injection of several doses, and poorer countries may have difficulties in adopting their widespread use. Therefore one group is engineering harmless bacteria to express papillomavirus capsid proteins and generate virus-like particles. A vaccine based on live bacteria would be inexpensive to produce and could be administered orally. If effective, such a vaccine could eventually replace the more expensive injected vaccines and contribute to the prevention of a major cancer worldwide.

KEY TERMS

Apoptosis

Basal cells

Condyloma

Heparan sulfate

Integrin

Keratinocytes

Mucosal epithelium

Pap test

Proteasome

Proteoglycan

Raft culture

Retinoblastoma (Rb)

Squamous cell carcinoma

Tumor suppressor protein

Ubiquitin

Ubiquitination

FUNDAMENTAL CONCEPTS

• Papillomaviruses cause warts on skin and in oral or genital mucosa.

• Low risk human papillomaviruses such as types 1, 2, 6, and 11 cause benign warts

• High risk human papillomaviruses such as types 16, 18, and 31 can cause cervical cancer and other malignancies.

• Productive papillomavirus infections require differentiation of infected keratinocytes.

• Oncogenic papillomaviruses cause cancer through the combined action of two early viral proteins: E6 and E7.

• These viral oncoproteins inactivate critical cellular tumor suppressor pathways mediated by p53 and Rb proteins.

REVIEW QUESTIONS

1. What are the two major receptors utilized during infection of human mucosal epithelium by papillomaviruses?

2. Which papillomavirus protein has DNA helicase activity?

3. Human cervical cancers caused by papillomaviruses do not have mutated p53 like most human cancers; how do papillomaviruses target p53?

4. Which two papillomavirus proteins help to immortalize HeLa cells?

5. Explain the concept of "plasmid replication".

6. What role does the viral E7 protein play in replication?

LARGER DNA VIRUSES OF EUKARYOTES

This section covers DNA viruses with genomes ranging from 25 kb to over 1000 kb, larger than the genomes of some cellular organisms! This is a very diverse group of viruses, and our understanding of the largest DNA viruses is still at a very primitive level.

The smallest members of this group are the adenoviruses, whose linear DNA genomes code for 30 to 50 proteins. Adenoviruses incorporate up to ten different proteins in their icosahedral capsids. Their "early" genes, transcribed immediately upon entry of the genome into the nucleus, carry out many of the same functions as the early genes of the small DNA viruses, with the same result: induction of the cellular machinery for DNA replication. Several viral gene products interfere with cellular transcription or translation and inactivate cellular defenses, therefore increasing the chance that progeny viruses are made in large numbers. Study of the expression of adenovirus "late" genes led to the discovery of RNA splicing.

Herpesviruses, poxviruses, and baculoviruses have genomes of intermediate sizes up to several hundred kb, and each can code for 100 or more proteins. Herpesviruses are known for their ability to remain latent in the host organism until an external trigger induces lytic growth, and the mechanisms of latency are being intensively studied. Vaccinia virus, a poxvirus, was the world's first known antiviral vaccine, used against the deadly smallpox virus. Unlike most DNA viruses of vertebrates, poxviruses replicate exclusively in the cytoplasm, importing all the transcription machinery they need within the virion core. Baculoviruses are insect viruses that are used for insect control, expression of useful eukaryotic proteins, and production of vaccines.

Viruses of algae are among the largest viruses known, with genomes ranging up to 500 kb. Most have very big icosahedral capsids, with internal lipid envelopes and a core containing the genome. Many have only been recently discovered, and their importance in controlling algal blooms and in influencing global levels of oxygen and carbon dioxide has led to a lot of recent research on these viruses.

Finally, mimivirus is a real giant among viruses, with a genome of 1.2 Mb. It replicates in an amoeba, and its huge size originally fooled scientists into believing it was a small bacterium. Its mysteries are only now being unraveled.

Adenoviruses

Philip Branton
Richard C. Marcellus

Adenoviridae

From Greek *adenos* (gland), for site of first isolation, adenoids

VIRION

Naked icosahedral capsid (T = 25) with 20 triangular faces.
Diameter 70–90 nm.
Knobbed fibers protruding from each of 12 vertices.
Eleven proteins in virion.

GENOME

Linear, double-stranded DNA, 30–36 kb.
Short, inverted terminal repeat sequence.
Both 5' ends covalently bound to a virus-coded terminal protein.

GENES AND PROTEINS

Codes for up to 50 proteins.
Most regions of both DNA strands are transcribed by cellular RNA polymerase II.
Six early and two delayed early transcription units: E1A, E1B, E2A, E2B, E3, E4; IVa2, IX.
One major late transcription unit: five subclasses of late mRNAs, L1 through L5.
Each transcription unit gives rise to multiple mRNAs by RNA splicing.
Two small VA RNAs transcribed by RNA polymerase III.

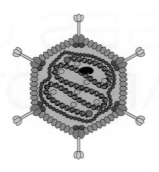
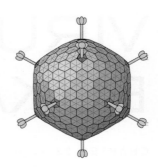

VIRUSES AND HOSTS

Fifty human serotypes in six subgroups (A through F).
Other adenoviruses in cattle, mice, birds, etc.

DISEASES

Respiratory syndromes including pneumonia, but not common colds.
Eye and gastrointestinal infections.

DISTINCTIVE CHARACTERISTICS

Multiple mRNAs arise by extensive alternative splicing.
Late transcripts are made from a single strong promoter and five alternative polyadenylation signals.
DNA synthesis is primed by terminal protein, uses viral DNA polymerase.
Can cause tumors in experimental animals but are not known to cause tumors in humans.
Adenoviruses are widely used as gene therapy and anticancer virus vectors.

Adenoviruses cause respiratory and enteric infections in humans

Adenoviruses were first isolated by Wallace Rowe in 1953 from tissue culture cell lines derived from tonsils and **adenoid** tissues of sick children. Shortly afterward, Maurice Hilleman and colleagues identified the same agent as the cause of a respiratory epidemic in army recruits in the United States. Originally (and incorrectly) postulated to be a cause of the common cold, adenoviruses were subsequently shown to be responsible for a number of other respiratory syndromes including pneumonia; they can also cause eye and gastrointestinal infections. The current adenovirus name and nomenclature were adopted in 1956, acknowledging the original source of the virus from adenoid tissue.

Adenoviruses can be oncogenic, but do not cause cancer in humans

Human adenovirus type 12 can induce malignant tumors in hamsters. This was the first human virus shown to be oncogenic. Although no evidence has linked adenoviruses to human cancer, these viruses have been used as models for understanding the molecular biology of cancer. Furthermore, their study has helped uncover basic mechanisms of gene expression in mammalian cells. In 1977, the laboratories of Philip Sharp and Richard

Roberts discovered RNA splicing by studying adenovirus late mRNAs, and were jointly awarded the Nobel prize for this discovery. In 1979, Robert Roeder used the adenovirus major late promoter to demonstrate specific initiation of transcription by RNA polymerase II *in vitro*, and he and others subsequently discovered many of the protein factors involved in transcription in mammalian cells by using this model system. Human adenoviruses have recently become of great importance as gene therapy vectors and are one of several promising new oncolytic virus anticancer agents.

The 100 known adenoviruses include 50 human virus serotypes, distinguished by their sensitivity to neutralizing antibodies. The human viruses are further classified into six subgroups (A to F) based on their ability to agglutinate red blood cells. This chapter will deal with the best-studied human adenoviruses: types 2 and 5, non-oncogenic members of group C, and type 12, a highly oncogenic member of group A.

Virions have icosahedral symmetry and are studded with knobbed fibers

Adenovirus virions have a remarkable resemblance to a regular icosahedron, with 20 equilateral triangular faces and 12 vertices (Figures 23.1 and 23.2). This handsome capsid contains a total of 11 viral proteins, but is constructed from two basic building blocks, the **hexon** and the **penton base** (Figure 23.2).

Hexons consist of trimers of the hexon protein (polypeptide II); each hexon has six nearest neighbors, hence the name. A total of 240 hexons cover the 20 faces

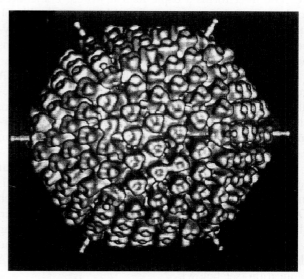

Figure 23.1 Adenovirus virion. A computer-generated image of adenovirus type 2 virion, viewed along a threefold symmetry axis. Reconstructed from cryoelectron microscopy images.

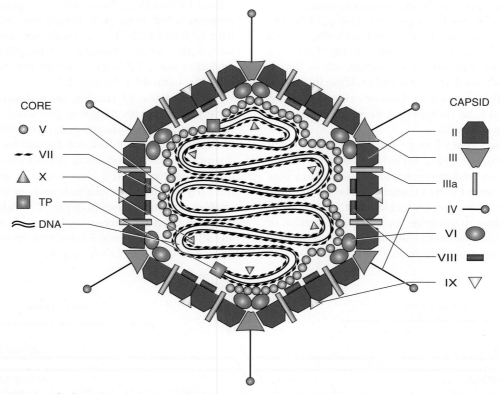

Figure 23.2 Structure of adenovirus virion. A schematic cross-sectional view of a human adenovirus type 5 virion, with components shown as indicated.

of the T = 25 icosahedral capsid. Penton base subunits, each made of five molecules of the penton protein (polypeptide III), lie at the 12 vertices of the icoashedron. Projecting from the penton bases like antennae on a space satellite are fibers made of trimers of polypeptide IV. The fiber forms a triple helical structure anchored to the penton base at one end and has a terminal knob at the other end. A number of "glue" proteins (polypeptides IIIa, VI, VIII, and IX) mediate interactions between hexons and pentons, and with the viral DNA.

The 30- to 36-kb linear, double-stranded viral DNA genome is organized into coiled domains by the histone-like polypeptide VII. Basic polypeptides V and X are also bound to the viral DNA. A 55-kDa virus-coded **terminal protein** is covalently linked via a serine residue to each of the 5' ends of the viral DNA (Figure 23.3).

The viral DNA contains **inverted terminal repeat** sequences of approximately 100 bp (Figure 23.3). This means that the first 100 nt at the 5' and 3' ends of each DNA strand are complementary and can base-pair, allowing single DNA strands to circularize, forming a short panhandle end. Both the terminal protein and the inverted terminal repeats are involved in viral DNA replication (see below).

Fibers make contact with cellular receptor proteins to initiate infection

Adenoviruses attach to specific cellular receptors via the knob of the fiber protein. Most adenovirus serotypes, as well as Coxsackievirus B3 (a picornavirus; see Chapter 11) utilize the same receptor on human cells, termed *Coxsackievirus and adenovirus receptor (CAR)*. CAR is a cell surface protein of unknown function, whose amino acid sequence shows that it is a member of the **immunoglobulin family** (Figure 4.2). Rodent cells also possess a closely related CAR protein that allows their infection by human adenoviruses.

Adenoviruses types 2 and 5 also utilize certain cell surface **integrin** proteins, including integrins αVβ3 and αVβ5, as **secondary receptors**. The penton base binds to these integrins via a domain similar to that used by extracellular adhesion molecules that interact with integrins. Binding to integrins leads to the migration of virus-receptor complexes to clathrin-coated pits, where they are taken up into endosomes (see Chapter 4).

Internalization within an endosomic vesicle occurs quickly (within 10 minutes), after which the virion escapes to the cytosol in a process dependent on the acidic pH of the endosome. Virus particles are transported toward the nucleus using interactions between the hexon protein and microtubules. They localize at nuclear pores where they are disassembled in a sequential manner, and the DNA is released into the nucleus. There the viral DNA associates with the nuclear matrix via the terminal protein, and transcription of viral genes commences.

Expression of adenovirus genes is controlled at the level of transcription

Although the adenovirus gene map (Figure 23.4) seems complicated at first, when broken down into its component parts it is easier to understand. Viral genes are arranged in a well-conserved set of transcription units. Eight transcriptional promoters are used by cellular RNA polymerase II to initiate transcription, and twelve polyadenylation signals define the 3' ends of mRNAs.

Transcription units can be divided into three classes, depending on the timing of their use during the infectious cycle. The six *early* transcription units are named E1A, E1B, E2A, E2B, E3, and E4. Note that E2A and E2B use a common promoter but different polyadenylation sites. The two *intermediate* transcription units are named IX and IVa2. Note that these genes use distinct promoters but share polyadenylation sites in common with E1B and E2B. There is only one *late* transcription unit, with a single promoter but five different polyadenylation sites. This gives rise to five families of late mRNAs, named L1 through L5.

Each of the 13 families of viral genes contains a variety of often overlapping protein-coding regions. Alternative RNA splicing generates different mRNA species that together code for a grand total of about 50 distinct adenovirus proteins. Adenovirus genomes can be seen as constructed from modular units similar to the early and late transcription units of the smaller and simpler polyomaviruses (Chapter 21). Adenovirus transcription units are interspersed in such a way that virtually all of the viral genome is utilized for coding proteins on one or the other DNA strand.

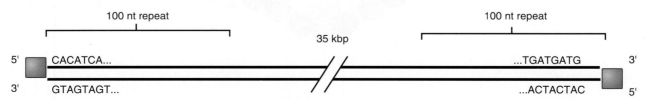

Figure 23.3 Terminal inverted repeats on adenovirus DNA. Linear, double-stranded adenovirus type 5 DNA showing the first 8 nt of the inverted terminal repeats. Terminal protein shown as an orange square.

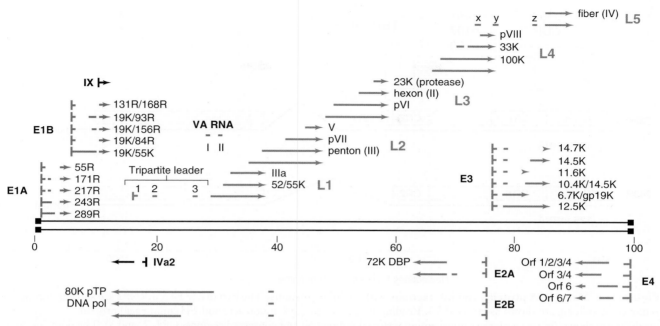

Figure 23.4 Transcriptional map and gene products of adenovirus type 5. Families of early (E, in blue), intermediate (in black), and late (L, in orange) mRNAs are shown in relation to the viral genome, divided into 100 map units (~350 nt per map unit). Rightward reading transcripts shown at top, leftward reading transcripts shown at bottom. Vertical bars show 5' ends of RNAs, corresponding to transcriptional promoters; arrowheads represent 3' ends of mRNAs, corresponding to polyadenylation sites. Introns are shown as interruptions in mRNAs. Names of proteins coded by mRNAs are shown adjacent to the corresponding mRNAs.

Adenovirus genomes also contain one or two copies of VA genes, transcribed at high levels by cellular RNA polymerase III, the enzyme used by the cell to transcribe tRNA and other small RNA genes. The small, highly base-paired VA (for "virus-associated") RNAs are involved in translational control by inhibiting activity of the cellular, double-stranded, RNA-dependent protein kinase **(PKR)**, one of the pathways by which **interferon** acts (see Chapter 33).

E1A proteins are the kingpins of the adenovirus growth cycle

E1A is the first viral gene to be transcribed. The E1A promoter, located near the left end of the genome (Figure 23.4), is driven by a transcriptional enhancer that requires only cellular transcription factors. Viruses that contain mutations in E1A genes make strongly reduced amounts of other early gene products, and therefore little DNA or progeny virions.

Two major E1A RNAs, denoted 13S and 12S (for their slightly different sedimentation rates in sucrose gradients), encode proteins of 289 and 243 amino acid residues, respectively (289R and 243R; Figure 23.5). These two proteins are identical except for a 46-amino acid region in the 289R protein that is missing in the 243R protein because of removal of a longer intron by

alternative splicing. Alignment of E1A proteins from different adenovirus serotypes reveals three regions with highly conserved amino acid sequences, termed CR1 (for "conserved region 1"), CR2, and CR3 (Figure 23.5); the 243R protein is missing CR3.

The two major E1A proteins are powerful transcriptional activators of a number of viral and cellular genes. The natural target cells of adenoviruses are terminally differentiated, non-dividing epithelial cells. Adenoviruses rely on a variety of host proteins for optimal DNA replication and production of adequate levels of precursor deoxynucleotides. The E1A proteins help provide these required cellular proteins by inducing non-dividing cells to enter the S, or DNA synthesis, phase of the cell cycle. An unintended result is the ability of adenoviruses to transform non-permissive cells and to induce tumor formation in experimental animals, thus making E1A an **oncogene**.

E1A proteins bind to the retinoblastoma protein and activate E2F, a cellular transcription factor

How do E1A proteins induce cells to enter the S phase of the cell cycle? The best-understood mechanism involves an interaction between E1A proteins and the **retinoblastoma (Rb)** family of cellular tumor

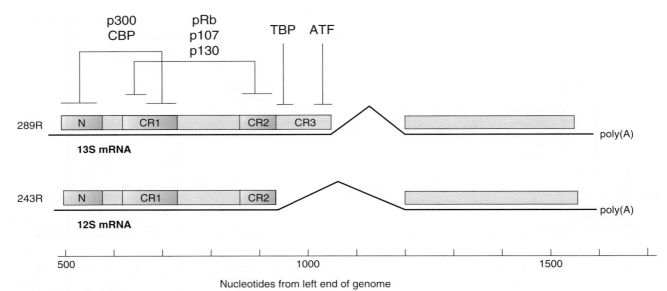

Figure 23.5 **Major E1A proteins and interactions with cellular proteins.** The two major E1A mRNAs are shown; introns removed by splicing are shown as inverted V's. Reading frames for the 289-amino acid and 243-amino acid protein products are shown (boxes) with the critical amino terminal region (N) and conserved regions (CR1, 2, and 3). The domains involved in binding by histone acetyl transferases (p300/CBP), Rb family members, and cellular transcription factors TATA-binding protein (TBP) and ATF are indicated.

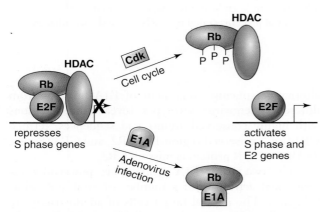

Figure 23.6 **S-phase gene activation by E1A proteins.** A complex between Rb and E2F binds to cellular and viral genes, and bound histone deacetylase (HDAC) represses transcription by altering chromatin structure. This complex is disrupted either during the cell cycle, by phosphorylation of Rb by cyclin-dependent kinase (Cdk; above), or during infection, by binding of adenovirus E1A proteins (below), allowing E2F to transcriptionally activate cellular and viral genes.

suppressor proteins (Figures 23.5 and 23.6). Rb was first identified as a gene that is deleted or mutated in human retinoblastomas and other cancers. Proteins of the Rb family bind to and inactivate a family of cellular transcription factors, termed E2F. E2F was first described as a transcription factor regulating the adenovirus E2 promoter (hence its name), but it also regulates promoters of many genes coding for cellular proteins involved in DNA synthesis and cell-cycle progression factors.

Rb and its family members p107 and p130 contain a *pocket domain* that binds to both the E2F and E1A proteins (Figure 23.6). **Histone deacetylases** (HDACs) bound to Rb are targeted to E2F-regulated genes and induce transcriptional repression by compressing chromatin structure, reducing access of proteins that activate transcription. In the G1 phase of normally cycling cells, Rb becomes phosphorylated by **cyclin-dependent protein kinases** (Cdks), and the phosphorylated Rb no longer binds to E2F. Thus free of repression, E2F can activate expression of S-phase genes, allowing progression into S phase.

The E1A proteins activate E2F by a similar mechanism: they bind to the Rb pocket, releasing E2F from the embrace of Rb. E2F is therefore free to activate transcription of its target genes (Figure 23.6). At the same time, E2F activates both viral genes E2A and E2B by binding to a site near their common promoter. This leads to production of three early viral proteins that direct adenovirus DNA replication.

E1A proteins also activate other cellular transcription factors

The E1A 289R protein independently activates transcription of adenovirus E3 and E4 genes by interactions between its unique CR3 domain and cellular transcription factors. CR3 contains two regions involved in transactivation: a zinc finger that interacts with the TATA-binding protein (TBP) of the basal transcription machinery, and a promoter-targeting region that activates cellular transcription factors such as ATF (Figure 23.5).

These interactions stabilize the transcription complex and stimulate viral gene expression.

Finally, the E1A amino terminus and part of CR1 engage in a complex with a distinct set of cellular proteins, the **histone acetyl transferases**. These enzymes, including p300 and the cyclic AMP-response element-binding protein, CBP (Figure 23.5), play an important role in the transcriptional activation of large blocks of cellular genes. This is carried out via the acetylation of nucleosomal histones and other proteins, resulting in a more relaxed chromatin structure ("**chromatin remodeling**") that is more accessible to specific transcription factors and is more easily transcribed. Thus these proteins reverse the transcriptional repression effects of the histone deacetylases referred to above. Binding of E1A proteins to histone acetyl transferases increases histone acetylation and thus promotes entry into the cell cycle by enhancing expression of cell-cycle genes.

E1A proteins indirectly induce apoptosis by activation of cellular p53 protein

Expression of E1A proteins leads indirectly to stabilization and activation of the cellular **tumor suppressor protein**, p53 (Figure 23.7; see also Figure 22.6). p53 is a transcription factor that both activates and represses transcription, and functions to police the natural progression of the cell cycle. Activation of p53 serves as a warning signal to the cell that something is wrong, and activated p53 induces growth arrest and cell death by **apoptosis**.

Apoptosis is a regulated form of cell suicide invoked when cells undergo extensive DNA damage, virus infection, or other insults. p53 activity is regulated in part by at least two cellular proteins that control its stability in opposite ways. Mdm-2 binds to p53 and targets it for **ubiquitin**-mediated degradation, whereas p14^{Arf} binds to Mdm-2 and blocks its binding to p53, thus stabilizing p53.

E1A proteins stabilize and activate p53 by two mechanisms (Figure 23.7). First, by binding to Rb (Figure 23.6) they activate expression of p14^{Arf}, which is encoded by an E2F-inducible gene. As a result, increased levels of p14^{Arf} bind to Mdm-2 and reduce its level in the cell. Second, the association of E1A with p300/CBP prevents p53 from activating transcription of the Mdm-2 gene, further reducing the level of Mdm-2. The resulting accumulation of p53 could induce the early death of infected cells by apoptosis and severely limit virus production were it not reversed by other adenovirus gene products, particularly two E1B proteins.

E1B proteins suppress E1A-induced apoptosis and target key proteins for degradation, allowing virus replication to proceed

The E1B region also produces multiple mRNAs by alternative splicing. The two best characterized proteins, denoted 19K and 55K (for their molecular weights), are produced from a single 2.2-kb mRNA by initiation of translation in two different reading frames.

The 19K protein is a functional homologue of the cellular protein Bcl-2, which is a suppressor of apoptosis. Oligomers of another Bcl-2 family member, called Bax, stimulate the **caspase** cascade and lead to apoptotic cell death (see Figure 33.4). Bcl-2 reverses this effect by binding to Bax, preventing oligomer formation. Both Bcl-2 and Bax are associated with mitochondrial membranes in a complex that regulates apoptosis. The adenovirus 19K protein mimics Bcl-2 by binding to Bax and preventing oligomer formation, thereby blocking apoptosis (Table 23.1).

The 55K protein binds directly to p53 and blocks transcription from p53-dependent promoters, therefore inhibiting p53-dependent apoptosis. More importantly, 55K and a protein coded by the adenovirus E4 transcription unit, called E4orf6 (for "open reading frame 6"), form an E3 ubiquitin ligase complex, involving members of the cellular Cullin protein family and Elongins B and C, which targets bound proteins for degradation in the proteasome. p53 bound to this complex is ubiquinated, transferred to the proteasome, and degraded (Table 23.1). A growing list of other key cellular proteins are also targeted for degradation by this complex.

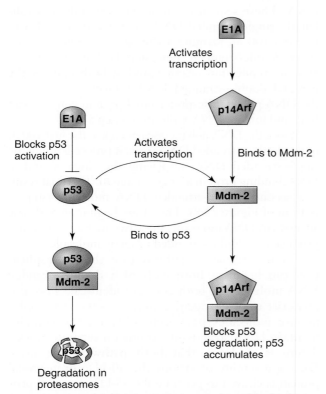

Figure 23.7 Degradation pathway of p53 is blocked by E1A proteins. See text for details.

Table 23.1 Functions of adenovirus E1B proteins

Protein	Function	Cellular proteins involved
19K	• Suppresses apoptosis	• Binds to Bax (homologue of Bcl-2)
55K	• Suppresses apoptosis	• Binds to p53 (transcriptional repression)
	• Induces p53 turnover (with E4orf6)	• Binds to p53 and E4orf6 in complex with cellular Cullin-Elongin B/C E3 ubiquitin ligase complex
	• Promotes transport, stabilization and translation of viral mRNAs (with E4orf6)	• Requires E4orf6 and E3 ubiquitin ligase complex
	• Host cell shut-off (with E4orf6 and L4-100K)	• eIF4E, eIF2α, eIF4G

Thus the two major E1B proteins protect infected cells from both p53-dependent and p53-independent apoptosis, reversing activities stimulated by E1A proteins. This allows the cell to survive long enough to produce high levels of progeny adenovirus virions before eventually succumbing to the effects of virus infection.

The preterminal protein primes DNA synthesis carried out by viral DNA polymerase

Once E1A has induced the cell to enter S phase and the early viral gene products have accumulated, viral DNA replication begins. Three essential proteins made by E2 genes carry out viral DNA replication. A single-stranded DNA binding protein (ssDBP) is produced from the E2A region, while both a viral DNA polymerase and an 80-kDa preterminal protein (pTP) are encoded by E2B (Figure 23.4).

DNA replication can begin at either end of the linear DNA molecule (Figure 23.8). The replication origins, including the first 50 nt of each inverted terminal repeat, contain binding sites for a complex between the adenovirus DNA polymerase and preterminal protein. Binding by cellular transcription factors called nuclear factor I and nuclear factor III (also known as Oct I) enhances the efficiency of initiation of DNA replication, perhaps by increasing the affinity of the preterminal protein/DNA polymerase complex for the replication origin.

The first step in DNA replication is unusual. Instead of an RNA primer (used by cells and by most DNA viruses) or a DNA primer (used by parvoviruses), adenoviruses use the preterminal protein as a primer for replication. This protein contains a serine residue whose hydroxyl group forms a phosphodiester bond with deoxycytidine, in a reaction that cleaves dCTP to dCMP. This covalently bound dC residue is then used as a primer by the viral DNA polymerase for the further addition of nucleotides complementary to the template DNA strand (step 2, Figure 23.8).

As the DNA polymerase extends the growing DNA chain, it displaces the other strand of parental DNA, which is then coated with the single-stranded DNA binding protein. Synthesis proceeds to the end of the linear DNA, producing a double-stranded progeny DNA containing a parental and a daughter strand, and a displaced parental single-stranded DNA molecule (step 3).

Single-stranded DNA is circularized via the inverted terminal repeat

At this point the double-stranded progeny DNA can serve as a template for further replication, but in each round only one double-stranded DNA molecule is formed, and there would therefore be no net accumulation of viral DNA. To solve this problem, we come back to the inverted terminal repeat sequences on adenovirus DNA. These sequences allow the two ends of a full-length, single-stranded DNA molecule to anneal over the first 100 nt, forming a panhandle structure with a short, double-stranded stem (step 4). This stem recreates the replication origin found at both ends of the parental double-stranded DNA molecule. Therefore, the DNA polymerase-preterminal protein complex can bind and initiate DNA replication (step 5).

Once the DNA polymerase has copied the first 100 nt of the double-stranded DNA stem, it proceeds along the single-stranded DNA, displacing the single-stranded, DNA-binding protein as it goes (step 6). The final result is a second double-stranded DNA molecule (step 7, bottom of Figure 23.8). Therefore, beginning with one adenovirus DNA molecule, these two pathways together produce two double-stranded progeny molecules.

In an alternative pathway (not shown), replication can begin at both ends of a double-stranded DNA molecule, as both ends are identical. This also gives rise to two progeny genomes, as the DNA polymerase molecules pass each other on the complementary DNA strands like trains on separate tracks. There is evidence that both pathways are used. During assembly of virions the 80-kDa preterminal protein is cleaved to generate the 55-kDa terminal protein, which remains covalently attached to the genome after packaging.

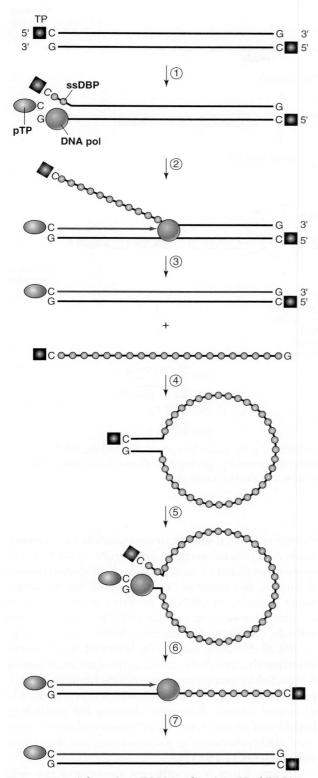

Figure 23.8 Adenovirus DNA replication. Viral DNA is replicated by three viral proteins: preterminal protein (pTP, blue oval), DNA polymerase (DNA pol, orange circle), and single-stranded DNA binding protein (ssDBP, small green circles). The cleaved, 55 kDa terminal protein (TP) is represented by a black square. Newly synthesized DNA is shown in blue. See text for details.

The major late promoter is activated after DNA replication begins

More than half of the adenovirus genome is dedicated to coding for at least 18 different late mRNAs. These mRNAs encode structural proteins, scaffolding proteins, a protease, and other products required for packaging of progeny virus. All late mRNAs are produced from a single primary transcript that extends 29 kb in the rightward direction from the highly active "major late promoter", at 16.5 map units (Figures 23.4 and 23.9). During the early phase this promoter has a low activity, and the rare transcripts terminate prematurely, so that only low levels of L1 mRNAs are made. Transcription of late genes is activated after the onset of viral DNA replication, and late transcripts become the major RNAs made in the infected cell.

At least two factors control the increase in late transcription. First, newly replicated viral DNA interacts with a different set of DNA binding proteins, thus altering DNA structure to make the major late promoter more accessible to cellular transcription factors. Second, polypeptide IVa2, which is encoded by an intermediate gene, binds to viral DNA at a site shortly downstream of the position where late transcription begins. IVa2 cooperates with factors bound to an upstream enhancer to activate transcription. Together, these steps activate the major late promoter by a factor of 100 or more.

Five different poly(A) sites and alternative splicing generate multiple late mRNAs

There are five families of late mRNAs, denoted L1 to L5, each defined by a unique polyadenylation site (Figures 23.4 and 23.9). Each primary transcript from the major late promoter is cleaved at only one of these sites, and the resulting polyadenylated RNA is spliced in a complex fashion to generate one of the 18 possible late mRNAs. Three short exons in the first 4 kb of these transcripts are spliced together to form the noncoding 200-nt "tripartite leader". This leader is then spliced to an mRNA "body"; multiple alternative 3' splice sites give rise to several different mRNAs for each of the five late gene families. As a result, all late mRNAs have the same tripartite leader at their 5' ends. Each of these mRNAs is translated into a unique protein by use of reading frames beginning at different AUG initiation codons within the mRNA body.

The tripartite leader ensures efficient transport of late mRNAs to the cytoplasm

What are the advantages to adenovirus in expressing its late mRNAs in such a fashion? First, all late mRNAs are made by turning on a single, highly active promoter, allowing coordinate expression of late proteins when

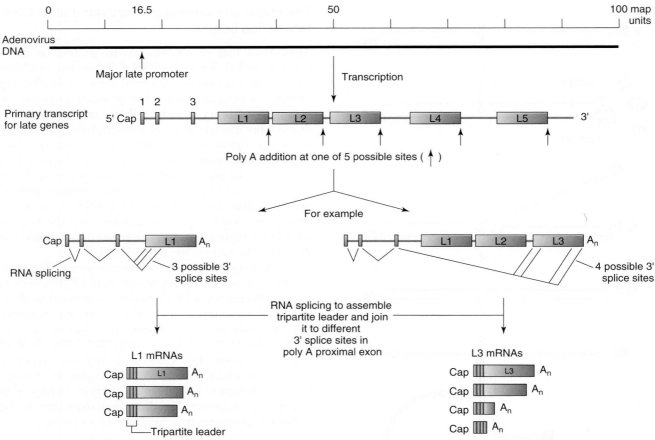

Figure 23.9 Production of late adenovirus mRNAs. Transcription initiates at the major late promoter, and the RNA is cleaved and polyadenylated at one of 5 possible sites. The tripartite leader is formed by splicing 3 small exons; it is spliced to the mRNA body at one of several alternative 3' splice sites. Exons are shown as shaded rectangles.

they are needed, after DNA synthesis begins. Second, all late mRNAs contain the same leader sequence at their 5' ends. This allows the mRNA transport and translation machinery to concentrate on these RNAs.

The Cullin-based E3 ubiquitin ligase complex discussed above, containing the E1B 55K protein and the E4orf6 protein, is responsible for selective transport of late mRNAs to the cytoplasm (Table 23.1). E4orf6 targets the 55K protein to the nucleus, but both 55K and E4orf6 also contain nuclear export signals that are involved in shuttling these complexes out of the nucleus. It is still uncertain whether direct shuttling, targeted protein degradation, or both are responsible for the selective transport and stabilization of viral mRNAs. Nevertheless, the effect is that late adenovirus mRNAs predominate in the cytoplasm during the late phase of infection.

The tripartite leader directs efficient translation of late adenovirus proteins

The cellular translation initiation factor eIF4F normally binds to the 5' cap structure on mRNAs and enables the 40S ribosomal subunit to scan mRNAs for an appropriate AUG initiation codon (see Figure 11.5a). EIF4F is dephosphorylated by an adenovirus host shutoff protein during the late phase of infection, and it can no longer direct scanning of mRNAs, resulting in the inhibition of translation of most cellular mRNAs. However, viral mRNAs containing the tripartite leader continue to be translated efficiently, because the leader allows **ribosome shunting** that avoids the scanning mechanism. Shunting is directed by complementary regions between the tripartite leader and the 18S RNA, which is part of the 40S ribosomal subunit. Ribosome shunting has parallels to initiation of protein synthesis at "internal ribosome entry sites" (IRES elements) of picornaviruses and flaviviruses (see Chapters 11 and 12).

Adenoviruses make different amounts of the various late mRNAs (and therefore of their protein products) by controlling which polyadenylation sites and 3' splice sites are recognized most efficiently. This system, using a single transcription unit and producing multiple mRNAs by different RNA processing reactions, may be a simpler and more economical means than having distinct transcriptional enhancers and promoters for each of the many late mRNAs needed.

Adenovirus-induced cell killing

Two viral proteins have been implicated in cell killing and virus release. One is the E3 11.6-kDa "death protein". Although it is nominally an early protein, the 11.6-kDa protein is expressed at high levels during the late phase when its mRNA is made using the major late promoter and an E3 polyadenylation site. The death protein is a membrane glycoprotein found in the Golgi membranes, endoplasmic reticulum, and most predominantly in the nuclear membrane; how it kills cells is not known.

The E4orf4 protein also can induce cell death, via a p53-independent mechanism. When expressed alone, E4orf4 kills a wide range of cancer cells, but it has little effect on normal human cells. This specificity of killing suggests that E4orf4 could be of use therapeutically in the treatment of human malignancies (see below). Although it has not been proven to be involved in the death of adenovirus-infected cells, E4orf4 could play an important role.

Cell transformation and oncogenesis by human adenoviruses

Adenoviruses (and other DNA tumor viruses) can oncogenically transform cells as a result of their ability to induce active cycling in cells that are normally in a resting state. Transformed cells are not subjected to normal control of the cell cycle, and therefore grow and divide in an unregulated fashion.

Adenovirus infection of human cells usually results in cell death, and therefore no transformed cells can survive. The human cell line 293 was generated *in vitro* by transfection of fetal embryonic kidney cells with a fragment of adenovirus DNA containing only the E1A and E1B genes, allowing survival and growth of transformed cells. Although rodent cells can be readily infected by human adenoviruses, they are non-permissive for viral replication and therefore survive infection. Virtually all human adenoviruses are capable of transforming rodent cells in culture, and such cells usually form tumors when injected into newborn or immunologically compromised rodents. However, only group A adenoviruses yield tumors by direct injection of virus into animals.

Therapeutic applications of human adenoviruses

In the near future we will almost certainly see adenoviruses used in a variety of clinical applications. Their ability to infect a wide range of cells and to express genes at high levels make them ideal tools to fight disease in humans and other mammals. Adenoviruses can be readily manipulated to allow expression of almost any gene by insertion of the gene under control of a suitable promoter into the viral genome.

Vaccines and gene therapy. As discussed in Chapter 37, human adenoviruses are already being used experimentally in gene therapy. In this application they function primarily as vectors to deliver genes that correct genetic deficiencies associated with human diseases. Most of the commonly used adenovirus vectors are missing E1A and E1B genes (and often many more) and therefore cannot replicate in normal cells. Replication-competent adenovirus vectors could also be used as vaccines to enhance the immune response against any desired target. Numerous genes have been inserted into adenovirus vectors for experimental vaccination, including genes encoding hepatitis B surface and core antigens, herpes simplex glycoprotein, rabies virus glycoprotein, respiratory syncytial virus F protein, and several HIV-1 proteins.

New cancer therapies. In a novel approach currently being tested in clinical trials, a replication-competent adenovirus mutant is injected into human tumors. The idea is that the virus will replicate in and spread through the tumor tissue, killing the cancerous cells. The virus strain, ONYX-015, is a mutant that does not express the E1B 55K protein, which plays a vital role in inactivating and degrading p53.

It was hypothesized that in normal cells infected with this virus p53 will remain active, leading to poor replication of the virus, little spread, and only limited pathological consequences. However, more than 50% of human tumors lack p53; in such cells, inactivation of p53 by the 55K protein is not required for virus replication, and therefore the ONYX-015 virus replicates actively in tumor cells and kills them selectively. While ONYX-015 does seem to kill cancer cells selectively, the basis for this effect now seems more complex than simply the presence or absence of p53. This virus is still being tested in clinical trials.

Clinical use of other adenovirus functions. Viral proteins have evolved to modulate many aspects of cell growth, apoptosis, and the immune system. It should be possible to take advantage of these viral proteins to develop new therapies. Such proteins could be expressed in target cells using viral vectors. In addition, understanding their mechanism of action could identify new targets for the development of drugs that mimic the action of the viral protein. Both E1B 19K and 55K proteins are potent inhibitors of apoptosis, and could be of value in diseases characterized by excess or unwanted apoptotic cell death, such as certain neurological disorders (e.g., Alzheimer's disease) or stroke. The E4orf4 protein kills cancer cells selectively and could be applied to the treatment of human malignancies. Finally, several E3 proteins reduce the immune response, and could be used in immune suppression therapy. Disease-causing viruses such as human adenoviruses may soon be developed into valuable new therapeutic agents.

In human adenoviruses, E1A is the oncogene that stimulates uncontrolled cell-cycle progression. Expression of E1A alone in rodent cells usually produces abortive transformation. In this process, transformed cells begin to proliferate, but die soon afterward due to E1A-induced apoptosis. Formation of stable trans-formants requires coexpression of E1A and either E1B 19K or E1B 55K proteins, preferably both. As discussed previously, these E1B proteins blunt the pro-apoptotic effects of E1A, and inhibit p53-induced growth arrest and apoptosis. This system has provided valuable tools to understand the more general roles of the Rb and p300/CBP families, cell-cycle progression genes, and p53.

KEY TERMS

Adenoid

Apoptosis

Caspase

Chromatin remodeling

Cyclin-dependent protein kinase

Hexon

Histone acetyl transferase

Histone deacetylases (HDACs)

Immunoglobulin gene superfamily

Integrins

Interferon

Inverted terminal repeat

Oncogene

Penton base

PKR

Retinoblastoma (Rb)

Ribosome shunting

Secondary receptor

Terminal protein

Tumor suppressor protein

Ubiquitin

FUNDAMENTAL CONCEPTS

- Many human adenoviruses bind to both a primary receptor (CAR) and a secondary receptor (integrin), which leads to uptake in endocytic vesicles.

- Adenovirus genes are organized in six early, two intermediate, and five late families.

- E1A proteins bind to cellular Rb protein, dissociating it from cellular E2F protein and stimulating transcription of other viral genes and numerous cellular genes that bring non-dividing cells into the cell cycle.

- E1A proteins also engage cellular histone acetyl transferases, resulting in relaxed chromatin and increased transcription of many cellular genes.

- E1A proteins indirectly induce apoptosis by increasing levels of cellular p53 protein.

- E1B proteins reverse E1A-induced apoptosis by reducing p53 levels.

- Viral DNA synthesis begins at either end of the linear genome with a protein primer covalently bound to a nucleotide.

- Eighteen different late viral mRNAs are made from a single transcription unit by use of five polyadenylation signals and multiple splicing sites.

- Some adenoviruses can induce cancer in experimental animals but not in humans.

REVIEW QUESTIONS

1. Discuss the principal methods used by adenoviruses to generate approximately 50 distinct adenovirus proteins by using a very limited number of transcriptional promoters.

2. How do the adenovirus E1A proteins induce apoptosis via p53?

3. DNA replication initiating on a single-stranded DNA template requires a primer. What is the primer for initiating adenovirus DNA replication?

4. Adenovirus inhibits a cellular factor that binds to the 5' cap structure on mRNA and enables the 40S ribosomal subunit to begin scanning the mRNA for an appropriate AUG initiation codon. Nevertheless, late adenovirus mRNAs are efficiently translated; by what mechanism?

Herpesviruses

Bernard Roizman
Gabriella Campadelli-Fiume
Richard Longnecker

Herpesviridae

From Greek *herpein* (creep), referring to spreading of lesions

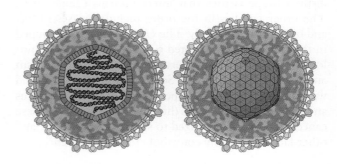

VIRION

Enveloped icosahedral capsid (T = 16).
Diameter 125 nm.
Diameter of enveloped virion 200 nm.
Capsid contains six proteins, envelope contains at least ten glycoproteins.
Material between capsid and envelope, called the "tegument", contains at least 14 viral proteins.

GENOME

Linear, double-stranded DNA, 125–250 kb.
Contains unique regions, inverted repeat elements, and terminal direct repeat sequences.

GENES AND PROTEINS

Herpes simplex virus type 1: 90 different transcriptional units.
Transcribed by cellular RNA polymerase II.
Most mRNAs are unspliced.
Most transcriptional units encode only one protein.
Genes are expressed in three temporal classes:
- α *(immediate early): early gene activation (six genes)*
- β *(early): DNA replication, late gene activation*
- γ *(late): virion proteins, regulatory proteins*

VIRUSES AND HOSTS

Three subfamilies: alpha, beta, and gammaherpesvirus.
Nine human herpesviruses, including herpes simplex virus, varicella-zoster virus, Epstein–Barr virus, cytomegalovirus.

Over 100 known herpesviruses from many animal species.

DISEASES

Chickenpox, mononucleosis, pneumonia, hepatitis, encephalitis.
Recurrent eye, mouth, and genital lesions.
Kaposi's sarcoma, Burkitt's lymphoma, nasopharyngeal carcinoma.
Life-threatening infections in immune-suppressed individuals.
Neonatal infections, birth defects.

DISTINCTIVE CHARACTERISTICS

Tegument proteins function in early steps of virus replication: activation of transcription and host shutoff.
Virion formation begins by budding of nucleocapsids through the inner nuclear membrane.
Many viral genes counter host defenses against virus infection.
Most herpesviruses become latent in the body and can be reactivated months or years after primary infection.

Herpesviruses are important human pathogens

Human herpesviruses are best known as the cause of discomfort from recurring "cold sores", lesions of the mucosa of the mouth and lips. Genital herpes lesions are a common sexually transmitted disease, and chickenpox is a common childhood disease that can be serious particularly when people are first exposed as adults. These three relatively mild but often annoying human diseases are caused, respectively, by herpes simplex virus types 1 and 2, and by varicella-zoster virus. However, under conditions of reduced immune response found in newborns, transplant recipients, and AIDS victims, herpesviruses can cause serious and often fatal disease, including **encephalitis**, pneumonia, and **hepatitis**.

These viruses belong to the family *Herpesviridae*, one of the 3 families of herpesviruses that comprise the order *Herpesvirales*. This family includes hundreds of viruses of mammals, birds, and reptiles. Among these are nine distinct herpesviruses that infect humans (Table 24.1). The other families in the order include viruses of fish and bivalves, respectively. The *Herpesviridae* are grouped in three subfamilies denoted α, β, and γ, based on biological and physical properties including cell **tropism** and genome organization. Some cancers are associated with infection by the γ-herpesviruses Epstein–Barr virus and Kaposi's sarcoma virus (human herpesvirus 8). These cancers are rare or localized to specific populations, and other factors probably contribute to their incidence.

Most herpesviruses can establish latent infections

One of the most intriguing characteristics of herpesviruses is their ability to establish **latent infections**. After an initial infection, viral DNA is harbored in a latent state in neurons, **B** or **T lymphocytes**, or other cell types. Latently infected individuals can remain without symptoms for months or years, or even their entire life. Reactivation of latent virus can lead to recurrent disease, such as repeated outbreaks of labial or genital herpes, or herpes zoster ("shingles"), a localized rash that can occur years after the initial chickenpox infection. People whose immune systems are impaired can suffer reactivation of cytomegalovirus or Kaposi's sarcoma virus with dire results. The mechanisms by which herpesviruses establish latent infections and reactivate are presently under intense study.

This chapter will concentrate on two well-studied human herpesviruses: herpes simplex virus type 1 and Epstein–Barr virus. Although both these viruses infect humans via the oral mucosa, they have very different lifestyles and cause distinct diseases. This is mainly due to the cell types that they target for establishment of long-term latency: neurons for herpesvirus type 1 and B

lymphocytes for Epstein–Barr virus. The characteristics and disease implications of these latent infections will be described in detail after a discussion of the strategy of lytic infection employed by the best-studied model, herpes simplex virus.

HERPES SIMPLEX VIRUS

Herpes simplex virus genomes contain both unique and repeated sequence elements

All herpesviruses contain a double-stranded linear DNA genome, which is between 120 and 230 kb in length depending on the virus species. The 150-kb herpes simplex virus genome has an unusual structure; it consists of two covalently linked components, long (L) and short (S) (Figures 24.1 and 24.2). Each of these components contains a stretch of unique sequences, U_L and U_S, flanked by inverted repeats. The sequence repeat at the left end of U_L is designated *ab*, and the internal inverted repeat is designated *b'a'*. The inverted repeats flanking U_S are designated *a'c'* and *ca*. In shorthand, the linear genome sequence can be written $a_n b$-U_L-$b'a'_n c'$-U_S-*ca*, where the subscript *n* designates one or more copies of the *a* sequence and the prime designates complementary sequences.

The *a* sequence varies from 200 to 500 bp in length in virus isolates, and contains no open reading frames. It contains signals for packaging viral DNA in capsids, and enhances recombination between genomes. The *b* sequence contains four open reading frames. Since the genome contains two identical copies of the *b* sequence, these four viral genes are therefore present in two copies per genome, albeit in inverted orientation. The *c* sequence contains a single open reading frame, also present in two copies per genome. Most viral genes are present in only one copy per genome; the U_L region contains 65 protein coding sequences, and the U_S region contains 14. Thus herpes simplex viruses contain at least 84 different genes.

Table 24.1 **Human herpesviruses**

Subfamily and characteristics		Members	Diseases
α	Short growth cycle, rapid spread Latent in sensory neurons.	Herpes simplex virus 1 & 2 Varicella-zoster virus	Labial and genital lesions Chickenpox, zoster (shingles)
β	Long growth cycle, slow spread, restricted host range. Latent in secretory glands and lymphoreticular cells.	Cytomegalovirus Human herpesvirus 6A, 6B Human herpesvirus 7	Mononucleosis, hepatitis, fetal abnormalities Roseola Roseola
γ	Growth primarily in epithelial cells and B or T lymphocytes. Latent in lymphocytes.	Epstein-Barr virus Human herpesvirus 8	Mononucleosis, Burkitt's lymphoma, nasopharyngeal carcinoma Kaposi's sarcoma

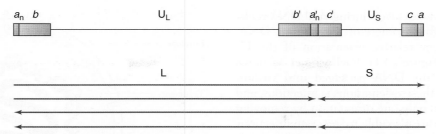

Figure 24.1 The genome of herpes simplex virus. Top: organization of unique and repeated sequence elements. Repeated sequence elements are shown in small letters (a, b, c) and unique sequence elements are shown in capital letters (U_L, U_S). Elements shown with a prime are inverted. Below: four isomeric forms of viral DNA with different orientations of the long (L) and short (S) components due to frequent recombination at the repeated sequence elements.

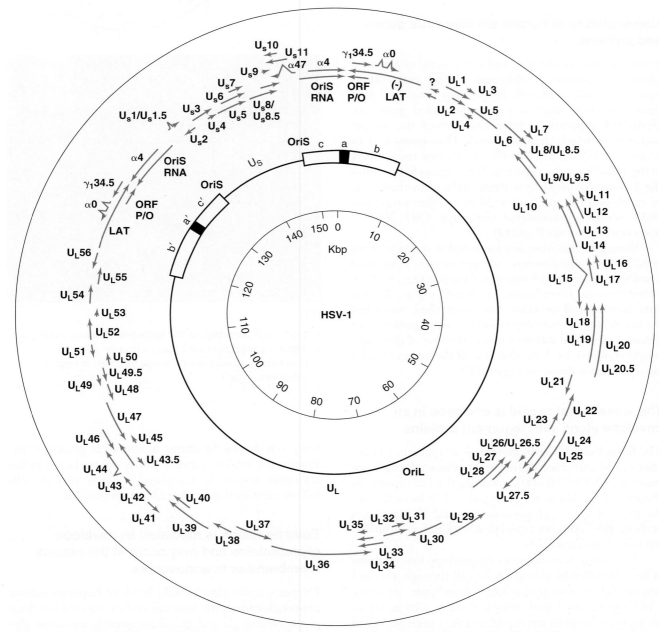

Figure 24.2 Map of herpes simplex virus genes. Genes are shown on the circularized 152-kb map of herpes simplex virus type 1 DNA. Arrows indicate direction of transcription and location of polyadenlyation sites. Introns are shown as interruptions connected by a V. The positions of the three DNA replication origins (two OriS and one OriL) are shown on the map.

Recombination takes place during viral DNA replication to generate four isomers of the genome DNA, each with a different relative orientation of the U_L and U_S segments (Figure 24.1). Each isomer accounts for one-quarter of the DNA packaged into virions. DNA in any one of these orientations is infectious, and genomes frozen into only one orientation by removing the inverted repeats are also able to generate infectious virus, but exhibit reduced virulence in experimental animal systems. The genomes of other herpesviruses show variations on the theme of interspersed unique and repeated sequences.

Nomenclature of herpes simplex virus genes and proteins

Partly because of the size and complexity of the viral genome, a number of different naming conventions have been used to identify herpesvirus genes and proteins. Presently, most herpes simplex viral genes are numbered sequentially from one end of the unique sequence in which they are found. Thus genes U_L1 to U_L56, and U_S1 to U_S12 ($\alpha47$), are located respectively in the L and S unique segments. Some genes lying within the b and c segments have names related to their time of expression ($\alpha0$, $\alpha4$, and $\gamma_1 34.5$) or other properties (LAT = latency-associated transcript; ORF P/O = open reading frames P and O).

Some proteins were first identified as *virion proteins* (e.g., VP5) or *glycoproteins* (e.g., gp), or were designated as *infected cell proteins* found in cell extracts (e.g., ICP25). Other proteins, first identified by function (e.g., Vhs for viral host shutoff protein), have kept their functional names. A number of herpes simplex virus proteins are denoted by several different names because of these historical conventions. A circular map of the herpes simplex virus genome is shown in Figure 24.2.

The icosahedral capsid is enclosed in an envelope along with tegument proteins

The herpes simplex virus genome is wrapped in a bilayered capsid constructed from six proteins, of which the major one, VP5 (150 kDa), makes up the 162 capsomers (Figure 24.3). These are arranged as 150 hexamers on the faces of a T = 16 icosahedron, and 12 pentamers at the vertices. Another capsid protein, VP26, is located at the tips of the capsomers.

The capsid is enclosed in an envelope that contains at least 10 different glycoproteins (gB through gM) and several other non-glycosylated membrane proteins. Unlike most enveloped viruses, there is a large amorphous mass between the capsid and the envelope, called the **tegument**, which contains about 14 virus-coded proteins. Tegument proteins are introduced into the cell upon infection by herpesviruses, and some of them

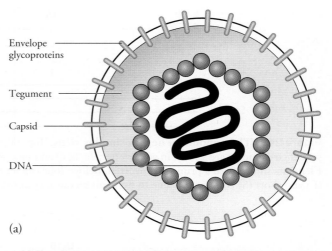

(a)

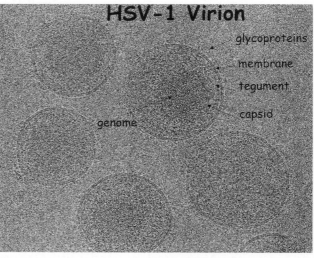

(b)

Figure 24.3 Structure of the herpesvirus virion. (a) Diagram of herpesvirus particle with major elements denoted. (b) Cryoelectron microscopy image of virion of herpes simplex virus type 1.

carry out crucial functions in the virus growth cycle. Even some viral messenger RNAs are packaged in the tegument and can be translated in newly infected cells before most viral genes have been transcribed.

Entry by fusion is mediated by envelope glycoproteins and may occur at the plasma membrane or in endosomes

Herpes simplex virus initially binds to **heparan sulfate proteoglycans** at the host cell surface via viral envelope glycoproteins gB and gC. Subsequently, envelope glycoprotein gD binds to at least two alternative receptors: **nectin** 1, a member of the nectin family of intercellular adhesion molecules; and herpesvirus entry mediator

(HVEM), a member of the **tumor necrosis factor (TNF)** receptor family. The two receptors are differently distributed in human cells and tissues.

Receptor binding triggers fusion of the viral envelope with a cellular membrane. This requires three conserved glycoproteins: gB, gH, and gL. Depending on the target cell, fusion takes place at the plasma membrane or in endosomes.

Fusion releases the nucleocapsid and tegument proteins into the cytoplasm. Nucleocapsids, along with some of the tegument proteins, are transported to the nuclear pores by dynein motors along the microtubular network of the cell. At the nuclear pore the DNA is released and enters the nucleus. Empty capsids remain at the cytoplasmic side of nuclear pores for several hours until they disintegrate. Viral DNA in the nucleus is circularized, either by direct ligation of the ends or by recombination between the *a* sequences at the ends. The circularized DNA is localized near nuclear structures known as nuclear domain 10 (ND10), where early transcription takes place. Steps along the entry pathway are shown in Figure 24.4.

Viral genes are sequentially expressed during the replication cycle

Like the smaller DNA viruses of eukaryotes discussed in this book, transcription of herpesvirus genes is carried out in the nucleus by cellular RNA polymerase II. Viral genes can be classified in at least four groups based on the time of their expression during the replication cycle. Transcription of the *immediate early* or α genes takes place during the first several hours after infection. Some α gene products are required for the subsequent transcription of *early* or β genes, which are expressed between about 4 and 8 h after infection. The β gene products are mostly involved in viral DNA replication.

The vast majority of viral genes are *late* genes, designated γ_1 and γ_2. Genes in the γ_1 group begin to be transcribed early but their transcription is stimulated several-fold after the onset of viral DNA replication; expression of γ_2 genes begins only after DNA replication starts. Although β and γ genes are spread throughout the genome of herpes simplex virus, it is notable that all six α genes are localized within or near the boundaries of the inverted repeat sequences *b* and *c* (Figure 24.5). Furthermore, three out of the six α genes contain introns, but only one of the 78 β and γ genes contains an intron.

Tegument proteins interact with cellular machinery to activate viral gene expression and to degrade cellular messenger RNAs

The precise functions of only a few of the tegument proteins are known. Although the α genes can be transcribed at a low level by basal cell transcription factors, expression is not optimal. A viral tegument protein known as α-transinducing factor (α-TIF) or VP16 forms a complex with at least two key cellular proteins, Oct-1

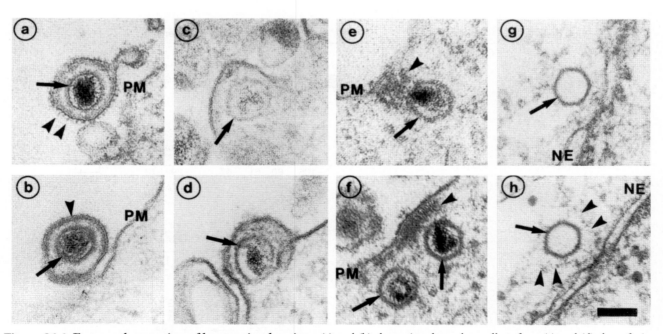

Figure 24.4 Entry and uncoating of herpes simplex virus. (a) and (b) show virus bound to cell surface; (c) and (d) show fusion of the viral envelope with the plasma membrane; (e) and (f) show nucleocapsids (arrows) detaching from the tegument (arrowhead) at the plasma membrane; (g) and (h) show empty capsids (arrows) at the nuclear pores (fibers shown by arrowheads) after release of DNA into the nucleus. PM = plasma membrane; NE = nuclear envelope.

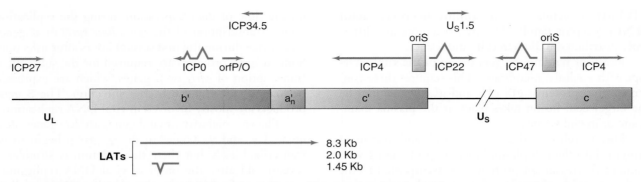

Figure 24.5 Transcriptional map of the b and c regions of herpes simplex virus. Above: transcripts of the six α genes (ICP0, 4, 22, 27, 47, and U$_S$1.5) and three other genes (ICP34.5, ORF P/O) are shown as blue arrows. Introns are shown as interruptions connected by a V. The two copies of oriS are shown flanked by α genes. The remainder of the U$_S$ region, containing γ genes, is not shown. Below: latency-associated transcripts (LATs).

and HCF-1. This complex binds to specific response elements in viral DNA that have a central consensus sequence TAATGARATT. These elements, along with binding sequences for other cellular transcription factors, are found upstream of the promoters of all α genes (Figure 24.6). The bound protein complex acts as transactivator to enhance the transcription of these genes—the first set of viral genes expressed in infected cells. Thus optimum α gene expression depends on the presence of a tegument protein that is introduced into the cell at the time of infection.

Another tegument protein known as *virion host shutoff* (Vhs) protein, a product of the U$_L$41 gene, interacts with several cellular proteins and mediates the selective degradation of both cellular and viral mRNAs during the first few hours after infection. Viral mRNA accumulates faster than it is degraded, but cellular mRNA synthesis is inhibited by other herpesvirus proteins; as a consequence, viral protein synthesis predominates. Vhs is also responsible for the transition from α to β and γ protein synthesis by degradation of α gene mRNAs. Interestingly, the RNA degradation activity of Vhs made after the onset of viral DNA synthesis is suppressed by binding to newly synthesized VP16 and VP22, and this inactive complex is incorporated into progeny virions. This mechanism allows large amounts of late viral mRNAs to be made and translated even though Vhs is being made at the same time.

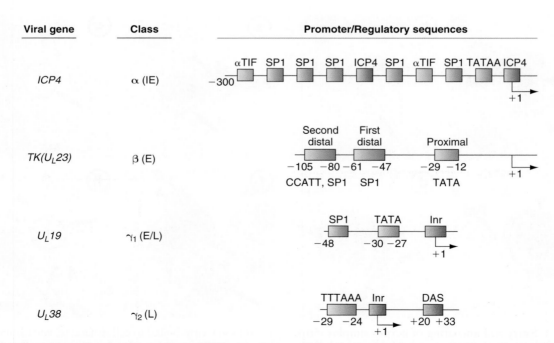

Figure 24.6 Structure of promoter regions of different classes of herpes simplex virus genes. A gene from each class (α, β, γ$_1$, and γ$_2$) is shown. +1 = transcription start site. Boxes indicate TATA boxes or sequence elements that interact with cellular (SP1, CCAAT) or viral (αTIF, ICP4) proteins. Inr = initiator element; DAS = downstream sequence activator element.

Immediate early (α) genes regulate expression of other herpesvirus genes

A major function of the α genes is to set up a regulatory cascade that leads to properly timed expression of other viral genes. The products of five of the six α genes (Table 24.2) are involved in the regulation of viral and cellular gene expression. Mutations in the genes that code for ICP0 and ICP4 strongly reduce expression of β and γ genes.

ICP0 is best described as a promiscuous activator protein, because its expression activates transcription of many genes introduced into cells by either infection or transfection. ICP0 activates transcription of β and γ genes by relieving the silencing of these genes, whereas ICP4 recruits cellular transcription factors to their promoters (Figure 24.6). ICP0 also activates transcription of β genes by a number of other pathways, not fully understood, involving interactions with a variety of cellular proteins.

The decrease in the rate of synthesis of α gene products at 4 hours postinfection involves several proteins in addition to the tegument protein Vhs. ICP4 can repress transcription of its own gene by direct binding to the ICP4 promoter region (Figure 24.6). Another α gene product, ICP27, colocalizes with spliceosomes in the nucleus and blocks RNA splicing. This not only inhibits expression of cellular genes, most of which make spliced mRNAs, but also inhibits the expression of the three α genes (ICP0, ICP22, and ICP47) that contain introns.

Table 24.2 Functions of herpes simplex virus α proteins

Protein	Activities
ICP0	Promiscuous activator of gene transcription (viral and cellular). Acts as ubiquitin ligase (E3). Localizes to nuclear domain 10 structures.
ICP4	Major regulatory protein, required for all β and γ virus gene expression. Binds to viral DNA; interacts with cellular transcription factors. Regulates transcription both positively and negatively.
ICP22 U_S1.5	Together with U_L13, enhance the expression of a subset of late genes through activation of cyclin-dependent cdc2.
ICP27	Early: colocalizes with spliceosomes and blocks RNA splicing. Late: shuttles between nucleus and cytoplasm to transport viral mRNAs.
ICP47	Binds to TAP1/TAP2 transporter, blocking presentation of antigenic peptides to immune system via MHC class I protein.

β gene products enable viral DNA replication

The function of most β genes is to replicate the viral DNA. Seven of these proteins work together to carry out essential steps in DNA replication (Table 24.3). Like other large DNA viruses, herpesviruses make their own DNA polymerase, which is directed by other viral proteins to begin DNA synthesis at any one of three viral origins of DNA replication on the herpes simplex virus genome. One origin (oriL) is in the middle of the U_L region, and the other two (oriS) are in the c inverted repeats flanking the U_S region (Figure 24.2).

It is not clear why the virus has three origins, since any two can be inactivated and the virus can still replicate in cultured cells. The three origins share extensive sequence homology and contain a central palindromic sequence of 144 bp (oriL) or 45 bp (oriS) that can form a stem structure on single-stranded DNA.

All three origins are flanked by genes whose promoters direct divergent transcription away from the origin, reminiscent of the replication origins of polyomaviruses. Interestingly, oriL is flanked by genes coding for the viral single-stranded DNA binding protein (U_L29) and the viral DNA polymerase (U_L30), both intimately connected to DNA replication. The two copies of oriS are flanked by the important regulatory α genes coding for ICP4, ICP22, and ICP47 (Figure 24.5).

DNA replication initially proceeds in a bidirectional fashion from a replication origin

The mechanism of herpesvirus DNA replication can be reconstructed only from knowledge of the proteins involved and in comparison with better-studied systems, because it has so far been impossible to reconstitute origin-dependent herpesvirus DNA replication *in vitro*. Replicating DNA in infected cells is found as high molecular weight, branched head-to-tail multimers called **concatemers**, suggesting that DNA is made by a **rolling circle mechanism**, which can produce linear multimers from a circular template. The branching could result from initiation of replication at several origins on the concatemers, as well as intermolecular recombination between replicating DNAs. However,

Table 24.3 Herpes simplex virus β proteins that direct viral DNA replication

Protein	Activity
U_L9	Binds to replication origins
U_L29 (ICP8)	Binds to single-stranded DNA
U_L5 + U_L8 + U_L52	Helicase-primase complex
U_L30	DNA polymerase
U_L42	Processivity factor

herpesvirus initially sets up bidirectional replication on the circular viral DNA, and only later shifts to rolling circle replication.

An origin-binding protein (U_L9) binds to specific 10 bp sequences in oriS or oriL and unwinds the two DNA strands with the help of the single-stranded DNA binding protein ICP8 (Figure 24.7). A DNA helicase-primase complex made from three other β proteins (U_L5, U_L8, and U_L52) then binds to the single-stranded DNA and synthesizes RNA primers, the first step in the initiation of DNA replication. This is similar to the mechanism by which polyomaviruses and eukaryotic cells initiate their DNA replication, but polyomaviruses use the cellular DNA polymerase α/RNA primase.

At this point the herpesvirus DNA polymerase, product of the U_L30 gene, binds to the RNA primers and extends them as DNA chains. U_L42, a viral protein that is bound to the DNA polymerase, increases the **processivity** of the polymerase, allowing it to extend growing DNA strands over considerable distances without dissociating. These two proteins have functions similar to those of the cellular proteins DNA polymerase δ and PCNA (see Chapter 21). A structure is set up in which two replication forks progress in opposite directions from the origin (Figure 24.7).

Rolling circle replication subsequently produces multimeric concatemers of viral DNA

At some point one of the strands in the parental DNA is nicked, resulting in rolling circle replication (Figure 24.8). In this mode, only one of the leading DNA strands is extended, as the DNA polymerase copies the remaining intact circular template strand. As it progresses, the DNA polymerase encounters the other end of the same DNA strand and displaces it from the template as it continues around the circular DNA. The DNA helicase-primase complex, along with DNA polymerase, fills in the complementary lagging-strand on the extruded DNA to produce a double-stranded, linear concatemer.

In addition to the proteins listed above, herpes simplex virus encodes other proteins concerned with DNA metabolism. Thymidine kinase (U_L23), ribonucleotide reductase (U_L39/U_L40), and deoxyuridine triphosphatase (U_L50) are involved in producing nucleoside triphosphates used as building blocks for DNA replication, while uracil N-glycosylase (U_L2) is involved in DNA repair. These genes are essential for virus replication in non-dividing cells (e.g., neurons) but not in dividing cells, which can make cellular analogues of these enzymes.

DNA replication leads to activation of γ_1 and γ_2 genes

Although γ_1 genes can be transcribed in the absence of viral DNA replication, their optimal expression requires viral DNA synthesis, presumably because of the increase in the number of template DNAs available for transcription. In contrast, the transcription of γ_2 genes is totally dependent on viral DNA synthesis. These genes code for viral stuctural proteins, proteins involved in the assembly of capsids and packaging of viral DNA, and many other proteins whose functions are only beginning to be discovered.

Most γ_2 gene promoters contain few or no binding sequences for cellular transcription factors upstream of their transcription start sites. However, downstream activation sequences, located beyond the transcription start site, are found in a number of γ_2 genes (Figure 24.6).

The expression of late genes is tightly regulated by three α proteins (ICP4, ICP22, ICP27) and a protein kinase encoded by U_L13. ICP4, ICP22, and ICP27 are localized in replication/transcription factories that appear in cell nuclei after DNA replication has begun. ICP22 and U_L13 activate the mitotic **cyclin-dependent kinase** cdc2, which in turn forms a complex with the viral DNA polymerase processivity

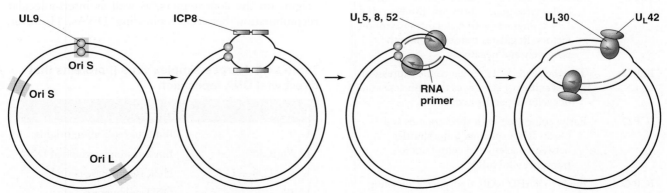

Figure 24.7 Establishment of bidirectional DNA replication. Replication is shown beginning at one of the three possible origins. RNA primers are blue; newly synthesized DNAs are orange.

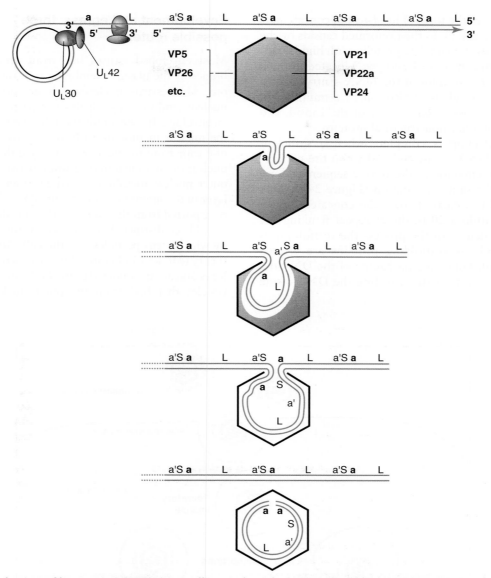

Figure 24.8 Packaging of herpesvirus DNA. Top: rolling circle replication mediated by herpesvirus DNA polymerase (U_L30) and processivity factor (U_L42) generates head-to-tail concatemers of herpesvirus DNA. **a** = terminal *a* repeats; L and S = long and short components; *a'* = internal *a* repeats (*b* and *c* regions are not shown). Below: precapsids made from VP5 and other capsid proteins are formed around scaffolding proteins (VP21, VP22a, VP24, shown in gray). Precapsids package viral DNA under the direction of packaging proteins that recognize the terminal *a* repeats and cleave off genome-length, linear DNA fragments. Scaffolding proteins are extruded from the capsids in the process.

factor and **topoisomerase** IIα. Activation of cdc2 leads to increased expression of a subset of γ_2 genes, but the exact mechanism is not understood. It has been postulated that rolling circle replication produces DNA knots that need to be resolved for efficient transcription. In addition, the carboxy-terminal domain (CTD) of RNA polymerase II is phosphorylated by a reaction mediated by cdc9, ICP22, and both U_S2 and U_L13 protein kinases. Coincident with the onset of synthesis of late proteins, ICP27 begins to shuttle between the nucleus and cytoplasm, exporting viral mRNAs.

Viral nucleocapsids are assembled on a scaffold in the nucleus

An immature capsid lacking DNA is assembled by accumulation of the major capsid protein VP5 and several other proteins around a scaffold (Figure 24.8). This is reminiscent of the way in which bacteriophage λ constructs and fills its capsid (Chapter 8). Three late proteins—VP21, pre-VP22a, and VP24—form the scaffold. VP24 is actually a viral protease that cleaves itself from a precursor protein, producing VP21 as another cleavage product.

Viral DNA in the form of head-to-tail concatemers is cleaved and stuffed into the preformed capsids by the interaction of several viral late proteins with highly conserved packaging sites (pac1 and pac2) located in the *a* sequence near the termini of the DNA. Entry of viral DNA into the capsids is carried out by a multimer of U_L6 protein located at one vertex of the capsid, and involves several other nonstructural proteins.

DNA packaging is complete when a sufficient length of the DNA is inserted and when the packaging machinery encounters the next *a* sequence along the DNA in the same orientation (Figure 24.8). The packaged DNA is cleaved from the concatemer at a specific site within a 20-nt direct repeat flanking the terminal *a* sequence. In the process the scaffolding is dismantled and the scaffolding proteins are ejected from the capsid. Following packaging of the DNA the capsid changes conformation, sealing the DNA inside the capsid.

Envelopment and egress: three possible routes

Most enveloped viruses bud from cytoplasmic membranes or the plasma membrane to form infectious virions. Herpesvirus nucleocapsids are assembled in the nucleus, and need to exit this compartment and acquire an envelope. It is generally thought that herpesviruses use the nuclear membrane to form an envelope as a means of escape from the nucleus. They bud through the inner nuclear membrane into the **lumen** between the inner and outer nuclear membranes, acquiring an envelope and a layer of tegument proteins (Figure 24.9). Virions are then transported from the lumen to the outside of the cell.

Three distinct theories account for the mechanism by which virus particles exit the cell. According to one theory (Model 1 in Figure 24.9), virions retain their envelopes and are transported to the Golgi membranes within vesicles that bud from the outer nuclear membrane.

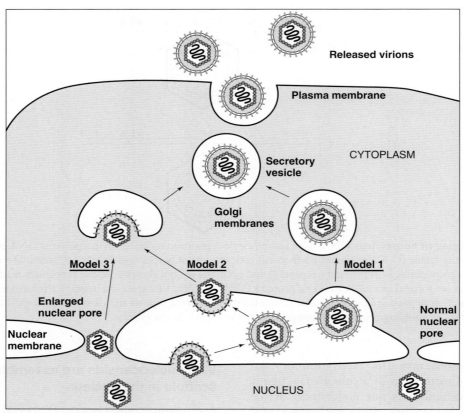

Figure 24.9 Envelopment and egress of herpesvirus virions. Nucleocapsids are assembled in the nucleus. Three possible routes of acquiring an envelope and escaping from the cell are shown. The "single nuclear envelopment" model (1) predicts that after budding through the inner nuclear membrane, virions are enclosed in vesicles derived from the outer nuclear membrane, and are transported to the plasma membrane without further envelopment. The "envelopment/de-envelopment" model (2) predicts that capsids acquire an envelope by budding through the inner nuclear membrane, become de-enveloped at the outer nuclear membrane, and are subsequently re-enveloped at Golgi membranes. The "single cytoplasmic envelopment" model (3) predicts that nuclear pores become enlarged and allow translocation of capsids to the cytoplasm, where they become enveloped at Golgi membranes. In all three models the transport vesicles fuse with the plasma membrane to release virions into the extracellular space.

According to another theory (Model 2 in Figure 24.9), virions lose their envelope by fusion with the outer nuclear membrane, releasing nucleocapsids into the cytoplasm. These nucleocapsids subsequently reacquire an envelope by budding into Golgi membranes. In either case, the envelope proteins would be glycosylated within the Golgi, and further transport of mature virions to the extracellular space is via secretory vesicles.

A third theory is based on the observation that nuclear pores become enlarged and the Golgi membranes are fragmented during viral replication. According to this hypothesis, nucleocapsids exit through the enlarged nuclear pores without an envelope (Model 3 in Figure 24.9), and become enveloped at multivesicular bodies derived from fragmented Golgi membranes.

Many viral genes are involved in blocking host responses to infection

Virus replication in cells cultured *in vitro* requires a relatively small number of genes—at most 37 out of the 84 identified to date. Among the genes dispensable for replication *in vitro* are a number that block host responses to infection. This is an active subject of research and is important for understanding how herpesviruses can survive and thrive in the host animal.

Vhs and ICP27. Stress responses activated during the initial stages of infection result in the transcription of several hundred cellular genes (Chapter 33). Most of these newly synthesized cellular mRNAs, as well as pre-existing cellular mRNAs, are degraded by Vhs, an endoribonuclease with a substrate specificity similar to that of ribonuclease A. Among the targets of Vhs are proteins associated with activation of the **interferon** pathway, a major host response to infection. ICP27 inhibits splicing of mRNAs during the early stages of infection, also contributing to blocking host responses. In addition, at late stages of infection transcription of cellular genes is also grossly reduced, possibly as a result of changes in the chromatin structure of cellular DNA.

ICP47. Herpes simplex virus blocks the presentation of antigenic peptides to the immune system (see Chapter 34). A fraction of newly synthesized viral proteins are normally degraded to peptides in **proteasomes**, especially as a consequence of misfolding. These peptides are translocated to the endoplasmic reticulum and are ultimately presented to the immune system by **major histocompatibility (MHC)** class I proteins at the cell surface. A viral protein, ICP47, binds to the transporter of antigenic peptides (TAP1/TAP2) and blocks the translocation of peptides into the endoplasmic reticulum, therefore inhibiting antigen presentation.

ICP34.5. Herpesvirus genes are tightly packed and are transcribed from both DNA strands. During the late phase of infection, polyadenylation signals are frequently misread, resulting in extended downstream transcription and the accumulation of large amounts of complementary RNAs that can anneal to each other. The resulting double-stranded RNA activates a cellular protein, double-stranded **RNA-dependent protein kinase (PKR)**. Activated PKR phosphorylates the a subunit of eukaryotic initiation factor 2 (eIF-2α), which in turn shuts off protein synthesis (see Chapter 33). Activated PKR can also lead to activation of the important cellular transcription factor, **NFκB**.

ICP34.5 binds to protein phosphatase 1α and redirects it to efficiently dephosphorylate eIF-2α, thereby reversing the effect of the activated PKR. The carboxy-terminal domain of ICP34.5 is homologous to a corresponding domain of a cellular protein, GADD34. This suggests that herpes simplex virus acquired part of a cellular gene whose function is to control protein synthesis in infected cells. PKR is a key factor in determining host susceptibility to infection. In fact, herpesvirus mutants lacking the $\gamma_1$34.5 gene are highly attenuated. These mutants, however, become fully virulent in mice lacking components of the interferon pathway that control PKR.

ICP0. Viral DNA introduced into cells during infection accumulates at nuclear bodies known as ND10 or PML structures, as they contain the promyelocytic leukemia protein, PML. Newly made ICP0 also localizes in ND10 structures and performs multiple functions. ICP0 dissociates histone deacetylases from chromatin in ND10 structures, therefore relieving transcriptional silencing that occurs as a result of strong DNA binding by deacetylated histones. ICP0 also acts as an E3 **ubiquitin ligase** to degrade several components of ND10, including the cellular proteins PML and SP100, and it causes a dispersal of the components of ND10. Cells lacking PML appear to be unable to respond to exogenous interferon.

Anti-apoptosis genes. Programmed cell death, or **apoptosis**, results from a complex cascade of events in which a cell can self-destruct (Figure 33.4). Viral replication is one stimulus that can induce apoptosis. Recent observations suggest that apoptosis is a primary line of host defense and that herpes simplex virus, like many other viruses, has evolved a number of mechanisms designed to block this host response to infection, allowing infected cells to survive long enough to produce progeny virus.

Wild-type herpes simplex virus blocks apoptosis caused by viral gene expression or by exogenous agents such as osmotic shock, thermal shock, and Fas ligand interaction. However, at least four different viral mutants have been shown to allow apoptosis to proceed. These include mutants that (1) lack glycoprotein D, an envelope protein; (2) have a temperature-sensitive U_L36 tegument protein that does not properly assemble virions; (3) lack ICP4 or ICP27, important viral regulatory proteins, and (4) lack the major subunit of ribonucleotide reductase, encoded by U_L39. Analysis of these mutants led to the discovery that the virus encodes at least three genes whose function, in part, is to block apoptosis. Cell death induced by mutants lacking the gD gene is blocked by expressing either glycoprotein D or glycoprotein J in infected cells. Apoptosis induced by a mutant lacking the gene encoding ICP4 is blocked by a protein kinase encoded by the U_S3 gene. This kinase prevents apoptosis at both pre- and post-mitochondrial stages; its target is not known.

Herpes simplex virus establishes latent infection in neurons

Latency is a hallmark of all known herpesviruses in their natural hosts. In humans and experimental animal systems, herpes simplex virus replicates in skin or mucosal cells at the portal of entry, producing infectious virus. The progeny virions, and more rarely the virus present in the initial inoculum, infect nerve endings and undergo **retrograde transport** along axons to the cell bodies of **dorsal root neurons** innervating the site of viral entry. Viral DNA enters the nucleus of the neuron, is circularized, and remains dormant; transcription is repressed and there is no DNA replication.

Several factors contribute to the lack of expression of viral genes in neurons. As described above, expression of herpesvirus α genes is activated by a complex between viral tegument protein VP16 and two cellular transcription factors, Oct-1 and HCF-1. This complex binds to α gene promoters and leads to their transcription. Because VP16 is a tegument protein that enters neurons with the infecting virus, little VP16 actually makes it to the nucleus of the neuron, which is a long distance along the nerve axon from the site of infection. Furthermore, neurons contain two transcription factors named Luman and Zhangfei, which bind to HCF-1 and prevent formation of the active transcription factor complexes. In the absence of α gene transcription, the rest of the genome remains silent, with the exception of a class of RNAs called latency-associated transcripts (LATs).

Latency-associated transcripts include stable introns

Herpes simplex virus latency-associated transcripts are a family of RNAs (8.3, 2, and 1.45 kb) that map to the *b* repeat sequences and are antisense to ICP0 mRNA (Figure 24.5). A surprising finding was that the 2-kb and 1.45-kb species are stable, unresolved lariat structures that accumulate in the nucleus. Lariat structures are generated from introns during RNA splicing; these two LATs are therefore stable introns. It is very unusual that introns accumulate in cells, because they are usually rapidly degraded after splicing occurs. The 2-kb and 1.45-kb RNAs are probably spliced from the 8.3-kb LAT transcript, but very little full-length transcript accumulates. Although they contain substantial open reading frames, there is no evidence that LAT RNAs express any proteins, and their accumulation as introns in the nucleus argues against a role as mRNAs.

The role of LATs remains controversial. It has been suggested that they inhibit ICP0 and ICP34.5 expression by an antisense mechanism involving microRNAs, therefore inhibiting viral gene expression. Extension of transcription beyond the polyadenylation site could also produce RNA antisense to ICP4 mRNA. LATs are not essential for establishment of latency, although the presence of the LAT transcriptional unit can increase the efficiency of latent infection in some model systems, perhaps by reducing productive infection. LATs have been shown to protect cells against apoptosis, perhaps contributing to the survival of neurons that express low levels of viral proteins.

In humans, local or systemic stimuli—such as physical or emotional stress, hyperthermia, exposure to ultraviolet light, menstruation, and so on—induce reactivation of herpes simplex virus from latency. Low levels of viral transcripts are made in sensory neurons, leading to formation of small amounts of progeny virions. Virus is transported along the axon to mucosal sites where the virus can replicate. This causes the appearance of local lesions, whose spread is usually limited by host inflammatory and immune responses.

EPSTEIN–BARR VIRUS

Epstein–Barr virus was discovered in lymphomas in African children

Epstein–Barr virus is a very successful pathogen, present in over 95% of the adult population. Most primary infections occur in childhood, resulting in largely asymptomatic or unrecognized infections. Infection in adolescence and adulthood can cause infectious mononucleosis, a disease whose symptoms are caused by proliferation of infected B lymphocytes and their subsequent destruction by activated T lymphocytes.

Denis Burkitt was an Irish physician pursuing a medical career in Kampala, Uganda. In 1957, Burkitt saw a young boy with swelling of both sides of his jaws that proved to be a lymphoma. This lymphoma is the most common cancer in African children and is often fatal upon spread to other parts of the body. Burkitt became fascinated with the tumor and he eventually embarked on what he called his "long safari", documenting the incidence of the lymphoma in sub-Saharan Africa. Burkitt found that the geographic distribution of patients with this lymphoma followed closely the incidence of malaria, suggesting that a transmissible infectious agent was involved.

Anthony Epstein attended a lecture by Burkitt at Middlesex Hospital in London in 1961. Epstein and his colleague Yvonne Barr immediately began collaboration with Burkitt by studying biopsies sent from Uganda. By 1964 it became clear that cells cultured from the tumor biopsies contained herpesvirus-like particles; this new herpesvirus was named Epstein–Barr virus.

In 1967, Werner and Gertrude Henle, working in Philadelphia, discovered that Epstein–Barr virus is the agent responsible for infectious mononucleosis. Several other diseases, including nasopharyngeal carcinoma and Hodgkin's lymphoma, have been linked to infection with Epstein–Barr virus. The association of Burkitt's lymphoma with malaria is now believed to result from a weakened immune system in malaria victims, allowing development of lymphomas in children infected with Epstein–Barr virus.

Epstein–Barr virus infects mucosal epithelial cells and B-lymphocytes

The genome of Epstein–Barr virus is approximately 180 kb in length, similar to that of herpes simplex virus. It contains five unique sequence elements (U1 to U5) interspersed with one major repeated sequence element (IR1) and three smaller internal repeats (IR2 to IR4) (Figure 24.10). The major internal repeat element contains six to twelve tandem repeats of 3 kb, depending on

the virus strain. There are three replication origins; two of these (Ori-lyt) are used during productive infection and one (Ori-P) is used during latent infection. Like herpes simplex virus, Epstein–Barr virus DNA circularizes upon entering the cell via a terminal repeat (TR) sequence.

Epstein–Barr virus replicates transiently in epithelial cells of the oral mucosa, releasing progeny virus into the saliva. This productive cycle is important both for spread of the virus to other individuals and for subsequent infection of resting B-lymphocytes in the oral cavity (Figure 24.11). In contrast, infection of B-lymphocytes leads to a long-term, latent infection, with expression of a limited class of latency genes, and little or no production of progeny virus.

The ability of Epstein–Barr virus to infect B lymphocytes is determined by its binding to two key proteins found on the surface of B-lymphocytes: CD21, a **complement** receptor, and HLA class II (also known as MHC class II), which is involved in immune recognition (see chapter 34). Viral envelope protein gp350/220 binds to CD21, leading to uptake of the virus into endocytic vesicles. Viral envelope protein gp42 binds to HLA class II and helps direct virus entry. gp42 is complexed with viral envelope glycoproteins gH and gL, and this complex (along with gB) is believed to mediate fusion with the membranes of the endocytic vesicles, releasing the nucleocapsid into the cytoplasm.

Interestingly, gp42 is underrepresented in virions made from B lymphocytes, probably because it binds to HLA class II protein on the surface of virus-producing cells and therefore is less likely to be incorporated into the viral envelope. Virus produced by B lymphocytes has reduced infectivity for B lymphocytes but enhanced infectivity for epithelial cells, allowing productive infection and virus transmission.

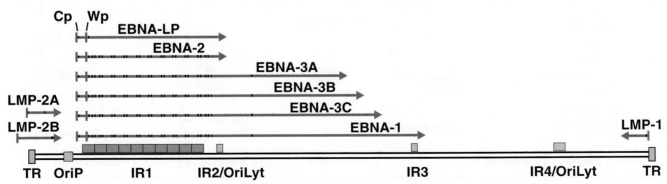

Figure 24.10 Linearized genome of Epstein-Barr virus and transcripts expressed during latent infections. Upon infection the genome circularizes; a linear genome is shown for convenience. EBNA mRNAs are generated by alternative splicing and 3' processing of long transcripts that initiate at promoters Cp or Wp. LMP-1 and LMP-2A genes are transcribed from a bidirectional promoter. LMP-2B is transcribed from a promoter just upstream of the LMP2A initiation site and contains an additional exon near its 5' end. TR = terminal repeats; IR = internal repeats; OriP and OriLyt are replication origins used during latent and lytic infection, respectively. RNA transcripts are shown in blue; vertical lines show 5' ends and arrows show 3' ends. Black parts of lines show introns; open reading frames are shown in orange.

Epstein–Barr virus expresses a limited set of proteins in latently infected B lymphocytes

Two sets of viral proteins are produced in latently infected B lymphocytes: *Epstein–Barr nuclear antigens* (EBNAs) and *latent membrane proteins* (LMPs). Six distinct EBNAs (EBNA-1, -2, -3A, -3B, -3C, and LP) are made from mRNAs differentially spliced from extraordinarily long primary transcripts that initiate near Ori-P and can extend halfway around the 180-kb genome of Epstein–Barr virus (Figure 24.10). EBNA-LP is made from a multiply-spliced transcript whose exons are coded within the multiple repeats of IR1; the other five EBNAs are made from coding regions downstream of IR1. Three LMPs (LMP-1, -2A, and -2B) are made from coding regions extending in either direction from the terminal repeat element of the circularized genome.

Three patterns of gene expression have been detected in infected B-lymphocytes. These are termed latency I, II, and III (Figure 24.11). Primary infection of B lymphocytes results in latency III; the six EBNAs, three LMPs, and several small RNAs coded by the virus genome are expressed in these cells. Under these conditions, B lymphocytes undergo rapid cell division, and the B-lymphocyte population expands. Cultured human B lymphocytes immortalized by infection with Epstein–Barr virus *in vitro* also exhibit the latency III pattern of gene expression.

The expression of viral proteins on the surface of infected B lymphocytes leads the infected host to produce cytotoxic T lymphocytes (see Chapter 34) that attack and kill most of the infected B-lymphocytes. The expansion of the B-lymphocyte population and their killing by cytotoxic T lymphocytes can lead to the symptoms of infectious mononucleosis. Some of the surviving infected B-lymphocytes have a latency II expression pattern. These cells express only EBNA-1, the three LMPs, and the small viral RNAs. Cells derived from nasopharyngeal carcinomas also show the latency II gene expression pattern.

Latently infected B lymphocytes eventually exit the cell cycle and differentiate into long-lived memory B lymphocytes with a latency I gene expression pattern. In these cells, only the small viral RNAs and, in some cases, EBNA-1 are expressed. By limiting expression of highly immunogenic viral proteins, the virus is able to avoid immune surveillance and therefore can survive within these latently infected cells. In humans infected with Epstein–Barr virus, approximately one to ten memory B lymphocytes per million B lymphocytes in the blood harbor viral DNA. Cells derived from Burkitt's lymphomas also show the latency I gene expression pattern.

Occasionally, latently infected cells undergo a productive replication cycle as they traffic in the oral cavity and differentiate into plasma B cells. The released virus can infect the oral epithelium, giving rise to infectious virus that can spread to other individuals via saliva. Most latently infected individuals continually produce small amounts of virus throughout their lifetime.

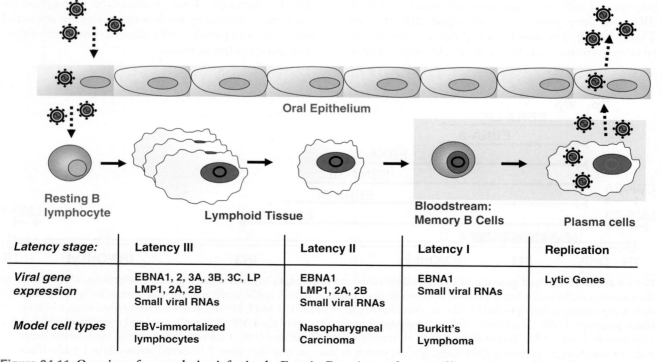

Latency stage:	Latency III	Latency II	Latency I	Replication
Viral gene expression	EBNA1, 2, 3A, 3B, 3C, LP LMP1, 2A, 2B Small viral RNAs	EBNA1 LMP1, 2A, 2B Small viral RNAs	EBNA1 Small viral RNAs	Lytic Genes
Model cell types	EBV-immortalized lymphocytes	Nasopharyngeal Carcinoma	Burkitt's Lymphoma	

Figure 24.11 Overview of events during infection by Epstein–Barr virus and stages of latency. See text for details.

Epstein–Barr virus nuclear antigens direct limited replication of the viral genome and activate viral and cellular genes

EBNA-1 binds to multiple sites within oriP and directs replication of the viral genome, which usually replicates only once per cell division during latent infections. EBNA-1 also acts to segregate the daughter genomes into each of the two progeny cells during cell division, presumably by binding viral DNA to cellular chromosomes. EBNA-1 binding within oriP also enhances expression of the latency genes.

EBNA-2 and EBNA-LP together are responsible for transcriptional activation of viral LMP genes and a number of cellular genes. EBNA-2 carries out this activation by binding to the promoters of viral or cellular genes via a cellular DNA-binding protein called RBP-Jk. EBNA-LP binds to EBNA-2 and enhances its activation activity, as do a number of general cell transcription factors, leading to recruitment of RNA polymerase II to these genes and their transcription. One of the major target cellular genes activated is c-myc, which in turn activates many other cellular genes and can lead to unchecked cellular growth and division. Interestingly, most Burkitt's lymphomas have permanently activated c-myc genes due to translocation of the c-myc gene to actively transcribed regions. During latency II, EBNA-2

expression is strongly reduced, leading to loss of expression of the viral LMP genes and the cellular genes that promote cell growth and division.

Latent membrane proteins mimic receptors on B lymphocytes

LMP-1 is an integral membrane protein that mimics the cellular protein CD40 (Figure 24.12a), which normally responds to the presence of the CD40 ligand (CD40L). CD40L is typically expressed on activated T lymphocytes and provides a key costimulatory signal for B-lymphocyte maturation when combined with signals through the B-cell receptor (BCR). Signaling by CD40 turns on a cascade of events leading to activation of the important cellular transcription factor, NFκB. This is one of the steps that lead to B-lymphocyte proliferation during an immune response. LMP-1 constitutively activates NFκB by binding to several intermediary proteins, including TRAFS and TRADD, whose binding is normally dependent on CD40L-activated CD40.

LMP-2A also mimics a cellular receptor, but in this case it is the B-cell receptor (BCR), which when paired with LMP-1 can drive resting B lymphocytes to proliferate and differentiate into memory B cells. LMP-2A binds to the cellular tyrosine kinases Lyn and Syk, as does an activated B-cell receptor (Figure 24.12b). These

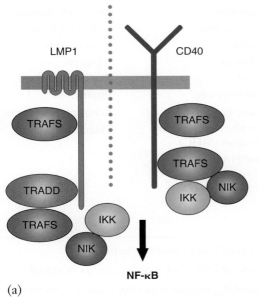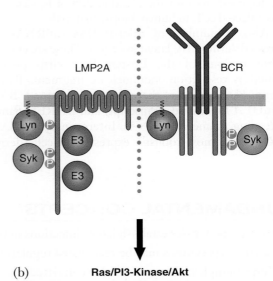

Figure 24.12 LMP-1 and LMP-2A mimic functions of normal B-lymphocyte proteins. (a) LMP-1 (left) resembles a constitutively active CD40 receptor (right) by stably interacting with proteins normally found bound to an activated CD40 receptor. These include tumor necrosis factor receptor-associated factors (TRAFs), tumor necrosis factor receptor-associated death domain protein (TRADD), inhibitory κB kinase (IKK), and NFκB inducing kinase (NIK), leading to activation of NF-κB. (b) LMP-2A (left) mimics the functions of an activated B-cell receptor (BCR; right). The B-cell receptor, consisting of surface immuglobulin, associates with membrane proteins that recruit tyrosine kinases such as Lyn and Syk following activation. LMP-2A binds to these two kinases and to E3 ubiquitin ligases (E3), all of which are important in modulating LMP-2A signaling. The combined signaling of LMP-1 and LMP-2A drives B-cell proliferation and maturation, allowing Epstein–Barr virus to establish a long-lived latent infection in memory B lymphocytes.

kinases are activated when the B-cell receptor binds to an antigen, leading to a cascade of events that eventually generates plasma cells, which produce specific antibodies (see Chapter 34). LMP-2A has been shown to block signaling through the B-cell receptor, perhaps by sequestering Lyn and Syk and therefore preventing their activation; one result is blockage of the activation of Epstein–Barr virus lytic replication. LMP-2A also allows activated B lymphocytes to survive (anti-apoptosis) by activating the Ras/PI3-Kinase/Akt pathway (Figure 24.12b).

Small, untranslated viral RNAs expressed during latent infections target host defense mechanisms

Two additional classes of viral genes are expressed during latency, but their products are RNAs that do not code for proteins. The Epstein–Barr encoded-RNAs (EBERs) are two small, untranslated RNAs that are transcribed by RNA polymerase III and resemble adenovirus VA RNAs (Chapter 23). As do VA RNAs, EBERs may be involved in modulating the cellular interferon response by inhibiting activity of PKR, a protein kinase that can interrupt protein synthesis. EBERs may also provide a survival function during latent infection by acting on cellular proteins involved in protein synthesis such as L22 and La, thereby enhancing cell growth and division. Since EBERs are abundantly expressed in latently infected cells, their presence is useful as a diagnostic tool for the identification of Epstein–Barr virus-infected cells in human biopsy material.

Also, a number of **microRNAs** (miRNAs) are expressed during EBV latent infection. These are encoded within regions of the Epstein–Barr virus genome defined by restriction endonuclease fragments BamHI A and H, and have therefore been named BART (*BamHI A Rightward Transcript*) and BHRF1 (*BamHI H Reading Frame 1*) miRNAs. Interestingly the BART miRNAs are most robustly expressed in cells derived from nasopharyngeal carcinomas, whereas the BHRF1 miRNAs are not detected in nasopharyngeal cells but are expressed in B lymphocytes.

The targets that have been identified for the miRNAs are diverse and include viral genes such as DNA polymerase and LMP-1. Cellular targets include a number of genes that are important for host immune responses and cell survival. These include stress-induced natural killer (NK) cell ligand (MICB), which is important for recognition and killing by NK cells; interferon-inducible T-cell attracting chemokine (CXCL-11/I-TAC), which regulates the host interferon response; and p53 up-regulated modulator of apoptosis (PUMA), which is a pro-apoptotic protein. By repressing the expression of these viral or host proteins at appropriate times in the viral life cycle, the virus may gain an advantage in establishing latency and persisting in humans.

KEY TERMS

Apoptosis	Nectin
B lymphocyte	NFκB
Complement	PKR
Concatemer	Processivity
Cyclin-dependent kinase	Proteasomes
Dorsal root neurons	Proteoglycans
Encephalitis	Retrograde transport
Heparan sulfate	Rolling circle mechanism
Hepatitis	T lymphocyte
Interferons	Tegument
Latency	Topoisomerase
Latent infection	Tropism
Lumen	Tumor necrosis factor
Major histocompatibility complex (MHC)	Ubiquitin ligase
microRNA	

FUNDAMENTAL CONCEPTS

- Most herpesviruses establish latent infections in their hosts after an initial primary infection.
- Latent herpesviruses may be reactivated regularly, as in cold sores, or only after many years, as in shingles.
- Herpes simplex virus genomes have inverted repeat sequences flanking unique sequence elements, as well as direct terminal repeated sequences. The inverted repeat sequences contain a small number of viral genes, which are therefore present in two copies per genome.
- The herpesvirus capsid is surrounded by an amorphous region called "tegument" that contains numerous viral proteins that function upon entry of the virion into the cell.
- Herpesviruses enter the cell by fusion of their envelope with cell membranes; attachment and fusion are directed by several envelope glycoproteins.
- After fusion, herpesvirus nucleocapsids are transported to the cell nucleus where they release their viral DNA into the nucleus via the nuclear pores.

- Viral DNA is circularized once it enters the nucleus.
- All herpesvirus mRNAs are made by cellular RNA polymerase II; transcription factors regulate the transcriptional program into several temporal classes called α (immediate early), β (early), γ_1 and γ_2 (late).
- A tegument protein called Vhs degrades host cell as well as viral messenger RNAs.
- α gene products regulate viral and cellular gene expression.
- β gene products direct viral DNA replication.
- A rolling circle mechanism produces multimeric concatemers of viral DNA that are cleaved into genome-length fragments during packaging into preformed capsids.
- Herpesvirus nucleocapsids appear to bud through the inner nuclear membrane to acquire an envelope, but the precise mechanism by which they acquire their final envelope and exit the cell is uncertain.
- A number of herpesvirus gene products suppress host cell defenses, including apoptosis.
- Herpes simplex viruses establish latent infections in neurons, where they are transcriptionally silent except for some RNAs that represent unusually stable introns.
- Epstein–Barr virus causes mononucleosis and, in certain regions of the world, malignant lymphomas.
- Epstein–Barr virus infects B lymphocytes and establishes a variety of types of latent infections.
- Epstein–Barr virus encodes two small RNAs, transcribed by cellular RNA polymerase III, which inhibit PKR and enhance cell growth.

REVIEW QUESTIONS

1. Explain why only 25% of the genomes of herpes simplex virions are identical. Are all viral genomes infectious?

2. Herpesvirus virions contain a large amorphous mass between the envelope and the capsid. What is this material and what is its composition?

3. For most viruses, packaging the viral genome is under tight control. How does herpes simplex virus ensure packaging of full-length genomes?

4. What are the two receptors that Epstein–Barr virus uses to infect B cells?

5. How is the cellular gene c-myc targeted by Epstein–Barr virus?

Baculoviruses

Eric B. Carstens

Baculoviridae

From Latin *baculum* (stick), referring to shape of virion

VIRION

Enveloped rod-shaped nucleocapsid.

Diameter of nucleocapsid 30-60 nm, length 300 nm.

Material between the capsid and the envelope is called "tegument".

Two types of virion are made: budded and occlusion-derived virions.

GENOME

Circular, double-stranded DNA, 80–180 kb.

Arranged in virions as a cylindrical core.

Condensed with a small basic virus-coded protein.

GENES AND PROTEINS

Autographa californica MNPV codes for over 150 proteins.

Most mRNAs are unspliced.

Genes are expressed in four temporal classes:

- Immediate early: *activation of early genes*
- Early: *DNA replication, activation of late genes (viral RNA polymerase)*
- Late: *assembly of budded virions (capsid and envelope proteins)*
- Very late: *assembly of occluded virions (polyhedrin)*

VIRUSES AND HOSTS

Four genera:

- Alphabaculovirus: *infect lepidopterans (moths and butterflies), enclose virions in polyhedra*

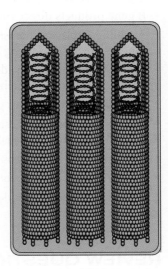

- Betabaculovirus: *infect lepidopterans, enclose virions in granules*
- Gammabaculovirus: *infect hymenopterans (sawflies, wasps, bees, and ants)*
- Deltabaculovirus: *infect dipterans (flies, mosquitoes)*

DISEASES

Important pathogens of commercial silkworms and larvae of butterflies, moths, and other insects.

DISTINCTIVE CHARACTERISTICS

Occlusion bodies contain virions in a form that allows insect-to-insect transmission.

Used as biological insecticides.

Widely used for high-level expression of foreign proteins including vaccine production.

Currently being investigated as gene therapy vectors.

Insect viruses were first discovered as pathogens of silkworms

Throughout the ages, economic pressures have helped to establish the direction of basic research. Before viruses were recognized as infectious agents, scientific efforts were focused on establishing a cause for a particular disease. Following the arrival of silk production in Europe in the thirteenth century, most likely as a result of smuggling silkworms out of China, the rearing of silkworms became a huge industry, especially in Italy, and outbreaks of disease in silkworms were financially devastating to the silk industry. Early accounts of silkworm diseases are found in both ancient Chinese and Western literature. Marco Vida of Cremona, an Italian bishop of the sixteenth century, described in poetic form an infection of silkworm that is reminiscent of baculovirus pathogenesis. In 1856, A. Maestra and Emilio Cornalia independently observed polyhedral bodies in silkworm larvae and made the connection between the presence of polyhedra and the symptoms in the diseased silkworms.

Stanislaus Von Prowzek showed in the early 1900s that jaundice of the silkworm is caused by a filterable agent, but he believed the agent to be a bacterium and the occlusion bodies to be by-products of the disease. Shortly thereafter, R. Glaser and J. W. Chapman (in gypsy moths) and C. Acqua (in silkworms) independently established the viral etiology of polyhedrosis. In 1943, Glaser and Wendell Stanley concluded, based on biochemical analysis, that polyhedra contain nucleoproteins. In 1947, Gernot von Bergold identified virus-like, rod-shaped particles in polyhedral bodies by electron microscopy, and subsequently found that they contained nucleic acids and proteins. These viruses were named baculoviruses because of their shape (*baculo* means "stick" or "rod" in Latin).

Baculoviruses are used for pest control and to express eukaryotic proteins

Baculoviruses replicate only in invertebrates, primarily insects of the order *Lepidoptera* (butterflies and moths). Other insects, including mosquitoes (*Diptera*), sawflies (*Hymenoptera*), and caddis flies (*Trichoptera*), can be infected by baculoviruses (Table 25.1). Baculovirus-like particles have been found in shrimp (*Decapoda*), but there are no equivalent baculoviruses that infect vertebrates, plants, bacteria, or archaea. The important role played by baculoviruses in insect diseases suggested that they might be harnessed to control insect pests. Rather than a disease agent that needed to be eradicated, baculoviruses were recognized as potentially beneficial.

The first practical application of an insect virus is credited to Edward Steinhaus, for the control of the alfalfa caterpillar in California in the 1940s. In 1950, the Insect Pathology Research Institute in Sault Ste-Marie, Ontario, successfully controlled sawfly populations with applications of baculovirus. Animal tests demonstrated that mice and guinea pigs were unaffected by exposure to baculoviruses; experimental or natural exposure of humans to baculoviruses caused no adverse health effects or signs of infection. For example, a survey of naturally occurring baculoviruses on market cabbage demonstrated that over 110 million virus-containing polyhedra might be consumed in one serving of coleslaw. Therefore, in contrast to research on animal viruses, where the goals are to cure or prevent infection, research on baculoviruses eventually focused on enhancing viral pathogenesis and virulence.

Much of the earlier work on baculoviruses centered on their potential use as species-specific insecticides. Each virus was named (using the Latin binomial) according to the insect from which it was isolated, leading to the description of hundreds of different baculoviruses isolated from as many species of insects (see Table 25.1). However, few of these viruses have been studied in detail. Exceptions are viruses that infect insects of some economic importance and that were readily adapted to cell culture. One virus in the latter group was originally isolated from an alfalfa looper (*Autographa californica*) and was named *Autographa californica* multicapsid nucleopolyhedrovirus (AcMNPV). Try to say that before brushing your teeth in the morning!

AcMNPV replicates well in cells from the cabbage looper (*Trichoplusia ni*) and the fall armyworm (*Spodoptera frugiperda*, Sf) cultured *in vitro*. Since the late 1970s, the AcMNPV-Sf virus-cell system has become the model for the study of baculoviruses. The results of basic research with AcMNPV led to a renewed interest in using genetically modified baculoviruses for insect pest control, as well as the use of baculoviruses and insect cells as an efficient eukaryotic gene expression system. As more detailed information has accumulated about baculovirus gene structure and function, baculoviruses have also emerged as important potential vectors for human gene therapy.

Baculovirus virions contain an elongated nucleocapsid

Virions of baculoviruses consist of rod-shaped nucleocapsids 30 to 60 nm in diameter and 250 to 300 nm in length, surrounded by a lipid-containing envelope (Figure 25.1). The space between the envelope and the nucleocapsid is denoted the **tegument**. The nucleocapsid contains a single molecule of circular, double-stranded DNA arranged as a cylindrical core. The two ends of the nucleocapsid are capped asymmetrically, giving polarity to the virion.

Table 25.1 **Some representative baculoviruses**

Genus	Virus	Host
Alphabaculovirus	*Autographa californica* multinucleocapsid nucleopolyhedrovirus (AcMNPV)	Alfalfa looper
	Bombyx mori nucleopolyhedrovirus (BmNPV)	Silkworm
	Lymantria dispar multinucleocapsid nucleopolyhedrovirus (LdMNPV)	Gypsy moth
Betabaculovirus	Cydia pomonella granulovirus (CpGV)	Codling moth
Gammabaculovirus	Neodiprion lecontei nucleopolyhedrovirus (NeleNPV)	Redheaded pine sawfly
Deltabaculovirus	Culex nigripalpus nucleopolyhedrovirus (CuniNPV)	Mosquito

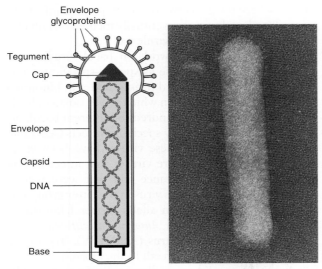

Figure 25.1 Budded baculovirus virion. At right, an electron micrograph of a negatively stained *Autographa californica* MNPV budded virion. At left, a diagram showing structural elements.

The core contains a small, highly basic protein, P6.9 (also known as Ac100), which is bound to the DNA. This protein likely helps to condense the genome DNA during packaging into capsids. The nucleocapsid contains a major capsid protein, VP39 (Ac89). Two other proteins, PP78/83 and VLF-1, together may form a capsid end structure. Capsids contain at least eight other proteins of unknown functions. The tegument includes at least one identified protein, GP41 (Ac80).

One of the consequences of having a tubular nucleocapsid is the potential for packaging DNA longer than the viral genome. Virions of various lengths have been observed by electron microscopic examination of AcMNPV mutants that produce genomes of different sizes. This characteristic has been used to incorporate extra DNA into virions when using baculovirus as an expression vector. However, there is no information on how the length of the nucleocapsid is regulated during production in the nucleus.

Baculoviruses produce two kinds of particles: "budded" and "occlusion-derived" virions

The baculovirus infectious cycle is unique because two types of virions are produced: **budded virions** and **occlusion-derived virions**. Budded virions (Figure 25.1) consist of single nucleocapsids surrounded by an envelope. They are formed by budding of nucleocapsids through the plasma membrane, which is modified by incorporation of a virus-encoded fusion protein. In contrast, occlusion-derived virions (Figure 25.2) are formed in the nucleus where they become enveloped

and are then occluded in a crystalline protein matrix. Two kinds of **occlusion bodies** have been described: polyhedra 0.15 to 15 μm in diameter, which can contain many virions, are made by nucleopolyhedroviruses. Ovocylindrical occlusion bodies 0.3 × 0.5 μm and containing only one virion are made by granuloviruses.

The occlusion bodies consist largely of a single viral-encoded polypeptide called **polyhedrin** or **granulin**. Polyhedrins and granulins from different baculoviruses have highly conserved amino acid sequences. The occlusion bodies are very characteristic of baculovirus infections and represent an obvious cytopathic effect easily seen by light microscopic examination of infected insect cells.

Because occluded and budded virions are enveloped at distinct sites in the cell, there are differences in the protein composition of the two kinds of virions. For example, the tegument of occluded virions contains a 41-kDa glycoprotein, GP41, not present in budded virions; however, GP41 may be involved in the egress of nucleocapsids from the nucleus prior to the formation of budded virions. Occlusion-derived virions initiate infection in the insect gut epithelium (see below).

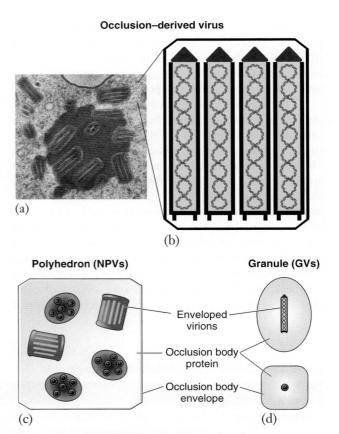

Figure 25.2 Occluded baculoviruses. (a) Electron micrograph of a thin section through a polyhedron in which baculovirus particles are occluded. (b) A diagram showing occluded virions. (c) and (d) Diagrams showing a polyhedron (nucleopolyhedrovirus) and a granule (granulovirus) with occluded virions.

The virion envelope has a trilaminar structure typical of lipoprotein unit membranes. The loose-fitting envelope of budded virions of AcMNPV (Figure 25.1) contains terminal spikes composed of a single virus-coded glycoprotein, GP64. This protein is required for infection of cell cultures via fusion of the viral envelope and the cell membrane. Occlusion-derived virions are several hundred-fold less infectious in cultured cells than budded virions, and do not contain GP64. However, at least six other viral proteins are known to be associated with the envelope of occlusion-derived virions. These differences in envelope composition reflect the different target cells that budded and occlusion-derived virions infect in the insect.

Baculoviruses have large, circular DNA genomes and encode many proteins

The baculovirus genome is a covalently closed circular, double-stranded DNA molecule. Most physical maps are now compared with that of AcMNPV (Figure 25.3), whose complete nucleotide sequence (133,894 bp) has been determined. Based on analysis of open reading frames and transcriptional promoters, it has been estimated that AcMNPV encodes about 152 proteins. Some of the immediate early genes including *ie-0*, *ie-1*, *ie-2*, and *pe38* are clustered within a 7-kbp region, but there do not appear to be other significant groupings of genes by function or time of expression.

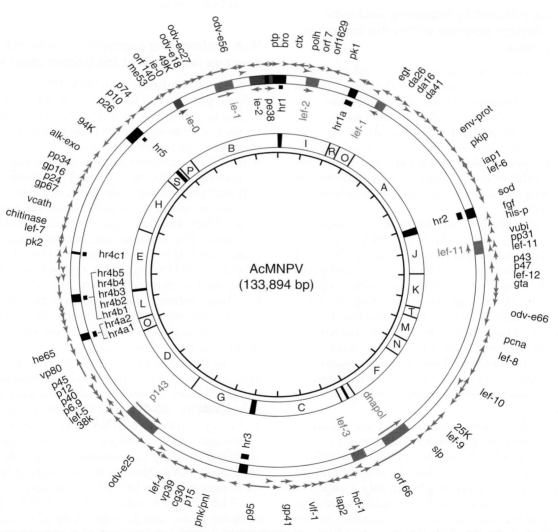

Figure 25.3 Genetic map of *Autographa californica* MNPV. The inner circle shows the locations of *Eco*RI restriction sites; the 20 largest *Eco*RI fragments are labeled alphabetically, according to size. The locations of homologous repeat sequences (hrs) usually associated with multiple *Eco*RI sites are labeled on the outer circle and shown in black. The locations of genes essential for viral DNA replication are shown in orange, and immediate early genes are shown in green, in the same circle. Blue arrows in the periphery show the location and orientation of identified coding regions. Known genes are labeled.

The sequences of 51 other baculovirus genomes, ranging from 81 to 179 kbp, are also known. Baculovirus genomes carry repeated nucleotide sequences (hrs) that are transcriptional enhancers and may also function as initiators of DNA replication. Computer analysis of baculovirus genome sequences revealed a common set of about 29 homologous core genes that are present in all species. These genes are probably the minimum set required for virus replication and form a pool that are the essence of what constitutes the baculovirus family. However, other genes are unique to particular species within the family; these genes likely determine individual species adaptation including host range or tissue tropism, virulence factors, and differences in virus morphology.

Insects are infected by ingesting occlusion bodies; infection spreads within the insect via budded virions

In natural populations, insects become infected with baculoviruses by feeding on cadavers of infected insects or on leaves contaminated with occlusion bodies. The occlusion bodies dissolve in the alkaline conditions of the insect gut (the acid pH of the human gut does not dissolve occlusion bodies). The released virions pass through the **peritrophic membrane** and invade the columnar epithelial and regenerative cells of the midgut by attachment and fusion of the viral envelope with the cell **microvilli**.

Once inside the cell, nucleocapsids are transported to nuclear pores. Either the viral DNA is released directly into the nucleus, or the nucleocapsids first enter through the nuclear pore and then are uncoated in the nucleus. The eclipse phase of the replication cycle follows, resulting in the appearance of a ring-like **virogenic stroma** in the nucleus, and production of viral nucleocapsids. These nucleocapsids exit the nucleus and travel to the basal cell plasma membrane, where they form virions by budding. These virions subsequently infect the midgut connective tissue, tracheal epidermal cells and tracheoblasts, initiating a second round of replication.

The budded virions produced by these infections are released into the **hemocoel** and spread to infect other tissues including hemocytes, fat body, and epidermis. In these tissues, infection initially results in release of budded virus but there is subsequently a switch in virus production, carried out by an unknown mechanism. Assembled nucleocapsids are retained in the nucleus of infected cells where they become associated with membranes, eventually becoming enveloped either singly or as groups of multiple nucleocapsids. In nucleopolyhedroviruses, these enveloped virions then become associated in the nucleus with crystallizing polyhedrin protein, forming the distinctive occlusion bodies or polyhedra associated with infection by these viruses. Maturation of the polyhedra involves the deposition of an outer envelope (**calyx**) over the surface of the occlusion body, which may be facilitated by fibrillar structures composed of a very late gene product called P10.

The end result of virus infection is death and liquefaction of the insect from two to nine days after infection, depending on virus dosage and insect host. The final stages of granulovirus infection are different and not as well understood as for nucleopolyhedroviruses. During granulovirus infection, the cytoplasm merges with the nucleus and a distinct nuclear membrane structure is no longer present. Normally, single granulovirus virions are occluded within small occlusion bodies called granules. It is not clear if an outer envelope (calyx) is present on granules.

Viral proteins are expressed in a timed cascade regulated at the transcription level

In Sf21 cells infected with AcMNPV, several virus-induced proteins can be distinguished by 3 hours after infection, well before viral DNA synthesis begins. The number of virus-specific proteins synthesized dramatically increases beginning at about 6 hours post-infection, and the synthesis of host cell proteins gradually decreases from that time onward. By about 24 hours, only viral proteins are made, indicating that baculovirus infection results in the shutoff of host cell protein synthesis. These kinds of experiments defined the temporal phases of baculovirus infection: the immediate early phase is from 0 to 3 hours; the early phase is from 3 to 6 hours; viral DNA synthesis begins between 6 and 8 hours; the late phase is from 6 to 24 hours; and the very late phase continues until the death of the cell. Different viral proteins are synthesized during these specific time frames, although some viral proteins expressed early continue to be synthesized late.

The timing and levels of baculovirus gene expression are transcriptionally controlled. Immediate early and early genes are transcribed by the host cell RNA polymerase II, but late and very late genes are transcribed by a virus-coded RNA polymerase. These RNA polymerases can be distinguished based on their differential susceptibility to α-**amanitin**, a specific inhibitor of cellular RNA polymerase II; α-amanitin inhibits early transcription but not late or very late transcription.

The RNA polymerases recognize DNA promoter elements in the viral genome. Many of the early genes carry promoters and transcription start sites that mimic normal host insect gene promoters. These include TATA-box-like elements and initiator motifs related to the sequence ATCA(G/T)T(C/T). However, the late and very late AcMNPV gene promoters usually contain the motif (A/G/T)TAAG, which serves as the transcription initiator for the virus-coded RNA polymerase.

Immediate early gene products control expression of early genes

The viral genes responsible for temporal control of baculovirus gene expression were discovered by using reporter genes whose transcription was driven by specific baculovirus promoters. Insect cells were cotransfected with plasmids containing a variety of specific viral DNA fragments and a reporter plasmid driven by an early promoter. One of the viral DNA fragments enabled transactivation of the reporter plasmid, and was found to contain an immediate early viral gene called *ie-1*. The IE-1 protein is responsible for transactivating other early genes and is considered the major regulator of early baculovirus gene expression. Other immediate early baculovirus genes, including *ie-2*, *pe38*, and *ie-0*, were identified based either on their ability to transactivate other viral genes or on the time of their expression in infected cells. IE-0 is the only baculovirus protein known to be made as a result of RNA splicing; a small exon located about 4 kb upstream of the *ie-1* start codon is spliced to the *ie-1* open reading frame, producing a protein identical to IE-1 except for an additional 54 amino acids at its N-terminus.

Early gene products regulate DNA replication, late transcription, and apoptosis

A large number of early baculovirus proteins are made. These are responsible for three major activities: replication of viral DNA, expression of late and very late viral genes, and modification of cell and host physiology via cellular signaling pathways. Much remains to be learned about the detailed functions of these proteins. Table 25.2 lists some early viral proteins known to be involved in viral DNA replication. It is notable that baculoviruses make their own DNA polymerase, as do most large DNA viruses (including adenoviruses, herpesviruses, and poxviruses), even though baculoviruses replicate their DNA in the cell nucleus where cellular DNA polymerases are located. The exact mechanism of baculovirus DNA replication is not well understood, although recent data suggest that homologous recombination may be involved.

DNA replication. Baculoviruses contain multiple copies of highly homologous DNA sequences located in several different specific regions of the genome. In AcMNPV, these homologous regions (hrs, see Figure 25.3) consist of multiple copies of a 60 to 70 bp repeat carrying an imperfect inverted repeat sequence of 30 bp

Table 25.2 Early *Autographa californica* MNPV proteins involved in DNA replication

Protein	Mol. wt. (kDa)	Function
LEF-1	31	Primase for DNA replication
LEF-2	24	Primase accessory factor, associates with LEF-1
LEF-3	45	Single-stranded DNA binding protein/ helicase transporter
P143	143	Helicase, host range factor, transcription factor
IE-1	67	May bind to replication origins
DNAPOL	114	DNA polymerase

with an *Eco*RI restriction site at its center. Several groups have shown that hrs act as transcriptional enhancers and as initiators of DNA replication, at least in transient assays in cell cultures.

Replication assays rely on the ability of the restriction endonuclease *Dpn*I to cleave methylated DNA but not unmethylated DNA. Plasmid DNA produced in bacteria is methylated, but when it is introduced into eukaryotic cells and undergoes replication, the DNA is no longer methylated. Therefore, *Dpn*I can be used to distinguish between input plasmid DNA (methylated) and progeny unmethylated DNA. Plasmids carrying either AcMNPV hrs or early promoter regions have been shown to replicate by this assay when transfected into insect cells that are subsequently infected with AcMNPV. Therefore, there is a strong link between induction of early viral transcription and initiation of viral DNA replication.

Apoptosis. Baculoviruses also code for proteins that inhibit **apoptosis** (programmed cell death). These inhibitors are divided into two major groups called P35 and inhibitors of apoptosis (IAP). P35 has only been found in AcMNPV and its close relatives, whereas *iap* genes are apparently present in most baculoviruses and they are often present as multiple copies. The expression of these genes inhibits the normal process of apoptosis, allowing for greater production of budded virus and development of the very late stage of virus infection.

Acquired cellular genes. A number of baculovirus-coded proteins such as the viral proliferating cell nuclear antigen (PCNA), **ubiquitin** (vUBI), and superoxide dismutase (SOD) share a high degree of amino acid sequence similarity with homologous cellular genes. This suggests that these genes have been acquired during evolution by the virus from the host cell genome. Although these genes may not be essential for virus replication, they likely enhance some stage of virus multiplication in their natural hosts. In addition, some baculoviruses have amplified copy numbers of certain genes, or carry additional unique genes that are not similar to any other known genes. This partly accounts for the variety of baculovirus genome sizes.

Late genes are transcribed by a novel virus-coded RNA polymerase

Once DNA replication begins, another set of baculovirus genes is expressed. The late genes code for viral proteins involved in assembly of budded virions, while the very late genes are involved in the formation of occluded virions. The virus-coded RNA polymerase is responsible for transcription of late and very late genes. Additional sequences downstream of the late initiator motif appear to be responsible for the high transcriptional activity of very late genes such as polyhedrin and P10. The product of the *vlf-1* gene, very late factor, is involved in the high-level expression of very late genes.

The development of *in vitro* transcription assays led to the identification of viral proteins required for transcription from late and very late promoters (Table 25.3). The viral RNA polymerase contains four virus-coded subunits. These proteins were also implicated in the regulation of late and very late transcription by transient *in vivo* expression assays (Figure 25.4). A chloramphenicol acetyl transferase reporter gene under the control of a baculovirus late promoter was transactivated by a combination of viral genes expressed on different transfected plasmids. This assay is based on a set of overlapping clones (A through M) that together represent the entire AcMNPV genome. Omission of certain clones (B, C, D, E, F, and I) had no effect on reporter gene activity (shown by presence of a band of acetylated chloramphenicol, labeled AC). However, omission of other clones (A, G, H, J–M) resulted in the absence of chloramphenicol acetyl transferase activity (band representing unmodified chloramphenical, labeled C). Therefore, these clones must code for factors essential for transcription of late genes.

By substitution of individual clones with smaller fragments, a total of 19 specific genes (late expression factors or *lefs*) were identified that were capable of transactivating the expression of late but not early promoters. Proteins coded by these genes include many of those shown in Tables 25.2 and 25.3. A library of plasmids, each carrying one of these genes, is sufficient to support expression of a transfected reporter gene under the control of an AcMNPV late or very late promoter.

Table 25.3 *Autographa californica* MNPV proteins that regulate late transcription

Protein	Mol. wt. (kDa)	Function
P47	47	Viral late RNA polymerase subunit
LEF-4	54	Viral late RNA polymerase subunit, RNA capping
LEF-8	102	Viral late RNA polymerase subunit
LEF-9	59	Viral late RNA polymerase subunit
LEF-7	27	Stimulatory for transient DNA replication
LEF-5	31	Transcription initiation factor?
LEF-6	20	mRNA transporter?
LEF-10	9	?
LEF-11	13	Essential for DNA replication *in vivo*
LEF-12	21	May not be essential when all other genes are present
39K/PP31	31	Phosphorylated DNA-binding protein?

As discussed above, several late expression factors are also required for replication of transfected reporter plasmids, supporting the hypothesis that DNA replication is required for efficient late gene expression. It is not known whether proteins required for viral DNA replication also play some role in regulating late or very late gene transcription. Once viral structural proteins are synthesized, assembly of nucleocapsids occurs in the nucleus.

Baculoviruses are widely used to express foreign proteins

In the early 1980s, several lines of evidence indicated that the very late gene product, polyhedrin, is not essential for virus replication in cell culture. Viruses missing the *polyhedrin* gene could be easily distinguished from polyhedrin-positive virus by microscopic examination of plaques in cell monolayers. G. Smith and Max Summers came up with the ingenious idea that the very strong polyhedrin promoter could be used to drive the high-level synthesis of a foreign protein by infecting cells with recombinant viruses.

The production of the recombinant virus occurs in two stages (Figure 25.5). First, the foreign gene is cloned adjacent to the polyhedrin promoter in a plasmid

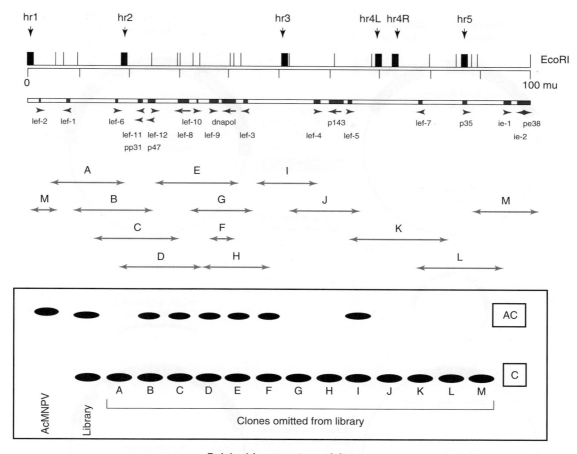

Polyhedrin promoter activity

Figure 25.4 Identification of baculovirus late expression factors (LEFs) by a transient expression assay. Top: locations of hrs superimposed on the EcoRI fragment map. Middle: locations of the *Autographa californica* MNPV late expression factors and a set of overlapping cloned DNA fragments (A–M) used to transfect cells. Bottom: radioactive chloramphenicol is incubated with extracts of transfected cells; acetylated chloramphenicol (AC) is then separated from unacetylated chloramphenicol (C) by thin-layer chromatography. Data adapted from Passarelli and Miller, *J. Virol.* 67, 3481–3488 (1993).

Recombinant baculoviruses as gene therapy vectors

In 1985, Lois Miller's group demonstrated that foreign gene expression from recombinant baculoviruses is cell type-dependent. Baculoviruses can attach to and penetrate non-permissive cells from a wide variety of tissues and species, and the viral genome is transported to the nucleus, but viral promoters are not recognized in these non-permissive cells, so no viral gene expression occurs. However, reporter genes can be expressed in mammalian cells if they are driven by mammalian promoters engineered into the baculovirus genome.

In 1995, Michael Strauss reported the development of a baculovirus vector that could be used for efficient gene transfer into mammalian hepatocytes *in vitro*. Nearly 100% of human cells exposed to baculoviruses can express the desired gene (provided high multiplicities of infection are

used), and expression of transduced genes under the control of mammalian promoters is high.

The advantages of using baculoviruses in human gene therapy include (1) the known capacity of baculoviruses to tolerate large inserts of foreign DNA, (2) the inability of baculoviruses to replicate in vertebrates, (3) a history of biosafety in vertebrates including a lack of toxic or adverse immune responses, (4) ease of manipulation in insect cell cultures, and (5) production of high virus titers. Current efforts in this area are directed at enhancing baculovirus stability in the human body and extending the life span of the viral genome in non-permissive cells. Continued research on baculovirus gene regulation and gene function, and the basis of host range specificity, will provide important information for future applications of baculoviruses in biotechnology.

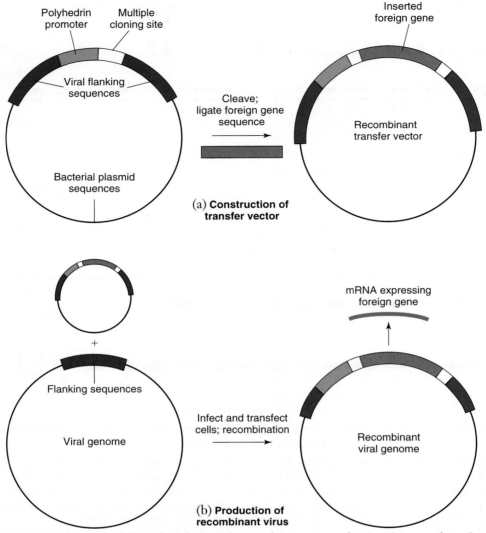

Figure 25.5 **Preparation of a recombinant baculovirus designed to express a foreign gene product.** See text for details.

Recombinant baculoviruses and VLPs for vaccine production

Many viruses construct their capsids and envelopes by spontaneous self-assembly of multiple copies of a small number of distinct protein subunits. Because of this property of self-assembly, expression of viral structural proteins in the absence of the viral genome often leads to the production of artificial virus-like particles (VLPs). These VLPs exhibit excellent antigenic properties and often stimulate strong protective immune responses when used as vaccines. In contrast to inactivated or attenuated virus preparations, VLPs contain no viral genetic information, so they are not infectious and cannot replicate.

Scientists have developed the baculovirus/insect cell expression system to make VLPs corresponding to a variety of human pathogenic viruses. One of the two recently licensed vaccines against oncogenic human papillomaviruses contains VLPs produced in insect cells infected by baculoviruses expressing human papillomavirus L1 capsid proteins. A number of clinical trials using baculovirus/insect cell-derived VLPs corresponding to such diverse viruses as human immunodeficiency virus and influenza virus are currently under way.

transfer vector, replacing the polyhedrin-coding region (step a). Second, insect cells are cotransfected with infectious, wild-type viral DNA and the transfer vector carrying the foreign gene (step b). During replication of the viral DNA, homologous recombination can occur between sequences flanking the polyhedrin gene on the transfer vector and the identical sequences on viral DNA, generating recombinant genomes with the foreign gene in place of the polyhedrin gene. These recombinant genes are incorporated into progeny virions. Recombinant viruses can be easily identified by screening for their polyhedra-negative phenotype. Since their invention, recombinant baculoviruses have become one of the most widely used vehicles for the high-level expression of foreign proteins.

KEY TERMS

α-amanitin	Occlusion body
Apoptosis	Occlusion-derived virions
Budded virions	Peritrophic membrane
Calyx	Polyhedrin
Granulin	Tegument
Hemocoel	Ubiquitin
Microvilli	Virogenic stroma

FUNDAMENTAL CONCEPTS

- The family *Baculoviridae* consists of a large number of different species of rod-shaped enveloped viruses with double-stranded circular DNA genomes varying from 80 to 180 kb pairs.
- Baculoviruses infect only invertebrates, notably insects, with a very high degree of host species specificity.
- Two phenotypes of virions are produced in infected insects:
  1. budded virus, which is responsible for systemic infection of many tissues within one host (horizontal transmission) and
  2. occlusion-derived virus, which is responsible for spreading of the virus from one host to another (vertical transmission).
- Baculoviruses undergo a temporally regulated replication cycle in insect cell cultures, producing both budded and occlusion-derived virus.
- Some baculoviruses are used as effective biological pest control agents, both in agriculture and forestry.
- Baculoviruses are widely used as efficient eukaryotic expression vector systems, taking advantage of the highly active polyhedrin gene promoter to drive foreign gene expression.
- Baculoviruses engineered to express caspid proteins of other viruses can result in the synthesis of large amounts of virus-like particles (VLPs) in infected insect cells. These VLPs have been used as vaccines against viruses such as human papillomaviruses, Ebola virus, and influenza viruses.

REVIEW QUESTIONS

1. What were the first two applications of baculoviruses as insecticides?

2. Baculoviruses produce two types of virus particles; one particle is crystalline-like and able to survive in the environment for extended periods of time. Discuss the composition of this particle.

3. *Autographa californica* MNPV early and late genes are transcribed by different polymerases recognizing different promoter elements. What are the two polymerases and the sequences of the elements they recognize?

4. What are virus-like particles (VLPs) and how were they used to produce the human papillomavirus vaccines?

Poxviruses

Richard C. Condit

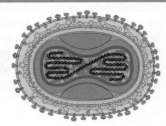

Poxviridae

From English *pocks* (pox), referring to blistering skin lesions

VIRION

Complex, ovoid- or brick-shaped particles, 310 x 240 x 140 nm
Surface ridges or "tubules".
No typical symmetry elements.
Internal core and lateral bodies.
Virions exist in two infectious forms:
- *Mature virus (MV): one lipid membrane*
- *Extracellular virus (EV): two lipid membranes*

GENOME

Linear, double-stranded DNA, 150–250 kb (vaccinia virus: 200 kb).
Covalently closed hairpin ends: no free 3' and 5' ends.
10-kb inverted terminal repeats.

GENES AND PROTEINS

150–250 genes (vaccinia virus: 200).
Each gene has its own transcriptional promoter.
No spliced mRNAs (no introns).
Genes are distributed on both DNA strands.
Genes are expressed in three temporal classes with distinct functions:
- Early: *dissolve core, direct DNA replication and intermediate gene expression, combat host defenses*
- Intermediate: *direct late gene expression, combat host defenses*

- Late: *virion structural proteins, virion transcription enzymes*

VIRUSES AND HOSTS

Two subfamilies:
Chordopoxvirinae: infect vertebrates:
- Humans: *variola, vaccinia (vaccine strain), molluscum contagiosum*
- *Viruses that infect a variety of birds and mammals: monkeys, cattle, etc.*
Entomopoxvirinae: infect insects (beetles, butterflies, flies, etc.).

DISEASES

Smallpox: high fatality rate, now eradicated.
Molluscum contagiosum: the only existing natural poxvirus infection of humans; relatively rare but more common in immunocompromised patients.
Monkeypox, cowpox, tanapox are animal poxviruses that occasionally infect humans.

DISTINCTIVE CHARACTERISTICS

An unusually large and complex virus.
Virus-coded enzymes for transcription and RNA processing are packaged in the virion core.
Early RNAs are made in the intact core and extruded into the cytoplasm.
DNA genome replicates in the cytoplasm using exclusively virus-coded enzymes.
Internal envelope is not formed by budding but is assembled *de novo*.

Smallpox was a debilitating and fatal worldwide disease

Smallpox, caused by the poxvirus variola, embodies simultaneously one of humankind's greatest trials and one of its greatest triumphs. The human disease was called "*small*pox" in fifteenth-century France and England to distinguish it from "large pox", or syphilis, which caused larger lesions (**pocks**) and infected older (larger) people. Once a gruesome and lethal disease worldwide, smallpox has now been eradicated from the planet in a magnificent and unprecedented international collaboration of science and society.

Smallpox plagued humans for at least 3000 years. Scarred Egyptian mummies provide evidence for the existence of smallpox as early as the second millennium B.C. Written records suggest that smallpox was **endemic** in regions of Egypt and India by the first century, from where it spread throughout China, southwestern Asia, Europe, and northern Africa by the tenth century A.D. European exploration and colonization in the sixteenth, seventeenth, and eighteenth centuries brought the disease to North and South America, southern Africa, and Australia, so that by the eighteenth century, smallpox was common in virtually all densely populated areas worldwide. Smallpox influenced the course of history, killing rulers, disabling attacking armies, and decimating native populations infected by advancing conquerors.

The human toll of smallpox is almost unimaginable today. Although in rural areas the incidence of smallpox may have been relatively low, in large cities in seventeenth century Europe smallpox probably behaved like a typical childhood disease, so that virtually every child could be expected to contract it. Smallpox strains varied in mortality between 1–40%, but the average case fatality rate in endemic countries was approximately 25%.

The infection was horrific in its presentation. The disease usually began as a respiratory infection, transmitted by virus contained in nasopharyngeal secretions, pus, or scabs from an actively infected person, or by virus-contaminated bedding or clothing. The infection progressed though an asymptomatic phase lasting approximately 12 days, followed by sudden onset of high fever, splitting headache, and vomiting lasting four days. After a slight improvement, the characteristic smallpox skin rash developed and fever resumed, lasting approximately 12 days. Painful, isolated, pus-filled blisters covered the entire body as well as the mouth and airway, making breathing difficult and eating and drinking virtually impossible. In severe cases, the rash became confluent and/or hemorrhagic.

Most survivors were scarred for life, many were blinded, and some suffered debilitating limb deformities. Comte de la Condamine, an eighteenth-century French mathematician and scientist, wrote: "No man dared to count his children as his own until they had had the disease".

Variolation led to vaccination, which has eradicated smallpox worldwide

Early attempts at smallpox prevention, called **variolation**, took advantage of the fact that infection with variola virus by an unnatural route resulted in a disease of reduced severity and decreased mortality. In China in the tenth century A.D., practitioners pulverized pock scabs from individuals undergoing mild smallpox infections, and blew the powder through a tube into the nostrils of the recipient, resulting in a generalized but nevertheless relatively mild infection. In India at about the same time, variolation was accomplished by seeding freshly scratched skin of a recipient with pus from an active smallpox infection. This cutaneous route of variolation resulted in primary and satellite lesions at the site of inoculation, but a much milder generalized infection, and an overall mortality rate of approximately 1%, significantly lower than the natural infection.

The practice of cutaneous variolation spread through southwest Asia and northern Africa. By the late eighteenth century variolation became well established in Europe and North America, reducing significantly the impact of smallpox where practiced, and setting the stage for the discovery of vaccination at the turn of the nineteenth century.

Immunization against smallpox by inoculation with cowpox was introduced by Edward Jenner, an English physician, in 1796. Jenner's bold experiment represented a synthesis of three concepts, all of which were fairly common knowledge at the time: (1) cutaneous variolation produced resistance to smallpox, (2) milkmaids were rarely pockmarked and often resistant to variolation, and (3) milkmaids often acquired pock-like lesions on their hands, called cowpox disease, presumably from milking cows bearing similar lesions on their teats.

Reasoning that the apparently protective effects of cowpox could be artificially induced using the procedures of variolation, Jenner inoculated eight-year-old James Phipps with cowpox from the hand of a milkmaid, Sarah Nelmes, on May 14, 1796. The procedure excited a limited cowpox disease on Phipps's arm, but otherwise "the indisposition attending it was barely perceptible". Most importantly, six weeks later Jenner challenged Phipps with smallpox by variolation, and found that he was resistant. To distinguish his procedure from variolation, Jenner named it **vaccination** (*vacca* means "cow" in Latin), and within 10 years vaccination against smallpox had been adopted worldwide.

Despite vigorous individual efforts over the next 150 years to control smallpox with vaccination, in 1958 the disease was still endemic in numerous countries in South America, Africa, southwestern Asia, the Indian subcontinent, and China. In that year, the World Health Organization launched the Global Smallpox Eradication Program, a massive, internationally coordinated program of surveillance and vaccination, which ultimately lead to the eradication of smallpox worldwide, certified in 1979. Vaccination for smallpox has now been discontinued, since the small chance of serious complications from vaccination exceeds the negligible risk of contracting smallpox.

Poxviruses remain a subject of intense research interest

Although smallpox has been eradicated, interest in poxviruses persists, and with good reason. Poxviruses encode and package in virions a full complement of enzymes required for viral mRNA synthesis and modification, and thus provide an ideal model system for studying the basics of mRNA metabolism. Poxviruses also encode dozens of proteins that have evolved to confound the host inflammatory and immune responses. This allows researchers to study host responses to viral infections, and may lead to development of useful immunotherapeutic and anti-inflammatory reagents. Foreign genes can be spliced into poxviruses and expressed during virus infection, providing research vectors for eukaryotic gene expression as well as promising recombinant vaccines. Finally, recent fears that smallpox might be used as a biological weapon have spawned renewed interest in development of both improved vaccines and antipoxvirus drugs.

The two subfamilies of poxviruses are based on host range: *Chordopoxvirinae* infect vertebrates and *Entomopoxvirinae* infect insects. Most of the vertebrate poxviruses cause blistering skin diseases, while the insect poxviruses cause a generalized necrosis of infected larvae. Some poxviruses are transmitted among different animal species, resulting in sometimes misleading nomenclature. For example, cowpox virus actually uses rodents as a reservoir host, cattle being only an occasional and accidental host for the virus. A list of some poxviruses is shown in Table 26.1.

Vaccinia virus, used for vaccination against smallpox, is used widely in research laboratories. Because vaccinia virus is the vaccine strain, it is relatively safe to work with and has been studied in clinical and laboratory situations for over 200 years. Although it was originally isolated as a poxvirus that infected cows and/or dairy workers, vaccinia virus is distinct from cowpox and is now found in the wild only as "escaped" vaccine virus that has recently emerged in buffaloes in India and in cattle in Brazil, in both cases resulting in transmission to humans. Vaccinia virus may be a variant of cowpox virus that has drifted significantly during two centuries of artificial propagation, or it may represent a species of poxvirus that has become extinct since its initial isolation.

Linear vaccinia virus genomes have covalently sealed hairpin ends and lack introns

The genome of vaccinia virus is a 200-kb double-stranded DNA molecule that contains 10-kb inverted repeats at each end (Figure 26.1a). Several genes lie within these repeats and are therefore present in two copies per genome. The last 4 kb of the repeats at each end lack coding sequences, but consist of about 200 short tandem sequence repeats of unknown function (Figure 26.1b). The genome, although linear, contains no free 5' and 3' ends; each terminus is sealed in a hairpin structure. The extreme terminal 100 bp exist in two related forms (Figure 26.1c). Each of these regions is imperfectly base paired, and in their unfolded, single-stranded form, they represent inverse complements of each other (Figure 26.1d). The hairpins and immediate flanking sequences have important functions in DNA replication.

Poxvirus genes do not contain introns and therefore viral mRNAs are not spliced. Each gene contains its own transcriptional promoter, and genes are closely packed; intergenic regions larger than 100 bp are rare, and coding sequences sometimes overlap. By convention, genes are numbered sequentially within individual Hind III restriction endonuclease fragments, and designated as being transcribed rightward or leftward (Figure 26.2). Thus gene D8L is the eighth gene from the left end of the Hind III D DNA fragment, and is transcribed in a leftward direction.

Genes near the ends of the genome are usually transcribed toward the nearest end, while genes in the center of the genome can be transcribed in either direction. Approximately 50 "housekeeping" genes, involved in processes common to all poxviruses such as transcription, DNA replication, and morphogenesis, are conserved among the poxviruses and are generally clustered in the center of the genome (Figure 26.2). Genes located near

Table 26.1 Some poxviruses and their hosts

Virus	Reservoir host	Other hosts
Variola (smallpox)	Humans	None
Molluscum contagiosum	Humans	None
Vaccinia	Unknown	Humans, cows, buffaloes
Monkeypox	Squirrels, rodents	Humans, monkeys
Cowpox	Rodents	Humans, cows, cats, zoo animals
Orf	Sheep	Humans, various ruminants
Fowlpox	Birds	Humans (as vaccine vector)
Myxoma	Rabbits	None
Entomopox	Insects	None

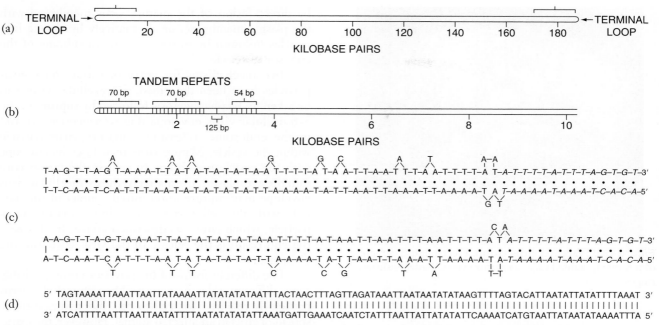

Figure 26.1 Vaccinia virus genome structure. From the top down are shown: (a) the double-stranded DNA genome, showing terminal loops; (b) an enlargement of the inverted terminal repeat, emphasizing the tandem repeats within the terminal repeat; (c) the two hairpin loops (nonitalic font) as they are joined to the terminal repeats (italic font); (d) the duplex structure formed at a concatemer junction, consisting of the unfolded hairpins base-paired to each other.

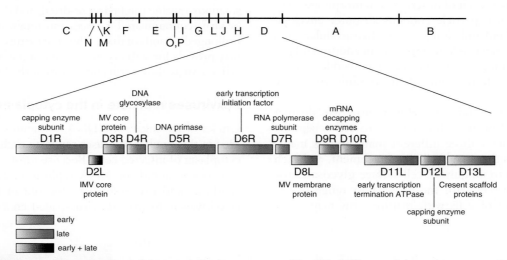

Figure 26.2 Functional organization of the vaccinia virus genome. Top: Hind III map of the entire genome. Below: exploded map of the 16-kb Hind III D fragment. Individual genes are represented as rectangles (direction of transcription: R: rightward, L: leftward). Early and late genes are indicated by shading, and a function for each gene, where known, is indicated.

the termini vary among poxviruses, and they encode functions that are distinct for each virus such as host range and pathogenesis.

Two forms of vaccinia virions have different roles in spreading infection

Vaccinia virions are large ($310 \times 240 \times 140$ nm), enveloped, ovoid or brick-shaped particles (Figure 26.3)

containing as many as 100 proteins. Thin sections of virions (Figure 26.4b) reveal a bioconcave core flanked by lateral bodies. The lateral bodies are trypsin-sensitive and therefore contain protein, but otherwise are of unknown composition and function. The core is surrounded by a bilayer "core wall" structure of unknown composition. Within the core is a tubular structure folded into three sections. The core contains the virus DNA, presumably within the tubular structure. Virion

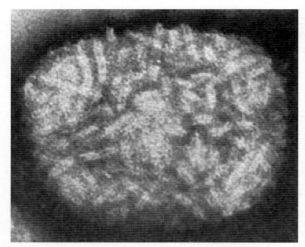

Figure 26.3 Vaccinia virus virion. Electron micrograph of negatively stained vaccinia mature virus, showing surface ridges.

cores also contain a spectacular array of virus-coded enzymes required for early mRNA synthesis and modification.

Virions exist in two forms, called mature virus (MV) and extracellular virus (EV), which differ in membrane composition as a result of differences in morphogenesis. The mature virus particle contains an outer envelope composed of lipid and viral proteins. Extracellular virus is a mature virus particle wrapped in an additional lipid bilayer containing a distinct set of embedded virus-coded proteins. Both forms of the vaccinia virion are infectious.

Mature virus and extracellular virus use distinct viral structural proteins for binding to cell receptors. There are at least three different mature virus membrane proteins that may facilitate attachment, and each of these binds to ubiquitous cell surface **glycosaminoglycans**, consistent with the broad host range of the virus. Infection of cells with mature virus may occur

by direct fusion of the mature virus membrane with the plasma membrane, or alternatively by uptake into vesicles followed by fusion with the membrane of the endosomal vesicle.

Two mechanisms for entry of extracellular virus particles have been described. Extracellular virus can be taken up by phagocytosis followed by rupture of the extracellular virus envelope due to a decrease in pH in the endosome, releasing a mature virus particle within the vesicle. Mature virus then fuses its envelope with the vesicle membrane to release the virion core into the cytoplasm. Alternatively, the extracellular virus envelope may rupture upon initial contact of the particle with the cell surface, releasing a mature virus particle, which enters as described above. In all cases, the end result is the release of the virus core into the cytoplasm.

The different forms of the poxvirus virion probably serve different roles in nature. The extracellular virus particle is required for cell-to-cell and tissue-to-tissue spread of virus in an infected animal. However, the extracellular virus membrane is fragile, and probably does not survive well in the environment. In contrast, the mature virus particle is stable, and is likely the form responsible for transmission of infections between individuals. Poxviruses are notoriously stable in the environment; the smallpox vaccine can be freeze-dried, and dried pox scabs or virus-contaminated surfaces can retain infectivity for prolonged periods of time. Thus the mature virus particle may provide for stability in the environment while extracellular virus allows efficient spread in the body.

Poxviruses replicate in the cytoplasm

Unlike other eukaryotic DNA-containing viruses covered in this book, poxviruses replicate exclusively in the cytoplasm of infected cells. Because most of the cellular enzymes required for DNA replication and for RNA synthesis and processing are located in the nucleus, poxviruses must provide virus-coded enzymes to carry

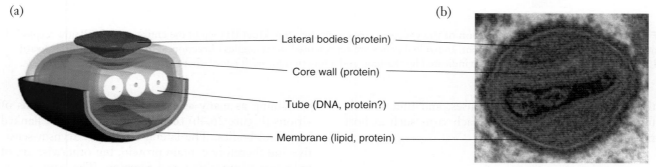

Figure 26.4 Model of the structure of vaccinia virus virions a. A cutaway model of the mature virus (MV) particle, showing internal structures. The membrane has been removed from the upper half of the virion, the near end has been removed, and the core wall has been rendered transparent, thus revealing the multiple layers, the concavities in the core, the lateral bodies, and the tubular internal structure. b. Electron micrograph of a thin section of vaccinia mature virus, showing the envelope, lateral bodies, the core wall and the internal tube. To view an online dynamic model of vaccinia virus virions, see www.vacciniamodel.com.

out these processes. Their large genome size enables poxviruses to code for a variety of enzymes involved in RNA and DNA metabolism.

Vaccinia virus genes required for DNA replication are expressed immediately following infection, and are therefore called early genes. Early genes are expressed within virus cores, which are released into the cytoplasm after fusion of the virion envelope with cellular membranes during virus entry (Figure 26.5). Early messenger RNAs are extruded from cores into the cytoplasm, where they are translated into early virus proteins. Some of these early proteins dissolve the core and release viral DNA into the cytoplasm, where further transcription and DNA replication take place. Genes required for virion formation are expressed after DNA replication has begun, and comprise two classes, intermediate and late. Regulation of gene expression is accomplished primarily at the level of transcription.

Poxvirus genes are expressed in a regulated transcriptional cascade controlled by viral transcription factors

The timing of poxvirus gene expression is dictated by the sequential synthesis of virus-coded transcription initiation factors (Figure 26.5), which bind to specific promoter elements on viral DNA. Initiation factors required for transcription of early genes (virus early transcription factors, VETF) are packaged in infecting virions and act within the core upon its release into the cytoplasm. Certain early genes encode distinct intermediate transcription factors (VITF) that trigger intermediate gene expression. Late transcription factors (VLTF) are in turn encoded by some of the intermediate genes. Finally, late genes encode early transcription factors, which are packaged into virions in preparation for a subsequent round of infection. This organization of gene expression assures that sufficient newly replicated virus genomes will accumulate before synthesis of structural proteins begins. If viral DNA replication is inhibited, expression of both intermediate and late genes is prevented.

Roughly one-half of vaccinia virus genes are controlled by early transcriptional promoters, the other half being controlled by intermediate or late promoters. Some genes have compound promoters, allowing for their expression throughout infection. Vaccinia virus promoters are approximately 35 nucleotides in length regardless of gene class. Mutational analysis has defined optimum consensus DNA sequences that distinguish early, intermediate, and late promoters.

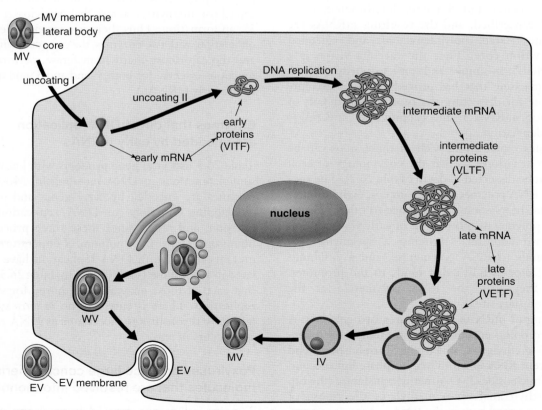

Figure 26.5 The vaccinia virus life cycle. See text for description. IV: immature virus; MV: mature virus; WV: wrapped virus; EV: enveloped virus; VETF: viral early transcription factors; VITF: viral intermediate transcription factors; VLTF: viral late transcription factors.

Table 26.2 Enzymes packaged in vaccinia virions

Function	Protein	No. of subunits	Mol. wt. (kD)
RNA synthesis	RNA polymerase	8	410
Transcription initiation	Early gene specificity factor	1	94
	Early gene initiation factor	2	166
mRNA 5' end capping	Cap guanylyltransferase	2	130
	Ribose-methyltransferase	1	39
Transcription termination	Cap guanylyltransferase	2	130
	DNA-dependent ATPase	1	72
mRNA 3' end polyadenylation	Poly(A) polymerase	2	94
Nucleic acid topology	RNA helicase	1	77
	DNA helicase	1	56
	DNA topoisomerase I	1	37
Protein modification	Protein kinase I	1	34
	Protein kinase II	1	52
	Protein phosphatase	1	20

Virus-coded enzymes packaged in the core carry out early RNA synthesis and processing

When purified vaccinia virions are permeabilized *in vitro* with a detergent and incubated with ribonucleoside triphosphates and S-adenosylmethionine, early genes are transcribed, and the resulting mRNAs are capped, methylated, polyadenylated, and extruded from the virion core. Thus the vaccinia virion is a "mini-nucleus", a simple and easy-to-purify mRNA synthesis machine that has proved to be of extraordinary value for dissecting the biochemical details of transcription, mRNA 5' end capping, and mRNA 3' end polyadenylation.

Enzymes representing each of these steps can be purified from virions, so that the enzymology of each process can be dissected in detail (Table 26.2). The vaccinia virus RNA polymerase is a complex, eight-subunit enzyme that bears some resemblance to other eukaryotic RNA polymerases. Initiation of early transcription requires two additional proteins: a specificity factor that binds to RNA polymerase, and a 2-subunit initiation factor (VETF) that binds to the promoter. Together they load the RNA polymerase at the transcription start site.

Capping of mRNAs also requires two enzymes. A 2-subunit capping enzyme (the cap **guanylyltransferase**) adds 7-methylguanosine to the 5' triphosphate end of the nascent RNA in a 5'-5' triphosphate linkage, and a distinct (nucleoside-2'-O-)-methyltransferase (the cap **ribose-methyltransferase**) methylates the ribose of the first transcribed base in the RNA.

Transcription termination occurs in response to a sequence element, U_5NU, in the growing RNA chain, and requires two protein factors: a DNA-dependent ATPase and the same cap guanylyltransferase that modifies the mRNA 5' end. Polyadenylation of freshly terminated mRNA 3' ends is catalyzed by a 2-subunit poly(A) polymerase, one subunit of which is the same cap ribose-methyltransferase that participates in mRNA 5' cap formation. Thus there is a remarkable economy of function, with two enzymes, the cap guanylyltransferase and the cap ribose-methyltransferase, performing dual functions essential for both 5' end and 3' end maturation of the early mRNA.

Enzymes that direct DNA replication are encoded by early mRNAs

Table 26.3 lists viral gene products with known or presumed functions in DNA metabolism. Most of these enzymes are encoded by early genes and many have overlapping functions in DNA replication, recombination, and repair since these three processes have numerous steps in common. Temperature-sensitive mutants defective in DNA replication have been isolated in six of these genes (noted in Table 26.3), proving that each of those enzymes is required for viral DNA replication. There is currently no *in vitro* system that faithfully reconstitutes vaccinia virus DNA replication as it occurs *in vivo*.

Poxviruses produce large concatemeric DNA molecules that are resolved into monomers

The presently accepted model describing poxvirus DNA replication (Figure 26.6) is fashioned after the "rolling hairpin" model of DNA replication in parvoviruses (see

Table 26.3 Enzymes of DNA metabolism encoded by vaccinia virus

Function	Protein	No. of subunits	Mol. wt. (kD)
DNA replication, repair, recombination	DNA polymerase*	1	116
	DNA pol. processivity factor	1	49
	DNA primase*	1	90
	Topoisomerase I	1	36
	ssDNA binding protein	1	34
	DNA ligase*	1	63
	Holliday junction resolvase	2	42
	Protein kinase (BAF antagonist)*	1	34
	Multifunctional "scaffold" protein*	?	22
	Uracil DNA glycosylase*	1	25
	dUTPase	1	16
	dsDNA break repair	1	50
Nucleotide metabolism	Thymidine kinase	1	19
	Thymidylate kinase	1	23
	Ribonucleotide reductase	2	121

* Genes coding for these proteins have been shown to be required for carrying out viral DNA replication.

Chapter 20). This model is based on the structure of the virus genome and on two additional observations: (1) replication appears to initiate within or near the hairpin termini; and (2) head-to-head and tail-to-tail **concatemers** (oligomers of genome-length units) accumulate during infection.

Replication is initiated by a site-specific nick in one of the terminal hairpins, revealing a 3' end that primes DNA synthesis (step 1). Extension from this 3' end (step 2) copies the sequence of the melted hairpin to its 5' end, approximately 100 nt. This nascent DNA chain then folds into a new hairpin (step 3) that can prime further DNA synthesis (step 4). Synthesis continues through the entire length of the genome (200,000 nt) using one DNA strand as template, then through the distal hairpin, and back through the entire length of the genome using the complementary strand as template (step 5), to generate a head-to-head dimer. This process forms a "concatemer junction" that contains the two complementary hairpin ends in a perfectly base-paired duplex (see Figure 26.1d).

Further replication by the same process yields higher-order concatemers (not shown). Resolution of dimers or concatemers into unit-length molecules (step 6) requires a virus-coded **resolvase**. The resolvase makes a staggered double-stranded break by nicking the complementary strands at each end of the concatemer junction, whereupon each strand of the junction can refold into a hairpin end as the concatemer is resolved. Joining of the adjacent 5'

and 3' ends by DNA ligase would then produce mature poxvirus DNA molecules.

This rolling hairpin model for vaccinia virus DNA replication does not formally require a DNA primase, as DNA strand synthesis occurs by elongation of nicked parental DNA strands rather than initiation by RNA primers. Therefore the recent discovery of a viral DNA primase that is required for DNA replication implies that the actual mechanism of vaccinia virus DNA replication may deviate significantly from this model.

Postreplicative mRNAs have 5' end poly(A) extensions and 3' end heterogeneity

Intermediate and late genes are called postreplicative genes, because their transcription begins only after DNA replication has started. They use distinct transcriptional promoters and initiation factors, but are otherwise similar and can be considered together. Postreplicative RNA synthesis, capping, and polyadenylation are carried out by the same enzymes used for early mRNA synthesis (see Table 26.2), freshly synthesized from early mRNAs.

Initiation of transcription of intermediate genes requires two virus-coded factors (VITF) synthesized from early viral mRNA, and one cellular factor. The virus factors are the cap guanylyltransferase referred to above, and one of the smaller RNA polymerase subunits. The cellular factor is a stress granule protein implicated

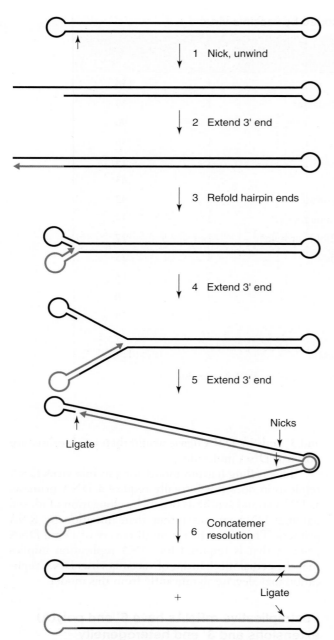

Figure 26.6 A model for vaccinia virus DNA replication. See text for details. Newly synthesized DNA is shown in orange.

in cellular mRNA metabolism. Initiation of transcription of late genes requires at least three viral protein factors (VLTF) produced from intermediate mRNAs, and may require in addition an early gene product and a cellular factor.

Intermediate and late mRNAs share two unique structural features that distinguish them from early mRNAs. First, 5'-terminal "poly(A) heads" are added to these mRNAs via a slippage mechanism during the initiation of transcription. Both intermediate and late

promoters contain the sequence TAAA at the RNA start site, usually located just upstream of the translation initiation ATG for a given gene. The viral RNA polymerase presumably synthesizes the initial AAA sequence, then backs up while still gripping the nascent RNA, adds another AAA sequence, then backs up and repeats the process several more times before clearing the promoter and continuing downstream. The resulting RNA 5' end contains 30 to 50 A residues upstream of the first AUG codon. This extra 5' noncoding region may aid in recognition of the AUG by ribosomes.

Second, these mRNAs have heterogeneous 3' ends, resulting from a transcription termination mechanism that is fundamentally different from the process used during early transcription. The RNA polymerase ignores the early termination sequence U_5NU, and instead terminates inefficiently at multiple sites far downstream, yielding a family of transcripts 2 to 4 kb in length with heterogeneous 3' ends. At least three viral genes that influence the length at the 3' end of postreplicative mRNAs have been identified. One of these genes codes for the same ribose-methyltransferase that participates both in mRNA 5' cap formation and in 3' end polyadenylation.

Just in case you lost track, the cap guanylyltransferase participates in mRNA capping, early mRNA transcription termination, and intermediate gene transcription initiation; and the cap ribose-methyltransferase participates in mRNA capping and polyadenylation, and in postreplicative gene transcription elongation or termination. Although the details of this remarkable multifunctionality remain to be resolved, it appears that each protein has multiple independent activities rather than multiple uses of a single enzymatic active site.

Mature virions are formed within virus "factories"

Assembly of vaccinia virions is a unique and complex process that provides fascinating insights into intracellular protein trafficking and organelle biogenesis. Virion assembly is studied by electron microscopic examination of cells infected with wild-type or mutant viruses (Figure 26.7); viral structural and assembly proteins are localized using labeled antibodies.

Viral DNA replication results in the formation of cytoplasmic structures called "DNA factories", recognizable as large areas of uniform electron density, cleared of normal cellular organelles. The first evidence of assembly of virus particles is the appearance within factories of rigid, crescent-shaped membrane structures (Figure 26.7; see lower-right corner of Figure 26.5). These contain both lipids and viral proteins, and are destined to become the mature virus envelope. Crescents are remarkable in that they comprise a single

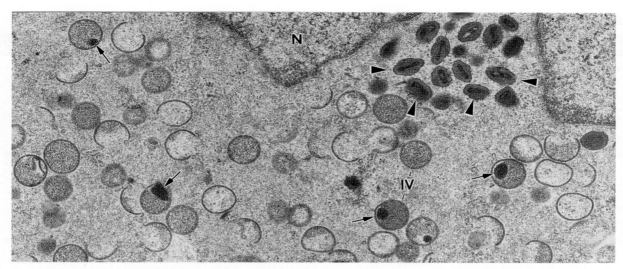

Figure 26.7 Electron micrograph of a vaccinia virus-infected cell. N: nucleus; IV: immature virions; arrows: condensing nucleoids within immature virions; arrowheads: mature virions.

lipid bilayer that is at no time directly connected to a preexisting intracellular membrane.

Crescent formation is directed by viral proteins that are synthesized on endoplasmic reticulum, which may be present within factories. Lipids are delivered to crescents from the endoplasmic reticulum in a novel form, perhaps as small vesicles or **micelles** containing viral envelope proteins. The crescents ultimately develop into spheres called "immature virions", and in the process enclose **viroplasm**, a substance of intermediate electron density that is known to contain several viral core proteins.

At a later stage, viral DNA enters the spheres to form a discrete electron dense nucleoid, perhaps a precursor to the tubular structure in the interior of the core in mature virions. Immature virions then develop the distinguishing internal core and lateral body structures by rearrangement and assembly of the material captured inside spheres, and in the process acquire the final brick shape to become mature virus (MV). In a normal infection the vast majority of progeny virions remain intracellular mature virus.

Extracellular virions are extruded through the plasma membrane by actin tails

A small fraction of mature virus undergoes further maturation to become extracellular virus (EV), which is specifically designed for export and infection of neighboring cells. Mature virus particles are wrapped in Golgi-derived cisternae containing specific viral extracellular virus proteins (center left of Figure 26.5). The resulting wrapped virus (WV) therefore consists of a mature virus particle encased in two additional bilayer membranes.

Wrapped virus is transported to the cytoplasmic side of the plasma membrane, where the outermost membrane of the wrapped virus fuses with the plasma membrane. This results in an extracellular virus (EV) particle that is attached to the cell membrane. Actin tails are assembled within the cell beneath the extracelluar virus particles, resulting in the formation of **microvilli** extending from the cell into the extracellular space, with extracellular virus particles perched at their tips. These microvilli can contact neighboring cells, infecting them with extracellular virus. Alternatively, extracellular virus particles can be released from the cell surface and infect distant cells.

Poxviruses make several proteins that target host defenses against invading pathogens

As obligate intracellular parasites, viruses have no choice but to confront host defenses (see Chapter 33 and 34). Cells respond to infection by synthesizing and secreting **interferons** and other **cytokines**. These are polypeptide hormones that communicate with other cells in the organism, inducing an antiviral state or stimulating the innate and adaptive immune systems to fight the infection. Poxviruses have evolved a number of genes that cleverly combat host defenses against infection (Table 26.4).

Interferon response. The first line of defense against virus infections is the interferon response (Chapter 33), triggered by the synthesis of double-stranded RNA (dsRNA) in infected cells. A significant quantity of poxvirus dsRNA is produced during the late phase of infection as a result of synthesis of transcripts with

Table 26.4 Poxvirus genes that interfere with host defense systems

System	Virus gene	Activity	Protection against host defense
Interferon	K3L	Homologue of eukaryotic initiation factor 2α (eIF-2α)	Pseudosubstrate for interferon-induced protein kinase (PKR)
	E3L	Double-stranded RNA binding protein	Sequesters dsRNA, an activator of interferon-induced PKR and 2′,5′-oligo A synthetase
	B8R	Secreted, soluble interferon-γ receptor	Binds to and inhibits interferon-γ
Interleukin-1	B14R	Serine protease inhibitor (serpin)	Inhibits proteolytic activation of interleukin-1, decreasing inflammatory response
	B16R	Soluble interleukin-1 receptor	Binds to and inhibits interleukin-1β
Tumor necrosis factor (TNF)	B28R, A53R	Soluble TNF receptor	Binds to and inhibits TNF-α and β
Complement	C3L	C4B and C3B binding protein	Inhibits classical and alternative complement pathways
Chemokines	B29R/C23L	Chemokine binding protein	Interferes with leukocyte migration

---

Medical uses of poxviruses

Poxviruses hold potential for medical use in two areas, recombinant vaccines and treatment of cancer. For use in vaccines, poxviruses can be engineered to express foreign proteins under control of viral transcriptional promoters (see below). Recombinant poxviruses can be constructed that express antigens of another pathogen, for example, the rabies virus surface glycoprotein. This recombinant virus can be used to vaccinate an animal, which will become immune both to the poxvirus and to rabies virus. Recombinant poxvirus vaccines are now used widely in

veterinary practice. While there are currently no recombinant poxvirus vaccines licensed for use in humans, some are being evaluated in clinical trials, in particular for immunization against HIV.

Some engineered vaccinia viruses as well as myxoma virus, a rabbit poxvirus, have been found to replicate preferentially in human tumor cells and are effective in selectively reducing tumors in animal models. With further development, use of such viruses as anticancer agents in humans can be envisioned.

---

heterogeneous 3′ ends, made on opposite DNA strands from convergent viral promoters.

Poxviruses synthesize two proteins that inhibit the interferon-induced antiviral state. The K3L gene product, a homologue of eukaryotic protein synthesis initiation factor eIF-2α, competitively inhibits the dsRNA-activated protein kinase **PKR**, relieving the inhibition of protein synthesis induced by interferon. The E3L gene product binds to and sequesters dsRNA, interrupting both the PKR and the parallel **2′, 5′-oligoadenylate synthetase** antiviral pathways.

Cytokine signaling. Virus infection triggers synthesis of a large number of cytokines, several of which are important in coordinating the innate immune response, including **interleukin-1, tumor necrosis factor (TNF)**, chemokines, and interferons, including interferon γ. Poxviruses have at least two different strategies for interfering with cytokine function. In one method, the virus produces a *ser*ine *p*rotease *in*hibitor (**serpin**), which specifically blocks a proteolytic cleavage required

for activation of interleukin-1β. In another strategy, the virus produces soluble receptors for tumor necrosis factor, interleukin-1, chemokines, and interferons, which bind to these cytokines and prevent them from interacting with their intended receptors. Importantly, while all these cytokines influence primarily the innate response, interferon γ also activates critical components of the adaptive immune response.

Complement. **Complement** plays a role in both the innate and adaptive immune responses (Chapter 34). In both cases, interaction of virus surface components with elements of the complement system triggers a cascade of complement reactions that results ultimately in **phagocytosis** of the virus particle. The innate ("alternative") pathway involves a nonspecific interaction of the complement system with the virion surface, while the adaptive ("classical") pathway involves a reaction of the complement system with antibody-coated virus particles. The vaccinia virus C3L gene product, synthesized and secreted during infection, binds to and

The vaccinia virus-T7 system

In an intriguing technological perversion of nature, vaccinia virus has been adapted for use as an efficient and versatile tool for expression of foreign proteins in eukaryotic cells. The system is composed of two recombinant vaccinia viruses. The first contains a copy of the bacteriophage T7 RNA polymerase gene (see Chapter 7) under transcriptional control of a vaccinia virus promoter, and the second contains the coding sequence for the gene of choice to be expressed, cloned downstream of a T7 RNA polymerase promoter.

Coinfection of cells with the two viruses results in synthesis of T7 RNA polymerase from the first virus, which then can transcribe the gene of choice from the T7 promoter in the second virus. The same outcome can be achieved more simply by infecting cells with the virus that expresses T7 RNA polymerase, followed by transfection with a DNA molecule containing any coding sequence cloned downstream from a T7 promoter.

The scheme has several advantages as an expression system: (1) recombinant vaccinia viruses are relatively easy to prepare by homologous recombination between an infecting vaccinia virus genome and a transfected DNA that contains the foreign gene embedded within a nonessential virus gene; (2) the entire process takes place in the cytoplasm, obviating the need for RNA splicing or nuclear-cytoplasmic transport of any of the component RNAs or proteins; (3) the system is very specific and efficient, owing to the remarkable activity and promoter specificity of the T7 RNA polymerase and to the amplification of expression by synthesis of numerous T7 RNA polymerase molecules, which efficiently and repeatedly transcribe the gene of choice.

The system has found utility in two areas: over-expression of foreign proteins for purification and biochemical analysis, and expression of foreign proteins for assay of their intracellular function. An interesting use of the system is the construction of recombinant vesicular stomatitis virus (see Chapter 15). In this example, cells are infected with vaccinia virus expressing T7 RNA polymerase, then are transfected simultaneously with four plasmid DNAs. One DNA contains the vesicular stomatitis virus genome sequence downstream from a T7 promoter. The other three DNAs contain the coding sequences of the three vesicular stomatitis virus proteins (P, L, and N) required for viral genome replication, each cloned downstream from a T7 RNA polymerase promoter. The T7 RNA polymerase supplied by vaccinia virus transcribes all four genes simultaneously, synthesizing the RNA genome of vesicular stomatitis virus and messenger RNAs for all three replication proteins. This allows replication, expression, and packaging of the genome of vesicular stomatitis virus.

---

inactivates two components of the complement system, C3B and C4B, which are critical to both the innate and adaptive complement pathways.

In summary, even a partial list of the many poxvirus immune defense functions reveals an ingenious multi-faceted attack on the host defense system. Some have likened the poxvirus immune defense to "star wars". Extending this metaphor, the virus carries a MIRV (multiple independently-targeted reentry vehicle) warhead, and launches this missile in a preemptive strike on the host's first lines of defense. Importantly, the virus probably has evolved to target the most critical points of the host's intrinsic and immune defense systems, and thus has much to tell us about the relative priorities of the different elements of these systems.

KEY TERMS

2', 5' oligoadenylate synthetase	Phagocytosis
Complement	PKR
Concatemer	Pock
Cytokine	Resolvase
Endemic	Ribose methyltransferase
Glycosaminoglycan	Serpin
Guanylyltransferase	Tumor necrosis factor (TNF)
Interferons	Vaccination
Interleukin-1	Variolation
Micelle	Viroplasm
Microvilli (singular: microvillus)	

FUNDAMENTAL CONCEPTS

- Immunization against smallpox (1796) was the first successful vaccination attempt.
- Smallpox is the only human infectious disease that has been totally eradicated.
- Poxviruses replicate exclusively in the cytoplasm although they have DNA genomes.
- Poxvirus cores contain a full complement of viral enzymes that direct transcription of early genes and processing of viral transcripts, which are extruded into the cytoplasm.

- Poxvirus DNA can replicate only after the core has been dissolved and the DNA genome is released into the cytoplasm.
- Poxvirus envelopes are assembled *de novo* within virus factories in the cytoplasm; they are not formed by budding from pre-existing cellular membranes as are most animal virus envelopes.
- To economize on restricted genome size, viruses often encode multifunctional proteins.
- Poxviruses have evolved elaborate mechanisms for evading the host immune system.
- Recombinant poxviruses can potentially be used for vaccination against many diseases.

REVIEW QUESTIONS

1. What are two likely origins for the smallpox vaccine strain, vaccinia virus?

2. Poxviruses replicate in the cytoplasm, but splicing of cellular RNAs takes place in the nucleus. Are poxvirus RNAs spliced, and if so, where?

3. How do the two forms of poxvirus virions differ, and are they both equally infectious?

4. Discuss two unique structural features that distinguish poxvirus intermediate and late mRNAs from early mRNAs.

5. Describe a "DNA factory".

6. Certain poxviruses can interfere with two major players of host antiviral defenses. What are they?

CHAPTER 27

Viruses of Algae and Mimivirus

Michael J. Allen
William H. Wilson

Phycodnaviridae
From Greek *phykos,* referring to seaweed, and *DNA,* referring to the nature of the genome

Mimiviridae
From "*mi*crobe *mi*micking", because it stains gram-positive, like some bacteria

VIRION
Regular icosahedral capsids, diameter 100–500 nm.
DNA is enclosed within an internal lipid membrane.
Some phycodnaviruses have an additional external envelope.
Complex virions may contain over 100 proteins.

GENOME
Linear or circular double-stranded DNA, 160–1200 kb.
Often contain unique repetitive regions, inverted repeats and terminal repeat sequences.

GENES AND PROTEINS
Paramecium bursaria chlorella virus-1 (PBCV-1): 366 genes.

Emiliania huxleyi virus-86 (EhV-86): 472 genes.
Acanthamoeba polyphaga mimivirus: 911 genes.
Most genes (~80%) are of unknown function.

VIRUSES AND HOSTS
Six genera of *Phycodnaviridae: Chlorovirus, Coccolithovirus, Phaeovirus, Prasinovirus, Prymnesiovirus,* and *Raphidovirus.*
Genera are determined by host range of viruses.

DISEASES
Some phycodnaviruses are instrumental in termination of algal blooms.

DISTINCTIVE CHARACTERISTICS
Associated with aquatic environments.
Among the largest viruses currently known.
A highly diverse family of viruses, reflected in a wide variation of encoded proteins.

Aquatic environments harbor large viruses

The *Phycodnaviridae* (literally translated as "DNA viruses that infect algae") comprise a genetically diverse, yet morphologically similar, family of icosahedral viruses that infect marine or freshwater eukaryotic algae. Phycodnaviruses have particle diameters from 100 to 220 nm, and contain double-stranded DNA genomes ranging from 160 to 560 kb.

Algae are a diverse group of photosynthetic eukaryotic organisms that are ubiquitous in marine, freshwater and terrestrial habitats. Often characterized by the nature of their photosynthetic apparatus, the products of multiple and complex **endosymbiotic** events, the algae do not represent a single evolutionary group. The number of algal species (mostly single-celled microalgae) has been estimated to be as high as several million. It is likely that viruses infect all of these species, although not all of these viruses will be assigned to the *Phycodnaviridae* family. It should be noted that the "blue-green algae" are prokaryotic cyanobacteria (the evolutionary source of algal chloroplasts) and their viruses are not considered here.

Given the number of potential hosts, it is surprising that so few phycodnaviruses have been isolated to date (see Table 27.1 for some examples). This is perhaps a reflection of the low importance with which they have been regarded in the past. However, 50% of the planet's oxygen is produced by marine microalgae, and algae play a pivotal role in global carbon fixation from atmospheric CO_2. Because of the increasing interest in global climatic changes that are currently being observed, it is likely that research into algae (and their viruses) will become increasingly fashionable and essential in the future.

Viruses play a fundamental role in cycling nutrients in the global ecosystem, as they kill or lyse up to 20% of the microbial biomass in the earth's oceans each day. It could be argued that if we lived in a world without the well-studied human viruses, we would all live longer; yet if we lived in a world without the poorly studied algal viruses, we would all certainly be dead! With only approximately 150 formally identified viruses, and 100 or so others mentioned in the literature, it is clear that most phycodnaviruses, containing an enormous reservoir of genetic diversity, remain to be discovered.

Phycodnaviruses are diverse and probably ancient

Members of the *Phycodnaviridae* are currently grouped into six genera (named after the hosts they infect): *Chlorovirus, Coccolithovirus, Phaeovirus, Prasinovirus, Prymnesiovirus*, and *Raphidovirus* (Table 27.1). Phylogenetic analysis of sequenced representative genomes places the *Phycodnaviridae* within the Nucleo-Cytoplasmic Large DNA Viruses (NCLDVs), a "superfamily" of diverse viruses that also includes the *Poxviridae, Iridoviridae, Asfarviridae*, and *Mimiviridae*. The herpesviruses are also distant cousins to this group.

However, as genomic information has become available for more algal viruses, it has become clear that the phycodnaviruses themselves are a very diverse group. Some newly isolated large, double-stranded DNA viruses of algae are currently not officially assigned to the *Phycodnaviridae* because of their high divergence from the original founder members of the family. This divergence is most likely a reflection of the ancient origins of the phycodnaviruses and the highly diverse nature of their hosts. Whereas terrestrial plants "only" emerged about 470 million years ago, algae (and presumably their viruses) were already well established by then. Indeed, the ancestors of phycodnaviruses could be between 2 and 2.7 billion years old and could have witnessed the divergence of prokaryotes and eukaryotes.

It is no surprise that "rogue" phycodnavirus members lying on their own distinct branches of the NCLDV tree are beginning to appear. Indeed, the two algal virus species *Pyramimonas orientalis* V01 and *Chrysochromulina ericina* V01 (infecting *P. orientalis* and *C. ericina*, respectively) actually are closer phylogenetically to the largest known virus, *Acanthamoeba polyphaga* mimivirus. This "giant" virus replicates in the amoeba for which it is named, and therefore is not a virus of algae. However, mimivirus will also be discussed in this chapter because of its unusually large size and its similarity to the phycodnaviruses, both with regards to structure and genomics, and our still very incomplete understanding of its biology.

Phycodnavirology: a field in its infancy

Reports of probable viruses infecting at least 44 different species of eukaryotic algae have appeared since the early 1970s. Initially these were based upon incidental observations of virus-like particles in electron micrographs. It was not until 1979 that the first phycodnavirus

Table 27.1 Phycodnaviruses and their hosts

Virus genus,	Type species	Algal host	Description
Chlorovirus	PBCV-1	*Paramecium bursaria* associated *Chlorella*	Freshwater single-celled green algae, symbiotically associated with the protozoan *Paramecium bursaria*.
Coccolithovirus	EhV-86	*Emiliania huxleyi*	Single-celled red algae with global abundance in the marine environment, known for their calcified shells and bloom formation.
Phaeovirus	EsV-1	*Ectocarpus siliculousus*	Multicellular filamentous brown algae (seaweed).
Prasinovirus	MpV-SP1	*Micromonas pusilla*	Dominant single-celled green algae found in both warm and cold marine waters.
	OtV-1	*Ostreococcus tauri*	Single-celled green algae famous for being the smallest free-living eukaryote. Ubiquitous in the marine environment.
Prymnesiovirus	CbV-PW1	*Chrysochromulina brevifilum*	Single-celled marine algae.
Raphidovirus	HaV01	*Heterosigma akashiwo*	Single-celled, free-swimming marine algae that tend to produce toxic surface aggregations known as harmful algal blooms.

was isolated; that virus infects the marine unicellular alga *Micromonas pusilla*. However, this report was largely ignored until the early 1990s, when high concentrations of viruses in aquatic environments were being described.

In the early 1980s a group of viruses that infect freshwater unicellular, chlorella-like green algae were characterized; they were called "chloroviruses". These reports were followed in the early 1990s by research into viruses of marine filamentous brown algae. Thereafter, use of genetic markers such as virus-encoded DNA polymerases, a core gene present in all NCLDVs, revealed that phycodnaviruses are a diverse and ubiquitous component of aquatic environments. The field of phycodnavirology is now firmly established and is rapidly expanding.

Conserved structure, diverse composition

Despite the wide host range displayed by the *Phycodnaviridae*, all family members, as well as mimivirus, have similar structure and morphology. The virions are large, layered structures between 100 and 500 nm in diameter, with a double-stranded DNA-protein core surrounded by a lipid bilayer, which is itself enclosed in a capsid with icosahedral symmetry. Variations around the core structural theme include an outer lipid layer in coccolithoviruses (a consequence of the budding exit mechanism through the cell membrane), a star-shaped surface structure on mimivirus (associated with entry of the viral genome into the cell), and the short filaments on the capsid of some phycodnaviruses as well as the long "hairy" appendages of mimivirus.

Yet, despite the conserved structure, the phycodnaviruses are probably the most genetically diverse of any virus family. In a comparison of the first three phycodnavirus genomes to be fully sequenced (PBCV-1, EhV-86, and EsV-1), a mere 14 genes out of a total of more than 1000 genes were found to be common to all three viruses. This outstanding display of biodiversity is mirrored in genome structure, size and, crucially, life style, and creates a myriad of problems for authors aiming to produce a chapter summarizing the general properties of such a diverse family! This is further exacerbated by an extensive lack of knowledge of the fundamental workings of these interesting, but mysterious, viruses. As a general rule, if it can vary, it inevitably does vary within the *Phycodnaviridae*.

CHLOROVIRUSES

Known chloroviruses replicate in *Chlorella* isolated from symbiotic hosts

The chloroviruses are the best studied of all phycodnaviruses. Their hosts, *Chlorella*, are small, unicellular, nonmotile, asexual green algae with a global distribution. While most chlorella species are free-living, many species have symbiotic relationships with aquatic organisms

from different classes in the animal kingdom including *Rhizopoda*, *Ciliata*, *Hydrozoa*, and *Turbellaria*.

To date, the only described chloroviruses infect symbiotic chlorella, often referred to as zoochlorellae, such as those associated with the protozoan *Paramecium bursaria* (Figure 27.1), the freshwater coelenterate *Hydra viridis*, and the heliozoon *Acanthocystis turfacea*. These organisms apparently profit from the photosynthetic capacities of the chlorella, which they harbor, and like chlorella themselves, they become green! Surprisingly, these viruses have not been observed to replicate in chlorella living within their symbiotic hosts, but they replicate well (and can form plaques) in these algae when cultured *in vitro*.

The best-studied chlorovirus is *Paramecium bursaria* chlorella virus-1 (PBCV-1), which infects *Chlorella* NC64A, a strain isolated from *P. bursaria*. Complete genome sequences are also available for five other chloroviruses.

The linear genomes of chloroviruses contain hundreds of genes, and each virus species encodes some unique proteins

The PBCV-1 genome (Figure 27.2) is a linear, 330-kb, double-stranded DNA molecule. Its termini consist of 35-nt-long covalently closed hairpin loops flanked by identical 2221-bp **inverted repeats**. The terminal loops are incompletely base paired, as are similar terminal loops in the genomes of poxviruses (see Figure 26.1). The predicted 366 PBCV-1 protein-encoding genes are evenly distributed on both strands and, with one exception, intergenic space is minimal. The exception is a 1788-nucleotide

Figure 27.1 *Paramecium bursaria* **filled with the symbiotic alga *Chlorella* NC64A,** which is a host for chloroviruses.

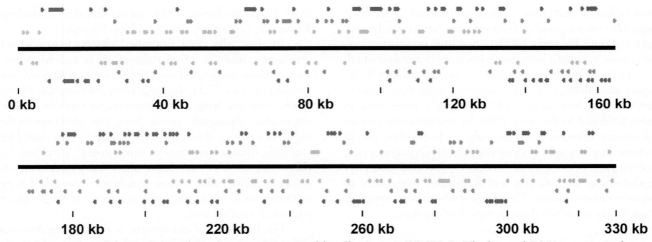

Figure 27.2 Map of the genome of *Paramecium bursaria* chlorella virus-1 (PBCV-1). The linear dsDNA genome is shown in a two segments for convenience. Blue arrows: early genes; orange arrows: early/late genes; green arrows: late genes. Arrows indicate direction of transcription of genes. Rightward-reading genes are shown above each DNA segment, and leftward-reading genes are shown below.

sequence near the middle of the genome, which encodes 11 tRNA genes. Unsurprisingly, viruses infecting the same host are most similar, and overall approximately 80% of the genes are found in all six sequenced genomes. Despite the high similarity in genome content, each of the six sequenced chloroviruses contains genes that encode unique proteins and functions. These include genes encoding dTDP-D-glucose 4,6 dehydratase, ribonucleotide-triphosphate reductase, mucin-desulfatating sulfatase, an aquaglyceroporin, an alkyl sulfatase, a potassium ion transporter, **ubiquitin**, a calcium-transporting ATPase, and a Cu/Zn superoxide dismutase.

Chlorovirus capsids are constructed from many capsomers and have a unique spike

The PBCV-1 virion (Figure 27.3) has a diameter of 190 nm and consists of an icosahedral outer capsid (T=169) surrounding an internal lipid bilayer, within which is enclosed the viral DNA and many viral proteins. The 54-kDa major capsid protein (Vp54) is a glycoprotein that comprises approximately 40% of the total protein in virions, the remainder composed from at least 110 different virus-encoded proteins. Vp54 folds into two **jelly-roll β-barrels** (Figure 27.4a), structures commonly found in capsid proteins of many other viruses, including bacterial and animal viruses (e.g., adenovirus).

The virus capsid is formed from 1680 trimeric capsomers of Vp54 (Figure 27.4b, c), which are arranged in groups as 20 curved triangular facets (**trisymmetrons**; 66 capsomers each) and 12 curved pentagonal facets (**pentasymmetrons**; 30 capsomers each) on the surface of the virion (Figure 27.3). At the center of each pentasymmetron lies a pentameric capsomer formed from a distinct capsid protein. This is a common pattern for the capsids of many phycodnaviruses, and it is likely

that trisymmetrons and pentasymmetrons are assembly intermediates used during the construction of complete capsids. One of the pentasymmetrons has at its center a unique spike that projects 25 nm outward from the virion, and this is likely the organelle with which virions contact the cell wall of host cells and initiate its degradation (see below).

The carbohydrate portion of Vp54 (Figure 27.4a), which is oriented to the outside of the particle, contains seven sugars: glucose, fucose, rhamnose, galactose, mannose, xylose, and arabinose. Six oligosaccharide chains are attached to the protein. However, the four glyscolyated asparagine residues are not located in typical eukaryotic consensus sequences, suggesting that PBCV-1 encodes most, if not all, of the machinery

Figure 27.3 Structure of PBCV-1 virion. The fivefold averaged cryo-EM structure of PBCV-1 viewed down a quasi-twofold axis. Hexagonal arrays of major capsomers form trisymmetrons (green, violet, and blue) and pentasymmetrons (yellow). The unique vertex with its spike structure is at the top. Capsomers in neighboring trisymmetrons are related by a 60° rotation, giving rise to the boundary between trisymmetrons.

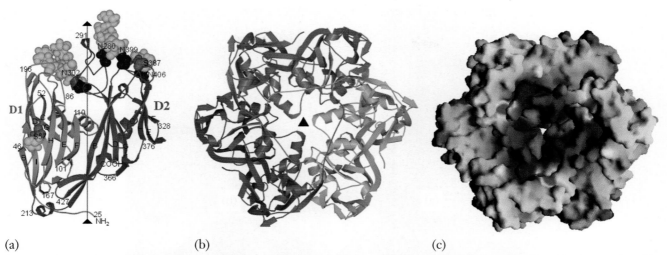

(a)　　　　　　　　　　　　(b)　　　　　　　　　　　　(c)

Figure 27.4 Structure of PBCV-1 capsid protein Vp54. (a) Structure of the Vp54 monomer with strategic amino acids labeled. The carbohydrate moieties (yellow-green) and glycosylated Asn and Ser residues (blue and gray, respectively) are shown as space-filling atoms. (b) The Vp54 trimer viewed from the inside of the virus, with each monomer in a different color. (c) Surface of the Vp54 trimer also viewed from the inside of the virus, colored according to charge distribution (positive is blue, negative is red).

required to glycoslate its major capsid protein. Indeed, glycosylation of Vp54 probably occurs independently of the host endoplasmic reticulum–Golgi system.

Virus entry begins by binding to and degradation of the host cell wall

PBCV-1 initiates infection by attaching to the chlorella cell wall by its spike; attachment is followed by degradation of the host wall at the point of contact by virus-packaged enzymes (Figure 27.5). These enzymes may reside in a small pocket between the spike and the internal membrane. Chloroviruses encode several proteins involved in polysaccharide degradation that may be involved in this process. Following host cell wall degradation, the internal membrane of the virion fuses with the host cell membrane, resulting in entry of the viral DNA and virion-associated proteins into the cytoplasm. An empty virus capsid is left attached to the cell wall (Figure 27.5b).

This process triggers a rapid depolarization of the host cell membrane and the rapid release of potassium ions from the cell, probably directed by a virus-encoded potassium channel located in the virus internal membrane. This depolarization is likely to serve two purposes: it may prevent infection by a second virus, and it lowers the turgor pressure within the cell, which aids entry of viral DNA.

Transcription of viral genes is temporally controlled and probably occurs in the cell nucleus

Circumstantial evidence indicates that the viral DNA and probably DNA-associated proteins quickly move to the nucleus, where early transcription is detected

within 5 to 10 minutes post infection. No viral RNA polymerase has been detected, either by examining the known virus-coded protein sequences or by enzymological methods. Therefore it appears that a host RNA polymerase must be responsible for transcribing the viral genome, and host RNA polymerases are localized in the nucleus. However, it is not understood how the large viral genome is transported through the cytoplasm, nor how it enters the nucleus.

Shortly after infection, host chromosomal DNA begins to be degraded, presumably to aid in inhibiting host transcription and/or to provide a readily available source of nucleotides for viral DNA replication. Viral DNA replication begins 60 to 90 minutes after infection. Of the 365 protein-coding genes of PBCV-1, 227 are expressed before viral DNA synthesis begins. Of these, 127 are considered "early" genes (blue in Figure 27.2), because their RNA transcripts disappear before initiation of viral DNA synthesis, and 100 are considered "early/late" genes (orange), because their RNAs are detected at later times as well. Most of the remaining genes are transcribed only after DNA synthesis begins, and are therefore denoted "late" genes (green). These data were determined by hybridizing copies of polyadenylated RNAs from virus-infected cells to probes corresponding to each of the 365 viral genes.

Progeny virions are assembled in the cytoplasm

Assembly of virus particles takes place in localized regions in the cytoplasm, called virus assembly centers (Figure 27.5c), starting at 2 to 3 h after infection. This poses a problem: if the infecting viral DNA is first transported to the nucleus, where it is transcribed to generate

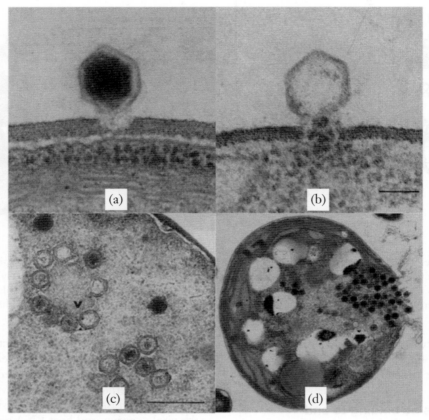

Figure 27.5 Infection of Chlorella strain NC64A by PBCV-1. (a) Attachment of PBCV-1 to the algal cell wall and digestion of the wall at the point of attachment. (b) Viral DNA entering the host cell. (c) Assembly of viral capsids in the cytoplasm. (d) Virion release following cellular lysis.

viral mRNAs, how then does it exit to the cytoplasm to be packaged into virus particles? And where does viral DNA replication occur? The answers to these questions are presently not clear. However, studies with another member of the NCLDV superfamily, African swine fever virus, suggest that viral DNA replication begins in the nucleus, and these replicated DNA molecules are then transported to the cytoplasm, where further replication and packaging take place. The chloroviruses may follow the same pathway. By 5 to 6 h, the cytoplasm becomes filled with infectious progeny virus particles and by 6 to 8 h, localized lysis of the host cell releases progeny virions (Figure 27.5d).

Small and efficient proteins

Many enzymes encoded by PBCV-1 are either the smallest or among the smallest known proteins encoded by phycodnaviruses. Their small sizes and the finding that many virus-encoded proteins are "user-friendly" have resulted in the biochemical and structural characterization of several PBCV-1 enzymes.

Examples include the smallest eukaryotic ATP-dependent DNA ligase; the smallest type II DNA **topoisomerase** (capable of cleaving double-stranded DNA 3 to 50 times faster than the human homologue);

the first RNA **guanylyltransferase** (RNA capping enzyme) whose atomic structure was resolved; a small prolyl-4-hydroxylase that converts proline incorporated into polypeptide chains into hydroxyproline in a sequence-specific fashion; a dCMP deaminase that is also capable of deaminating dCTP (usually enzymes from two different protein families are used for these reactions); and a small (94 amino acid) potassium ion channel protein; ATCV-1 has an even smaller version with only 83 amino acids.

A virus family with a penchant for sugar metabolism: hyaluronan and chitin

The chloroviruses are unusual because they encode many enzymes involved in sugar metabolism. For example, PBCV-1 encodes the enzymes glutamine:fructose-6-phosphate aminotransferase, UDP-glucose dehydrogenase, and hyaluronan synthase, all of which are involved in the synthesis of **hyaluronan**. Hyaluronan is a polymer of disaccharides, themselves composed of D-glucuronic acid and D-N-acetylglucosamine, linked together via alternating β-1,4 and β-1,3 glycosidic bonds. All three genes are transcribed early in PBCV-1 infection and hyaluronan accumulates on the external surface of the infected chlorella cells. However, some chloroviruses have a **chitin** synthase instead of, or in addition to, the hyaluronan synthase. Chitin is an insoluble linear homopolymer of D-N-acetylglucosamine linked by β-1,4 glycosidic bonds. Chlorella infected with these viruses produce either chitin or a chitin/hyaluronan on the surface of infected cells.

Intriguingly, a few chloroviruses appear to lack both genes and produce no extracellular polysaccharides during infection. The functional relevance of these energetically costly biosynthesis pathways remains to be elucidated; however, it is interesting to note that the chloroviruses also encode genes for enzymes involved in sugar degradation, which are essential for viral infection. PBCV-1 encodes two chitinases, a chitosanase, a glycanase, gluconase, and glucouronic lyase. One of the chitinases and the chitosanase are found packaged into the virion and are involved in cell wall digestion, allowing entry of the viral DNA, following attachment of the capsid (see above); these enzymes are also probably involved in release of progeny virus.

COCCOLITHOVIRUSES

Viruses that control the weather

There are not many viruses that have a clear link to controlling the weather, yet the coccolithoviruses ("cocco", derived from Greek *kokkis*, meaning berry or grain, referring to their shape; and "lith", from Greek *lithos*, meaning stone) are able to make this bold claim. Coccolithoviruses infect the cosmopolitan and ecologically important **coccolithophore** *Emiliania huxleyi*, a tiny (5 μm diameter) marine alga (Figure 27.6) that floats freely in the ocean. It is a eukaryotic cell that has an elaborate armor of calcareous (chalk) plates called **coccoliths**.

When conditions are right, *E. huxleyi* can grow in huge numbers to form what is known as a "bloom". Blooms typically contain around 10,000 to 100,000 algae/mL of seawater over areas up to 100,000 km^2, which would give a total of 10^{20} to 10^{21} organisms in the top 10 cm of seawater! Light reflected by coccoliths in these massive and impressive blooms can even be seen from space (Figure 27.7); blooms can range from the size of a small country to a whole continent. Blooms can only be seen from space when the cells are dying and the chalky shell is released into the surrounding sea. When this happens the sea looks a milky white color. The chalk of the White Cliffs of Dover, epitomized in Dame Vera Lynn's famous war-time song, is formed from the coccoliths of algae like *E. huxleyi* killed by viruses over geological time.

Typically, these blooms collapse over a period of 2 to 3 days, releasing a biogenic sulfur gas called dimethyl sulfide (DMS) into the atmosphere, a bit like an oceanic sneeze! DMS is the smell commonly associated with the sea, but has a multitude of functions in algae. Its precursor, dimethly sulfoniopropionate (DMSP) is thought to act as an **osmolyte**, regulating osmotic pressure within the algae. Cleavage of DMSP by the membrane-bound enzyme DMSP lyase during virus infection produces DMS, which is thought to attract certain protozoan-like grazers that engulf the algae. It would seem that the virus-infected cells make a more appealing snack for grazers! High concentrations of DMS can also prevent further infection of *E. huxleyi* by coccolithoviruses, although the mechanisms by which this antiviral property works are not understood.

In the upper atmosphere, oceanic DMS is oxidised into acidic particles (methanesulfonate, sulfur dioxide, and sulfuric acid) that eventually form cloud condensation

Figure 27.6 Electron micrograph of cells of the coccolithophore *Emiliania huxleyi* and a bound virion of EhV-86.

Figure 27.7 True color satellite image of a milky *E. huxleyi* bloom. This bloom occurred in the English Channel, south of Plymouth, UK on July 30, 1999. At this point the bloom was effectively "dead", and up to 1 million *E. huxleyi*-specific coccolithoviruses per mL of water were found in the middle of the high reflectance water.

nuclei (Figure 27.8). The increased cloud cover reflects heat and sunlight back out to space. This is a process that believers of the "Gaia hypothesis" (which postulates that planet Earth is a self-regulating "organism") can extol, since something as small as a virus can control the weather.

Many genes looking for a function

With almost 500 genes and a double-stranded DNA genome of 408 kb, the coccolithovirus *E. huxleyi* V-86 (EhV-86) is the largest algal virus genome sequenced to date (Figure 27.9), but a virus that infects *Pyramimonas orientalis*, a marine microalga, has an even larger genome, estimated at 560 kb. Proteins encoded by coccolithovirus genomes are remarkably different from those of other viruses, as 80% of their protein-coding sequences have no matches in current protein databases. Such degree of novelty provides a tantalizing glimpse of the potential benefits locked within, but many of the gene secrets still remain a mystery.

Of the genes that do have database matches, most are similar to core genes of viruses in the NCDLV group. For example, EhV-86 contains 21 of the core set of 40 to 50 conserved virus genes found in NCLDVs; these genes encode some of the principal features of virion structure, genome replication and expression. The presence of the RNA polymerase holoenzyme (6 subunits of RNA polymerase), together with a family of novel promoter sequences, indicates that coccolithoviruses have their own transcription machinery. This distinguishes these viruses from the chloroviruses, which do not code for RNA polymerase, and therefore must depend on the cell's transcription machinery, which is normally located in the nucleus. Expression of many coccolithovirus transcripts may therefore occur in the cytoplasm rather than the host nucleus. It is not known whether these viruses also have a nuclear phase in their life cycle, as the chloroviruses and some other members of the NCLDV group apparently have.

Expression of coccolithovirus genes is temporally regulated

Microarray analysis of transcribed genes in infected cells showed that a set of genes located between kb 200 and 300 on the genome are specifically transcribed at very early times (within 1 h) after infection. Most of these "early" genes are preceded by a repeated GTTCCC(T/C)AA sequence that may be part of a promoter for RNA polymerase. Interestingly, none of these early genes have homologs in protein sequence databases, and most of the reading frames are modest in length. It is not presently known whether these genes are transcribed by a virus-coded RNA polymerase, which could be packaged within the virion, or cellular RNA polymerases, which reside in the nucleus.

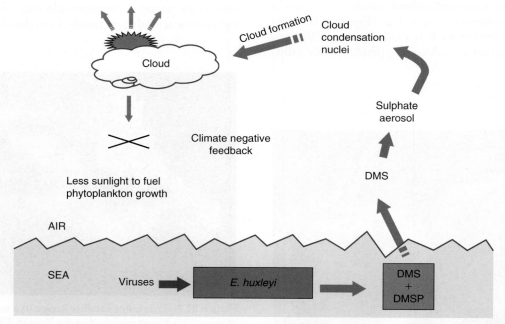

Figure 27.8 Interaction of algae and their viruses with climate. The Gaia hypothesis states that the Earth is a self-regulating organism. This may seem plausible when the activity of coccolithoviruses is taken into consideration. They kill continent-sized blooms of their host organism *E. huxleyi* to produce a massive flux of dimethylsulfide (DMS) into the atmosphere, whose oxidation products subsequently seed cloud formation, blocking the vital fuel of phytoplankton growth, sunlight.

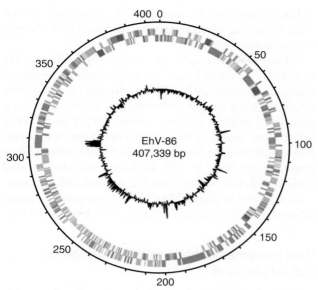

Figure 27.9 Map of the genome of Emiliania huxleyi virus-86 (EhV-86). The scale is numbered clockwise in kbp. The two colored circles show protein-coding regions on the two DNA strands. Genes are color-coded by putative function: red, information transfer; sky blue, degradation of large molecules; yellow, central or intermediary metabolism; light blue, kinases; pink, virus-specific; light green, no known function; dark green, no known function but contain transmembrane helices; gray, miscellaneous. Inner circle: G + C content.

Cheshire Cat dynamics: sex to avoid virus infection

The Red Queen's race in Lewis Carroll's book, "Alice's Adventures in Wonderland", is a common metaphor for an evolutionary arms race. A good example can be seen in predator–prey dynamics, particularly virus–host interactions in which hosts must rapidly evolve immune mechanisms to allow survival of the species: "It takes all the running you can do to stay in the same place". "Red Queen dynamics" can also help explain the bloom–bust cycles of *E. huxleyi*, which seem to be controlled by the coccolithovirus. However, *E. huxleyi* has adopted a novel sex strategy to avoid coccolithovirus infection. This has been dubbed the "Cheshire cat" escape strategy, after the disappearing antics of another famous character in Lewis Carroll's book.

The microalga may respond to coccolithovirus attack by switching from its usual diploid life stage (where it is susceptible to infection) to a haploid cell, changing its physical appearance and making it impenetrable ("invisible") to the coccolithovirus. This would have to occur immediately upon infection before the virus genome has had a chance to replicate (i.e., when only one copy of the genome is present, ensuring that one haploid progeny cell must be virus-free and therefore can survive). As diploids, *E. huxleyi* are non-motile,

coccolith-bearing cells, perhaps ironic given some early thinking that coccoliths actually played an antiviral role. The motile, flagellated, haploid cells lose their chalk armor and are instead covered by organic scales, which may act as a physical barrier to infection. This is potentially a clever antiviral strategy and will create a reservoir of resistant haploid cells that can have sex and possibly create a huge variety of new *E. huxleyi* genotypes. Eventually, some of the fitter genotypes will succeed in generating new blooms when the environmental conditions are favorable. Alternatively, there may be extremely low levels of haploid cells present naturally at all times, which have the opportunity to proliferate when the more vigorous diploid community becomes decimated by viral attack.

This genetic landscape that is constantly changing in response to coccolithovirus infection may be critical in buffering the effects of rapid climate change. Microorganisms such as these algae will probably be the first organisms to react and adapt to a changing ocean. Increased levels of algae will fix increased amounts of CO_2 and have a buffering effect on the concentrations of this "greenhouse gas". Selection for sexual reproduction as an antiviral mechanism will also maintain a high diversity, ensuring that there will always be a genetic variant that can adapt to a particular environmental condition.

Survival of the fattest: the giant coccolithovirus genome encodes sphingolipid biosynthesis

There are several surprises in the coccolithovirus genome, including a group of genes involved in sphingolipid biosynthesis that is thought to be responsible for the production of the sphingolipid, **ceramide**. These genes have never before been found in a virus; they are more commonly seen in animals and plants. Sphingolipid production may be crucial for the generation of lipid raft membrane structures that could aid virus exit from the cell by budding. However, ceramide can also control a cell death mechanism termed **apoptosis**. Typically, apoptosis is used as a defense against virus infection and is controlled by signaling within the host. Premature death of infected cells prevents virus replication and therefore acts as an antiviral mechanism, preventing spread of progeny virions (see Chapter 33).

Coccolithoviruses may have circumvented this defense by acquiring the ability to control the timing of the apoptotic cascade during the infection process, essentially short-circuiting host-controlled apoptosis. Virus-induced **caspase** activation has also been observed during infection (caspases are proteases essential for apoptosis), and, intriguingly, caspase cleavage sites are also found in many EhV proteins, providing further evidence for the manipulation of this pathway during viral infection. For an invading virus, the ability to control when your host will die and to ensure your own survival is an effective propagation strategy. Perhaps as a result of this delay of cell death, coccolithoviruses have burst sizes as large as 1000 progeny virions per algal cell.

PRASINOVIRUSES

Small host, big virus

We now will discuss viruses that infect the world's smallest free-living eukaryote: the green alga *Ostreococcus* (Figure 27.10). *Ostreococcus tauri* (a marine prasinophyte) has a diameter of less than 1 micron, it has no cell wall but instead a naked plasma membrane, and it lacks scales and flagella. Typically containing just a single mitochondrion, chloroplast, and Golgi body, this tiny marine alga has a global distribution and is found at a wide range of ocean depths. Distinct genotypes can be distinguished within the algae in the *Ostreococcus* genus; these genotypes correlate not with geographic location but with their ability to grow in environments with high or low ambient light. That is, variation among these algae appears to correlate with depth in the ocean and not with "horizontal" geographic location.

Two sequenced species of *Ostreococcus*, *O. tauri*, and *O. lucimarinus*, have genomes of 12.6 and 13.2 Mb, respectively. Despite the small size of these algae and their genomes, twenty viruses have so far been

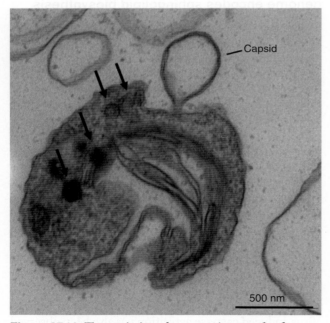

Figure 27.10 **Transmission electron micrograph of a thin section of *Ostreococcus* tauri** (strain OTH 95) infected by OtV-1, eight hours post infection. The empty capsid of the infecting OtV-1 virion is still attached to the host cell membrane. Virions at different stages of assembly can be seen (arrows).

isolated from both *O. tauri* and *O. lucimarinus*. All of these viruses are phycodnaviruses, assigned to the family *Prasinoviridae*.

The prasinoviruses are morphologically similar to other phycodnaviruses, with icosahedral capsids of diameter approximately 120 nm. To date, the genomes of three virus strains infecting *O. tauri* have been sequenced in their entirety: strains OtV-1 and OtV-5 infect algae found in high light conditions, and strain OtV-2 infects algae found in low light conditions. Their genomes are 180- to 190-kb linear, double-stranded DNAs, and code for between 232 and 268 different proteins and four to five tRNAs. OtV-2 contains 42 unique protein coding sequences not found in OtV-1 or OtV-5.

Viral genomes contain multiple genes for capsid proteins

Among the genes of novel function found are N-myristoyl-transferase, 3-dehydroquinate synthase, multiple glycosyl- and methyl-transferases, prolyl 4-hydroxylase, and 6-phosphofructokinase. Other functions encoded include enzymes for DNA replication, recombination, and repair; nucleotide metabolism and transport; transcription; protein and lipid synthesis, modification and degradation; signaling; sugar metabolism; and eight major capsid proteins.

The presence of so many major capsid proteins in each of these viruses is, so far, unique among the phycodnaviruses. Five of these capsid proteins have closest similarity to capsid proteins of *Pyramimonas orientalis* virus-1 (PoV-1), two have closest similarity to capsid proteins of *Heterosigma akashiwo* virus-1 (HaV-1), and one has closest similarity to a capsid protein encoded by *Paramecium bursaria* chlorella virus NY-2A, all viruses in different genera of the phycodnavirus family. The presence and diversity of so many homologous genes coding for major capsid proteins (also observed in Chlorella viruses) raises interesting questions of the potential host range of these viruses, although to date host range appears restricted specifically to *Ostreococcus* species. The ability to switch the major structural component of the virus capsid could have profound implications for future potential host interactions, although this is an avenue of research yet to be explored.

It works both ways

In common with the genetic promiscuity of their relatives, there is evidence of horizontal gene transfers between host and virus in all three sequenced OtV genomes. OtV-1, OtV-2, and OtV-5 contain 11, 14, and 6 genes respectively with close homology to genes of the host, *O. tauri*. The majority of these genes are of unknown function and the direction of transfer is yet to be determined, although to date the vast majority of horizontal gene transfers between the phycodnaviruses and their hosts are believed to have gone in the host-to-virus direction. However, OtV-2 appears to have transfered genes encoding DNA topoisomerase II and ribonucleotide reductase to its host, as these genes are conserved among the NCLDV family.

Not much room for maneuver

As the smallest free-living eukaryote, *Ostreococcus* represents a significant challenge for the replication of a large virus. A virus 120 nm in diameter is replicating inside a cell with a diameter of less than 1000 nm; it can be estimated there is physically room for no more than 100 virions within a cell. When the space required for the nucleus, the chloroplast, ribosomes, and other essential intracellular structures is taken into consideration, this makes for a very crowded infection process! Experiments reveal a typical burst size of 6 to 15 virus particles per cell. Following virus uptake, genome replication begins by 2 hours post infection, and virions are assembled in the cytoplasm from 6 hours until 20 hours post infection, after which cellular lysis occurs. The host cell nucleus (and the chromosomes contained within), mitochondria, and the chloroplast remain intact throughout this period.

PHAEOVIRUSES

Seaweed viruses

Phaeoviruses are latent viruses that infect the brown algae, more familiarly known as seaweeds. As multicellular, filamentous macroalgae, the brown algae differ from the unicellular, microalgal hosts of the other *Phycodnaviridae*. Another major difference is that during infection, the virus genome is integrated into the host cell genome; the virus genome is then inherited in a Mendelian fashion and virus particles only appear in the reproductive organs of the infected algae. This lifestyle resembles that of temperate viruses such as bacteriophage lambda, which integrates its DNA into the host cell genome and only replicates under certain conditions.

Seaweeds predominantly inhabit temperate near-shore marine environments around the world. They can be significant components of fouling communities that occur on marine structures such as docks, buoys, and the exterior surfaces of ships. Viruses are known to infect members in the genera *Ectocarpus*, *Feldmannia*, *Hincksia*, *Myriotrichia*, and *Pilayella*.

Phaeoviruses have a temperate life cycle and integrate their genomes into the host

The life cycle of the filamentous brown alga, *Ectocarpus*, is perhaps the best studied, and it serves as a model organism for studying the molecular genetics of brown algae. *Ectocarpus* has a complicated sexual life cycle comprising alternating generations. Male and female (haploid) gametophytes produce swimming gametes. Phaeoviruses only infect these free-swimming gametes.

Following virion attachment, the viral genome and associated protein is released into the cytoplasm and moves to the nucleus within 5 minutes. The viral genome is then integrated into the host cell genome, and is transmitted during cell division to all the cells in the developing host thallus. The virus genome remains latent in the vegetative filamentous cells until the formation of algal reproductive organs in the mature host. This differentiation triggers viral DNA replication in the reproductive organs, followed by breakdown of the nuclear membrane, virion assembly, and release of virions from the cells. Since viruses displace the normal reproductive organs, the algae are rendered sterile; however, this does not affect their vegetative growth.

PRYMNESIOVIRUSES AND RAPHIDOVIRUSES

The lesser-known *Phycodnaviridae*

These genera of phycodnaviruses are thought to be widespread in the oceans since they infect bloom-forming species of phytoplankton. However, very little research has been conducted on them. Of historical interest, one of the earliest reports of a phycodnavirus was that of a putative prymnesiovirus observed in thin sections of a *Chysochromulina* species in 1974; this was well before the importance of algal viruses had been appreciated.

Prymnesioviruses infect phytoplankton from the algae class *Prymnesiophyceae*, in the division *Haptophyta* (generally referred to as haptophytes). These algae have a global distribution and they are often associated with large-scale blooms. To date, prymnesioviruses have been isolated from members of the genera *Chrysochromulina* and *Phaeocystis*. Large monospecific blooms of *Chrysochromulina* are rare; typically they are present in low concentrations. This property has led to speculation that viruses that infect *Chrysochromulina* only require a low host density for propagation and that these viruses may even prevent bloom formation.

In contrast, a lot of information exists on the biogeochemical impact on the marine ecosystem by members of the genus *Phaeocystis*. They form dense, spatially and temporally extensive, monospecific blooms consisting of a mixture of colonial cells (within a gelatinous matrix) and unicellular cells that collapse suddenly in a virus-induced crash. This crash leads to a rapid shift in the composition of the bacterial community due to the massive flux of released organic nutrients. Intense *Phaeocystis* blooms can lead to anoxia and impressive foam formation on beaches during their decline, hence their label as harmful algal blooms. Like *E. huxleyi*, *Phaeocystis* species play important roles in CO_2 and sulfur cycling, which ultimately have major implications on global climate; infection by viruses is known to affect these processes.

Raphidoviruses infect algae from the class *Raphidophyceae*, which are often associated with toxic red tides and subsequent fish kills, particularly in the aquaculture industry. This class of algae includes important bloom-forming species found in coastal and subarctic regions of the world's oceans, although freshwater species also exist. To date, the only viruses reported in this genus infect the red tide-forming species *Heterosigma akashiwo*; these viruses are referred to as *Heterosigma akashiwo* virus (HaV).

*A*lthough not usually associated with human illness, blooms of *H. akashiwo* have caused massive fish kills, typically of caged, farmed fish such as salmon and yellowtail in places like Japan and Peru. The susceptibility of *H. akashiwo* to HaV differs among clonal strains in the laboratory. Marine field surveys and cross-reactivity tests between *H. akashiwo* host strains and HaV clones suggest that this strain-dependent infection plays an important role in determining clonal composition and maintaining intraspecies diversity in natural *H. akashiwo* populations. This diversity probably contributes to the success of *H. akashiwo* as a ubiquitous and problematic bloom-former in coastal regions.

MIMIVIRUS

The world's largest known virus

In a textbook aiming to portray the individual personalities of viruses, the mimivirus stands out as possibly the most personality-filled virus currently under study. From the humblest of origins (it was originally isolated from a water cooling tower in Bradford, England), it has since taken the world by storm, and currently holds the world record for being the largest known virus. In a research field often dominated by the claim "my virus is bigger than yours!" the mimivirus stands head and shoulders above its peers as the largest. It is only because so little is known about it that it has not been assigned a chapter of its own.

Yet the study of the mimivirus from a virological perspective got off to a slow start. It was originally thought to be a small *Legionella*-like bacterium, and was tentatively named *Bradfordcoccus* because it appeared to be a gram-positive bacterium. However, research progress was hindered by the repeated failure to amplify ribosomal RNA sequences. Electron microscopy ultimately revealed a polyhedral structure, similar to many virions of the nucleo-cytoplasmic large DNA viruses, but substantially larger.

It was then shown to be a virus that replicates in an amoeba, and so was named after its host, *Acanthamoeba polyphaga* mimivirus. "Mimivirus" was chosen as a shortened version of *mi*crobe *mi*micking virus. The classic definition of a virus is an obligate intracellular parasite that lacks the core protein synthesizing apparatus (ribosomes and associated factors) and does not have systems to supply chemical energy or metabolic needs. The mimivirus was one of the first viruses to blur these boundaries.

Mimivirus is unquestionably a virus

Easily visualised by light microscopy, it is clear to see why this hairy 750-nm monster was mistaken for a bacterium at first glance (Figure 27.11). The mimivirus is currently the virus with the largest fully sequenced genome: linear, double-stranded DNA of 1,181,404 base pairs and a staggering 911 predicted protein-encoding genes plus six tRNA genes. It has a coding capacity that eclipses numerous bacteria and archaea (such as *Chlamydia*, *Rickettsia*, *Mycoplasma*, and *Nanoarchaeum*).

This huge genome size is one of the reasons why the study of the mimivirus has triggered a plethora of articles debating viral evolution and (more controversially) the very origins and definitions of life. However, despite the controversy, the mimivirus is a *bona fide* virus. Although it codes for a number of transfer RNAs and proteins involved in protein synthesis, it has no ribosomes of its own. Furthermore, like all viruses, it uncoats and replicates its genome within a host cell, then packages it within a virion. Phylogenetic analysis based on conserved genes in the NCLDV superfamily places mimivirus as a member of this group of viruses, closely related to the *Phycodnaviridae* family. Some

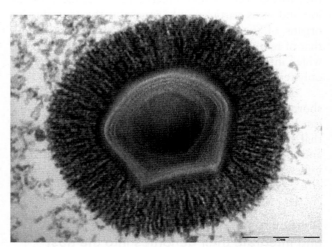

Figure 27.11 Transmission electron micrograph of a mimivirus particle in a thin section. A heavy fringe of fibers surrounds the capsid, within which several internal layers and the core can be seen.

newly isolated phycodnaviruses have shown even closer phylogenetic relationships with the mimivirus lineage, further justifying its inclusion in this chapter.

Why such a large genome?

The mimivirus exhibits the genome compaction typical of DNA viruses, as a full 90.5% of its DNA is occupied by protein-coding sequences, with an average distance between open reading frames of only 157 nucleotides. The main difference between the mimivirus and all other viruses is sheer size. It is the blue whale of viruses. This raises the obvious questions: how and why did it get this big?

The large genome size is clearly not due to the accumulation of noncoding or junk DNA, as in many eukaryotic organisms. Both DNA strands encode approximately the same number of genes (450 genes on the "R" strand versus 465 genes on the "L" strand). The obvious conclusions are that the mimivirus is either a serial acquirer of genetic information (by horizontal gene transfer) or it has a propensity for duplication of its own genes. However, horizontal gene transfer from the host has been predicted to contribute a minimal component of the mimivirus genome, although a lateral gene transfer has been suggested for at least one gene (see later).

Clues are beginning to emerge which suggest that self-duplication is the main mechanism by which the genome size has expanded. There appears to have been a segmental 200-kb duplication of the telomeric regions at both ends of the linear genome; furthermore, tandem duplications are common throughout the genome. Of particular note are two families of **paralogous** gene families (genes that apparently arose by duplication) within the mimivirus genome. One family contains 66 members that encode **ankyrin** domains, and another family consists of 12 members located in perfect tandem sequence from L174 to L185. Indeed, it has been suggested that more than one-third of mimivirus genes have at least one homologue on the genome.

A third hypothesis to explain the large size of the mimivirus genome is that it has evolved from a free-living cellular organism that became an intracellular parasite and lost many of the requirements for cellular existence. This theory is not easy to prove or disprove.

Mimivirus has a unique mechanism for releasing its core

Little is known of the precise molecular workings of the mimivirus infection cycle, but data are rapidly accumulating that allow a rough outline to be pieced together. It is likely that the mimivirus tricks the amoeba into internalizing it by having the appearance of an innocuous bacterial food source. It is taken up within the host cell

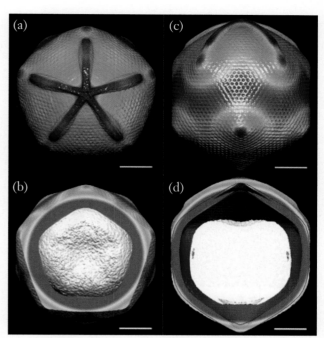

Figure 27.12 Stargate structure on the mimivirus capsid. Particle reconstructions from cryoelectron micgrographs of mimivirus particles, using fivefold symmetry averaging. Images are colored red at smaller diameters through yellow and green to blue at greater diameters. (a) Stargate structure (blue) seen down the unique fivefold vertex. (b) Section through the edge of a particle in the same orientation after removing the stargate structure, showing a depression in the core (in white). (c) Stargate structure seen from one side (threefold symmetry axis). (d) Section through the center of a particle in the same orientation as in (c) shows a depression in the core (white) below the stargate vertex.

phagosome, which these amoeba use to scavenge bacteria and detritus from their environment. The outer fibril layer becomes partially digested within the phagosome, allowing interaction of the capsid surface with the phagocytic vesicle membrane.

Mimivirus virions feature a unique pentagonal star-shaped structure at a single vertex of their icosahedral capsids (Figure 27.12). The five edges of the capsid that radiate from this vertex are decorated with a starfish-like feature that opens gaps between the five adjacent faces of the capsid. This structure opens within the phagosome to create a large "stargate", revealing the virus core wrapped within a membrane. This membrane apparently fuses with the phagosome membrane, allowing passage of the virus core into the cytoplasm (Figure 27.13).

Virus replication occurs exclusively in the cytoplasm

The remaining replication cycle appears to take place entirely in the cytoplasm in virus "factories" that remain

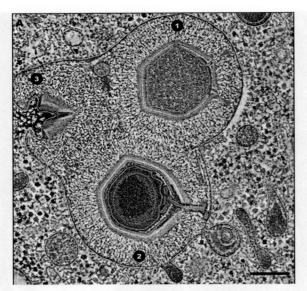

Figure 27.13 Core membrane of mimivirus fuses with phagosome membrane. Transmission electron micrographs of a late phagosome in a cell infected with mimivirus, showing three particles at different stages of uncoating. At the early uncoating stage (1), a partial opening of the inner protein shell at the stargate assembly is initiated; the stargate is denoted by two red arrows. Later (2), the opening of the stargate allows for the extrusion of the viral membrane toward the phagosome membrane. In the final uncoating stage (3), fusion between viral and phagosome membranes occurs, and the core is extruded into the cytoplasm. (Scale bar: 100 nm).

associated with the original virion core structure (Figure 27.14), similar to the life cycle of poxviruses (see Chapter 26, Figure 26.5 and 26.7). While the external fibril layer remains within the phagosome, small spherical compartments can be seen in the cytoplasm 2 hours post infection; these precisely match the size of the intact virion core containing the viral genome.

The mimivirus virion is highly complex and has been shown to contain at least 130 different virus-coded proteins, including a set of proteins homologous to the subunits of eukaryotic RNA polymerase. Presumably, this plethora of proteins is required immediately upon release of the virion core into the cytoplasm.

Mimivirus early transcription appears to be driven by a highly conserved AAAATTGA sequence in the region immediately upstream of the messenger RNA start site of about one-half of mimivirus genes. This sequence probably serves as a mimivirus-specific promoter similar to the familiar TATA box of higher eukaryotes. Presumably these transcripts are extruded from the intact core into the cytoplasm, where they are translated.

At some point, the mimivirus cores apparently release their viral DNA into the cytoplasm. Empty cores can be seen surrounded by a mass of DNA, which increases in size as the infection progresses and viral DNA is replicated. New virions can be seen assembling at the edges of these virus factories (Figure 27.14a). Eventually the amoeba undergoes lysis at approximately 14 hours post infection. A typical burst size is usually greater than 300 viruses per infected cell.

Genome replication

The linear 1,181,404-bp mimivirus genome contains no inverted terminal repeats, but does contain a nearly perfect 617-bp inverted repeat at positions 22,515 and 1,180,529. Interaction of these regions would lead to the creation of a "Q"-like circular form of DNA containing short and long tail extensions, 259 bp and 22,515 bp, respectively (Figure 27.15). This form could be involved in replication of the viral genome.

No coding sequences are found on the short tail; however, there are 12 protein-coding regions on the longer tail, 7 of which have predicted functions involved in DNA replication or binding. These include four proteins that may bind to the origin of replication, a chromosome-segregation ATPase, and two DNA-binding proteins. The origin of replication itself is predicted to be located 400,000 nt from the left end of the genome.

Genes coding for translation factors and DNA repair enzymes

With such a large genome size, it is no surprise that the mimivirus encodes a wide variety of proteins with a diverse array of potential functions. However, it is striking that for such a large virus, so far few proteins with truly novel metabolic functions have been identified on the mimivirus genome. Genes for four aminoacyl-tRNA synthetases, an mRNA cap-binding protein, translation initiation factors, tRNA methyltransferases, and peptide release factor are clearly of interest since they are all involved in protein synthesis. Perhaps the most interesting aspect is not their presence but their number; at least ten proteins have accessory roles in protein synthesis. It appears that the mimivirus has a heavy requirement for additional translation factors, which is perhaps no surprise given the potential number of proteins it encodes, yet these genes still contribute a mere ten proteins among the potential predicted total of 911 mimivirus proteins.

An interesting feature of the mimivirus genome is the presence of a range of DNA repair enzymes capable of correcting errors introduced by oxidation, ultraviolet light irradiation, and alkylating agents. These include three types of topoisomerase IIA (common to most large DNA viruses), topoisomerases IA and IB, and a DNA UV damage repair endonuclease. The mimivirus genome may be more prone to damage due to its large size. Long periods in between infections could result in prolonged exposure to the external environment and may make mimivirus

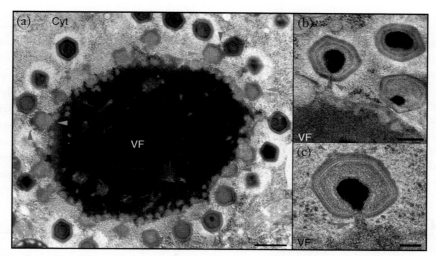

Figure 27.14 Mimivirus factories. Transmission electron micrographs of a cell infected with mimivirus. (a) Virus factories (VF) surrounded with virus particles at various stages of assembly. Fiber-less particles lie close to the edge of the virus factory (blue, purple, and red arrowheads); fiber-covered particles lie farther away (yellow arrowhead). The stargate can be seen on some particles (red arrowheads). (Scale bar, 100 nm). (b,c) Higher-magnification images of particles undergoing DNA packaging (green arrowheads) at a position opposite the stargate structure, seen clearly in c (red arrowheads).

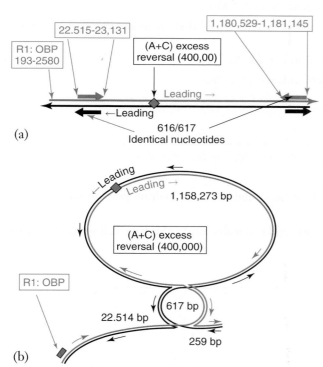

Figure 27.15 Schematic structure of the mimivirus genome. (a) Viral genome with inverted repeats (orange and black arrows), and a likely origin of replication (green diamond). A gene encoding a possible origin of replication DNA-binding protein (R1: OBP) is located at the extreme left end. (b) Putative circularized Q-like form obtained by pairing the two inverted repeats. Note that repeats are not drawn to scale, as they are only 617 nt out of a total of 1,181,404 nt.

susceptible to UV damage, whereas exposure to oxidizing conditions during envelopment in the phagocytic vesicle may trigger significant oxidative damage.

Ancestors of mimivirus may have transferred genes from bacteria to eukaryotes

MutS is an interesting DNA repair gene that provides a telling insight into the checkered history of the genome of the mimivirus. MutS is part of a DNA mismatch repair enzyme system that binds to regions on DNA where bases are mismatched as a result of DNA damage; it is highly conserved from bacteria to humans. Phylogenetic analysis shows that the mimivirus MutS protein is most closely related to MutS homologues in the bacterium *Nitratiruptor* (a nitrate-reducing bacterium found in deep-sea hydrothermal vents) and the eukaryote *Octocorallia* ("soft corals", marine organisms that form colonial polyps).

The *Octocorallia* homologue of MutS, known as Msh1, is encoded on the mitochondrial genome of all octocorals, but is not present in mitochondria of even the closely related hexacorals. Therefore, it has been suggested that an ancestral mimivirus was involved in transfer of this gene from marine bacteria to the ancestors of the *Octocorallia* lineage, perhaps 450 million years ago. This has spawned the "melting pot" hypothesis for amoebae and other scavengers: by phagocytosis they concentrate a variety of "food" organisms and force interactions between their genomes, thus facilitating horizontal gene transfers between organisms in different domains of life.

Conclusion

The phycodnaviruses and mimivirus represent an incredibly diverse group of viruses with very large genomes but a common virion architecture. With far-reaching implications for the function of the global ecosystem, these understudied giants of the virus world have only recently begun to reveal their tantalizing secrets. Full of surprises, novelty, and mystery, they will continue to enthrall researchers into the distant future. The future is bright for the phycodnaviruses; the future is green!

KEY TERMS

Ankyrin	Hyaluronan
Apoptosis	Inverted repeat
Caspase	Jelly-roll β-barrel
Ceramide	Osmolyte
Chitin	Paralogous
Coccolithophore	Pentasymmetrons
Coccoliths	Topoisomerase
Endosymbiotic	Trisymmetrons
Guanylyltransferase	Ubiquitin

FUNDAMENTAL CONCEPTS

• Phycodnaviruses are large, double-stranded DNA viruses infecting algae.

• The existence of phycodnaviruses has been known for a little over three decades.

• Phycodnaviruses are ancient and, despite their conserved physical appearance, have incredibly diverse lifestyles.

• Horizontal gene transfer is a recurring theme in phycodnavirus and mimivirus evolutionary history.

• Coccolithoviruses can affect the weather by causing the demise of coccolithophore blooms and triggering the release of climate-active gases into the atmosphere.

• Phycodnavirus genomes harbor genes that encode the smallest known versions of many proteins.

• The mimivirus is the world's largest sequenced virus, with a genome of 1.2 Mb.

• Mimivirus extrudes its core into the cytosol via a "stargate" structure in its capsid that opens up within phagosomes in infected amoebae.

• Mimivirus appears to replicate exclusively in the cytoplasm of infected cells.

• The phycodnaviruses and mimivirus are full of novel genes of unknown function.

REVIEW QUESTIONS

1. Why would you predict that the phycodnaviruses may play a role in understanding global warming?

2. How diverse are the genomes of the phycodnaviruses?

3. The Vp54 protein that makes up the capsid of the PBCV-1 virion contains a carbohydrate portion unexpectedly containing what?

4. Plant viruses require a physical break to infect a cell because of rigid cell walls. Algae also have cell walls; how does the chlorovirus PBCV-1 enter a cell?

5. How do the coccolithoviruses control the weather?

VIRUSES THAT USE A REVERSE TRANSCRIPTASE

The viruses covered in this section use a reverse transcriptase, an enzyme that produces a DNA copy of an RNA molecule, during their life cycle. These viruses may be relics of the transition between the RNA world and the DNA world; reverse transcriptase was probably responsible for generating DNA copies of the RNA molecules postulated to have been the genomes of primitive cells.

Retroviruses are the best-known viruses that use reverse transcriptase; the enzyme was first discovered in these viruses. Upon entering the cell, retroviruses use reverse transcriptase packaged within the virion core to make a fully double-stranded DNA copy of their single-stranded RNA genome. This DNA copy is then transported to the nucleus where another packaged viral enzyme, integrase, directs the insertion of the viral DNA into the cellular chromosome at random sites. At this point, the viral genome has quite simply become a new set of cellular genes, which can be transmitted indefinitely by the cell during DNA replication and cell division.

In response to appropriate signals, the integrated retrovirus genome can be transcribed by cellular RNA polymerase II, giving rise to full-length viral genome RNA and to spliced versions that serve as messenger RNAs for viral structural and enzymatic proteins. Certain retroviruses have acquired copies of genes that encode cellular signaling proteins, and in some cases these viruses can transform normal cells into cancer cells; their study has led to great advances in the understanding of oncogenesis.

The discovery of human immunodeficiency virus (HIV) brought to the fore public awareness of retroviruses. This sexually transmitted virus caused an ongoing worldwide epidemic of acquired immune deficiency syndrome (AIDS) that has killed millions of people and remains unchecked despite the availability of effective anti-HIV chemotherapy. AIDS is caused by loss of cellular immunity due to depletion of T lymphocytes; illness and death is a result of infection by a variety of pathogens that can no longer be repulsed by the weakened immune system.

Hepatitis B virus also uses reverse transcriptase during its replication, but it packages a DNA genome within its virions. This DNA genome is transcribed in the cell nucleus to generate viral mRNAs as well as a pre-genome RNA transcript that is packaged in viral cores along with reverse transcriptase. This virus infects liver cells and can cause a lifelong infection, which can give rise to liver cancer or degeneration; like HIV, it is sexually transmitted. An effective vaccine is available. Some plant viruses also have a double-stranded DNA genome that is made from an RNA by reverse transcriptase.

Retroviruses

Alan Cochrane

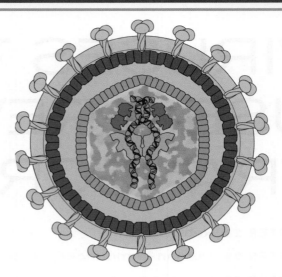

Retroviridae

From *reverse transcription*, characteristic of these viruses

VIRION

Spherical enveloped particle.
Diameter 100 nm.
Envelope assembled at plasma membrane.
Icosahedral or conical capsid.
Contains reverse transcriptase.

GENOME

Linear, single-stranded RNA, positive sense.
Two identical molecules, 7–10 kb, packaged in virions.
5′ cap, 3′ poly(A) tail.

GENES AND PROTEINS

mRNAs are transcribed from integrated provirus DNA by RNA polymerase II.
Three major genes, each translated as a polyprotein and cleaved into mature proteins:
- gag: *nucleocapsid proteins*
- pol: *reverse transcriptase, protease, and integrase*
- env: *envelope proteins*

Gag and Pol proteins are made from unspliced mRNA.
Env proteins are made from singly spliced mRNA.
Lentiviruses and human T-cell leukemia viruses have additional regulatory genes expressed from multiply spliced mRNAs.

VIRUSES AND HOSTS

Seven genera are recognized: *Alpha*- to *Epsilonretrovirus, Lentivirus, Spumavirus.*

- Simple retroviruses: *Rous sarcoma virus (chickens), mouse mammary tumor virus (mice), Jaagsietke sheep retrovirus.*
- Lentiviruses: *human, feline, bovine, and simian immuno- deficiency viruses*
- Spumaviruses: *simian foamy virus.*

DISEASES

Humans: AIDS, leukemia.
A variety of cancers in monkeys, mice, cats, sheep, birds, etc.

DISTINCTIVE CHARACTERISTICS

Two identical copies of the genome per virion.
Genome RNA is converted to DNA by reverse transcription.
A cellular tRNA is used as a primer to initiate reverse transcription.
A DNA copy of the genome (provirus) is integrated into host cell DNA.
Retrovirus populations exist as quasispecies in their hosts because of frequent errors made by reverse transcriptase.
Can cause cancer by expressing mutated cellular regulatory genes or by integrating into or nearby cellular regulatory genes.

Retroviruses have a unique replication cycle based on reverse transcription and integration of their genomes

The first members of the retrovirus family were described at the beginning of the twentieth century, when a virus responsible for equine infectious anemia was isolated. Shortly afterward, in 1911, Peyton Rous showed that the Rous sarcoma virus induces tumor formation in chickens. In neither case was the nature of these viruses understood until much more recently.

The discovery of **reverse transcriptase** independently in the laboratories of Harold Temin and David Baltimore in 1971 demonstrated that retroviruses integrate a DNA copy of their RNA genome into the chromosomes of infected cells. Shortly thereafter the laboratory of Michael Bishop and Howard Varmus showed that the ability of Rous sarcoma virus to induce tumors results from the acquisition of an altered cellular gene (an **oncogene**) by the virus. These discoveries were all awarded Nobel prizes, although Peyton Rous had to wait 55 years between his discovery and his prize!

Retroviruses have been isolated from numerous animal species, including chickens, mice, cattle, horses, cats, monkeys, and humans. Retroviruses are presently grouped into seven genera (Table 28.1), based on differences in morphology and genome organization. A previous but related classification was based on the pathology associated with infection: *oncoviruses* (many viruses in the *Alpha-* to *Epsilonretrovirus* genera) are tumor-inducing viruses; *lentiviruses* induce slowly progressing, wasting disease; and *spumaviruses* ("foamy" viruses) induce persistent infection without any associated pathology.

The discovery of human immunodeficiency virus (HIV), a lentivirus, in the mid-1980s (see Chapter 29)

led to an explosion of work on this newly emerged human pathogen, responsible for a worldwide epidemic of acquired immunodeficiency syndrome (AIDS).

Viral proteins derived from the gag, pol, and env genes are incorporated in virions

Retroviruses are roughly spherical and approximately 100 nm in diameter (Figure 28.1). The envelope contains the external surface protein (SU), bound by noncovalent interactions to the transmembrane protein (TM), which traverses the lipid bilayer. Coating the inner surface of the membrane is the viral matrix protein (MA). The capsid protein (CA) forms an icosahedral or conical core, depending on the virus strain. Three virus-coded enzymes—a protease (PR), an integrase (IN), and a reverse transcriptase (RT)—are associated with the virus core. The viral strucural proteins are often identified by their glycosylation status and their molecular weights; for example, the human immunodeficiency virus SU protein is called gp120, TM is gp41, and CA is p24.

Table 28.1 Retrovirus genera

Genus	Examples of virus	Host
Alpharetrovirus	Rous sarcoma virus	Chickens
Betaretrovirus	Mouse mammary tumor virus	Mice
Gammaretrovirus	Murine leukemia virus	Mice
Deltaretrovirus	Human T-cell leukemia virus type 1	Humans
Epsilonretrovirus	Walleye dermal sarcoma virus	Fish
Lentivirus	Human immunodeficiency virus type 1	Humans
	Simian immunodeficiency virus	Monkeys
	Feline immunodeficiency virus	Cats
Spumavirus	Simian foamy virus	Monkeys

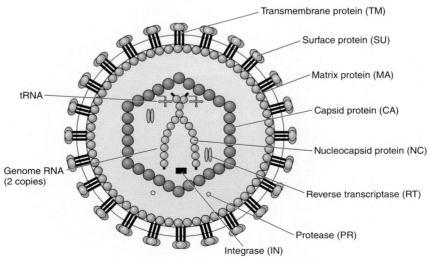

Figure 28.1 A typical retrovirus virion. Schematic diagram showing relative positions of structural proteins and RNAs in the virion.

The virus genome (Figure 28.2) is a positive-strand RNA 7 to 10 kb long, complexed with the nucleocapsid protein (NC). Genome RNAs are capped at their 5' termini and polyadenylated at their 3' termini, as are eukaryotic mRNAs. A 150- to 200-nt repeated sequence (R) is present at both the 5' and 3' ends of the genome RNA. Adjacent to the repeated sequences at either end are unique regions designated U5 (80–200 nt) and U3 (240–1200 nt).

Unlike most other viruses, retroviruses package two identical copies of the genome in each virion. Within the virion the two RNAs exist as a dimer, held together in a head-to-head configuration by interaction of sequences known as a "kissing loop" located in the U5 region. A specific cellular transfer RNA is bound to the genome RNA by base pairing between the primer binding sequence (PBS), located just downstream of U5, and 18 nucleotides at the 3' end of the tRNA. Different virus strains bind different cellular tRNAs. Viral proteins are generated from three genes designated gag (for "group-specific antigen"), pol (for polymerase), and env (for envelope proteins).

Retroviruses enter cells by the fusion pathway

The retrovirus life cycle takes place in two phases. During the early phase (Figure 28.3) the virus enters the cell, makes a DNA copy of its RNA genome, and inserts this copy into the host cell genome. Retroviruses can only infect certain species and types of cells, a property based mainly on the interaction of SU with receptor proteins on the cell surface. Examples of virus receptors are:

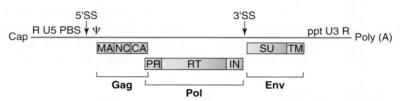

Figure 28.2 Structure of retrovirus RNA. Beginning at the left end, features are: methylated cap; repeat region (R); untranslated 5' sequence (U5); primer binding sequence (PBS); 5' splice site (5' ss); psi (ψ) packaging sequence; gag, pol, and env reading frames for viral structural genes; 3' splice site (3' ss); polypurine tract (ppt) used during reverse transcription; untranslated 3' sequence (U3); repeat region (R), poly(A) tail.

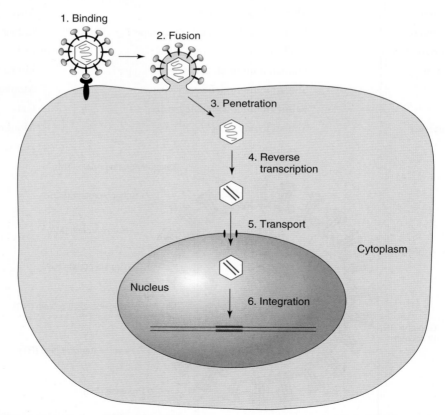

Figure 28.3 Early phase of retrovirus life cycle. RNA genomes are shown as curved light blue lines; they are converted to double-stranded DNA (provirus), shown as parallel straight dark blue lines, by reverse transcription.

- CD4 and the **chemokine** receptors CCR5 and CXCR4, for HIV-1 (see Chapter 29)
- the amino acid transporter EcoR/Rec-1, for murine leukemia virus
- the phosphate transporters Pit-1/Glvr-1 and Pit-2/ Glvr-2, for the gibbon ape leukemia virus
- Tva, a homologue of the low-density lipoprotein receptor, for avian sarcoma leukosis virus type A

Following interaction of SU with its receptor, the viral envelope either fuses directly with the plasma membrane, or the virus undergoes receptor-mediated endocytosis and fusion is initiated by a decrease in pH within the endosome. Fusion is induced by a conformational shift in the SU protein that exposes the hydrophobic amino terminus of the TM protein, which is then inserted into the cell membrane. Fusion releases the virion core into the cytoplasm.

Viral RNA is converted into a double-stranded DNA copy by reverse transcription

In the cytoplasm, deoxynucleoside triphosphates penetrate into the core, and the process of conversion of viral RNA to double-stranded DNA begins. Reverse transcriptase, the enzyme responsible for this conversion, exists as a dimer that has two distinct activities: an RNA/DNA-dependent DNA polymerase activity, and a **ribonuclease H** activity. Ribonuclease H is an enzyme that selectively hydrolyses the RNA part of an RNA-DNA hybrid, but does not degrade unhybridized RNA. The cellular tRNA molecule that is bound to the primer-binding site (PBS) on the viral RNA serves as a primer for the initiation of synthesis of DNA complementary to the genome RNA. Conversion of RNA to DNA is a complex process that proceeds as follows (Figure 28.4):

1. *Synthesis of minus-strand strong-stop DNA.* DNA synthesis initially extends from the tRNA primer up to the 5′ end of the viral RNA, resulting in a short (100–400 nt) DNA molecule that is a copy of the U5 and R regions. This is called "minus-strand strong-stop DNA" because it is a complementary copy of part of the positive-strand genome RNA, and DNA synthesis stops abruptly at the end of the RNA template during *in vitro* reactions.

2. *Removal of template RNA.* RNase H digests the RNA part of the newly made DNA-RNA hybrid (but not the RNA-RNA hybrid at the primer binding site), removing the RNA template and exposing the DNA copy of the U5 and R sequences.

3. *Strand transfer.* The DNA copy of the R sequence hybridizes with the R sequence at the 3′ end of the genome RNA. This is called the "first strand transfer", since the strong-stop DNA strand is transferred from the 5′ end to the 3′ end of the genome RNA.

The actual geometry of how this transfer takes place within the virus core during reverse transcription is not clear. Although the resulting complex is drawn as a linear molecule in Figure 28.4, it could remain circular.

4. *Copying of full-length genome.* The strong-stop DNA is extended by reverse transcriptase to copy the entire remaining part of the genome RNA up through the primer-binding site, making a full-length minus-strand copy of the genome. Because of the strand transfer, this DNA molecule contains a copy of the R sequence sandwiched in between the U3 sequence, originally at the right end of the viral RNA, and the U5 sequence, originally at the left end. The U3-R-U5 sequence is called a "long terminal repeat" (LTR).

5. *Removal of template RNA.* As it is copied into DNA, most of the genome RNA is digested by RNAse H. However, an RNA sequence just to the left of U3, consisting of a polypurine tract (ppt), is resistant to digestion, and remains hybridized to the newly synthesized minus-strand DNA. This residual RNA serves as a primer for the subsequent synthesis of plus-strand DNA by reverse transcriptase.

6. *Synthesis of plus-strand strong-stop DNA.* Synthesis is initiated by the ppt RNA primer and extends through the U3-R-U5 long terminal repeat just formed, and on through the 18 nucleotides of the tRNA that were initially hybridized to the primer binding site on genome RNA. The short DNA intermediate made is designated "plus-strand strong-stop DNA".

7. *Removal of tRNA and ppt primer.* RNase H digestion of the 3′ end of the tRNA, now part of an RNA-DNA hybrid, removes the tRNA from the minus-strand DNA copy and exposes the primer binding site on the plus-strand strong-stop DNA. The ppt primer is also removed.

8. *Second strand transfer.* This exposed primer binding site (PBS) can hybridize with its complementary PBS' sequence at the other end of the newly synthesized minus-strand DNA. This is called the "second strand transfer" since, like the first strand transfer, the plus-strand strong-stop DNA is transferred from one end of the template to the other end. Again, this is drawn as a linear molecule in Figure 28.4 but could involve a circular intermediate, with the plus-strand strong-stop DNA acting as a bridge between the two ends of the minus-strand DNA.

9. *Extension of both DNA strands.* Both the minus and plus DNA strands are then extended by reverse transcriptase to the ends of their respective template strands. This results in a linear, double-stranded DNA with long terminal repeats at both ends. This DNA is called **proviral DNA**.

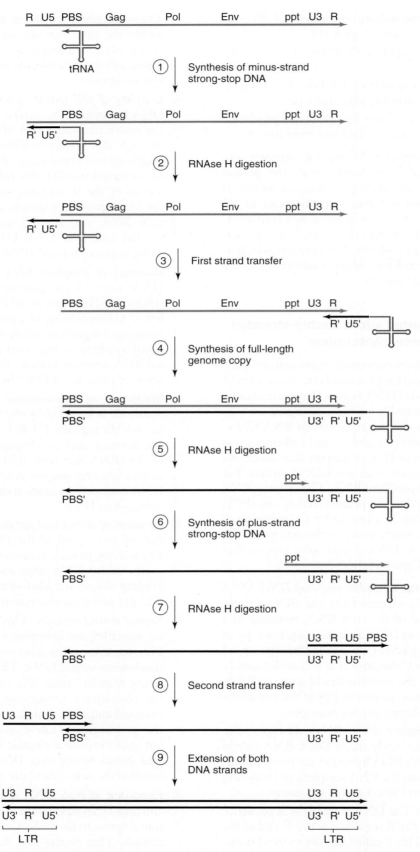

Figure 28.4 Reverse transcription. Blue lines correspond to RNA, black lines correspond to DNA. Arrowheads denote 3' ends. See text for details.

The reverse transcriptase of retroviruses lacks the 3'-to-5' exonuclease activity that cellular DNA polymerases use for proofreading. This exonuclease activity is used by cellular DNA polymerases to remove regions that contain mispaired bases, allowing their repair. Because this activity is missing, between 1 and 10 nucleotide errors are introduced during synthesis of each proviral DNA molecule. As a result, retrovirus populations cannot be considered uniform but rather are a collection of variants or **quasispecies**.

The linear double-stranded viral DNA resulting from reverse transcription remains associated with components of the virus core in a **preintegration complex**. The large size of this complex prevents its entry through the nuclear pores, and most retroviruses must await the breakdown of the nuclear envelope that occurs during cell division. For this reason, these retroviruses can only productively infect cells that undergo mitosis. However, lentiviruses code for proteins that mediate the active transport of the viral DNA complex through the nuclear pore (shown in Figure 28.3; see also Chapter 29).

A copy of proviral DNA is integrated into the cellular genome at a random site

Insertion of the proviral DNA into the host cell chromosome is catalyzed by the viral integrase, present in the core of the infecting virion. The integrase binds to the two ends of linear viral DNA and brings them together and in close proximity to cellular DNA.

The reaction proceeds as follows (Figure 28.5):

1. The integrase removes the two 3'-terminal nucleotides of each strand of the linear viral DNA.

2. Viral DNA is inserted into host DNA by a concerted cleavage and ligation reaction. The integrase brings the two 3'-OH ends of viral DNA close to two phosphodiester linkages 4 to 6 nt apart on the host DNA, and joins viral to cellular DNA strands. This leaves a 4- to 6-nt single-stranded gap on the target host DNA, and a 2-nt unpaired region on the viral DNA.

3. Host enzymes carry out repair synthesis of the gap, simultaneously removing the terminal 2 unpaired nucleotides of the viral DNA. This generates a direct repeat of host DNA (4–6 bp depending on the virus) and results in the loss of the terminal 2 bp of viral DNA. Note that the loss of viral sequences at the ends of viral DNA has no effect on progeny virus because the ends are not used in the synthesis of viral RNA (see below).

Integration sites appear to be distributed randomly over the host genome. Once integrated, the proviral DNA becomes part of a host cell chromosome and is replicated along with host DNA, just like any cellular

gene. Consequently, spread of the infection within an animal can be achieved either by infection of new cells with progeny virus or by multiplication of cells already containing proviral DNA. Furthermore, virus infections can be transmitted from parent to offspring if an egg or sperm cell becomes infected and contains integrated proviral DNA. The stable integration of the viral DNA also renders "curing" of a retrovirus infection difficult, requiring the destruction of all cells containing proviral DNA, whether or not it is actively expressed.

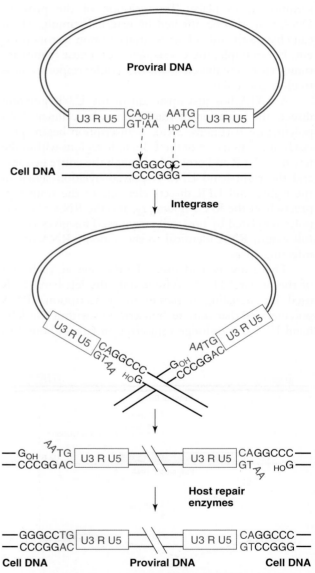

Figure 28.5 Integration of proviral DNA into host cell DNA. Viral integrase cleaves 2 nucleotides from the 3' ends of the viral DNA and then joins these 3' ends to host DNA in a concerted cleavage-ligation reaction at sites 4–6 nt apart on the host DNA (dashed arrows). Host cell enzymes repair the resulting gapped structure. Proviral DNA is blue; host DNA is black.

Sequence elements in the long terminal repeats direct transcription and polyadenylation by host cell enzymes

With integration, the early phase of retrovirus replication is completed. The late phase involves expression of viral RNA, synthesis of viral proteins, and assembly of virions. Proviral DNA can remain unexpressed in the host cell for long periods of time, like any other cellular gene, if the cell does not contain the appropriate set of transcription factors. The U3 region contains transcriptional enhancers that interact with cellular transcription factors and determine in what cell type and to what extent transcription takes place. Transcription of the provirus DNA may also be affected by external stimuli, which can change the mix of active transcription factors present. For example, the U3 sequence of mouse mammary tumor virus contains elements that confer responsiveness to glucocorticoids.

A TATA box just upstream of the U3/R junction directs the initiation of transcription by cellular RNA polymerase II (Figure 28.6). Transcription begins precisely at the junction of the U3 and R regions within the left-hand LTR and proceeds through the entire genome and the right-hand LTR. A polyadenylation signal in the right-hand LTR directs cleavage of the transcript precisely at the R/U5 boundary, and the RNA 3′ end is polyadenylated by host cell enzymes. This gives rise to full-length RNA identical to the genome RNA of the infecting virus.

There are two identical LTRs, one at each end of the proviral DNA. Why is only the left-hand LTR used for signaling initiation of transcription? RNA polymerases that initiate transcription within the left-hand LTR can dislodge transcription factors bound to the right-hand LTR as they pass through that region on the proviral DNA, inactivating its ability to initiate transcription. This is called **promoter occlusion**. This concept is supported by the observation that mutation or deletion of the left-hand LTR can lead to initiation of transcription in the right-hand LTR, resulting in transcription of cellular genes downstream of the proviral DNA (see below).

Similarly, why is only the right-hand LTR used for signaling cleavage and polyadenylation of the transcript? The cellular cleavage and polyadenylation machinery recognizes the highly conserved AAUAAA polyadenylation signal in RNA. Some retroviruses position this signal within the U3 region so that only the copy present in the right-hand LTR is transcribed. However, the AAUAAA is located within the R region of other retroviruses, and therefore it is transcribed and present at both the 5′ and 3′ ends of viral RNA. In some of these viruses, the U3 region contains sequences that enhance recognition of the polyadenylation signal near the 3′ end of viral RNAs, ensuring that cleavage and polyadenylation take place only at that end. Other viruses have sequence elements adjacent to the polyadenylation signal near the 5′ end that repress recognition of that AAUAAA.

Differential splicing generates multiple mRNAs

Transcription produces genome-length RNA identical to that found in the virion. How are the numerous viral proteins, some of which are encoded in different reading frames, made from this RNA? This problem is solved in part by splicing of the primary transcript (Figure 28.6). All retroviruses make at least two mRNAs: unspliced RNA is used for synthesis of the Gag and Gag/Pol proteins, and a singly spliced form, from which the Gag/Pol reading frames have been removed, is used for synthesis of the Env proteins.

In some retroviruses, more complex splicing patterns generate additional mRNAs. For example, Rous sarcoma virus (see Figure 28.8) has an alternative 3′ splice site downstream of the env gene that permits the expression of the Src protein. For human T-cell leukemia virus and members of the lentivirus subfamily (Chapter 29), viral RNA can undergo multiple splicing events to generate mRNAs encoding regulatory proteins.

The Gag/Pol polyprotein is made by suppression of termination and use of alternative reading frames

The unspliced, full-length RNA is used to synthesize Gag and Gag/Pol **polyproteins**, the precursors of the MA, CA, NC, PR, RT, and IN proteins (Figure 28.2). The generation of these two polyproteins from a single mRNA requires modification of the normal translation

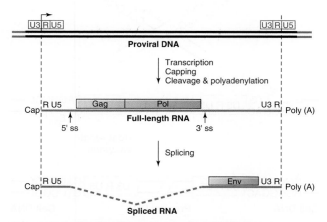

Figure 28.6 Production of retrovirus RNAs. Transcription initiates at junction of U3 and R, and RNA is cleaved and polyadenylated at junction of R and U5. Full-length RNA serves as mRNA for Gag and Gag/Pol and is packaged in virions. A fraction of genome-length RNA is spliced, removing the Gag/Pol reading frames, and serves as mRNA for Env.

process. For viruses of the *Gammaretrovirus* genus, suppression of translation termination is used. The termination codon UAG separating the gag and pol reading frames is correctly recognized 19 times out of 20, resulting in synthesis of only the Gag polyprotein. However, occasionally a glutamine-tRNA misreads the UAG codon as CAG and inserts glutamine at that site. This allows translational readthrough, permitting the generation of the Gag/Pol polyprotein. The misreading process is stimulated by a particular secondary structure called a **pseudoknot**, located just beyond the termination codon in the mRNA.

Avian retroviruses and human immunodeficiency virus type 1 (HIV-1) employ a different strategy. Not only is there a termination codon between the gag and pol regions, but the two proteins are also in different reading frames. This barrier is overcome by inducing the ribosome to shift its reading frame at a precise position within the RNA prior to the termination codon.

Ribosomal frameshifting is induced by the presence of two sequence elements within the RNA: a heptamer sequence, where the ribosome stalls, and a secondary structure downstream of the heptamer that induces ribosome stalling (Figure 28.7). Examples of such heptamer sequences are A AAA AAC (HTLV-1), A AAU UUA (Rous sarcoma virus), and U UUU UUA (HIV-1). Base pairing between the codons on these mRNA sequences and the anticodon sequences of the two tRNAs bound to the ribosome is similar in two different reading frames. The stalled ribosome is able to equilibrate between these two states because of the similar strength of base pairing interactions. Therefore the translating ribosome occasionally shifts the reading frame back one nucleotide, and resumes translation of the sequence (Figure 28.7). In the new reading frame, the gag termination codon is no longer recognized and the pol region is translated, resulting in the generation of the Gag/Pol protein.

Why have these viruses evolved such a complex scheme for the generation of the Gag/Pol polyprotein? This process ensures the synthesis of Gag and Gag/Pol in a particular ratio, so that the required quantities of each protein necessary for proper virus assembly are present. Only a few molecules of the enzymes reverse transcriptase, protease, and integrase are needed, but many molecules of the structural proteins encoded by Gag are required to form a single virion.

Virions mature into infectious particles after budding from the plasma membrane

The Env protein is translocated directly into the **lumen** of the endoplasmic reticulum as it is being synthesized. It is subsequently transported through the Golgi apparatus and the endosome compartment and finally arrives at the plasma membrane. In the course of these events, the protein undergoes glycosylation and cleavage by host enzymes to generate the mature forms of SU and TM. In contrast to Env, both Gag and Gag/Pol proteins are released into the **cytosol** upon translation. The Gag and Gag/Pol proteins interact with each other to initiate assembly of the virus core.

Two different assembly pathways have been elucidated. For mouse mammary tumor virus and Mason–Pfizer monkey virus ("B-" and "D-type" viruses), the core is assembled within the cytoplasm, generating spherical structures designated "A-type particles" visible by electron microscopy. These cores bud through the plasma membrane in regions where the SU and TM proteins have accumulated, acquiring an envelope. For most other mammalian and avian retroviruses ("C-type" viruses), as well as lentiviruses, assembly of the core occurs directly at the plasma membrane during budding. As the particle is assembled, the core is extruded outward, resulting in a C-shaped bulge on the surface of the plasma membrane, hence the designation "C-type".

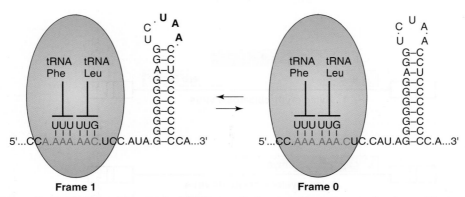

Figure 28.7 Ribosome frameshifting. A base-paired RNA stem induces stalling upstream by the translating ribosome (green oval) at the heptamer sequence, shown in orange. While stalled, the ribosome can shift from reading frame 1 to frame 0. Translation in frame 0 does not recognize the UAA termination codon (bold letters).

The Gag protein is targeted to the plasma membrane by the fatty acid **myristate**, linked post-translationally to its N-terminal amino acid. The NC portion of Gag binds to a **packaging signal** designated psi (ψ), located near the 5' end of viral genome RNA (Figure 28.2). Selective encapsidation of full-length viral RNA is achieved by positioning the psi sequence downstream of the 5' splice site, so that only unspliced viral RNA contains the packaging signal. The Gag/Pol polyprotein is incorporated into the assembling virus core by interaction between its CA region and the corresponding region of Gag. Finally, the tRNA primer is incorporated into the assembling core via binding to the RT and NC portions of the Gag/Pol protein.

As virions are assembled and extruded from the cell, the viral protease becomes activated. The protease then cleaves the Gag and Gag/Pol polyproteins into the individual structural (MA, CA, and NC) and enzymatic (PR, RT, and IN) proteins, and they rearrange to form the electron-dense core characteristic of mature virions. It is only at this stage that virions become infectious; therefore, the protease is an important target of antiviral chemotherapy directed against retroviruses (see Chapter 36).

Acute transforming retroviruses express mutated forms of cellular growth signaling proteins

Study of the mechanisms by which retroviruses transform cells in culture and induce tumors in animals has provided important insights into the causes of cancer. There are two classes of oncogenic retroviruses: the *acute transforming* viruses, which transform cultured cells rapidly and induce cancer within a short time after inoculation into animals; and the *nontransforming* viruses, which induce cancer only after a long latency period.

Tumor formation by **acute transforming retroviruses** is directed by proteins that are not required for virus replication. An example is one of the first retroviruses discovered, Rous sarcoma virus (Figure 28.8). A viral gene was identified by mapping mutations that inactivated tumor formation by Rous sarcoma virus in chickens, but that did not affect virus replication. This gene was designated *src*, for sarcoma.

Src was one of the first oncogenes identified, and its presence in a virus lent credence to the theory that certain cancers were caused by oncogenes of viral origin. However, the viral src gene was subsequently shown to be nearly identical to a cellular gene, called *c-Src*, that codes for a protein tyrosine kinase involved in cellular signaling pathways. The previous assumption that src was an oncogene of viral origin was proved incorrect. It became clear that Rous sarcoma virus had acquired a mutated copy of the c-Src gene, probably by integrating its proviral DNA nearby the c-Src gene and incorporating that gene into the viral genome during subsequent transcription from the LTR.

Numerous other acute transforming retroviruses contain a variety of oncogenes that are derived from cellular genes. These cellular genes (**proto-oncogenes**) code for growth factors, receptor proteins, protein kinases, or transcription factors involved in pathways that regulate cell growth and division. Comparison of viral oncogenes with their cellular counterparts revealed that the viral forms have mutations or deletions that alter the activity of the proteins. Consequently, cells expressing these mutated viral forms of the protein behave as if they are receiving a growth stimulus independent of

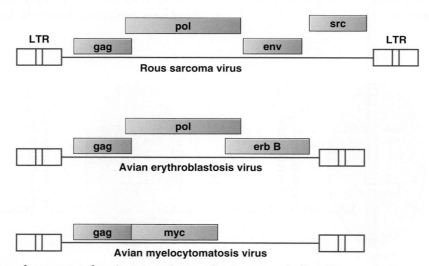

Figure 28.8 Genomes of acute transforming retroviruses. Oncogenes are indicated in orange. Rous sarcoma virus is replication-competent because it contains all structural genes; however, most transducing retroviruses have deletions of structural genes and often express their oncogenes as fusion proteins.

the external environment of the cell, and show uncontrolled growth and division. Many of the mutations found within viral oncogenes are similar to those found in altered genes associated with tumor formation in animals or humans, suggesting common tumor-inducing mechanisms.

Additional mechanisms for tumor promotion that do not rely on a mutant form of a host gene have been described for particular retroviruses. For example, the transforming potential of the mouse retrovirus spleen focus forming virus (SSFV) is due to mutations in the viral env gene that allow it to mimic the activity of erythropoietin and to stimulate growth of cells that respond to this hematopoietic growth factor. The outgrowth of these cells renders them susceptible to subsequent random mutations that result in their transformation. Human T-cell leukemia virus type 1 (HTLV-1) codes for a protein designated Tax, which augments viral gene expression by activation of cellular transcription factors. As a result other cellular genes (IL-2 and IL-2 receptor) are also induced, resulting in cell proliferation due to an autocrine feedback that stimulates growth of infected cells.

Rous sarcoma virus turns out to be a unique example of a transforming retrovirus that is both tumor-inducing and able to replicate on its own. Most transforming retroviruses that have picked up cellular genes have also lost large segments of viral genes by deletion, rendering them unable to replicate on their own; they depend on coinfection with a replication-competent retrovirus. Two examples of such deleted viral genomes are shown in Figure 28.8. The nontransforming "helper virus" provides the structural and enzymatic proteins necessary for producing virions. These virions incorporate the RNA genomes of either the helper virus itself or the defective transforming virus. The defective viral RNA must at the minimum contain a packaging signal (psi) recognized by the Gag protein of the helper virus, as well as the U3, R, U5, and PBS sequences required for reverse transcription and integration into the cellular genome.

Retroviruses lacking oncogenes can transform cells by insertion of proviral DNA near a proto-oncogene

In contrast to acute transforming viruses, nontransforming retroviruses lack oncogenes. However, the replication of a retrovirus can result in mutagenesis of the host cell because it involves integration of proviral DNA into the host genome. Integration takes place at random sites within the cellular genome. If viral DNA is inserted adjacent to or within a cellular proto-oncogene, the altered pattern of expression of the gene can lead to tumor formation if the infected cell is not killed by the resident retrovirus.

One mechanism of insertional activation of cellular proto-oncogenes is called "promoter insertion". A viral LTR is inserted just upstream of a cellular gene, and transcription that initiates within the LTR progresses through the adjacent cellular gene. This can happen if the left-hand LTR is inactivated by mutation or deletion, allowing the right-hand LTR to initiate transcription (Figure 28.9a). Alternatively, the integrated proviral DNA may undergo a deletion that removes most of the genome, leaving the left-hand LTR intact and adjacent to a cellular proto-oncogene (Figure 28.9b). In both cases, transcription of the adjacent cellular gene is now regulated by the viral LTR rather than its own control elements, and over-expression of the cellular gene may result in inappropriate cell growth and division.

Another mechanism involves insertion of a retrovirus polyadenylation signal into a cellular gene. Many mRNAs contain signals within their 3' untranslated region that regulate translation or RNA stability. Insertion of a provirus DNA into this region of a gene can result in the use of the polyadenylation signal within the left-hand LTR, generating an mRNA missing its normal 3' untranslated region. Consequently, the mRNA is no longer regulated in the appropriate fashion and expression of the protein can be altered.

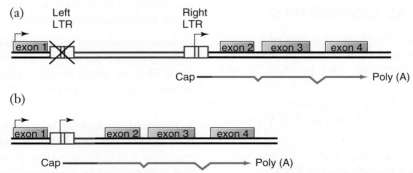

Figure 28.9 Promoter insertion near proto-oncogenes can lead to cell transformation and tumor formation. Transcription of the proto-oncogene (exons shown as filled boxes) initiates within the viral LTR, either as a result of inactivation of the left-hand LTR (a) or deletion of the right-hand LTR (b). Result is a mixed viral/cellular mRNA that expresses the proto-oncogene under control of viral promoter/enhancer.

Memories of infections past and the enemy within: endogenous retroviruses and retrotransposons

The stable insertion of proviral DNA into the host chromosome means that, even in the absence of expression, the viral genome will be replicated each time the infected cell divides. If present in germ line cells, the viral genome (and therefore virus) can be inherited, and will persist by "vertical transmission". Sequencing of genomes of a many animal species has revealed that a significant fraction of the genome (8% in humans) is composed of retrovirus DNAs; these are called "endogenous retroviruses". In most instances, the viral DNA cannot produce infectious virus because of mutations that disrupt or delete reading frames or gene expression control sequences.

Comparative sequencing data show that retroviral DNA has been inserted into germ line genomes at different times during evolution, some occurring after the division of humans and apes. Some genes of endogenous retroviruses have even been adapted by the host organism for specialized functions. An example is the env gene of human endogenous retrovirus W, called syncytin-1. Expressed during fetal development, Syncytin-1 can induce membrane fusion, and is active in the formation of the fused trophoblast layer of cells surrounding the fetus.

An even larger fraction of most genomes (41% in humans) is composed of retrovirus-like elements termed **retrotransposons**. These elements are related to the retroviruses by their mechanism of replication; they express an RNA that is converted into DNA by reverse transcriptase and is subsequently integrated into the genome. They can therefore spread within the cell genome, but do not code for proteins that allow them to escape the cell. Retrotransposons may have been the precursor of retroviruses.

The two major types of human retrotransposons are designated LINE (long interspersed nuclear element) and SINE (short interspersed nuclear element). Only LINEs encode reverse transcriptase. SINEs are dependent on the reverse transcriptase provided by LINEs for their replication.

The large number of endogenous retroviruses and retrotransposons present in the genome have enormous mutagenic potential. Hosts have evolved multiple mechanisms to suppress their expression; these include DNA methylation to block transcription, interfering RNA to silence translation of mRNAs, and APOBEC proteins (see Chapter 29) that sequester RNA into particles within the cytoplasm. However, the capacity of these transposable elements to alter genome structure and gene function also makes them a strong driving force in evolution. Balancing the negative and positive effects of retrotransposition is key in maintaining the viability of the host.

KEY TERMS

Acute transforming
   retroviruses

Chemokine

Cytosol

Lumen

Myristate

Oncogene

Polyprotein

Preintegration complex

Promoter occlusion

Proto-oncogene

Proviral DNA

Pseudoknot

Quasispecies

Reverse transcriptase

Ribonuclease H

Ribosomal
   frameshifting

FUNDAMENTAL CONCEPTS

• Retroviruses package two identical copies of a positive-strand RNA genome along with the viral enzymes reverse transcriptase, integrase, and protease in an enveloped virion.

• Upon entry into the cell, the RNA genome is reverse transcribed to double-stranded proviral DNA within the viral nucleocapsid.

• Reverse transcription is initiated on a primer provided by a cellular tRNA bound to virion RNA.

• Reverse transcriptase contains a ribonuclease H activity that degrades RNA in RNA/DNA hybrids.

• Proviral DNA is imported into the nucleus and integrated into the host cell chromosome at random sites.

• RNA polymerase II of the host cell transcribes the integrated viral DNA genome.

• Retroviruses express three set of genes called gag (capsid proteins), pol (viral enzymes) and env (envelope proteins) on unspliced and singly spliced mRNAs.

• Viral proteins are synthesized as polyproteins, which are subsequently cleaved by viral and cellular proteases to the individual viral proteins.

• Some retroviruses have acquired cellular genes within their genomes during integration and transcription; some of these genes produce altered cellular signaling proteins that can transform infected cells into tumor cells.

• Retroviruses lacking oncogenes can also transform cells if their proviral DNA is inserted nearby a gene encoding a cellular signaling protein whose transcription is stimulated by the retrovirus promoter/enhancer.

REVIEW QUESTIONS

1. The genome of retroviruses contains a sequence known as the "kissing loop". What is the "kissing loop"?

2. What is the purpose of the cellular tRNA incorporated into every retrovirus particle?

3. The retrovirus proviral DNA is larger than the genome RNA. Why?

4. Lentiviruses are able to replicate in resting cells, but other retroviruses cannot. Why?

5. Retroviruses are the only RNA viruses whose genome is replicated by the host cell enyzmatic machinery. What cellular enzymes accomplish this genome replication?

Human Immunodeficiency Virus

Alan Cochrane

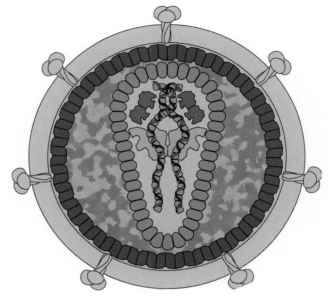

Lentivirus

From Latin *lentis* (slow), for slow progression of disease

VIRION

Spherical enveloped particle.
Diameter 100 nm.
Conical capsid with probable icosahedral symmetry.

GENOME

Linear single-stranded RNA, positive sense, 9.3 kb.
Two identical genome RNAs in each virion.
Cellular tRNA^{lys3} molecules packaged in virions used as primers for reverse transcription.

GENES AND PROTEINS

Four capsid proteins: matrix (MA), capsid (CA), nucleocapsid (NC), p6.
Three enzymes: protease (PR), reverse transcriptase (RT), integrase (IN).
Two envelope proteins: surface (SU) and transmembrane (TM).
Six regulatory proteins: Vif, Vpu, Vpr, Tat, Rev, Nef.

VIRUSES AND HOSTS

Human immunodeficiency virus types 1 and 2 (HIV-1, HIV-2).
Simian immunodeficiency virus.
Equine, bovine, feline immunodeficiency viruses.

DISEASES

Acquired immune deficiency syndrome (AIDS) was first described in 1981.

A major global pandemic today (more than 33 million people infected).
HIV replicates in and kills lymphocytes and macrophages.
Infection results in depletion of CD4+ T cells, rendering the host immune-incompetent.
As a result, opportunistic infections by other pathogens are often fatal.
HIV is transmitted through sexual contact and blood exchange.

DISTINCTIVE CHARACTERISTICS

Proviral DNA can enter nucleus without requirement for cell division.
Lentiviruses make a complex set of singly and doubly spliced mRNAs.
Six regulatory proteins control virus production and pathogenesis.

Human immunodeficiency virus type 1 (HIV-1) and acquired immunodeficiency syndrome (AIDS)

In 1981, the term *acquired immunodeficiency syndrome (AIDS)* was coined to describe a condition in a group of previously healthy young males within the Los Angeles/San Francisco area who showed a marked depletion of their immune **CD4-positive T lymphocytes**, rendering them immune-incompetent. As a consequence, they suffered from a number of **opportunistic infections** (the most prevalent being pneumonia) that were often fatal. Subsequent epidemiological studies suggested that the syndrome was due to a transmissible agent that was acquired through sexual contact or blood exchange; hemophiliacs, recipients of blood transfusions, and intravenous drug users were also affected.

In 1983, a retrovirus isolated from the blood of individuals with AIDS was characterized by groups led by Luc Montagnier in Paris and Robert Gallo in Maryland. This virus, subsequently named human immunodeficiency virus type 1 (HIV-1), was demonstrated (despite much controversy and debate) to be the causative agent of AIDS. HIV-1 is characteristic of a subfamily of retroviruses named the lentiviruses (Table 29.1), so named because of the slow progression of diseases caused by lentiviruses.

HIV-1 has caused major global pandemic that affects males and females equally. The World Health Organization estimated in 2009 that 25 million people have died of AIDS and that more than 33 million people worldwide were living with HIV-1 infections, most of these being inhabitants of Africa and South-East Asia. With no effective vaccines and limited access to drug treatments, the number of infected people continues to grow as the pandemic expands into other developing countries. Despite intensive information programs and the availability of treatment, the number of infected people also continues to increase in many developed countries, people between the ages of 16 and 40 being most at risk. In response to this health crisis considerable

Table 29.1 Lentiviruses

Virus	Host
Human immunodeficiency virus type 1	Humans
Human immunodeficiency virus type 2	Humans
Simian immunodeficiency virus	Apes and old world monkeys
Feline immunodeficiency virus	Cats
Equine infectious anemia virus	Horses
Caprine arthritis-encephalitis virus	Goats
Visna-maedi virus	Sheep

resources have been committed to understanding HIV-1 biology and pathology, rendering it perhaps the most intensively studied disease agent of our time.

HIV-1 was probably transmitted to humans from chimpanzees infected with SIVcpz

Many species of African monkeys and apes are hosts for specific strains of simian immunodeficiency virus (SIV), closely related to HIV. These viruses have been isolated and their nucleotide sequences compared. As a result of these studies, it has been determined that HIV-1 is a zoonotic infection likely transmitted in the early 1900s from butchered chimpanzees infected with the chimpanzee strain of SIV (SIVcpz) to humans in West-Central Africa. In turn, chimpanzees likely acquired SIVcpz from red-capped mangabeys and greater spot-nosed monkeys, as SIVcpz appears to be a recombinant between the viruses that infect these two monkey species.

Despite high levels of viremia, old world monkeys infected with SIV do not succumb to the same clinical syndromes seen in humans infected with HIV-1, suggesting that these monkeys have adapted to SIV through selection over a long period of time. However, chimpanzees living in the wild and infected with SIVcpz appear to suffer from significant disease, loss of fertility, and early death, the higher pathogenicity perhaps reflecting a relatively recent acquisition of this virus.

HIV-1 infection leads to a progressive loss of cellular immunity and increased susceptibility to opportunistic infections

In the majority of cases, HIV-1 is transmitted upon exposure to mucous membranes, usually during sex or ingestion of breast milk. Direct infection of cells lining the gut or vagina has not been detected, indicating that the virus must overcome this physical barrier to establish the infection. Transit of HIV-1 across the mucosa can be facilitated by local abrasions, inflammation, ulcerations or interaction with **dendritic cells**. Dendritic cells collect virus by the binding of the HIV-1 SU protein (gp120) to DC-SIGN, a cell-surface protein. These cells then move to the subepithelial space and present bound virus to CD4-positive T cells under conditions that greatly enhance infection of the T cell. Following amplification of the virus in the local lymph node, virus is released into the circulation, seeding the infection into all the lymphatic tissues of the body.

The subsequent course of the infection can be roughly divided into three phases: (1) acute infection; (2) a subsequent clinically latent phase of variable duration; and finally (3) AIDS (Figure 29.1). Two to six weeks following exposure to the virus, individuals can develop a mononucleosis or influenza-like syndrome (fever, malaise, lethargy, nausea, diarrhea, headaches,

(a)

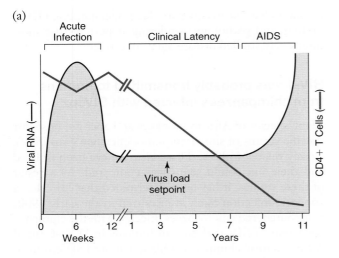

(b)

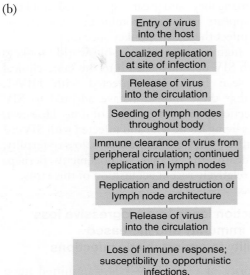

Figure 29.1 Events associated with progression to AIDS.
(a) Changes in viral RNA and CD4-positive T cell levels in
the peripheral circulation from acute infection to end-stage
disease. (b) Stages of HIV-1 infection.

stiff neck, or swelling of lymph nodes) that requires
hospitalization in a minority of individuals.

The first organ system to be affected by the infection
is the gut associated lymphoid tissue (GALT). There is a
significant depletion of CD4-positive T cells in this lymphoid tissue during acute infection, which is not restored
even after resolution of the initial viremia. Given the role
of the GALT in regulating intestinal flora, it is has been
suggested that its depletion may result in release of bacterial products such as lipopolysaccharide into the circulation, inducing a state of chronic immune activation
known to persist in the chronic phase of HIV disease.

Increased numbers of T cells from HIV-infected individuals undergo spontaneous apoptosis even though they
are not directly infected. Within 1 to 2 months, the primary infection resolves but there is little recovery of the
GALT. Virus titers drop as the immune system responds
with both **cytotoxic T lymphocytes** and **antibodies** (see
Chapter 34). However, infection within the lymph nodes
persists throughout the course of the disease.

Several courses of disease following acute infection
have been documented in the absence of treatment:

1. *Rapid progressors* (10–15% of infected individuals)
   develop late stage symptoms in 2 to 3 years.

2. *Slow progressors* (70–80%) develop late stage
   symptoms in 8 to 10 years.

3. *Long-term non-progressors* (5%) show no decline
   in CD4-positive T-cell levels.

Several factors help predict clinical outcome. After
resolution of the acute infection, varying levels of viral
RNA genomes ("virus load setpoint", Figure 29.1) can
be detected in the blood of different individuals. The
higher the basal level of viral RNA, the more rapidly
patients progress to full-blown AIDS. Also important
is the nature of the virus itself. For example, patients
infected with a strain of virus that lacks the viral nef
gene progress to AIDS very slowly if ever.

The nature of the immune response is also crucial.
Patients who develop a predominantly cytotoxic T-cell–
based immune response have a better chance of long-
term survival. The diversity of the **epitopes** recognized
by the immune system also appears to play a role; recognition of a limited number of epitopes is associated with
a poor prognosis.

Over the course of clinical latency, virus replication
persists in the lymph nodes, resulting in a gradual depletion in the level of circulating CD4-positive T cells and
destruction of the lymph node architecture. Individuals
progressing toward end-stage disease have high levels of
virus in the blood, indicative of the failure of the immune
system to contain the infection. Patients exhibit chronic
fever, night sweats, diarrhea, a number of infections such
as cytomegalovirus, pneumonia, oral thrush, herpes
simplex, neoplasms such as Kaposi's sarcoma and lymphomas, and neurological syndromes including dementia and
neuromuscular disorders. Neurological symptoms correlate with virus replication in the central nervous system.
While current therapies delay or prevent disease progression, they do not eradicate the infection. Consequently,
withdrawal from therapy results in reemergence of the
virus and continued disease progression.

Antiviral drugs can control HIV-1 infection and prevent disease progression, but an effective vaccine has yet to be developed

At present, recommended treatment involves use of a
combination of (usually three) drugs targeted at the viral
enzymes reverse transcriptase and protease, although
drugs targeting other aspects of the viral life cycle (virus
entry and integration of proviral DNA) are also available.

However, treatment is costly and requires a high degree of compliance (>95% adherence to medication schedule) to contain the infection. Drug side effects (nausea, diarrhea, fatigue, bone loss, body fat redistribution, heart problems due to high blood lipids, diabetes) are also common, with the result that patients often do not complete treatment regimens or delay treatment until significant CD4-positive T-cell depletion has occurred.

Despite considerable effort, therapeutic and/or preventive vaccines remain a distant hope, as the most straightforward strategies have yielded very limited success. This is due in part to the high degree of variation in the virus, both within one individual and within the population. Also complicating HIV treatment is the rapid emergence of drug-resistant strains, which currently represent approximately 10% of new infections in developed countries.

HIV-1 is a complex retrovirus

The basic replication pattern of lentiviruses is identical to that of other retroviruses. However, sequence analysis of the genome of HIV-1 revealed it to be considerably more complex than many other retroviruses. In addition to the standard gag, pol, and env genes, six additional reading frames were identified: vif, vpr, vpu, tat, rev, and nef (Figures 29.2 and 29.3, Tables 29.2 and 29.3). Since all of these additional reading frames are conserved among various clinical isolates of the virus, they must play important roles in the virus life cycle. The following discussion of the replication of HIV-1 will emphasize the unique features of this virus, including the roles of these additional viral proteins.

In contrast to simpler retroviruses, which make only two mRNAs (unspliced and singly spliced),

Table 29.2 HIV-1 structural proteins

Name	Abbreviation	Alternative name (M. Wt. in KDa)
Matrix	MA	p17
Capsid	CA	p24
Nucleo**c**apsid	NC	p7
Protease	PR	p14
Reverse **t**ranscriptase	RT	p66/51
Integrase	IN	p32
Surface protein	SU	gp120
Trans**m**embrane protein	TM	gp41
Virion **protein R**	Vpr	p15

Table 29.3 HIV-1 nonstructural proteins

Name	Abbreviation	Alternative name (M. Wt. in KDa)
Viral **i**nfectivity **f**actor	Vif	p23
Virion **p**rotein **u**nique to HIV-1	Vpu	p16
Transactivator of **t**ranscription	Tat	p15
Regulator of **e**xpression of **v**irion proteins	Rev	p19
Negative **e**ffector	Nef	p27

splicing of the HIV-1 primary transcript generates more than 25 mRNAs that fall into three size classes (Figure 29.3):

1. the unspliced 9-kb full-length RNA, used to produce Gag and Gag-Pol proteins
2. the singly spliced 4-kb class of RNAs, which encode Vif, Vpr, Vpu, or Env
3. the doubly spliced 2-kb class of RNAs, which encode Tat, Rev, or Nef.

In each class of spliced RNA there are multiple species, generated by the presence of several different 3' and 5' splice sites.

HIV-1 targets cells of the immune system by recognizing CD4 antigen and chemokine receptors

The **CD4 antigen,** found on the surface of both helper T lymphocytes and **macrophages,** is the primary receptor for the virus. Thus HIV-1 attacks at the very heart of the immune system, as both CD4-positive T cells and

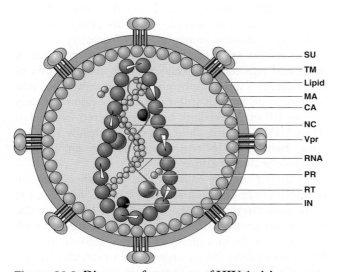

Figure 29.2 Diagram of structure of HIV-1 virion. Components of the virion are labeled; see Table 29.2. For the sake of clarity, only one of the two genome RNA molecules is shown covered by NC protein.

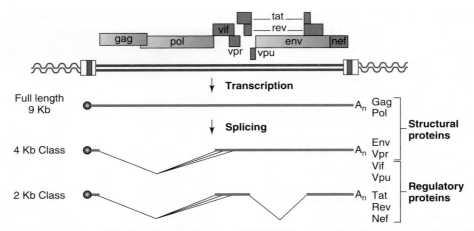

Figure 29.3 Genome structure and RNA splicing pattern of HIV-1. Above: a map of the integrated proviral DNA framed by two long terminal repeats, and the reading frames for viral proteins. Below: the pathway by which the primary transcript is converted into more than 25 different mRNAs in three size classes by alternative splicing. Right: viral proteins produced from the different classes of viral mRNA.

macrophages are vital for development of both humoral antibody and cell-mediated immunity (see Chapter 34). However, the presence of CD4 alone on the cell surface is not sufficient for infection. Infection also requires the presence of either the CCR5 or CXCR4 **chemokine** receptor. The ability of **ligands** for these receptors (Mip1α, Mip1ß, and Rantes for CCR5; SDF-1 for CXCR4) to block virus entry demonstrates their significant role in the infection process.

Viruses using CXCR4 are designated X4 (or T cell-tropic) viruses, whereas those using CCR5 are termed R5 (or macrophage-tropic) viruses. These strain differences depend on variations in the SU protein (gp 120), particularly in a region that undergoes a high rate of evolution, designated variable domain 3. Although both forms of HIV-1 may be present in an inoculum, only the R5 strain is sexually transmitted. Natural resistance to HIV-1 infection is associated with a mutation in CCR5 that results in a loss of cell surface expression of the protein. In contrast to CD4+ T cells, HIV-1 infection of macrophages results in a low level of virus replication, with the bulk of the virus accumulating in intracellular vacuoles. Release of the virus occurs upon fusion of the vacuoles with the plasma membrane.

Binding to CD4 induces a conformational change in gp120 that exposes regions that interact with the chemokine receptor. This binding triggers a conformational change in the envelope protein TM (gp41) that induces the fusion of the viral envelope with the plasma membrane, allowing release of the nucleocapsid into the cytoplasm (Figure 29.4).

Virus mutants arise rapidly because of errors generated during reverse transcription

Once released into the cytoplasm, the viral nucleocapsid partially breaks down to permit access to nucleotide pools within the cell. The viral capsid associates with cellular microfilaments and reverse transcription of the viral genome begins. Experiments in several species have identified host proteins (Ref1, Lv1) that can block this stage of the infection. Designated "restriction factors", they either accelerate capsid breakdown or block transport of the viral **preintegration complex** into the nucleus. Old world monkeys experimentally infected with HIV-1 express Trim5α, a protein that promotes rapid degradation of the capsid before reverse transcription can occur. Unfortunately, the human Trim 5α homolog fails to recognize HIV-1.

Reverse transcriptases do not have efficient proof-reading mechanisms to remove incorrectly incorporated nucleotides in proviral DNA. The resulting error rate during reverse transcription (1–10 mutations per proviral DNA molecule synthesized) renders HIV-1 infections a moving target for both the immune system and drug treatment strategies. Approximately 2×10^9 virus particles are made and destroyed every day in an untreated infected individual. Only a small fraction of these particles successfully integrates proviral DNA into infected cells, but this still translates into approximately 10^4 to 10^5 virus variants being generated every day. Thus HIV-1 has the ability to evolve rapidly to overcome barriers to its replication, be they an immune response or drug treatment.

Current therapy calls for the simultaneous use of multiple drugs affecting several steps of the virus life cycle in an attempt to generate a higher barrier for the virus to overcome. Although this approach has met with great success, the emergence of viruses resistant to combinations of three drugs suggests that even this barrier may not be sufficient to control the infection over the long term. The same issues apply to vaccine design. Immunization with one or a small group of viral antigens will generate an immune response to only a limited

(a)

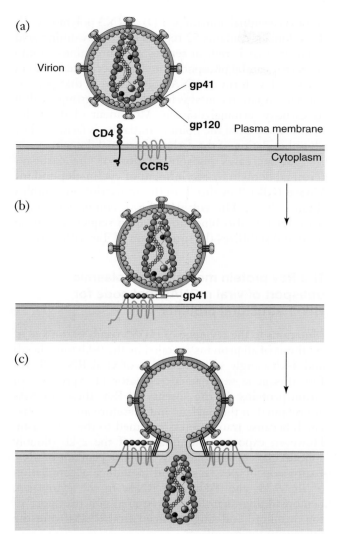

(b)

(c)

Figure 29.4 Model of HIV-1 entry. (a) gp120 on the virion binds to CD4 on the cell surface. (b) gp120 undergoes a conformational change that exposes regions that bind to chemokine receptors (CCR5 or CXCR4). Changes in gp120 conformation expose the fusion domain of gp41, resulting in its insertion into the cell membrane. (c) The close proximity of the viral and cellular membranes results in their fusion, releasing the virion core into the cytoplasm.

number of the possible variants of the virus, something that the virus can easily overcome.

Unlike other retroviruses, HIV-1 directs transport of proviral DNA into the cell nucleus

Most retroviruses cannot productively infect non-dividing cells because the preintegration complex, containing proviral DNA, is unable to enter the intact nucleus (see Chapter 28). Integration of proviral DNA must therefore await the disintegration of the nuclear membrane when the cell divides. In contrast, HIV-1 is able to transport the preintegration complex into the nucleus

via the nuclear pores, and can therefore infect cells that are not actively dividing.

Three virus-encoded proteins in the preintegration complex facilitate this transport: MA, Vpr, and IN. MA encodes a classical nuclear import signal that interacts with the **importin** family of proteins. In contrast, both Vpr and IN interact directly with components of the nuclear pore. However, in resting T cells the preintegration complex remains in the cytoplasm. Only after T-cell activation (see Chapter 34) is the preintegration complex transported into the nucleus and viral DNA integrated into the cellular genome.

This situation presents a problem for the immune system. To generate an immune response, T cells are activated by presentation of antigen fragments during contact with dendritic cells within lymph nodes. However, dendritic cells collect HIV-1 virions on their surface. Consequently, T cells that would support an immune response against HIV-1 are exposed to virus while interacting with dendritic cells; these conditions enhance the ability of the virus to replicate by facilitating entry of proviral DNA into the nucleus, and eventually lead to productive infection and death of the T cells.

Latent infection complicates the elimination of HIV-1

After integration of the HIV-1 proviral DNA into the host cell genome, there are two possible outcomes: **latent infection**, in which viral RNA is not expressed, or active infection, leading to production of progeny virus. Up to 1% of the cells within the lymph nodes and circulation of recently infected individuals contain integrated HIV-1 provirus, but only about 1% of those infected cells actively express viral RNA.

Although latently infected cells do not directly contribute to disease, they do represent a significant hurdle to treatment and a cure. Because they do not express viral proteins, these cells cannot be readily distinguished from uninfected cells by the immune system. Removal of patients from multiple antiviral drug therapy invariably results in reemergence of the infection. The virus that appears is not drug-resistant, and is probably generated from the latently infected cells. Many of these cells are memory T cells that have a very long lifetime in the body. Therefore, unless latently infected cells can be activated in some fashion to render them recognizable to the immune system, treatment for HIV-1 must be continued for life if the disease is to be held in check.

Virus latency is regulated by transcriptional control elements present in the HIV-1 LTR. Binding sites for transcription factors **NFκB** and NFAT, whose activities are sensitive to external stimuli, are present in the U3 region of the LTR. NFκB is held in the cytoplasm in an inactive state by association with the inhibitor **IκB**. A variety of stimuli induce the degradation of IκB, resulting in the transport of NFκB into the nucleus where it can activate transcription. One such stimulus is activation of T cells via the **T-cell receptor** (see Chapter 34); others are virus infections or treatment with various **cytokines** (see Chapter 33). The U3 region contains binding sites for other cellular transcription factors such as Sp1, LEF, Ets-1, and USF, which also play a role in modulating HIV-1 transcription.

The Tat protein increases HIV-1 transcription by stimulating elongation by RNA polymerase II

Tat is a highly basic 86-amino acid protein produced by doubly spliced HIV-1 mRNA; it is localized in the cell nucleus. Expression of this protein dramatically increases the amount of viral RNA produced in infected cells, thus its name, abbreviated from "*trans*activator of *transcription*".

Key to the function of Tat is the presence of a sequence element located just after the HIV-1 transcription start site designated the *Ta*t-responsive element (TAR) (Figure 29.5). Tat binds to the TAR sequence on nascent (growing) RNAs, but not to the same sequence on proviral DNA. This was the first demonstration that transcription by RNA polymerase II could be modulated by an RNA element on the nascent transcript.

TAR forms a stem-loop structure in viral RNA, and two regions are essential for its function: (1) a bulge in the stem, which forms the recognition element for Tat binding; and (2) the nucleotide sequence within the loop. Tat increases viral RNA abundance not by increasing the rate of initiation of transcription by RNA polymerase, but rather by increasing the elongation efficiency of RNA polymerase molecules (Figure 29.5).

Elongation by RNA polymerase II is stimulated by increasing the extent of phosphorylation of the carboxy-terminal domain (CTD) of RNA polymerase II. This domain contains 52 repeats of a seven–amino acid sequence that is rich in serines and threonines, amino acids that can be phosphorylated by specialized protein kinases. The form of RNA polymerase II that participates in initiation contains few phosphate groups within its carboxy-terminal domain. Movement of the polymerase away from the point of initiation is facilitated by phosphorylation of the carboxy-terminal domain.

Tat increases phosphorylation of the carboxy-terminal domain by recruitment of **cyclin-dependent kinase (cdk)-9/cyclin** T to the transcription complex (Figure 29.5). The cdk 9/cyclin T complex interacts with Tat but also binds to the loop sequences within TAR and stabilizes the interaction of Tat with TAR.

The Rev protein mediates cytoplasmic transport of viral mRNAs that code for HIV-1 structural proteins

Mutations in the rev gene of HIV-1 result in the loss of synthesis of all proteins encoded by the 9-kb (unspliced) and 4-kb (singly spliced) classes of viral RNA; therefore this gene was named *r*egulator of *e*xpression of *v*irion proteins. In the absence of Rev, these mRNAs are retained in the nucleus and therefore are not translated, because translation is confined to the cytoplasm. However, export and translation of the 2-kb (doubly spliced) class of viral mRNAs is not dependent on Rev. Expression of Rev was found to induce accumulation of the 9-kb and 4-kb viral RNAs in the cytoplasm, demonstrating that Rev is required for transport of the unspliced and singly spliced viral RNAs from the nucleus to the cytoplasm.

Regulation of mRNA transport by Rev requires two classes of sequences on viral RNA: (1) *cis*-acting

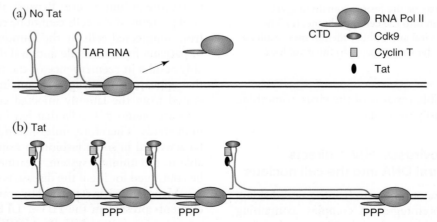

Figure 29.5 Mechanism of Tat function. (a) In the absence of Tat, RNA polymerase II molecules that initiate transcription at the HIV-1 promoter lack processivity and most are released from the template before reaching the polyadenylation site. (b) Tat recruits cyclin T and cyclin-dependent kinase-9 (Cdk9) to the transcription complex shortly after initiation, resulting in hyperphosphorylation (P) of the carboxy-terminal domain of RNA polymerase II. This increases processivity of the polymerase and generates increased levels of full-length transcripts. CTD: carboxy-terminal domain of RNA polymerase II.

repressive sequences (CRS) present in the gag, pol, and env regions inhibit the transport of RNAs to the cytoplasm; (2) a 240-nt sequence within env, termed the *Rev response element (RRE)*, is a target for Rev binding and required for Rev action (Figure 29.6). The 116-amino acid Rev protein contains both a nuclear localization signal and a nuclear export signal, indicating that this protein can shuttle continuously between the nucleus and cytoplasm. Transport to the cytoplasm involves binding of the nuclear export signal on Rev to the cellular protein **exportin** 1, which mediates the docking of Rev to the nuclear pore. If Rev is simultaneously bound to an mRNA via its RRE, the export of Rev results in the export of the mRNA.

Messenger RNAs that contain *cis*-acting repressive sequences cannot leave the nucleus unless they contain an RRE, in which case Rev supplies a pathway for their export. Since the doubly spliced 2-kb RNAs (and most cellular mRNAs) do not contain a *cis*-acting repressive sequence, their export does not require Rev. Once HIV-1 mRNAs are in the cytoplasm, Rev can remain bound and can enhance their translation by a mechanism that is not yet understood. Transport of Rev back into the nucleus requires dissociation of the Rev/RNA complex. Subsequently, Rev binds to importin β, which docks Rev to the cytoplasmic face of the nuclear pore (Figure 29.6). Rev is returned to the nucleus, where the cycle is repeated.

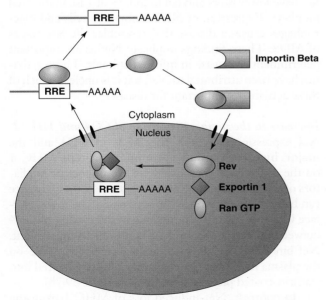

Figure 29.6 Mechanism of Rev function. Rev is transported into the nucleus via nuclear pores by direct interaction with importin beta. Within the nucleus, Rev binds to the Rev-responsive element (RRE) in viral RNAs, and recruits exportin 1 and Ran GTP to generate a complex capable of exporting the RNA to the cytoplasm. The complex dissociates in the cytoplasm, releasing the mRNA and allowing Rev to return to the nucleus and repeat the cycle.

Together, the Tat and Rev proteins strongly upregulate viral protein expression

Tat and Rev are essential for virus replication because they regulate HIV-1 transcription and transport of mRNAs that code for viral structural proteins. Their action results in the expression of viral proteins in two stages. Following integration of proviral DNA and its transcription at a basal level, only the doubly spliced 2-kb RNAs (Figure 29.3) are transported to the cytoplasm. This permits the synthesis of Tat, Rev, and Nef. Both Tat and Rev are then transported to the nucleus where they act to augment transcription of provirus DNA (Tat) and the transport of viral mRNAs to the cytoplasm (Rev). Expression of proteins encoded by the 9-kb and 4-kb classes of mRNAs (Gag, Gag/pol, Env, Vif, Vpr, and Vpu) can then take place.

Mutational studies of viral genes determined that vif, vpr, vpu, and nef are not essential for production of infectious HIV-1 in many cell culture systems *in vitro*. However, the conservation of the reading frames for these proteins in most clinical isolates of the virus suggests that they play important roles during infection within the host. These roles are described below.

The Vif protein increases virion infectivity by counteracting a cellular deoxcytidine deaminase

Vif (*viral infectivity factor*) is a 193-amino acid protein found in the cytoplasm of infected cells and incorporated at low levels in virions via an interaction with viral genome RNA. Deletion of the vif gene reduces infectivity of HIV-1 in cell cultures and in animal models used to test pathogenicity. Virus lacking Vif enters cells normally but generates a lower level of proviral DNA than wild-type virus. The absence of Vif within infecting virions cannot be compensated by expressing it in the cells being infected.

Vif is required to overcome the action of a host cell protein, APOBEC3G. APOBEC3G is a member of a family of cellular proteins that deaminate cytidines (in RNA) or deoxycytidines (in DNA) to either uridine or deoxyuridine. The original member of this family, APOBEC1, was so named because it deaminates a specific cytidine residue in the mRNA coding for *Apo*lipoprotein B, allowing two proteins to be made from the same gene using either the native or edited messenger RNA.

APOBEC3G is incorporated into virions during assembly, and can induce deamination of multiple deoxycytidine residues in the DNA product of reverse transcription made during a subsequent infection. This leads to mutations in viral structural and regulatory proteins, with a corresponding reduction in infectivity. APOBEC3G therefore functions to defend the cell against infection by mutating the DNA copy of the

HIV-1 genome. Vif acts by binding to APOBEC3G and inducing **ubiquitination** and degradation of this protein by **proteasomes**. This prevents its incorporation into virions, and therefore counteracts this cellular antiviral defense mechanism.

However, mutants of APOBEC3G that lack deoxycytidine deaminase activity retain some antiviral activity. Recent studies have shown that APOBEC3G also directly impairs the reverse transcription reaction, significantly reducing the yield of proviral DNA.

The Vpr protein enhances HIV-1 replication at multiple levels

Vpr (*virion protein R*) is a 100-amino acid protein that is recruited into virions (10–100 molecules per virion) by virtue of its interaction with the carboxy-terminal region of Gag. One of its major effects is to permit the infection of non-dividing cells by serving as a signal for the active transport of the preintegration complex into the cell nucleus (see above).

Vpr also facilitates packaging within the virion of the cellular enzyme uracil DNA glycoslase. This enzyme can remove deoxyuridine residues incorporated into viral DNA during reverse transcription because of high levels of dUTP in the cell. Other lentiviruses block effects of high dUTP levels by encoding and packaging a dUTPase, which hydrolyzes dUTP to dUDP, rendering it unavailable for the synthesis of viral DNA.

Vpr can also arrest infected cells in the G2 stage of the cell cycle. Vpr may do this by targeting for degradation cellular proteins that are needed to pass from the G2 phase to mitosis. Arresting or delaying cells in G2 may be beneficial for virus replication since transcription directed by the HIV-1 LTR is the most active at this stage of the cell cycle.

The Vpu protein enhances release of progeny virions from infected cells

Vpu (*virion protein unique to HIV-1*) is an 81-amino acid protein that is inserted into membranes via its amino-terminal domain. This protein accumulates in the Golgi apparatus and the endosome compartment of the cell. No homologs have been identified in related lentiviruses such as HIV-2 and simian immunodeficiency virus. Vpu has two known activities within the cell.

Degradation of CD4. The cellular CD4 protein is a receptor for HIV that interacts with gp120 at the cell surface (see above). However, both CD4 and gp160, the precursor of gp120, are made in the endoplasmic reticulum, and they can bind to each other at that intracellular site. The aggregate that forms retains gp160 inside the cell and therefore reduces gp120 incorporation into the virions released (Figure 29.7a). Vpu acts by binding

to CD4 and to the cellular protein ß-TrCP. This induces the ubiquitination of CD4 and its subsequent degradation by proteasomes, thus releasing gp160 and increasing surface expression of its cleavage products, gp 41 and gp 120.

Enhancement of virus release from the plasma membrane. This activity is dependent upon the transmembrane portion of Vpu. In the absence of Vpu, virions accumulate on the cell surface in a partially budded state. Expression of Vpu results in enhanced release of virus from the cell surface. Remarkably, this effect is not restricted to HIV-1; Vpu also enhances the release of other, unrelated viruses. Recent work has determined that Vpu induces degradation of tetherin, a host cell protein located on the plasma membrane that is believed to interact with budding virions and promote their endocytosis.

The Nef protein is an important mediator of pathogenesis

Nef (*negative effector*) is a 210-amino acid protein that is localized at the inner face of the plasma membrane via a residue of **myristate** added to its N-terminal amino acid. Infection of monkeys with simian immunodeficiency virus containing point mutations in the Nef gene resulted in the rapid generation of revertant viruses with a functional nef gene, showing that there is considerable selection pressure for an active Nef protein. Simian immunodeficiency virus mutants with deletions of the Nef gene are viable, but have lower titers and fail to induce disease in infected monkeys. Expression of Nef in mouse T cells and macrophages causes a disease that resembles the late stages of AIDS. These findings implicate Nef as an important determinant of disease in infected animals. Three activities have been attributed to Nef, but it is unclear which of these activities is important for disease.

Decrease in the surface expression of CD4 and MHC 1. Nef expression decreases the levels of CD4 and the **major histocompatibility complex protein** MHC 1 on the cell surface. Because these are important mediators of immune responses (see Chapter 34), their absence can be important in disease progression. The loss of surface CD4 is caused by an increase in the cycling of CD4 between the cell surface and the endosome compartment. Nef binds to CD4 and to the adaptor protein AP2 on the plasma membrane, and this complex is recruited into clathrin-coated pits and endosomes (Figure 29.7b).

In contrast, Nef-induced loss of MHC 1 from the cell surface is due to a block in trafficking of MHC 1 from the Golgi apparatus to the plasma membrane. This activity of Nef requires its interaction with another adaptor protein, AP1. The loss of cell surface MHC 1 means that the cell cannot present viral antigens to circulating cytotoxic T lymphocytes, thus masking the infection from the immune system.

(a)

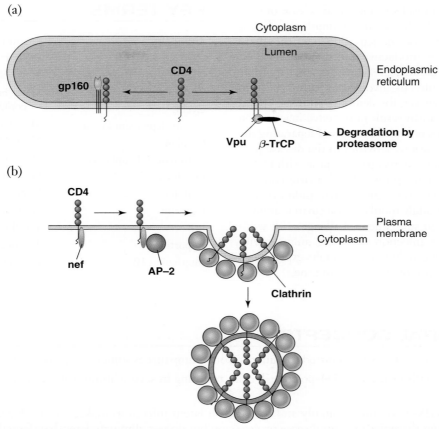

Figure 29.7 Down-regulation of CD4 expression. (a) In the absence of Vpu, CD4 interacts with gp160 in the endoplasmic reticulum, resulting in formation of a complex that blocks transport to the cell surface. Vpu binds to the cytoplasmic domain of CD4 and through interaction with β-TrCP directs CD4 to the proteasome degradation pathway. (b) Nef binds to the cytoplasmic tail of CD4 at the inner surface of the plasma membrane, and via interaction with AP-2 increases uptake of CD4 into clathrin-coated pits, lowering its concentration on the cell surface.

Enhancement of virus infectivity. HIV Nef mutants make virus particles that have reduced capacity to infect cells. This effect cannot be reversed by expression of Nef in cells infected with Nef-minus virus. Nef may enhance infectivity by modification of virion structure; however, no difference in the structure of virions produced in the presence or absence of Nef has been detected. Virions produced in the absence of Nef appear to have a reduced ability to complete proviral DNA synthesis upon infection of target cells. There has been a suggestion that the presence of Nef might facilitate the passage of the nucleocapsid from virions into the cytoplasm by altering the structure of the actin matrix that lines the inner surface of the plasma membrane, where nucleocapsids are released after fusion.

Modification of cell signaling. Nef has been implicated in alterations of signaling in T cells, resulting in general activation of the cell and promoting virus replication. However, unlike T cells activated by antigens, T cells activated by Nef cannot effectively mount an immune response. Activation of T cells by Nef is caused by modulation of the activity of a number of cellular protein kinases involved in signaling pathways.

One such target is the host serine/threonine protein kinase Pak2, known to be involved in stimulation of cell growth and inhibition of apoptosis. Nef also prevents apoptosis of the infected cell by inhibition of Ask-1, a kinase that links death receptors with cellular caspases involved in initiating apotosis. By stimulating PI-3 kinase, Nef also inactivates Bad, a pro-apoptotic factor. Expression of FasL on the surface of the infected cell is also increased by Nef, inducing apoptosis of surrounding uninfected CD8-positive and CD4-positive T cells by interaction between FasL and Fas. In so doing, Nef acts to kill cells that could otherwise help to clear the infection, prolonging the lifetime of the infected cell and maximizing production of new virus.

Nef also interacts via its proline-rich domain with members of the **Src** family of tyrosine protein kinases and alters their activity. Activation of the kinase Hck results in increased expression of interleukin-6, **tumor necrosis factor-α**, and **interleukin-1β**, all activators of HIV-1 replication. This also results in increased release of chemokines from infected cells, recruiting uninfected T cells to the sites of virus infection and activating them, thus making them susceptible to productive infection by HIV-1.

At present, it is unclear which of the multiple activities of Nef are directly involved in pathogenesis in infected animals. However, the demonstration that point mutations within Nef can result in selective inactivation of particular functions provides a means of separating the contributions of the various activities to the disease process. The initial failure to detect pathogenesis with HIV-1 lacking Nef resulted in the proposal such a virus could be used as an attenuated virus vaccine. Initial studies yielded promising results in adult monkeys; vaccinated animals displayed resistance to subsequent challenge with wild-type virus. However, infection of young animals with virus lacking Nef induced disease, and adults experienced complications over a prolonged period of time.

KEY TERMS

Antibodies	Latent infection
CD4 antigen	Ligand
CD4-positive T lymphocytes	Macrophage
Chemokines	Major histocompatibility
Cyclin	complex (MHC) protein
Cyclin-dependent kinase	Myristate
Cytokines	NF-κB (nuclear factor-κB)
Cytotoxic T lymphocytes	Opportunistic infections
Dendritic cells	Preintegration complex
Epitopes	Proteasomes
Exportin	Src-related kinases
IκB	T-cell receptor
Importin	Tumor necrosis factor
Interleukin 1β	Ubiquitination

FUNDAMENTAL CONCEPTS

- HIV-1 is a lentivirus that was transmitted to humans from chimpanzees sometime in the early 1900s.
- HIV-1 infects and kills primarily CD4-positive T cells, resulting in severe immune deficiency for >90% of people infected.
- The capacity of HIV-1 to mutate rapidly and to establish latent infections makes curing this infection extremely difficult. Current drug therapies prevent disease progression but do not eliminate latently infected cells.
- HIV-1 codes for multiple auxiliary proteins that overcome barriers to virus replication.
- Tat stimulates elongation by RNA polymerase II molecules that have initiated transcription of HIV-1 proviral DNA.
- Rev directs export of most HIV-1 mRNAs from the nucleus to the cytoplasm.
- Vif inactivates the host protein APOBEC-3G, which inhibits reverse transcription and increases mutations in proviral DNA.
- Vpr retains infected cells in the G2 phase, increasing virus replication.
- Vpu enhances release of HIV-1 virions from the cell surface.
- Nef has multiple effects on the cell, maximizing virus replication while reducing the effectiveness of the host immune response.

REVIEW QUESTIONS

1. The genetic makeup of an individual's immune system can be an important determinant in the development and progression of AIDS. Why?

2. Discuss the three size classes of mRNAs generated by HIV-1, in contrast to the two size classes generated by simpler retroviruses.

3. At the molecular level, how is HIV latency controlled?

4. A number of cellular proteins, including APOBEC3G, are incorporated into HIV virions. What role does APOBEC3G play in the virions?

5. The HIV protein Nef is able to induce a loss of MHC 1 from the surface of virus-infected cells. How does this "hide" the virus from the immune system?

CHAPTER 30

Hepadnaviruses

Christopher Richardson

Hepadnaviridae

From *hepa*titis and *DNA*

VIRION

Spherical enveloped particle.

Diameter 42 nm.

Icosahedral capsid, T = 4.

Abundant smaller spherical and filamentous forms lack nucleocapsid and are not infectious.

GENOME

Circular, double-stranded DNA with a single-stranded gap on one strand, 3.2 kb.

GENES AND PROTEINS

mRNAs are transcribed by cellular RNA polymerase II from five promoters.

Four overlapping reading frames, seven viral proteins.

- *Three surface proteins: LS, MS, SS*
- *Core proteins: C, E*
- *Polymerase (reverse transcriptase) protein: P*
- *Regulatory protein: X*

VIRUSES AND HOSTS

Hepatitis B viruses of human, chimpanzee, duck, gibbon, gorilla, ground squirrel, heron, orangutan, snow goose, woodchuck, woolly monkey.

DISEASES

Hepatitis: incubation period 30–180 days.

Transmission by blood and sexual contact.

Acute disease can be mild or severe.

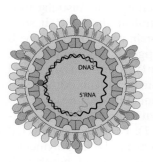 

Chronic or associated diseases: cirrhosis, liver cancer, serum sickness.

Treatment: pegylated interferon, entecavir, telbivudine, tenefovir, adefovir, lamivudine.

Prevention:
- *vaccination with recombinant hepatitis B surface antigens*
- *injection with anti-hepatitis B immunoglobulin after exposure to virus.*

DISTINCTIVE CHARACTERISTICS

Unusual partly single-stranded, partly double-stranded circular genome results from incomplete genome replication in cell.

Pregenome RNA is both an mRNA and a template for synthesis of genome DNA.

Reverse transcriptase (polymerase protein) generates genome DNA from pregenome RNA within capsid during virus assembly.

Makes large amounts of non-infectious spherical and filamentous particles.

Has a satellite virus: hepatitis delta virus (see Chapter 31, Viroids).

At least seven distinct viruses cause human hepatitis

Hepatitis (infection of the liver) constitutes a major worldwide public health problem. Among the many agents that cause hepatitis are a number of unrelated viruses. Before the viruses causing hepatitis were identified, these pathogens were classified by modes of transmission and epidemiology. The terms "hepatitis A" and "hepatitis B" were first introduced by MacCallum in 1947 to differentiate between infectious (**enteric**) and serum hepatitis. Type A hepatitis was predominantly transmitted via the fecal–oral route while type B hepatitis was transmitted primarily through blood contact. Eventually these terms were adopted by the World Health Organization Committee on Viral Hepatitis. Hepatitis A virus is now known to be a member of the *Picornaviridae* (Chapter 11). A new bloodborne hepatitis virus, distinct from A and B, was named C, and is a member of the *Flaviviridae* (Chapter 12). At least four other hepatitis viruses, named D (or delta) to G, are presently known.

The discovery of hepatitis B virus

In 1963, Baruch Blumberg discovered a previously unknown protein, which he denoted the Australia antigen, in the blood of an Australian aborigine. It soon became apparent that this protein was related to type B hepatitis, and the Australia antigen is now known as the hepatitis B surface antigen. In 1973, David Dane found virus-like particles in the serum of patients with type B hepatitis; these particles subsequently became known as Dane particles and were shown to be virions of hepatitis B virus (HBV) (Figure 30.1). Paul Kaplan detected endogenous RNA/DNA-dependent DNA polymerase (reverse transcriptase) within the core of hepatitis B virions. The hepatitis B virus DNA genome was eventually cloned and sequenced by Pierre Tiollais, Patrick Charny, Pablo Valenzuela, and William Rutter in 1979.

Hepatitis B virus is distinct in the world of viruses with its very compact 3.2- kb circular DNA genome, extensive use of overlapping reading frames, production of a pregenome RNA, and dependence on a reverse transcription step for replication. Because of its use of reverse transcriptase, hepatitis B virus is compared here with the retroviruses, which were the first viruses known to use reverse transcriptase for replication. Although they share reverse transcription as an obligatory step in their replication cycles, these viruses differ in an important way: retroviruses package an RNA genome that is reverse transcribed into DNA once it enters the host cell, whereas hepadnaviruses package a DNA genome that is reverse transcribed from a pregenome RNA in the cell during virus assembly (see below).

Human hepatitis B virus became the archetype of the *Hepadnaviridae* family. Similar viruses are found in a number of animals (see chapter opener), and some of these viruses have served as models for the study of hepadnavirus replication and pathogenesis. Although the hepatitis B surface antigen has been found in other primates, humans remain the primary reservoir for this virus. Cauliflower mosaic virus, a plant virus, also has circular DNA, has overlapping reading frames, and uses a reverse transcriptase during its replication. It is believed to be a distant cousin of hepadnaviruses (see Table 3.6).

The World Health Organization estimates that over 2 billion people worldwide have been exposed to hepatitis B virus, and around 350 million of these individuals are chronic carriers. Transmission is primarily through blood (transfusions, needle sharing) and sexual contact. The virus can also be passed from mother to newborn babies during birth. A satellite virus known as hepatitis delta virus (see Chapter 31) is associated with hepatitis B virus infection. Hepatitis delta virus can replicate only in cells infected with hepatitis B virus, since it borrows the hepatitis B surface proteins to package and form an envelope around its own capsid proteins and genome RNA. However, hepatitis delta virus resembles viroids and is quite different from the hepadnaviruses.

Dane particles are infectious virions; abundant non-infectious particles lack nucleocapsids

The hepatitis B virion, or Dane particle (Figure 30.1), has a diameter of 42 nm. Its envelope contains hepatitis B surface (S) proteins and lipids; however, electron microscopy does not reveal a typical unit membrane structure, and the envelope contains a lower proportion of lipids than most membranes and viral envelopes. The envelope surrounds the nucleocapsid or core, which is composed of 180 molecules of hepatitis B core (C) protein arranged with T = 4 icosahedral symmetry. The core contains at least one molecule of reverse transcriptase (P), as well as the viral genome.

High concentrations of non-infectious particles can be found in the serum together with Dane particles during the acute phase of infection. These particles are composed primarily of lipids and hepatitis B surface proteins. Both filamentous and spherical non-infectious particles, each having a diameter of 22 nm, are found in the serum. The spheres contain only small and middle hepatitis B surface proteins (SS and MS), whereas the filaments also contain large hepatitis B surface proteins (LS). These particles contain no core proteins, reverse transcriptase, or viral genome DNA. Since non-infectious particles contain the viral surface antigens, they can induce an immune response in the infected host. Production of massive amounts of non-infectious particles may provide a "stealth" mechanism for evading the immune response; they can divert the immune attack by binding to antibodies and **complement**, allowing the less abundant virions to traverse the bloodstream undetected.

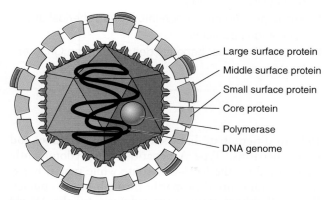

Large surface protein
Middle surface protein
Small surface protein
Core protein
Polymerase
DNA genome

Figure 30.1 **Structure of virions of hepatitis B virus.** Genome is shown as an orange wavy line bound to polymerase (orange sphere). Capsid, shown as an icosahedron, is constructed from core protein. Envelope, containing three forms of surface protein, is shown as orange subunits surrounding the capsid.

Eight major strains with different genome sequence variations (genotypes) and four major antigenic strains (serotypes) of hepatitis B virus are found worldwide. The genotypes, denoted Group A through Group H, are distributed in specific regions of the world. For example, genotypes A and D are found in Europe, D and E in Africa, B, C, and D in Asia, B in Alaska, C in Australia, and F, G, and H in Central and South America. Genotypes vary in the severity of disease symptoms, in the ability to induce chronic liver disease, and in their response to **interferon** or chemotherapy.

Serotypes are defined by antibody recognition sites on the small surface protein. A common **epitope** present on the surface protein of all known hepatitis B strains is called determinant "a". The four distinct serotypes contain two pairs of mutually exclusive epitopes named "d" or "y" and "w" or "r". Determinant d has a lysine at position 122 on the small surface protein, while y has an arginine at that position; w has a lysine at position 160 while r has an arginine at that position. Thus, a common strain of hepatitis B virus might be classified as adw or adr.

The viral genome is a circular, partly single-stranded DNA with overlapping reading frames

Hepatitis B virus has one of the smallest viral genomes known: the circular DNA is 3.2 kb long. This DNA has the unusual property of being partly double-stranded, with a single-stranded region of variable length (Figure 30.2a). Furthermore, unlike other circular DNA viral genomes, neither DNA strand forms a covalently closed circle. The (full-length) minus DNA strand is joined at its 5' end to a tyrosine residue in the P protein, a virus-coded DNA polymerase with reverse transcriptase activity. The complementary plus DNA strand is joined to a short, capped RNA at its 5' end.

The plus-strand is shorter than the minus-strand because its synthesis is interrupted before completion, giving rise to the single-stranded region. The ends of the DNA are held together as a circle by hydrogen bonding between short repeated sequences called DR2, near the 5' ends of both strands. These peculiar features arise from the mechanism of generation of the genome DNA via reverse transcription (see below). *In vitro* incubation of disrupted virions with deoxyribonucleoside triphosphates allows the virion-associated DNA polymerase to extend the plus DNA strand, filling in the single-stranded gap (Figure 30.2b).

Coding regions in the genome of hepatitis B virus are organized in a highly efficient fashion (Figure 30.3). Four partly overlapping open reading frames, designated C (for capsid protein), P (for polymerase), S (for surface protein), and X (for a regulatory protein), are translated to yield seven different viral proteins. Every base pair in the genome of hepatitis B virus is involved

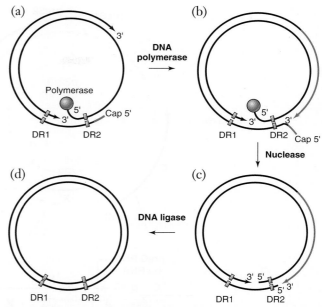

Figure 30.2 Structure of hepatitis B virus genome, and its conversion to a covalently closed, circular DNA. (a) The 5' ends of minus-strand (inside circle) and plus-strand (outside circle) genome DNA are joined by short complementary regions named DR2 (gray boxes). The 5' end of the minus-strand is attached to the viral polymerase protein (orange sphere) and the 5' end of the plus-strand is attached to a short, capped RNA (blue line). The plus-strand is incomplete. (b) DNA polymerase extends the plus-strand, filling in the ss gap (orange). (c) Nucleases remove the 5' ends of both strands. (d) DNA ligase joins the 3' and 5' ends of each strand, forming a fully covalently closed double-stranded circular DNA.

in coding for at least one viral protein! The genome also contains two enhancer elements (Enh 1 and 2) that regulate levels of transcription, a polyadenylation signal, a packaging signal (ε), and direct repeats (DR1 and DR2) that are involved in reverse transcription. These sequence elements necessarily overlap with one or more coding regions (see inset in Figure 30.3). Few other viruses have so many overlapping functions within such a compact genome.

Nucleocapsids enter the cytoplasm via fusion and are transported to the nucleus

Replication of hepatitis B virus is most efficient in liver cells (hepatocytes), although other cell types, including monocytes, epithelial cells, and endothelial cells, have been found to support minimal viral replication. Hepatocytes freshly explanted from duck liver support duck hepatitis B virus infection, but there are currently no *in vitro* cultivated cell lines that can be productively infected with hepatitis B virus. Consequently, initial steps of virus entry are poorly understood. However,

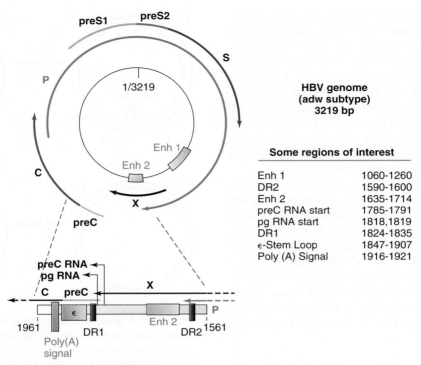

Figure 30.3 Coding and signaling regions on the hepatitis B genome. The four partially overlapping open reading frames (C, P, S, and X) are shown with arrowheads pointing in the direction of transcription and translation. Translation of the C and S open reading frames starts at two or three alternative sites (preC or C and preS1, preS2, or S). The overlap of control sequences and coding sequences in a 400-nt stretch of the genome is shown in the inset.

several cell lines are capable of supporting hepatitis B virus DNA synthesis upon transfection with cloned viral DNA. These cell systems have helped to elucidate much of the hepadnavirus life cycle.

Attempts to define a receptor for hepatitis B virus have yielded various candidates, including apolipoprotein-H, fibronectin, interleukin-6, annexin V, and an unknown 80-kDa protein. Which of these proteins are functional cellular receptors remains to be established.

It is believed that the large hepatitis B surface protein undergoes proteolytic cleavage as it interacts with the cellular membrane, exposing a fusion peptide. This results in the fusion of the virion envelope and the host cell membrane and the release of the nucleocapsid into the cytoplasm. The nucleocapsid is then transported to the nuclear membrane. Studies with duck hepatitis B virus have suggested that release of the viral genome into the nucleus occurs by interaction of the core with the nuclear pore complex.

The viral genome is converted within the nucleus into a covalently closed, circular form (Figure 30.2). This process involves extension of the plus-strand DNA across the single-stranded gap region to form a fully double-stranded DNA, removal of the terminal primers (the P protein on the minus-strand, and the short, capped RNA primer on the plus-strand), and covalent

ligation of the resulting 3′ and 5′ ends of each DNA strand (Figure 30.2b, c, and d). The P protein may take part in the extension of the plus-strand, but inhibitor studies have shown that all of these steps (strand extension, primer removal, and ligation) can be carried out by host cell enzymes in the nucleus.

Transcription of viral DNA gives rise to several mRNAs and a pregenome RNA

In contrast to retroviruses, hepadnavirus DNA is not immediately integrated into the host cell genome, but remains in a free, circular form. This circular viral DNA uses cellular RNA polymerase II to direct the synthesis of a set of five messenger RNAs, one of which also serves as a pregenome RNA (pg RNA). Transcription is influenced by two enhancer elements (Enh I and Enh II) (Figure 30.3), which are binding sites for several cellular transcription factors. Five different promoter elements (preC, pregenome, preS1, S, and X) direct initiation of transcription by RNA polymerase II (Figure 30.4). The preC and pregenome promoters are closely spaced, only 10 to 15 nucleotides from each other (see insert, Figure 30.3). A single polyadenylation signal specifies the 3′ ends of all viral RNAs. This gives rise to five classes of RNA transcripts, ranging in size from 900 to 3500 nucleotides (Figure 30.4).

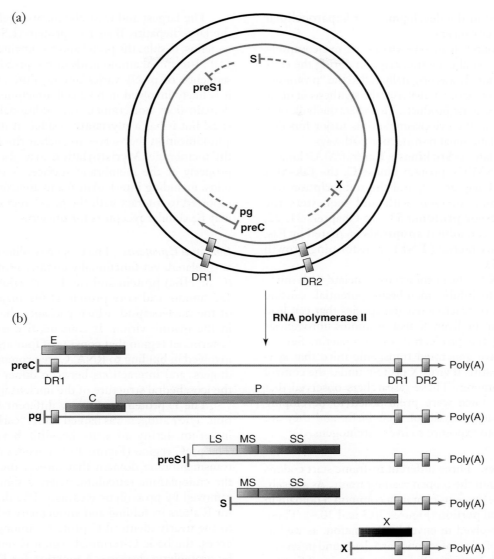

Figure 30.4 Transcription of hepatitis B virus DNA. (a) Circular DNA is transcribed by host RNA polymerase II, which recognizes five promoters (preC, pg, preS1, S, and X) to generate five classes of capped mRNAs that share a common 3′ polyadenylated end. (b) Open reading frames and translation products are shown above each RNA. pg RNA also serves as a template for virus replication.

The smallest mRNA encodes the X protein. Transcripts initiated at the S promoter encode both the small (SS) and medium (MS) surface proteins, while those initiated at the preS1 promoter encode the large surface protein (LS). Transcripts initiated at the pregenome promoter have a dual role. They can serve as messenger RNAs for the C and P proteins, and can also serve as pregenome RNA (pgRNA), an intermediate in the synthesis of progeny viral genome DNA. Transcripts initiated at the nearby preC promoter encode the E protein.

The 3.5-kb pg and preC transcripts are exceptional in that they extend more than once around the circular DNA genome. Their promoters lie upstream of the polyadenylation signal, but RNAs initiated at these promoters are not cleaved on first passage of the polyadenylation signal, which lies very close to the promoter region. This may be due in part to the sequence of the polyadenylation signal, UAUAAA, which differs from the consensus AAUAAA found in most genes in vertebrate cells.

The roles of hepatitis B virus proteins

X protein. The smallest open reading frame encodes the 154–amino acid X protein. This protein is aptly named, because its function remains elusive; many contradictory findings have been made over the years. X has been implicated in the development of **hepatocellular carcinoma** (liver cancer).

X is not found in mature virions or nucleocapsids, but it stimulates viral gene transcription during the early phases of infection. It was originally called the "promiscuous transactivator" since it increases the synthesis of many viral and cellular gene products. However, the bulk of X protein resides in the cytoplasm, and its major function may be to regulate signal transduction pathways.

X up-regulates c-**Src kinase**, **Ras**/raf/MAP kinase, stress kinase (SAPK), protein kinase C, the **Jak–Stat pathway** (see Chapter 33), and the transcription factor **NF-κB**, and interacts with and sequesters the **tumor suppressor protein** p53 (see Chapters 21, 22, and 23). It also can inhibit **apoptosis** mediated by Fas, **tumor necrosis factor (TNF)**, transforming growth factor β, or p53.

Recently X has been shown to associate with mitochondria and to inhibit membrane potential, causing the generation of reactive oxygen species that can lead to mutations in the liver. X also promotes fibrogenesis by stimulating the production of fibronectin. Several laboratories have generated transgenic mice that synthesize the hepatitis B virus X protein under the control of its own promoter. These researchers observed that the transgenic mice were predisposed to developing liver tumors within a year of their birth, and they are more sensitive to exposure to liver carcinogens.

Surface proteins. Three different in-frame start codons are utilized within the S open reading frame. As a result, all three surface proteins contain a common S domain in their C-terminal portion (Figures 30.3 and 30.4). These proteins are involved in envelope formation, as well as the formation of non-infectious particles found in serum.

The small hepatitis B surface protein (SS, also known as SHBs), originally designated as Australia antigen, is the smallest and most abundant surface protein. SS has four α-helical hydrophobic regions. It contains 14 cross-linked cysteine residues, providing for high stability, and it can be glycosylated at asparagine-146. SS has at least two transmembrane regions. Prior to DNA sequence analysis, different strains of hepatitis B virus were classified based on the antigenic epitopes contained within SS (see above).

The second most abundant membrane protein produced is the middle hepatitis B surface protein (MS, or MHBs). The MS protein contains an additional 55 N-terminal amino acids encoded by the preS2 domain (Figure 30.3). These additional amino acids are primarily hydrophilic and are believed to reside on the external face of hepatitis B virions. The preS2 domain also contains a glycosylation site at asparagine-4 that is always glycosylated in the MS protein. However, viruses that lack the ability to express MS protein still retain the ability to produce infectious virions.

The largest and least abundant envelope protein is the large hepatitis B surface protein (LS, or LHBs). It contains, besides the preS2 and S domains, an additional 119 N-terminal amino acids in the preS1 domain. The sequence of preS1 varies among different strains and it is likely involved in host cell attachment. The preS1 domain does not contain any additional glycosylation sites, but it has a **myristate** residue at its N-terminus, a modification that serves to anchor the N-terminus in the membrane. **Myristoylation** may also help LS fold properly at the membrane surface. A variety of proteins, including interleukin-6 and annexin V, have been reported to interact with the preS1 region of LS; these may be cellular receptors for the virus.

Core and E proteins. The C open reading frame encodes two related, yet functionally distinct proteins: the core (C or HBc) protein and the E (or HBe) protein. The 185-amino acid core protein is the major component of the nucleocapsid, which packages the viral genome in the mature virion. It contains five α-helices and a C-terminal region that consists of four arginine clusters involved in binding to RNA. Core proteins readily form dimers, and interactions between dimer pairs produce the icosahedral structure of the nucleocapsid.

The E protein has a very different fate and function. The *e* antigen was named for its "early" appearance in serum during an acute hepatitis B virus infection. The preC region (Figure 30.3) encodes a hydrophobic transmembrane domain that directs the E protein to the endoplasmic reticulum, where a signal sequence is removed by proteolytic cleavage. The distinct location for E alters its folding and antigenicity when compared to the nearly identical C protein. Among other differences, the basic C-terminal region is removed from E by proteolytic cleavage. A function for E has not been established. One theory is that high levels of E protein may suppress the host immune system and prevent it from eliminating cells that contain hepatitis B virus. In support of this hypothesis, woodchuck hepatitis virus variants that lack E are incapable of establishing persistent viral infections.

Polymerase protein. The largest open reading frame in the genome encodes the hepatitis B DNA polymerase protein (P, or HBp). This 90 kDa protein functions as an RNA/DNA-dependent DNA polymerase (reverse transcriptase). The polymerase plays a critical role in hepadnavirus genome replication and pregenome RNA encapsidation. P has four characteristic domains: an N-terminal domain for priming minus-strand DNA synthesis; a spacer domain; a reverse transcriptase domain, which occupies roughly 40% of the protein; and a **ribonuclease H** domain. The ribonuclease H activity degrades the pregenome RNA during the process of genome replication.

The pregenome RNA is packaged by interaction with polymerase and core proteins

The pregenome RNA (pgRNA), in addition to serving as mRNA for C and P proteins, is used as a template to produce the DNA genome of hepatitis B virus. This occurs in a "packaged" RNA-protein complex including both the C and P proteins. The viral replication machinery targets the pregenome RNA for packaging by recognition of a region known as the epsilon (ε) stem-loop (Figures 30.3 and 30.5).

All viral transcripts have a copy of ε near their 3′ ends, just upstream of the polyadenylation site; however,

1. Polymerase binds to 5' ε-stem-loop on pg RNA

2. Initiation of RT activity

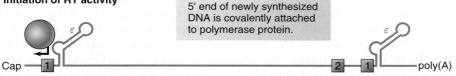

5' end of newly synthesized DNA is covalently attached to polymerase protein.

3. Polymerase-DNA transfers to 3' DR1 site

4. Elongation of minus-strand DNA; RNA template degradation

5. Capped RNA fragment remains at 5' end of template

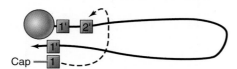

Capped RNA fragment translocates to DR2↓ and acts as primer for plus-strand DNA synthesis.

6. Initiation of plus-strand DNA synthesis

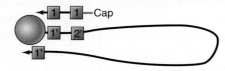

7. Transfer of plus-strand DNA to 3' DR1 site on minus-strand DNA; elongation of plus-strand DNA

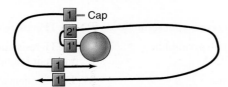

Polymerase extends plus-strand DNA along minus-strand DNA template, forming circular ss/ds DNA.

Figure 30.5 Reverse transcription of pregenome RNA to form hepatitis B virus genome DNA. Boxes labeled 1 and 2 represent DR1 and DR2 sequence elements. ε stem-loops are shown at both ends of pregenome RNA (pg RNA). Orange sphere represents reverse transcriptase (P protein). 3′ ends of growing DNA chains (black) are shown as arrows. RNA (blue) degraded by RNAse H is shown as dashed line. See text for details.

only pregenome and preC RNAs have a second copy of the ε stem-loop near their 5′ ends. It appears that only the 5′ ε stem-loop is functional in packaging, and only in the pregenome RNA. The slightly longer preC RNA is not packaged, apparently because translation of the preC open reading frame (which overlaps with the ε stem-loop) interferes with packaging. The pregenome RNA starts just downstream of the beginning of the preC open reading frame and therefore this region is not translated in pg RNA (Figure 30.3).

Analysis of the ε region sequence predicts a series of inverted repeats that fold into a three-dimensional stem-loop structure (Figure 30.5). This secondary structure is conserved among all hepadnaviruses despite many differences in the primary sequence. The polymerase protein is believed to recognize and interact directly with the ε stem-loop, initiating both encapsidation and reverse transcription of the pregenome RNA. The C-terminus of the polymerase protein interacts with the core protein to begin encapsidation only when the polymerase is bound to the ε stem-loop.

Genome replication occurs via reverse transcription of pregenome RNA

The generation of the hepatitis B virus DNA genome from pregenome RNA is formally similar to the synthesis of retrovirus proviral DNA from genome RNA (Chapter 28). In both cases, a reverse transcriptase initiates DNA synthesis near the 5′ end of the template RNA, then jumps to a repeated sequence at the 3′ end, and copies the remainder of the RNA, degrading the template as it progresses. However, many details of the process are different, as described below (Figure 30.5 and Table 30.1).

Step 1. Initiation of the reverse transcription process begins when P binds to the 5′ ε stem-loop. Retrovirus reverse transcriptases use a cellular tRNA bound to their genome RNA as a primer to initiate synthesis of a DNA copy (Chapter 28). Hepatitis B virus, in contrast, initiates DNA synthesis on the pregenome RNA template by covalently linking the first nucleotide of the DNA chain to the hydroxyl group of a tyrosine residue in the polymerase protein itself. This has similarities to the mechanism by which adenoviruses initiate replication of their linear, double-stranded DNA genomes (see Chapter 23), but in the case of hepadnaviruses the polymerase plays the roles of both the adenovirus terminal protein and the DNA polymerase. The P protein remains linked to DNA throughout the process, so presumably the active site extends the 3′ end of the growing DNA molecule, while the 5′ end remains attached to another part of the same molecule of P, forming a circular complex.

Steps 2 and 3. Reverse transcription progresses for only a few nucleotides beyond the 5′ ε stem-loop, after which the enzyme and the newly synthesized DNA are translocated onto an identical sequence in the direct repeat region 1 (DR1) near the 3′ end of the pregenome RNA. This step is reminiscent of the transfer of the retrovirus "minus strong-stop" DNA to the repeat region at the 3′ end of the retrovirus genome RNA.

Table 30.1 Comparison of reverse transcription by retroviruses and hepadnaviruses

Step	Retroviruses	Hepadnaviruses
Template for DNA synthesis	Plus-strand virion RNA	Plus-strand pregenome RNA
Site of DNA synthesis	Residual capsid after virus entry (cytoplasm)	Newly assembled capsid before virion formation (cytoplasm)
Primer for initiation of minus-strand DNA synthesis	Cellular tRNA bound near 5′ end of genome RNA	Polymerase protein bound near 5′ end of pregenome RNA
First switch: to 3′ end of RNA template	Via terminal repeat (R) sequence	Via subterminal DR1 sequence
RNA template degradation	RNAse H activity in reverse transcriptase	RNAse H activity in Polymerase
Initiation of plus-strand-DNA synthesis	RNAse H-resistant polypyrimidine tract in genome RNA	RNAse H-resistant capped 5′ RNA fragment bound to DR2 sequence
Second switch: to 3′ end of minus-strand DNA	Via primer-binding site	Via DR1 sequence
Final product	ds linear DNA with long terminal repeats	Partly ss/ds circular DNA
Fate of DNA	Integrated into cellular chromosomal DNA	Incorporated into virions

Steps 4 and 5. The P protein then makes a complete (minus-strand) copy of the remainder of the pregenome RNA by extending the DNA chain to the 5′ end of the RNA template. As reverse transcription proceeds, the RNA template is degraded by the ribonuclease H activity of P. A short RNA fragment at the capped 5′ end, including the DR1 region, remains resistant to ribonuclease H degradation. This fragment is translocated to the identical DR2 region near the 5′ end of the newly synthesized minus-strand DNA, where it forms a short RNA/DNA hybrid.

Steps 6 and 7. The translocated RNA fragment serves as a primer for subsequent synthesis of plus-strand DNA. As in step 2, the polymerase begins copying the template minus-strand DNA near its 5′ end, making a short DNA product with the capped RNA fragment at its 5′ end. This nascent DNA chain then jumps to the 3′ end of the same minus-strand DNA molecule by annealing to the DR1 sequence, and the polymerase proceeds to copy the minus-strand. The newly synthesized plus-strand DNA remains hydrogen-bonded to the minus-strand, forming a double-stranded DNA.

However, plus-strand DNA synthesis is usually not completed by the time the virion is released from the host cell, and stalls when the dNTP pools within the virion are exhausted. This explains the region of single-stranded DNA found within the genomes of purified virions. The resulting molecule, shown in step 7 of Figure 30.5, is identical to the molecule shown in Figure 30.2a. Table 30.1 and Figure 30.6 summarize in schematic form the similarities and differences between the reverse transcription mechanisms and the replication cycles of retroviruses and hepadnaviruses.

Virions are formed by budding in the endoplasmic reticulum

Because more S mRNA is made than preS1 mRNA, larger amounts of the SS and MS proteins are made than LS protein. These proteins are inserted into the endoplasmic reticulum as they are translated (Figure 30.7). By interaction between their transmembrane regions, S proteins aggregate in specific regions of endoplasmic reticulum membranes and exclude host membrane proteins. Regions rich in S proteins bud into the lumen of the endoplasmic reticulum in the absence of cores to produce the abundantly secreted 22-nm diameter spheres and filaments (Figure 30.7). These non-infectious particles contain little or no LS protein, since its presence favors the retention of surface proteins within the endoplasmic reticulum.

The assembled core associates with areas of the Golgi membranes rich in these surface proteins. LS is believed to interact with the core protein at the cytoplasmic face of the Golgi membranes, pulling the

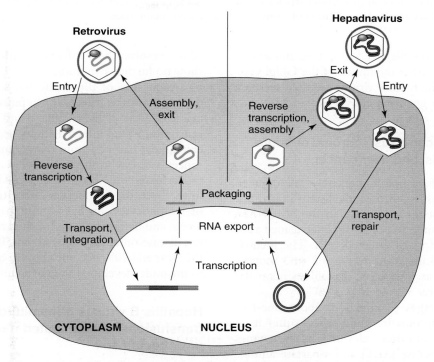

Figure 30.6 Schematic comparison of replication cycles of retroviruses and hepadnaviruses. RNAs are shown in blue and DNAs in black. Capsids are shown as hexagons, and the virus envelope is shown as a gray circle. Reverse transcriptase is shown as orange sphere. Numerous details in the replication cycles are omitted.

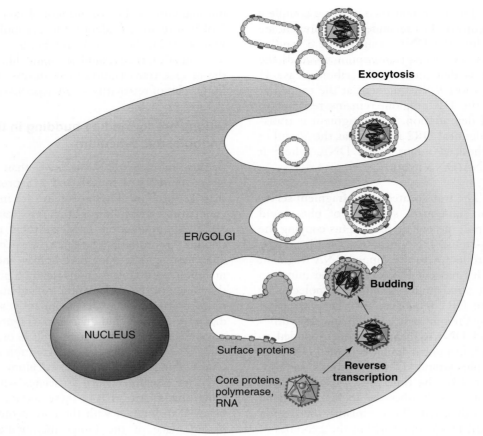

Figure 30.7 Assembly of nucleocapsids and virions of hepatitis B virus. Surface proteins are inserted into membranes of the endoplasmic reticulum and Golgi apparatus. Nucleocapsids are assembled in the cytosol and reverse transcription generates the DNA genome. Virions and non-infectious particles are formed by budding into the lumen of endoplasmic reticulum and Golgi membranes, and are released upon fusion of vesicles with the plasma membrane.

nucleocapsid into a vesicle that forms a 42-nm enveloped Dane particle within the lumen of the Golgi vesicle (Figure 30.7). The mature virions are then secreted into the extracellular environment via **exocytosis**.

Hepatitis B virus can cause chronic or acute hepatitis, cirrhosis, and liver cancer

Hepatitis B virus does not immediately kill infected cells but instead leads to chronic, persistent infections. It appears that most tissue damage in the host is the result of a host immune response directed against viral antigens on the surface of infected cells. The incubation period for viral infection averages 2 to 3 months but can be as long as 6 months, depending upon the level of virus exposure.

In the initial stages of infection, about one-half of patients are asymptomatic, while others suffer from flu-like symptoms, **jaundice**, elevated liver enzymes (alanine aminotransferase [ALT] and aspartate aminotransferase [AST]) in the blood, and **serum sickness**, a hypersensitivity response to large amounts of viral antigen in the serum. Fortunately, about 90% of infected

adults resolve their infection within six months and show no further symptoms.

About 1% of those infected suffer from a severe disease known as **fulminant hepatitis**, which results in massive levels of liver damage and often is fatal. The remaining 9% of infected adults become chronic carriers, almost half of whom may eventually develop liver cancer. Other chronically-infected individuals progress to **cirrhosis**, which is characterized by fibrin deposition in the liver, the production of **ascites** fluid, infiltration of lymphocytes, flu-like symptoms, high serum ALT levels, and hemorrhaging. Infected children have different disease outcomes; only about 50% manage to clear the virus and 40% become chronic carriers, perhaps due to the underdeveloped state of a child's immune system.

Hepatitis B virus is transmitted by blood transfusions, contaminated needles, and unprotected sex

Transmission of hepatitis B virus is primarily through blood or sexual contact. Hepatitis B virus can also be passed on to babies at birth. There are presently no drugs or treatments that ensure clearance of hepatitis B virus from infected individuals. However, there are several ways to minimize the likelihood of contracting the virus.

Risky activities to avoid include intravenous drug use and unprotected sexual intercourse with multiple partners. In professions where the virus might be expected to be in the environment, care when handling potentially contaminated surfaces, fluids, and sharp objects will minimize accidental exposure. Members of these professions are encouraged to be vaccinated. Infection with hepatitis B virus nevertheless remains a problem for certain occupational groups such as morticians, doctors, dentists, dialysis workers, hemophiliacs, mental care facility workers, tattoo-parlor workers, firefighters, police officers, daycare workers, and barbers.

A recombinant vaccine is available

In 1986, Pablo Valenzuela and William Rutter of Chiron Corporation produced the first recombinant hepatitis B vaccine in yeast (Recombivax), and this is the basis of the vaccines still used today. In a new approach, Charles Arntzen, formerly of the Boyce Thompson Institute at Cornell University and now at Arizona State University, is currently testing edible hepatitis B vaccines engineered into bananas! Another method of prevention is through **immunoprophylaxis**, used primarily on newborns of mothers who are carriers of hepatitis B virus. In this procedure, anti-hepatitis B immunoglobulins are injected into the exposed individual to neutralize circulating virus and prevent it from entering target cells. In spite of the availability of a vaccine, millions of individuals still are chronically infected with the virus.

Antiviral drug treatment has real success

Currently, six antiviral drugs are effectively used to treat chronic carriers of human hepatitis B virus. One is pegylated interferon (see Chapter 33), an immune system modulator that works by stimulating cytokine synthesis and by enhancing host antiviral responses. Lamivudine (3TC), a nucleoside analogue that was originally developed for the treatment of HIV (see Chapter 36) was also shown to be active against hepatitis B virus by Dr. Lorne Tyrrell (University of Alberta) and Biochem Pharma (Montreal) in 1994. 3TC interferes with the DNA polymerase activity of the P protein. Since then four other effective nucleoside analogues have been approved. These include entecavir, telbivudine, tenofovir, and adefovir.

Development of viral resistance against 3TC was initially a problem but a combination drug approach alleviated this problem substantially. Antiviral treatments can reduce the viral load to base levels in 48 weeks and are generally administered indefinitely to prevent reemergence of the virus. Interferon can have some secondary effects, such as nausea and fever, which makes this treatment uncomfortable. The new nucleoside analogues are more easily tolerated. More effective drugs with shorter treatment durations are under development.

There is still no cure for individuals infected with hepatitis B virus, and infection remains a major worldwide health problem, particularly because it can be a persistent, chronic infection that can eventually lead to serious liver disease. Chronic hepatitis can be controlled by the nucleoside inhibitors to a large extent, but treatments are expensive and only available in developed nations. Availability of the vaccine worldwide, especially in poorer countries, is still a major problem. More effective vaccines and antiviral agents are currently being developed by a number of laboratories and pharmaceutical companies.

Some interesting facts about hepatitis B virus infection

- You should never use someone else's toothbrush. It might transmit hepatitis B virus to you!
- The first hepatitis B virus vaccine was prepared from the blood of paid chronic carriers.
- Hepatitis B virus can linger on drying surfaces like door handles for over a week and still remain infectious.
- There have been several reports of joggers contracting hepatitis B from contaminated brambles or thorns on their running paths.
- Most people who are acutely infected with hepatitis B virus do not even realize they have it.
- The hepatitis B virus vaccine is effective in at least 90% of those who are vaccinated with the three-injection regimen.
- Two billion people worldwide have been exposed to hepatitis B virus.
- Some laboratories have been working on generating an edible hepatitis B virus vaccine by expressing HBs proteins in potatoes, carrots, and bananas.

KEY TERMS

Apoptosis	Jak–Stat pathway
Ascites	Jaundice
Cirrhosis	Myristate
Complement	Myristoylation
Enteric	NF-κB (nuclear factor-κB)
Epitope	Ras
Exocytosis	Ribonuclease H
Fulminant hepatitis	Serum sickness
Hepatitis	Src kinases
Hepatocellular carcinoma	Tumor necrosis factor (TNF)
Immunoprophylaxis	
Interferons	Tumor suppressor protein

FUNDAMENTAL CONCEPTS

- Hepatitis B virus, like retroviruses, encodes a DNA polymerase with reverse transcriptase activity.
- Virus replication begins in the nucleus using host RNA polymerase II, which produces viral mRNAs and a full-length pre-genome RNA template from circularized viral DNA.
- Hepadnaviruses package the RNA pre-genome template in the cytoplasm and use viral reverse transcriptase to generate the partially double-stranded viral DNA genome.
- Hepatitis B virus can cause cirrhosis and is the leading cause of hepatocellular carcinoma (liver cancer).
- The X protein is a multifunctional oncogenic protein that can initiate hepatocarcinogenesis.
- Immune attack by cytotoxic T cells and inflammatory cytokines (Fas, TNF) is responsible for liver damage.
- Recombivax vaccine is recombinant surface antigen expressed in and purified from yeast.
- New nucleoside analogue polymerase inhibitors are effective in reducing viral loads and reversing the symptoms of hepatitis B.
- Woodchucks, ground squirrels, ducks, and transgenic mice containing the viral genome have been used to study the molecular virology and immunology of hepatitis B.

REVIEW QUESTIONS

1. High concentrations of noninfectious hepatitis B virus particles can be found in the serum of patients; these were once used as a source of material for preparation of a vaccine. Speculate why this practice was subsequently discontinued.

2. What is meant by the statement "the genome of hepatitis B virus is said to be an example of extreme efficiency"?

3. Which of the hepatitis B virus proteins is associated with the ability of the virus to cause cancer? How does it accomplish this?

4. When the hepatitis B vaccine was introduced into the United States, children were one of the first groups targeted for vaccination; why?

5. Explain why the best worldwide treatment for hepatitis B infections is an effective vaccination program.

VIROIDS AND PRIONS

CHAPTER 31 Viroids and Hepatitis Delta Virus
CHAPTER 32 Prions

This section covers two categories of very small infectious agents that are not viruses. Viroids are small, single-stranded RNAs that do not code for any proteins, yet can replicate and cause disease in plants. Prions are normal cellular proteins that under certain circumstances can undergo structural changes and become pathological, causing neurological diseases in humans and other animals. We study them in a course on virology because they are small, noncellular infectious agents.

Viroids are circular, single-stranded RNAs 250 to 400 nt long that are not encapsidated. Because they encode no proteins, viroids depend on cellular RNA polymerases for replication. Viroids trick cellular RNA polymerases that normally transcribe DNA genomes into replicating the viroid RNA by making long RNA oligomers. These oligomers self-cleave and then ligate to reform circular viroid RNAs. Viroids are therefore "ribozymes," or RNAs with catalytic activity. Viroids are transmitted among plants by sucking insects or by mechanical means, like many plant viruses. They can cause plant diseases, but the basis of their pathogenicity is not well understood.

Hepatitis delta virus shares some characteristics of viroids: it has a small, single-stranded circular RNA genome with self-cleavage activity. However, it also has a single protein-coding gene. The RNA genome is encapsidated by the surface protein of hepatitis B virus, a required "helper" virus.

Prions are infectious proteins; they do not contain any detectable informational nucleic acid. The normal cellular protein, called PrP^C, is a copper-binding protein of uncertain function. A rare event can lead to its misfolding into a beta-helix configuration, resistant to protease digestion. In this form, ingestion or introduction into the central nervous system can lead to propagation of the misfolded form and neurological disease, dementia, and death.

Scrapie is a naturally occuring prion disease of sheep; kuru was a human disease transmitted among members of a New Guinea tribe by handling prion-infected human tissue. Mad cow disease is a recently discovered prion disease, transmitted by feeding cattle with meal made from discarded carcasses of prion-infected cattle and other animals. A small number of people in Britain have succumbed to a neurological disease believed to have arisen from ingestion of tissues of prion-infected cattle.

Viroids and Hepatitis Delta Virus

Jean-Pierre Perreault
Martin Pelchat

Viroids
Name means "virus-like" particles

VIROIDS

Autonomously replicating pathogens.
Unencapsidated, circular single-stranded RNA, 250–400 nt.
Extensive internal base pairing.
Do not encode proteins.
Some possess catalytic activity (hammerhead self-cleaving structure).

VIROID-LIKE PLANT SATELLITE RNAS

Require helper virus for replication.
Circular, single-stranded RNA, 240–500 nt.
Highly base-paired structure like viroids.
Do not encode proteins.
Unlike viroids, are encapsidated by helper virus proteins.
All possess catalytic activity (hairpin or hammerhead self-cleaving structure).

HEPATITIS DELTA VIRUS (HDV)

The only viroid-like satellite RNA known to infect humans and animals.
Depends on hepatitis B virus for its replication.
Circular single-stranded RNA, ~1680 nt.
Extensive internal base pairing.
Encodes two proteins from one open reading frame by RNA editing.
Packaged by hepatitis B virus S protein.
RNA is capable of self-cleavage (hepatitis delta virus ribozyme).

DISEASES

Viroids are transmitted between plants via mechanical breaks and insects.
Symptoms of viroid infections include leaf discoloration, dwarfing, enhanced fruit production, orange spotting, decline in fruit yield, stunting, etc.
Hepatitis delta virus is a blood-borne pathogen that replicates in the liver.
Hepatitis delta virus frequently causes fulminant hepatitis in both primate and non-primate hosts.
Hepatitis delta virus is only detected in the presence of hepatitis B virus.

DISTINCTIVE CHARACTERISTICS

Viroids are the only known infectious, pathogenic RNAs that are not packaged and do not code for any protein product.
They use cellular RNA polymerases to replicate RNA.
They replicate via rolling circle forms, RNA cleavage, and ligation.
Some viroids, and hepatitis delta virus, have enzymatic activity ("ribozymes": self-cleavage).
They may be relicts of the ancient RNA world.

Viroids are small, circular RNAs that do not encode proteins

Until the early 1970s, viruses were believed to be the smallest agents of infectious diseases. The discovery of small, self-replicating RNA pathogens changed this view. With genomes ranging from 240 to 1700 nucleotides, these infectious RNAs cause various diseases in plants and animals. They are divided into four major groups: viroids, satellite RNAs, viroid-like plant satellite RNAs, and hepatitis delta virus.

Viroids were first discovered as the causative agent of potato spindle tuber disease. This novel pathogen consists solely of a small, unencapsidated single-stranded RNA, for which the term "viroid" ("virus-like") was proposed. Since then, over 30 different viroids have been identified as the causative agents of a number of diseases that affect plants of economic importance. Viroids are composed only of a circular, single-stranded RNA molecule 250–400 nucleotides in length, a size significantly smaller than the genomes of the smallest viruses. Viroids differ from viruses not only by their small size and lack of encapsidation of their RNA; viroids do not code for any proteins and therefore must rely on host enzymes for their replication. It is believed that they cause disease by direct interaction between the viroid RNA, or other viroid-specific RNAs generated in the course of the infection, and cellular targets. **RNA interference** (RNAi) may be also be involved in viroid pathogenesis. However, the details of the molecular mechanisms of pathogenesis are still unknown.

The two families of viroids have distinct properties

Viroids can be classified into two families based on comparative sequence analyses and whether they possess a central conserved region (Table 31.1). There are currently only four members in the avocado sunblotch viroid (ASBVd) group, also known as the *Avsunviroidae*

family or group A. In this family, both plus- and minus-strand viroid RNAs have the ability to self-cleave their RNA multimers (see below). All other viroids identified to date fall into the potato spindle tuber viroid (PSTVd) group, also known as the *Pospiviroidae* family or group B. The RNAs of viroids in this family possess a highly conserved central region (C), and are not capable of self-cleavage.

Sequence analysis indicates that RNAs of *Pospiviroidae* family members are characterized by five domains: T_L (terminal left), P (pathogenic), C (central), V (variable), and T_R (terminal right) (Figure 31.1b). The boundaries between these domains are defined by sharp changes in degree of sequence homology among different viroids. Initially, these structural domains were presumed to have specific functional roles; for example, the P domain was associated with pathogenicity. However, the situation is more complex, and expression of disease symptoms is now thought to be controlled by discrete determinants located within different domains. Members of this family are divided into five subfamilies

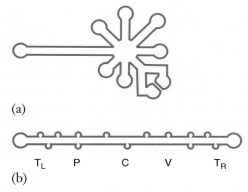

(a)

(b) T_L P   C   V T_R

Figure 31.1 Structure of viroid RNA. (a) Branched structure proposed for peach latent mosaic viroid, a member of group A. (b) Rod-like structure proposed for group B viroids, showing the five domains (TL: terminal left; P: pathogenic; C: central; V: variable; TR: terminal right).

Table 31.1 Classification of viroids

Family	Characteristics	Subfamily	Examples
Avsunviroidae (group A)	Self-cleaving RNA; self-ligating; replicate in chloroplasts	Avsunviroid	Avocado sunblotch viroid (ASBVd)
		Pelamoviroid	Peach latent mosaic viroid (PLMVd) and chrysanthemum chlorotic mottle viroid (CChMVd)
		Unassigned	Eggplant latent viroid (ELVd)
Pospiviroidae (group B)	RNAs do not self-cleave; replicate in nucleus	Pospiviroid	Potato spindle tuber viroid (PSTVd) and 8 others
		Cocadviroid	Coconut cadang-cadang viroid (CCCVd) and 3 others
		Hostuviroid	Hop stunt viroid (HpSVd)
		Apscaviroid	Apple scar skin viroid (ASSVd) and 11 others
		Coleviroid	Coleus blumei-1 viroid (CbVd) and 4 others

(Table 31.1) based mainly on the presence of highly conserved sequences within the C domain, and in part on the presence of sequence homologies in the other domains.

Viroid RNAs have a high degree of base pairing throughout the molecule. Members of the *Pospiviroidae* family fold into a rod-like secondary structure that includes short, double-stranded regions separated by small, single-stranded loops (Figure 31.1b). Although avocado sunblotch viroid is also believed to adopt a rod-like secondary structure, two other members of the *Avsunviroidae* family, peach latent mosaic viroid and chrysanthemum chlorotic mottle viroid, have complex secondary structures (Figure 31.1a), in which several stem-loops (hairpins) emerge from a central core. The actual structure of viroids inside the cell remains undetermined, and it is possible that these RNAs adopt several alternative conformations during the different steps of their life cycles.

Viroids replicate via linear multimeric RNA intermediates

By convention, the plus polarity is assigned to the most prominent viroid RNA species, and the minus polarity to its complementary strand; since viroids do not code for proteins, these designations are arbitrary. Viroids replicate by a **rolling circle replication** mechanism that generates linear multimers from a circular RNA template (Figure 31.2). A host RNA polymerase copies plus-strand circular RNA (top of Figure 31.2) into single-stranded linear, complementary minus-strands (step 1). Because there is no strong termination signal for

the RNA polymerase on the template RNA, transcription can continue around the circle for several cycles, producing a multimeric linear RNA (step 2).

For *Avsunviroidae*, these RNAs are subsequently cleaved to unit-length minus-strands (step 3), which circularize (step 4). The circular minus-strand RNAs then serve in the same fashion as templates for the synthesis of multimeric, linear plus-strand RNAs, which are cleaved and circularized (steps 5–8). This cycle can generate several new circular plus-strand RNA daughter molecules from each plus-strand template, because transcription continues for multiple rounds at both steps 1 and 5. This strategy is called "symmetric" replication, since both plus- and minus-strand circles are replicated in the same fashion.

Members of the *Pospiviroidae* family follow a similar scheme but directly copy the linear, minus-strand multimers made in step 2 into linear, plus-strand RNA multimers, therefore skipping steps 3 through 6. These multimers are subsequently cleaved and circularized by host enzymes into plus-strand progeny circles. This is called "asymmetric" replication.

Three enzymatic activities are needed for viroid replication

Rolling circle replication as depicted in Figure 31.2 requires three distinct enzymatic activities for completion: (1) an RNA-dependent RNA polymerase, to generate the multimeric linear strands; (2) an RNA-cleaving activity (endoribonuclease), to process them to unit-length linear molecules; and (3) an RNA ligase, to circularize the linear monomers.

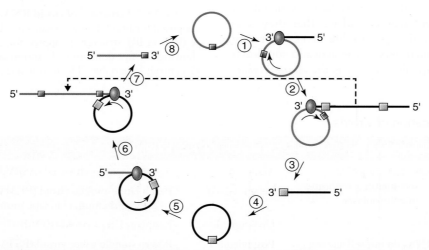

Figure 31.2 Rolling circle replication of circular RNAs. Steps 1 to 8 show symmetric replication, in which both plus- and minus-strand circles are formed and used as templates for RNA synthesis. During asymmetric replication, the multimeric minus-strand linears resulting from step 2 are directly copied into mutimeric plus-strand linears (dashed horizontal line with vertical arrow), thus skipping steps 3–6. Plus-strands are light blue, minus-strands are black. RNA polymerase is shown as an orange oval. Arrows inside circles show direction of rotation of the circular RNA through the RNA polymerase, which is shown here as stationary. Rectangles on the RNAs show cleavage/ligation sites. 3' and 5' ends of linear RNAs are labeled. See text for details.

Polymerization. Cellular DNA-dependent RNA polymerase II appears to be involved in the replication of *Pospiviroidae*, since their replication can be inhibited by low concentrations of **α-amanitin**, a known inhibitor of this enzyme. The polymerization step for *Avsunviroidae* is clearly different since it is unaffected by high levels of α-amanitin. This is probably related to the distinct sites where viroids in the two groups are localized. Viroids in the *Pospiviroidae* family are found in the cell nucleus, whereas viroids in the *Avsunviroidae* family are found within chloroplasts in the cytoplasm of plant cells. Chloroplast RNA polymerase resembles bacterial RNA polymerase in its resistance to α-amanitin. Therefore these viroids probably replicate in chloroplasts using the chloroplast RNA polymerase.

Cellular RNA polymerases normally transcribe only double-stranded DNA templates into RNA. It is not clear how viroids recruit these RNA polymerases and induce them to use their single-stranded RNA genomes as templates to be copied into complementary RNAs. Replication initiation sites have recently been identified in both *Avsunviroidae* and *Pospiviroidae*, and these should help determine how host RNA polymerases are recruited.

Cleavage. Viroids in the *Avsunviroidae* family can cleave their multimeric linear RNAs *in vitro* in the absence of any proteins, yielding monomeric linear RNA molecules. *Self-cleavage* is a term used for the intrinsic property of an RNA to cleave itself; such RNAs are **ribozymes**, because they are RNA molecules that have enzymatic activity in the absence of proteins.

Self-cleavage of *Avsunviroidae* RNAs is carried out by a **hammerhead structure** that directs cleavage at a specific phosphodiester bond in the RNA (Figure 31.3a). This structure is so named because it resembles the head of a hammer. Other self-cleaving RNA structures have also been identified in viroid-like satellite RNAs (**hairpin structure**), (Figure 31.3b) and hepatitis delta virus RNA (**delta structure**) (Figure 31.3c). All of these structures align the target cleavage site by base pairing with other regions of the RNA molecule. These structures have been shown to be responsible for the production of monomers both *in vivo* and *in vitro*. Self-cleavage occurs by nucleophilic attack by the 2'-hydroxyl of the nucleotide at the cleavage site, generating a 2', 3'-cyclic phosphate at one side of the cleavage site, and a 5'-hydroxyl terminus at the other side (Figure 31.3d).

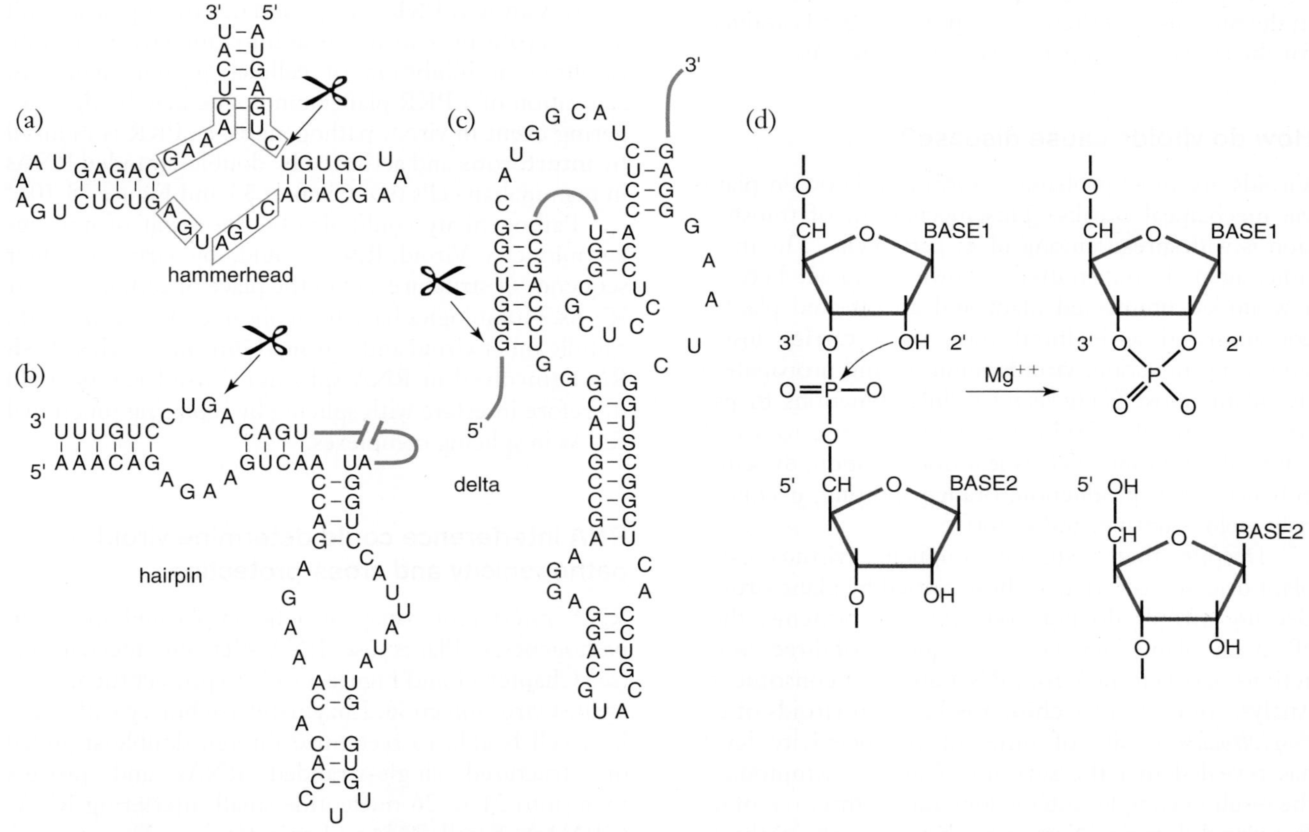

Figure 31.3 Self-cleaving RNA structures and the cleavage reaction. (a) Hammerhead self-cleaving structure of peach latent mosaic viroid. The conserved nucleotides essential for catalytic activity are boxed. (b) Hairpin self-cleaving structure of satellite tobacco ringspot virus satellite RNA. (c) Delta self-cleaving structure of hepatitis delta virus. (d) The self-cleavage reaction involves a Mg^{2+}-catalyzed phosphoryl transfer reaction that cleaves the RNA, producing 2'-3' cyclic phosphate (top right) and 5'-hydroxyl (bottom right) termini. The arrows and scissors indicate the cleavage sites in (a)–(c).

It is unclear how the multimeric plus-strands of *Pospiviroidae* are cleaved to give monomers, because a self-cleavage activity has not been found. It has been proposed that host cell enzymes carry out cleavage of these RNA molecules. Some effort has been spent in searching for a plant cell enzyme that can catalyze cleavage of viroid RNA multimers, but none has yet been found.

Ligation. Ligation of peach latent mosaic viroid (an *Avsunviroidae* family member found in the chloroplast) can occur without the participation of host proteins. Linear monomers resulting from hammerhead-mediated self-cleavage have been shown to self-ligate *in vitro*, producing 2'-5' phosphodiester bonds. It therefore seems likely that members of the *Avsunviroidae* family need only a host cell RNA polymerase activity for their replication, since they provide their own self-cleavage and self-ligation activities.

In contrast, *Pospiviroidae* require cellular enzymes for all three steps: polymerization, cleavage, and ligation. A wheat germ RNA ligase known to participate in splicing of tRNA introns can circularize linear unit-length potato spindle tuber viroid RNAs. This enzyme, which produces 3'-5' phosphodiester bonds, is localized in the nucleus, and therefore appears a logical candidate for the activity that ligates *Pospiviroidae* RNAs.

How do viroids cause disease?

Viroids are most probably transmitted between plants via mechanical breaks. This mechanism of transmission is widespread among plant pathogens. The infectious agent is transmitted following contact between a wounded uninfected plant and an infected plant, a contaminated agricultural tool, or a carrier insect. Following infection, viroids replicate and propagate in the plant, most likely using cellular junctions to pass from cell to cell. Viroid infections give rise to a wide range of symptoms such as leaf discoloration, dwarfing, enhanced fruit production, orange spotting, decline in fruit yield, stunting, and so forth.

Despite their extreme simplicity, viroids cause plant diseases as varied as those caused by plant viruses. Because viroids do not code for any proteins, their effects on plants must be a consequence of direct interactions between the viroid RNA and host constituents. Analysis of molecular **chimeras** between viroids of the *Pospiviroidae* family of different pathogenicity levels has revealed that the severity of disease symptoms is the result of complex interactions among three out of the five viroid domains. Sequence changes in the pathogenicity (P) domain can affect infectivity and the severity of symptoms. For example, some mutations that increase base-pairing interactions in this domain decrease severity of symptoms.

Interaction of viroid RNA with cellular RNAs or proteins may disrupt cell metabolism

The molecular mechanisms by which viroids cause disease in plants remain mysterious. Both host nucleic acids and proteins have been proposed to be the initial viroid target. Some viroids have a region complementary to certain cellular RNAs. This suggests that disease could result from inhibition of the function of these RNAs, or their cleavage directed by viroid RNA (e.g., *trans*-acting hammerhead cleavage). For example, a portion of potato spindle tuber viroid has sequence similarities to the U1 RNA of mammals (involved in RNA splicing), and some viroids contain sequences that could base pair with 7S ribosomal RNA. However, it is difficult to explain how changing a few nucleotides can convert a severe variant into a mild one, since these changes generally are not located in the regions proposed to base pair.

Alternatively, there may be host proteins that recognize and interact with different viroid structures. Mammalian **double-stranded RNA-dependent protein kinase (PKR)** is activated by potato spindle tuber viroid, as its secondary structure resembles that of double-stranded RNA. The level of activation has been correlated to the severity of plant disease symptoms. Activated PKR phosphorylates the alpha subunit of eukaryotic protein synthesis initiation factor 2 (eIF2), resulting in inhibition of cellular protein synthesis. Activation of a PKR plant homologue may be the triggering event in viroid pathogenesis, as PKR is induced by **interferons** and activated by double-stranded RNAs in mammalian cells (see Chapter 33 and Figure 33.10).

Pathogenicity could also be the result of molecular mimicry. Viroid RNAs could, by virtue of their sequence or structure, act in the place of certain cellular RNAs. Homologies have been observed between potato spindle tuber viroid and a group I intron, as well as U3B RNA (involved in RNA splicing). Viroid RNAs could therefore interfere with splicing by replacing functional RNAs in splicing complexes.

RNA interference could determine viroid pathogenicity and cross-protection

RNA interference may also be implicated in viroid pathogenesis. Plants use RNA-silencing mechanisms (see Chapter 11 and Figure 33.11) to protect themselves against virus infection. Enzymatic machinery within the host cell is able to recognize foreign double-stranded or structured single-stranded RNAs and process them into 21 to 26 nucleotide small interfering RNAs (siRNAs). Small RNAs identical to viroid sequences have been detected in infected plants, and it has been shown that these siRNAs were produced by the cellular enzymatic machinery following viroid infection. The development of symptoms associated with viroid infections could result from the activation or down-regulation of targeted genes by siRNAs, but no specific targets have yet been identified.

Cross-protection is known to occur between strains of the same viroid, and between different viroids that are very similar in sequence. Plants infected with one viroid will not allow replication and pathogenesis by a related viroid. This is reminiscent of viral interference, in which the presence of one virus inhibits replication by a superinfecting virus. The mechanism of cross-protection in viroids is not understood. One hypothesis is that a limiting host factor is required for viroid replication, transport, or accumulation. Differential affinity of the viroid RNA for this factor could determine which of two coinoculated viroids prevails, and the interaction of viroid RNA with the host factor could determine pathogenicity.

Circular plant satellite RNAs resemble viroids but are encapsidated

Plant satellite RNAs are small, single-stranded RNAs that depend on a **helper virus** for replication and encapsidation, but show little or no sequence similarity with their helper virus. Moreover, satellite RNAs are not essential to helper virus replication, and most do not encode any protein. Satellite RNAs can be either linear or circular. For a discussion of linear satellite RNAs of cucumber mosaic virus, see Chapter 10. Circular satellite RNAs, like viroids, have a highly base-paired structure. Unlike viroids, they are encapsidated and can therefore be transmitted more efficiently from plant to plant. Furthermore, like *Avsunviroidae*, all viroid-like circular plant satellite RNAs studied so far have a hairpin or hammerhead structure that catalyzes cleavage of multimeric RNA to monomers during rolling circle replication.

A number of different RNA-containing plant viruses support the replication of satellite RNAs, presumably by using their own RNA-dependent RNA polymerases to replicate the satellite RNA, and their capsid proteins to encapsidate it. The presense of satellite RNAs can affect the replication of the corresponding helper virus and modulate disease symptoms caused by the virus. Given their similairities, it is likely that viroids and viroid-like satellite RNAs have evolved from a common ancestor.

Hepatitis delta virus is a human viroid-like satellite virus

Hepatitis delta virus (HDV) is unique among human pathogens in that it shares several characteristics with both plant viroids and viroid-like plant satellite RNAs. It has been proposed that hepatitis delta virus evolved from a primitive viroid-like RNA through the capture of a cellular transcript. This blood-borne pathogen replicates in the liver and frequently causes **fulminant hepatitis** in both primate and nonprimate hosts. Hepatitis delta virus has also been associated with the development of liver cancer. It is detected only in the presence of hepatitis B virus (see Chapter 30), and it uses the hepatitis B virus envelope protein (S antigen) to package its RNA genome. Because hepatitis delta virus needs a viral protein for packaging, it is considered an RNA satellite of hepatitis B virus. With greater than 15 million people affected worldwide, hepatitis caused by hepatitis delta virus infection is a serious public health problem.

Hepatitis delta virus may use two different cellular RNA polymerases to replicate

The genome of hepatitis delta virus (Figure 31.4) is a 1682-nt single-stranded circular RNA molecule that can be folded into a rod-like structure by virtue of extensive internal base pairing. It possesses a single protein-coding open reading frame, complementary to the genome RNA. An 800-nt polyadenylated messenger RNA is synthesized by cellular RNA polymerase II and is translated into proteins called hepatitis delta antigens. The 3' end of this mRNA is specified by the classical polyadenylation signal, AAUAAA, which directs RNA cleavage by cellular enzymes and the addition of poly A to the cleaved 3' end of the RNA.

The right terminal domain (about 360 nt) of the genome RNA of hepatitis delta virus is reminiscent of viroids, and includes self-cleaving motifs used on both genome and complementary antigenome RNAs (delta motif; see Figure 31.3c). Hepatitis delta virus replicates in the nucleus via a symmetric rolling circle mechanism as shown for viroids in Figure 31.2.

The production of antigenome RNA from genome RNA templates (steps 1 and 2) may use cellular DNA-dependent RNA polymerase I, because it is resistant to α-amanitin, which inhibits RNA polymerase II. This could explain two differences between transcription (mRNA synthesis) and replication of genome RNA. First, there is evidence that transcription and replication initiate at different sites, implying the use of different polymerases. Second, the machinery responsible for genome replication does not recognize the cleavage/polyadenylation signal that leads to mRNA 3' end formation. It is known that RNA polymerase I, used by the cell to transcribe ribosomal RNA genes, does not interact with the cellular factors used for cleavage and polyadenylation of RNA polymerase II transcripts. This would explain why multimeric antigenome RNAs are generated by rolling-circle replication using circular genome RNA as template, but are not cleaved at the polyadenylation signal.

An alternative interpretation of experimental results suggests that RNA polymerase II is used for both transcription and replication. This model proposes that the polyadenylation signal is not always recognized by the cellular cleavage factors, allowing production

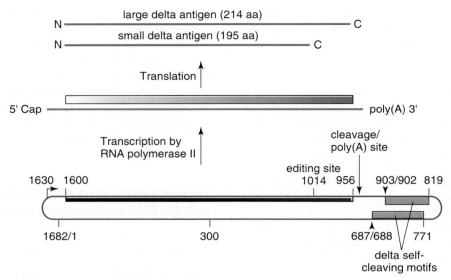

Figure 31.4 Hepatitis delta virus genome, mRNA, and proteins. Bottom: viral genome. The viroid-like domain necessary for rolling circle replication is at the right. Green shaded regions indicate self-cleaving motifs on the genome RNA (nt 688–771) and antigenome RNA (nt 819–902); arrowheads indicate the cleavage sites. The complement of the single open reading frame encoding both forms of delta antigen is shown as a black rectangle. The transcriptional start site at nt 1630 and the cleavage/polyadenylation site are noted. Middle: viral messenger RNA, shown in light blue, with the open reading frame shown above. Top: viral proteins (N = amino terminus, C = carboxy terminus). The larger form of delta antigen results from editing (A–G) at position 1014 on genome RNA.

of multimeric antigenome templates as well as 800-nt mRNAs by the same cellular RNA polymerase. However, insensitivity of antigenome synthesis to α-amanitin remains unexplained by this model.

The second stage of replication, formation of genome RNA from antigenome templates (steps 5 and 6 in Figure 31.2), is sensitive to α-amanitin, and therefore presumably uses RNA polymerase II. There is no polyadenylation signal on the antigenome RNA, allowing multigenome-length transcripts to accumulate by rolling-circle replication. Monomeric circular antigenome and genome RNAs are formed by self-cleavage and ligation of multimeric linear RNAs (steps 3, 4 and 7, 8). The small delta antigen and several other cellular factors are also involved in genome replication.

RNA editing generates two forms of hepatitis delta antigen

A fraction of hepatitis delta virus antigenome RNAs are edited during replication, resulting in the deamination of a specific adenosine residue (at nucleotide 1014) to an inosine. This editing event is carried out by a cellular enzyme known as adenosine deaminase (ADAR1) acting on RNA. The modified residue forms a Watson–Crick base pair with cytosine rather than uridine during subsequent replication, changing the original adenosine to a guanosine. The specificity of editing is likely determined by the primary and secondary structural characteristics of the hepatitis delta virus RNA.

As a result of editing, the UAG termination codon that normally ends the open reading frame at amino acid 195 is changed to a UGG codon, which codes for tryptophan. Edited antigenome RNAs give rise to genome RNAs by replication; these genome RNAs are then transcribed by RNA polymerase II to form modified messenger RNAs. Up to 30% of hepatitis delta virus messenger RNA molecules contain this modified termination codon. On these altered messenger RNAs, translation progresses to amino acid 214 before encountering a stop codon. Thus hepatitis delta virus makes two forms of the delta antigen: small (195 amino acids) and large (214 amino acids) (Figure 31.4).

Both the small and large forms of this protein contain nuclear localization signals and RNA-binding regions. The small delta antigen is required for viral RNA replication, while the large delta antigen is required for genome packaging and is an inhibitor of viral RNA replication. Because the two delta antigens are expressed at different stages of viral infection, regulation of editing levels is essential; the mechanism by which this regulation is accomplished is still being worked out.

Conclusion: viroids may be a link to the ancient RNA world

The various small self-replicating RNA pathogens share a considerable number of characteristics and may well have evolved from a common ancestor. To date, viroids have only been found in the plant

Engineering ribozymes for novel uses

The plant pathogenic RNAs, and the more complex hepatitis delta virus, have provided us with most of the naturally occurring self-cleaving RNAs known to date. The hammerhead, hairpin, and delta self-cleaving RNA motifs (Figure 31.3a–c) have been studied extensively. These motifs can be formed by annealing two different RNA fragments in vitro, such that one RNA fragment acts as the ribozyme and the other as the substrate (Figure 31.5). If the complementary region on the two RNAs is short enough, the cleavage products will dissociate from the ribozyme, thereby permitting the binding of more substrate molecules for cleavage. Via successive rounds of binding and cleavage, a single ribozyme molecule can therefore cleave many substrate molecules, establishing a classic enzyme/substrate relationship. Furthermore, by changing the binding sequence between the ribozyme and its substrate, it is possible to create a ribozyme with a new substrate specificity.

Novel ribozymes with catalytic profiles different from those of the naturally occurring molecules have been made by designing oligonucleotides synthesized in vitro or by mutagenesis of existing ribozymes. A wide variety of RNAs can be targeted for cleavage by such engineered ribozymes. Because of their ability to interact directly with RNA, ribozymes are currently being developed as potential therapeutic agents for a wide range of diseases, including virus infections (human immunodeficiency virus, viral hepatitis, etc.), cancer, diabetes, cardiovascular disease, and osteoporosis.

Hammerhead and hairpin ribozymes derived from plant RNA pathogens can cleave many RNAs and inhibit virus replication. However, the capacities of these two ribozymes are limited by their requirement for high magnesium concentrations; mammalian cells contain one-tenth the concentration of magnesium needed for optimal activity of these ribozymes. The delta ribozyme is therefore extremely interesting, because it can function in the environment of human and animal cells.

Viroids and related satellite RNAs have been a rich source of ribozymes to date, and there is considerable promise of more to be found. In the coming years, altered forms of these versatile molecules will surely emerge as a new class of drugs.

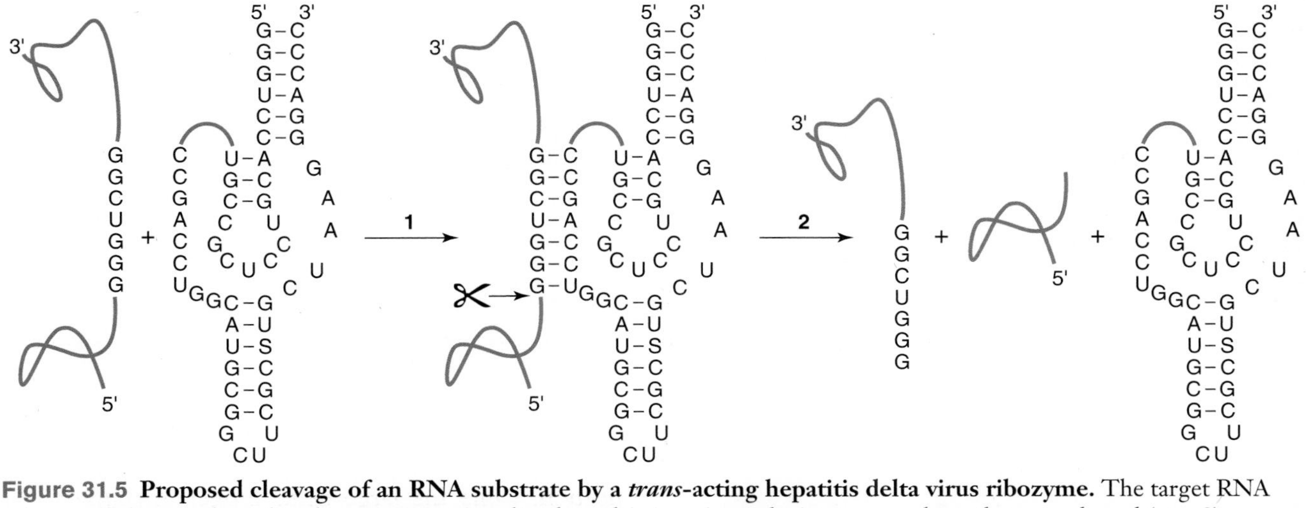

Figure 31.5 Proposed cleavage of an RNA substrate by a *trans*-acting hepatitis delta virus ribozyme. The target RNA substrate (left) anneals to the ribozyme (step 1) and is cleaved (scissors), producing two products that are released (step 2). The ribozyme (right) is then ready to process a new molecule of substrate RNA.

kingdom, although hepatitis delta virus, a viroid-like RNA, replicates in humans. Because viroids can replicate and possess enzymatic activities (RNA cleavage and ligation), they are considered as a model living fossil from a precellular RNA world. They are used to study the evolutionary connection between RNA and DNA genomes. They are also ideal biological molecules with which to study RNA structure/function relationships.

KEY TERMS

α-amanitin

Chimera

Delta structure

Fulminant hepatitis

Hairpin structure

Hammerhead structure

Helper virus

Interferon

PKR (double-stranded, RNA-dependent protein kinase)

Ribozyme

RNA interference

Rolling circle replication

FUNDAMENTAL CONCEPTS

- Viroids, plant satellite RNAs, and hepatitis delta virus are small, circular, single-stranded infectious RNAs.
- Viroids and plant satellite RNAs do not encode any proteins.
- Viroids infect plants and require no helper virus; viroid-like satellite RNAs require a helper virus to replicate.
- Hepatitis delta virus infects hepatocytes in humans and other animals; its replication requires coinfection with hepatitis B virus.
- Hepatitis delta virus RNA and certain viroids possess a self-cleaving structure essential for their replication.
- Viroids and hepatitis delta virus RNA are replicated by host cell RNA polymerases that normally transcribe DNA.
- The mechanisms by which viroids cause disease are not fully understood, but may include RNA interference.
- These small, circular, replicating RNAs may be a relict from the ancient RNA world.

REVIEW QUESTIONS

1. The two families of viroids are differentiated from each other based on what factors?

2. Compare and contrast the two types of genome replication used by the two families of viroids, referred to as symmetric and asymmetric replication.

3. What are ribozymes and how do the three types of ribozymes discussed in this chapter differ from one another?

4. How might RNA interference contribute to the pathogenicity of viroids?

5. The only known human viroid-like agent, hepatitis delta virus, has an RNA genome but does not encode its own RNA polymerase. How is it replicated?

CHAPTER 32

Prions

Dalius J. Briedis

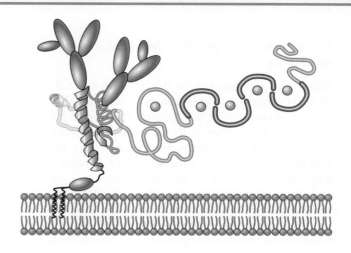

Prions

From *pr*oteinaceous *i*nfectious particles

PROPERTIES

Small, filterable infectious particles that contain protein but no detectable nucleic acid.

Prion proteins (PrP^C) are encoded by the host genome.

PrP^C is found in neuronal synapses, binds copper, has unknown function.

Prion proteins become infectious and pathogenic (PrP^{Sc}) as a result of protein conformational changes.

PrP^{Sc} can catalyze its own formation from PrP^C in animals.

PrP^{Sc} aggregates and accumulates in diseased brain.

DISEASES

Chronic, progressive, invariably fatal central nervous system degeneration.

Brain pathology is spongiform encephalopathy–large vacuoles in cortex and cerebellum give brain a sponge-like appearance.

Affected areas contain microscopic insoluble amyloid fibrils and macrocrystalline arrays known as amyloid plaques.

No signs of a host immune response.

Can arise spontaneously or by ingestion of infected tissue.

Affects wild and domesticated ruminants (sheep and goats: scrapie; cattle: mad cow disease); mink, cats.

Experimentally transferred to mice, hamsters, chimpanzees.

Human diseases: Creutzfeldt–Jakob disease (CJD), kuru.

DISTINCTIVE CHARACTERISTICS

Proteinaceous infectious agent that contains no nucleic acid and consists only of a single species of protein called PrP.

A new kind of infectious agent that can transmit a disease and replicate itself without the intervention of informational nucleic acids.

Prions are proteins that cause fatal brain diseases

The concept that a protein can transmit an infectious disease from one individual to another without the intervention of a nucleic acid genome has caused considerable surprise and controversy within the scientific community, but has now become widely (if not universally) accepted. A proteinaceous infectious agent called a **prion** ("pree-on" or "pry-on") appears to contain no nucleic acid, and only a single species of protein called PrP (prion-associated protein).

All viruses use nucleic acid as a repository for the information necessary for their replication. A number of viruses package in their virions proteins that are necessary for replication once the genome enters the host cell. These include virion-associated RNA or DNA polymerases (retroviruses, negative-strand viruses, reoviruses), and tegument proteins of herpesviruses. However, these proteins are encoded by information in the viral nucleic acid genome, and their activities serve to express and replicate the genome. Prions appear to represent a new kind of infectious agent that can transmit a disease (**prion disease**) and replicate itself without the intervention of informational nucleic acids.

Prion diseases were first detected in domestic ruminants

Sheep and goat herders in Britain have for hundreds of years observed a disease in their animals that they termed **scrapie**. Affected animals develop behavioral changes, tremor (especially of head and neck), and uncoordinated movements that progress to prostration and death without associated fever. Scrapie got its name from the habit of affected animals to itch, rub, and scrape their skin. In severe cases, the sheep's wool is rubbed off their sides and back.

Central nervous system tissue from such animals reveals large vacuoles in the cortex and cerebellum, giving the brain the appearance of a sponge. This condition is therefore termed **spongiform encephalopathy**. There are no signs of brain inflammation, suggesting that there is no infectious organism present. However, sheep and goat herders have long recognized that scrapie behaves like an infectious disease. Isolated, unaffected flocks appeared never to develop the disease unless they were introduced to animals from affected flocks. Under such circumstances previously healthy animals would develop scrapie, albeit with an incubation period of from 2 to 5 years.

Bovine spongiform encephalopathy ("mad cow disease") developed in Britain and apparently spread to humans

Cattle have increasingly been fed meat and bone meal prepared from animal sources as the beef industry has evolved. This started in the United Kingdom as early as 1926, but the volume of recycled animal protein increased rapidly over the twentieth century. In the 1970s and early 1980s, a number of changes were made in the rendering process for transforming diseased and otherwise unusable animal carcasses into meat, bone meal, and animal feed. Continuous instead of batch processing led to a greater risk of cross-contamination between feeds from different species. In addition, the falling price of tallow (animal fat) led to a decrease in the amount of expensive processing using organic solvents.

A disease similar to scrapie, called bovine spongiform encephalopathy (**BSE**) or **mad cow disease**, spread throughout cattle herds in Britain during the 1980s, presumably via an infectious agent present in their feed. It is presently unclear whether this disease crossed the species barrier from sheep and goats to cattle, or whether it is a rare, preexisting disease of cattle that was propagated by feeding remains of diseased cattle to livestock. During the subsequent 20 years, millions of cows in the United Kingdom and parts of continental Europe were slaughtered in attempts to control this disease (Figure 32.1).

BSE has since apparently itself crossed the species barrier and has infected humans. The first human case of an apparent BSE-derived disease was diagnosed in the United Kingdom in 1994. Since then, more than 100 individuals have been affected by this disease, called **new variant Creutzfeldt-Jakob disease (nvCJD)**, with an incidence in the United Kingdom of 10 to 15 cases per year.

Human prion diseases can be either inherited or transmitted

During the 1950s, an apparently transmissible spongiform encephalopathy of humans was identified among the Fore people in the mountainous interior of Papua New Guinea. This disease was called **kuru** ("shivering" or "trembling" in the Fore language). Kuru was characterized by cerebellar movement disorders and a shivering-like tremor progressing to total muscular incapacity, dementia, and death within one year of the onset of symptoms (Figure 32.2). The disease had a very high yearly incidence rate of 1%.

Early investigators were able to explain the high frequency of transmission of the disease by ritual funeral preparation of victims, including ritual cannibalism of their brains. D. Carleton Gajdusek proved the transmissibility of this human spongiform encephalopathy, transferring it by direct inoculation of diseased human brain tissue into chimpanzee brains. Gajdusek was awarded the Nobel Prize in Medicine in 1976 for his work on spongiform encephalopathies.

A number of central nervous system disorders in humans were shown to have similar symptoms and progression, and to share the spongiform pathology of the brain; these include **Creutzfeldt–Jakob disease (CJD)**, Gerstmann–Sträussler–Scheinker syndrome, and fatal familial insomnia (Table 32.1). Many of these cases were

Figure 32.1 Incineration of British cattle infected with mad cow disease.

shown to have a familial distribution, and are therefore linked to inherited genes. However, other cases were labeled "sporadic" and have no obvious genetic component.

In some cases, Creutzfeldt–Jakob disease was shown to be transmitted from human to human by (1) corneal allografts, (2) reusable intracerebral electrodes used for electroencephalograms, (3) grafts of **dura mater** (covering of the brain) from cadavers, and (4) human growth hormone purified from pituitary glands of cadavers.

Figure 32.2 Fore victims of kuru. Five women and one girl, all victims of kuru, who were still ambulatory, assembled in 1957 in the South Fore village of Pa'iti. The girl shows the spastic strabismus, often transitory, which most children with kuru developed early in the course of the disease. Every patient required support from the others in order to stand without the aid of the sticks they had been asked to discard for the photograph.

Similarities between symptoms and brain pathology in these diseases and kuru and scrapie led to extensive efforts to identify a causative infectious agent.

The infectious agent of prion diseases contains protein but no detectable nucleic acid

Initial efforts to identify a conventional infectious agent in affected nervous tissue were fruitless. However, direct inoculation of affected brain tissue from kuru and Creutzfeldt–Jakob disease victims into the brains of chimpanzees, and scrapie brain tissue into mice and hamsters, eventually led to transmission of the disease. Despite incubation periods of months before symptoms appeared, it was now possible to detect and to attempt to identify the infectious agent.

It was quickly found that prion infectivity is unusually resistant to inactivation by ionizing or ultraviolet irradiation, which damage nucleic acids. The target size for inactivation is several orders of magnitude lower than had ever previously been found for a virus, meaning that if there is a nucleic acid genome, it is unusually small. Infectious fractions of scrapie-infected brain contain no demonstrable nucleic acid even when analyzed by highly sophisticated techniques. Thus the ratio of nucleic acid to infectious particles would have to be much lower than ever previously encountered in viruses.

Scrapie infectivity was found to be resistant to treatment with agents that were usually effective inactivators of nucleic acids (psoralen cross-linking agents, nucleases, high concentrations of zinc ions, ammonium hydroxide) while sensitive to agents that were usually effective inactivators of proteins (strong alkali, phenol, sodium dodecyl sulfate). Purification of infectious material resulted in preparations highly enriched for a 27- to 30-kd protein not detected in similarly purified fractions of uninfected brain tissue. This ultimately led Stanley

Table 32.1 Species affected by prion diseases

Species	Disease	Abbreviation
Sheep	Scrapie	
Elk, mule deer	Chronic wasting disease	CWD
Mink	Transmissible mink encephalopathy	TME
Cattle	Bovine spongiform encephalopathy	BSE
Cats	Feline spongiform encephalopathy (probable cross-species spread from BSE)	FSE
Zoo ruminants	Transmissible spongiform encephalopathy of zoo ruminants	TSE
Humans	Creutzfeldt-Jakob disease	CJD
	(New) variant Creutzfeldt–Jakob disease	vCJD
	Gerstmann-Sträussler-Scheinker syndrome	GSS
	Fatal familial insomnia	FFI
	Kuru	

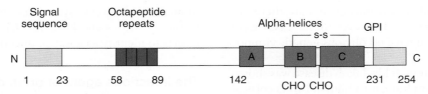

Figure 32.3 Domains and modifications of the prion protein. Amino acid numbers are shown below the map. N-terminus is at left, C-terminus at right; the mature protein begins at amino acid 23 and ends at amino acid 230. CHO: carbohydrate residues added to two asparagine residues; GPI: glycophosphatidyl inositol membrane anchor; S–S: putative disulfide bond; A, B, C: α-helical regions in PrP^C.

Prusiner in 1982 to propose that scrapie was caused by a novel *pr*oteinaceous *in*fectious particle for which he coined the term "prion" (Pr); the scrapie prion protein was named PrP^Sc. His proposal was very controversial at the time, but has gained acceptance and eventually led to his being awarded the Nobel Prize in 1997.

PrP^Sc is encoded by a host cell gene

Soon after the purification of PrP^Sc, its N-terminal amino acid sequence was determined. This led to the demonstration that PrP^Sc is encoded by a cellular gene. Uninfected animals express a cellular isoform (PrP^C) of the prion protein. PrP^C (Figure 32.3) is a 207-amino acid glycoprotein normally expressed on the surface of neurons, lymphocyctes, follicular **dendritic cells**, and other tissues. During synthesis and transport to the plasma membrane via the endoplasmic reticulum and the Golgi membranes, a 22-amino acid N-terminal signal peptide and 24 C-terminal amino acids are cleaved from the precursor 253-amino acid polypeptide. Oligosaccharide chains are added at one or two asparagine residues toward the carboxy terminal end of the protein. A carboxy-terminal phosphatidylinositol glycolipid anchors PrP^C in the plasma membrane.

Although PrP^C is highly conserved among mammalian species, attempts to identify its normal function by genetic means have proven frustrating. Knockout mice lacking PrP^C remain healthy throughout development, suffering only minor electrophysiological defects. A number of octapeptide amino acid repeats in the N-terminal region bind copper ions via histidine residues. PrP^C has been implicated in copper regulation, copper buffering, and copper-dependent signaling. PrP^C is presumed to function in transmembrane transport or signaling. A possible functional role involves shuttling or sequestration of copper ions at nerve synapses via binding to the N-terminal octapeptide residues.

Differences between PrP^C and PrP^Sc

Proteinase K treatment of PrP^C present in brains of normal animals results in complete digestion of the protein, while similar treatment of PrP^Sc from diseased animals gives a proteinase-resistant 27- to 30-KDa core

fragment (PrP 27–30) that retains infectivity. This fragment is missing N-terminal amino acids 23 to 89, but is otherwise identical in amino acid sequence and glycosylation pattern to PrP^C. Although PrP^C is a soluble protein, PrP^Sc tends to form oligomers and aggregates that can be detected as fibrils in purified preparations or in infected brains (Figure 32.4a).

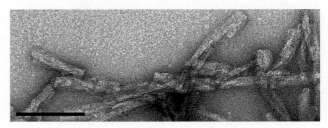

(a)

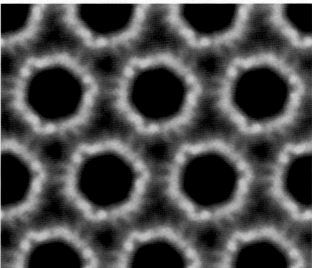

(b)

Figure 32.4 Prion fibrils and two-dimensional crystals. Negatively stained preparations of PrP 27–30 were observed by electron microscopy. (a) A typical prion rod or fibril (black bar = 100 μm). (b) Crystallographic averaging of a two-dimensional crystal shows threefold symmetry. Each hexagon is believed to be formed from trimers of PrP 27–30 dimers. Further analysis led to the model shown in Figure 32.6.

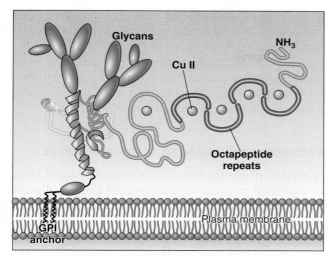

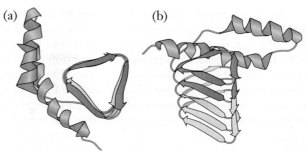

Figure 32.6 β-helical model of PrP 27–30. (a, b) Top and side views, respectively, of PrP 27–30 modeled with the N-terminal portion of the protein as a left-handed β-helix (ribbon arrows displayed in a triangular barrel). The structure of the α-helical part (ribbon helices) was derived from the known structure of the C-terminal region of PrP^C.

Figure 32.5 The structure of the prion protein. The prion protein is shown inserted into the plasma membrane via the glycoinositol phospholipid anchor. The carbohydrate groups (glycans) are shown as blue ovals at top left. The three α-helical regions in the C-terminal region are at the left, and at the right is the flexible N-terminal region containing the octapeptide repeats (blue half-circles) with bound copper ions (green spheres).

The structure of PrP^C has been determined by nuclear magnetic resonance and x-ray crystallography (Figure 32.5). PrP^C contains three α-helices in the C-terminal half of the molecule, while the N-terminal region, containing octapeptide repeats that bind to copper ions, is relatively unstructured. Because of its tendency to aggregate, PrP^{Sc} has not been crystallized, but physical measurements show that it contains less α-helical content and substantial amounts of β strands, and therefore has undergone a significant conformational change.

Two-dimensional crystals of PrP^{Sc} have been analyzed by electron microscopy (Figure 32.4b) and a model of the structure of this protein has been proposed (Figure 32.6). In this model, the N-terminal region of PrP^{Sc} is folded into a highly stable, parallel β-helix. Therefore the conformational change from PrP^C to PrP^{Sc} could involve formation of this β-helix structure.

The prion hypothesis: formation of infectious and pathogenic prions from normal PrP^C

Prusiner's original statement of the prion hypothesis proposed that a misfolded form of a protein (PrP^{Sc}) can catalyze the refolding of properly folded native protein molecules (PrP^C) into a similar misfolded conformation. In one model, formation of PrP^{Sc} is a nucleation-dependent oligomerization process (Figure 32.7). This model states that the conformational change leading to conversion of PrP^C into PrP^{Sc} is a reversible reaction,

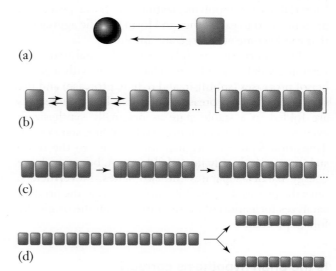

Figure 32.7 A mechanism for prion propagation. The reaction is broken into four stages. In (a), there is an equilibrium between the PrP^C state (black disc) and a rare monomeric precursor of PrP^{Sc} (green square). (b) Monomeric precursors can interact with each other, weakly and reversibly, until a stable infective seed is formed (shown as a bracketed structure). Once this critical size has been acquired, then monomeric precursors of PrP^{Sc} can add irreversibly to allow growth of the PrP^{Sc} particle (c). The infective particle can then "reproduce" by breakage to smaller, stable particles (d).

but PrP^{Sc} monomers are less stable than PrP^C, resulting in little or no accumulation of PrP^{Sc}.

Aggregation of PrP^{Sc} would promote the conversion of PrP^C by binding to and stabilizing the otherwise unfavored PrP^{Sc} conformation. Thus massive conversion of PrP^C to PrP^{Sc} would require an initial nucleation process. Infection with exogenous PrP^{Sc} would introduce a "seed" that initiates aggregation. Genetically inherited prion diseases would result from an increased stability of mutated PrP^{Sc} or an increased tendency of the protein to undergo aggregation and thus stabilize the altered conformation.

In an alternative proposed mechanism, the PrP^{Sc} form is postulated to be inherently more stable than PrP^{C}, but its formation is highly improbable because PrP^{C} folds in a relatively stable conformation upon its synthesis in the cell. In this case, exogenous PrP^{Sc} would catalyze the rearrangement of a molecule of PrP^{C}, or of a partially destabilized intermediate, to the more stable PrP^{Sc} conformation. The presence of the product PrP^{Sc} would then catalyze more transformation of PrP^{C} to PrP^{Sc}. If this model is correct, genetically inherited prion diseases would result from mutations in PrP^{C} that increase the population of unstable intermediates, or the rate of spontaneous conversion of PrP^{C} to PrP^{Sc}.

The nucleation-polymerization model (Figure 32.7) resembles the well-studied processes of crystal growth, tubulin polymerization, viral capsid assembly, and bacterial flagellar polymerization. In these processes, proteins bind to each other and form specific aggregates that can become very large and insoluble.

The alternative template-assisted, catalyzed conversion model is similar to what is seen with a number of proteases (subtilisin and α-lytic protease) and the **serpin** family of protease inhibitors. These molecules are folded in a stable state as originally synthesized, even though a distinct, more stable folding state exists. Transition from the original folded state to the more stable state is a highly improbable event, because a large energy barrier separates the two states. Nevertheless, once the process begins, it is autocatalytic, as the product catalyzes formation of more of itself from the originally folded protein.

Is the prion hypothesis correct?

Although generally accepted, the prion hypothesis has not convinced all scientists in the field. It remains possible that a small virus might transmit prion disease by infecting and replicating in brain neurons, leading by some change in the infected cell to the conformational change in PrP^{C} that causes disease. The most highly purified preparations of PrP^{Sc} contain 100,000 protein molecules per infectious unit. This could reflect inefficient infection of the host, as with some viruses. However, it could also mean that an unknown virus remains hidden among the protein in the prion preparations. If this were the case, it is not clear why the virus should purify along with the prion protein, nor why the virus should remain resistant to treatments that normally destroy nucleic acids. Also, a virus hypothesis would not easily explain genetically inherited prion diseases.

Prion proteins can be produced by recombinant DNA technology in bacteria; these proteins can be folded into states that resemble the native PrP^{c} and the altered PrP^{sc} conformations. Recently, it was shown that infectious prions could be generated from purified recombinant prion protein. However, this required mixture of the recombinant protein with cellular extracts and prolonged and repeated cycles of incubation under conditions that favor protein misfolding. If, as seems likely, the prion hypothesis is correct, then this result means that other cellular factors may be required to help catalyze the conformational switch. Efforts are continuing to provide definite proof of the prion hypothesis.

Pathology and diagnosis of prion diseases

A number of human diseases are caused by production of misfolded proteins. In diseases such as cystic fibrosis and α1-antitrypsin deficiency, degradation and/or mistrafficking of misfolded proteins causes loss of protein function. A second class of misfolding disease includes systemic amyloidoses, type II diabetes, and several common neurological diseases. These are caused by toxicity resulting from the aggregation of specific misfolded proteins into highly ordered structures made up of fibers known as **amyloid**. These fibers can be detected by their characteristic binding of the dye, Congo red.

A growing list of diseases in which fibrillar accumulation causes cellular damage in the central nervous system includes Alzheimer's disease, in which neurodegeneration is associated with deposition of the amyloid β protein; Parkinson's disease, which is associated with deposition of α-synuclein; and Huntington's disease, in which an unusually long tract of polyglutamine in a protein leads to brain lesions. All of the proteins involved in these diseases undergo a conformational alteration to a common structure in the amyloid fibril, a "cross β" repeat structure in which β strands are aligned perpendicular to the axis of the amyloid fiber. As with prion diseases, these amyloidoses can be initiated by inherited mutations in the gene coding for the responsible protein.

It is now clear that a growing number of intracellular proteins are prone not only to pathological aggregation but can also be released and "infect" neighboring cells. Therefore, many complex diseases may obey a simple model of propagation in which the penetration of "seed proteins" into hosts determines spatial spread and disease progression. Some have termed these proteins "prionoids," since they seem to infect neighboring cells (and possible other hosts) just as do prions.

All tissue abnormalities of prion diseases are confined to the central nervous system. Affected areas contain microscopic insoluble amyloid fibrils (Figure 32.4a) and macrocrystalline arrays known as amyloid plaques. The prion protein accumulates selectively and abnormally in neurons in the central nervous system during the course of the disease. Until recently, diagnosis of prion disease in individuals with suspicious clinical presentations was made by histological examination of brain biopsies or, more commonly, autopsy specimens. Recent advances have been made in concentrating and detecting abnormal PrP in blood and lymphoid tissues such as tonsils. Immunodetection using western blots or direct fluorescent detection before and after proteinase digestion have allowed detection of as little as 10^{-8} µg (10^6 molecules) of abnormal PrP.

Proteins of yeast and other fungi can form self-propagating states resembling prions

Reed Wickner proposed that an inheritable yeast phenotype [URE3] is the result of a prion-like, self-propagating change in the conformation of the nitrogen metabolism protein Ure2. Amyloid-like formations formed by other chromosomally encoded yeast and fungal proteins, such as Sup35, Rnq1, HET-s, and Swi1 have been shown to result in adoption of distinct and inheritable "prion states." Prion states in these simple organisms have been studied extensively and have been shown to result from aggregation of cellular proteins into self-propagating amyloid fibrils whose transmission occurs naturally through cell division.

Fibrils formed *in vitro* from recombinant yeast prion proteins and then introduced into yeast cells have been shown to result in inheritable prion states, providing final evidence that proteins alone can act as self-propagating infectious agents. Yeast prions also exhibit other properties of their mammalian counterparts, exhibiting both strain variability and barriers to interspecies transmission.

Prion proteins in yeast and other fungi do not share sequence similarity with mammalian prion proteins. The prion domains of certain yeast proteins have an exceptionally high content (approximately 40%) of two "amyloidogenic" amino acids, glutamine and asparagine.

Formation of prions in fungi results in phenotypic changes that vary with the aggregating protein. The best-characterized yeast prion state is associated with Sup35, a protein involved in translation termination. Conversion of this protein into amyloid fibrils depletes active Sup35 monomer from the cell, and the resultant phenotype is defective in protein synthesis.

Not all fungal prion states, however, are deleterious. For example, the [Het-s] prion state of the fungus *Podospora anserina* results in accumulation of amyloid fibrils made of the protein HET-s, and these fibrils appear to serve a functional role. The [Het-s] state has been shown to mediate heterokaryon compatibility and prevents exchange of nuclei between incompatible yeast genotypes.

Genetics of prion diseases: mutations in the prion gene can increase occurrence of disease

PrP^C is encoded by a single exon of a unique gene (two copies per diploid genome). There are at least 20 known mutations in the PrP^C gene that are associated with prion disease, presumably because of a high rate of spontaneous PrP^{Sc} formation. A number of families that suffer from genetic Gerstmann–Sträussler–Scheinker syndrome have a common double mutation in PrP^C at amino acids 178 and 200 (D178N; E200K), while other such families have a mutation of leucine to proline at amino acid 102 (P102L). Mice with a **transgene** containing the P102L mutation spontaneously develop a lethal scrapie-like disease. On the other hand, sheep homozygous for a PrP gene encoding substitution of arginine for glutamine at codon 171 (R171Q) are entirely resistant to scrapie.

Humans who are homozygous for valine at position 129 are predisposed to both acquired and sporadic Creutzfeldt–Jakob disease. All victims of new variant Creutzfeldt–Jakob disease in the United Kingdom up to the present have been found to be homozygous for methionine at position 129. Homozygosity at this locus was also present among most victims of kuru. Taken together, this information suggests that heterozygosity at position 129 protects against both inherited and infectious prion disease.

In fact, heterozygosity in the PrP gene, particularly at position 129, is present at unexpectedly high levels in humans worldwide. This implies that there has been selective pressure for maintenance of heterozygosity in humans, suggesting that prion diseases similar to kuru may have been commonplace over the past 500,000 years of human evolution. It has been speculated that this may be the result of widespread cannibalism among humans in the past. Many of our ancestors may have succumbed to prion disease as a result of consumption of human tissues.

Prion diseases are not usually transmitted among different species

To a large extent, prion diseases respect species barriers; prions from one animal cannot usually infect animals of a different species. This suggests that prion infectivity depends on sequence similarity between donor and recipient prion proteins. This requirement may mean that interactions of PrP^{Sc} that lead to its stabilization or aggregation are stronger between molecules of the same amino acid sequence and therefore the same three-dimensional conformation.

However, the species barrier is not absolute. Sheep scrapie can be used to infect mice and hamsters, a discovery that eventually led to the purification and identification of prions. Sheep are susceptible to infection with bovine spongiform encephalitis (BSE) at the same frequency and incubation time as scrapie, although sheep PrP differs from bovine PrP at amino acids 98, 100, 143, 146, 155, 168, 198, and 218. In contrast, cattle are very resistant to infection with sheep scrapie but are susceptible to BSE. This raises the question of whether BSE was acquired during the epidemic in cattle by ingestion of remains not of scrapie-infected sheep, but of cattle that had been infected by a rare, spontaneous case of BSE sometime in the past.

Mice are resistant to infection with hamster prions, which have 16 differences in amino acid sequence from mouse PrP. A striking finding was made when knockout mice lacking the PrP gene were generated: these mice are resistant to prion disease caused by inoculation with PrP^{Sc} from infected mice, and cannot propagate infectious prions. Introduction of a hamster or human prion transgene into such mice restored susceptibililty to infection with hamster or human prions, respectively. These results are consistent with the prion hypothesis: pre-existing PrP^{C} is required to propagate infectious prions and to cause prion disease.

Strain variation and crossing of the species barrier

Distinct strains of prions can produce distinct and reproducible patterns of disease, incubation time, and glycosylation and protein cleavage patterns of PrP^{Sc}. In some cases, these strains are attributed to different conformational structures of PrP^{Sc}. For example, two strains of transmissible mink encephalopathy exist: hyper (HY) and drowsy (DY). These two strains reproducibly transmit different behavioral changes, with different neuropathology and incubation periods. Proteinase K treatment of the PrP^{Sc} species corresponding to HY and DY gives rise to digested proteins with distinct amino termini, although the parent PrP^{C} proteins have the same amino acid sequence. This shows that amino acid sequence differences are not necessary to produce different strains.

Human cases of new variant Creutzfeldt-Jakob disease (nvCJD) appear to be linked to the BSE epidemic in the United Kingdom. New variant CJD is distinct from sporadic CJD; it has a distinct pathology, has a far younger age of onset than sporadic CJD, and presents with very different initial symptoms. Intracerebral inoculation of BSE-infected brain extract into macaque monkeys produced disease and pathology very similar to those of new variant CJD patients. It seems possible, therefore, that new variant CJD is caused by a different prion strain that is less restricted by the species barrier than that of scrapie. CJD and new variant CJD show different proteinase K-resistant PrP^{Sc} species, and different patterns of di-, mono-, and nonglycosylated PrP^{Sc} are revealed by electrophoresis. The protein pattern of new variant CJD is similar to that of BSE-infected animals, but quite different from that of sporadic or acquired CJD. A proteinase K-resistant diglycosylated species is particularly prominent among the new variant CJD cases.

The nature of the prion infectious agent

Prions challenge our traditional concepts of infectious agents. The prion hypothesis proposes that prions transform a pre-existing normal cellular protein into an infectious, pathogenic agent. We are used to the concept that information contained in nucleic acids can be transferred from cell to cell or from organism to organism, as with viruses. Virus genomes code for proteins that enable their replication and transmission, and thus viruses are self-propagating.

Similarly, computer viruses transfer information from computer to computer, propagating themselves, although their "program" is carried by electronic bytes rather than by a nucleic acid. Nevertheless, computer viruses are analogous to real viruses, because their sequence of bytes (analogous to nucleic acid) codes for programs (analogous to enzymes that direct nucleic acid replication) that enable them to be replicated, *de novo*, using elements provided by a computer (analogous to the cell). Both viruses and computer viruses can cause disease or damage to their host cell or computer.

The prion hypothesis proposes that information coded in the three-dimensional structure of a protein, and therefore ultimately in its amino acid sequence, can be transferred from organism to organism and propagate itself, without the intervention of a nucleic acid. To be sure, the amino acid sequence of prions is encoded in nucleic acid genes of the host organism, but only the protein, not the prion gene, is transmitted and infectious. It may prove useful to modify the concept of an "infectious agent" to include any transfer of information between cells or organisms that results in repetitive self-propagation.

Richard Dawkins, in his best-selling 1976 book, *The Selfish Gene*, postulated that evolution occurs whenever information is copied over and over again with introduction of variations and with selection of some variants over others. The most familiar unit of evolving information is the gene, but evolution can occur based on any mode of information transfer. Human ideas, for example, can be transmitted by human communication from generation to generation. The copying of ideas from one person to another is imperfect, just as the copying of genes from parent to child is sometimes inaccurate. Agriculture may be thought of as an idea that first arose in the ancient Middle East and then spread and evolved among humans for thousands of years.

Prions are transmissible, replicable, and variable disease-causing agents that are distinct from viruses. Whether we define them as "living" or "nonliving" or as an "infectious enzyme," we do know the following about them: (1) they have arisen in organisms during evolution; (2) they are able to propagate themselves and the diseases they cause; and (3) they appear to be able to evolve and to adapt themselves to different hosts.

Can prion diseases be prevented or cured?

At the moment, no proven effective treatments exist for the prevention, cure, or even delay of progression of prion diseases. All are uniformly fatal. Success in inhibiting prion replication in cultured cells has been obtained with a variety of agents including polyanions, sulfonated dyes, cyclic tetrapyrroles, quinacrine, quinoline, acridines, and phenathiazines, as well as other amyloidophilic compounds such as pyridine dicarbonitriles, peptide aptamers, and β-sheet breakers. However, none of these agents has yet been shown to have significant efficacy in reducing or preventing prion disease in animals. Anti-prion protein antibodies and antibody fragments can block prion formation in cultured cells in a dose-dependent fashion. Because some variants of prion proteins seem to protect against prion disease, vaccination or gene therapy are other possible solutions to curing or preventing prion diseases.

KEY TERMS

Amyloid

BSE

Creutzfeldt–Jakob disease (CJD)

Dendritic cells

Dura mater

Kuru

Mad cow disease

New variant Creutzfeldt–Jakob disease (nvCJD)

Prion

Prion disease

Scrapie

Serpin

Spongiform encephalopathy

Transgene

FUNDAMENTAL CONCEPTS

- Prions are proteins that can cause fatal brain diseases.
- Human prion diseases can be transmitted or inherited.
- The infectious agent of prion diseases contains protein but no detectable nucleic acid.
- Human prions are encoded by a host cell gene.
- The prion hypothesis proposes that misfolded forms of a cellular protein can catalyze the refolding of properly folded native protein molecules into similarly misfolded conformations.
- Prion diseases of humans may have much in common with more familiar ailments such as Alzheimer's and Parkinson's diseases.
- Misfolded proteins of yeast and other fungi can form self-propagating states that resemble prions.
- Prion diseases are not usually transmitted among different species, but bovine spongiform encephalopathy ("mad cow disease") developed in Britain and apparently spread to humans.
- The existence of prions suggests that information coded in the three-dimensional structure of a protein can be transferred from organism to organism and propagate itself without the intervention of a nucleic acid.

REVIEW QUESTIONS

1. Describe how prion diseases in humans are both infectious and inheritable, and discuss how they are transmitted.
2. What is the prion hypothesis, and how does it explain prion diseases?
3. How does new variant Creutzfeldt–Jakob disease differ from sporadic Creutzfeldt–Jakob disease?
4. Are prions capable of evolution?

HOST DEFENSES AGAINST VIRUS INFECTION

Organisms have coexisted with viruses for most of evolutionary time. Many viruses damage or kill their host cells and therefore cause disease and death. As a result, host organisms have acquired a variety of means to protect themselves against infection and the pathological effects of infection. These defenses can be conveniently separated into two categories: intrinsic cellular defenses and host immune responses.

Intrinsic cellular defenses are rapid responses by infected cells to invading pathogens. They begin with molecular detection systems such as toll-like receptors, which recognize a variety of molecules that are specific to viruses or other pathogens. For viruses, the targeted molecules include glycoproteins and nucleic acids, but in particular double-stranded RNA, which is a byproduct of the expression of many viral genomes.

Detection of virus infection leads to transcriptional activation of genes that code for interferons, other cytokines, and inducers of apoptosis. Some of these proteins are secreted and act on nearby cells, inducing synthesis of antiviral proteins that reduce or inhibit virus replication. Interferons and cytokines also help induce longer-term host immune responses.

Plants and simpler eukaryotes use RNA interference as a defense mechanism against virus infections. Viral double-stranded RNAs are cleaved into small interfering RNAs, which direct cleavage and therefore inactivation of viral mRNAs or genome RNAs.

Immune responses are mediated by specialized cells that circulate in the blood or lymph and that are localized in the thymus, lymph nodes, spleen, and gut. Natural killer cells can recognize and kill infected cells that are lacking certain cell surface proteins. Viral proteins and peptides derived from them are recognized by receptors on B and T lymphocytes, activating those cells to differentiate into plasma cells, which produce antibodies, or into cytotoxic T cells, which kill virus-infected cells. Cells that mediate the adaptive immune response can be long-lived, and the organism is therefore protected against future infections by the same pathogen. Viruses have developed a number of strategies to counter both intrinsic cellular defenses and adaptive immune responses.

Intrinsic Cellular Defenses Against Virus Infection

Karen Mossman
Pierre Genin
John Hiscott

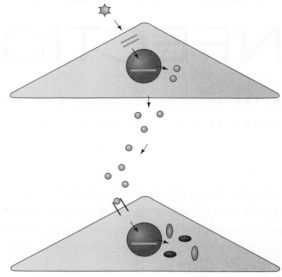

DETECTION OF VIRUS INFECTION

Toll-like receptors located on the cell surface or in endosomal membranes recognize conserved pathogen structures.

Various cytoplasmic receptors recognize viral nucleic acids.

Virus entry into host cells is recognized by an unknown mechanism, independent of nucleic acid.

HOST CELL RESPONSE TO VIRUS INFECTION

Double-stranded RNA is a potent stimulator of signal transduction cascades.

Interferon regulatory factors (IRFs) are activated by phosphorylation.

Nuclear factor κB (NFκB) is activated upon phosphorylation and degradation of its inhibitor.

IRFs and NFκB induce synthesis of interferons, pro- and anti-inflammatory cytokines, and chemokines.

Virus infection induces both extrinsic and intrinsic pathways of apoptosis, leading to the activation of cysteine proteases called caspases.

INTERFERONS: STRUCTURE AND SYNTHESIS

Interferons-α, β, and λ have antiviral activity and are produced by most cell types.

Interferon-γ is an immune modulator that is produced by lymphocytes and macrophages.

Interferon mRNAs are unstable, leading to a short burst of interferon synthesis upon induction.

Interferons are secreted into the extracellular medium.

INDUCTION OF ANTIVIRAL ACTIVITY

Interferons interact with specific receptors on the surface of target cells.

Binding leads to dimerization of receptors, phosphorylation, and activation of Jak kinases.

Jak kinases phosphorylate Stat proteins, which dimerize and transit to the nucleus.

Stat proteins bind to interferon-stimulated genes in the nucleus and activate transcription.

More than 1000 cellular genes are activated by the Jak–Stat pathway.

Interferons-α, β, and λ primarily induce genes involved in blocking virus replication.

Interferon-γ primarily stimulates components of the adaptive immune response

VIRAL DEFENSES AGAINST INTERFERON RESPONSES

Homologues of interferon-regulatory factors block transcription of interferon genes.

Viral proteins inhibit activation of interferon-stimulated genes.

Small viral dsRNAs block activation of dsRNA-dependent protein kinase (PKR).

Viral proteins that bind to dsRNA reduce activation of PKR.

Soluble homologues of cytokine receptors block cytokine production and inhibit B-cell activation and antibody production.

RNA INTERFERENCE

Plants, invertebrates and fungi do not produce interferon, but use RNA interference as a host defense mechanism.

Viral double-stranded RNA is cleaved by Dicer into small interfering RNAs (siRNAs) that are 21 to 23 nt long.

siRNAs become part of a RISC complex that cleaves viral RNAs with homologous sequences.

Mammals use a similar, yet distinct, strategy to produce microRNAs (miRNAs), which either degrade or inhibit the translation of messenger RNAs.

INTRODUCTION

To stay healthy and survive, organisms must defend themselves against invading pathogens, including viruses. Physical and chemical barriers such as the skin, mucous membranes, tears, and acidic or basic environments are the first line of defense. Gut and respiratory tract secretions contain enzymes and other molecules that prevent attachment of viruses and infection of exposed cells.

If an invading pathogen is able to breach these barriers, cells exposed to the pathogen respond with **intrinsic cellular defense** systems. These are systems that act rapidly, within minutes of the initial infection of a cell, and locally, within the infected cell or in nearby cells. Intrinsic cellular defense systems are deployed in three stages: (1) pathogen recognition, (2) signal amplification, and (3) pathogen control.

A variety of specialized cellular proteins can detect invading pathogens; the best-characterized pathogen-detecting proteins are the **toll-like receptors**. Once these proteins detect the presence of a pathogen, they set off a highly coordinated response that involves receptor-mediated signaling, activation of gene transcription, secretion of proteins, and cell-to-cell communication. This response leads to the secretion of **cytokines**, including **interferons**, and the activation of enzyme systems that inhibit replication of the infecting pathogen within cells. Thus an infected cell responds both by setting up its own intracellular defensive measures and by sending signals to nearby cells that in turn activate their defenses against infection.

If the infection spreads, an organism-wide immune response is activated. The immune response utilizes a variety of specialized immune cells in the circulatory system, and can be divided into two phases, the **innate** and the **adaptive immune responses**. The innate immune response also acts quickly, and involves the action of sentinel cells, including **macrophages**, **dendritic cells**, and **natural killer (NK) cells**, as well as **complement**. The adaptive immune response takes days to weeks to deploy its maximum effects. It involves recognition of foreign antigens by **B** and **T lymphocytes**, leading to the production of antibodies and the generation of **cytotoxic T lymphocytes** that kill infected cells.

This chapter focuses on intrinsic cellular defenses and the signaling responses by infected cells, particularly the production of cytokines. The following chapter focuses on sentinel cells involved in the innate immune response, and on the deployment of the adaptive immune response.

DETECTION OF VIRUS INFECTION BY HOST CELLS

Host cells sense virus infection with toll-like receptors and a variety of other molecular detection systems

Organisms have evolved multiple ways of recognizing different aspects of virus infection, including virus attachment and entry, virus replication, and release of virus components from infected cells. Binding of virion structural proteins to cell surface receptors can initiate antiviral pathways. For example, when non-enveloped viruses such as adenoviruses bind to cell surface receptors like integrins, the cell responds with the production of several cytokines, including **interleukin-1α**.

The entry of enveloped viruses into cells also triggers an antiviral response. This response is nonspecific, as enveloped viruses from a variety of different families can elicit a similar response by the infected cell. Instead of producing cytokines, cells respond to entry of enveloped viruses directly by inducing a small subset of **interferon-stimulated genes**, most of which encode intracellular proteins that block virus transcription and translation. How cells recognize entry of these enveloped viruses is not yet understood.

Toll-like receptors (TLRs), located either on the cell surface or in endosomal membranes, are specialized proteins that recognize a variety of conserved molecular structures typically found in pathogens ("pathogen associated molecular patterns") (Table 33.1). Toll-like receptors are highly conserved from flies to humans and share structural and functional similarities. They were originally discovered as proteins that are important for defending against fungal infections in the fruit fly, *Drosophila melanogaster*. Most mammals have between 10 and 15 toll-like receptors, with humans and mice having 10 and 13, respectively.

TLR-2 and TLR-4, present on the surface of many cell types, recognize envelope glycoproteins from

Table 33.1 Toll-like receptors and molecules they recognize

Toll-like receptor	Binds to	Examples of viruses detected
TLR-1	Lipopeptides	Hepatitis C virus
TLR-2	Lipoproteins, glycoproteins	Herpesviruses, measles virus, hepatitis B and C virus
TLR-3	Double-stranded RNA	Respiratory syncytial virus, West Nile virus, influenza virus, herpesvirus
TLR-4	Lipopolysaccharide, glycoproteins	Vesicular stomatitis virus, respiratory syncytial virus, murine mammary tumor virus
TLR-5	Flagellin	
TLR-6	Diacyl lipopeptides	Vaccinia virus, hepatitis C virus
TLR-7, TLR-8	Single-stranded RNA	Human immunodeficiency virus, vesicular stomatitis virus, influenza virus, hepatitis C virus
TLR-9	Unmethylated CpG DNA	Herpesviruses, vaccinia virus, hepatitis B virus
TLR-10	Unknown	

viruses such as respiratory syncytial virus, measles virus, herpes simplex virus, and vesicular stomatitis virus (Figure 33.1). Binding of these glycoproteins to TLR-2 or TLR-4 begins a signaling cascade that eventually leads to production of cytokines, including interferons, which help to inhibit virus replication and spread.

Several toll-like receptors recognize viral nucleic acids

Viral nucleic acids are the most potent inducers of intrinsic and innate antiviral responses. Toll-like receptors located within endosomes recognize both single- and double-stranded RNA and DNA (Figure 33.1).

Many DNA viruses transcribe both strands of certain regions of their DNA genomes, giving rise to self-complementary RNAs that can anneal to form double-stranded RNA. Viruses with single-stranded RNA genomes replicate by synthesizing full-length complementary RNA strands that can form double-stranded RNA by hybridizing to the genome RNA. Furthermore, a number of viruses have double-stranded RNA genomes. Thus the presence of double-stranded RNA in a cell is perhaps the most reliable indicator that a replicating virus is present, and it makes sense that the cell recognizes and responds to this molecule.

TLR-3 recognizes double-stranded RNAs. The expression of TLR-3 is widespread and includes dendritic cells, macrophages, fibroblasts, and epithelial cells. Binding of double-stranded RNA by TLR-3 only occurs at low pH, indicating that binding occurs within endosomes. TLR-3 is essential for the recognition of infection by multiple viruses, including encephalomyocarditis virus, respiratory syncytial virus, herpes simplex virus, influenza viruses, and West Nile virus.

The related toll-like receptors TLR-7 and TLR-8 recognize guanosine- and uridine-rich single-stranded

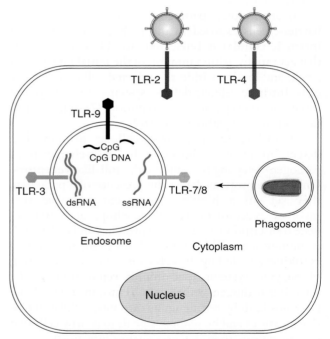

Figure 33.1 Toll-like receptors that recognize virus components. Cell surface toll-like receptors (TLR-2 and TLR-4) recognize virion glycoproteins; endosomal toll-like receptors (TLR-3, TLR-7, TLR-8, and TLR-9) recognize different forms of viral nucleic acids. Phagosomes can deliver viral components to endosomes, where they are recognized by toll-like receptors.

RNAs from viruses such as human immunodeficiency virus, vesicular stomatitis virus, and influenza virus. TLR-9 recognizes both single-stranded and double-stranded viral DNA from various members of the herpesvirus family as well as unmethylated CpG dinucleotides in bacterial DNA. The expression of TLR-7, TLR-8, and TLR-9 is largely restricted to plasmacytoid dendritic cells, which are major interferon-producing cells within the blood, and B lymphocytes. Interferon production in plasmacytoid dendritic cells does not require virus replication. It is thought that virus particles taken up in these cells by endocytosis are retained within endosomes until they are degraded, and the released nucleic acids then bind to toll-like receptors.

Autophagy also plays a role in recognition of viral nucleic acids by toll-like receptors. Autophagy is an intrinsic cellular process that increases cell survival during periods of stress, such as nutrient starvation and infection. Stressed cells degrade much of their contents by forming specialized membrane compartments called **phagosomes**, which are similar to lysosomes. Autophagy has been shown to be required for TLR-7-mediated recognition of vesicular stomatitis virus particles in the cytoplasm, by delivering viral RNA to vesicles containing TLR-7 (Figure 33.1).

Other cellular proteins are also involved in recognition of viral RNAs

Although toll-like receptors were the first pattern recognition receptors identified, alternative mechanisms for sensing virus infection were subsequently discovered. The RNA helicase, *r*etinoic acid-*i*nducible *g*ene I (RIG-I), and a related protein, *m*elanoma *d*ifferentiation-*a*ssociated gene 5 (MDA5), play important roles in intrinsic antiviral responses induced by double-stranded RNA. Both RIG-I and MDA5 contain a C-terminal RNA helicase domain and two N-terminal caspase recruitment domains (CARDs).

RIG-I and MDA5 appear to recognize RNAs from different viruses, with the recognition depending on both RNA structure and length (Figure 33.2). RIG-I preferentially recognizes short double-stranded RNA molecules and single-stranded RNA molecules with a 5'-triphosphate group. MDA5, on the other hand, preferentially recognizes long double-stranded RNA molecules and the synthetic double-stranded RNA mimic polyinosinic-polycytidylic acid (polyI:C).

RIG-I plays an essential role in the recognition of single-stranded RNA viruses, including paramyxoviruses, influenza viruses, vesicular stomatitis virus, and Japanese encephalitis virus, while MDA5 is required for the recognition of picornaviruses. Unlike the membrane-bound toll-like receptors, these are soluble, cytosolic proteins, and can therefore bind to nucleic acids localized in the cytosol.

Viral double-stranded DNAs in the cytoplasm are recognized by at least three different cellular proteins

In addition to recognizing double-stranded RNA, a form of nucleic acid not routinely produced in uninfected

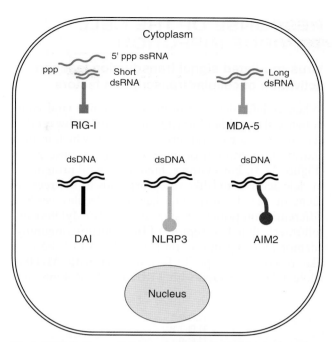

Figure 33.2 Cytosolic sensors that recognize viral nucleic acids. The cytoplasmic receptors RIG-I and MDA-5 detect single-standed or double-stranded viral RNA; DAI, NLRP3, and AIM2 bind to cytoplasmic, double-stranded DNA.

cells, host cells have evolved mechanisms of recognizing double-stranded DNA in the cytoplasm (Figure 33.2). This is also a sign of infection, because most DNA is usually restricted to the nucleus in healthy, uninfected eukaryotic cells. *D*NA-dependent *a*ctivator of *i*nterferon regulatory factors (DAI) is expressed in most cell types, and its levels dramatically increase following interferon signaling. DAI binds to bacterial, viral, and mammalian DNA, and has been shown to mediate cellular responses to herpesviruses and poxviruses. Recognition of cytosolic DNA by DAI leads mainly to the production of interferon.

Two additional cellular proteins recognize viral double-stranded DNA within the cytoplasm, namely *a*bsent *i*n *m*elanoma 2 (AIM2) and (hold your breath for this name!) *n*ucleotide-binding domain, *l*eucine-*r*ich repeat-containing (NLR) family, *p*yrin domain-containing 3 (NLRP3) protein. Both AIM2 and NLRP3 contain a pyrin domain that mediates protein-protein interactions. Proteins that are involved in inflammation and **apoptosis** often contain pyrin domains.

NLRP3 is one of over 20 human NLR proteins. These proteins are best known for recognizing microbial components such as peptidoglycans and toxins, and endogenous stress-related molecules such as uric acid and ATP. AIM2 and NLRP3 induce production of **pro-inflammatory cytokines** following infection with poxviruses and adenoviruses, respectively. It was recently found that NLRP3 also mediates innate immunity to influenza virus via recognition of viral RNA.

RESPONSE OF THE CELL TO VIRUS INFECTION

Virus-mediated signal transduction leads to activation of cellular transcription factors

When cellular detector proteins recognize viral components, they initiate signal transduction pathways that activate a variety of transcription factors, which in turn activate transcription of many cellular response genes (Figure 33.3). For example, binding of viral proteins or nucleic acids to toll-like receptors leads to the recruitment of the downstream adaptor molecules *m*yeloid *d*ifferentiation primary response gene *88* (MyD88) and *t*oll/*i*nterleukin-1 *r*eceptor (TIR)-domain-containing, adaptor-inducing *i*nterferon-β (TRIF). All of the toll-like receptors except TLR-3 interact with MyD88, while TLR-3 interacts with TRIF, and TLR-4 interacts

with both adaptor molecules. The toll-like receptors bind to their adaptor proteins via TIR domains.

Through a series of intracellular associations with additional signaling molecules, these adaptors recruit **protein kinases**, including the IκB kinase (IKK) enzyme complex. This enzyme complex leads into the NF-κB signal transduction cascade. The NF-κB heterodimer is bound in the cytoplasm to an inhibitor, IκBα; when this inhibitor is phosphorylated by IKK, it releases NF-κB, which then translocates to the nucleus.

Another kinase that is activated is *T*ank-*b*inding *k*inase 1 (TBK-1), which phosphorylates two *i*nterferon *r*egulatory *f*actors, IRF-3 and IRF-7 (Figure 33.3). Phosphorylated IRF-3 and IRF-7 form dimers that are translocated to the nucleus. Once in the nucleus, NF-κB, IRF-3, and IRF-7 act as transcription factors and induce the expression of interferons, other cytokines, and interferon-stimulated genes (ISGs).

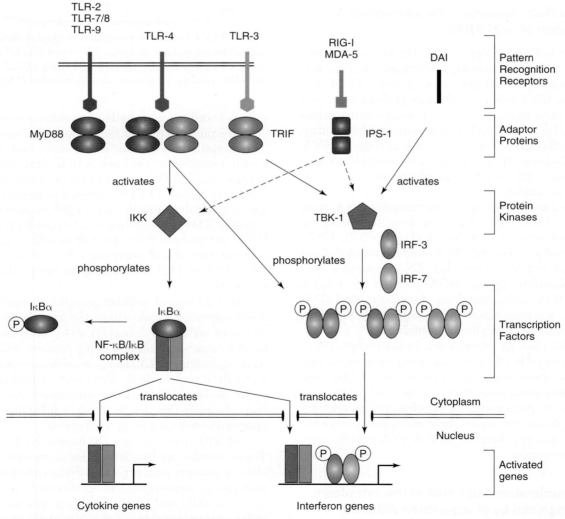

Figure 33.3 Signal transduction following virus recognition. Toll-like receptors signal through the adaptor proteins MyD88 and TRIF to activate kinases such as IKK and TBK-1. These kinases then activate transcription factors such as NF-κB, IRF-3 and IRF-7. Once activated, these transcription factors induce the production of cytokines and interferons, which are then secreted and act on neighboring cells.

The cytosolic detector proteins RIG-I and MDA5 recruit the adaptor protein *i*nterferon-β *p*romoter *s*timulator 1 (IPS-1) upon activation by viral RNAs (Figure 33.3). IPS-1 contains a caspase recruitment domain in its N-terminal region, and self-dimerization can take place by association of these domains. Interaction of IPS-1 with additional cellular proteins leads to the activation of IKK and TBK-1 and the nuclear translocation of NF-κB, IRF-3, and IRF-7.

Virus infection can lead to the activation of other signal transduction pathways, resulting in the activation of additional kinases. For example, herpesviruses rapidly activate c-*J*un *N*-terminal *k*inase (JNK), *e*xtracellular *r*egulated *k*inase (ERK), *p*hospho*i*nositide *3*-kinase (PI3 kinase), and *m*itogen-*a*ctivated *p*rotein *k*inase (MAPK). These kinases then phosphorylate a number of transcription factors, leading to their activation. These cellular signal transduction events lead to many changes in the transcriptional profile of an infected cell, inducing synthesis of a wide array of cytokines, chemokines, and stress-related molecules.

Cellular recognition of virus infection leads to production of cytokines

One of the first responses of an organism to virus infection is the secretion of cytokines (Table 33.2), which are proteins that stimulate protective reactions in the infected host. Cytokine secretion is a hallmark of both the intrinsic cellular defense system and the innate immune response, and cytokines mediate communication between the innate and adaptive immune systems (see Chapter 34). The production and secretion of cytokines can be induced by cellular recognition of virus binding, entry, uncoating, or replication; by the accumulation of viral nucleic acids; and by virus-induced stress responses. While several cellular transcription factors are involved in producing cytokines, NF-κB is a key factor in this response.

Cytokines that are produced within a virus-infected cell are secreted and can act either on nearby cells or at distant sites. Cytokines bind to specific cell surface receptors and initiate cellular signal transduction pathways, leading to activation of transcription of numerous cellular genes. Many of the proteins produced are involved in protecting the cell from virus infection or activating the immune system. There are four main types of cytokines: interferons, pro-inflammatory cytokines, **chemokines**, and **anti-inflammatory cytokines** (Table 33.2).

1. *Interferons.* One of the first cytokines produced by virtually all virus-infected cells is interferon. Interferons are major players in host defense against most viruses, and are discussed in detail below.
2. *Pro-inflammatory cytokines.* Other cytokines that are rapidly made following virus infection include

Table 33.2 Cytokines and their functions

Type	Examples	Function
Interferons	Interferon-α/β	Antiviral defense; dendritic cell maturation; natural killer cell activation
	Interferon-λ	Antiviral defense
	Interferon-γ	Macrophage activation; inhibition of interleukin-6; antiviral defense
Pro-inflammatory cytokines	Interleukin-1	Costimulator of T cells; induces fever, acute-phase proteins (liver) and metabolic wasting
	Interleukin-6	Stimulates B cell growth; costimulator of T cells; induces acute-phase proteins (liver)
	Interleukin-12	Stimulates growth of natural killer cells and T cells; induces production of interferon-γ and tumor necrosis factor-α
	Tumor necrosis factor-α	Activates neutrophils, induces fever, acute-phase proteins (liver), and metabolic wasting; lyses infected cells
Chemokines	Interleukin-8	Recruits neutrophils; promotes angiogenesis
	RANTES	Recruits T cells, eosinophils, and basophils; activates natural killer cells
	CXCL10 (IP-10)	Recruits monocytes, macrophages, T cells, natural killer cells, and dendritic cells
Anti-inflammatory cytokines	Interleukin-4	Inhibits macrophage function; promotes Th2 cells; stimulates B cells
	Interleukin-10	Inhibits macrophage function; stimulates B cells
	Transforming growth factor-β	Inhibits T cells and macrophages; induces production of immunoglobulin A

*t*umor *n*ecrosis *f*actor alpha (TNFα), *inter/*eukin 6 (IL-6), and interleukin 12 (IL-12). These three pro-inflammatory cytokines activate immune cells within the circulatory system.

3. **Chemokines.** These are "chemo-attractant" cytokines that serve to recruit immune cells such as lymphocytes and antigen-presenting cells to the site of infection. The recruited immune cells limit virus replication and stimulate the adaptive immune response, which helps to clear virus and protects against a future infection by the same virus.

4. **Anti-inflammatory cytokines.** Many immune cells produce anti-inflammatory cytokines, which suppress the activity of the pro-inflammatory cytokines and help to return the immune system to a resting state. Sustained secretion of pro-inflammatory cytokines and their movement throughout the circulatory system lead to the characteristic hallmarks of a virus infection, including sleepiness, lethargy, fever, nausea, appetite suppression, and/or muscle pain. It is important that the immune response not be prolonged beyond what is necessary to fight the present infection, and these anti-inflammatory cytokines do just that.

Recognition of virus infection can trigger death of infected cells

Many cells also respond to a virus infection by inducing a self-destruction process called apoptosis, or programmed cell death. This can be an effective host defense system, because the premature death of virus-infected cells reduces the spread of the infection within the organism.

Apoptosis, an important process in both development and differentiation of multicellular organisms, is controlled by signals that monitor cell-cycle progression, metabolism, growth regulation and stress. Cell death results from the activation of **caspases**, which are members of a family of *c*ysteine proteases that cleave after *asp*artate residues. There are twelve known caspases in humans. These proteases are normally present as inactive "procaspases."

Two main apoptosis pathways are known: the *extrinsic* pathway, triggered by a cell surface receptor, and the *intrinsic*, or mitochondrial, pathway, triggered by a variety of stress responses within the cell. Both of these pathways have many components, and only a simplified outline is shown in Figure 33.4. The extrinsic pathway can begin with the cytokine tumor necrosis factor α (TNFα), which binds to its receptor on the cell surface. This leads to trimerization of the receptor, recruitment of adaptor molecules, and assembly of a protein complex called the "death-inducing signaling complex" or DISC. DISC activates caspase 8, which in turn activates caspases 3 and 7. These caspases cleave a

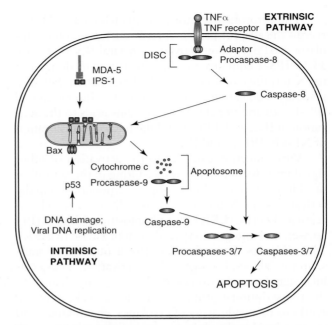

Figure 33.4 Intrinsic and extrinsic pathways of apoptosis. The extrinsic pathway (upper right) begins by binding of a ligand such as TNF α to a cell surface receptor; adaptor proteins form a "death-induced signaling complex" (DISC) with procaspase-8, leading to the activation of caspase-8. The intrinsic pathway (lower left) responds to developmental signals and internal stress such as virus infection by activating cellular p53, which leads to the binding of oligomerized proteins such as Bax to the mitochondrial membrane, release of cytochrome C, and activation of caspase-9 by the "apoptosome." Both caspases 8 and 9 activate caspases-3 and -7, which are the "executioner" caspases that lead to cell death. Cytosolic receptors such as MDA5 (upper left) recruit IPS-1 upon activation, also leading to release of mitochondrial proteins and apoptosis.

number of cellular proteins that lead directly to cellular destruction.

The intrinsic pathway can be induced in response to DNA damage or viral DNA replication, which activate the cellular protein p53. This leads to the formation of oligomers of Bax and similar proteins at the outer mitochondrial membrane. These oligomers create holes in the membrane, resulting in the leakage of cytochrome c and other mitochondrial proteins into the cytosol. These proteins in turn form a complex, the "apoptosome," which leads to activation of caspase 9; caspase 9 in turn activates caspases 3 and 7, as above.

Viruses can also induce apoptosis by binding to the cell surface, altering the organization of cellular membranes, modifying the cytoskeleton, disrupting the cell cycle, or activating cellular signal transduction pathways (see Chapters 21 and 23). Furthermore, proteins that detect virus infection, including TLR-3, RIG-I, and MDA-5, can induce apoptosis. Binding of double-stranded RNA to TLR-3 and recruitment of the adaptor protein TRIF (see Figure 33.3) can lead to apoptosis through the activation of caspase-8. RIG-I and MDA5 recruit the adaptor protein IPS-1 upon detection of viral RNAs (Figure 33.3). IPS-1 dimerizes and binds to the outer mitochondrial membrane, and its recruitment leads to activation of caspase-3, as shown in Figure 33.4, triggering apoptosis. Mice deficient for IPS-1 are much more susceptible than wild-type mice to infection with a variety of RNA and DNA viruses, suggesting that this pathway plays a critical role in defending against virus infection.

Activation of IRF-3 by pathways shown in Figure 33.3 can also induce apoptosis, in part by transcriptional induction of proteins such as the death domain-associated protein Daxx, and in part by causing the translocation of proteins to the mitochondria and induction of the mitochondrial apoptosis pathway.

Other antiviral signal transduction pathways involve "inflammasomes"

A subset of caspases is known as the "inflammatory caspase" subfamily. These enzymes are mostly involved in the regulation of cytokine maturation. One member of this subfamily, caspase-1, can be part of a large protein complex, known as the "inflammasome."

Caspase-1 is rapidly activated following recognition of viral RNA or double-stranded DNA by the inflammasome components AIM2 and NLRP3 (Figure 33.5). Both of these DNA sensors contain pyrin domains, allowing for interaction with the pyrin domain of the adaptor molecule, *a*poptosis-associated *s*peck-like protein (ASC). The ASC adaptor then recruits procaspase-1 through its caspase activation domain. Activated caspase-1 causes

the maturation and secretion of pro-inflammatory cytokines such as interleukin-1β and interleukin-18.

Recent work has found that infections by bacteria or viruses can lead to a new pathway toward cell death called **pyroptosis**. Unlike apoptosis, pyroptosis is dependent on caspase-1 and is characterized by the release of pro-inflammatory cytokines.

INTERFERONS

Virus-infected cells secrete interferons, which protect nearby cells against virus infection

Of the virus-induced cytokines, interferons possess the greatest antiviral activity. Interferons released from virus-infected cells bind to receptors on nearby uninfected cells and stimulate transcription of a set of genes encoding proteins with antiviral activities, resulting in protection of those cells against viral infection (Figure 33.6).

Interferon was discovered in 1957 by Alick Isaacs and Jean Lindenmann. These scientists were studying the phenomenon called **viral interference**, in which infection of cultured mammalian cells by one virus can interfere with a subsequent infection of the cells by the same or another virus. Isaacs and Lindenmann found that the culture medium from cells infected with influenza virus could transmit antiviral protection to cells that had not previously come in contact with the virus; they named the substance present in the medium "interferon." Interferons were subsequently shown to

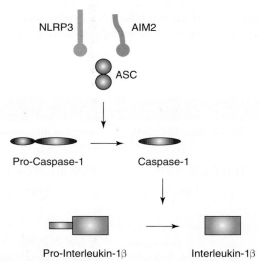

Figure 33.5 Cellular sensor proteins activate caspases and cytokines. As part of the "inflammosome," AIM2 and NLRP3 signal through the adaptor ASC to activate caspase-1, which cleaves pro-interleukin-1β to mature interleukin-1β.

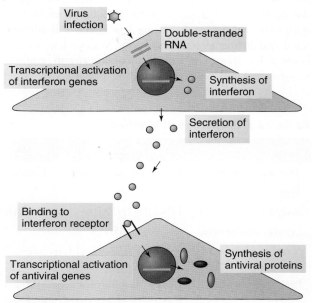

Figure 33.6 The interferon system. Virus infection leads to production of double-stranded RNA, which activates transcription of interferon genes. Interferons are synthesized and secreted, then bind to receptors on nearby cells, inducing expression of genes that make antiviral proteins.

be proteins that could be purified and used to protect cells against virus infection.

Their discovery stimulated an enormous amount of research whose underlying hope was that interferons could be used to prevent and cure virus infections in humans. It was soon found that interferon induced by one kind of virus can inhibit replication of many different virus types. However, the antiviral activity of interferon is generally limited to the animal species in which the interferon is produced. Three major types of interferon, designated types I (interferons α and β), II (interferon-γ) and III (interferons λ₁, λ₂, and λ₃) have been described. Besides their antiviral effects, interferons inhibit cell growth and division and stimulate the immune system.

Interferons are a first line of host defense against viruses; however, therapeutic use has been limited

A gene coding for interferon-α was cloned by Charles Weissmann and collaborators in 1980, allowing the protein to be produced in large amounts and opening up the possibility of therapeutic use. Recombinant interferons α and β are presently used to treat chronic infections by hepatitis B and C viruses, cancers such as melanoma, multiple myeloma, renal cell carcinoma and Kaposi's sarcoma, and the neurodegenerative disease, multiple sclerosis. However, their original promise as major antiviral drugs has not been fulfilled.

Therapeutically administered interferon can lead to unpleasant side effects and its antiviral effects are short-lived. Interferons are most effective in reducing viral spread during the initial stages of virus infection, before most symptoms are detected, but are less effective if administered late during an acute infection. Nonetheless, we now know a lot about the induction of interferon synthesis in virus-infected cells, the signaling pathways by which interferons stimulate the expression of antiviral genes, and the mechanisms by which antiviral proteins inhibit virus replication. Interferons play a major role in the "rapid reaction" first line of defense against viral infections.

Interferons α, β, γ, and λ are made by different cells, bind to different receptors, and have distinct functions

The interferons are classified into three groups based on amino acid sequence homology and the receptors they use (Table 33.3). Type I interferons include interferon-α and interferon-β. Thirteen different interferon-α genes have been identified and are clustered on the short arm of human chromosome 9. Both human and mouse interferon-α genes lack introns and share a high degree of homology at the nucleotide sequence level, suggesting that the gene cluster was derived from a common ancestor gene by successive duplications. A single gene encodes interferon-β; it also lacks introns and is more distantly related to the interferon-α family.

Interferon-α, previously called leukocyte interferon, is produced by peripheral blood leukocytes and many other cell types in response to viral infection or treatment with double-stranded RNA. The different interferon-α subtypes all have the same number of amino acids and share a high degree of amino acid sequence homology; differences in molecular weights are due only to different levels of glycosylation. Interferon-β, also called fibroblast interferon, is a glycosylated protein produced mainly by fibroblasts and epithelial cells.

Type II interferon, called interferon-γ or "immune interferon," is only distantly related to the type I interferons. Interferon-γ is encoded by a single gene containing three introns. It is mainly produced by activated T lymphocytes and natural killer (NK) cells. Secretion of interferon-γ is induced by mitogenic stimuli, antigens,

Table 33.3 Properties of interferons

	Interferon-α	Interferon-β	Interferon-λ	Interferon-γ
Alternative name	Leukocyte interferon	Fibroblast interferon	Interleukins 28A, 28B, 29	Immune interferon
Type	I	I	III	II
Receptors	IFNAR1; INFAR2c	IFNAR1; INFAR2c	IL28Rα; IL10Rβ	IFNGR1; IFNGR2
Length, amino acids	165	165	196-200	146
Number of genes in humans	13	1	3	1
Principal producer cells	Leukocytes, fibroblasts, macrophages, epithelial cells	Fibroblasts, epithelial cells	Leukocytes, fibroblasts, macrophages, epithelial cells	T lymphocytes, macrophages, natural killer cells
Inducing agents	Viruses, double-stranded RNA			Mitogens, antigens, interleukin-2
Major activity	Antiviral action			Immune stimulation

and cytokines such as interleukin-1, interleukin-2, or interferon-γ itself. Although interferon-γ possesses antiviral activity, it is primarily an immune modulator rather than an antiviral agent.

Type III interferons were discovered in 2003 by two independent groups using computational and bioinformatic methods. Three human type III interferon genes are clustered on chromosome 19: interferon-λ_1, interferon-λ_2, and interferon-λ_3, also known as interleukin-29, interleukin-28A, and interleukin-28B, respectively. Interferon-λ_2 and interferon-λ_3 share 96% amino acid sequence identity but they share only 81% amino acid sequence identity with interferon-λ_1.

Transcription of interferon genes is activated by virus infection or double-stranded RNA

Virus infection is the best natural inducer of interferon expression, although other infectious agents can also induce interferon production. Most RNA- and DNA-containing viruses induce interferon synthesis; however, double-stranded RNA viruses, such as reovirus and blue tongue virus, are the best inducers. Synthetic and natural double-stranded RNAs are also potent inducers of interferon synthesis, and double-stranded RNA may represent the natural signaling molecule by which virus infection leads to interferon production (Figures 33.1, 33.2, and 33.3).

Furthermore, lysis of virus-infected cells can release viral double-stranded RNA into the extracellular medium. Unlike single-stranded RNA, double-stranded RNA is stable and has been detected within the lung fluid of individuals infected with viruses such as influenza virus. Indeed, even before the discovery of interferon, it was appreciated that double-stranded RNA acts as a "viral toxin" and leads to many of the symptoms associated with a virus infection.

Transcriptional activation occurs by binding of transcription factors to interferon gene enhancers

Induction of interferon synthesis requires transcriptional activation of interferon genes, which are strongly repressed in uninfected cells. Signaling pathways shown in Figure 33.3 lead to activation of cellular transcription factors. These activated transcription factors bind to regulatory sequences called positive regulatory domains (PRD), which lie 200 nucleotides upstream of the transcription start sites on both interferon-α and interferon-β genes (Figure 33.7). The binding of transcription factors to the interferon-β gene enhancer has been carefully studied and serves as a paradigm for the interaction between transcription factors and the general transcriptional machinery.

Four different transcription factors are known to bind to the virus-inducible enhancer (VRE) within the interferon-β promoter (Figure 33.7). Interferon regulatory factors IRF-3 and IRF-7 bind to positive regulatory domains I and III, NF-κB binds to positive regulatory domain II, and ATF-2/c-Jun binds to positive regulatory domain IV. Finally, the high mobility group protein HMG-1, a small histone-like protein associated with chromatin, binds to the minor groove of DNA within positive regulatory domains IV and II, promoting the binding of ATF-2/c-Jun and NF-κB. Together with the coactivator CBP/p300, this collection of DNA-bound proteins assembles to form an active transcription complex called the "enhanceosome."

Although not identical, the interferon-α enhancer also contains positive regulatory domains. While interferon-β can be induced without IRF-7 by the constitutively expressed IRF-3 (along with NF-κB and ATF-2/cJun), interferon-α induction requires IRF-7. Type III interferons are produced in a similar fashion to type I interferons. Induction of interferon-λ_1 and interferon-$\lambda_{2/3}$ resembles induction of interferon-β and interferon-α, respectively.

There are presently nine known members of the IRF family of transcription factors, of which five are thought to contribute to type I and/or type III interferon production. IRFs function in a number of biological processes such as response to pathogens, cytokine signaling, regulation of cell growth, and **hematopoiesis**. All IRFs share a high degree of homology in their N-terminal DNA-binding domains, and they bind to the consensus DNA sequence GAAANNGAAANN (where N can be any

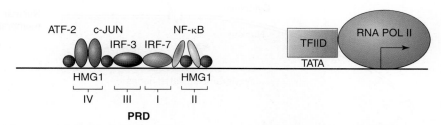

Figure 33.7 Transcriptional activation of the interferon-β gene. Virus infection leads to activation of cellular transcription factors that bind to positive regulatory domains (PRD) upstream of the TATA box and start site on the interferon-β gene, forming an "enhanceosome."

nucleotide). The C-terminal portion is unique to each specific IRF member. Viral versions of IRFs are produced by human herpesvirus 8, the virus responsible for Kaposi's sarcoma. These viral IRFs block the interferon response, allowing more efficient replication of human herpesvirus 8.

Once transcription of interferon genes is activated, interferon mRNAs are transported to the cytoplasm and interferons are synthesized and secreted into the extracellular medium (Figure 33.6). Interferon mRNAs contain AU-rich elements that render them unstable, and the transcriptional activation of interferon genes is short-lived. As a result, infected cells give rise to a burst of interferon that is sent out to neighboring cells, but synthesis is soon halted.

Interferon signal transduction is carried out via the Jak–Stat pathway

Cytokines and interferons are intercellular messengers that induce many important biological responses in target cells. Signaling by interferons is mediated by a pathway that includes the *J*anus tyrosine *k*inases (Jak kinases) and the *s*ignal *t*ransducers and *a*ctivators

of *t*ranscription (Stat proteins). These proteins also mediate cellular responses to many other cytokines and growth factors such as growth hormone, prolactin, and erythropoietin. This pathway is activated by a cascade of phosphorylation events on tyrosines located in the receptor molecules, in the intermediary Jak kinases, and in the Stat proteins that activate transcription of a variety of target genes (Figure 33.8).

Signaling begins by the binding of the cytokine to a cell-surface receptor. Interferon-α and -β signal through the same receptor complex, which contains the *interferon alpha receptor* 1 (IFNAR1) and IFNAR2c proteins. Interferon-λ signals through a different receptor complex, which contains the *interleukin-28 receptor* α (IL28Rα) and the *interleukin-10 receptor* β (IL10Rβ) proteins. These four receptor components are transmembrane proteins with a binding domain on the extracellular side and an intracellular domain that binds to one of the Jak kinases, either Jak1 or Tyk2 (Figure 33.8, top left).

While the interferon receptor proteins IFNAR1 and IFNAR2c are expressed in virtually all cell types throughout the body, IL28Rα is only expressed in a few cell types, most notably epithelial cells. This restricted

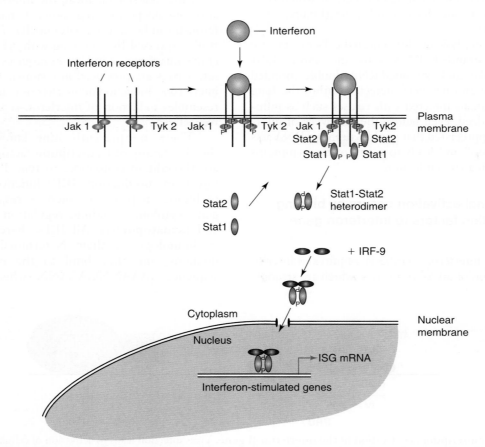

Figure 33.8 Interferon signal transduction. Binding of interferon to receptors at the surface of target cells activates the Jak–Stat pathway and leads to transcriptional activation of interferon-stimulated genes (ISG). See text for details.

expression of the type III interferon receptor complex may allow the host to rapidly eliminate viruses at the major sites of virus entry before the virus can establish an infection, without activating additional immune responses.

Binding to interferon brings two or more receptor molecules together, and their associated Jak kinases phosphorylate tyrosine residues on each other as well as on the cytoplasmic domains of the receptor molecules (Figure 33.8, top center). The phosphorylated Jak kinases are now activated. Subsequently, Stat1 and Stat2 proteins are recruited by binding to the phosphorylated receptor, and they in turn are tyrosine-phosphorylated by the activated Jak kinases, leading to the dimerization of the Stat proteins (Figure 33.8, top right). After these Stat dimers are released, they bind to cytoplasmic IRF-9 and the trimeric complex translocates to the nucleus, where it binds to sequences within the promoter regions of a variety of interferon-stimulated genes to activate their transcription.

The **Jak–Stat pathway** represents an extremely rapid signaling system between the plasma membrane and the cell nucleus, and explains at least part of the basis for the specificity of signals that are induced by different cytokines. A remarkable feature of this system is that newly induced Stat DNA binding activity can be detected in the nucleus within minutes of cytokine binding. This timing reflects the rapidity of the activation of these emergency response genes.

Multiple cellular responses affecting cell growth and differentiation, resistance to a wide range of pathogens, and modulation of immune functions are induced by interferon via the Jak–Stat pathway. These responses are carried out by proteins expressed from the *interferon-stimulated genes* (ISGs). More than 1000 such genes have been identified to date; a partial list is presented in Table 33.4.

Antiviral activities induced by interferons

The ability of interferons to confer an antiviral state to uninfected cells is their defining activity. Interferons provide an early line of defense against viral infections, and are part of the host's intrinsic cellular and innate immune responses to infection. Adaptive immunity, provided by humoral antibodies and specific B- and T-lymphocyte responses, is much slower to respond (days to weeks). The vital role of interferons in antiviral defense has been demonstrated by studies in which animals that cannot mount an interferon response succumb to viral infections more rapidly and at a lower infectious dose than normal animals.

There is evidence that the earliest stages (virus entry and uncoating) and the final stages (assembly and release) of virus infection can be inhibited by interferon treatment of cells. However, the best-characterized antiviral effects are directed against viral mRNA synthesis and stability, and viral protein synthesis. Three

Table 33.4 Selected cellular proteins induced by interferons

Protein	Function	Induced by interferon type
dsRNA-dependent ser/thr protein kinase	Phosphorylates eIF-2α, inhibits protein synthesis	α/β
2', 5'-oligo(A) synthetase	Activates ribonuclease L	$\alpha/\beta/\lambda$
Ribonuclease L	mRNA degradation	α/β
Myxovirus-inhibiting protein	GTPase; inhibits replication of negative-strand RNA viruses	$\alpha/\beta/\lambda$
Interferon regulatory factor-1	Transcription factor	$\alpha/\beta, \gamma$
Interferon regulatory factor-7	Transcription factor	$\alpha/\beta/\lambda$
Interferon-stimulated gene protein-15	Cytokine immunomodulator	$\alpha/\beta/\lambda$
Interferon-stimulated gene protein-56	Inhibition of translation	$\alpha/\beta/\lambda$
Cytokine-responsive gene protein-2	Chemokine	$\alpha/\beta, \gamma$
Class I major histocompatibility complex antigen	Antigen presentation	$\alpha/\beta, \gamma$
Class II major histocompatibility complex antigen	Antigen presentation	γ
Rev-binding protein-27	Inhibits Rev-dependent HIV activation	$\alpha/\beta, \gamma$
Immunoglobulin G-Fcγ receptor	Binds to IgG Fc fragment	γ
Nitric oxide synthase	Macrophage activation	γ
Mn-dependent superoxide dismutase	Superoxide scavenger	γ
RING-12, proteasome component	Protein degradation	γ

interferon-induced systems that are understood in some detail are (1) the Mx proteins; (2) **2′, 5′-oligo(A) synthetase** and **ribonuclease L**; and (3) **PKR (double-stranded RNA-dependent protein kinase)**.

1. *The Mx proteins.* Mx (for "myxovirus") proteins are interferon-inducible, 70- to 80-kDa proteins with the ability to hydrolyze GTP. Expression of Mx affects virus replication by interfering with the transcription of influenza virus and other RNA viruses. The cellular localization of Mx plays a large role in determining its antiviral activity. For example, nuclear Mx1 in rodents blocks the replication of viruses such as orthomyxoviruses, which replicate within the nucleus, whereas cytoplasmic Mx2 in rodents blocks the replication of viruses such as rhabdoviruses and bunyaviruses, which replicate in the cytoplasm. Interestingly, human MxA, which is found in the cytoplasm, has antiviral activity against a range of viruses, regardless of the compartment where the virus replicates.

    Mx proteins inhibit the activity of the viral RNA polymerase, thus blocking viral mRNA production. They also interfere with the transport of influenza virus ribonucleoprotein complexes from the cytoplasm to the nucleus, where they are transcribed and replicated. It is not clear exactly how the Mx proteins block these steps, and the role of the GTPase activity of Mx proteins in their function is not understood.

2. *2′, 5′-oligo(A) synthetase and ribonuclease L.* The enzyme 2′, 5′-oligo(A) synthetase is produced in interferon-treated cells, and is activated by binding to double-stranded RNA in virus-infected cells. This enzyme has the unique ability to produce oligomers of ATP via a 2′, 5′ linkage, in contrast to the normal 3′, 5′ linkage found in natural RNAs (Figure 33.9). These oligomers contain up to 10 adenylate residues. They in turn bind to another interferon-induced protein, ribonuclease L, and activate its ribonuclease activity. Ribonuclease L then degrades both host and viral mRNAs, leading to inhibition of virus replication by suppression of viral protein synthesis. Since host mRNAs are also cleaved, the cell may die if sufficient ribonuclease L activity is present. Furthermore, ribonuclease L cleavage of cellular RNAs can result in the accumulation of small RNA species that stimulate RIG-I-like receptor pathways, leading to amplification of the cellular interferon response.

3. *Double-stranded RNA-dependent protein kinase (PKR).* PKR is a serine-threonine protein kinase that is present in normal cells at low levels; its expression is increased by interferon treatment. Like 2′, 5′-oligo(A) synthetase, PKR is activated by double-stranded RNA. When two molecules of

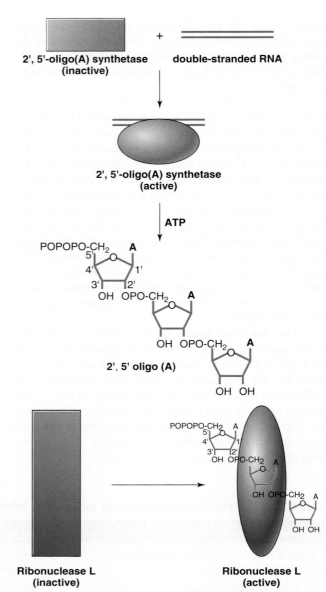

Figure 33.9 Antiviral activity directed by 2′, 5′-oligo(A) synthetase and ribonuclease L. Double-stranded RNA activates 2′, 5′-oligo(A) synthetase, which produces oligomers of ATP joined by 2′, 5′ linkages. These oligomers bind to and activate ribonuclease L, which degrades viral and cellular mRNAs.

PKR bind to the same double-stranded RNA molecule, they phosphorylate each other, inducing a conformational change that greatly increases their serine-threonine kinase activity (Figure 33.10). Activated PKR then phosphorylates a variety of cellular proteins.

The best-known substrate is the alpha subunit of translation initiation factor eIF-2, which is phosphorylated by PKR at serine-51. Phosphorylation results in the formation of an inactive complex that involves eIF-2 bound to guanosine diphosphate

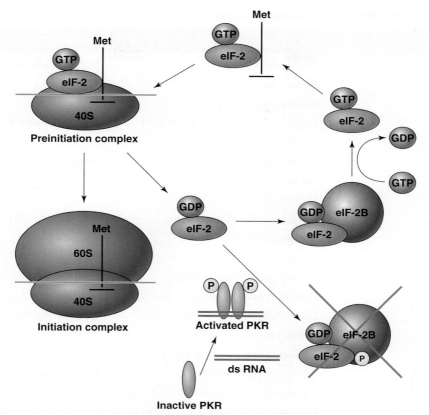

Figure 33.10 Mechanism of antiviral activity directed by PKR. When activated by double-stranded RNA, PKR phosphorylates the eIF-2α subunit of translation initiation factor eIF-2 (bottom right), blocking the recycling of GDP with GTP normally carried out by eIF-2B, and ultimately blocking initiation of protein synthesis. 40S and 60S: small and large ribosomal subunits.

(GDP) and the recycling factor eIF-2B. Normally, eIF-2B catalyzes the exchange of GDP for GTP, and the regenerated eIF-2-GTP complex binds to initiator methionine tRNA and forms a translation initiation complex with mRNA and a 40S ribosomal subunit. Since the GDP-GTP exchange reaction is blocked, insufficient met-tRNA-eIF-2-GTP complex is formed, and translation initiation can no longer take place.

Apoptosis may also play a role in the antiviral effect of PKR, since its over-expression results in programmed cell death. PKR is believed to play an important role in regulation of cellular protein synthesis in the absence of virus infection.

Interferons have diverse effects on the immune system

Interferon-γ, or immune interferon, is produced by a restricted set of immune cells (T lymphocytes, macrophages, and natural killer cells) in response to immune and/or inflammatory stimuli, and its major function is to stimulate the development and actions of immune effector cells. Interferons-α and -β play a more limited role in

modulating the adaptive immune response to viral infection. Interferons affect the immune response in dramatic fashion, and this will be discussed in Chapter 34.

Viruses have developed numerous strategies to evade the interferon response

Viruses fight back against the host interferon response by making proteins specifically designed to interfere with these host defense systems. Table 33.5 gives some examples of a growing list of viral gene products that interfere with the interferon response. Other examples are discussed below.

An important target for many viruses is the double-stranded, RNA-activated protein kinase (PKR). Adenovirus produces high levels of VA I RNA, which is a small (160 nt) RNA transcribed from the adenovirus genome by RNA polymerase III. By virtue of self-complementary sequences, it has a structure similar to double-stranded RNA. VA I RNA binds to PKR as do other double-stranded RNAs, but fails to activate the protein kinase, perhaps because it is too small to bind to two PKR molecules at the same time (see above). Therefore it occupies the dsRNA binding site and

Table 33.5 Inhibition of the interferon system by viruses

Target	Virus (gene product)	Effect
Interferon induction	Vaccinia virus (E3L)	Sequesters double-stranded RNA
	Ebola virus (VP35)	
	Influenza virus (NS1)	
	Hepatitis C virus (NS3/4A, NS5A)	Inhibits toll-like receptor signaling
	Vaccinia virus (A52R)	
	Paramyxovirus (V)	Inhibits RNA helicase signaling
	Hepatitis A virus (3ABC)	
	Hepatitis C virus (NS3/4A)	
	Herpes simplex virus (ICP0)	Inhibits function of IRF-3
	Rotavirus (NSP1)	
	Rabies virus (P)	
Interferon signaling	Human papillomavirus (E6)	Inhibits/degrades Jak/Tyk2
	Human cytomegalovirus	
	Paramyxovirus (V)	Inhibits/degrades Stats
	Rabies virus (P)	
	Human cytomegalovirus (IE1)	
Function of products of interferon-stimulated genes	Adenovirus (VAI RNA)	Blocks activation of PKR
	Epstein–Barr virus (EBV small RNA)	
	Human immunodeficiency virus (Tat)	
	Herpes simplex virus (ICP34.5)	
	Adenovirus (E1A)	Inhibits function of interferon-stimulated genes
	Influenza virus (NS1)	
	Crimean Congo hemorrhagic virus (L)	

inhibits PKR activation, allowing viral protein synthesis to take place even in cells that express high levels of PKR induced by interferon. Epstein–Barr virus (a herpesvirus) encodes a small RNA that performs a similar function.

Both reovirus σ3 capsid protein and vaccinia virus SK-1 protein block PKR activation by binding strongly to double-stranded RNA in the infected cell. These proteins therefore sequester the double-stranded RNA and do not allow it to come in contact with PKR, so that activation cannot occur. The HIV transcriptional transactivator Tat and the hepatitis C virus NS5A protein appear to inhibit PKR antiviral activity by binding directly to PKR. In poliovirus-infected cells, PKR is proteolytically cleaved.

Adenovirus E1A protein inhibits the production and action of interferon at the level of transcription. E1A sequesters p300/CBP, a coactivator required for transcription of the interferon genes. Human herpesvirus 8 produces viral homologues of the interferon regulatory factors, called v-IRFs, which block interferon gene activation by inhibiting, rather than stimulating, transcription when bound to interferon genes.

Some viruses evade apoptosis or the inflammatory response by synthesizing proteins that interfere with signaling pathways. For example, poxviruses produce proteins called CrmA, CrmB, and CrmC that bind to caspases, resulting in a delay in apoptosis and allowing the virus to replicate. Poxviruses also produce soluble forms of cytokine receptors ("viroceptors") that perform the dual functions of blocking the production of antiviral cytokines and inhibiting B cell activation, thus delaying the onset of the humoral immune response.

RNA INTERFERENCE

Small interfering RNAs are involved in combating virus infections in plants and invertebrates

In mammals, the primary response to virus infection is the production of interferon. However, plants, invertebrates, and fungi do not produce interferon. Instead, they use **RNA interference** (RNAi) as an important antiviral defense mechanism (see Chapter 10). RNA interference is initiated by the double-stranded RNA-specific ribonuclease III enzyme called Dicer (Figure 33.11).

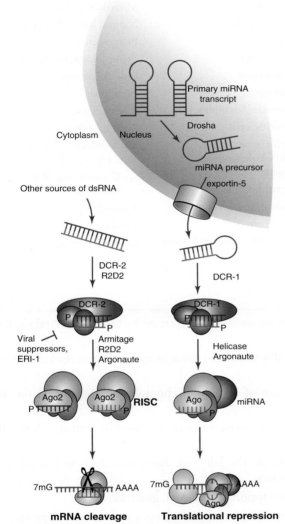

Figure 33.11 RNA interference and microRNA. Left: Plants use RNA interference as an antiviral defense mechanism. Viral double-stranded RNA is cleaved by the enzyme Dicer (DCR-1 and DCR-2) to make small interfering RNAs (siRNAs). A host RNA-dependent RNA polymerase (not shown) can amplify siRNAs. SiRNAs are bound by the RISC complex, which includes the protein Argonaute (Ago). There siRNAs are unwound and a single strand is hybridized to viral genome RNA or mRNAs, which are subsequently cleaved. Right: Mammalian cells use similar enzymatic machinery to generate microRNAs (miRNAs). Imperfect RNA hairpins in cellular transcripts are recognized and cleaved by the enzyme Drosha and are exported from the nucleus. Once in the cytoplasm, these RNA molecules are further processed by Dicer to short miRNAs. MiRNAs are incorporated into the RISC complex where they are hybridized to cellular or viral mRNAs. Perfect matches lead to mRNA degradation, whereas imperfect matches block protein synthesis.

Dicer cleaves double-stranded RNAs resulting from virus infection into **small interfering RNAs (siRNAs)** that are 21 to 23 nucleotides in length, and have 2- to 3-nt 3′ overhangs at both ends. These siRNA molecules are then bound by a collection of proteins known as the ***RNA-induced silencing complex* (RISC)**. One of the small RNA strands is used as a "guide" that hybridizes to complementary sequences within the viral genome RNA or messenger RNAs. A ribonuclease within the silencing complex then cleaves the hybridized viral RNA molecule.

Not surprisingly, plant and invertebrate viruses have evolved strategies to block RNA interference, including the production of proteins known as suppressors of RNA silencing (see Chapter 10). For example, both tomato bushy stunt virus and flock house virus encode proteins that directly bind double-stranded RNA, thus sequestering the RNA and making it unavailable to the RNA interference machinery.

MicroRNAs are used to control gene expression in vertebrates

The RNA interference machinery is also present in vertebrates, including mammals, but instead of defending against virus infection, this machinery regulates gene expression via either degradation of mRNA or inhibition of mRNA translation. This regulation occurs by the production of **microRNAs**, also known as **miRNAs**. These are similar in length to small interfering RNAs, but are produced by a different mechanism.

Unlike siRNAs, which are produced from long and perfectly complementary viral double-stranded RNA molecules, miRNAs are typically made from imperfect RNA hairpins in transcripts synthesized by cellular RNA polymerase II (Figure 33.11, right side). These hairpin structures are typically found within the introns of capped and polyadenylated transcripts ("primary miRNAs"). Primary miRNA stem-loop structures typically have an imperfectly paired region of approximately 30 base pairs.

Within the nucleus, the ribonuclease III enzyme Drosha cleaves the primary miRNA into a smaller miRNA precursor, which is then exported to the cytoplasm. Once in the cytoplasm, miRNA precursors are further processed to miRNAs by Dicer, the enzyme involved in the generation of siRNAs. The processed miRNAs then associate with the RISC complex. If an miRNA is only partially complementary to its messenger RNA target, it may inhibit translation of the mRNA. If, however, the miRNA is perfectly complementary to its target, the mRNA is degraded, as occurs with siRNAs.

So far, more than 500 different human miRNAs have been discovered, and many are expressed in a developmental or tissue-specific manner. Even though mammals encode the RNAi machinery, to date siRNAs complementary to viral RNAs have not been found within virus-infected cells. Furthermore, there is no evidence that mammalian viruses encode suppressors of RNA silencing, which might be expected if this were a major antiviral mechanism available to mammalian cells.

However, many viruses do encode miRNAs. Of particular interest, viruses that have been found to make miRNAs are all nuclear DNA viruses, including members of the herpesvirus, adenovirus, and polyomavirus families. Viral miRNAs target both viral and cellular messenger RNA species.

Simian virus 40 and Epstein–Barr virus both encode miRNAs that target early viral messenger RNAs. These miRNAs accumulate during the late phase of infection, when early viral proteins are no longer needed for virus replication but can become targets of the host adaptive immune response. Besides "autoregulating" synthesis of their own proteins, viral miRNAs target cellular mRNAs that encode proteins involved in apoptosis or host antiviral immune responses. For example, Epstein–Barr virus encodes miRNAs that target PUMA, a pro-apoptotic protein, and CXCL-11, a potent T-cell chemokine.

KEY TERMS

2', 5'-oligo(A) synthetase	Interleukin-1α
Adaptive immune responses	Intrinsic cellular defense
Anti-inflammatory cytokines	Jak–Stat pathway
Apoptosis	Macrophages
Autophagy	MicroRNAs (miRNAs)
B lymphocytes	Natural killer (NK) cells
Caspases	Phagosomes
Chemokines	Pro-inflammatory cytokines
Complement	Protein kinases
Cytokines	Pyroptosis
Cytotoxic T lymphocytes	Ribonuclease L
Dendritic cells	RNA interference
Double-stranded, RNA-dependent protein kinase (PKR)	RNA-induced silencing complex (RISC)
Hematopoiesis	Small interfering RNAs (siRNAs)
Innate immune responses	T lymphocytes
Interferons	Toll-like receptors
Interferon-stimulated genes	Viral interference

FUNDAMENTAL CONCEPTS

• Toll-like receptors and other cellular pattern recognition receptors sense viral infections by binding to viral nucleic acids and glycoproteins.

• Recognition of viral nucleic acids, particularly double-stranded RNA, leads to the activation of transcription factors, including interferon regulatory factors (IRFs) and NFκB.

• Activated IRFs and NFκB induce the production and secretion of cytokines, chemokines, and interferons by virus-infected cells.

• Interferons-α, -β, and -λ are expressed by most cell types, whereas interferon-γ is expressed by activated NK cells and T lymphocytes.

• Upon binding to receptors on the cell surface, interferons activate the Jak–Stat signal transduction cascade and induce transcription of a large number of interferon-stimulated genes.

• Proteins coded by interferon-stimulated genes block virus transcription, translation, and spread.

• In response to viral infection, cytokine signaling and sensing of intracellular stress can lead to apoptosis, or programmed cell death.

• Plants and invertebrates do not express interferon, but instead use RNA interference to fight against viral infections.

REVIEW QUESTIONS

1. Organisms from flies to humans have a variety of primary "sensors" on their cells that can detect foreign molecules. What are these sensors called and what types of molecules do they "see"?

2. Infection by herpesviruses can lead to the activation of a signal transduction pathway that results in the activation of four cellular kinases. What are these and what are the consequences to the cell?

3. Viruses can induce apoptosis of a virus-infected cell by binding of viral RNA to one of three detection proteins. What are these proteins and how do they induce apoptosis?

4. Why are the Type I interferons known as antiviral interferons?

5. Plants and invertebrates do not produce interferon in response to virus infection but do produce Dicer. What is Dicer and how does it contribute to antiviral defense?

Innate and Adaptive Immune Responses to Virus Infection

Malcolm G. Baines
Karen Mossman

THE INNATE IMMUNE RESPONSE

Innate immune responses are rapid but non-specific responses of the host to infection.

Complement is a set of serum proteins that recognize foreign macromolecules on pathogens or cells and mark them for destruction.

The inflammatory response is triggered by cytokine production by infected cells.

Circulating leukocytes and monocytes migrate to regions producing cytokines and phagocytose pathogens.

Tissue macrophages are activated by interferon-γ to kill microbes and infected cells by producing toxic oxygen and nitrogen compounds.

Natural killer (NK) cells recognize virus-infected cells lacking major histocompatibility complex (MHC) molecules and target them for death by apoptosis.

THE ADAPTIVE IMMUNE RESPONSE

Antigens are macromolecules that give rise to specific immune responses that lead to production of:
* *antibodies*
* *cytotoxic T lymphocytes*

B and T lymphocytes arise in the bone marrow and thymus (primary lymphoid organs).

Immune responses occur in the spleen, lymph nodes, and gut (secondary lymphoid organs).

B-cell receptors are surface antibodies that can bind to a specific epitope on intact antigens.

T-cell receptors are surface molecules that can bind to small peptide epitopes bound to MHC proteins.

A great variety of specific B- and T-cell receptors are created by somatic recombination.

Antibodies (immunoglobulins) are formed from dimers of heavy chains and light chains.

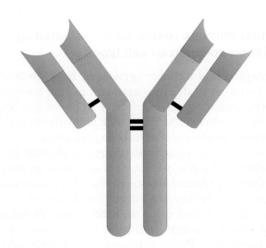

The variable parts of heavy and light chains form antigen-binding sites

Antigen-presenting cells such as dendritic cells and macrophages digest antigens and present digested fragments in association with MHC molecules to T cells.

T cells that recognize and bind to peptides on MHC molecules differentiate into helper and cytotoxic T cells.

B cells that recognize and bind to antigens differentiate into plasma cells, which secrete antibodies.

EFFECTS OF INTERFERONS ON THE IMMUNE RESPONSE

Interferons stimulate antigen processing via the proteasome and stimulate antigen presentation via MHC proteins.

Interferons modulate production of cytokines that favor development of Th1 cells and a cytotoxic T-cell response.

Interferon-γ stimulates macrophage activity.

Interferons can influence immunoglobulin isotype switching.

VIRUS STRATEGIES TO COUNTER HOST DEFENSES

Viruses express mutated versions of cellular genes encoding cytokines and cytokine receptors.

These genes encode virokines, which mimic cytokines, and viroceptors, which mimic cytokine receptors.

Virokines can alter the host immune response and protect infected cells from detection.

Viroceptors can bind to host cytokines and inhibit host responses to infection.

Viruses evade complement by down-regulating, sequestering, or degrading critical complement components.

Viruses make homologues of MHC proteins that prevent NK cell recognition of infected cells.

Viruses make proteins that interfere with intracellular antigen processing and presentation.

Viruses evade adaptive immune responses via antigenic drift and shift.

The host immune response is mediated by circulating specialized cell types

All complex multicellular organisms have evolved a broad spectrum of defenses against infection by viruses and other microorganisms. These defenses are organized in a multilevel strategy that helps the organism resist infection and combat infectious diseases.

The first level of defense comprises physical barriers, including the skin and mucosal secretions in the gut and respiratory tracts that non-specifically prevent access of viruses to cells they can infect. Should the virus succeed in infecting host cells, the second level of defense against infection comes into play: these responses are called **intrinsic cellular defenses** and are discussed in detail in Chapter 33. Intrinsic cellular defense systems respond rapidly (within minutes) to an infection, induce a set of intracellular antiviral defense mechanisms within infected cells and their immediate neighbors, and direct the production and secretion of cytokines, including interferons, that signal the presence of an infection within the organism.

If the infection begins to spread within the organism, the third level of host defense, the **immune response,** comes into play. This is the subject of the present chapter; most of the information presented here refers to the mammalian immune system, particularly that of man. The immune response is mediated by a variety of specialized cell types that originate primarily in bone marrow, and circulate through the body in the blood and lymphatic systems as well as in most tissues. Each of these cell types carries out specific functions that are listed in Figure 34.1, and they will be encountered as we discuss the different elements of the mammalian immune response. As obligate intracellular pathogens, viruses can remain almost invisible to most of the cells and molecules of the host immune systems; this chapter will explain how the problem of "seeing" inside the cell is dealt with by the immune system.

Innate immune responses are rapid but non-specific; adaptive immune responses are slower but long-lasting and highly specific

Immune responses can be classified as *innate* or *adaptive*. Innate immunity is exercised by a variety of cells and serum proteins that non-specifically recognize foreign pathogens or infected cells. Like intrinsic cellular defenses, innate immune responses can be rapid, and they usually act locally at the site of infection. Acute or innate responders to viral infection include the **complement** system and inflammatory cells such as **polymorphonuclear leukocytes** and **natural killer (NK)** cells, **monocytes,** and **macrophages** (Figure 34.1).

The slower adaptive immune response is mediated by populations of specifically activated **B lymphocytes,** which synthesize virus-specific **antibodies,** and **T lymphocytes,** which differentiate into **cytotoxic T lymphocytes** that kill virus-infected cells. These cell types interact with each other, and both antibody production and cytotoxic lymphocytes are maintained by long-lived "memory" cells that in many cases provide lifelong immunity against reinfection by a pathogen. Whereas the cells responsible for nonspecific innate immunity are broadly distributed in the blood, tissues, and organs throughout the body, pathogen-specific immune B and T lymphocytes are mostly localized in specialized organs of the lymphatic system.

THE INNATE IMMUNE RESPONSE

Complement proteins mark invading pathogens or infected cells for destruction

The term *complement* refers to a self-regulating cascade of over a dozen proteins that were originally shown to "complement" the specific antibacterial activities of antibodies by a **classical activation pathway.** A number of

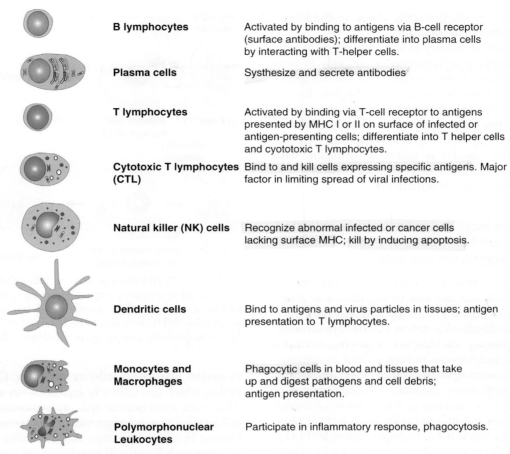

Figure 34.1 Cells involved in the innate and adaptive immune responses.

serum proteins, including **ficolins**, which bind to N-acetyl glucosamine, and **collectins**, which bind to mannose residues expressed on microbial membrane carbohydrates and lipopolysaccharides, activate complement by the **lectin receptor-activated pathway**. Finally, an **alternate pathway** that spontaneously activates complement and normally serves to eliminate senescent cells also can amplify complement activation by the classical and lectin pathways.

All mechanisms for activating complement result in the proteolytic cleavage of the critical third complement component, C3, into fragments C3a and C3b, and the attachment of C3b to the target cell or pathogen. C3b bound to a virus particle is recognized by receptors on the surface of inflammatory cells such as macrophages, increasing the efficiency of phagocytosis and leading to engulfment and destruction of the virus particle.

C3b also has an enzymatic activity that triggers the activation of the remaining cascade of complement proteins (C5–C9), resulting in the formation of a **membrane attack complex**. This creates a pore in the viral envelope that permits serum proteases and nucleases to enter the virus and destroy infectivity. Complement pore-forming activities may also create a hole in the surface of infected host cells or microorganisms to which antibody has bound.

The cleavage of complement components C3 and C5 also generates small complement fragments C3a and C5a, which are called **anaphylatoxins**. These complement fragments trigger tissue **mast cells** to release **histamine** and other mediators that augment blood flow and vascular leakage, and attract leukocytes to the site of infection. Some viruses such as Epstein–Barr virus directly interfere with complement-dependent functions.

The inflammatory response is mediated by cytokines and migrating leukocytes

The accumulation of leukocytes at sites of infection, the local activation of their antimicrobial functions, and the production of cytokines are collectively called the **inflammatory response**, because these activities lead to local inflammation of tissues. Nonspecific **phagocytosis** and cell-mediated cytotoxicity during an inflammatory response are important resistance mechanisms. Polymorphonuclear leukocytes and monocytes (Figure 34.1) can spontaneously phagocytose free virus particles or debris from virus-infected cells. These cell types, primarily found in the blood, are acute responders to infection.

→ Leukocytes + monocytes

Phagocytosis by leukocytes is greatly augmented by binding of complement fragments or antibody to virus particles; phagocytic cells have receptors that recognize these components on the surface of pathogens. By this means the host can recognize and destroy virus particles against which the host organism has already produced circulating antibody from a previous exposure. Uptake into phagocytic vesicles is followed by fusion with lysosomes and activation of lysosomal proteases and nucleases that degrade virus particles.

Macrophages localized in tissues are activated by infection and kill viruses or infected cells using toxic oxygen compounds

Specific antiviral antibodies bound to virus particles and virus-infected cells can also lead to engulfment or attack by macrophages (Figure 34.1), leading to cell lysis, often referred to as **antibody-dependent cellular cytotoxicity**. In contrast to phagocytic leukocytes, macrophages primarily reside in the body's tissues and organs, and are activated *in situ*. The activation of macrophages by interferons, particularly **interferon-γ**, induces a vigorous respiratory burst resulting in the generation of toxic forms of oxygen that kill microbes and infected cells: these include nitric oxide (NO), superoxide (O^-_2), hydrogen peroxide (H_2O_2), singlet oxygen (O·), and hydroxyl radical (OH·). These highly reactive oxygen species can oxidize envelope lipids and cause strand breaks in viral nucleic acids. Activated macrophages also release interleukin-12 (a mediator of T-cell differentiation), soluble protein **lymphotoxins**, and **tumor necrosis factors**, which induce apoptosis via specific cellular receptors (see Chapter 33, Figure 33.4).

(TNF)

Natural killer cells recognize virus-infected cells and kill them via apoptosis pathways

Natural killer (NK) lymphocytes are capable of selectively and rapidly killing virus-infected cells and cancer cells either directly or by recognition of "cell-surface bound antibody". NK cells constitute a minority population of circulating large, granulated lymphocytes. The activity of NK cells is augmented by **interferon-α**, which is one of the first cytokines produced by host cells in response to viral infection. NK cells appear to constitute a very important early defense mechanism for limiting the severity of viral disease.

NK cells can detect abnormalities in the expression of cell-surface proteins that may be altered by viral infection. NK cells carry several *activating receptors* that bind to cell-surface proteins expressed by proliferating cells (Figure 34.2). Binding to these proteins activates NK cells and can lead to a cytotoxic response. NK cells also express several dominant *inhibitory receptors* that bind

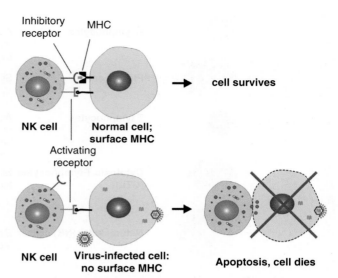

Figure 34.2 Cell killing by natural killer (NK) cells. NK cells recognize activating and inhibitory receptors on the surface of other cells. The absence of inhibitory receptors, for example MHC proteins, triggers release of NK cell factors which induce apoptosis in the target cell.

to **major histocompatibility complex (MHC)** molecules, which are normally expressed on all cell types. Therefore, most normal cells are protected against NK cell attack if they are expressing MHC proteins. It is the balance between activation and inhibition that decides whether an NK cell will attack and kill a particular cell within the body.

The major function of MHC proteins is to bind to peptides produced by degradation of antigens and to display them on the cell surface, leading to the activation of specific antiviral immune responses (see below). Infection by many viruses leads to reduced expression of MHC proteins on the cell surface, thereby inhibiting cellular immune responses. However, this reduced expression of MHC proteins can be recognized by NK cells, which are no longer inhibited from attacking the cell. NK cells will then secrete cytotoxic proteins, which attack and kill the infected cell.

One such cytotoxic protein is **perforin**, a pore-forming protein with some similarity to complement component 9. Perforin makes holes in the cell surface membrane as well as intracellular membranes. This facilitates the entry of a second cytotoxic protein, **granzyme B**, which activates **apoptosis**, or programmed cell death. Granzyme B is a protease that can activate both caspase-3, the "executioner caspase," as well as the pathway leading to release of mitochondrial proteins and the formation of the apoptosome (see Figure 33.4). Many viruses carry genes whose expression can suppress apoptosis in infected cells, thereby prolonging the virus replication cycle. The fate of the infected cell depends partly on the outcome of a battle between cellular mechanisms that lead to apoptosis and viral mechanisms that suppress it.

THE ADAPTIVE IMMUNE RESPONSE

Viruses and other infectious agents produce a variety of proteins and other macromolecules that are recognized by the host as foreign; these molecules can give rise to specific immune responses and are termed **antigens**. Viral structural proteins are often the most effective antigens, because they are displayed on the surface of the virion and are easily accessible to the protective elements of the immune system that prevent disease. Many nonstructural viral proteins are expressed in infected cells, and these can also be recognized by the immune system.

The immune response generates specific antibodies that recognize and bind to viral antigens. Circulating antibodies can bind to proteins on the surface of virus particles within the body and neutralize their infectivity. The immune response also produces clones of cytotoxic lymphocytes that recognize specific viral antigens expressed in infected cells and kill those cells, limiting and controlling the spread of the infection. This section will briefly describe how the adaptive immune system recognizes and responds to viral infections.

Primary and secondary organs of the immune system harbor B and T lymphocytes

The cells responsible for producing the immune response are lymphocytes found in blood, lymph nodes, spleen, bone marrow, and thymus (Figure 34.3). The bone marrow contains multipotent stem cell precursors for both the red and white blood cell populations, including lymphocytes. The bone marrow and thymus are **central or primary lymphoid organs** responsible for the production of naïve B lymphocytes (B for bone marrow) and T lymphocytes (T for thymus), respectively.

B and T lymphocytes circulate in the blood and populate the **peripheral or secondary lymphoid organs** (spleen, lymph nodes, gut), where the immune responses subsequently occur. The spleen primarily responds to microbial antigens in the central blood circulatory system, whereas the lymph nodes respond to infectious agents that enter the body through the skin and mucous membranes lining the oral cavity, the respiratory system, the digestive system, and the urogenital system. These infectious agents spread via extracellular fluids into the **lymphatic system**, a set of vessels that transports cells and microbes to the lymph nodes (Figure 34.4). Within the lymph nodes, B lymphocytes respond to foreign antigens by differentiating into **plasma cells**, which produce specific antibodies. A subpopulation of T lymphocytes within lymph nodes differentiates into cytotoxic cells that kill virus-infected cells.

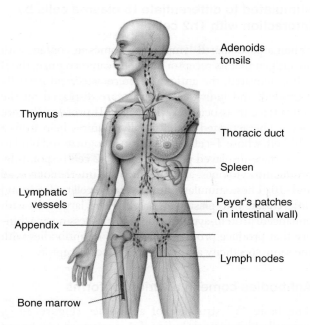

Figure 34.3 Anatomy of the human immune system. The primary lympoid organs are the bone marrow and thymus; the secondary lymphoid organs are the spleen, lymph nodes, Peyer's patches, adenoids, and tonsils.

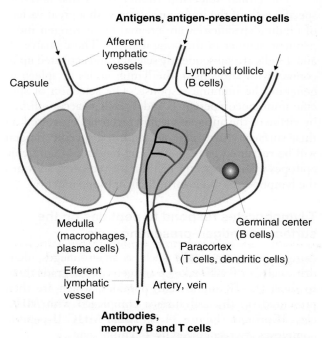

Figure 34.4 Structure of a lymph node. Lymph nodes are a major site of interaction between antigens and cells of the immune system. Naïve lymphocytes enter from blood via capillaries. Antigens and antigen-presenting cells enter via the afferent lymphatic vessels. Activation of B cells leads to formation of a germinal center and the production of plasma cells, which secrete antibodies. Antibodies and memory lymphocytes leave via the efferent lymphatic vessel.

B and T cells have specific surface receptors that recognize antigens

Both B and T cells express unique surface receptor molecules that can recognize and bind to specific parts of antigens called **epitopes**. These receptor molecules are produced by genes that rearrange during lymphocyte development to generate millions of different binding specificities; each individual B- or T-cell expresses many copies of a surface receptor that has a single binding specificity.

The **B-cell receptors** are membrane-bound antibodies. These antibodies bind to epitopes on intact proteins or other antigens with which B cells come in contact in the circulation. The B-cell receptor on each cell represents the specific antibody that this cell will produce upon stimulation after encountering an antigen to which it can bind.

The **T-cell receptors** bind not to intact proteins but to small peptide epitopes produced by intracellular digestion of proteins. These peptides are complexed with specialized presenting molecules called major histocompatibility complex (MHC) proteins, which are located on the surface of all cell types in the body (except red blood cells). Like the B-cell receptors, each T-cell receptor has binding specificity for a particular antigenic determinant.

The **clonal selection theory** states that antigen-specific cells of the immune system with a great variety of binding specificities are created in an antigen-independent manner in the bone marrow. These "naïve" B and T cells are subsequently selected and activated upon contact with antigens in the lymph nodes or spleen to generate the immune response. The selection of specific immunocompetent B and T cells is accomplished by virtue of the antigen-specific receptors expressed on their surface. Only cells that bind to a specific antigen will be recruited to respond. The binding of antigenic epitopes with the specific cell surface receptors activates the lymphocytes to perform their immune function.

T lymphocytes respond to peptides on the surface of antigen-presenting cells

Antigen-presenting cells such as macrophages, dendritic cells, or B cells take up antigens and degrade them to short (12–18 amino acid) peptides, which are then presented on the cell surface complexed with MHC class II protein (Figure 34.5). These MHC II-peptide complexes are recognized by T lymphocytes.

When a T-cell receptor comes in contact with an MHC II protein carrying a peptide to which it can bind, the T-cell will be stimulated to differentiate into active T-helper cells of two kinds, called Th1 and Th2 cells, or into precursors of cytotoxic T-cells, called Tc cells. Th1 cells help activate cell-mediated immunity, and Th2 cells help activate antibody production (see below).

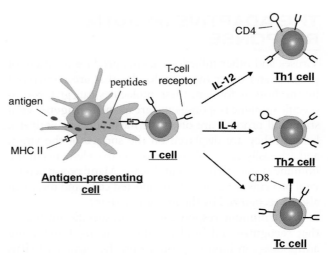

Figure 34.5 Generation of helper (Th) and cytotoxic (Tc) T lymphocytes. An antigen-presenting cell ingests proteins and displays peptide fragments on the cell surface in association with MHC II protein. Immature T cells whose T-cell receptor recognizes these peptides are activated and differentiate into Th1, Th2, or Tc cells, expressing CD4 or CD8 surface proteins.

Th cells express an MHC II recognition protein called CD4 on their surface, while Tc cells express an MHC I recognition protein called CD8. CD4 and CD8 proteins direct these different T-cell populations to interact with specific target cells displaying either MHC type II or type I proteins on their surface.

B lymphocytes respond to antigens and are stimulated to differentiate to plasma cells by interaction with Th2 cells

When a particular B lymphocyte comes in contact with an antigen that its receptor antibody can recognize, the B cell is activated, the antigen is internalized and partially degraded, and individual peptides are displayed on the cell surface in association with MCH II proteins (Figure 34.6). The activated B cell usually requires help from a Th2 cell, whose T-cell receptor can recognize and bind to the surface-displayed peptide. The Th2 cell responds by producing cytokines, among which are interleukins-4, -6 and -10. These stimulate the activated B cell to divide and differentiate into plasma cells, which are large cells with enhanced protein-synthesizing and processing machinery that produce prodigious amounts of antibodies with the same specificity as the original B-cell receptor.

Antibodies come in a variety of forms

The basic "Y" structure of antibodies (Figure 34.7), also called immunoglobulins (Ig), consists of two linked dimers, each containing a light chain and a heavy chain. The antigen-binding sites lie at the tips of the upper arms of the Y; these are the variable regions of both light

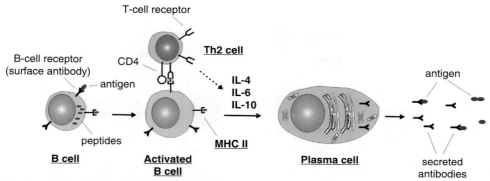

Figure 34.6 Antibody production by B cells. Proteins that bind to antibodies on the surface of immature B cells are taken up in the cell and digested; peptide fragments are presented on the surface of the B cell in association with MHC II protein. These cells are stimulated to differentiate into antibody-producing plasma cells when they come in contact with activated Th2 helper cells whose T-cell receptor shares specificity for the displayed peptide.

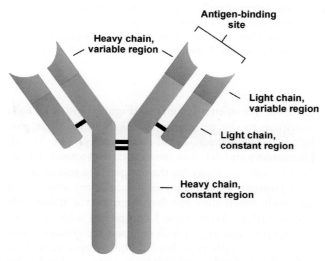

Figure 34.7 Structure of the immunoglobulin G molecule. Two dimers of a heavy chain and a light chain, joined by a disulfide bond, are themselves joined by disulfide bonds to form an IgG molecule. The antigen-binding site is constituted by the N-terminal variable regions of the heavy and light chains. The constant region of the heavy chains determines the immunoglobulin isotype, and is responsible for binding to complement, macrophages, etc.

and heavy chains, which are responsible for the millions of distinct binding specificities of antibodies. The antigen-binding site is capable of binding to an antigenic determinant with a molecular weight of between 500 and 1500 Da. Therefore the binding site can accommodate an epitope or antigenic determinant of about six sugar molecules or thirteen amino acid residues. The remaining parts of both light and heavy chains have invariant or constant amino acid sequences.

Five different types of immunoglobulin molecules (IgM, IgD, IgG, IgE, and IgA) are synthesized by separate clones of B lymphocytes; each of these **isotypes** has its own individual functions in the adaptive immune response.

The large IgM macroglobulin is the first antibody produced in an immune response. IgM is short-lived, with a half-life of about 3.5 days. It is most effective at agglutinating particulate or cellular antigens in the serum, activating complement for cell lysis, and binding to antigens to mark them for phagocytosis and phagosomal degradation. IgM is a pentamer of the basic immunoglobulin structure, and contains a 15-kDa J chain that joins the ringed structure.

The IgD molecule appears on the surface of immature and naïve B-lymphocytes simultaneously with IgM, and IgD may act as an antigen receptor.

The IgG molecule is the most abundant immunoglobulin. It is long-lived, with a half-life of about 25 days. IgG is involved in agglutination of cellular antigens, neutralization of toxins and viruses, and binding to bacteria and virus particles, leading to their uptake and destruction by phagocytosis. There are actually four subclasses of IgG.

IgA is involved in providing immune defenses in the mucous membranes of the gut, respiratory and urogenital tracts. It is a dimer of the basic immunoglobulin "Y" structure, with the addition of a joining J chain and a secretory protein (molecular weight 60,000) added during its secretion through the mucosal epithelium.

IgE mediates allergies by virtue of its ability to attach to tissue basophils known as mast cells. The combination of antigen with IgE activates the mast cells, causing them to degranulate and release histamine, heparin, and arachidonic acid metabolites that cause the symptoms of allergies.

The enormous diversity of antibody specificities

Antibody specificity is a property of the N-terminal variable amino acid sequences of the immunoglobulin proteins. How does the immune system provide for antibody diversity? Immunoglobulin diversity is

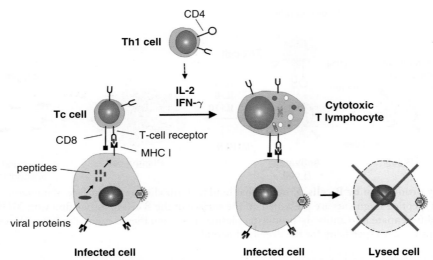

Figure 34.8 Generation of cytotoxic T lymphocytes and killing of virus-infected cells. Peptides derived from digested viral proteins are displayed on the surface of virus-infected cells in association with MHC I protein (left). When Tc cells come in contact with infected cells, they are activated if their T-cell receptor recognizes these peptides. Activated Th1 cells (top) help to activate Tc cells by production of interleukin-2 and interferon-γ. The result is cytotoxic T lymphocytes (right), which recognize and kill infected cells displaying the peptide in question.

explained by the existence of a large library of variable region exons within the genes that code for both the heavy chain and light chains. These exons are randomly selected and combined by a process of somatic recombination during lymphocyte differentiation and development. Somatic mutation of these regions during immune responses further increases the specificity and binding affinity of the antibodies.

When a specific lymphocyte clone is activated with the help of Th2 cells (Figure 34.6), it first produces IgM, and later the other immunoglobulin isotypes. Th2-cell dependent isotype switching in B cells requires recombination of the assembled specific variable gene sequence of the heavy chain from one constant gene sequence to the next (e.g., IgM to IgG to IgE to IgA). This occurs by cleavage of the DNA, removal of the region coding for the constant region of the heavy chain, and joining of the upstream segment coding for the variable region to another downstream constant region. Isotype switching is also essential for increasing the affinity of antibodies (via further somatic mutations in the variable regions) and for production of long-term immune memory cells.

Cytotoxic T cells are generated upon interaction of Tc cells with MCH I-bound peptides

Cell-mediated immune reactions are carried out by T lymphocytes and target host cells infected by intracellular bacteria and viruses. Tc cells are generated by the encounter between an antigen-presenting cell and a naïve CD8-positive T cell (Figure 34.5). Tc cells are primed to respond when their T-cell receptor binds to a peptide displayed in association with a MCH I protein

on the surface of an infected cell (Figure 34.8). Th1 cells contribute to the maturation of Tc cells via their production of interleukin-2 and interferon-γ. The Tc cell differentiates and develops into an active cytotoxic T lymphocyte with specificity for the antigen first encountered.

Cytotoxic T lymphocytes have the capacity to kill cells with which they come into contact if those cells express surface MHC type I complexed with the specific peptide that their T-cell receptor recognizes. Thus virus-infected cells that display viral peptides on their surface in a complex with MHC type I proteins will be killed when they encounter cytotoxic T lymphocytes that recognize those peptides. Cytotoxic T lymphocytes also cause the rejection of organ transplants, can kill cancer cells, and can participate in autoimmune reactions that cause diseases such as thyroiditis, diabetes, systemic lupus erythematosus, and perhaps multiple sclerosis. Some viruses such as herpes simplex virus type 1 and cytomegalovirus can interfere with MHC-I expression and peptide presentation, effectively blocking viral antigen expression on infected cells and preventing recognition by Tc cells.

EFFECTS OF INTERFERONS ON THE IMMUNE RESPONSE

Interferons stimulate antigen processing and presentation

Interferon promotes protective immune responses by its ability to increase the expression of major histocompatibility complex proteins. All interferon family members share the ability to enhance the expression of MHC class I proteins and promote the development of CD8-positive cytotoxic T cells. In addition, interferon-γ is uniquely able to induce the expression of MHC class II proteins and therefore to promote enhanced CD4-positive T-helper cell responses leading to enhanced production of cytotoxic Tc cells.

Interferons also play an important role in antigen processing by regulating the expression of proteins required to generate antigenic peptides. Interferon modifies the activity of several components of **proteasomes**, which digest viral proteins and make peptides available to MHC class I proteins. This action enhances the immunogenicity of viral proteins by increasing the quantity and repertoire of peptides displayed in association with MHC class I proteins.

Interferons and the development of CD4-positive helper T cells

The distinct helper T-cell populations are defined by the cytokines they produce after stimulation. Th1 cells synthesize interferon-γ and interleukin-2, and promote cell-mediated immunity and the cytotoxic response (Figure 34.8). Th2 cells synthesize interleukins-4, 5, 6, and 10, and thereby facilitate antibody production by B cells (Figure 34.6). Interferon has an important effect on Th1 cell development and plays a dual role in this process. First, it facilitates Th1 cell production by enhancing the synthesis of interleukin-12 in antigen-presenting cells (Figure 34.5). Second, it blocks the development of Th2 cells by inhibiting interleukin-4 production, which is required for Th2 cell formation. Thus interferon directs the immune system to generate a cytotoxic response, which is most effective in suppressing viral infections, as infected cells are killed and eliminated.

The role of interferon in macrophage activation and cellular immunity

Macrophages function as key effector cells in innate and adaptive immune responses. To carry out these functions, macrophages must first become activated, a process that enhances cytocidal activities. Interferon-γ alone is able to stimulate macrophage activation. Activated macrophages use a variety of interferon-induced mechanisms to kill microbial targets. Two of the most important killing mechanisms involve the production of reactive oxygen and reactive nitrogen intermediates. For example, nitric oxide is generated in macrophages as a result of the interferon-dependent induction of the inducible form of nitric oxide synthase (iNOS).

Effects of interferons on antibody production

Interferons play complex roles in regulating antibody production since they exert their effects either indirectly, by regulating the development of specific T-helper cell subsets (as described above), or directly at the level of B cells. The best-characterized action of interferons on B-cell activation is their ability to influence immunoglobulin (Ig) isotype switching. For example, interferon-γ can switch production from IgM to IgG2a and IgG3 in stimulated B cells. In this fashion, interferons modulate the immune response and increase host defenses against certain intracellular bacteria and viruses.

VIRUS STRATEGIES TO COUNTER HOST DEFENSES

Viruses make proteins that mimic cytokines and cytokine receptors and interfere with host defenses

Several large DNA viruses, including herpesviruses and poxviruses, possess genes coding for viral proteins that mimic cytokines or cytokine receptors. These proteins are called **virokines** and **viroceptors**, and they combat host immune responses against these viruses. Genes for virokines and viroceptors are almost certainly derived from host cell genes that were incorporated into viral genomes and evolved by mutation and selection to confer an advantage in virus replication, spread or survival. Th2 virokines can alter the host Th1 immune response by preventing the synthesis or release of Th1 cytokines from infected cells, competing for host cytokine receptors, interfering with cellular signal transduction pathways, or inducing host responses that are advantageous to virus replication.

Viral epidermal growth factor (vEGF) is a virokine produced by vaccinia virus, a member of the poxvirus family. vEGF has amino acid sequence homology to epidermal growth factor (EGF) and transforming growth factor alpha (TGF-α). It binds to the cellular epidermal growth factor receptor, and plays an important role in stimulating the growth of cells, resulting in increased virus replication.

Other examples of virokines include homologs of interleukin-6, 8, 10, and 17, and secreted serine proteases. Two other virokines are the G protein of respiratory syncytial virus, which mimics the chemokine fractalkine, and the Sis protein of simian sarcoma virus, which is a homolog of platelet-derived growth factor.

MT7 is a viroceptor produced by myxoma virus, a rabbit poxvirus; it is a homolog of the interferon-γ receptor. MT7 is rapidly secreted in high levels from virus-infected cells. It binds to and sequesters interferon-γ, and therefore inhibits binding of interferon-γ to host cell receptors, thus reducing the Th1 immune response to viral infection. Poxviruses and herpesviruses also make G protein-coupled receptors that bind to a variety of chemokines and prevent the recruitment of immune cells to the site of infection. Collectively, these proteins are known as **virulence factors**, as they are important for the virus to establish and maintain an infection within its host.

Viruses evade innate immune responses

Viruses have evolved many strategies to block innate immune proteins, pathways, and cells. As one of the first lines of defense against incoming viruses, the complement system is a centerpiece of innate immunity. Viruses have developed many ways to evade complement actions. One strategy is to prevent complement binding to antibody-antigen complexes. Several herpesvirus glycoproteins cause the shedding or internalization of viral protein-antibody complexes, while the spike protein of some coronaviruses displays receptor activity for the Fc (constant) portion of immunoglobulins. Herpesviruses and poxviruses make a number of proteins that are similar to complement control proteins. These viral mimics bind and sequester, down-regulate or increase the decay of critical host complement activators such as C3b, C4b, and C5b.

Viruses have also learned to protect themselves by "stealing" complement-regulating proteins that healthy mammalian cells use for protection against complement. Members of the poxvirus, herpesvirus, retrovirus, and togavirus families incorporate complement-regulating proteins such as CD55 (C3b decay accelerating factor), CD59 (Protectin, an inhibitor of the assembly of the membrane attack complex) and CD46 (a cofactor that increases proteolysis and degradation of complement components) into their envelopes. The herpesvirus cytomegalovirus causes infected cells to express increased CD55 and CD46 on their surface.

In addition to evading the antiviral effects of complement, viruses inhibit the activity of cells involved in innate resistance, such as natural killer (NK) cells and dendritic cells. NK cells express both inhibitory and activating receptors to recognize and kill virus-infected cells and avoid killing healthy cells (Figure 34.2). Cytomegalovirus produces a non-functional MHC class I decoy that binds to the inhibitory receptors on NK cells, preventing NK-cell activation. Cytomegalovirus also makes proteins that either down-regulate or sequester host signaling molecules that activate NK cells. Poxviruses evade activating receptors of NK cells by secreting soluble competitive inhibitors of these receptors.

Human immunodeficiency virus is capable of infecting and killing both NK cells and CD4-positive T cells. Since CD4-positive T cells produce interleukin-2, a cytokine required for NK cell function, the loss of these cells indirectly affects NK cell antiviral activity. Effective antiviral responses also require interactions between NK cells and dendritic cells. Human immunodeficiency virus and herpesviruses prevent these interactions by (1) blocking NK cell function, (2) preventing the expression of cytokines that activate dendritic cells, (3) killing specific functional subsets of NK cells, and (4) changing the expression of specific NK cell receptors and coreceptors. Hepatitis C virus blocks interactions between NK cells and dendritic cells by inhibiting the production of dendritic cell proteins that are required for binding to NK cells.

Herpesviruses directly evade the antiviral function of dendritic cells by interfering with antigen presentation and processing. In human cytomegalovirus, at least six proteins interfere with antigen presentation by preventing the processing of viral proteins, by down-regulating or sequestering MHC class I and II proteins, or by blocking the loading of viral peptides onto MHC proteins. Human immunodeficiency virus takes a different approach; it prevents the maturation of dendritic cells and induces the expression of interleukin-10, a cytokine with immune suppressive function.

Viruses evade adaptive immune responses

Viruses evade both the humoral and cell-mediated arms of the adaptive immune response. Viruses often escape humoral immunity by changing the proteins recognized by antibodies. In a process called **antigenic drift**, a number of viruses can change epitopes on viral structural proteins with mutations in regions normally recognized by antibodies. It is well known that human immunodeficiency virus uses this strategy. If serum and virus are taken from an HIV-infected individual, the antibodies in the serum cannot neutralize the virus that is currently present. However, these antibodies can neutralize virus samples harvested several months previously. Filoviruses modify the sugars on virion glycoproteins, preventing recognition by B-cell and T-cell receptors.

Due to its segmented genome, influenza virus can evade humoral immunity via **antigenic shift**. Two different strains of influenza virus replicating within the same cell can recombine genome segments, in a process called **reassortment**, to form a new strain that contains a mixture of the genome segments from the two original strains. The reassortant virus can therefore express a new surface antigen that is not recognized by antibody in its natural host (see Chapter 18).

Viruses can escape the cell-mediated immune response by blocking almost every stage of MHC class I antigen presentation. For example, herpesvirus proteins with repeats of acidic amino acids escape processing by the proteasome, a cytoplasmic processing unit that generates peptides for MHC I presentation. Multiple herpesvirus proteins also block the transporter associated with antigen processing (TAP) complex, preventing peptide transport into the endoplasmic reticulum, where peptide loading of MHC-1 occurs. Although the adenovirus protein E3-19K binds to TAP, it does not prevent peptide transport. Instead, it prevents formation of the peptide-loading complex within the endoplasmic reticulum.

A cowpox virus protein, V203, has an endoplasmic reticulum-retention signal and is able to retain MHC I within the endoplasmic reticulum. The human immunodeficiency virus Nef protein disrupts the trafficking of MHC I. In addition, many viral proteins cause the down-regulation or degradation of cell-surface proteins such as MHC class I and II, CD4, and CD8.

KEY TERMS

Alternate pathway

Anaphylatoxins

Antibodies

Antibody-dependent cellular cytotoxicity

Antigenic drift

Antigenic shift

Antigens

Apoptosis

B lymphocytes

B-cell receptors

Central or primary lymphoid organs

Classical activation pathway

Clonal selection theory

Collectins

Complement

Cytotoxic T lymphocytes

Epitopes

Ficolins

Granzyme B

Histamine

Immune response

Inflammatory response

Interferon-α

Interferon-γ

Intrinsic cellular defenses

Isotype

Lectin receptor-activated pathway

Lymphatic system

Lymphotoxins

Macrophages

Major histocompatibility complex (MHC)

Mast cells

Membrane attack complex

Monocytes

Natural killer (NK) cells

Perforin

Peripheral or secondary lymphoid organs

Phagocytosis

Plasma cells

Polymorphonuclear leukocytes

Proteasomes

Reassortment

T lymphocytes

T-cell receptors

Tumor necrosis factors

Virokine

Viroceptor

Virulence factors

FUNDAMENTAL CONCEPTS

• Hosts respond to viral infections with four layers of defense: physical barriers, intrinsic cellular defenses, innate resistance, and immune responses.

• The host immune response is composed of a rapid, non-specific innate response, and a slower, highly specific adaptive immune response.

• The complement cascade consists of a set of serum proteins that recognize and bind to pathogens, helping antibodies and phagocytic cells to clear virus or virus-infected cells from the body.

• Cytokines produced by infected cells work in concert with leukocytes, macrophages, and natural killer cells to recognize and kill virus-infected cells.

• B and T lymphocytes are produced in primary lymphoid organs (bone marrow and thymus) and carry out their adaptive immune functions in secondary lymphoid organs (spleen, lymph nodes, gut).

• Dendritic cells, macrophages, and B cells engulf antigens, process the proteins and present the peptides to T lymphocytes on cell surface MHC membrane proteins.

• Within infected cells, viral proteins are processed into small peptides, which are presented on the surface of infected cells by MHC proteins for recognition by B and T lymphocytes.

• Activated T lymphocytes can secrete cytokines that stimulate B lymphocytes to develop into plasma cells.

• Plasma cells produce antibodies, or immunoglobulins, which bind to viral proteins and either prevent the virus from infecting cells or target the virus for degradation or phagocytosis.

• Activated T lymphocytes can develop into cytotoxic T cells that recognize and kill virus-infected cells.

• Viruses have developed many different strategies to counteract innate and adaptive immune responses.

REVIEW QUESTIONS

1. For phagocytic cells to "see" a foreign organism they must recognize specific molecules. What are some examples of these molecules, and how are they recognized?

2. Although antibodies can be effective in destroying some invading viruses, a cellular response mediated by cytotoxic T cells is usually required. Which cell–cell interaction is required for generation of antigen-specific T cells, and how does this interaction proceed?

3. What is meant by the statement that B cells "see" antigens and T-cells "see" peptides?

4. Naïve B cells express two immunoglobulin isotypes on their surface that may serve as antigen receptors. Which immuglobulin isotypes are these, and what is their structure?

5. Compare antigenic shift and antigenic drift.

ANTIVIRAL AGENTS AND VIRUS VECTORS

These three chapters deal with practical applications of virology that protect humans or animals against virus infection, help the organism recover from an infection, or modify viruses to be used as vectors that express specific genes for gene therapy or as antitumor agents.

Vaccines are now more than 200 years old, as the first successful antiviral vaccine was shown by Jenner in 1796 to prevent smallpox. Vaccine technology has progressed rapidly, and there are now dozens of effective antiviral vaccines. Chapter 35 discusses the history and the technology of vaccine development, and outlines the advantages and problems of different kinds of viral vaccines: live attenuated viruses, killed whole viruses, subunit vaccines, DNA vaccines, and recombinant vaccines. Vaccination campaigns against diseases that are highly transmissible and are therefore public health risks can be unpopular, and the ethics of requiring vaccination with products that are never 100% safe are discussed.

Antiviral chemotherapy is a comparatively recent science that has progressed rapidly with the discovery of effective agents against HIV, once thought to lead inexorably to disease and death. Scientific understanding of steps in viral replication and the availability of automated tests to detect antiviral activity of thousands of compounds have led to significant breakthroughs. Compounds that block virus attachment, uptake, uncoating, DNA or RNA replication, assembly, and release are now available for use against specific viruses.

A number of viruses have been adapted for use as vectors. This involves deleting certain viral genes and inserting into the viral genome an expression cassette that contains control sequences and foreign genes to be expressed. Propagation of the resulting vector usually requires coinfection with a wild-type virus or expression of missing viral proteins by the infected cell. Chapter 37 discusses vectors made from adenoviruses, retroviruses, and adeno-associated viruses. Gene replacement therapy for genetic diseases has seen some recent success, and a variety of virus vectors directed against human cancers are being developed.

Antiviral Vaccines

Brian Ward

Vaccine

From Latin *vacca* (cow), because vaccinia virus was thought to be cowpox virus

Vaccination is the most effective means available for prevention of viral infections.

Smallpox and rabies vaccines were groundbreaking developments in medical history.

Serious adverse effects following immunization are rare but sometimes important.

Requiring vaccination when no vaccine is 100% safe raises ethical issues.

MODERN VACCINE PRODUCTION

Virus production in cultured cells or embryonated eggs.

Sophisticated virus purification and detection methods.

Genetic engineering to modulate virus pathogenicity, infectivity, or immunogenicity.

Production of viral proteins in bacteria, yeast, or eukaryotic cell expression systems.

Production of viruses, virus-like particles, or viral proteins in plants.

TYPES OF ANTIVIRAL VACCINES

Live wild-type viruses that infect other animals but are attenuated in humans.

Live attenuated human viruses with mutations that reduce virulence.

Inactivated viruses: retain immunogenicity but are no longer infective.

Subunit vaccines: contain one or more viral proteins but no viral genome.

DNA vaccines: express immunogenic viral proteins but cannot generate virus.

Recombinant viruses engineered to incorporate one or more immunogenic proteins

Chimeric vaccines: benign or attenuated microbes (viruses or bacteria) that express specific immunogenic viral proteins.

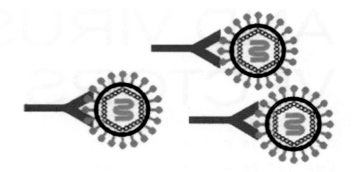

Nanoparticle vaccines delivering one or more viral proteins in virus-like particles, virosomes, proteosomes, etc

LIVE ATTENUATED VIRUS VACCINES

Naturally stimulate both cell-mediated and humoral immunity.

Are among the most successful antiviral vaccines.

Can give long-lived immunity after even a single dose.

Can occasionally cause serious illness in certain individuals.

Can be expensive to make and store.

INACTIVATED VIRUS VACCINES

Often stimulate only humoral immunity if administered without adjuvants.

Can induce immunity to many different antigens simultaneously.

Usually require multiple doses to achieve protection.

Are incapable of replication and therefore do not cause disease if properly made.

Are expensive to manufacture and may require the handling of highly pathogenic viruses

Are less sensitive to storage conditions than live attenuated viruses.

Are safer for immunocompromised individuals.

SUBUNIT VACCINES

Usually stimulate only humoral immunity if administered without adjuvants.

Can induce immunity to many different antigens simultaneously.

Usually require multiple doses to achieve protection.

Present no danger of inadvertent infection with live virus.

Are cheaper to manufacture and test, and can be very stable.

Avoid delivery of "irrelevant" and potentially immunopathologic viral antigens.

Are safer for immunocompromised individuals.

DNA VACCINES

Can be engineered to stimulate both cell-mediated and humoral immunity.

Share other advantages of subunit vaccines.

May involve risks because foreign genes are expressed in the body.

A BRIEF HISTORY OF ANTIVIRAL VACCINES

Viral pathogens have been serious threats to health and society throughout the evolution of *Homo sapiens*. The horror of watching an individual die from rabies virus encephalitis or Ebola virus hemorrhagic fever is surely the same today as it was for our hunter-gatherer ancestors. However, these deadly **zoonotic** (animal-borne) virus infections have had relatively little impact on human civilization due to their local distribution and their low capacity to spread from person to person.

In contrast, widespread virus **epidemics** that are spread directly from person to person and are highly contagious, such as smallpox, influenza, and measles, have been limited only by population size and the speed of transportation. The incubation periods of these pathogens are all relatively short (7–10 days) and the diseases run their course to recovery or death quite rapidly. When transportation of humans was restricted to foot, hoof, and oar, and population centers were small and dispersed, travelers infected in one city were either dead or had recovered and were immune by the time they reached the next city. With population growth, the great trading cities of the Middle Ages became large enough to maintain foci of these diseases year round. Also, faster sailing vessels ensured periodic delivery of unwanted viral passengers to all parts of the known world.

From the middle of the thirteenth century until recent times, human population growth has been slowed and occasionally reversed by waves of destruction wrought by viruses such as measles, yellow fever, and smallpox. As much as 50% of the population of Europe is estimated to have been killed by these agents between 1300 and 1700 A.D. The indigenous populations of the Americas were decimated at least in part by epidemics of viral diseases in the 100 years following the arrival of Europeans in the mid-sixteenth century. The terrible reality of these viruses and the impact of vaccines are readily apparent from the data shown in Table 35.1.

Edward Jenner and his European counterparts are widely credited with discovering vaccination in the late 1700s (see Chapter 26). However, attempts to confer active immunity using microbial products have been documented in many cultures extending back at least one to two centuries before Jenner's birth. For example, Chinese physicians of the sixteenth century attempted active immunization against smallpox using dried scabs in various forms, soiled undergarments, and even powdered fleas fed on scabs!

Many cultures in the Middle East have long recognized the durability of immunity to leishmania parasites after natural infection. Children were intentionally inoculated with fully virulent organisms at inconspicuous places on the body (e.g., lower back, buttocks) to prevent subsequent, and less controlled, naturally occurring leishmania infections in more visible locations. Jenner and his contemporaries were nevertheless the first to recognize the full potential of active immunization and to systematically study vaccines and vaccination.

Table 35.1 Impact of antiviral vaccines on disease frequency: pre-vaccination era vs. 1998 (USA)

Disease	Dates	Annual reported cases: Pre-vaccination	Post-vaccination (1998)	Decrease (%)
Smallpox	1900–04	48,164	0	100
Poliomyelitis (paralysis)	1951–54	1,314	1*	100
Measles	1958–62	503,282	89	100
Mumps	1968	152,209	606	99.6
Rubella	1966–68	47,745	345	99.3
Congenital rubella syndrome	1958–62	823	5	99.4

Single case of vaccine-associated paralytic poliomyelitis.
From *Morbidity and Mortality Weekly Report 1999*, **48**, *243–248.*

The fruits of their labors have fundamentally changed human history. Some milestones in vaccine development are presented in Table 35.2.

Early vaccine technology was crude but effective

Like all medical interventions, the development of new vaccines has been limited by the availability of appropriate technologies (Table 35.3). Nowadays we may recoil at the idea of spraying powdered scabs up the nose, but the inevitability and destructiveness of smallpox went a long way toward easing any squeamishness about the purity or the origins of early vaccines. It would be hard to imagine people lining up today to be inoculated with crudely purified pus from cowpox lesions or dried spinal cord from rabbits infected with rabies.

Early in the history of vaccine development there were occasional setbacks due to contaminated vaccine lots (e.g., infections from bacterial contamination of vaccinia-induced pustules) and serious vaccine-associated side effects (e.g., encephalomyelitis following rabies vaccine). Many of the technologically crude vaccines made from these "natural products" actually worked well, and their popularity spread. However, public acceptance was far from universal.

Modern vaccine developers could learn a great deal from the showmanship of vaccine pioneers like Louis Pasteur, who arranged public demonstrations of his anthrax and rabies vaccines in the early 1880s. Pasteur's most potent ally in convincing Parisians to try vaccines was fear. Injection of spinal cord material from a rabid rabbit would certainly be less intimidating to someone who had just been bitten by a rabid wolf. Dried pus from a smallpox or cowpox lesion would likewise seem quite benign in the setting of a smallpox epidemic with 20 to 25% mortality.

Table 35.2 Milestones in antiviral vaccine development

Pre-1700s	Chinese doctors use powdered smallpox scabs to "immunize" intranasally. Mediterranean-area doctors use directed leishmania-infected sandfly bites to induce long-term protection from reinfection.
1721	Lady Montagu brings concept of variolation (inoculation with pus from recovering smallpox victim) from Turkey to England.
1798	Jenner publishes *Variolae Vaccinae*, the use of cowpox inoculation to protect against smallpox.
1885	Pasteur and collaborators introduce air-dried rabbit spinal cord as rabies vaccine.
1900	Walter Reed demonstrates that yellow fever is caused by a filterable virus.
1930–45	Introduction of vaccines for Japanese B encephalitis (1930), yellow fever (1935), and influenza (1936).
1946–75	Introduction of vaccines for polioviruses types 1–3 (Sabin attenuated strains and Salk inactivated virus); measles, mumps, and rubella viruses; tick-borne encephalitis virus; mouse brain, duck embryo, and tissue culture vaccines for rabies virus; inactivated influenza A and B viruses; and adenoviruses.
1975–present	Introduction of vaccines for hepatitis B virus, hepatitis A virus, varicella zoster virus (chickenpox), live, cold-adapted influenza virus, rotavirus, and human papillomavirus.

Table 35.3 Technological advances supporting antiviral vaccine development

1850	Germans add glycerin to "cow lymph" to stabilize and decontaminate smallpox vaccine.
1879	Galtier succeeds in passaging rabies from rabbit to rabbit.
1884	Metchnikoff descibes role of phagocytes in immunity.
1885	Pasteur and colleagues fix biological characteristics of rabies by passage in rabbit brains and discover that virulence in spinal cord is lost with drying.
1887	Petri describes overlapping glass plates to use for sterile cultures.
1897	Loeffler and Frosch describe "filterable viruses" (foot-and-mouth disease virus).
1931	Goodpasture first uses embryonated eggs to culture viruses.
1936	Stanley isolates nucleic acids from tobacco mosaic virus.
1950	Enders, Weller, and Robbins launch systematic study of cell culture.
1953	Description of first immortalized cell line (HeLa).
1960	Production of reassortant influenza A viruses in eggs.
1972	First DNA cloning performed.
1974	First recombinant proteins expressed in bacteria.
1985	Hepatitis B surface antigen gene cloned and expressed.

Embryonated chicken eggs and cell culture played major roles in vaccine development in the twentieth century

The technological leaps that had the greatest impact on antiviral vaccine development in the twentieth century were (1) serial passage of human viruses in animals; (2) the adaptation of viruses for growth in embryonated chicken eggs; and (3) the growth of viruses in vertebrate cells cultured *in vitro*.

As late as the 1880s, pioneering microbiologists were restricted to growing "germs" either in whole animals or on potato slices and in jars containing broth. In 1881, Robert Koch introduced gelatin and agar-based culture systems and, in 1887, R. Petri invented the overlapping glass plates that have immortalized his name. These innovations set the stage (in 1931) for the first use of eggs to culture viruses; an egg can be thought of as a sterile tissue culture flask with a calcium shell. Early vaccines made possible by growth in embryonated eggs include the yellow fever virus strain 17D (1932) and influenza virus (1936).

In the early 1950s, John Enders, Thomas Weller, and Frederick Robbins began the first systematic study of cell culture, using tissues from children who had died at birth. The use of such **primary cell cultures**, and the subsequent establishment of immortalized cell lines such as HeLa, dramatically accelerated the development of antiviral vaccines in the late 1950s and early 1960s. The achievements of this era include improved vaccines for smallpox and rabies as well as poliovirus, measles, mumps, and rubella vaccines.

Production of vaccines against avian influenza strains has been problematic

Nevertheless, the following story illustrates how technological limitations persist to modern times. Prior to the sudden emergence of the porcine H1N1 pandemic influenza virus in Mexico in mid-2009, the principal focus of the world's influenza experts had been avian influenza. Between 1997 and 2009, many parts of the world experienced a major **panzootic** (widespread epidemic in animals) of H5N1 avian influenza strains. These viruses were pathogenic in domestic and wild birds and led to hundreds of millions of bird deaths in South and East Asia either naturally or through culling of infected stock. Although relatively few cases of H5N1 disease occurred in humans, human mortality rates were very high (33–70%).

During this decade of active spread of H5N1 virus in animals and birds, there was increasing concern that mutation of these viruses could lead to human-to-human spread and a major influenza **pandemic** (widespread human epidemic). When vaccine manufacturers attempted to make the first H5N1 vaccines, they ran into a serious technical problem. Influenza vaccine production depended on a single technology: the creation of reassortant vaccine-strain virus (see Chapter 18), and

its amplification in embryonated eggs. Unfortunately, the H5N1 viruses were highly pathogenic for avian cells and rapidly killed the embryos before the required reassortant viruses could be produced. As a result of this technological limitation, it was initially quite difficult to make candidate vaccines against these dangerous viruses.

Had a virulent and transmissible H5N1 virus "emerged" while the manufacturers were struggling to overcome this problem, the results could have been devastating. At the time of writing, the major influenza vaccine manufacturers are still heavily dependent on embryonated eggs. However, several vaccines produced with alternate technologies are either available or rapidly approaching the market. These include the production of virus or noninfectious virus-like particles in tissue culture cells.

TYPES OF ANTIVIRAL VACCINES

Antiviral vaccines can be grouped into several categories (Table 35.4 and Figures 35.1 and 35.2):

1. *Live wild-type viruses.* These are the simplest vaccines to make and are among the most effective. A virus that infects one animal species can sometimes

Table 35.4 Classification of currently available human antiviral vaccines

Live wild-type viruses	Vaccinia (cowpox)[a]
Live attenuated viruses	Adenovirus[b]
	Influenza A (cold adapted)
	Measles
	Mumps
	Polio (Sabin)
	Rotavirus (human-recombinant)
	Rubella (german measles)
	Varicella (chickenpox)
	Yellow fever
Whole inactivated viruses	Hepatitis A
	Influenza A
	Influenza B
	Polio (Salk)
	Rabies
	Tick-borne encephalitis
Subunit vaccine	Hepatitis B surface antigen
Virus-like particles	Human papillomavirus
Chimeric virus	Rotavirus (human-bovine)

[a]*Smallpox eradicated (supply of vaccine tenuous).*
[b]*Military use only.*

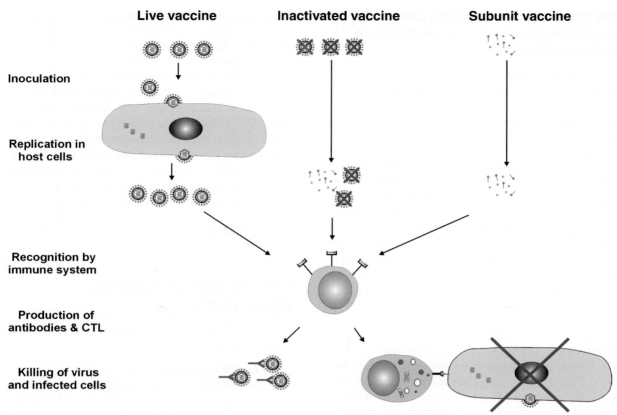

Figure 35.1 Major types of antiviral vaccines. Live attenuated viruses, inactivated whole viruses, or purified viral proteins ("subunits") are inoculated into the host. Live viruses replicate in host cells and produce progeny virus. Both progeny virus and viral proteins made within infected cells are recognized by the immune system, which generates specific antibodies and cytotoxic T lymphocytes (below). Inactivated virus and subunit vaccines do not replicate in the host, but are recognized by the immune system. Because they do not replicate, these vaccines are usually less efficacious, are often administered with adjuvants, and may require booster doses.

be used to infect another species to protect against a closely related pathogen. Viruses are often adapted to replicate well only in their natural host; they can infect other species, but replicate poorly and have limited or no pathogenic effects. However, these viruses may share **immunogenic** determinants with the target virus, resulting in an effective immune response against that virus. At this point, vaccinia virus is the only such viral vaccine licensed for use in humans, and in its present form it is no longer being used because smallpox has been eradicated.

2. *Live attenuated viruses.* Serial passage of pathogenic viruses in animals, eggs, or cells grown *in vitro* often results in the acquisition of a variety of mutations that reduce the pathogenic properties of the virus in its natural host. Although there is often little knowledge of the mechanisms of attenuation in these "black box" viruses, these are among the most successful vaccines currently used. These attenuated viruses replicate in humans and therefore can often induce protective immune responses as effectively as the parent viruses, while causing little or no disease. Table 35.5 illustrates how two such vaccines were produced.

The recently introduced cold-adapted influenza A virus vaccine is a "second-generation" live attenuated virus that was designed with specific characteristics. Cold-adapted influenza vaccines are produced by combining the H and N genes of the targeted virus strain with six other gene segments from mutant viruses known to have restricted growth at 37°C. The H and N genes code for the envelope glycoproteins hemagglutinin and neuraminidase, which are the most immunogenic influenza virus proteins. The reassortant viruses produced cannot replicate in the lung at core body temperature, but grow well in the cooler nasal mucosa, where they elicit an excellent immune response without endangering the vaccinated individual.

3. *Inactivated viruses.* This is a common route for production of vaccines when a live virus vaccine is not available. Pathogenic viruses are grown in eggs or in cultured cells and purified. The purified virus is then subjected to inactivation by treatment with formaldehyde, organic solvents, detergents, or high temperature. When carefully controlled, these treatments preferentially inactivate the virus genome and

Recombinant vaccines

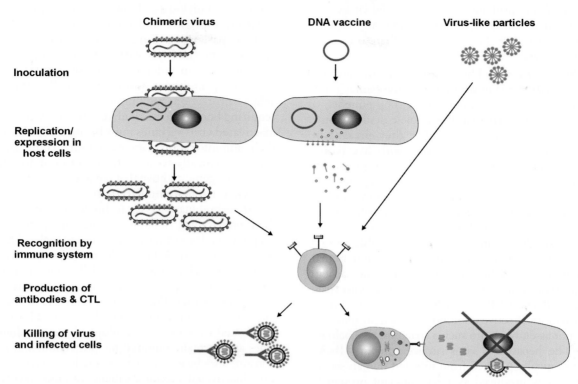

Figure 35.2 Recombinant antiviral vaccines. A genome fragment (green) encoding an immunogenic viral protein is engineered into the genome (red) of a nonpathogenic virus or into a plasmid DNA. The recombinant virus or plasmid is inoculated into the host, where the viral protein is expressed and induces an immune response. Recombinant viruses or expression vectors can be used to make virus-like particles (right) containing an immunogenic viral protein: these virus-like particles are inoculated into the host where they induce an immune response against the targeted virus.

Table 35.5 Passage histories for two vaccine-strain viruses

Edmonston-B and Moraten measles vaccine strains
• Isolated by Enders & Peebles in 1954 from patient (D. Edmonston)
• Twenty-four passages (35°–36°C) in primary human kidney cells
• Twenty-eight passages in primary human amnion cells
• Six passages in chicken embryos
• Marketed as Edmonston B vaccine in 1963
• Seven passages (36°C) in chicken embro fibroblasts
• Forty passages (32°C) in chicken embryo fibroblasts
• Marketed as "further attenuated" measles vaccine (Moraten strain) in 1968
Oka-strain varicella virus
• Isolated by Takahashi et al. in 1974 from three-year-old Japanese boy (family name Oka)
• Initial growth in human lung fibroblast cells
• Twelve passages in primary guinea pig fibroblasts
• Two passages in WI-38 cells (human diploid fibroblasts)
• Three to six passages in MRC-5 cells (human diploid lung cells)
• Licensed for use in Japan and Korea in late 1980s, and in USA in 1995

therefore virus infectivity, while retaining the antigenicity of viral proteins. Such inactivated viruses can retain their ability to induce an immune response but are incapable of replication and therefore do not cause disease.

4. *Subunit vaccines.* These consist of purified viral proteins that are known to be immunogenic. These vaccines do not contain other components of the virus, in particular the nucleic acid genome. Their major advantage is that they present no danger of inadvertent infection with live pathogenic virus, which can be a problem with the other vaccine categories. They may also be significantly cheaper to manufacture, because they do not require expensive virus growth in cell culture systems and testing of the virus produced.

Presently, subunit vaccines are available for only two human viruses, hepatitis B virus and human papillomavirus. These vaccines exploit the capacity of purified viral proteins to self-assemble into virus-like particles (VLP). Virus-like particles tend to be much more effective immunogens than proteins that are not assembled into such a structure (see below).

The hepatitis B virus surface antigen (HbsAg) was originally purified from infected human serum, but is now produced as a recombinant protein by expressing the cloned surface antigen gene in yeast or mammalian cells, where it can form virus-like particles. The human papillomavirus (HPV) L1 protein is produced using recombinant technologies and is expressed in either yeast or transfected insect cells, where it also spontaneously forms virus-like particles.

Advantages and drawbacks of vaccine types

The immune response elicited by live wild-type or attenuated vaccines comes closest to that observed in natural infection, because these viruses multiply within the vaccinated individual. Immunity after even a single dose of these vaccines is typically long-lived (at least 10 years in most cases). Some live virus vaccines can be taken orally or nasally, which simplifies their delivery.

However, these vaccines are expensive to manufacture and are relatively unstable, requiring careful attention to refrigeration via the **cold chain**, a problem in developing countries. Live vaccines may also be subject to inactivation by preformed maternal antibodies when administered in early childhood. For example, maternal antibodies can inactivate measles vaccine-strain virus in some children younger than 12 months of age. Live virus strains can occasionally cause serious illness and even death in subjects with abnormal immune systems.

Inactivated whole virus vaccines, like live virus, can generate immune responses to a wide range of viral antigens. These vaccines are generally less sensitive

to storage conditions and are safer for immunocompromised individuals. However, their production is relatively expensive and requires the manipulation of virulent pathogens. Furthermore, because inactivated viruses cannot multiply within the host, they tend to be less successful in inducing a robust immune response. Therefore, multiple doses are usually required to achieve protection, and immunity often fades with time, requiring booster doses. Unlike some live virus vaccines, inactivated virus vaccines must be injected, which makes their administration somewhat more complex. These vaccines typically require the addition of an **adjuvant** (a non-antigenic additive that enhances immunogenicity) to elicit a strong immune response.

If recombinant DNA technologies can be used in their preparation, subunit vaccines are potentially the simplest and least expensive vaccines to make. These vaccines can be remarkably stable and their use avoids the delivery of irrelevant and potentially immunopathologic viral antigens responsible for vaccine-associate adverse events (see below). These vaccines can also be used without risk in immunocompromised subjects. However, like inactivated virus vaccines, multiple doses of subunit vaccines are normally required for protection, and periodic boosting may be necessary to ensure long-term immunity. Like inactivated vaccines, subunit vaccines also typically include an adjuvant to enhance immunogenicity.

New categories of antiviral vaccines

Several new types of vaccine (Figure 35.2) targeting various viruses are newly licensed or under late-stage development.

1. *Chimeric vaccines* are based on live organisms that are either nonpathogenic in humans (e.g., strains of the bacterium *Lactobacillus*, and a variety of attenuated viruses) or known and "trustworthy" (e.g., attenuated *Salmonella* Ty21a, and vaccine-strain measles virus). These organisms can be genetically manipulated to express one or more viral proteins, and their intrinsic characteristics can be used to deliver vaccine antigens to desired locations (e.g., gut mucosa, **dendritic cells**). The newly introduced rotavirus vaccines based on recombinant human and bovine chimeric viruses are good examples.

2. *DNA vaccines* are based on DNA plasmids that encode one or more viral proteins in a form that can be expressed in host cells. Injection of these plasmids into a human or animal results in transient expression of the viral proteins in host cells (e.g., myocytes, immune cells). Such plasmids can be effective mimics of viral infection and can elicit broader immune responses than soluble proteins, because the viral proteins are synthesized in cells within the host and thus are more efficiently recognized by the immune system.

3. *Virus-like particle (VLP) vaccines* include "nanoparticles" bearing viral antigens that can be produced using a wide range of technologies. Although relatively few viral proteins can self-assemble like the hepatitis B virus sAg and the human papillomavirus L1 protein, recombinant proteins that combine a multimerization domain from one protein with an antigenic viral protein can create a family of novel VLPs. Other nanoparticle vaccines can be generated by mixing viral antigens with either lipids (e.g., liposomes, nano-emulsions) or proteins that tend to aggregate (e.g., **proteosomes**). Finally, VLPs that mimic the enveloped viruses can be generated in a number of prokaryotic and eukaryotic expression systems, including plants.

4. *Multivalent peptide vaccines* with hundreds and even thousands of peptide epitopes are being developed to address the problems associated with highly mutable RNA viruses such as HIV and hepatitis C virus.

Among these newer approaches, chimeric live vaccines share the strengths and weaknesses of the live attenuated vaccines. Nanoparticle vaccines such as virus-like particles can elicit very powerful immune responses, particularly when viral antigens are combined with immunostimulatory molecules such as ligands for pathogen-recognition receptors (see Chapter 33). More than a decade of frustrating efforts with DNA vaccines that work beautifully in animal models but less well in humans has led to the concept of sequential, multiple-formulation vaccination strategies referred to as "prime-boost" (beginning with a DNA vaccine "prime", followed by a live attenuated vaccine "boost"). These strategies take advantage of the characteristics of each component vaccine to generate specific patterns of response (more antibody, more cell-mediated response, or balanced humoral and cellular responses).

HOW DO ANTIVIRAL VACCINES WORK?

In the current environment of molecular sophistication, it is sometimes hard to believe that products developed half a century ago are still in use. Until very recently, the process of attenuation was essentially "black-box" science. Pathogenic viruses were repeatedly passaged through eggs or tissue culture cells under varying conditions until they acquired an attenuated phenotype. Astonishingly by today's ethical standards, the appearance of the attenuated phenotype was typically demonstrated by inoculation of children! Several of our most successful vaccines, including yellow fever, measles, mumps, rubella, and Sabin polio, were generated in this way.

We have learned a great deal about the molecular characteristics of some of these products in recent years. The three Sabin attenuated poliovirus vaccine strains very occasionally revert to neurovirulence, and we now know the sites on the viral genome of many of the

mutations that affect virulence (Table 35.6). These data help to explain the observation that vaccine-associated paralytic polio is most often ascribed to pathogenic revertants of types 2 and 3 Sabin vaccine-strain viruses. Most of these vaccines continue to be used effectively even though we remain ignorant of the precise mechanisms of attenuation and protection against infection.

Is worldwide eradication of poliovirus possible?

Many of the complexities involved in national and international vaccine program decisions are exemplified by the ongoing global effort targeting poliovirus. Like smallpox and measles, humans are the only known host for the polioviruses. As a result, polio can be targeted for eradication.

Unlike smallpox and measles, there are two good vaccines for poliovirus: both the inactivated (Salk) and live attenuated (Sabin) vaccines are highly effective in preventing polio. However, they each have distinctive strengths and weaknesses. The live vaccine can be administered orally, increasing compliance and avoiding needle-related injuries. Live vaccine-strain viruses are also shed in the stool of vaccinated children and this can immunize close contacts (e.g., family members) or boost their immunity. In children with suboptimal immunity, virus shedding can persist for months or years. All of the oral poliovirus vaccine strains also have the capacity to persist in sewage for weeks, potentially infecting and therefore immunizing or boosting immunity in community members exposed to contaminated water.

This capacity to spread within families and communities is a significant advantage in regions with low rates of vaccine coverage. However, community spread of live vaccine strains also explains the fact that two-thirds of those who suffer from vaccine-associated paralytic polio are close contacts of vaccinated children. Several water-based community outbreaks of flaccid paralysis caused by vaccine-derived revertant polioviruses (including Type 1) have been described.

In countries that can achieve and sustain high rates of vaccine coverage, the inactivated vaccine can provide excellent protection without any risk of disease from revertant viruses. However, this vaccine must be delivered by needle, with attendant risks for the vaccinee, the vaccinator and the community (e.g., inadequate disposal of used needles). Problems with virulent revertants led most wealthy countries to switch from the Sabin live vaccine to the Salk inactivated vaccine in the 1990s. However, the live vaccine is still preferred in resource-poor settings with lower rates of vaccine coverage.

The persistence and shedding of live attenuated viruses in some hosts, the appearance of virulent revertants, and their capacity to spread through contaminated water represent major challenges for the eradication effort. It is likely that a complex global program that evolves from use of the live vaccine to the inactivated vaccine and continues vaccination long after cases of polio disappear, will be needed before eradication can be achieved.

Table 35.6 Attenuated oral polioviruses: reversion to neurovirulence

Virus strain	Number of mutations	Number of amino acid changes	Number of mutations required for return to neurovirulence
Sabin strain 1	57	23	>10
Sabin strain 2	23	5	5–6
Sabin strain 3	~6	3	1–2

From Minor, P. D. (1992). *The molecular biology of poliovaccines.* Journal of General Virology 73, 3065–3077.

The role of the immune system in fighting viral infections

Cellular immune responses (see Chapter 34) mediate recovery from the vast majority of viral illnesses. Individuals who lack the ability to make soluble antibodies via the **humoral immune response** recover from most viral infections in spite of their defect, showing that antibody formation is not required for recovery. However, individuals with defects in cell-mediated immunity are highly susceptible to severe, protracted, and frequently fatal infections with a wide range of viral pathogens. Thus killing of infected cells by activated **cytotoxic lymphocytes** appears to be the major mechanism by which the immune system clears viral infections from the body. This makes sense, as infected cells can be recognized and killed before they produce large amounts of virus, which could then spread throughout the body. The only exception to this rule appears to be the enterovirus family, including poliovirus and the echo and Coxsackie viruses. Control of these pathogens requires an intact humoral response.

However, antibodies still have a major role in protecting against viral infections. High titers of preformed **neutralizing antibodies** can confer immunity against most viral pathogens. Antibodies acquired from the mother near the end of pregnancy are critically important for the protection of infants between birth and 6 to 9 months of age. Antiviral antibodies are also important for limiting reinfection with the same virus and can sometimes provide partial or complete protection against closely related viruses. Therefore, the humoral response is vital for protection against infection by viral pathogens, while cellular immunity is important for recovery from viral infections.

Adjuvants play an important role in vaccination with inactivated or subunit vaccines

Because they are able to replicate in the body, live attenuated virus vaccines induce both cellular and humoral immunity, as do most natural viral infections. However, inactivated whole virus or subunit vaccines cannot replicate and therefore are less efficient in inducing immune responses. In order to stimulate the immune response, these vaccines are often mixed with adjuvants such as aluminum hydroxide or aluminum phosphate salts. Until recently, aluminum salts were the only adjuvants licenced for use in humans.

Aluminum salts preferentially stimulate **type 2 T helper lymphocytes** (Th-2 cells), which secrete **cytokines** that favor the production of antibodies (see Chapter 34). However, aluminum-based adjuvants do not efficiently induce cellular immune responses. Early vaccine scientists showed that the levels of neutralizing antibodies correlate well with protection against many virus infections, and therefore this was one of the primary goals of vaccine development. The viruses targeted by most vaccines developed up to the present time either cause acute infections by one prevalent serotype (measles, rubella, smallpox) or can be neutralized by antibodies before reaching their target tissues (poliovirus, hepatitis B, rabies, human papillomaviruses).

Vaccines that stimulate cell-mediated immunity are being developed

However, it is now clear to vaccine developers that the generation of antibodies is not always sufficient for an effective antiviral vaccine. Viruses with the capacity to establish latent infections (herpesviruses), viruses with multiple serotypes and limited cross-neutralization (respiratory syncytial virus, dengue virus), and those that rapidly mutate their structural proteins (influenza A virus, human immunodeficiency virus, hepatitis C virus) present much more difficult targets for vaccines that generate predominantly antibody responses.

Furthermore, antibodies are not often of much use against viruses that have already established a chronic infection (human immunodeficiency virus, hepatitis B virus, papillomaviruses, herpesviruses). Also, a weak or non-neutralizing antibody response generated by a vaccine can sometimes lead to immunopathology (respiratory syncytial virus, atypical measles) or even facilitate virus entry into target cells (dengue virus).

Presently, a major focus of vaccine designers is the harnessing of cell-mediated immunity to create vaccines for the above-mentioned elusive and complicated viral targets, which remain serious threats to human health. In the last decade, it has become clear that the cellular immunity generated by some live attenuated vaccines can be very long-lived (>50 years for smallpox vaccine).

However, protection by other live attenuated vaccines such as measles and mumps may be lost over time in the absence of periodic boosting by either revaccination or exposure to natural disease.

NEW DEVELOPMENTS IN ANTIVIRAL VACCINES

New approaches to vaccine development show great promise

In the last 50 years, spectacular advances in molecular biology, immunology, biotechnology, and protein chemistry have been the driving forces behind antiviral vaccine development. The modern era of vaccine development was foreshadowed by the introduction of the serum-based hepatitis B surface antigen (HBsAg) vaccine in 1981, and was fully inaugurated with the cloning and expression of the HBsAg gene in yeast and mammalian cells and the licensing of a recombinant vaccine in 1986. Several of the most promising advances in vaccine development (Table 35.7) are discussed here.

New adjuvants are being developed

Until very recently, only aluminum salts were used as adjuvants in human vaccines. Although aluminum-based adjuvants help the immune system to respond to many viral antigens, they do not stimulate **type 1 T helper lymphocytes** (Th-1 cells), involved in the cellular immune response. The dominance of alum is now being challenged by a rapidly expanding repertoire of adjuvants that manipulate the immune response in increasingly sophisticated ways. Several commercial vaccines now contain oil-in-water-based adjuvants that can both enhance and broaden the antibody response. One such adjuvant contains an emulsion of squalene (a fat-soluble hydrocarbon isolated from fish oil) ± vitamin E (also mostly hydrocarbon) in water. Variants of this adjuvant have been used successfully for inactivated human influenza vaccines (including pandemic vaccines) in Europe and Canada.

Other adjuvants that are being tested interact with **toll-like receptors** (TLRs), crucial signaling molecules that elicit strong innate responses to pathogens and also help to stimulate adaptive immune responses (see Chapter 33). For example, CpG elements are short, unmethylated DNA motifs not commonly found in the human genome but often present in bacterial genes. DNA containing these motifs binds to toll-like receptor 9 (TLR-9). Injection of CpG DNA can elicit innate cellular Th-1-like responses and therefore increase the host response to antigens.

New delivery systems for viral antigens

Chimeric molecules that contain cytokines, **lectins**, or **monoclonal antibodies** fused to viral antigens can help to direct the antigen to an appropriate target cell or enhance immune responses by interacting with cells in the immune system. Dendritic cells, which present antigens to the immune system, can be isolated from an individual, exposed to a viral antigen, and reintroduced into the body. This can lead to enhanced antigen presentation, which may result in clearance of chronic viral infections.

A variety of auto-assembling vesicles or nanoparticles, within which viral antigens can be trapped, are being tested as delivery systems for subunit vaccines. These include proteosomes, virosomes, liposomes, and virus-like particles (VLPs). Viral antigens trapped on or in these particles can promote both humoral and cellular immune responses. An alternative to introduction of viral antigens is the insertion of their genes into expression plasmids, which are then introduced into cells in the body. These DNA vaccines can produce the protein of interest intracellularly, leading to establishment of both humoral and cellular immunity, without the dangers inherent in live or inactivated virus vaccines.

Historically, the options for mechanical delivery of vaccines into the human body have been remarkably limited, based for the most part on injection with needles, contributing to many people's fear of vaccines. Aerosols, eye drops, and suppositories are being developed to direct vaccines to specific body locations. Genetic engineering of plants can now produce a variety of viral antigens in natural foods such as bananas. Direct consumption of antigen-containing food can sometimes result in acquisition of immunity to the targeted viruses. Plants can also be engineered to produce virus-like particles that can be purified and used as vaccines. Finally, a variety of devices are being tested to deliver tiny particles coated with either antigens or plasmid DNA at high velocity into superficial tissues and cells. In the case of the so-called "**gene guns**," this can be a quick and efficient system for localized delivery of DNA vaccines or other antigens.

Vaccination with defined proteins

With the exception of the vaccines for hepatitis B virus and human papillomavirus, which contain single viral proteins, all antiviral vaccines presently in use are based on either live attenuated, live chimeric, or inactivated whole viruses. As more is learned about the biology of viral pathogens, it should be possible to create a new generation of "minimalist" vaccines that contain only the antigens or peptides that elicit protection, while avoiding those that contribute to vaccine-associated adverse events (see below). Many such products are currently under development or in field trials for a range of viral pathogens including measles, respiratory syncytial virus, influenza, HIV, dengue, and several herpesviruses. Other advantages of these simpler vaccines include ease of manufacture and the possibility of combining several antigens in multivalent formulations to protect against a range of pathogens.

However, the utility of some of these narrowly targeted vaccines may eventually prove to be limited, because

Table 35.7 New technologies for human antiviral vaccines

A. Adjuvant systems	
Oil-in-water emulsions	Elicit strong humoral response without major toxicity.
Cholera toxin B subunit	Elicits strong humoral response without toxicity.
Toll-like receptor ligands	Elicit strong humoral and cellular immune responses.
• CpG motifs	Short, unmethylated DNA motifs that bind to TLR-9 and elicit potent, innate and cellular (Th-1-like) responses.
• Imiquimod/resiquimod	TLR-2 ligand that elicits strong humoral and cellular responses
• Detoxified lipopolysaccharide	TLR-4 ligand that mimics bacterial infection and stimulates strong humoral and cellular responses
Pulsed dendritic cells	Presentation of antigens to dendritic cells outside of the body and their reintroduction can lead to enhanced response to antigens.
Conjugates	A variety of molecules such as cytokines, monoclonal antibodies, and lectins can be conjugated to antigens to enhance immune reactions.
B. Delivery systems	
Liposomes	Antigens are trapped inside vesicles made from synthetic lipids, enhancing immune reactions. Stability and consistency have been major problems.
Immune stimulating complexes (ISCOM)	More complex liposomal preparations that retain the capacity to elicit strong immune responses, but are more stable.
Proteosomes	Auto-assembling vesicles containing bacterial outer membrane proteins from Neisseria that can trap antigens and promote strong humoral and cellular responses.
Plasmid DNA	An expression plasmid containing the gene that encodes a viral antigen is introduced into the body. Expression of this gene *in vivo* can lead to potent induction of both humoral and cellular responses.
Viral or bacterial vectors	Genes that encode viral antigens of interest can be inserted into non-pathogenic viral vectors (avian poxviruses, Semliki Forest virus) or bacterial vectors (attenuated Salmonella typhi) that retain some useful characteristics of the vector organisms (binding and entry, tissue tropism).
Virus-like particles	Viral proteins, alone or in combination, that can either spontaneously self assemble or be coaxed into assembling with lipids to form nanoparticles resembling viruses.
C. Delivery technologies	
Aerosol	A variety of ingenious mechanical devices are being studied for aerosolization of antiviral vaccine preparations for delivery either to the nasal or the lower respiratory mucosa.
Eye drops, troches, suppositories	Formulations that deliver vaccine antigens to conjunctival, vaginal, or gastrointestinal mucosae.
Foods	Genetic manipulation of plants has made it possible to consider the delivery of viral and other antigens in foods such as bananas.
Gene gun	Accelerates gold particles carrying plasmid DNA to high velocities for penetration of skin to the desired depth, targeting dermal layers or muscle.

viruses can undergo mutation of the targeted gene and therefore escape the immune response directed against a single antigen. Because of this, vaccine developers are focusing on highly conserved, essential viral proteins and are considering vaccination with mixtures or "cocktails" of defined proteins. Plasmid-based DNA vaccination may be particularly well suited for the delivery of either single or small groups of viral proteins to the immune system.

Use of live viruses with defined attenuation characteristics

The first intentional exposure of healthy subjects to a live vaccine-strain virus can make both vaccinator and vaccinee quite nervous, particularly when the vaccine has been produced by random passage. Such black-box vaccines continue to be introduced. For example, the

Oka-strain varicella (chickenpox) vaccine was licensed in North America in the late 1990s, and a live attenuated rotavirus vaccine was licensed only in the mid-2000s.

However, by using genetic engineering methods, vaccine developers now are able to create molecularly defined, attenuated vaccine-strain viruses. Among the best examples of this approach are the cold-adapted influenza A vaccine (described above) and one of the chimeric rotavirus vaccines. In the latter, the surface glycoprotein genes of human rotaviruses have been inserted into a non-pathogenic bovine rotavirus.

Use of live vectors and chimeric viruses

Because of many years of experience with hundreds of millions of vaccinated people, the advantages and dangers of live virus vaccines such as vaccinia and measles are now well known. We have simultaneously gained considerable experience with live bacteria used as vaccines, such as Bacille Calmette–Guerin (BCG, against tuberculosis) and attenuated *Salmonella typhi* (against typhoid fever). Most importantly, advances in molecular biology have given us the capacity to modify these organisms to carry foreign genetic material. The safe track records of these viral and bacterial agents make them attractive vectors for the delivery of antigens from unrelated organisms for which vaccines are not yet available. Examples of such chimeric products include genetically modified vaccine-strain measles virus carrying the structural proteins of respiratory syncytial virus, and attenuated *Salmonella typhi* expressing rotavirus glycoproteins.

A chimeric vaccinia virus expressing the major glycoprotein of rabies virus is currently in use in the northeastern part of North America to control a major rabies **epizootic** in raccoons. Other viruses are also the subject of intense investigation as possible vectors for foreign genetic material. Examples include canarypox viruses that express HIV sequences, and alphaviruses such as Semliki Forest virus that express paramyxovirus proteins. As noted above, both human–human and human–bovine chimeric rotavirus vaccines have recently been licensed, and many other conceptually similar vaccines are currently in clinical trials. All seek to exploit the capacity of the vector organism to express the targeted proteins within host cells, eliciting a cytotoxic T-cell response.

Vaccines that can break tolerance

Until very recently, vaccine development efforts have focused almost exclusively on products capable of preventing or modifying acute disease. Many hundreds of millions of people around the globe are already infected with viruses such as hepatitis B and C, papillomaviruses, and human immunodeficiency virus. In many cases,

these viruses persist in the body because of the establishment of some degree of **tolerance** by the human immune system. Several of the new approaches outlined in Table 35.7 have the proven capacity to break tolerance in animal models. Among the most promising include DNA vaccination, pulsed dendritic cell vaccination, and the use of potent Th1-biasing adjuvants such as ligands for toll-like receptors. Several of these strategies are already in advanced-stage clinical trials.

The changing vaccine paradigm

Just as there can be no one-size-fits-all for the development of antiviral vaccines, there is no single optimal vaccination strategy for all viruses. There is still an urgent need for vaccines that can protect against viruses that cause significant illness and death, such as respiratory syncytial virus, HIV, dengue virus, and hepatitis C virus. However, development efforts are not limited to these viruses. Viruses with more limited capacity to do harm, such as agents of mild childhood diarrhea and the common cold, are currently being targeted. Vaccine developers are also targeting ubiquitous and persisting viruses such as the herpesviruses Epstein–Barr virus and cytomegalovirus. However, the consequences for human biology of vaccinating against these types of viral agents are currently unknown.

Finally, many of the scientific advances outlined above have to some extent democratized the process of vaccine development, which was once the exclusive purview of the major pharmaceutical companies. Presently, many small academic and biotechnology company laboratories around the world are actively pursuing vaccine development programs. The range of viruses against which vaccines are currently under development is impressive (Table 35.8). These developments could lead to both favorable and unfavorable outcomes. On the one hand, more humans throughout the world will likely enjoy protection against more viral diseases than ever before. On the other hand, the unbridled use of vaccines for everything from tooth decay to cancer has the potential to tarnish the relatively good name that vaccines have enjoyed for almost half a century.

ADVERSE EVENTS AND ETHICAL ISSUES

Vaccine-associated adverse events

Although the currently available antiviral vaccines are among the safest pharmaceutical products ever introduced, none is 100% safe. The history of vaccine development is troubled by several unfortunate incidents in which vaccinated individuals have suffered severe "adverse events" (Table 35.9). In some cases, these incidents can be explained by the biological properties of the product. For example, vaccine-associated paralytic polio due to reversion of Sabin polio vaccine-strain virus to neurovirulence is understood in terms of the mutations required to maintain attenuation (see above). However, due to incomplete knowledge about many immunologically mediated reactions, most adverse events following immunization are poorly understood and therefore are difficult to foresee.

Table 35.8 Some antiviral vaccines currently under development

Virus family	Virus species	Most promising strategies
Arenavirus	Junin	Live attenuated
	Lassa fever	Recombinant vaccinia virus
Calicivirus	Norwalk	Recombinant capsid "ghosts"
Filovirus	Ebola/Marburg	DNA
		Recombinant vesicular stomatitis virus
Flavivirus	Hepatitis C	Recombinant envelope glycoprotein, peptide, polypeptide
	Dengue	Tetravalent attenuated, DNA
Hepadnavirus	Hepatitis B	DNA, recombinant surface antigen including pre-S1/S2 regions; curative hepatitis B vaccine: DNA with CpG motifs
Herpesvirus	Epstein–Barr	Whole inactivated, subunit (gp350)
	Herpes simplex 1, 2	Recombinant polypeptide, vectored
Orthomyxovirus	Influenza A	Vectored, DNA, peptide
Paramyxovirus	Measles	Vectored, DNA
	Parainfluenza	Live attenuated
	Respiratory syncytial	Subunit G and fusion protein, DNA, vectored, peptide
Parvovirus	Parvovirus B19	Empty capsid with VP1 protein
Picornavirus	Enterovirus 71	Whole inactivated
Reovirus	Rotavirus	Live reassortants of simian or human origin
Retrovirus	HIV-1, 2	Recombinant polypetide, DNA, vectored, chimeric
Unclassified	Hepatitis E	Recombinant protein ORF2

Table 35.9 Severe adverse events associated with antiviral vaccine administration

Vaccine implicated	Event
Smallpox ("lymph" vaccines)	Sepsis due to bacterial contamination
Yellow fever (1942)	Vaccine lot contaminated with hepatitis B virus, leading to 28,000 cases of hepatitis B
Inactivated polio (1955)	Incomplete inactivation of virus, leading to 204 cases of paralytic disease (Cutter incident)
Inactivated measles (1960s)	Atypical (severe) disease upon exposure to natural measles infection
Attenuated polio (Sabin)	Vaccine-associated paralysis (~1 per 2.4 million persons vaccinated) due to reversion to more pathogenic strain
Live vaccines (vaccinia, polio, measles)	Dissemination and death in immunocompromised individuals
Attenuated live measles	Unexplained mortality in girls who received high titer formulations (relative risk of death doubled to age 5); more than 20 million doses of vaccine distributed
Inactivated influenza A	Apparent risk of Guillain Barré Syndrome (~1 per 1 million persons vaccinated), only in some years
Rotavirus (1999)	Intussusception—bowel folding and obstruction (vaccine withdrawn)

A great deal of energy is being expended internationally to refine the tools available for the surveillance of problems associated with (but not necessarily caused by) vaccination. These surveillance activities are essential not only to ensure the safety of new vaccines but also to protect the enormous societal investment already made in national and international vaccination programs. Even transient losses of public confidence in vaccination programs can have devastating effects. Examples of crises of public confidence in vaccines include the pertussis vaccine in Japan in the mid-1990s (a real association with fever and hyporesponsiveness), measles vaccine in the United Kingdom (a spurious association with autism), and poliovirus vaccine in northern Nigeria in 2004–2005 (a politically motivated claim of attempted genocide). This last situation resulted in the reintroduction of polio into many countries and has likely set back polio eradication efforts by many years.

It is important to understand that the safety demanded of a vaccine can change in different situations. For example, an occasional serious complication with a rotavirus vaccine might theoretically be tolerable in the developing world, where rotaviruses kill more than a million children every year. Such a problem would, however, be unacceptable in North America and Europe, where only a small number of children die from rotavirus infection annually. As we approach worldwide eradication of measles and poliovirus, there are increasing concerns about the safety of the currently used vaccines. In the 1990s, most developed countries replaced the live Sabin (oral) polio vaccines with inactivated vaccines in their national programs to avoid the small but measurable risk of reversion (approximately 1 per 2 million doses).

Ethical issues in the use of antiviral vaccines

Many of the viruses targeted by vaccine programs are among the most highly transmissible infectious agents known. As a result, public health authorities have great interest in maintaining high levels of vaccine coverage to protect the small number of individuals who are missed by the program, who have failed to respond to vaccination, who respond initially to vaccination but subsequently lose protection, who have a valid contraindication to vaccination, or who simply refuse vaccination. The protection afforded these individuals by high levels of vaccine coverage in the community is termed **herd immunity**.

The desire to establish herd immunity has led public health authorities either to enforce vaccination or to make avoidance of vaccination difficult. In the past, compulsory or heavy-handed vaccination programs for smallpox occasionally resulted in street riots. The debates surrounding some childhood vaccination programs are only slightly less vigorous today. Some individuals at each end of the spectrum are quite prepared to label those at the opposite extremes as "abusers" of children. Ironically, the great successes of the quasi-compulsory vaccination programs that have eliminated smallpox and nearly eliminated polio and measles have also contributed to the growing wave of anti-vaccination sentiment, because people currently feel themselves to be less at risk.

Another ethical problem arises from the concept of "altruistic" vaccination. For example, annual vaccination of healthcare providers against seasonal influenza virus strains can reduce the impact of influenza on hospitalized and institutionalized individuals, although the danger of influenza to the (younger and healthier) vaccinated individuals is not as great. Although many such institutions have programs to promote immunization, there has been no attempt to legislate or mandate annual influenza vaccination for healthcare workers.

Since no vaccination can ever be 100% safe, it seems appropriate that there should be little or no "cost" or penalty associated with a personal choice not to receive any particular vaccine. However, it is also reasonable that many jurisdictions require those who refuse measles vaccine to keep their children out of school in the event of a measles outbreak. A similar argument can be made to exclude from the workplace (without pay) healthcare personnel reluctant to be vaccinated in the event of a major influenza pandemic. Taking this argument further, it is also logical to ask if society should be asked to pay for the lifelong care of an individual who suffers a vaccine-preventable complication such as paralytic polio after refusing vaccination. Such issues are obviously controversial and highly charged.

This discussion would not be complete without mentioning the balancing responsibilities of the vaccine industry, governments, healthcare systems, and the healthcare providers to develop and administer the safest possible vaccines and to take the issue of vaccine safety very seriously. This commitment requires the support of adequate surveillance systems and a program to fairly compensate the rare individual who suffers a serious adverse event as a result of vaccination.

The story of the rabies vaccine

Human rabies is a devastating disease with periods of relative lucidity and apparent madness alternating over days to weeks in a downward spiral to inevitable death. Beginning with Zinke's observation in 1804 that rabies could be transmitted by the injection of saliva from a rabid animal, several investigators worked through the nineteenth century to develop a successful rabies vaccine. By 1879, Victor Galtier had succeeded in passaging the virus in rabbits and had demonstrated limited efficacy with a rabbit tissue vaccine in herbivores.

Simultaneously, Émile Roux and colleagues in Pasteur's group had been able to "fix" rabies virus (i.e., establish consistent biological properties) by serial passage in rabbit brains. They had also made the key discovery that viral virulence disappeared over a 15- to 20-day period of drying spinal cord from infected rabbits. Armed with this information, they were able to demonstrate that a series of injections with spinal cord material dried for variable periods of time could protect dogs.

In today's terms, it is stunning that pre-exposure prophylaxis had been demonstrated in only 50 to 60 dogs before the first human, a preteen boy named Joseph Meister, received this vaccine in 1885. Desperation and fear go some way toward explaining why Meister's parents and the investigators themselves conducted this experiment.

To everyone's relief (including Joseph's), the vaccine appeared to work and its popularity spread. However, with more widespread use it became apparent that this vaccine induced serious autoimmune demyelination of the central nervous system in as many as four per thousand vaccinees. The associated mortality was significant (30–50%) and many of those who recovered were left with permanent neurological sequelae. This disease is now called postinfectious encephalomyelitis, and is known to be mediated by autoreactive T cells. The myelin component of the early vaccine was eventually implicated.

Subsequent generations of rabies vaccines produced in suckling mouse brain and duck embryos had progressively less myelin and dramatically lower incidences of such complications (1/8000 and 1/32,000 respectively). The successful adaptation of rabies virus to the human diploid fibroblast cell line WI-38 in the early 1960s finally permitted the development of a rabies vaccine that is both effective and safe. To date, more than 10 million people have been inoculated with rabies vaccines grown in cultured cells, with seroconversion and protection rates approaching 100% and without a single proven case of vaccine-associated neurological damage.

KEY TERMS

Adjuvant

Autoimmune demyelination

Cellular immune responses

Chimera (adj: chimeric)

Cold chain

Cytokines

Cytotoxic T lymphocytes

Dendritic cells

DNA vaccines

Epidemic

Epizootic

Gene gun

Herd immunity

Humoral immune response

Immunogenic

Lectins

Monoclonal antibodies

Neutralizing antibodies

Pandemic

Panzootic

Primary cell cultures

Proteosomes

Tolerance

Toll-like receptors

Type 1 and type 2 T helper lymphocytes

Zoonotic

FUNDAMENTAL CONCEPTS

• No vaccine can ever be 100% safe, but vaccines are among the safest medical interventions ever introduced.

• Vaccines are generally held to a very high safety standard because they are administered to healthy subjects to prevent disease.

• Antiviral vaccines have had enormous positive impacts on global health: for example, eradication of smallpox, control of measles and polio.

• Vaccine science has evolved over time with advances in understanding of the immune response and developments in technology such as egg culture, tissue culture, and recombinant DNA technologies.

• Cellular immunity is critical for recovery from most viral infections.

• Antibodies can protect against many virus infections and prevent reinfection, but in most cases the humoral response does not play an important role in recovery from virus infections.

• Live attenuated virus vaccines, inactivated virus vaccines, and subunit vaccines have all been successful. All but live attenuated vaccines are administered with an adjuvant to augment the immune response.

• To date, only live attenuated vaccines reliably elicit both humoral and cellular immune responses

• Among the most promising new vaccine technologies include DNA vaccines, chimeric vaccines, nanoparticle vaccines, and virus-like particles.

• Loss of public confidence in vaccines can have devastating effects both for individuals who refuse vaccination and for populations.

REVIEW QUESTIONS

1. Given the large number of vaccine types available, discuss why a live attenuated virus vaccine may be the best choice.

2. A recent problem (2004–2005) with the campaign for the eradication of polio is a great example of an unexpected "adverse event." What was the adverse event?

3. The ability of some vaccines to induce "herd immunity" has long been thought of as a positive property. Discuss some of the ethical issues surrounding "herd immunity."

4. How would you construct a subunit chimeric vaccine against a human virus?

5. Subunit vaccines generally stimulate only humoral immunity. What modifications or additions to these vaccines could enable the stimulation of both humoral and cellular immunity in a vaccinated individual?

Antiviral Chemotherapy

Donald M. Coen

DISCOVERY OF ANTIVIRAL DRUGS

"Serendipity"—trying out compounds used for other
   purposes:
- *Amantadine (inhibits influenza virus uncoating)*
- *Acyclovir (inhibits herpesvirus DNA replication)*

Chemical modification of known active compounds:
- *Ganciclovir (inhibits herpesvirus DNA replication)*
- *Azidothymidine (inhibits HIV-1 reverse transcription)*

High throughput screening assays of many
   compounds:
- *Nevirapine (inhibits HIV-1 reverse transcription)*
- *Raltegravir (inhibits HIV-1 integrase)*

Rational design, often with the aid of three-dimensional
   structures of viral proteins:
- *Ritonavir (inhibits HIV-1 protease)*
- *Zanamivir (inhibits influenza neuraminidase)*

TARGETS OF ANTIVIRAL DRUGS

Inhibitors of virus attachment and entry:
- *Picornavirus capsid proteins*
- *HIV-1 envelope proteins and host chemokine receptors.*

Inhibitors of uncoating of viral genome:
- *Influenza M2 ion channel*

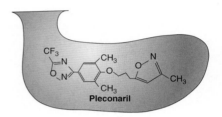

Inhibitors of viral genome replication:
- *Herpesvirus nucleoside/nucleotide kinases*
- *Virus DNA and RNA polymerases*
- *HIV-1 integrase*

Inhibitors of virion maturation:
- *HIV-1 protease*

Inhibitors of virus release from the cell:
- *Influenza neuraminidase*

The discovery and widespread use of antiviral compounds began relatively recently

Both bacteria and viruses (filterable infectious agents) were first shown to cause disease in the late nineteenth century. Antibiotics active against bacterial diseases began to be widely used in the 1940s. However, the clinical use of drugs to treat viral diseases in humans lagged behind, and the first antiviral drugs were introduced only in the 1960s.

A major factor accounting for this delay is that viruses replicate within host cells, using many elements of the host cell machinery. Therefore, there were fewer obvious targets for chemotherapy against viruses than against bacteria. For example, many antibiotics are directed against bacterial cell wall synthesis or bacterial protein synthesis, which are clearly distinct from their human counterparts. The development of compounds active against a broad spectrum of viruses is unlikely, because different viruses do not share common features

that distinguish them from their host cells. Each virus has its own array of specific proteins required for its replication, and therefore different antiviral drugs may have to be developed for each virus.

Because of these difficulties, pharmaceutical companies have not always been enthusiastic about investing resources in the development of antiviral drugs. Individual virus diseases often represent smaller potential markets than the wide variety of diseases caused by bacteria. Moreover, vaccination has been a very successful approach to controlling viral diseases, further dampening the interest of pharmaceutical companies to develop drugs against these diseases. Efforts to develop antiviral drugs have focused mainly on those viruses for which effective vaccines or public health measures are not available, and that cause widespread and easily diagnosed diseases whose treatment would earn substantial profits.

However, since 1980 many effective antiviral agents have been developed. This began with the success of **acyclovir**, a drug directed against certain herpesviruses.

The AIDS epidemic led to a pressing need for drugs active against human immunodeficiency virus (HIV) and opportunistic pathogens, and coincided with the development of expanding knowledge about virus genetics, molecular biology, enzymology, and protein structure. These factors have led to a rapid increase in the number and kinds of antiviral drugs available to treat human diseases. Indeed, many advances in **rational drug discovery** have come from efforts to develop antiviral drugs.

Although protein drugs such as **interferons** (see Chapter 33) and nucleic acid-based drugs such as oligonucleotides are of considerable interest for medicine and virology, for purposes of brevity only small-molecule drugs will be covered here. DISCLAIMER: The inclusion of certain drugs in this chapter should not be construed as a comment on their clinical merits compared with drugs that are not included. The spectrum of drugs available is changing rapidly, and the drugs described here are used only as examples of the mechanisms of a much larger group of compounds.

Antiviral drugs are useful for discoveries in basic research on viruses

Aside from their importance in treating viral diseases, antiviral drugs make excellent laboratory tools. As will be illustrated in detail below, drugs can block the functions of specific viral proteins that direct specific steps in virus infection. This is crucial for understanding both the timing of particular events in the virus life cycle and which proteins are needed for each step. Antiviral drugs, in conjunction with viral mutations that confer drug resistance, can help to identify the roles of viral proteins and dissect the details of how these proteins do their jobs. Drug resistance mutations can provide selectable markers for engineering interesting mutant viruses and performing viral genetics.

How are antiviral drugs obtained?

The first antiviral drugs were discovered by testing chemical compounds for their ability to inhibit virus replication in labor- and time-intensive assays of viral replication, such as assays of cytopathic effect or plaque assays. These compounds had usually been synthesized for some other purpose, and discovery of their antiviral activities was often serendipitous. Examples of drugs that were discovered this way are **amantadine** and acyclovir. When particular compounds were known to have antiviral activity, a variety of other compounds with similar structures were tested for activity. By use of this somewhat more rational procedure, effective antiviral drugs such as the nucleoside analogues **ganciclovir** and **azidothymidine** were discovered.

More modern and rational approaches have been used to discover some newer drugs. One approach is to find a way to assay virus replication that can be automated, using robots, and performed relatively rapidly (i.e., within one to a few days). An example of such an assay is the use of a dye that measures viability and proliferation of cells following virus infection. This assay is then used to screen large libraries of chemicals (hundreds of thousands of compounds) for their ability to inhibit virus replication in cultured cells. This approach is called a **cell-based high throughput screen**, because it involves applying chemicals to cells infected with the virus and assaying the effect on virus replication. Such an assay has been used to discover novel inhibitors of herpesvirus and HIV replication.

A related approach, called a **target-based high throughput screen**, starts with the identification of a viral gene product (almost always a protein) that would make a good drug target. In principle, the best targets are essential viral proteins that are sufficiently different from their cellular counterparts that the activity of the viral protein can be inhibited without affecting the function of similar cellular proteins. Nonessential viral proteins can also be targeted if they activate a drug, transforming it into an inhibitor of virus replication.

For high throughput screening, it is important that the target protein can be expressed and purified easily, and assayed rapidly (i.e., within hours or even minutes) using robots. The activity assay is then used to screen large libraries of chemicals. For example, HIV-1 reverse transcriptase can be expressed and purified, and its DNA polymerase activity can be easily assayed *in vitro*. This approach was used to discover non-nucleoside reverse transcriptase inhibitors such as **nevirapine**.

A still more rational approach is to exploit detailed knowledge about crucial viral proteins to design drugs that will inhibit their activity. Many viral proteins interact with other viral or cellular proteins in functional complexes, and these interactions can be targeted by drugs. Furthermore, knowledge of the three-dimensional structure of the protein can enable visualization of its active site and understanding of its mechanism of action. This information can then be used to design small molecules that bind to the active site and block activity of the protein. This approach has been used to discover anti-HIV protease inhibitors such as **ritonavir** and anti-influenza neuraminidase inhibitors such as **zanamivir**.

Antiviral drugs are targeted to specific steps of virus infection

As outlined in Chapter 1 and detailed in most chapters in this book, the infection of cells by viruses can be broken down into a common set of steps or stages (Figure 36.1). In principle, any step of viral infection can be targeted for inhibition. There are advantages to very early or late steps such as attachment, entry, and

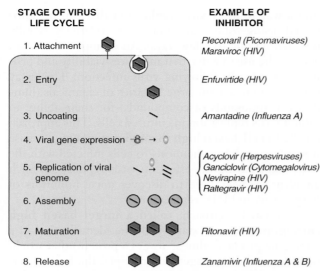

STAGE OF VIRUS LIFE CYCLE	EXAMPLE OF INHIBITOR
1. Attachment	*Pleconaril (Picornaviruses)* *Maraviroc (HIV)*
2. Entry	*Enfuvirtide (HIV)*
3. Uncoating	*Amantadine (Influenza A)*
4. Viral gene expression	
5. Replication of viral genome	*Acyclovir (Herpesviruses)* *Ganciclovir (Cytomegalovirus)* *Nevirapine (HIV)* *Raltegravir (HIV)*
6. Assembly	
7. Maturation	*Ritonavir (HIV)*
8. Release	*Zanamivir (Influenza A & B)*

Figure 36.1 Inhibitors useful at different stages in the virus replication cycle. Mature virions or capsids are shown as green hexagons, genomes as diagonal lines, viral mRNAs as blue lines, and viral proteins as orange ovals. Immature virions are shown as circles. For some viruses (e.g., HIV), maturation takes place after release of virions from the cell.

release, because inhibitors of these steps do not have to enter cells to exert activity.

Steps such as genome replication, assembly, and maturation require specific viral proteins, often acting as enzymes. Enzymes make good drug targets because they are usually at low concentrations inside cells (due to being catalysts), they are relatively well-understood mechanistically, and they normally interact with small-molecule substrates. Furthermore, pharmaceutical companies have a great deal of experience in developing enzyme inhibitors as drugs. It is therefore not surprising that most clinically useful antiviral drugs are enzyme inhibitors that target genome replication, maturation, and release. Nevertheless, there are antiviral agents that act at nearly every step of viral infection.

Drugs preventing attachment and entry of virions

Inhibition of attachment and entry prevents all subsequent steps in virus infection and permits clearance of virus particles by immune and other mechanisms. In principle, drugs can act by binding either to virions or to the cellular proteins to which virions attach (receptors). **Capsid-binding drugs** block attachment and/or entry of picornaviruses, especially rhinoviruses, a major cause of the common cold. There are many different serotypes of rhinoviruses, making vaccination impractical. There is already a large market for drugs that mitigate cold symptoms, and some think that market could be tapped by an effective antiviral agent, if it is very safe.

Capsid-binding drugs act by direct binding to virions. These drugs inhibit attachment and/or entry or even later steps depending on the drug-virus combination. Our understanding of these drugs comes largely from x-ray crystallographic studies of complexes between picornaviruses and such drugs. Figure 36.2 is a model showing how picornaviruses bind to their cellular receptor via the so-called "canyon" in the surface of the capsid. Beneath the canyon floor is the drug-binding site—a "pocket" in the VP1 capsid protein. This pocket is lined with hydrophobic amino acid side chains, and is ordinarily occupied by a lipid molecule. Binding of the cellular receptor molecule to the canyon is thought to release the lipid molecule from the pocket. This allows the capsid to undergo a conformational change that results in the eventual entry of the genome RNA into the cytoplasm.

With this mechanism of picornavirus attachment and entry in mind, how capsid-binding drugs act can be understood. Binding of the drug to the pocket displaces the lipid molecule. The presence of the drug can alter the structure of the canyon floor, thereby preventing proper interaction of the virion with the cellular receptor and preventing attachment. Alternatively, the drug

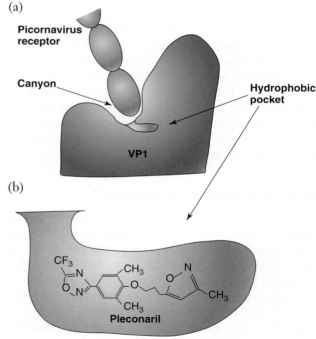

Figure 36.2 Receptor and drug binding to the picornavirus capsid. (a) Binding of the N-terminal domain of a picornavirus receptor to the "canyon" on the surface of the capsid. The hydrophobic pocket lies just below the canyon floor. (b) Binding of pleconaril, a capsid-binding drug, to the hydrophobic pocket displaces the bound lipid and interferes with attachment of virion to receptor and/or entry of viral genome into the cytoplasm.

can stabilize the capsid and prevent the conformational change that releases the viral RNA into the cell, thereby preventing uncoating and genome entry.

An example of a capsid- binding drug is **pleconaril**, which is active against many picornaviruses, including rhinoviruses. In clinical trials, oral pleconaril was found to reduce cold symptoms and virus shedding, but it was not approved by the U.S. Food and Drug Administration (FDA) due to safety concerns. A nasal spray formulation of pleconaril is currently under investigation.

An anti-HIV drug, **enfuvirtide**, which binds to HIV-1 virions and prevents virus entry, was approved by the FDA in 2003. Like all anti-HIV drugs, enfuvirtide is administered as part of a multi-drug cocktail, which reduces the chances of the development of drug resistance. This peptide drug blocks a conformational change in the viral envelope protein gp41, which is required for fusion between the viral envelope and the cell membrane. More recently, another anti-HIV drug, **maraviroc**, which blocks attachment by binding to a cellular receptor for HIV-1, was approved by the FDA (see text box).

Development of maraviroc

Maraviroc is an inhibitor of HIV-1 attachment and entry that was approved by the FDA in 2007. It is a highly unusual antiviral drug because it targets a host protein—CCR5—rather than a viral protein. HIV-1 virions initially attach to cells by binding to the host cell membrane protein CD4. However, HIV-1 does not enter cells unless its envelope glycoprotein gp120 subsequently binds to one of two host cell chemokine receptors, CXCR4 or CCR5. Binding to a chemokine receptor triggers fusion between the HIV-1 envelope and a cellular membrane, leading to virus entry. HIV-1 strains that are most frequently transmitted between individuals can only infect cells that express CCR5.

A crucial discovery came from studies of people who had been exposed repeatedly to HIV-1, yet did not develop AIDS. Certain of these people have a deletion in the gene encoding CCR5 and therefore lack this chemokine receptor. Despite its profound impact on susceptibility to HIV-1, the deletion has little negative impact on human health, making it an attractive target for intervention with a small molecule.

Drug companies performed target-based screens for compounds that could prevent binding of chemokines to CCR5. A compound identified in such a screen was then modified systematically, and eventually maraviroc was selected for clinical trials. Maraviroc blocks infection of HIV-1 strains that use CCR5 for attachment and entry, but not HIV-1 strains that use CXCR4. Because maraviroc is effective only in people who are infected with a virus strain that binds to CCR5, it is important to test patients to see which strains are present before beginning therapy with this drug.

Amantadine blocks ion channels and inhibits uncoating of influenza virions

Amantadine (Figure 36.3) was discovered by random screening in the 1960s. It has been most useful in certain settings, such as nursing homes, for prevention of infection among contacts of flu victims. Amantadine inhibits uncoating of most influenza virus strains. At first, amantadine, which is a weak base, was thought to inhibit the acidification of endosomes, the cellular compartment in which influenza uncoating occurs. However, the concentrations of drug required to inhibit endosomal acidification are much higher than those needed for a therapeutic effect.

The elucidation of the mechanism of action of amantadine represents a triumph of molecular genetic approaches using drug-resistant viruses. Alan Hay and collaborators at the National Institute for Medical Research in London (UK) isolated a set of influenza virus mutants resistant to amantadine. They then used influenza genetics to map the mutations to the M segment of the virus. Upon sequencing, all of the mutations were found to lie within the M2 gene (the other gene in the M segment is the M1 gene). The product of the M2 gene is a small protein that spans the virion envelope. It was noticed that the M2 protein resembles known channel-forming transmembrane proteins. Subsequent biochemical and electrophysiological experiments showed that the M2 protein is a channel for H^+ ions and that amantadine blocks this channel.

How would blocking an ion channel lead to inhibition of uncoating? A reasonable but not completely verified model is diagrammed in Figure 36.3. This starts with the fact that influenza virus enters cells by receptor-mediated endocytosis. As a result, the virions find themselves in endosomes whose pH is lowered. This permits fusion of the influenza envelope with the endosomal membrane, mediated by the altered conformation of the viral hemagglutinin at low pH (see Chapter 4). It is thought that M2 in the virion envelope functions to let H^+ ions enter the virion, and the low pH changes the conformation of the matrix protein, which lines the inside of the viral envelope. This results in liberation of the viral nucleocapsids from within the envelope, so that they can find their way to the nucleus for transcription. Amantadine would block the entry of H^+ ions, thereby preventing uncoating.

A key point about amantadine is that this drug and the drug-resistant mutants opened up a whole new area of virus biology. No one had imagined that a virus might encode an ion channel. Scientists are working on the possibility of discovering drugs against other viruses that encode similar proteins.

Nucleoside analogues target viral DNA polymerases

Virus genome replication is often performed by virus-coded enzymes that are essential for virus replication and

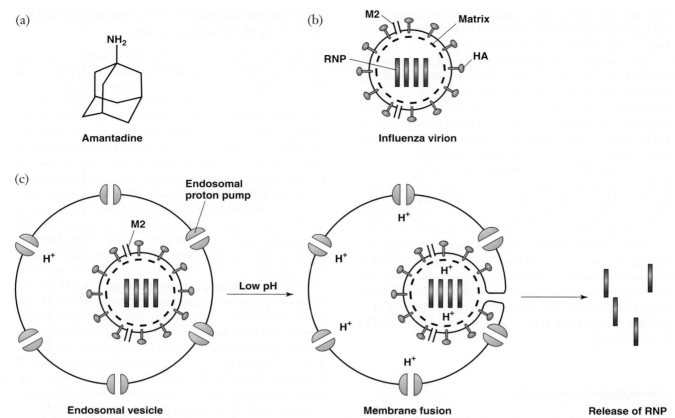

Figure 36.3 Proposed mechanism of action of amantadine on influenza virus uncoating. (a) Structure of amantadine. (b) Structure of influenza virus particle, showing HA and M2 proteins in envelope, matrix protein, and viral ribonucleoprotein (RNP). (c) Influenza virions in endosomes undergo fusion with the endosomal membrane upon a drop in pH induced by an endosomal proton pump. The M2 protein allows hydrogen ions to enter the virion, releasing the RNP from the matrix protein. Amantadine blocks the M2 channel, inhibiting this process.

differ substantially from host proteins. Therefore these viral proteins make good drug targets. Accordingly, more antiviral drugs target genome replication than any other stage of virus infection.

Many clinically useful antiviral drugs are nucleoside analogues that target viral DNA polymerases. Figure 36.4 shows three examples: acyclovir and ganciclovir, both active against herpesviruses, and azidothymidine (AZT), active against HIV-1. Upon entering cells, nucleoside analogues must first be activated by phosphorylation so that they mimic the natural substrates of DNA polymerases, deoxynucleoside 5'-triphosphates. These phosphorylated analogues can inhibit viral DNA polymerases in two ways. They can act as a competitive inhibitor, blocking binding of the natural nucleoside triphosphate to the active site of the enzyme. Alternatively, analogues can be incorporated into the growing DNA chain and prematurely terminate its elongation. Both of these events can be important for antiviral activity and for selectivity.

Nucleoside analogues will usually be toxic for host cells if they can be phosphorylated by host cell enzymes and if the phosphorylated form can inhibit host enzymes

such as DNA polymerases. Therefore, selectivity of these drugs depends on two factors. First, if virus-coded nucleoside kinases can phosphorylate nucleoside analogues more efficiently than their cellular counterparts, these drugs should be more active in infected than in uninfected cells. Second, if viral DNA polymerases are are more sensitive to inhibition by the phosphorylated analogues than are cellular DNA polymerases (or other cellular enzymes), replication of the viral genome in infected cells should be inhibited at drug concentrations low enough to avoid toxic effects on uninfected cells.

Acyclovir is selectively phosphorylated by herpesvirus thymidine kinases

Acyclovir exhibits a very high **therapeutic index** (TI = dose that exerts a 50% toxic effect divided by dose that exerts a 50% antiviral effect). Accordingly, it has been a very successful drug in treating both common herpes simplex virus infections such as genital herpes (many millions of doses of the drug have been administered) and less common but more serious effects of herpesvirus infections such as encephalitis.

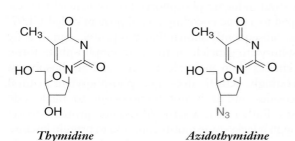

Deoxyguanosine   *Acyclovir*

Ganciclovir

Thymidine   *Azidothymidine*

Figure 36.4 Structures of selected nucleosides and nucleoside analogues. Altered parts are shown in orange.

The drug consists of a guanine base attached to what is essentially a broken sugar ring (Figure 36.4). It can be thought of as deoxyguanosine missing the CHOH moiety at the 3' position and the CH$_2$ moiety at the 2' position of ribose. This **acyclic** sugar-like molecule accounts for the name of the compound and for aspects of acyclovir action.

Figure 36.5 shows what happens to acyclovir in cells infected with herpes simplex virus or varicella zoster virus. Each of these viruses encodes a **thymidine kinase** that is able to phosphorylate not only thymidine and certain other nucleosides, but also a variety of nucleoside analogues, including some like acyclovir, which contain a purine instead of a pyrimidine base. No mammalian enzyme phosphorylates acyclovir as efficiently as do the thymidine kinases of these two herpesviruses, although some phosphorylation does occur in uninfected cells (1–3% of that found in cells infected with herpes simplex virus). Therefore, herpesvirus-infected cells contain a much higher concentration of phosphorylated acyclovir than do uninfected cells. This accounts for much of the antiviral selectivity of acyclovir.

Acyclovir is preferentially incorporated by herpesvirus DNA polymerases

Acyclovir is converted to the monophosphate by the viral enzyme, and is then converted to acyclovir diphosphate and acyclovir triphosphate probably exclusively by cellular enzymes. Acyclovir triphosphate is an inhibitor of herpesvirus DNA polymerases. Moreover, it inhibits viral DNA polymerases more potently than cellular DNA polymerases. The details of the inhibition of herpes simplex virus DNA polymerase by acyclovir triphosphate have been worked out with the purified enzyme *in vitro*. This fascinating mechanism, termed "induced substrate inhibition," involves three steps (Figure 36.6).

In the first step, acyclovir triphosphate acts as a competitive inhibitor of the incorporation of deoxyguanosine triphosphate (pppdG) into DNA, by competing with pppdG for binding to the DNA polymerase. High concentrations of pppdG can reverse inhibition at this stage. In the second step, acyclovir triphosphate serves as a substrate and is incorporated by the viral DNA polymerase into the growing DNA chain opposite a dC residue on the template strand.

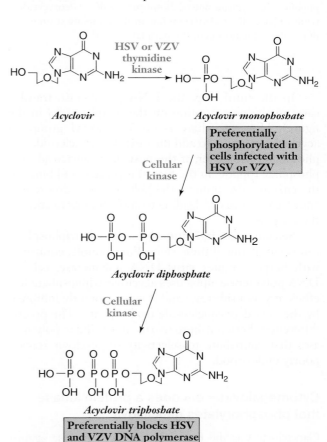

Figure 36.5 Phosphorylation of acyclovir. Shown are the enzymatic steps that result in the phosphorylation of acyclovir to the mono-, di-, and triphosphate forms, the last of which blocks viral replication. HSV: herpes simplex virus; VZV: varicella zoster virus.

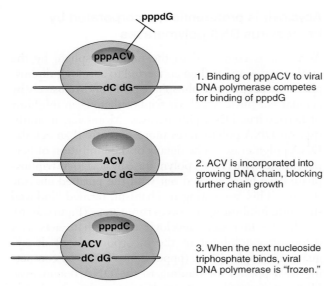

Figure 36.6 Mechanism of inhibition of herpes simplex virus DNA polymerase by acyclovir triphosphate. Large green oval is viral DNA polymerase, small gray oval is dNTP binding site, horizontal lines are template and growing daughter DNA strands. pppACV = acyclovir triphosphate; pppdG = deoxyguanosine triphosphate; pppdC = deoxycytidine triphosphate; dC = deoxycytidine in the template strand; dG = deoxyguanosine in the template strand.

1. Binding of pppACV to viral DNA polymerase competes for binding of pppdG

2. ACV is incorporated into growing DNA chain, blocking further chain growth

3. When the next nucleoside triphosphate binds, viral DNA polymerase is "frozen."

In the third step, the DNA polymerase translocates to the next position on the template (dG in the figure). However, there is no 3'-hydroxyl group on acyclovir on which to add the next deoxynucleoside triphosphate. Moreover, if the next deoxynucleoside triphosphate (pppdC in the figure) is present and binds to the enzyme, this induces the polymerase to freeze in a "dead-end complex," leading to apparent inactivation of the enzyme.

Selectivity of inhibition by acyclovir triphosphate can occur at any of these steps. For example, compared with herpes simplex virus DNA polymerase, cellular DNA polymerase alpha uses acyclovir triphosphate less efficiently as a substrate and it is less potently inhibited by the next deoxynucleoside triphosphate. The precise differences between herpesvirus and cellular polymerases that contribute to selectivity of acyclovir remain poorly understood.

Cytomegalovirus encodes a protein kinase that phosphorylates ganciclovir

Ganciclovir was the first drug approved for use against human cytomegalovirus, another herpesvirus. Cytomegalovirus infections are usually inapparent in normal adults, but can be life-threatening or sight-threatening in the immunocompromised. Ganciclovir is relatively efficacious in this setting. Its structure is shown in Figure

36.4. Like acyclovir, ganciclovir contains a guanine base linked to an acyclic sugar-like molecule, and lacks a 2' CH_2 moiety. However, it contains the 3' CHOH group that is missing in acyclovir. Thus, it more closely resembles the natural compound, deoxyguanosine, which may account for its greater toxicity.

Ganciclovir must be phosphorylated to act as an inhibitor, in a similar fashion to acyclovir (Figure 36.5). However, cytomegalovirus does *not* encode a homologue of herpes simplex virus thymidine kinase (which phosphorylates ganciclovir very efficiently). Despite this, cytomegalovirus infection induces a roughly 30-fold increase in intracellular phosphorylated forms of ganciclovir. What enzyme is responsible for this?

Biochemical experiments failed to uncover the correct answer, but, as was true for amantadine, molecular genetics succeeded in defining the drug target. A ganciclovir-resistance mutation that prevents cytomegalovirus from inducing phosphorylation of ganciclovir was mapped to a gene encoding a viral protein called UL97. It was subsequently shown that this protein can directly phosphorylate ganciclovir to the monophosphate form; cell kinases then generate the di- and triphosphates. Interestingly, UL97 does not phosphorylate natural nucleosides and it is not homologous to nucleoside kinases. Rather, it is a serine/threonine protein kinase, although it lacks some motifs common to protein kinases. How a protein kinase can phosphorylate a nucleoside analogue is an interesting problem in enzymology.

Ganciclovir triphosphate is a more potent inhibitor of cytomegalovirus DNA polymerase than of cellular DNA polymerases. Thus, as is the case for acyclovir and herpes simplex virus, ganciclovir has selective activity against cytomegalovirus at two steps—activation by phosphorylation and inhibition of DNA polymerase. However, selectivity at these two steps is not as great for ganciclovir as it is for acyclovir. Accordingly, ganciclovir is more toxic and is used only for very serious infections.

HIV-1 reverse transcriptase preferentially incorporates azidothymidine into DNA, leading to chain termination

Azidothymidine (AZT; zidovudine) was the first drug approved for treatment of acquired immune deficiency syndrome (AIDS) caused by HIV-1, based on clinical studies showing that it could prolong life of patients with AIDS. However, when used alone its benefits were rather limited, and, like all anti-HIV drugs, it is now used in combination with other drugs.

AZT, like acyclovir and ganciclovir, is a nucleoside analogue with an altered sugar moiety. In this case, a thymine base is attached to a ribose in which the normal 3' OH has been converted to an azido (N_3) group (Figure 36.4). Thus, like acyclovir, it leads to chain termination when incorporated into a growing DNA molecule.

The intracellular metabolism of AZT is very similar to that of thymidine (Figure 36.7). HIV-1 does not encode an enzyme that activates AZT by phosphorylation. Rather, AZT is a substrate for cellular thymidine kinase. Thus, unlike acyclovir and ganciclovir, activation of this drug is not selective; it occurs in uninfected as well as in infected cells.

AZT monophosphate is then converted to AZT diphosphate by cellular **thymidylate kinase**. However, AZT monophosphate is also an inhibitor of thymidylate kinase, which is essential for synthesis of thymidine triphosphate and therefore DNA. This likely accounts for at least some of the toxicity of this drug, which most often manifests itself clinically as bone marrow suppression.

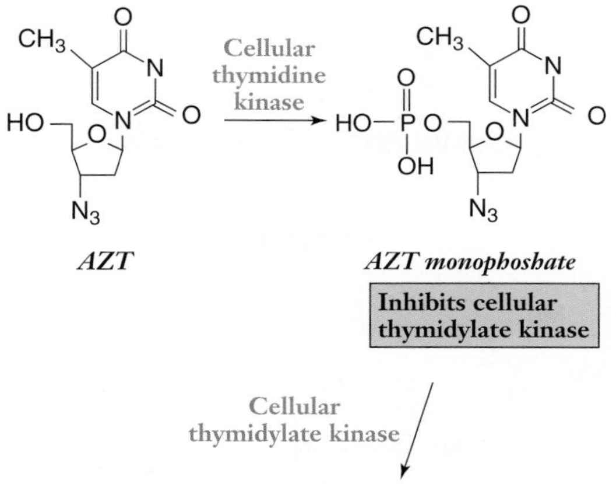

AZT

AZT monophoshate

> Inhibits cellular thymidylate kinase

Cellular thymidylate kinase

AZT diphoshate

Cellular nucleoside diphosphate kinase

AZT triphosphate

> Preferentially blocks HIV reverse transcriptase

Figure 36.7 Phosphorylation of azidothymidine (AZT). Unlike acyclovir (Figure 36.5), AZT is not selectively phosphorylated by a viral kinase, but follows the phosphorylation pathway of thymidine in the cell. As is the case with acyclovir, the triphosphate species exerts antiviral activity. However, the monophosphate species of AZT blocks the cellular enzyme, thymidylate kinase, contributing to toxicity.

AZT diphosphate is converted to the triphosphate by cellular **nucleoside diphosphate kinase**. AZT triphosphate is a more potent inhibitor of HIV reverse transcriptase than of human DNA polymerases; this accounts for the selectivity of AZT.

Since the development of AZT, several other nucleoside analogues have have been developed and approved for use against HIV. Like AZT, these compounds are phosphorylated by cellular kinases and their triphosphates inhibit HIV-1 reverse transcriptase. Some of these compounds exhibit less or different toxicity than AZT and some can be administered less frequently. Additionally, some of these compounds and other nucleoside analogues have been approved for use against infections by hepatitis B virus. Although virions of hepatitis B virus contain DNA, the viral genome is transcribed into a full-length RNA copy within the infected cell, and that RNA is subsequently reverse-transcribed into DNA. Following phosphorylation by cellular kinases, certain nucleoside analogue triphosphates are selective inhibitors of the viral enzyme responsible for this reverse transcription step.

Non-nucleoside inhibitors selectively target viral replication enzymes

Despite problems with AZT, it helped to validate the concept developed from viral genetics and biochemistry that HIV-1 reverse transcriptase is a useful drug target. Subsequently, several pharmaceutical companies developed target-based high throughput screens against HIV-1 reverse transcriptase, leading to the identification of a number of promising compounds that are not nucleoside analogues.

Interestingly, the various compounds are similar to each other in structure (Figure 36.8 shows one, nevirapine), and they act in a similar fashion. These **non-nucleoside inhibitors** bind to reverse transcriptase at a site close to the active site for DNA polymerase activity, and they drastically slow the rate of DNA polymerization. They are very selective—they don't even inhibit the reverse transcriptase of the closely related virus HIV-2—and their discovery makes a strong case for the validity of the target-based high throughput approach.

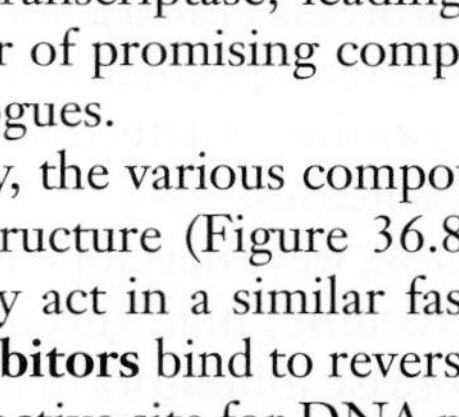

Nevirapine

Figure 36.8 Nevirapine, a non-nucleoside inhibitor of HIV-1 reverse transcriptase.

Due to the ability of the virus to rapidly develop resistance to these compounds, they are used in combination with other drugs. Using several different drugs simultaneously reduces the likelihood that a virus will become resistant to all of the drugs, and certain mutations that provide resistance against one drug may make the same protein more sensitive to a second drug or lessen the ability of the virus to replicate.

In 2007, the FDA approved an anti-HIV drug named **raltegravir**. This compound acts by inhibiting a viral enzyme, **integrase**. Integrase catalyzes a crucial step in HIV genome replication: integration of the DNA copy generated by reverse transcriptase into host cell DNA. Raltegravir came out of a drug discovery program that utilized a target-based assay of integrase.

Protease inhibitors can interfere with virus assembly and maturation

Many viruses encode proteases that are essential for viral replication, often at the stage of assembly and maturation. A great deal of effort has been made to discover drugs active against these enzymes. Much of the impetus for this effort came from the successes and the lessons learned from the development of anti-HIV-1 protease inhibitors. HIV-1 protease is an excellent target for antiviral chemotherapy because it is an enzyme that is essential for HIV replication; it cleaves the Gag precursor of viral capsid proteins within the assembled virion. Without these cleavages, the progeny virions are unable to infect cells and therefore the infection does not spread. HIV-1 protease is also a good target because it has an unusual substrate specificity; it cleaves between phenylalanine and proline, a site that is rarely if ever cleaved by human proteases.

Protease inhibitors were developed rationally rather than by random screening, high throughput, or otherwise. The use of protease inhibitors in combination with reverse transcriptase inhibitors led to strong reductions in the amount of HIV-1 circulating in patients and to dramatic improvements in the health and longevity of the patients.

Ritonavir: a successful protease inhibitor of HIV-1 that was developed by rational methods

One example of these drugs is ritonavir (structure in Figure 36.9), a **peptidomimetic** drug (it mimics a peptide). Its design began with one of the natural substrates of HIV-1 protease (Leu$_{165}$-Asn-Phe-Pro-Ile), a region of the Gag-Pol polyprotein that is cleaved to release reverse transcriptase.

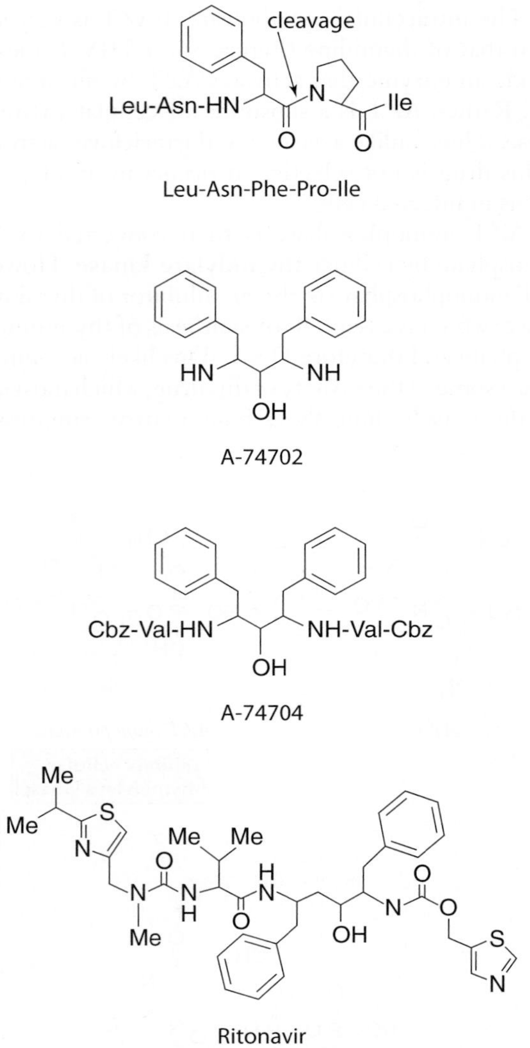

Figure 36.9 Steps in the development of ritonavir. Top: a natural substrate of HIV-1 protease. The peptide bond (arrow) between phenylalanine (phe) and proline (pro) is cleaved by the protease. Middle: two early symmetric inhibitors in which the peptide bond is replaced by a hydroxyethylamine. Bottom: the end result, ritonavir. Cbz = carboxybenzyl; Me = methyl.

A key insight was provided by the crystal structure of HIV-1 protease. The enzyme was found to be a symmetrical dimer of two identical subunits. The most closely related cellular proteases are not symmetrical dimers. The group that developed ritonavir therefore made a correspondingly symmetrical inhibitor in which the proline was replaced with phenylalanine, and the peptide bond between the two phenylalanines was modified by converting the normal C:O to CHOH, creating a hydroxyethylamine (A-74702 in Figure 36.9).

Hydroxyethylamine compounds not only mimic the original peptide, they also mimic the transition state of protease catalysis (i.e., the catalytic intermediate that binds most tightly to the enzyme). However, unlike the original peptide and the actual transition state, they cannot be cleaved. These features make them good inhibitors.

A-74702 is a very weak inhibitor, but adding symmetrical valine-carboxybenzyl groups at both ends (A-74704 in Figure 36.9) resulted in a 40,000-fold increase in potency. Further modifications increased potency by more than 100-fold, resulting in a compound that can inhibit the protease when present at picomolar (10^{-12} M) concentrations. The modifications also vastly improved the uptake and stability of the drug in the body (**pharmacokinetics**).

These improvements were facilitated by an iterative process of synthesizing an inhibitor and solving the structure of the protease bound to the inhibitor by x-ray crystallography. Then by looking at these structures, scientists made informed guesses about what groups to add or subtract to generate a new inhibitor. After many cycles of this process, the result was ritonavir.

HIV-1 infected cells treated with protease inhibitors continue to make viral proteins, but the Gag polyprotein is not cleaved. Therefore, virus particles bud from the infected cells, but they are immature and do not successfully infect other cells. The proof that these drugs act as expected is that drug-resistance mutations are located in the HIV-1 gene encoding the protease. These protease inhibitors are now used along with other drugs in some of the very successful triple therapy drug combinations that efficiently suppress HIV-1 replication and virus levels, and lead to healthier and longer lives for people infected with HIV-1.

Neuraminidase inhibitors inhibit release and spread of influenza virus

The inhibitors of influenza virus **neuraminidase** are perhaps the best examples of the success of a purely rational approach to drug design. These drugs inhibit virus release from infected cells. Most viruses do not require special mechanisms to be released from the cell surface, but influenza virus does. It initially attaches to cells by interactions between the virion hemagglutinin protein and cellular **sialic acid** (Figure 36.10a; also known as N-acetyl neuraminic acid), which is present on many membrane glycoproteins and **gangliosides.** This greatly facilitates virus entry.

However, at the end of the virus replication cycle, binding of virion hemagglutinin proteins to sialic acid

on the cell surface inhibits release of newly formed virus particles. To overcome this problem, influenza virus encodes another envelope protein, the enzyme neuraminidase. Neuraminidase cleaves sialic acid from the membrane glycoproteins, permitting release of the virus; without this enzyme, the virus remains stuck and cannot spread to other cells. Influenza neuraminidase differs sufficiently from cellular enzymes to make it an excellent drug target.

The design of the two neuraminidase inhibitors presently available, zanamivir and **oseltamivir** (Figure 36.10a), began when the three-dimensional atomic structure of the neuraminidase was determined by x-ray crystallography in 1983. Subsequently, the structure of the neuraminidase complexed with sialic acid was solved, showing that sialic acid occupies two of three well-formed pockets on the enzyme (Figure 36.10b). Zanamivir and oseltamivir were designed based both on this structure and that of neuraminidase bound to a transition state inhibitor. The idea was to maximize energetically favorable interactions with specific amino acid residues of the enzyme in the three pockets. The idea worked and highly potent inhibitors were derived. These neuraminidase inhibitors provide one of the best examples of rational drug design starting from the structure of a protein and a natural ligand.

Although these two compounds are potent *in vitro*, clinical trials showed that they do not rapidly alleviate flu symptoms. They can reduce the duration of fever by one or two days, but only if administered within 48 hours after the appearance of the first symptoms of infection. They have been marketed aggressively to patients, using, for example, actors from sitcoms to play the role of influenza virus. More recently, these compounds have shown effectiveness in reducing severe disease symptoms and deaths due to the emerging influenza H5N1 ("bird flu") and pandemic H1N1 ("swine flu") strains, which has led to stockpiling and more widespread use of these drugs.

Antiviral chemotherapy shows promise for the future

In summary, the field of antiviral chemotherapy is a very active and fertile area. Part of the excitement is that we are taking the knowledge we have gained over many years about viruses and are putting it to work to treat viral diseases. The successes with influenza virus, HIV-1, and herpesviruses lead to optimism that other viral diseases will be susceptible to antiviral therapies in the future. Furthermore, antiviral drugs provide a terrific avenue for learning more about viruses.

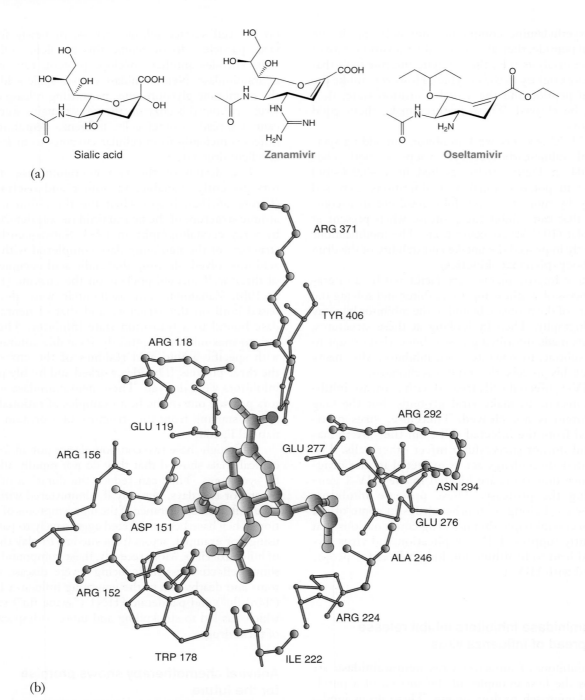

(a)

(b)

Figure 36.10 Rational design of neuraminidase inhibitors. (a) Structures of sialic acid, zanamivir, and oseltamivir. Ac = acetyl; Me = methyl. (b) Model of part of the active site of influenza neuraminidase bound to zanamivir, shown as ball-and-stick model in the center (small orange spheres = O; larger orange spheres = N). Amino acids near the binding site are labeled.

KEY TERMS

Acyclic

Acyclovir

Amantadine

Azidothymidine

Capsid-binding drugs

Cell-based high throughput screen

Enfuvirtide

Ganciclovir

Gangliosides

Integrase

Interferon

Maraviroc

Neuraminidase

Nevirapine

Non-nucleoside inhibitors

Nucleoside diphosphate kinase

Oseltamivir

Peptidomimetic

Pharmacokinetics

Pleconaril

Raltegravir

Rational drug discovery

Ritonavir

Sialic acid

Target-based high throughput screen

Therapeutic index

Thymidine kinase

Thymidylate kinase

Zanamivir

FUNDAMENTAL CONCEPTS

• Possible antiviral compounds are screened for activity by automated screening using infected cells or viral proteins as assays.

• Another approach is to design drugs based on knowledge of the active sites of the enzymes to be targeted.

• Once a promising candidate is discovered, chemists synthesize analogues that may have higher antiviral activity.

• An anti-HIV drug, maraviroc, blocks attachment of the virus by binding to a cellular receptor.

• Another anti-HIV drug, enfuvirtide, blocks a conformational change in viral envelope protein gp41 that is required for fusion and viral entry.

• Amantadine inhibits uncoating of influenza virus by blocking an ion channel protein that allows H^+ to enter virions within endosomes.

• Acyclovir is a nucleoside analogue that is preferentially phosphorylated by herpes simplex virus thymidine kinase and is incorporated into DNA by the viral DNA polymerase, blocking further DNA replication.

• Ganciclovir is phosphorylated by a cytomegalovirus protein kinase and preferentially inhibits activity of the viral DNA polymerase.

• A number of nucleoside analogues, beginning with azidothymidine, inhibit HIV reverse transcriptase by premature termination of growing DNA chains.

• Other inhibitors of HIV reverse transcriptase are not nucleoside analogues, but bind near the active site of the enzyme.

• A recent anti-HIV drug, raltegravir, inhibits viral integrase.

• Protease inhibitors active against HIV inhibit maturation of virus particles.

• Neuraminidase inhibitors suppress release of influenza virions from sialic acid residues on the surface of infected cells.

REVIEW QUESTIONS

1. What is the antiviral mechanism of pleconaril (a drug not currently approved by the FDA)?

2. Influenza virus enters cells via receptor-mediated endocytosis, yet replication of the virus is inhibited by drugs that block ion channels. How might such a drug target influenza virus replication?

3. Most drugs that inhibit herpesvirus DNA replication require activation by a viral nucleoside kinase. How is ganciclovir activated?

4. By means of rational drug design, zanamivir and oseltamivir were discovered as inhibitors of influenza virus. How do they combat influenza virus infections?

5. Pharmaceutical companies have not always been enthusiastic about investing resources in the development of antiviral drugs because the development of compounds active against a broad spectrum of viruses is unlikely. Discuss some of the reasons why broad-spectrum antibiotics have been produced against bacteria, but broad-spectrum antivirals are unlikely.

CHAPTER 37

Eukaryotic Virus Vectors

Rénald Gilbert
Bernard Massie

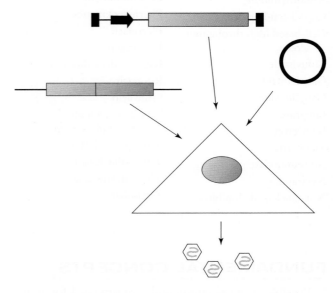

USES OF VIRUS VECTORS

For recombinant protein production.
As recombinant vaccines.
For gene therapy.
As anticancer agents.

CONSTRUCTION OF VECTORS

Deletion of some or all virus genes.
Retention of *cis*-acting replication and packaging
   sequences in the vector genome.
Insertion of gene(s) of interest into the vector genome.

PRODUCTION OF VECTORS

Transfection of genome into cell lines engineered to
   express virus genes deleted in the vector, or
Transfection of genome into cells that are co-infected
   with helper viruses that express missing genes.

ADENOVIRUS VECTORS

Are easy to produce at high titers.
Have been widely used in human trials.
Can induce immune responses that reduce their utility.

RETROVIRUS VECTORS

Can permanently introduce a foreign gene into target
   tissues by integration.

Are less immunogenic than first-generation
   adenovirus vectors.
Can be packaged with the G envelope protein of
   vesicular stomatitis virus to allow infection
   of many cell types.

ADENO-ASSOCIATED VIRUS VECTORS

Are considered safe because the parent virus is
   nonpathogenic for humans.
Are less immunogenic than other vectors.
Have limited capacity for insertion of large
   foreign genes.

Many viruses can be engineered to deliver and express specific genes

When viruses are genetically modified to deliver and express either their own or foreign genes, they are called **virus vectors**. Although DNA can be delivered to cells using a variety of physical and chemical methods, virus vectors are capable of much higher gene transfer efficiency. Virions are highly efficient gene transfer machines that interact with specific receptors on the cell surface and deliver their nucleic acid genomes to the appropriate compartment for optimal expression and replication (Chapters 2 and 4). This gives virus vectors an enormous advantage over other methods for gene transfer. The genomes of many different viruses of eukaryotes have been cloned and can be manipulated in bacterial plasmids and subsequently reintroduced into their natural host cells to produce recombinant virions.

Virus vectors have been designed for four major objectives: (1) the production of specific proteins in cultured cells; (2) vaccination of animals or humans against specific diseases; (3) gene therapy in animals or humans, usually to replace deficient genes; and (4) anticancer agents that specifically kill tumor cells. As Table 33.1 shows, vectors have been developed from a large variety of viruses whose genomes can be positive- or negative-strand RNA, single- or double-stranded DNA, ranging from 5 to 200 kb in size. Tens of thousands of research papers have been published on the subject in the past 20 years.

Virus vectors are used to produce high levels of specific proteins in cultured cells

The first virus used as a vector for protein production in mammalian cells was simian virus 40 (SV40), a polyomavirus. Richard Mulligan and Paul Berg expressed rabbit β-globin in monkey cells in 1979 using the SV40 genome as a vector. The 5.3-kb circular, double-stranded DNA SV40 genome was one of the first eukaryotic virus genomes to be cloned in a bacterial plasmid, and its complete nucleotide sequence was determined in the late 1970s. However, the small genome size was a handicap, because it limits the size of foreign DNA that can be packaged in the virion.

Table 37.1 Some eukaryotic virus vectors and their applications

Application	Virus family	Virus species
Protein production	Adenovirus	Adenovirus type 5
	Retrovirus	Human immunodeficiency virus type 1
	Baculovirus	*Autographa californica* nucleopolyhedrovirus
	Togavirus	Semliki Forest virus, Sindbis virus
	Adenovirus	Adenovirus types 2 and 5
Recombinant vaccines	Coronavirus	Transmissible gastroenteritis virus, mouse hepatitis virus, infectious bronchitis virus
	Flavivirus	Japanese encephalitis virus, yellow fever virus
	Herpesvirus	Bovine herpesvirus
	Orthomyxovirus	Influenza A and B viruses
	Paramyxovirus	Sendai virus, simian virus 5
	Parvovirus	Adeno-associated virus
	Picornavirus	Poliovirus
	Poxvirus	Vaccinia virus, canarypox virus, fowlpox virus
	Rhabdovirus	Rabies virus, vesicular stomatitis virus
	Togavirus	Semliki Forest virus, Sindbis virus, western equine encephalitis virus, Venezuelan equine encephalitis virus
Gene replacement therapy	Adenovirus	Adenovirus types 2 and 5
	Baculovirus	*Autographa californica* nucleopolyhedrosis virus
	Flavivirus	Kunjin virus, tick-borne encephalitis virus
	Herpesvirus	Cytomegalovirus
	Retrovirus	Human immunodeficiency virus type 1, simian immunodeficiency virus type 1, murine leukemia virus
	Paramyxovirus	Sendai virus, human parainfluenza virus
	Parvovirus	Adeno-associated virus, B19 virus
	Polyomavirus	Simian virus 40
	Togavirus	Semliki Forest virus, Sindbis virus
Antitumor agents	Adenovirus	Adenovirus type 5
	Herpesvirus	Herpes simplex virus
	Paramyxovirus	Newcastle disease virus, simian virus 5
	Parvovirus	Minute virus of mice, H-1 parvovirus, LuIII virus
	Poxvirus	Vaccinia virus, avian poxviruses
	Reovirus	Reovirus
	Rhabdovirus	Vesicular stomatitis virus

Both adenoviruses (35-kb genome) and baculoviruses (135-kb genome) proved to be more effective vectors for protein production because of their larger genome sizes and the ease of producing recombinant virions. They are still used for such purposes, and a number of vectors are available commercially as kits, along with selection systems for recombinant viruses and susceptible cell lines for propagation.

More recently, high-level protein production has been achieved with other vector systems, particularly those derived from togaviruses such as Semliki Forest virus and retroviruses such as human immunodeficiency virus type 1. Large-scale transfection of mammalian cells with plasmids has begun to supplant the use of recombinant viruses for the purposes of protein production.

Gene therapy is an expanding application of virus vectors

Vaccination and gene therapy are the most widely publicized uses of virus vectors. The idea of gene therapy arose in the late 1950s, when it was realized that viruses are able to transport their own genetic material into cells. Furthermore, tumor viruses such as polyomaviruses and retroviruses (Chapters 21 and 28) were shown in the late 1960s to integrate copies of the virus genome into the host cell during transformation. This led to the realization that viruses could be engineered to bring foreign genes (**transgenes**) into cells, and that those genes could in some cases be stably integrated into the cellular genome, replacing defective host cell genes.

Gene replacement therapy was initiated in the mid-1980s with the successful transfer of the adenosine deaminase gene to mouse bone marrow stem cells to correct a severe immune deficiency, using a retrovirus vector. But it was not until 1990 that the equivalent experiments were carried out in humans, with limited success. Developments in the past 20 years have led to more than 1500 clinical trials of gene therapy, most using virus vectors. However, gene therapy is still largely an experimental procedure, and much work remains to be done to make it available for general use in humans.

Gene transfer can be carried out *in vitro* by infecting cells isolated from an individual and then reintroducing cells expressing the desired gene into the host organism. Alternatively, genes can be transferred directly into cells in the body by injection of virus vectors. For gene therapy applications such as vaccination and anticancer therapy, transient but high-level expression of gene products in the target cells may be sufficient. Other applications such as gene replacement therapy require long-term expression, ideally for the lifetime of the individual.

With the advent of gene transfer technologies, new horizons are opening for the development of

biopharmaceuticals based on recombinant DNA technology. Presently, administration of therapeutic proteins is restricted to secreted and circulating proteins such as **antibodies**, serum proteins, **cytokines**, and hormones. In the future, it will be possible to synthesize therapeutic proteins directly in the cells of an individual. With this new drug delivery system, the possibilities are great: the delivery of genes encoding receptors for signaling molecules, membrane transport proteins, structural proteins, enzymes, transcription factors, and so on, can be used in a myriad of therapeutic applications.

Virus vectors are produced by transfection of cells with plasmids containing deleted genomes

For gene replacement, an ideal vector should be capable of delivering a transgene specifically to the appropriate target cells and expressing that gene sufficiently strongly to obtain a therapeutic effect. In most cases, this will require (1) efficient, targeted delivery, (2) control of both the level and duration of transgene expression, and (3) minimal toxicity and/or immune reaction to either the vector components or the transgene.

A typical virus vector is unable to replicate because of deletion of some or all essential viral genes (Figure 37.1). The vector retains the *cis*-acting sequence elements of the virus genome required for replication (ori) and packaging (Ψ), and the deleted viral genes are replaced

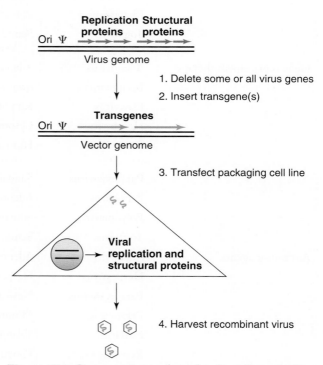

Figure 37.1 Construction and production of a typical virus vector. Ori = sequences required for genome replication; Ψ = *cis*-acting packaging sequences.

with cassettes controlling the expression of the foreign genes. Vectors are propagated in cell lines that express the deleted viral genes and can therefore provide the essential replication and packaging components normally provided by the intact virus. Development of an efficient producer cell line is often the most challenging requirement in the construction of a vector system. In some cases, viral proteins are provided by infection with a **helper virus**; however, this complicates production of recombinant vector stocks, because the helper virus must be removed.

Vectors based on viruses with RNA genomes are constructed in DNA plasmids containing copies of the virus genome made by reverse transcription. These plasmids are introduced into cells where they are transcribed either by endogenous cellular RNA polymerases or by bacteriophage T7 RNA polymerase, which is provided by specialized cell lines or by a vaccinia virus vector (see Chapters 7 and 26).

In most cases, both the 5' and the 3' ends of the genome RNA of the vector must be precisely reproduced by transcription from the plasmid DNA to enable replication of the transcripts by viral RNA polymerases. Various genetic engineering tricks are used to produce RNAs with precise 3' ends, including the use of T7 transcriptional terminator signals and self-cleaving **ribozymes** (see Chapter 31). The resulting virus genome RNAs are replicated by the viral RNA polymerase, which is usually provided by co-infection of the cells with an expression plasmid.

Virus vectors are engineered to produce optimal levels of gene products

The expression cassette (Figure 37.2) is one of the chief elements of a virus vector. It is a recombinant DNA molecule containing sequence information for efficient transcription of the desired gene and translation of its messenger RNA. A typical recombinant DNA used for gene expression in mammalian cells includes a strong **transcriptional enhancer** and promoter, appropriate sequences at the 5' end of the mRNA for efficient initiation of translation, RNA splicing signals, and finally a strong **polyadenylation signal**.

If two transgenes must be expressed coordinately, they can be cloned as two independent transcription units with their own enhancers, promoters, splice sites, and polyadenylation signals. Alternatively, they can be expressed on the same mRNA molecule but be controlled by independent translation initiation cassettes (internal ribosome entry sites or IRES; see Chapter 11). Expression cassettes for RNA virus vectors do not usually contain transcriptional promoters/enhancers, RNA splicing sites, or polyadenylation signals because they are expressed as RNA by viral RNA polymerases in the cytoplasm.

Although many different gene therapy vectors have been developed, this chapter will focus on three of the most widely used vectors: those derived from adenoviruses, retroviruses, and parvoviruses (Table 37.2).

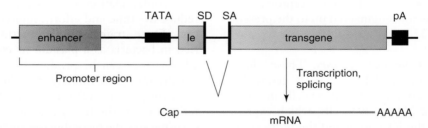

Figure 37.2 Typical mammalian expression cassette. Shown is a typical transgene with two exons. The first exon contains leader sequence (le) for efficient translation. Strong transcriptional enhancer and TATA box ensure high-level transcription from a defined site. Strong splice donor (SD) and splice acceptor (SA) sequences as well as a strong polyadenylation signal (pA) ensure efficient processing and export of mRNA.

Table 37.2 Properties of selected virus vectors

| Vector | Transgene capacity | Transduced cells | | Integrates transgene | Antigenicity |
		Dividing	Nondividing		
First-generation adenovirus	7 kb	Yes	Yes	No	Strong
Third-generation adenovirus	35 kb	Yes	Yes	No	Weak
Oncoretrovirus	8 kb	Yes	No	Yes	Weak
Lentivirus	8 kb	Yes	Yes	Yes	Weak
Adeno-associated virus	4.5 kb	Yes	Yes	Yes[1]/No	Weak

[1]*Efficient integration requires the presence of the Rep protein.*

Descriptions of vectors derived from bacteriophage T7, poxviruses, baculoviruses, flaviviruses, togaviruses, and bunyaviruses can be found in their respective chapters.

ADENOVIRUS VECTORS

Adenovirus vectors are widely used in studies of gene transfer and antitumor therapy

Adenoviruses (Chapter 23) have a double-stranded linear DNA genome of 25 to 45 kbp, depending on the serotype. Adenoviruses do not integrate their genome during productive infection, but the adenovirus genome can persist as a stable episome in some cell types for months or even years. Adenovirus vectors have been extensively used for gene transfer into mammalian cells since the mid-1980s. They are the best-characterized vectors, and the most common vectors used in preclinical and clinical trials. High-titer stocks can be produced easily, and high levels of transgene expression can be achieved with these vectors.

Replication-defective adenovirus vectors are propagated in complementing cell lines

The development of adenovirus vectors has involved the following genetic modifications: (1) eliminating the capacity of the vector to replicate on its own; (2) increasing its capacity to package exogenous genes; (3) reducing the host immune response to viral proteins; and (4) retargeting the vector to specific cell receptors. The various adenovirus vectors engineered up to the present are shown in Figure 37.3.

First-generation adenovirus vectors. These have undergone deletion of the E1A and E1B genes. E1 deleted adenoviruses can replicate only in cell lines such as human embryonic kidney 293 cells, which express integrated copies of the E1 region and therefore complement these functions. The E3 region is not required for virus production in cultured cells *in vitro* and its deletion increases the space available (total 7 kbp) for insertion of exogenous DNA.

Construction of first-generation vectors was routinely performed by transfecting 293 cells with two DNA fragments. One contains the wild-type adenovirus genome cleaved within the E1 region to remove the left-hand **inverted terminal repeat** and **packaging signals**. The second is a transfer plasmid that contains the left-hand end of the genome, the transgene cassette, and a homologous sequence to the right of the transgene cassette that allows for recombination to occur. This procedure, favored for its simplicity, is nevertheless inefficient. Recombination is a rare event and incompletely digested adenovirus genomes may give rise to nonrecombinant virus.

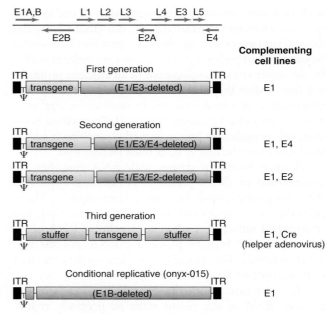

Figure 37.3 Production of adenovirus vectors. The adenovirus genome map showing major early (E) and late (L) transcription units is shown at top. Below are schematic structures of the genomes of five adenovirus vectors, and at the right, viral genes expressed in complementing cell lines. See text for details. Viral DNA: blue; transgene DNA: green; stuffer DNA: orange. ITR: inverted terminal repeat; Ψ: packaging signal; Cre: recombinase.

Other strategies have been developed to minimize contamination with nonrecombinant virus and to reduce the time and effort required for vector production. These strategies utilize a virus genome that is carried in bacteria as a plasmid or cosmid. Recombinant adenoviruses can be produced by either: (1) subcloning the transgene directly into the plasmid containing the virus genome; (2) using recombination within bacteria to insert the transgene into the virus genome; or (3) using specific recombinases such as Cre recombinase (see below) to insert the transgene.

Second-generation adenovirus vectors. First-generation adenovirus vectors can sustain transgene expression for years in post-mitotic cells of immunodeficient animals. However, expression persists for only a few weeks in immunocompetent animals, because the host immune system destroys transduced cells expressing viral proteins. To reduce immune rejection, adenovirus vectors were further crippled by deletion of genes in the E2 or E4 regions (Figure 37.3). The production of these second-generation vectors required the development of producer cell lines expressing the corresponding deleted genes, which are required for virus replication.

Deletions in E2A (single-stranded DNA binding protein), E2B (viral DNA polymerase/preterminal protein), and E4 were engineered. In most cases, *in vivo* studies showed significant reduction of the immune response, lower liver toxicity, and prolonged vector stability. However, adenovirus vectors lacking E4 are retained less stably in transduced cells and produce lower amounts of transgene protein. It appears that E4 is required for efficient expression of transgenes under the control of several promoters used in these vectors, particularly the cytomegalovirus and Rous sarcoma virus promoters.

Third-generation adenovirus vectors. Because second-generation adenovirus vectors do not fully escape the host immune response, and because there was a need for larger cargo space for some applications, the more drastic approach of eliminating all viral genes was undertaken. Fully deleted vectors have been given several names: third generation, helper dependent, high capacity, or "gutless." Those vectors offer two main advantages: (1) no viral genes are present, eliminating problems of immune response to viral proteins; (2) up to 35 kb are available for insertion of transgenes for expression.

However, these vectors are more difficult to produce; it is almost impossible to generate a stable cell line that complements all of the essential adenovirus genes, due to the toxicity associated with the over-expression of most of these proteins. The functions required for complementation of third-generation adenovirus vectors are provided by coinfection with a helper virus, usually a first-generation adenovirus vector. However, this leads to a further problem: the helper virus replicates at the same time as the vector, and it must be removed from the vector stock. Furthermore, recombination can occur between the DNA of the helper virus and the third-generation adenovirus vector, resulting in a hybrid population containing elements belonging to both vectors.

The most efficient strategy to eliminate the helper virus is to remove the packaging signal on the helper virus genomes made during propagation of the vector. This can be conveniently done by the use of the Cre recombinase, derived from bacteriophage P1. The helper adenovirus is engineered to contain *lox*P sequences flanking its packaging signal. In 293 cells expressing Cre recombinase, the packaging signal of the helper virus is deleted, resulting in DNA that still can produce viral proteins, but cannot be packaged into virus particles. Although cleavage by Cre recombinase is not complete and some helper viruses are still packaged, contamination of the vector with helper virus can be reduced to less than 0.1%.

Replication-competent adenovirus vectors are useful tools in antitumor therapy

The disadvantages of some adenovirus vectors, namely cytotoxicity and induction of an immune response, are assets in applications such as vaccination or cancer gene therapy. Adenoviruses that do not express E1B genes but are otherwise intact replicate poorly in normal cells but well in many tumor cells. This is a result of the complex interplay between adenovirus early gene products and cellular signaling pathways leading to **apoptosis** (see Chapter 23). The selective replication of E1B-deleted adenoviruses in tumor cells leads to death of tumor cells, and enables spread of virus progeny to nearby tumor cells, increasing the antitumor effect of the virus. Other strategies rely on tumor-specific promoters to drive E1A gene expression or expression of cytokine or suicide genes to further improve the antitumor effects of replicative adenovirus vectors.

Onyx-015 is a conditionally replicative adenovirus that lacks the E1B region and preferentially replicates in certain cancer cells. In recent clinical trials it was observed that injections of Onyx-015 directly into solid tumors, or its delivery through the intra-arterial or intravenous routes of administration, was safe but was ineffective at killing cancer cells. On the other hand, in conjunction with chemotherapy, the use of Onyx-015 resulted in complete or partial regression of tumors and to a significant extension of the life of the patients. This success is even more impressive considering that the treated individuals were affected by recurrent cancer of the head and neck, or by gastrointestinal cancer that had spread to the liver, two types of cancer with particularly poor prognosis. Onyx-015, also known as H101, has been commercialized in China for the treatment of head and neck cancer since 2005.

Interestingly, the first commercial gene therapy product to be approved was also an adenovirus used against cancer. This drug, referred to as Gendicine, is a first-generation adenovirus vector encoding the tumor suppressor p53, and was approved for use in China in 2003. P53 regulates apoptosis in normal cells. Many tumor cells have mutations in the gene for p53 that inhibit its capacity to induce apoptosis. Infection of tumor cells with an adenovirus encoding p53 restores its function and triggers cell death. As in the case of Onyx-015, Gendicine works better in combination with conventional treatments such as radiotherapy. Clinical data show that treatment with Gendicine can both extend and improve the quality of the patient's life.

Advantages and limitations of adenovirus vectors

Several features make first-generation adenovirus vectors attractive vehicles for gene transfer and protein production: (1) they can be easily produced at high titers; (2) they can transduce many cell types; (3) they do not require cell division for transgene expression; (4) their genome remains episomal, and therefore the target cell does not undergo permanent mutation. The high levels of recombinant proteins synthesized by first-generation adenovirus vectors in mammalian cells have only recently been surpassed by the lentivirus vector system. However, because adenoviruses do not integrate their genome into cellular chromosomes, cell division leads to dilution of the transgene and eventually to loss of transgene expression. Furthermore, first-generation adenovirus vectors are strongly immunogenic, causing the loss of transgene expression in immunocompetent hosts.

The use of second- and third-generation adenovirus vectors greatly reduces problems caused by host immune responses. However, vectors that lack the E4 region usually express lower levels of transgene proteins expressed from viral promoters. Furthermore, expression of these transgenes is often prematurely silenced. The good news is that long-term transgene expression without evidence of toxicity can be achieved with third-generation adenovirus vectors if nonviral promoters are employed. Because of their lack of toxicity, their weak antigenicity, their demonstrated long-term transgene expression, and their very large insert capacity, third-generation adenovirus vectors are very promising for gene therapy applications. The main hurdle limiting their use is the production of clinical-grade vectors due to the complexity of the production process and to the low levels of helper virus contaminating vector stocks.

RETROVIRUS VECTORS

Retrovirus vectors incorporate transgenes into the cell chromosome

Retroviruses (Chapter 28) use reverse transcriptase to generate a double-stranded DNA copy of their RNA genome and subsequently integrate this DNA into the cellular chromosome. They are therefore ideal vectors for gene replacement therapy, as transgenes introduced into target cells remain stably integrated.

In the case of retroviruses such as murine leukemia virus, integration of the proviral DNA occurs only in dividing cells. The integration complex of these retroviruses is unable to pass through the nuclear pores, and therefore has access to the cellular chromosomes only after the nuclear membrane dissassembles during mitosis. Lentiviruses such as human immunodeficiency virus are capable of transporting their integration complex into intact nuclei, and vectors based on lentiviruses are therefore useful for expressing transgenes in nondividing cells.

Packaging cell lines express retrovirus enzymatic and structural proteins

Retrovirus vectors are produced by packaging cell lines that express all essential viral proteins: Gag (capsid proteins), Pol (reverse transcriptase and integrase), and Env (envelope proteins). The plasmids used to produce these cell lines contain neither the long terminal

repeats (LTR) nor viral packaging signals (Ψ), ensuring that no virus genomes are made or packaged. Currently the best packaging cell lines are generated from split genome constructs (Figure 37.4a). These lines further ensure safety by reducing the possibility of generating replication-competent virus by recombination.

To produce a retrovirus vector, a plasmid called the "transfer construct" is introduced into the packaging cell line (Figure 37.4b). This plasmid includes the gene of interest (up to 8 kb) flanked by two long terminal repeats, and contains the primer binding site (PBS) for reverse transcription and the packaging signal, both near the left-hand LTR. Transcription by cellular RNA polymerase II begins in the left-hand LTR, using the enhancer and promoter sequences located in the U3

(a)

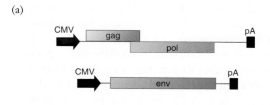

(b)

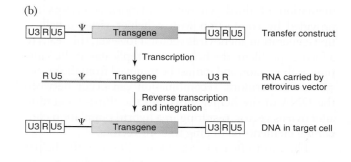

(c)

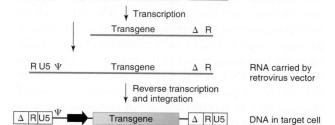

Figure 37.4 Production of retrovirus vectors. (a) Plasmid constructs used to establish a packaging cell line expressing gag, pol, and env gene products. (b) Production of a simple retrovirus vector by transfection of a transfer construct into the packaging cell line. The vector DNA integrated in the target cell is identical to the DNA in the transfer construct. (c) Production of a self-inactivating retrovirus vector. See text for details. CMV: cytomegalovirus promoter; U3, U5, and R, unique 3', unique 5' and repeat regions of the retrovirus long terminal repeat; C: packaging signal; pA: polyadenylation signal.

region, and ends in the right-hand LTR where the polyadenylation signal defines the RNA 3′ end. The enzymes necessary for reverse transcription and integration are produced by the packaging cell line and are incorporated along with the RNA into virions. In most cases, the cell is not lysed and continuously releases virions by budding into the extracellular medium.

The ability of a retrovirus to infect specific cell types (**tropism**) is determined by the envelope protein, which mediates attachment of the virion to the cell surface as well as entry by fusion or endocytosis. It is possible to modify the tropism of a retrovirus by replacing the envelope gene of the producer cell line with a gene derived from a different virus. This strategy is known as **pseudotyping**.

The G glycoprotein of vesicular stomatitis virus (VSV-G) is often used for pseudotyping retroviruses. Retroviruses pseudotyped with VSV-G can infect many different cell types and are stable. Unfortunately, VSV-G is toxic to cells in which it is made. For this reason, retroviruses pseudotyped with VSV-G are produced either by transient transfection of the producer cells with a plasmid encoding the VSV-G gene, or by using a producer cell line containing the VSV-G gene tightly regulated by an inducible promoter.

Strategies for controlling transgene transcription

Transgene expression in target cells can be controlled by the enhancer/promoter elements of the LTR or by using a second promoter that directly controls the transgene (Figure 37.4c). A common problem that arises when the transgene is controlled by the LTR is reduction or loss of transgene expression. This phenomenon, known as **silencing**, is most likely caused by methylation of DNA sequences located in the LTR, leading to loss of transcriptional activity. Some of the sequences within the LTR responsible for promoter silencing have been characterized and can be deleted to correct the problem.

An alternative approach is to insert a second promoter to control transgene expression (Figure 37.4c). However, interference between the LTR promoter and the second promoter (**promoter occlusion**) often occurs. This interference is thought to be caused by dislodging of transcription factors from the downstream promoter by RNA polymerases that initiate at the upstream promoter.

For this reason an elegant scheme has been designed to inactivate the transcriptional activity of the LTR after integration. Such vectors, referred to as self-inactivating retrovirus vectors, are safer because they cannot produce viral RNA after integration into their target cells. Furthermore, they cannot use the right-hand LTR to drive transcription of adjacent cellular genes, which could be problematic if the proviral DNA integrated

next to an oncogene (see Figure 28.9). A deletion is made within the U3 region of the right-hand LTR of the transfer construct, thus eliminating its transcriptional activity (Figure 37.4c). The left-hand LTR is normal, and drives transcription of the full-length RNA. After reverse transcription and integration, the U3 regions of both the left and right LTRs will be deleted and both LTRs will therefore be transcriptionally inactive (see the mechanism of reverse transcription in Chapter 28).

Lentivirus vectors are used for gene delivery to non-dividing cells

Lentiviruses have a major advantage over other retroviruses when used as vectors: the integration complex, containing the proviral DNA made by reverse transcriptase, is transported into the nucleus via the nuclear pores. This means that lentivirus vectors can transduce stationary cells as well as dividing cells. The best-characterized lentivirus is human immunodeficiency virus type 1, the causative agent of AIDS; most lentivirus vector systems used today are derived from this virus. The first experiment using a lentivirus as a vector was accomplished in 1996 by the group of Inder Verma at the Salk Institute in California. These researchers demonstrated that a replication-defective lentivirus could be used to stably express a reporter gene in mouse brain neurons, which are nondividing cells.

Production of lentivirus vectors requires additional *cis*-acting sequences

Lentiviruses are complex retroviruses that encode six accessory proteins (Vif, Nef, Tat, Rev, Vpr, and Vpu) in addition to the proteins coded by the gag, pol, and env genes (Chapter 29). Of these proteins, only Tat and Rev are essential for virus production in many cell lines. Tat is a powerful transcription factor that stimulates transcription from the LTR. Consequently, the presence of Tat is essential only if the wild-type LTR is used to control expression of the transgene or the RNA of the vector genome. The Rev protein promotes the efficient transport of unspliced or singly spliced viral RNA from the nucleus to the cytoplasm. Rev interacts with a sequence on viral RNA that is named the Rev-responsive element, or RRE. Hence, efficient production of lentivirus vectors requires the presence of both Rev and the RRE. The viral RNA must also contain the central polypurine tract (cPPT). This is a sequence of 118 nucleotides that significantly increases the transduction efficiency of lentivirus vectors in both dividing and nondividing cells.

Because some essential lentivirus proteins are toxic for cells, it is difficult to establish good producer cell lines for lentivirus vectors without tight regulation of expression of the packaging proteins. To circumvent this problem, vectors are often produced by simultaneously transfecting susceptible cells with three different plasmid constructs, referred to as packaging, envelope, and transfer constructs (Figure 37.5).

The packaging construct. This plasmid encodes all viral enzymes and structural proteins necessary for the production of lentivirus virions except for the envelope protein. The expression of these genes is placed under the control of a strong promoter such as the cytomegalovirus (CMV) immediate early promoter.

The envelope construct. The env gene of HIV-1 produces an envelope glycoprotein that targets the CD4 antigen and **chemokine** receptors, present on lymphocytes, **macrophages**, and certain other cell types. Because this restricted tropism severely limits cell types that can be transduced, the HIV-1 env gene is not used to generate lentivirus vectors. The use of a different envelope protein also improves the safety of the vector by reducing the likelihood of producing replication-competent HIV-1. The envelope protein most often used is that from vesicular stomatitis virus (VSV-G), as this enables the vector to infect many different cell types.

The transfer construct. The simplest form of this plasmid contains the transgene flanked by the two LTRs of HIV in addition to the other essential *cis*-acting elements, which are the packaging signal, the RRE, and the cPPT. The transgene carried by lentivirus vectors can be expressed from a strong promoter such as the CMV promoter, or from an inducible promoter that allows regulation of protein levels. In more recent constructs, the U3 region of the left-hand LTR, which contains the promoter and enhancer elements, has been replaced with a strong constitutive promoter such as CMV, or with the U3 region of another retrovirus such as Rous sarcoma virus (RSV). This modification permits transcription of viral RNA without the need of the Tat protein. An additional improvement is the deletion of the enhancer/promoter elements of the right-hand LTR, which generates self-inactivating vectors as described above.

Applications of retrovirus vectors: treatment of blood disorders

One of the most promising applications of retrovirus vectors is the treatment of blood disorders. Hematopoietic stem cells isolated from the affected individual are infected with a retrovirus containing a normal version of the faulty gene causing the disease. Cells expressing the corrected gene are then reintroduced into the body where they can differentiate and proliferate.

Such a strategy was successfully employed to treat children suffering from severe combined immunodeficiency disease (SCID-XI). This disease is caused by mutations in a cellular receptor (the γc cytokine receptor subunit) whose function is required for differentiation of T lymphocytes. Children affected with SCID-XI die at an early age because of the absence of T-cell-mediated and humoral immune responses. Children treated by retrovirus vector-mediated therapy demonstrated a complete recovery of their immune system; they were able to leave their protective isolation in the hospital and return home with their parents to enjoy a normal life. The result of this clinical trial made news in 2000 and is regarded as the first example of a successful treatment of a human disease by gene therapy.

Unfortunately some of the children that were treated with this retrovirus developed T-cell leukemia, a cancer resulting from the unrestricted proliferation of T lymphocytes. It was later demonstrated that there was a causative link between the leukemia and the gene therapy treatment. The retrovirus vector used to transduce the hematopoietic stem cells contained the wild-type LTR. It happened that the vector integrated itself near the T-cell proto-oncogene LMO2, whose transcription was then activated by the enhancer located within the LTR of the retrovirus vector. Activation of LMO2 promoted proliferation of T cells, resulting in leukemia.

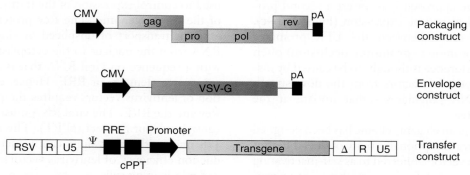

Figure 37.5 Production of lentivirus vectors. Cells are transfected with three different plasmids. The packaging construct provides the structural proteins and enzymes necessary for the assembly of the virion and processing of the RNA. The envelope construct encodes the gene for the envelope protein, in this example VSV-G. The transfer construct contains the transgene expression cassette and all the necessary *cis*-acting DNA sequences. RSV: Rous sarcoma virus U3 region (promoter); RRE: Rev-responsive element; cPPT: central polypyrimidine tract.

Two strategies could be used to reduce the risk of cancer in patients being treated with this vector. One strategy would be to decrease the vector dose in order to reduce the odds of an integration event near a cellular proto-oncogene. The other strategy would involve the use of a self-inactivating retrovirus vector, in which both LTRs would be inactivated after integration.

Applications of retrovirus vectors: treatment of neurological disorders

Interestingly, a recent gene therapy trial using a self-inactivating lentivirus vector for the transfer of the *ABCD1* transporter gene in autologous hematopoietic stem cells resulted in the successful treatment of adrenoleukodystrophy (ALD). ALD is a fatal brain disease, manifested by a malfunction in myelin turnover caused by mutations in the *ABCD1* transporter gene. This is the first successful gene therapy trial using a HIV-based vector and so far no side effects have been reported.

Because of their capacity to integrate their genome into the chromosomes of nondividing cells such as neurons, it should be possible to use lentivirus vectors to treat other neurological disorders such as Parkinson's disease. This progressive neurodegenerative disorder is caused by the death of neurons in the substantia nigra region of the brain. These neurons synthesize the neurotransmitter dopamine, which controls motor function and movement. A reduction of dopamine in the brain results in tremors, aberrant movements and rigidity, the main symptoms of Parkinson's disease.

One possible approach to treat Parkinson's disease would be to restore the production of dopamine by expressing the enzymes responsible for its synthesis. To test this hypothesis, a lentivirus vector expressing the three enzymes involved in the synthesis of dopamine was constructed and injected into the brains of animal models for Parkinson's disease. Neurons were transduced by the vector and an increase of dopamine production during several months was seen. When tested in non-human primates, the vector significantly reduced disease severity.

This vector is now being evaluated in patients with Parkinson's disease in a Phase I/II clinical trial. So far no toxicity has been associated with the vector injection, and there has been improvement of motor function and quality of life of the treated patients.

Advantages and limitations of retrovirus vectors

The main advantage of retrovirus vectors is their capacity to integrate their genome efficiently into the cell chromosome, which ensures the permanent presence of the transgene in transduced cells and their progeny. In addition, they are far less immunogenic than first-generation adenovirus vectors, because they do not encode any viral proteins.

The major advantage of lentivirus vectors over other retrovirus vectors is their ability to transduce nondividing cells, which include most cells in the body. Interestingly, lentivirus vectors are also less prone to silencing. As a consequence, transgene expression with lentivirus vectors can be proportionally increased by increasing the dose of infecting virus.

An additional problem with retrovirus vectors is the possibility of generating replication-competent virus during large-scale vector production. Even though retrovirus vectors do not encode any viral genes, possible recombination events between sequences of the transfer construct and viral genes integrated in the packaging cells may generate a replication-competent retrovirus. Because such retroviruses could cause cancer, their presence in preparations of retrovirus vectors intended for gene therapy applications is intensively scrutinized.

Another safety issue concerns the fact that the integration of proviral DNAs of retroviruses can happen at numerous sites in the cell genome. This raises the possibility that the vector might integrate near an oncogene and cause cancer by disturbing the function or expression of this oncogene. As described above, such an unfortunate event happened in some children who have undergone retrovirus gene therapy to cure severe immunodeficiencies.

Retrovirus virions are much less stable than virions of other vectors such as adenoviruses and adeno-associated viruses. Although retrovirus vectors can remain viable for several months at –80°C in the presence of cryoprotectants (glycerol, sucrose, or serum), the virion loses its infectivity within a few hours at 37°C or within a few days at 4°C. This has negative consequences for virus yield during vector production, which takes place at 37°C, and during vector purification, which is done at room temperature or at 4°C. It may be possible to increase stability of retrovirus vectors by introducing more stable versions of envelope proteins and reverse transcriptases into these vectors.

ADENO-ASSOCIATED VIRUS VECTORS

Adeno-associated virus vectors can insert transgenes into a specific chromosomal locus

Adeno-associated virus (AAV) has a small, nonenveloped virion 25 nm in diameter. Its 4.7-kb, single-stranded DNA genome contains only two genes: rep, which encodes replication and integration enzymes, and cap, which produces the capsid structural proteins (Chapter 20). Replication of AAV requires the presence of a helper virus, either an adenovirus (thus the name) or a herpesvirus. These helper viruses provide functions needed for efficient replication and transcription of AAV genomes and abundant synthesis of capsid proteins. Although the use of AAV as a gene transfer vehicle was first documented in 1984, its utility as vector for gene therapy applications has been exploited only since the mid-1990s.

AAV can infect dividing as well as nondividing cells, and its life cycle involves integration of its genome into the host chromosomes. In contrast to retroviruses, integration of AAV is site-specific. In humans, integration occurs preferentially on chromosome 19 at a site called *AAVS1*. Recombinant AAV vectors are deleted of all viral genes and consequently do not express the Rep proteins. In the absence of Rep proteins, integration of AAV is far less efficient and is no longer restricted to the *AAVS1* site.

Production of AAV vectors usually requires a helper virus

The essential genes that must be provided by helper adenoviruses include E1A and E1B, open reading frame 6 of early region 4 (E4orf6), E2A, and the small VA RNAs. One of the first strategies employed to produce AAV vectors was to cotransfect 293 cells, which express adenovirus E1A and E1B genes, with a transfer plasmid and a packaging plasmid (Figure 37.6). The transfer plasmid contains the transgene expression cassette flanked by the AAV inverted terminal repeats (ITRs), which include the viral origin of DNA replication and the DNA packaging signal. The packaging plasmid contains the rep and cap genes, regulated by their own endogenous promoters, but no ITRs. The remaining adenovirus gene functions were originally provided by infecting the cells with a first-generation adenovirus vector.

Although relatively efficient, this method has the disadvantage that vector preparations are contaminated by adenovirus virions. Fortunately, adenovirus can be partly removed by heat denaturation and by density gradient centrifugation. To avoid adenovirus contamination of AAV vector preparations, it is now current practice to produce AAV by transfecting the cells with a helper plasmid that codes for all essential adenovirus functions, eliminating the adenovirus vector (Figure 37.6). Although production of AAV by transfection of three plasmids can be scaled up by using suspension cultures of 293 cells, this production method is cumbersome.

To simplify AAV vector production, a method has been developed that is based on infection of suspension cultures of insect cells by recombinant baculoviruses. Three recombinant baculoviruses are used: one encodes the transgene expression cassette flanked

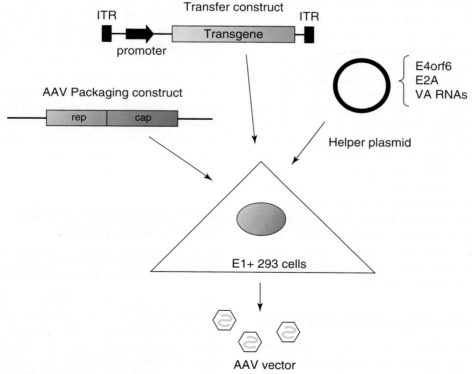

Figure 37.6 Production of adeno-associated virus (AAV) vectors. A cell line expressing the E1 region of adenovirus (usually 293 cells) is transfected with three plasmids: the transfer construct, the packaging construct, and the helper plasmid. The packaging construct provides AAV Rep and Cap proteins. The transfer construct provides the transgene expression cassette, which is replicated and packaged in AAV capsids because of the inverted terminal repeats (ITR). The helper plasmid provides adenovirus helper functions (listed on right).

by the AAV inverted terminal repeat sequence, and the other two carry the Rep and the Cap genes. The AAV transcriptional promoters were modified to function optimally in insect cells. Because insect cells can be grown in suspension, AAV vector production can be scaled up in bioreactors. Importantly, AAV vectors produced in insect cells are indistinguishable from those produced in mammalian cells.

The first AAV vectors constructed were derived from serotype 2 (AAV2). Today several serotypes (AAV1 to 12) that can infect primate cells have been described. The various serotypes have different transduction efficiencies for different tissues. For example, AAV serotypes 1, 6, 8, and 9 can transduce skeletal muscle very efficiently; AAV serotypes 8 and 9 work very well in liver; and AAV types 1, 4, 5, 7, and 8 can efficiently transduce neuronal cells. The transfer construct derived from AAV2 can be efficiently packaged into virions made of capsid proteins of different serotypes. Hence, it is possible to generate AAV vectors with different cell tropisms by changing the serotype of the Cap gene of the packaging construct.

Clinical trials using adeno-associated virus vectors

Transduction of the liver and skeletal muscle of laboratory animals with AAV vectors is nontoxic, relatively efficient, and provides long-term transgene expression. For these reasons, use of these two tissues in humans as a platform for systemic delivery of therapeutic proteins was considered a promising application of AAV vectors.

Hemophilia type B is a disease caused by a reduced level of coagulation factor IX in the blood. The current available treatment of this disease is infusion of factor IX concentrates. In two clinical trials, patients were treated with AAV vectors carrying factor IX either by direct injection into muscle or by infusion in the liver via the hepatic artery. In both of these trials no adverse event following AAV treatment was detected, demonstrating the safety of the vector. However, these clinical trials were disappointing because the level of factor IX produced was lower than predicted from preclinical studies using animal models. Moreover, production of factor IX was transient, and antibodies against AAV capsids were detected, which could prevent reuse of the same vector. AAV capsid components were apparently displayed as foreign antigens at the surface of the transduced liver cells, which were eliminated by the host immune system.

The most dramatic success with gene therapy employing recombinant AAV occurred in three trials for Leber congenital amaurosis, a genetic disorder progressively leading to blindness by age 10. Mutations of the RPE65 gene, involved in the retinoid cycle, are responsible for 15% of cases of Leber congenital amaurosis.

In these trials, patients received an intraocular injection of an AAV vector encoding the RPE65 protein. All the treated patients had already suffered significant visual loss. In all cases, treatment resulted in a general improvement of vision and increased sensitivity to light. Most importantly, no noticeable toxicity or immune response was observed. The success of these trials stems most likely from the fact that the eye is a small organ and therefore the vector dose needed is significantly lower than for treatment of a systemic disease such hemophilia. Moreover, the eye tends to tolerate the presence of foreign antigens such as the AAV capsid.

Advantages and limitations of AAV vectors

Because AAV does not cause disease, AAV vectors are considered to be very safe for administration to humans. AAV virions are stable; they are resistant to heat inactivation as well as large pH variations. Infection with AAV vectors does not trigger a strong cellular immune response, which allows for long-term transgene expression. This is partly due to the lack of expression of any viral genes by these gutless vectors. Furthermore, AAV vectors poorly infect professional antigen-presenting cells, whose function is to induce the cascade of events leading to the activation of the immune system (see Chapter 34).

The prolonged transgene expression observed after gene transfer *in vivo* using AAV vectors was first thought to be a consequence of their integration into the host chromosome. However, AAV DNA has been found in the form of stable, unintegrated double-stranded monomers and concatamers in the nucleus. The prolonged transgene expression observed with AAV vectors is most likely a consequence of the transcription of this episomal DNA. AAV DNA can integrate at random sites into the cell chromosome. However, this is an inefficient process, and the contribution of integrated AAV DNA to the production of the transgene protein is unlikely to be important.

The AAV genome is a single-stranded DNA molecule. A rate-limiting step during AAV infection is the conversion of the single-stranded DNA into double-stranded DNA. To bypass this step, researchers have constructed vectors in which the single-stranded DNA molecule can fold back on itself in such a way that it carries a double-stranded DNA copy of the transgene of interest. Although such vectors are more efficient at transducing cells, the disadvantage of double-stranded AAV vectors is a smaller capacity for foreign DNA sequences. Indeed, the major limitation of AAV vectors is their small insert size capacity of roughly 4.5 kb, which reflects the limited packaging capacity of the AAV capsid. This is sufficient to carry cDNAs from

Retargeting virus vectors.

The majority of gene therapy strategies using virus vectors target specific tissues or organs within the body. Many of the vectors used either have too broad a tissue tropism or are not able to infect the targeted tissue. Therefore significant research efforts are being made to target virus vectors by manipulating viral proteins responsible for attachment to receptors and uptake.

Adenovirus binding to cells is mediated by the fiber protein, which projects from the capsid. Modification of domains on the fiber protein involved in receptor binding can lead to the production of differently targeted vectors. An alternative is conjugation of the fiber protein with small-molecular-weight **ligands** that will bind to specific cellular receptors. Another alternative is to use the fiber protein of adenovirus serotypes that are known to infect target cells more efficiently. For example, the cellular receptor for adenovirus 35 is CD46, a protein located at the surface of many tumor cells. Vectors containing the fiber protein of adenovirus 35 transduce tumor cells expressing CD46 more efficiently than vectors made with the fiber protein of serotype 5, the most common serotype used. As discussed above, a similar strategy has been employed to modify the tropism of AAV vectors by means of capsids derived from different serotypes.

Most retrovirus and lentivirus vectors are pseudotyped with the vesicular stomatitis virus G glycoprotein, a protein conferring a very wide tropism. Methods have been developed to target retrovirus vectors to a specific cell type by modifying the gene encoding the viral envelope protein. Envelope proteins derived from viruses such as Ebola, Molokai, rabies, influenza, and respiratory syncytial virus have been successfully employed to alter the tropism of retrovirus vectors.

The role of many envelope proteins is twofold. They first bind to a cell surface receptor, and they then carry out fusion of the virion envelope with the cell membrane, releasing the capsid into the cytoplasm. It has been possible to separate these two functions for retargeting purposes. A lentivirus vector was constructed to express a glycoprotein whose sole function is to mediate fusion. Two examples of fusogenic glycoproteins are the influenza hemagglutinin (HA) and the glycoprotein of Sindbis virus. These proteins were mutated to abolish the interaction with their natural receptors. Retargeting was rendered possible by inserting into the envelope protein the binding sequence of an antibody targeted against a specific cell receptor. This approach was demonstrated to be effective for retargeting lentivirus vectors to B lymphocytes by using an antibody against CD20, a protein specifically expressed at the surface of these cells.

Recent progress in vector development

Chimeric viruses. The goal in designing chimeric viruses (also known as hybrid vectors) is to combine into one vector the advantageous properties of different viruses. An adenovirus/retrovirus **chimera** utilizes the ease of production, high titer, and wide tropism of adenoviruses in conjunction with the ability of retroviruses to integrate their proviral DNA. In this system, the transgene expression cassette flanked by retrovirus LTRs is delivered to cells using a first-generation adenovirus vector. Recombinant retroviruses are produced by coinfecting the cells with a second adenovirus carrying the retrovirus env, gag, and pol genes. The recombinant retrovirus vectors can infect nearby cells and integrate and express the transgene they carry.

Long-term maintenance of virus genomes. Two elements derived from the herpesvirus Epstein–Barr virus (see Chapter 24) have been useful for allowing long-term maintenance of vector DNA in dividing cells. Expression of the DNA-binding protein EBNA-1 permits the replication of circular DNA containing the latent origin of DNA replication (oriP) in dividing cells, thus maintaining the DNA in daughter cells. A circular template can be made from an adenovirus vector by inserting two recombination sites, such as loxP sites, in the vector backbone. Upon expression of the Cre recombinase, recombination between the two loxP sites will generate a circular DNA.

Site-specific integration. A major problem with integrative vectors derived from retroviruses is the fact that integration into the cell genome occurs at random locations, which can result in insertional mutagenesis with dire consequences. Zinc finger nucleases (ZFNs) can be used to solve this problem by creating lentivirus vectors with the ability to integrate the gene they carry at a predetermined chromosomal location. ZFNs are custom-made artificial nucleases that bind to a specific nucleotide sequence. Upon binding to its target DNA sequence, the ZFN creates a double-strand break that is quickly repaired by the cellular machinery by homologous recombination with the sister chromatid. If a gene expression cassette that is flanked by the same target DNA sequence is present, it may be integrated at this position by homologous recombination. Using this approach, it has been possible to develop lentivirus vectors that are able to integrate a gene expression cassette at a specific gene locus. To prevent integration of the vector by means of its natural integration mechanism, the lentivirus integrase was mutated to abolish its function.

small genes such as that coding for blood clotting factor IX, but not larger genes affected by several common genetic diseases.

The ability of the AAV genome to form concatemers by recombination across the inverted terminal repeats has been employed to increase its capacity for carrying foreign DNA. In this approach, a large cDNA (more than 4.5 kb) is split into two AAV vectors. The split junction of the cDNA is flanked by one half of an intron carrying a splice donor or acceptor sequence. Formation of head-to-tail concatemers via the inverted terminal repeat sequences of two AAV vectors, each carrying a different half of the split cDNA, regenerates the full-length cDNA with the intron at the junction of the two AAV vectors. The functional protein can then be produced after transcription and RNA splicing. Such split AAV vectors have been used successfully, albeit at lower efficiency than standard AAV vectors, to deliver the 6.5-kb cDNA coding for dystrophin, in which mutations cause a severe type of muscular dystrophy.

KEY TERMS

Antibody	Polyadenylation signal
Apoptosis	Promoter occlusion
Chimera	Pseudotyping
Chemokine	Ribozyme
Cytokine	Silencing
Helper virus	Transcriptional
Inverted terminal repeat	enhancer
Ligand	Transgene
Macrophages	Tropism
Packaging signal	Virus vectors

FUNDAMENTAL CONCEPTS

- Virus vectors are highly efficient means of expressing foreign genes within specific cell types.
- Vectors are produced by transfecting cells with plasmids containing a modified viral genome, usually lacking essential genes, and containing an expression cassette with the desired foreign gene.
- To make vectors, the transfected cells must either be engineered to express essential viral genes, or be infected with a helper virus that provides these functions.
- Expression cassettes used in DNA virus vectors are designed for maximal levels of expression by use of effective transcriptional enhancers, splice sites, and polyadenylation signals.
- Several different generations of adenovirus vectors have improved the capacity for incorporating foreign genes and reduced immune responses and toxicity of the vectors.
- Retrovirus vectors can integrate viral DNA containing a foreign gene into target cells and can be modified to infect specific cell types by expressing specific envelope glycoproteins.
- Retrovirus vectors have been used successfully to treat children suffering from severe combined immunodeficiency disease, but several of these children developed leukemia.
- Adeno-associated virus vectors can also integrate foreign genes into the cell chromosome; they are considered safer because these viruses cause no known human disease.
- Adeno-associated virus vectors have recently shown some success in treating a congenital disease that results in blindness in children.

REVIEW QUESTIONS

1. Describe two types of adenovirus vectors that have been used as antitumor agents.

2. Lentiviruses are unique among retroviruses in that they do not need to induce cell division in infected cells in order to replicate. Why?

3. Self-inactivating retrovirus vectors are much less likely to induce tumors in cells following integration of the proviral DNA into the cellular genome. Why?

4. A major problem with virus vectors has always been the host's immune response to the vector itself. How do adeno-associated virus (AAV) vectors circumvent this problem?

α-amanitin: a cyclic octapeptide produced by the fungus *Amanita phalloides*, which binds to and inhibits eukaryotic RNA polymerase II.

Abortive: an infection that does not lead to a complete virus replication cycle.

Actinomycin D: a tricyclic unsaturated hydrocarbon linked to two cyclic pentapeptides, made by a streptomyces species, that binds tightly to dG residues in DNA and blocks transcription by DNA-dependent RNA polymerases.

Acute transforming retroviruses: retroviruses that transform cells rapidly and efficiently by means of cellular oncogenes incorporated into the viral genome.

Acyclic: a version of a compound in which a group that is usually cyclic is broken or removed.

Acyclovir: nucleoside analogue used for treatment of herpes simplex and varicella zoster virus infections. It is selectively phosphorylated by viral thymidine kinase and, in its triphosphate form, selectively inhibits viral DNA polymerase.

Adaptive immune response: the host response mediated by activated B and T lymphocytes against specific antigens. Long-lived memory cells can provide life-long protection against re-exposure to a specific pathogen.

Adenoid: growth of lymphoid tissue in the upper part of the throat, behind the nose.

Adjuvant: a substance that enhances immune responses when administered along with antigens.

Adult T-cell leukemia: a malignant blood disease caused by T cell proliferation induced by HTLV-1 infection.

Alternate pathway: an arm of the complement system that induces formation of the membrane attack complex in the absence of pathogen-specific antibodies.

Amantadine: an antiviral agent that inhibits the activity of the M2 ion channel of influenza A viruses and blocks uncoating.

Ambisense: coding strategy used by arenaviruses and some members of the bunyavirus family, in which an RNA genome segment contains both positive- and negative-sense genes (*ambi* = both). Therefore some viral proteins are made from messenger RNAs copied from the genome RNA, while other proteins are made from messenger RNAs copied from the antigenome RNA.

Amyloid: insoluble proteins that accumulate in microscopic fibers and macrocrystalline arrays called plaques, usually resulting from a conformational change in a normal cellular protein.

Anaerobic methanogen: a microorganism that grows in the absence of oxygen and produces methane, often by reduction of carbon dioxide, to provide a source of energy. All known methanogens are archaea. They are found in anoxic environments such as marshes, the guts of animals, undersea sediments, volcanic springs, and deep underground.

Anaphase: phase of mitosis during which the chrosomes begin to migrate to opposite ends of the cell.

Anaphylatoxins: fragments of complement proteins that trigger inflammation.

Ankyrin: a family of adaptor proteins that mediate the attachment of integral membrane proteins to the spectrin-actin based membrane skeleton.

Antibodies: immunoglobulin proteins that recognize and bind to specific regions (epitopes) on a variety of types of molecules including proteins, carbohydrates, lipids, and nucleic acids.

Antibody-dependent cellular cytotoxicity: a mechanism by which macrophages or NK cells kill cells that express a foreign antigen on their surface by recognition of surface-bound antibodies.

Antibody-dependent enhancement: enhanced entry of certain viruses when the virions are complexed with antibody, resulting in an increased variety of host cells that can be infected.

Antigenic drift: a slow change in the antigenic specificity of a protein due to the accumulation of point mutations in the gene coding for the protein. RNA viruses tend to mutate rapidly because their RNA polymerases do not have proofreading mechanisms to eliminate mutations that occur during genome replication.

Antigenic shift: a rapid change in the antigenic specificity of a protein. For RNA viruses with multiple genome segments, this is usually due to reassortment of genome segments between two virus strains that infect the same cell.

Antigenome: a full-length complementary copy of a single-stranded (usually RNA) genome.

Antigens: molecules or parts of a molecule that can induce antibody formation by binding to a specific antibody or T-cell receptor.

Anti-inflammatory cytokines: cytokines that function to reduce inflammation.

Aphids: small insects that feed on plants using sucking mouthparts called stylets. They are common vectors of many plant viruses.

Aplastic crisis: a severe disease condition in which no red blood cells are produced.

Apoptosis: programmed cell death, involving a number of signaling pathways that lead to activation of caspases (proteases), fragmentation of chromosomal DNA, condensation of the nucleus, and blebbing of the cytoplasm.

Apoptosis: the process of programmed cell death.

Archaea: one of the three domains of life, distinct from Bacteria and Eukarya by virtue of phylogenetic relatedness (based on ribosomal RNA sequences). Differ from organisms in the other two domains by virtue of their cell surface proteins, the chemical nature of their lipids, and their translation, transcription, and replication machinery.

Arthropod-borne: transmitted (usually to mammals and birds) by arthropod vectors (e.g., mites, ticks, fleas, mosquitoes). Arthropod-borne viruses ("arboviruses") generally replicate both in the arthropod vector and in the vertebrate host.

Ascites: fluid that accumulates in the abdominal cavity as a result of impaired liver function or other pathologic conditions.

Asialoglycoprotein receptor: a liver-specific protein that binds to and internalizes glycoproteins that lack terminal sialic acid.

Attachment factor: a cell surface component involved solely in binding of virions to cells, but not in virus uptake.

Autocrine loop: self-stimulation of growth by a cell population that produces a growth-promoting molecule and responds to the same molecule.

Autoimmune demyelination: loss of myelin in the central nervous system due to an immune response against myelin introduced with an impure vaccine.

Autophagy: a cellular process responsible for the degradation and recycling of cellular components by ingestion into vesicles called autophagosomes, which fuse with lysosomes containing digestive enzymes.

Autoreactive T lymphocytes: T lymphocytes that recognize antigens on normal cells in the body, and destroy them.

Azidothymidine: nucleoside analogue used for treatment of human immunodeficiency virus infections. It is activated by cellular nucleoside kinases and, in its triphosphate form, selectively inhibits viral reverse transcriptase.

β-galactosidase: an enzyme that cleaves lactose to galactose plus glucose; often used as a reporter enzyme because artificial substrates such as Xgal produce colored cleavage products that are easy to detect and measure.

Bacteria: one of the threee domains of life, distinct from Archaea and Eukarya.

Bafilomycin A1: inhibitor of the vacuolar proton pump that acidifies endosomes.

Basal cells: cells in the edpidermis of mammalian skin that divide and differentiate into the outermost squamous cells of the epithelium.

B-cell receptors: transmembrane proteins on the surface of B lymphocytes that consist of a ligand binding domain (membrane-bound antibody) and an intracellular signaling domain.

B lymphocytes: white blood cells that originate in bone marrow and are involved in the adaptive immune response, especially in steps leading to the production of humoral antibodies.

Bronchitis: inflammation of the bronchi that results in coughing and excess sputum production.

BSE: bovine spongiform encephalopathy; prion disease in cattle.

Buckminsterfullerene: C_{60} molecule with icosahedral symmetry.

Budded virions: for baculoviruses, virions that are produced by budding from the plasma membrane (see **Occluded virions**).

Budding: the process of forming an enveloped virus particle by the folding of a cellular membrane around the virus capsid as it is extruded through the membrane.

Calyx: a carbohydrate-rich layer on the surface of occlusion bodies, sometimes called the outer or occlusion body envelope.

Capsid: the protein coat that encloses a viral genome.

Capsid-binding drugs: a class of drugs that bind to hydrophobic pockets on picornavirus capsids and inhibit viral attachment and/or entry.

Carboxylic ionophors: compounds that bind monovalent cations (including protons) and increase their membrane permeability.

Caspases: **c**ysteine-**asp**artic prote**ases** that are activated during the events leading to apoptosis.

Cathepsins: a family of intracellular proteases with a variety of specificities.

Caveolae: small flask-shaped invaginations of the plasma membrane that are enriched in cholesterol, glycosphingolipids, and the protein caveolin.

Caveolin: major membrane protein component of caveolae.

Cavins: proteins that bind to caveolins and regulate function of caveolae.

CD4 antigen: a protein on the surface of helper T lymphocytes that is involved in interactions with other cells in the immune system.

CD4-positive T lymphocytes: T lymphocytes that carry CD4 antigen on their surface.

Cell-based high throughput screen: method that uses automated assays for screening large numbers of drugs for their ability to inhibit virus replication or other viral processes in infected cells.

Cell-to-cell movement: movement of a plant virus from one cell to another. This is also called local movement.

Cellular immune responses: adaptive immune responses mediated by T lymphocytes. During virus infection, these responses are mediated primarily by cytotoxic CD8$^+$ T cells but also include CD4$^+$ T cell responses.

Central or primary lymphoid organs: lymphoid tissue (thymus and bone marrow) responsible for the generation of lymphocytes from immature progenitor cells.

Ceramide: type of lipid molecule composed of sphingosine and fatty acid.

Chemokines: secreted proteins that attract leukocytes and are active in inflammatory reactions.

Chimera: an organism or genome resulting from the artificial combination (usually by genetic engineering) of genomes of two or more distinct species. In Greek mythology, a chimera was a fire-breathing monster that had a lion's head, a goat's body, and a snake's tail. (adj.: **chimeric**.)

Chitin: long-chain polymer of N-acetylglucosamine (a glucose derivative) found in cell walls and exoskeletons.

Chloramphenicol acetyltransferase: an enzyme that transfers an acetyl group from acetyl-coenzyme A to chloramphenicol. This reaction can be measured by separating the acetylated from unacetylated chloramphenicol, and is used as a reporter assay to measure activity of genes.

Chlorosis: a lack of chlorophyll, causing a pale green, yellow, or white leaf. Many viruses cause chlorosis in plant leaves.

Chromatin remodeling: alteration of chromatin by modification (acetylation and methylation) of nucleosomal histones, changing their binding affinities to DNA and the access of chromatin to transcription factors.

Ciliated epithelium: epithelial cells that line body surfaces such as the respiratory tract and which are covered with cilia. Cilia are thin, elongated extensions of the plasma membrane containing microtubules; cilia move rhythmically to sweep up foreign particles and mucus.

Cirrhosis: severe damage to the liver, involving loss of normal structure, caused by environmental factors or infection by pathogens.

Cis-acting replication elements (cre): sequence elements located in viral RNA that are required for initiation of viral RNA replication. Cre probably interact with specific viral proteins.

CJD: see Creutzfeldt-Jakob disease.

Class I fusion proteins: envelope glycoproteins that form long alpha-helical coiled-coils, which undergo major conformational changes during fusion to bring the virus envelope and the target cell membrane in close proximity. Examples: orthomyxoviruses, paramyxoviruses, coronaviruses, filoviruses, retroviruses.

Class II fusion proteins: envelope glycoproteins that achieve fusion by a mechanism distinct from that of class I fusion proteins. Fusion involves rearrangement of multimers of protein subunits but only minor conformational changes in protein tertiary structure. Examples: flaviviruses, togaviruses, rhabdoviruses, hepadnaviruses.

Classical activation pathway: an arm of the complement system that is triggered by antibodies bound to antigens on pathogens.

Clathrin: a protein involved in the generation of endocytic vesicles; lines the cytoplasmic side of coated pits.

Clathrin-coated pits: invaginations of the plasma membrane involved in endocytosis, and covered on their cytosolic surface by a clathrin protein coat.

Clonal selection theory: a model of how B and T lymphocytes recognize and respond to antigens to protect the host from infections. In this model, immature lymphocytes have many different antigen receptors. Those that recognize host proteins are destroyed while those that recognize foreign antigen are activated and produce many clones of themselves.

Coccolithophore: single-celled haptophyte algae that produce calcium carbonate plates called coccoliths.

Coccoliths: individual plates of calcium carbonate formed by coccolithophores in the Golgi body, then excytosed and incorporated in the coccosphere.

Co-chaperone: a protein that associates with chaperones and stimulates their protein-folding and unfolding activities.

Coiled-coil: a protein structural element consisting of several alpha-helical chains coiled about each other, forming a rod-shaped domain.

Cold chain: a system to maintain the storage temperature for a temperature-sensitive product such as live vaccines all the way from the manufacturer to the site of administration. Many vaccines are susceptible to both excessive heat and freezing.

Collectins: proteins that contain C-type lectin domains, which bind to complex carbohydrates on microorganisms. Bound collectins are recognized by macrophages and other phagocytic cells and lead to uptake and destruction of the microorganism.

Complement: a large set of proteins involved in cascades of reactions stimulated by bacteria or antigen/antibody complexes, and leading to immune cytolysis and other protective mechanisms.

Concatemer: a linear or circular oligomer of nucleic acid genome sequences produced by rolling circle replication and/or recombination. Monomers can be arranged as direct repeats (head to tail) or inverted repeats (head to head; tail to tail) depending on how they are formed.

Condyloma: growth resembling a wart on the skin or a mucous membrane, usually of the genitals or anus, caused by infection with a human papillomavirus.

Confocal microscope: a microscope that uses focused lasers and computer imaging to visualize thin segments in living or fixed cells.

Core proteins: proteins associated with the viral genome inside the capsid.

CREB-binding protein (CBP): a histone acetylase that binds to a cellular transcription factor called cyclic AMP response element-binding protein (CREB). CREB binds to cyclic AMP response elements on DNA and activates transcription.

Crenarchaeota: one of the major phyla of Archaea, including most known hyperthermophiles, such as *Sulfolobus solfataricus*, but also many marine organisms that are not thermophiles.

Creutzfeldt-Jakob disease: prion disease in humans. The new variant CJD, suspected to be transmitted to humans by consumption of BSE-contaminated beef, strikes mostly young people.

Cryoelectron microscopy: a method for visualizing biological structures, including viruses, at high resolution (to about 1 nm) by suspending unstained objects in a thin sheet of noncrystalline ice at liquid nitrogen temperatures. Computer-assisted averaging is utilized to reconstruct three-dimensional images of the object.

Cucurbit: a plant in the family *Cucurbitaceae*. Common examples are squash, melons and cucumbers.

Cyclin-dependent kinases (CDKs): protein kinases that, in combination with cyclins, regulate the cell cycle by phosphorylating other proteins.

Cyclins: cellular proteins that regulate the cell cycle by associating with and activating cyclin-dependent kinases.

Cytokines: secreted proteins that act on cells of the innate and adaptive immune systems to stimulate or inhibit various immune functions.

Cytopathic: causes cell injury and death.

Cytoplasm: the portion of a euykaryotic cell enclosed by the plasma membrane, excluding the nucleus.

Cytosol: the soluble portion of the cytoplasm, excluding membrane-bound organelles and their contents.

Cytotoxic T lymphocyte: a class of lymphocytes, usually bearing the CD8 antigen, that attack and kill cells expressing foreign antigens presented by MHC class I molecules. In the context of virus infection, cytotoxic T lymphocytes are the primary effectors for killing virus-infected cells.

DC-SIGN (*d*endritic *c*ell-*s*pecific *i*ntercellular adhesion molecule-*g*rabbing *n*onintegrin): a type II transmembrane mannose-binding protein found on dendritic cells, involved in their interaction with T cells during antigen presentation.

Delta structure: structure of a ribozyme derived from hepatitis delta virus.

Demyelination: loss of the myelin sheaths covering axons in the central nervous system, leading to dysfunction and loss of muscle control.

Dendritic cells: cells that accumulate and present foreign antigen to lymphocytes in lymph nodes and elsewhere in the body.

Dicots: flowering plants that have two cotyledons, or seed leaves, which emerge from the germinating seed.

Discontinuous transcription: mechanism that copies and joins together nonadjacent sequences on an RNA or DNA template by displacement of the RNA polymerase along with its nascent RNA chain from one site to another.

Disseminated intravascular coagulation: systemic clotting of the blood; can be caused by activated neutrophils.

Divergent transcription: transcription of adjacent genes that progresses in opposite and diverging directions from a central point. RNAs are therefore transcribed from complementary genome strands.

DNA gyrase: a bacterial enzyme that overwinds double-stranded DNA by cleaving a phosphodiester bond and rotating one strand around the complementary strand using energy provided by ATP, to generate supercoiled DNA.

DNA vaccine: introduction into the body of plasmids that express specific proteins, usually of a selected pathogen, resulting in an immune response against those proteins.

DNA-dependent RNA polymerase: An enzyme that synthesizes RNA using a DNA template.

Domains of life: groups of related organisms at the base of a system of classification of all life forms, based on ribosomal RNA sequences and fundamental properties such as presence or absence of a nucleus, membrane composition, and the machinery for replication, transcription, and translation.

Dorsal root neurons: sensory neurons whose ends are in peripheral organs and whose cell bodies are grouped in dorsal root ganglia near the spinal cord.

Double-stranded RNA-dependent protein kinase: see **PKR**.

Dura mater: the tough outer membrane covering the brain and spinal cord (from Latin words for "hard" and "mother").

Dynactin: an adaptor protein complex that allows dynein to bind to its cargo.

Dynein: motor complex that typically carries cargo in the minus end direction (towards the center of the cell) along microtubules.

Early endosome: intermediate vesicle to which molecules bound to receptors during receptor-mediated endocytosis are first delivered.

Ectodomain: the part of a membrane protein that projects into the extracellular space.

Encephalitis: disease involving inflammation of the brain.

Endemic: natural to a specific animal or human population or place; indigenous; native. Refers to a disease or pathogen that is present and circulates among hosts in a specific location, but does not necessarily infect a large segment of the population.

Endoplasmic reticulum–Golgi intermediate compartment (ERGIC): a membranous cytoplasmic organelle that is continuous with the endoplasmic reticulum and the Golgi apparatus, and is involved in transport, processing, and modification of proteins.

Endosome: intermediate membrane vesicles in the receptor-mediated endocytosis pathway.

Endosymbiotic: relationship in which a unicellular organism is taken up into the cells of another organism and continues to reproduce, retaining at least minimal genetic information, eventually becoming dependent on the engulfing organism. Endosymbiosis is believed to be at the origin of a number of organelles such as mitochondria and chloroplasts, and is

therefore considered to be mutually advantageous to both the invading and the receiving organism.

Enfuvirtide: a peptide that blocks a conformational change in HIV-1 gp41 necessary for fusion of the viral envelope with the cell membrane.

Enteric: intestinal.

Envelope: a lipid bilayer membrane, usually derived from a cellular membrane, into which viral glycoproteins are inserted. Envelopes are used by many viruses as an external layer in which their nucleocapsids are wrapped.

Ependymal cells: epithelial cells that line the brain and spinal cord.

Epidemic: increase in infectious disease incidence over time and/or geographic range. Epidemics are larger than "outbreaks" (which are more geographically limited) and smaller than "pandemics."

Epitopes: short regions on antigens such as proteins (usually 10 amino acids or less) or carbohydrates, to which specific antibodies bind.

Epizootic: an epidemic in one or more animal species.

Erythema infectiosum: Latin for "infectious rash." A childhood viral disease with typical rash and fever, also known as "fifth disease."

Erythroid precursors: cells on the pathway of differentiation from hemopoietic stem cells to mature red blood cells.

Ethidium bromide: a fluorescent polycyclic aromatic molecule that intercalates between bases in DNA or RNA. Used to detect and quantitate nucleic acids.

Eukarya: one of the three domains of life; all eukarya, also called eukaryotes, have a nucleus, as well as other membrane-bounded organelles such as mitochondria or chloroplasts.

Euryarchaeota: one of the major phyla of Archaea, including hyperhalophiles, methanogens, thermophiles and some hyperthermophiles.

Exocytosis: the process of delivery of material within vesicles derived from the endoplasmic reticulum to the outside of the cell by their passage through the Golgi membranes and fusion with the plasma membrane.

Exportins: proteins that mediate the export from the nucleus to the cytoplasm of proteins or RNAs.

Family: in virology, a taxonomic grouping of genera whose members share common virion structure, genome organization and size, and replication strategy, and for which there is evidence of a common evolutionary origin by comparison of nucleotide and amino acid sequences.

Ficolins: a family of plasma proteins containing a collagen-like region and a C-terminal fibrinogen-like domain; these proteins function to recognize invading pathogens.

Filopodia: small, actin-containing projections from the cell surface.

Filterable virus: the name given to viruses when they were first distinguished from other microorganisms on the basis of their small size and their ability to pass through fine-pore filters.

Fission: see **Membrane fission**.

Folate receptor-α: a cellular membrane protein that binds to folic acid, a soluble B vitamin.

F-pili: filamentous protrusions from the surface of *E. coli*, coded by the F plasmid, that mediate bacterial conjugation.

Fulminant hepatitis: severe and rapidly progressive hepatitis often leading to hepatic failure and death.

Furin: a host cell endopeptidase, localized in the *trans*-Golgi compartment of the cell's secretory pathway, that cleaves proteins at arginyl residues.

Fusion: see **Membrane fusion.**

Fusion peptide: a short hydrophobic region within a viral envelope protein that is inserted into the target cellular membrane during virus-induced membrane fusion.

Ganciclovir: nucleoside analogue used for treatment of human cytomegalovirus infections. It is selectively phosphorylated by the viral UL97 protein kinase and, in its triphosphate form, selectively inhibits viral DNA polymerase.

Gangliosides: sialic acid–containing glycosphingolipids found in cellular membranes.

Gastroenteritis: inflammation of the gastrointestinal tract resulting in diarrhea, vomiting, and/or cramps.

Gene gun: apparatus that projects tiny particles (gold or other inert material) coated with DNA into cells at high speeds, allowing penetration of cell membrane and intracellular expression of proteins encoded by the DNA.

Genetic bottleneck: an event during which only a limited number of individuals of a species survive, resulting in a narrow amount of genetic diversity in a population.

Genetic drift: see **Antigenic drift.**

Genetic economy: principle that organisms with genomes of limited size use their protein-coding capacity economically, for example, by making capsids with multiple copies of a few different proteins.

Genetic shift: see **Antigenic shift.**

Genus: in virology, a grouping of closely related virus species. (Plural: **genera.**)

Glycolipid: lipid molecule with a polar head group that contains carbohydrate.

Glycoprotein: protein that has covalently bound carbohydrate residues.

Glycosaminoglycan: polysaccharides formed from repeating units made up of two different 6-carbon sugars, one of which is an amino sugar, usually N-acetyl glucosamine or N-acetyl galactosamine. These can be bound to proteins on the cell surface.

Gramineae: "true grasses," including cereals, bamboo, and many marsh and grassland plants.

Granulin: protein that forms ovoid occlusion bodies of granuloviruses.

Granzyme B: a cytolytic protein produced by cytotoxic T lymphocytes and natural killer cells; induces apoptosis in infected target cells.

Green fluorescent protein: a jellyfish protein that emits green light when exposed to ultraviolet light. Widely used in cell biology for intracellular localization of specific cellular proteins fused to green fluorescent protein by genetic engineering.

Guanylyltransferase: an enzyme that transfers a guanosine residue to the phosphorylated 5′ end of an RNA, creating an RNA "cap" structure.

Hairpin structure: a stem structure formed by base pairing in a nucleic acid that is internally self-complementary.

Hairy-cell leukemia: a malignant blood disease caused by proliferation of a type of B cell.

Hammerhead structure: secondary structure found in ribozymes of group A viroids. It can be drawn in a shape that resembles a hammerhead.

Hantavirus pulmonary syndrome: an often-fatal lung disease resulting from infection by hantaviruses such as sin nombre virus, transmitted by rodents.

Helical symmetry: symmetry defined by rotational and translational displacement about a single helix axis.

Helicase: an enzyme that unwinds DNA or RNA helices.

Helper virus: a virus that can provide functions necessary for the replication of another virus during infection of the same cell by both viruses.

Hemagglutination: the agglutination or clumping of red blood cells.

Hemagglutination assay: a method of titering viruses that agglutinate red blood cells.

Hemagglutinin: a protein that binds to receptor molecules on the surface of red blood cells, leading to their agglutination.

Hematopoiesis: the formation of the cellular components of the blood from stem cells found within the bone marrow.

Hemifusion: intermediate in the membrane fusion reaction where only the outer bilayer halves have fused.

Hemocoel: in arthropods, an open body cavity in which blood flows and bathes the tissues and organs.

Hemolytic anemia: a disease caused by shortage of red blood cells due to their fragility.

Hemorrhagic fever with renal syndrome: a disease caused by various hantaviruses involving internal bleeding and kidney pathology, which can be mild or severe with a high fatality rate.

Hemorrhagic fevers: diseases caused by a number of viruses, involving high fever, vascular damage leading to internal bleeding, and resulting shock, which if untreated can lead to organ failure and death.

Heparan sulfate: a sulfated polysaccharide (glycosamino-glycan) made from repeating disaccharides. These are covalently attached to core proteins to form proteoglycans, found on the cell surface.

Hepatitis: disease involving damage to and inflammation of the liver.

Hepatocellular carcinoma: malignant tumor of the liver.

Herd immunity: sufficiently high vaccine coverage in a population such that the capacity of viral pathogens to spread is limited, thus protecting the small number of unvaccinated individuals in the community.

Hexon: structural subunit of adenovirus virions, made from trimers of the hexon protein.

Hippocampus: a section of the brain that is located inside the temporal lobe and is part of the limbic system. The major functions are involved in memory and navigation.

Histone acetyl transferases: enzymes that acetylate nucleosomal histones and other proteins, reducing their binding affinity for DNA and resulting in a more relaxed chromatin structure that is more accessible to specific transcription factors and is more easily transcribed.

Histone deacetylases (HDACs): enzymes that remove acetyl groups from histones, therefore increasing their binding affinity for DNA and resulting in a more rigid chromatin structure.

Holin: a small, bacteriophage-coded protein that self-associates in the bacterial membrane under certain conditions of pH and ionic concentrations to open holes in the membrane, allowing phage endolysin to enter the periplasmic space and attack the bacterial cell wall.

HTLV-1 associated myelopathies (HAM): a disease similar to tropical spastic paraparesis.

Humoral immune response: all aspects of the response to foreign antigens by the adaptive immune system that result in production of antibodies.

Hyaluronan: an anionic, non-sulfated, usually large molecular weight glycosaminoglycan formed in the plasma membrane.

Hydrocephalus: swelling of the cerebral ventricles.

Hyperhalophile: an organism that grows in hypersaline waters with concentrations of sodium chloride above 1.5 molar.

Hypersensitive necrosis: a localized area of tissue death on a plant leaf that is induced in response to a pathogen, and is associated with limited spread.

Hyperthermophile: an organism that grows at temperatures above 80°C.

Hypoxemia: low oxygen level in the blood, leading to lack of oxygen supply to tissues and organs.

Icosahedral symmetry: symmetry of a regular icosahedron, with 12 axes of fivefold symmetry, 20 axes of threefold symmetry, and 30 axes of twofold symmetry.

IκB: an inhibitor that binds to the transcription factor NF-κB, maintaining it in an inactive state in the cytoplasm.

IκB kinase complex (IKK): a serine-threonine protein kinase that phosphorylates IκB, resulting in the dissociation and activation of NF-κB.

Immune response: the set of responses to foreign organisms or macromolecules that are used by an organism to protect against disease by mobilizing a variety of proteins and specialized cells in the circulation and in tissues that recognize, respond to, and destroy foreign pathogens and tumor cells.

Immunogenic: able to induce an immune response.

Immunoglobulin family: a family of membrane proteins that have characteristic repeated domains usually stabilized by disulfide bonds, first described in immunoglobulins.

Immunoprophylaxis: treatment of an infectious disease by injection of antibodies directed against the disease agent.

Importin (also called karyopherin): receptor protein that interacts with nuclear localization signals in proteins.

Inclusion bodies: aggregates of viral proteins or nucleic acid visible by light microscopy in infected cells by their staining characteristics.

Infectious subvirion particles (ISVPs): in reoviruses, particles that have lost part of the external capsid as a result of protease digestion in endosomal vesicles or the intestine.

Inflammatory response: induction of inflammation, mediated by cytokines and white blood cells; this response functions to remove injurious stimuli and to initiate the healing process.

Innate immune response: the part of the immune response that acts by rapid and relatively nonspecific detection of pathogens. Cellular proteins such as those in the complement pathway, or cytotoxic cells such as natural killer cells, bind to pathogens or infected cells and mark them for destruction. Distinguished from the adaptive immune response, which develops more slowly but is highly specific for particular antigens and usually results in development of both specific antibodies and cytotoxic cells.

Intasome: an assembly of the integrase and the host protein IHF on the phage lambda *attp* site; the intasome captures the bacterial *attb* site and carries out recombination between the two *att* sites.

Integrase: a retroviral enzyme that catalyzes the integration into cellular DNA of the double-stranded viral DNA produced by reverse transcription.

Integrins: a family of cell surface proteins (usually heterodimers) that interact with extracellular adhesion proteins and initiate intracellular signaling pathways.

Interferons: a family of cytokines that induce antiviral activity and protect cells against virus infection.

Interferon-stimulated genes: a collection of genes that encode antiviral and anti-proliferative properties and are induced in cells treated with interferon.

Interferon-α: a family of type I interferons that are rapidly produced by many cell types, particularly blood cells, that have antiviral and anti-proliferative activities, and that are an important component of the intrinsic cellular response and the innate immune response to viral infections.

Interferon-γ: a cytokine produced by activated T lymphocytes and natural killer cells that functions in the innate and adaptive arms of the immune response by activating cells such as macrophages.

Intergenic region: for polyomaviruses, a 400-nt region located on the viral genome between the early and late genes. Contains important transcription and replication control sequences.

Interleukin-1: a cytokine, secreted early in the immune response, that activates both T and B lymphocytes.

Interleukin-2: a cytokine that stimulates T-cell growth and proliferation.

Internal ribosome entry sites (IRES): regions on viral RNAs (particularly picornaviruses and flaviviruses) that bind to ribosomes and lead to cap-independent initiation of protein synthesis at internal sites distant from the 5′ end of the RNA. IRES contain extensive secondary and tertiary structures that interact with a variety of cellular proteins. Some cellular genes have been shown to contain IRES.

Intrinsic cellular defenses: rapid and usually local host responses by an organism to infection by a pathogen. Intrinsic defenses can protect the infected cell or nearby cells from damage by inducing a variety of protective systems within minutes to hours of the initial infection.

Inverted terminal repeat: a nucleotide sequence present at both ends of a nucleic acid molecule but in the opposite orientation; the two ends of a single-stranded DNA with such a repeat are complementary and therefore can anneal to each other.

Ionophores: small lipid-soluble molecules that bind specifically to certain ions and transport them across membranes.

Isotype: in immunology, the type of constant region in the heavy chain of an immunoglobulin. There are nine known immunoglobulin isotypes, each presumably with a different function in the immune response.

Jak–Stat pathway: signaling by interferons, other cytokines, and growth factors is mediated by a pathway that includes the Janus tyrosine kinases (Jak kinases) and the signal transducers and activators of transcription (Stat proteins), via a cascade of tyrosine phosphorylation events leading to the activation of transcription of a variety of target genes.

Jaundice: yellowing of the skin due to the presence of bile pigments in the blood, often caused by poor liver function.

Jelly-roll β-barrel: folding of a polypeptide chain into a number of parallel β strands, usually as two sheets, found in many virion capsid proteins.

Junctional adhesion molecule A (JAM-A): a cellular membrane protein found in tight junctions; member of the immunoglobulin superfamily of cell-adhesion molecules (CAMs).

Karyopherin: see **Importin**.

Keratinocytes: squamous epithelial cells that comprise skin and mucosa and provide a barrier between the host and the environment.

Kuru: prion disease of the Fore people in Papua New Guinea, caused by ingestion of prions during ritual cannibalism. Means "shivering" or "trembling" in the Fore language.

Laminin receptor: a cell surface protein that binds to laminin. Laminins are a family of proteins, each containing three polypeptides of about 820 kD, that are a component of the basal lamina, which lies underneath epithelial cells.

Latency: a state of inactivity in which the viral genome is present in the cell in a dormant state; little or no viral transcription or replication occurs.

Latent infection: an infection in which a pathogenic agent remains present in the organism but replicates slowly or not at all and produces few or no symptoms.

Lectin receptor-activated pathway: an arm of the complement system that begins with recognition of mannose residues on microbial carbohydrates.

Lectins: plant proteins that bind to specific oligosaccharides in cell membranes.

Leucine zipper domain: in proteins, an alpha-helical region with leucine residues at each 7th position, which enables dimerization between proteins.

Ligand: a specific molecule that is bound by a protein.

Lineage: a collection of individual organisms linked by close genetic relationships.

Lipid rafts: microdomains in plasma membranes of eukaryotic cells that are enriched in cholesterol, glycolipids, and sphingolipids and in certain membrane proteins. Lipid rafts are believed to be targets for binding of certain viruses during virus entry, as well as for assembly of certain enveloped viruses.

Lipopolysaccharide: A complex molecule containing fatty acids and oligosaccharides, found in the outer membrane of Gram-negative bacteria.

Long-distance movement: movement of a plant virus throughout a plant by way of the vascular tissue.

Lumen: the inner cavity of cellular vesicles, such as the endoplasmic reticulum.

Lymphatic system: a network of vessels and lymphoid tissues within the body that functions to transport extracellular fluids containing pathogens, macromolecules, and immune cells to lymphoid organs.

Lymphotoxin: a cytokine, also known as tumor necrosis factor β (TNF β), that kills infected cells by inducing holes in the plasma membrane.

Lysis: breaking open of a virus-infected cell, leading to its death and the release of virus particles within the cell.

Lysogen: a bacterial cell containing a prophage.

Lysosome: subcellular organelle in eukaryotic cells responsible for degradation of substances taken up by endocytosis.

Lysosomotropic agent: weakly basic compound that elevates the pH in acidic organelles of the cell.

Macrophages: cells present in a variety of tissues and organs that engulf and digest extracellular debris, cell fragments, and pathogens, and are involved in antigen presentation to cells of the adaptive immune system.

Macropinocytosis: uptake at the cell surface of relatively large amounts of extracellular fluid and particles within vacuoles 0.5 to 5 μm in diameter.

Mad cow disease: popular name for bovine spongiform encephalopathy (BSE).

Major histocompatibility complex (MHC) proteins: cell surface proteins that bind to antigens or antigen fragments and present them to cells involved in the immune response. In humans, some of these proteins are also called human leukocyte antigens, or HLA.

Maraviroc: an antiviral agent that binds to the cellular CCR5 chemokine receptor membrane protein, which serves as a co-receptor for HIV-1; binding of maraviroc can therefore block HIV-1 entry.

Mast cells: cells that reside in tissues and contain granules loaded with histamine and heparin; they respond to stimulation by degranulation, inducing inflammation.

Matrix protein: in virology, virus-coded proteins that lie just underneath the viral envelope and form a connecting bridge between the envelope and the viral nucleocapsid.

Membrane attack complex: a collection of complement proteins that forms a pore in the membrane of pathogens or infected cells, killing the pathogen or cell.

Membrane fission: the process by which one membrane divides or "buds" into two.

Membrane fusion: the process by which two membranes combine to form one.

Metalloproteinases: enzymes that degrade proteins by cleaving peptide bonds, and that use metal ions such as zinc in the catalytic mechanism.

Methyltransferase: an enzyme that transfers a methyl group from S-adenosyl-methionine to an acceptor molecule (e.g., a nucleotide base or ribose during RNA cap formation)

Micelle: small, water-soluble droplets of polar lipids.

MicroRNAs (miRNAs): short RNA molecules of approximately 22 nucleotides that bind to complementary sequences of target mRNAs, resulting in mRNA degradation or translational repression.

Microvilli: finger-like projections of the plasma membrane, containing a core of actin filaments, on the surface of various cell types such as those of the intestinal epithelium. Microvilli greatly increase the cell surface and play an important role in absorption of substances from the extracellular medium. (Singular: **microvillus**.)

Mitotic arrest-defective (MAD): proteins that control entry into anaphase, preventing separation of chromosomes on the mitotic spindle in dividing cells.

Mitotic spindle checkpoint: a control mechanism that prevents the onset of anaphase until chromosomes are aligned properly on a bipolar spindle; controlled by mitotic arrest-defective (MAD) proteins.

Monensin: monovalent **ionophore** that disrupts the passage of proteins through the Golgi complex, and has multiple effects on glycosylation.

Monoclonal antibodies: antibodies produced by cloned antibody-producing cells. Each monoclonal antibody therefore has a unique amino acid sequence and binds to a highly specific epitope.

Monocots: flowering plants that have a single cotyledon, or seed leaf, which emerges from the germinating seed.

Monocytes: white blood cells that phagocytose foreign particles or pathogens. When monocytes move from the blood into tissues, they differentiate into macrophages or dendritic cells.

Mosaic disease: a disease that causes symptoms on plant leaves that result in a mosaic pattern of light and dark green, or green and yellow.

Mucociliary flow: flow of mucus and debris along the epithelial lining of the respiratory tract modulated by the movement of cilia on the epithelial cells.

Mucosal epithelium: epithelial cells of the mucosal membrane.

Multinucleated giant cells: see **Syncytium.**

Multiplicity of infection (m.o.i.): the number of infectious virus particles inoculated per cell.

Multivesicular bodies: cytoplasmic vacuoles containing many small membrane vesicles and lysosomal degradative enzymes.

Myocarditis: disease involving inflammation of the muscular walls of the heart.

Myristate: a C_{14} linear, saturated fatty acid that can be joined via its carboxyl group to N-terminal glycine in a number of proteins; the fatty acid becomes embedded in a cellular membrane, linking the protein to the membrane.

Myristoylation: covalent binding of the 14-carbon fatty acid myristate to N-terminal glycine residues on proteins, leading to their association with lipid membranes.

Natural killer (NK) cells: lymphocytes that attack and kill abnormal cells, particularly cells infected with pathogens and cancer cells; part of the innate immune response.

Necrogenic: refers to a pathogen that causes necrosis, or death.

Nectins: a family of cell surface proteins, with an immunoglobulin-like structure, that function in cellular adhesion.

Negative (or minus) strand: for RNA viruses, the strand complementary to the RNA that contains protein-coding regions. Some viruses specifically package the negative RNA strand in virions.

Negative staining: staining of specimens for electron microscopy by use of heavy-metal salts that absorb electrons but do not bind to the specimen. The image is formed by exclusion of the stain by the specimen, therefore the image is "negative."

Nested set: a set of similar objects, each of which fits into the next larger object.

Neuraminidase: enzyme that cleaves terminal sialic acid (N-acetyl neuraminic acid) from glycolipids or glycoproteins.

Neutralizing antibodies: antibodies that inactivate or "neutralize" virus infectivity when they come in contact with the virus particle.

Nevirapine: non-nucleoside reverse transcriptase inhibitor.

New variant Creutzfeldt-Jakob disease (nvCJD or vCJD): prion disease in humans believed to be caused by transmission of prions from diseased cattle.

NF-κB (nuclear factor-κB): a heterodimeric transcription factor that regulates many aspects of immune responses, apoptosis, and a variety of signaling pathways. It is sequestered in an inhibitory complex in the cytoplasm, from which it is released upon receipt of a variety of extracellular signals; it travels to the nucleus and activates transcription of numerous target genes.

Non-nucleoside inhibitors: DNA polymerase inhibitors that are not nucleoside analogs.

Nonpermissive: describes a cell type that cannot support productive infection by a particular virus. Nonpermissive cells may block virus infection at many possible stages, including virion binding, uptake, gene expression, genome replication, and virion assembly.

Nucleocapsid: viral capsid containing the viral nucleic acid genome.

Nucleoside diphosphate kinase: an enzyme that phosphorylates nucleoside diphosphates (e.g., deoxyguanosine diphosphate) to nucleoside triphosphates.

Nucleosomes: complexes assembled from two copies each of the four cellular histones H2A, H2B, H3, and H4. Chromosomal DNA is condensed by wrapping around the exterior of cylindrical nucleosomes.

Obligatory intracellular parasite: an infectious element or organism that can replicate only within other cells.

Occlusion body: a crystalline protein matrix formed in the nucleus of cells infected with baculoviruses, consisting of enveloped viral nucleocapsids embedded in polyhedrin or granulin.

Occlusion-derived virions: virions embedded in occlusion bodies, formed in the nucleus.

2′, 5′-oligo(A) synthetase: an enzyme that produces oligomers of up to 10 ATP residues linked via a 2′, 5′ phosphodiesterase bond. Part of the interferon response, this enzyme is activated by double-stranded RNA. 2′, 5′-oligoadenylate binds to and activates ribonuclease L, which degrades mRNAs.

Oncogene: a viral or cellular gene whose expression can lead to cell transformation or tumorigenesis.

Oncolytic: able to kill or lyse tumor cells.

Oncotropism: the ability of a virus to replicate more efficiently in cancer cells than in normal cells.

Operator: a DNA or RNA binding site for a repressor protein that negatively regulates gene expression.

Opportunistic infections: infections that flourish in an immune-deficient individual but that are usually absent from healthy individuals.

Order: in virology, a rarely used taxonomic grouping of virus families whose members share distinctive characteristics of replication strategy, but which may diverge widely in host organism and genome size, and which may share little or no nucleotide sequence homology.

Oseltamivir: neuraminidase inhibitor used for treatment of influenza A and B virus infections.

Osmolyte: an organic compound that can affect osmosis, affecting cell volume and fluid balance.

Packaging signal: sequence on a DNA or RNA genome that is recognized by viral proteins involved in packaging the genome in the capsid.

Palmitoylation: covalent attachment of palmitic acid to cysteine sulfhydryl groups in a protein; increases affinity of the protein for lipid membranes.

Pandemic: increase in infectious disease incidence over time and a large geographic range spanning many countries and continents (e.g., the human influenza pandemic that spread worldwide in 1918–1920).

Panleukopenia: infectious disease (of cats) leading to shortage of white blood cells, among other symptoms.

Panzootic: increase in infectious disease incidence in animal species over time and large geographic range (e.g., panzootic of rabies in raccoons extending up eastern North America through the 1970s to the present).

Papain-like cysteine proteinases: enzymes that have structural homology to the proteinase papain, and that use cysteine as the nucleophile for hydrolysis of peptide bonds.

Pap test: a diagnostic test for cervical cancer at an early stage, by examination of exfoliated cells stained with Papanicolaou stain. It is the most successful test invented for cervical cancer prevention.

Paralogous: homologous gene sequences that presumably arose by a gene duplication event are called "paralogous" genes.

Pentasymmetrons: pentagonal capsid substructures composed of multiple trimeric or hexameric capsomers surrounding a central pentameric capsomer.

Penton base: structural subunit of adenovirus virions, made from pentamers of the penton protein and located at the 12 vertices of the icosahedral virion.

Peptidomimetic: drug that mimics a peptide, but is not itself a peptide.

Perforin: a cytolytic protein produced by cytotoxic T lymphocytes and natural killer cells. This protein forms a pore in the plasma membrane of a target cell, leading to apoptosis, often with the help of cytolytic proteins such as granzyme B.

Peripheral or secondary lymphoid organs: lymphoid tissue (e.g., spleen, lymph nodes, tonsils) where foreign or altered native antigens interact with lymphocytes.

Peritonitis: inflammation of the peritoneum (the lining surrounding the abdominal cavity).

Peritrophic membrane: chitinous protective sheath secreted around food in insect gut.

Permissive: a cell type is considered permissive for infection by a particular virus if it can be productively infected and give rise to progeny virus.

Persistent infection: an infection (in this case, by a virus) of cultured cells or of an animal during which the virus is not eliminated, but continues to replicate for a prolonged time at a low level, without killing the host.

Phagocytosis: the uptake of particulate material from the extracellular medium by vesicles in phagocytic cells.

Phagosome: a vacuole formed around a particle taken into a cell by phagocytosis.

Pharmacokinetics: study of the fate of a drug when administered to an organism, including its uptake, its distribution, its degradation, and its clearance.

Phyla (singular, phylum): a taxonomic rank, or grouping of related organisms, between "kingdom" and "class."

Picornavirus 3C-like proteinases: enzymes that have structural homology to the picornavirus 3C proteinase, which is involved in cleavage of picornavirus polyproteins into mature proteins.

PKR (double-stranded RNA-dependent protein kinase): a cellular serine-threonine protein kinase that inactivates protein synthesis by phosphorylating eukaryotic initiation factor 2B (eIF-2B), in response to virus infection or the presence of double-stranded RNA.

Plaque assay: a method for enumerating infectious virus particles by generation and counting of plaques, or focal areas of cell death, on a monolayer of cells. Each plaque usually arises from a single infectious virus particle.

Plaque-forming units (PFU): infectious virus particles as detected by a plaque assay. Usually expressed as PFU/ml.

Plasma cells: terminally differentiated B cells specialized in producing and secreting antibodies.

Plasmodesmata: membrane-lined connections between plant cells that allow small molecules to pass between cells.

Pleconaril: a capsid-binding drug active against rhinoviruses and other picornaviruses.

Pock: a localized blistering lesion on skin or mucosae caused by poxviruses or other agents.

Poliomyelitis: disease caused by poliovirus, involving central nervous system lesions that can lead to paralysis.

Polyadenylation signal: in eukaryotes, a consensus AAUAAA sequence in messenger RNAs that is recognized by proteins that bind to the RNA, cleave it downstream of the AAUAAA to create a new 3' end, and add a poly (A) tail to the 3' end.

Polyclonal: derived from a number of individual clones. Tumor cells or cells of the immune system are considered polyclonal if they have numerous different origins or antigenic specificities.

Polyhedrin: protein that forms the polyhedral occlusion bodies of nucleopolyhedroviruses.

Polymorphonuclear leukocytes: a collection of white blood cells, including neutrophils, eosinophils and basophils, that have granules within their cytoplasm and lobed nuclei.

Polyprotein: a precursor protein that is cleaved by proteolysis during or after synthesis to generate several mature functional proteins.

Positive (or plus) strand: the RNA strand that contains coding regions for proteins.

Positive staining: staining of specimens for electron microscopy by use of heavy-metal salts that bind to the specimen and impart electron density. Usually used for observation of thin sections of cells.

Preintegration complex: a complex of retrovirus capsid proteins and newly synthesized proviral DNA in a state ready to be integrated into the cellular genome.

Primary cell cultures: cells grown *in vitro* that are derived directly from an animal tissue. Cells from embryos or newborn animals are often used because of their ease of growth.

Primosome: a protein complex that synthesizes RNA primers on which DNA replication can be initiated.

Prion: pathogen that replicates and causes prion diseases, presumably a conformational variant PrP^{sc} of a normal cell protein, PrP^c. The term *prion* stands for "*p*roteinaceous *i*nfectious" particle. [Note: in biology, the name "prion" also refers to a group of 6 or 7 different species of small petrels (oceanic birds) restricted to the Southern Ocean surrounding Antarctica.]

Prion disease: neurodegenerative disease caused by the accumulation of prions in the brain, leading to spongiform encephalopathy, and characterized by accumulation of amyloid fibrils and plaques. Invariably fatal. Can be caused by inherited mutations in PrP gene or transmitted from infected animals, usually of same species.

Procapsid: a precursor to a viral capsid that usually does not contain the viral genome. Procapsids often change in structure or protein composition when they incorporate the viral genome to form a mature capsid.

Processivity: the ability of a polymerase to extend a growing chain without detaching from the template.

Progressive multifocal leukoencephalopathy (PML): a rare, fatal demyelinating disease of humans caused by the polyomavirus JC.

Prohead: term used frequently to designate procapsids of bacteriophages. See **Procapsid.**

Pro-inflammatory cytokines: cytokines that function to enhance inflammation.

Promoter occlusion: silencing the use of a transcriptional promoter, usually as a result of the presence of an active upstream promoter. Promoter occlusion may occur by dislodging of transcription factors from the downstream promoter by transcribing RNA polymerases that initiate upstream.

Prophage: a repressed bacteriophage chromosome that is integrated into the bacterial chromosome or replicates as a free plasmid.

Proteasomes: large specialized complexes that degrade proteins in the cell and make peptides available to MHC class I proteins for antigen presentation.

Protein kinases: enzymes that modify proteins by the covalent addition of phosphate groups.

Protein tyrosine kinase: an enzyme that phosphorylates proteins at specific tyrosine residues.

Proteoglycan: extracellular macromolecule composed of large negatively charged polysaccharide chains and protein.

Proteosomes: vesicles made from the outer membrane of *Neisseria* bacterial species, into which exogenous antigens are introduced. Injection of proteosomes into animals can induce immunity to these antigens.

Protomers: subunits used to build a larger structure.

Proto-oncogene: a cellular regulatory gene that, when mutated or overexpressed, can cause cellular transformation and oncogenesis.

Proviral DNA: a DNA copy of a retrovirus genome that is inserted into the chromosome of an infected host cell.

Pseudoknot: an RNA secondary structure in which bases in the loop at the end of a stem form hydrogen bonds with bases at a distant site on the same RNA molecule.

Pseudotype: see **Pseudotyping.**

Pseudotyping: production of a recombinant virus (pseudotype) that carries a foreign receptor-binding protein, allowing the virus to bind to and infect different cell types or species.

Pseudovirion: virions of polyomaviruses that encapsidate fragments of cellular chromatin instead of the viral genome.

Pyroptosis: a form of programmed cell death that occurs during inflammation and requires activated caspase 1.

Quasi-equivalent: in virus capsid structure, describes the symmetrical distribution of protein subunits on the surface of a capsid in such a way that they form similar but not identical bonds with each other.

Quasispecies: virus variants that differ by a small number of mutations.

Raft culture: *in vitro* culture system used to differentiate epithelial cells by growth at an air–liquid medium interface.

Raltegravir: an antiviral agent that acts by inhibiting HIV-1 integrase.

Ras: a cellular signaling protein that is activated by receptor tyrosine kinases and in turn activates the MAP kinase pathway, which itself induces transcription of cell-cycle and other genes.

Rational drug discovery: approaches for discovering new drugs that use basic knowledge from genetics, molecular biology, enzymology, and protein structure to choose drug targets and molecules to act on these targets.

Reassortant viruses: viruses that have packaged different genome segments from two (or more) parent viruses and therefore express some viral proteins from one strain and some viral proteins from the other strain. Reassortment can occur at high frequency with viruses that have fragmented genomes.

Reassortment: the exchange of genome segments among different viral strains in viruses with fragmented genomes.

Receptor tyrosine kinases: cell surface receptor proteins that respond to binding of signaling molecules by phosphorylating specific tyrosine residues on intracellular proteins, activating a variety of signaling pathways leading to activation or inactivation of numerous cellular genes.

Receptors (viral): cell surface proteins, lipids, or carbohydrates to which specific viruses attach and begin the process of entry into the cells. Viral receptors have their own cellular functions, and are only fortuitously exploited by viruses.

Receptor-mediated endocytosis: endocytic uptake of molecules from the extracellular space involving their specific binding to plasma membrane receptors.

Replication complex: for RNA viruses, a factory for viral RNA synthesis consisting of viral and cellular proteins associated with membranes in the cytoplasm of an infected cell.

Replicative intermediate: for RNA viruses, an RNA molecule on which one or several growing complementary RNA strands are being synthesized. The growing strands are usually base-paired to the template RNA only near their growing 3′ ends.

Replicon: a nucleic acid that, when transfected into cells, is capable of replication directed by proteins encoded in the nucleic acid. RNA virus replicons may lack coding sequences for viral structural proteins.

Reporter gene: a gene coding for a protein that can be detected within cells easily and quantitatively. Examples include **green fluorescent protein**, **beta-galactosidase**, and **chloramphenicol acetyltransferase**. The insertion of a reporter gene into a virus genome allows for the detection of virus infection and replication by assaying for the appearance of the reporter protein.

Reservoir: a population of organisms in which a pathogenic agent circulates in nature, not necessarily causing overt disease. Humans or domestic animals can be infected by coming in contact with members of this population, and they can suffer from disease, but do not necessarily transmit the agent efficiently to other members of their species.

Resolvase: an enzyme that makes staggered single-strand breaks at the ends of a concatemer junction, allowing formation of single-length DNA molecules with hairpin ends.

Retinoblastoma: a tumor of the retina. Can be caused by absence of retinoblastoma protein (pRb), a tumor suppressor protein.

Retrograde transport: transport of macromolecules within nerve axons in the direction from the nerve periphery to the nucleus, or cell body.

Reverse genetics: The process of engineering specific genetic changes in the genome of an organism and then looking for resulting phenotypic changes.

Reverse transcriptase: an enzyme that synthesizes DNA using an RNA template: also known as an RNA-dependent DNA polymerase.

Ribonuclease H: a ribonuclease that specifically degrades the RNA part of RNA/DNA hybrids.

Ribonuclease L: an enzyme that, when activated by binding of 2′, 5′-oligoadenylate, degrades cellular and viral mRNAs. Part of the interferon response.

Ribose methyltransferase: an enzyme that transfers a methyl group to the 2′ position of ribose during generation of a methylated cap structure at the 5′ end of an RNA.

Ribosomal frameshifting: a process by which ribosomes pause during translation of a coding region on a messenger RNA, move back one or two nucleotides on the messenger RNA, and recommence reading the messenger RNA in a different translational reading frame.

Ribosome shunting: a mechanism by which ribosomes are able to locate initiation codons for translation of messenger RNAs without "scanning" the RNA from its capped 5′ end by the classical pathway using initiation factor eIF4F.

Ribozyme: a *ribo*nucleic acid that acts as an *enzyme*. The RNA catalyzes a specific biochemical reaction in a manner that can transform several substrate molecules per ribozyme molecule. Some known ribozymes can cleave or ligate RNAs.

Ritonavir: protease inhibitor used for treatment of human immunodeficiency virus infections.

RNA helicase: an enzyme that unwinds helical structures (stems) in single-stranded RNA molecules, or unwinds the strands of double-stranded RNA molecules. This enzyme is usually required for transcription and replication of RNA.

RNA interference (RNAi): an RNA-dependent gene regulation system mediated by **miRNAs** and **siRNAs** that is important for development and defense against pathogens.

RNA world: a hypothetical prebiotic world in which self-replicating RNAs developed and evolved, preceding and perhaps leading to development of the DNA world that is universal in present-day life.

RNA-dependent DNA polymerase: an enzyme that synthesizes DNA using an RNA template; reverse transcriptase.

RNA-dependent RNA polymerase: an enzyme that synthesizes RNA using an RNA template. Viruses with RNA genomes must code for such an enzyme, which is not available in the host cell.

RNA-induced silencing complex (RISC): a multi-protein complex that uses one strand of an **siRNA** or **miRNA** as a template for recognizing complementary mRNA sequences to mediate RNA interference.

RNA polymerase II holoenzyme: a complex consisting of eukaryotic RNA polymerase II plus accessory proteins, which is active in transcription.

Rolling-circle replication: synthesis of a linear RNA or DNA from a circular template by an RNA or DNA polymerase that cycles around the template numerous times without detaching, resulting in an extended linear molecule that contains tandem repeats of the complement of the template sequence. These are usually cleaved into unit-length molecules.

Satellite nucleic acid: a nucleic acid genome that is able to replicate only in the presence of a helper virus, and which is packaged as a virion by capsid proteins coded by the helper virus. Genomes are usually less than 2 kb long and in some cases code for no proteins.

Satellite virus: a virus whose genome is able to replicate only in the presence of a helper virus, but which codes for its own capsid protein (and usually no other proteins).

Scaffolding proteins: proteins that are involved in the assembly of complex biological structures such as viral capsids, but are not retained in the final structure.

Scrapie: prion disease in sheep, characterized by rubbing or scraping of the skin.

Secondary receptor: cell-surface receptors that interact with viral capsid or envelope proteins to direct uptake, but only after the virus particle has bound to a primary receptor on the cell surface.

Self-assembly: ability of a structure such as a viral capsid to assemble spontaneously from its protein subunits, without the aid of other machinery.

Seroconversion: the appearance in serum of antibodies specific to a pathogen, indicating exposure to the pathogen but not necessarily overt disease.

Serpin: *ser*ine *p*rotease *in*hibitor. Can block cleavage of cytokines and therefore block their normal biological functions.

Serum sickness: a hypersensitivity response to large amounts of antigen in the circulatory system.

Shine-Dalgarno sequence: a consensus sequence located just upstream of translational start sites in bacterial mRNAs that facilitates initiation of protein synthesis by binding to 16S ribosomal RNA.

Shock: inability of circulatory system to maintain adequate blood supply to vital organs; most often caused by low blood volume due to loss of blood.

Sialic acid: a 9-carbon acidic sugar derived from mannose, found on many glycolipids and glycoproteins. Also called N-acetyl neuraminic acid.

Sickle cell disease: a hereditary form of anemia caused by an abnormal type of hemoglobin and characterized by unusual crescent-shaped red blood cells—"sickle cells."

Sieve elements: a cell type in the phloem, sieve elements combine and form tubes that are used to transport nutrients.

Signal peptidase: a host protease that cleaves the signal sequence from membrane or secretory proteins during or shortly after their transport into the lumen of the endoplasmic reticulum.

Signal sequence: amino acid sequence that directs insertion of the N-terminus of proteins into the endoplasmic reticulum during protein synthesis, and is cleaved from the protein in the process.

Silencing: reduction or elimination of the activity of a transcriptional promoter, often because sequences in the promoter region are methylated.

Small interfering RNAs (siRNAs): short, double-stranded RNA molecules of 20 to 25 base pairs that stimulate RNA interference and antiviral signaling pathways.

Sodium dodecyl sulfate: a 12-carbon ionic detergent that is used to denature proteins.

Solfataric: refers to sulfurous fumaroles or springs of volcanic origin. The word is derived from Italian or Latin words referring to sulfur ("solfa") and land ("terra"). A volcanic crater in the Campo Flegri near Naples is named Solfatara, and is the original source of the archaeon, *Solfolobus sofataricus*.

SOS genes: a collection of LexA-regulated genes that function to repair damaged DNA.

Spasticity: continuous contraction of certain muscles, usually resulting from damage to brain centers that control voluntary movement.

Species: in virology, a taxonomic grouping of virus isolates that share a high degree of nucleic acid and amino acid sequence homology in addition to having a common genome structure and size, common virion structure and replication strategy, and (often but not always) a limited host range.

Spliceosome: a complex made up of a number of small ribonucleoproteins bound to pre-messenger RNAs. Carries out cleavage and joining reactions leading to production of spliced messenger RNAs.

Spongiform encephalopathy: disease accompanied by large vacuoles in the cortex and cerebellum, giving the brain the appearance of a sponge.

Squamous cell carcinoma: a type of skin cancer arising from the upper layers of skin epidermis.

Src-related kinases: enzymes like c-src that phosphorylate tyrosines on specific target proteins and regulate numerous cellular signaling pathways.

Subacute sclerosing panencephalitis (SSPE): a progressive, fatal neurological disease that is a rare complication of measles virus infection. Only one in 300,000 measles victims are affected by SSPE, which seems to result from a slow, persistent infection and strikes up to 10 years after the acute illness.

Subgenomic mRNA: a messenger RNA that includes only part of the sequence of the genome of an RNA virus.

Supercoiled: double-stranded (usually circular) DNA molecules that are overwound and as a result are twisted into a coiled structure.

Surface lattice: a set of lines drawn on the surface of an object that connect points defining axes of symmetry.

Synchrotron: device used to accelerate electrons, which can then be used to create high-energy x-rays.

Syncytium: a group of cells that share a common cytoplasm with more than one nucleus. Can be formed by cell-cell fusion. (Plural: syncytia.)

Systemic mosaic: mosaic disease symptoms found throughout the infected plant.

Target-based high throughput screen: method for screening large numbers of drugs for their ability to affect biological activity of a purified protein by use of automated *in vitro* assays.

TATA box: a consensus sequence, often TATAAA, to which the TATA-binding protein (TBP), a part of the general transcription factor TFIID, binds. Usually located 25 to 30 nt upstream of eukaryotic transcription start sites.

Taxonomy: the science of classification of living organisms on the basis of genetic and evolutionary relatedness.

Tax-responsive elements (TRE): 21 bp repeated sequences in HTLV-1 proviral DNA that help form a complex with Tax and other proteins, recruiting histone acetylases to the DNA and increasing viral gene transcription.

T-cell lymphoma: a T-cell malignancy in which the proliferating T cells are localized mostly in lymphoid tissues rather than in the blood.

T-cell receptor: a receptor on the surface of T lymphocytes that binds to specific antigens in association with major histocompatibility proteins, leading to activation of T cells and production of antibodies or cytotoxic T lymphocytes.

Tegument: region between the envelope and the nucleocapsid in virions of certain viruses, notably baculoviruses and herpesviruses. Contains a number of virus proteins.

Telomerase: an enzyme that maintains and extends ends of linear chromosomes in eukaryotic cells by using a small template RNA to make repeated DNA sequences.

Temperate phage: a bacterial virus that can either replicate to produce progeny virions and cell lysis, or enter a repressed prophage state that can be harbored in a cell for multiple generations.

Terminal protein: an adenovirus-coded protein that is covalently bound to the 5′ ends of linear viral DNA. Directs and primes initiation of viral DNA replication.

Therapeutic index: the ratio of the dose of a drug that exerts a half-maximal toxic effect, to the dose that exerts a half-maximal therapeutic effect. Effective, safe drugs have a high therapeutic index.

Thermophile: an organism that grows at temperatures above 45 °C but below 80 °C.

Thrips: a minute insect of the family Thripidae (order Thysanoptera) commonly found on plants, some of which are vectors for tospoviruses. ("Thrips" is both singular and plural.)

Thymidine kinase: an enzyme that phosphorylates thymidine to thymidine monophosphate.

Thymidylate kinase: an enzyme that phosphorylates thymidine monophosphate (thymidylate) to thymidine diphosphate.

Tight junction: intercellular junction that brings plasma membranes of neighboring cells into very close proximity (1–2 nm); found in endothelial and epithelial cells.

T lymphocytes: white blood cells that mature in the thymus and are involved in the adaptive immune response, regulating development of B lymphocytes for antibody production, and generating cytotoxic T lymphocytes, leading to killing of target cells.

Tolerance: a state in which the immune system accepts as "self" organisms or cells of foreign origin. In the context of viral illness, tolerance can occur when infection occurs at a very young age or can be induced as a part of the biology of the infecting virus (HIV or human papillomaviruses).

Toll-like receptors: a set of cell-surface and endosomal proteins that recognize molecules specific to invading pathogens (e.g., lipopolysaccharide, CpG-containing DNA sequences, double-stranded RNA) and activate innate and adaptive immune responses via a protein kinase cascade.

Tonoplast: the membrane that surrounds the vacuole in plant cells.

Topoisomerase I: enzyme that cleaves a phosphodiester bond in a double-stranded DNA, forming a temporary covalent bond with the DNA, and allows one strand to rotate around the complementary strand, subsequently resealing the phosphodiester bond and leading to relaxation of overwound (supercoiled) DNA.

Topoisomerases: enzymes that relieve torsional stress in supercoiled DNA molecules by cleavage and resealing of phosphodiester bonds.

Toroidal coil: a spring-like coil that consists of a cylindrical form bent into a circle.

Transcription factors: proteins that bind to specific DNA sequences in transcriptional enhancers or promoters and alter the transcriptional activity of the gene by interacting with other transcription factors and RNA polymerases.

Transcriptional enhancer: a region that contains consensus-binding sequences for transcription factors and that enhances the transcription of specific genes. Often located immediately upstream of a gene, but can be found within the transcribed sequence or thousands of base pairs from the transcription start sequence.

Transgene: a foreign gene that is introduced artificially into a cell and expressed, sometimes as part of a virus vector.

Transmembrane anchor: a domain of 20 hydrophobic amino acids that is embedded in a lipid bilayer and therefore anchors a protein in the membrane.

Transovarial transmission: the transmission of a pathogen from mother to offspring via the egg.

Triangulation number: $T = h^2 + hk + k^2$, where h and k are integers; the number of protein subunits on the surface of viral capsids with icosahedral symmetry is 60 x T.

Trisymmetrons: triangular capsid substructures composed of multiple trimeric or hexameric capsomers.

Tropical spastic paraparesis (TSP): a neurological disease, caused by HTLV-1 infection, that leads to demyelination of motor neurons.

Tropism: the ability of a virus (or other pathogen) to infect and replicate in a particular tissue, organ, or species. Often depends on the initial interaction of the pathogen with cell surface receptors.

Tumor necrosis factor (TNF): a cytokine that activates the inflammatory response and can cause apoptosis.

Tumor suppressor protein: a protein, such as pRb or p53, whose absence or mutation can facilitate or cause tumor formation; therefore the wildtype protein suppresses tumor development.

Twofold symmetry axis: an axis around which an object rotated 180° appears identical to the original object.

Type 1 and type 2 T helper lymphocytes: T lymphocytes that secrete cytokines that contribute to the development of cellular immune responses (Th1 cells) or humoral immune responses (Th2 cells)

Type I transmembrane protein: an integral membrane protein oriented with its amino terminus facing the lumen of the endoplasmic reticulum or the extracellular space and its carboxy terminus facing the cytosol. Its single hydrophobic membrane-spanning anchor is usually near the carboxy terminus.

Type II transmembrane protein: an integral membrane protein oriented with its carboxy terminus facing the lumen of the endoplasmic reticulum or the extracellular space and its amino terminus facing the cytosol. Its single hydrophobic membrane-spanning anchor is usually near the amino terminus.

Ubiquitin: a 76-amino acid polypeptide that is covalently joined to lysine residues on proteins, which are then degraded in proteasomes.

Ubiquitination: covalent linkage of ubiquitin to a protein, marking it for degradation by proteasomes.

Ubiquitin ligase: an enzyme that catalyzes covalent binding of ubiquitin to proteins, marking them for degradation in proteasomes.

Uncoating: the removal of the protein coat, or capsid, from a viral genome within a cell, in preparation for its expression and replication.

Vaccination: a term coined by Edward Jenner and initially used to describe immunization against smallpox using cowpox. The term was generalized by Louis Pasteur in 1881, in honor of Jenner, to include artificial immunization against any disease.

Variolation: artificial inoculation with variola (smallpox) virus, designed to produce immunity to natural smallpox infection. Practiced in various forms from the tenth through the eighteenth centuries in several countries worldwide, variolation produced a disease of diminished severity while conferring lifelong immunity to smallpox. The practice was replaced by vaccination in the nineteenth century.

Vesicle transport: the transfer of vesicular membranes and their contents between two membrane compartments by coupled membrane fission and fusion.

Viral interference: the state of resistance of a cell to virus infection brought about by previous exposure of the cell to a similar or different virus.

Virion: a complete, infectious virus particle.

Viroceptor: a virus-coded protein that is a homologue of a cellular cytokine receptor. A viroceptor can bind to a cytokine and interfere with its activity, therefore reducing host cell defenses against the virus.

Virogenic stroma: electron-dense material that accumulates in baculovirus-infected nuclei, believed to be the site of viral replication and nucleocapsid assembly.

Viroid: a small (250–400 nt), circular RNA that can replicate in the absence of a helper virus by reprogramming cellular DNA-dependent RNA polymerases to copy its genome RNA. Viroids do not code for proteins and are not packaged into virions, but are probably transferred as naked RNAs from one host to another via insects or mechanical abrasion.

Virokine: a virus-coded protein that is a homologue of a cellular cytokine. By mimicking cytokines, virokines can augment virus replication or suppress host defenses.

Virological synapse: a small region of contact established between neighboring cells that allows passage of enveloped viruses from one cell to the next without exposure to the extracellular milieu.

Viroplasm: name given to localized regions in the cytoplasm of infected cells where poxvirus DNA accumulates and where virion assembly takes place.

Virulence factors: pathogen-encoded proteins that enable the pathogen to counteract aspects of the host immune response and therefore increase its own replication, usually to the detriment of the host organism.

Virus vectors: viruses that are genetically modified to express foreign genes and/or to infect cell types or species not originally targeted by the virus.

X-ray diffraction: the diffraction of x-rays by oriented molecular assemblies, usually in crystals, used to determine the atomic structure of the object.

Zanamivir: neuraminidase inhibitor used for treatment of influenza A and B virus infections.

Zoonosis: infection of an animal population with a pathogenic agent that can be transmitted to human beings.

Zoonotic disease: a disease that can be transmitted to human beings from animals.

Front Cover

Enterobacteria Phage Phi X174: McKenna, R., Xia, D., Willingmann, P., Ilag, L.L., Krishnaswamy, S., Rossmann, M.G., Olson, N.H., Baker, T.S., Incardona, N.L. "Atomic structure of single-stranded DNA bacteriophage phi X174 and its functional implications," *Nature* (1992) 355: 137–143. Image created by Jean-Yves Sgro with software Qutemol. **Human Rhinovirus 3:** Zhao, R., Pevear, D.C., Kremer, M.J., Giranda, V.L., Kofron, J.A., Kuhn, R.J., Rossmann, M.G., "Human rhinovirus 3 at 3.0 A resolution," *Structure* (1996) 4: 1205–1220. Image created by Jean-Yves Sgro with software Qutemol. **Simian Virus 40:** Stehle, T., Gamblin, S.J., Yan, Y., Harrison, S.C., "The structure of simian virus 40 refined at 3.1 A resolution," Structure (1996) 4: 165–182. Radially depth-cued image created by Jean-Yves Sgro with software VMD with ambient occlusion. *Qutemol:* http://qutemol.sourceforge.net/. Marco Tarini, Paolo Cignoni, Claudio Montani, "Ambient Occlusion and Edge Cueing for Enhancing Real Time Molecular Visualization," *IEEE Transactions on Visualization and Computer Graphics* (2006), Volume 12, Issue 5, pp. 1237–1244. ISSN:1077–2626. *VMD-Visual Molecular Dynamics:* http://www.ks.uiuc.edu/Research/vmd/. Humphrey, W., Dalke, A. and Schulten, K., "VMD - Visual Molecular Dynamics", *J. Molec. Graphics* (1996), vol. 14, pp. 33–38.

Chapter 1

Fig. 1.2a: Courtesy of Robley Williams, University of California at Berkeley and Harold Fisher, University of Rhode Island; **Fig. 1.2b:** Courtesy of John Finch, Cambridge University; **Fig. 1.2c:** Courtesy of The John Curtin School of Medical Research, ANU.

Chapter 2

Fig. 2.1a: Electron micrograph courtesy of Dr. Carla W. Gray, University of Texas, Dallas: reprinted from Gray, C.W., R.S. Brown, and D.A. Marvin, "Adsorption complex of filamentous fd virus," *Journal of Molecular Biology* 146 (1981): 621–627; **Fig. 2.1b, 2.1e–g:** Courtesy of Robley Williams, University of California at Berkeley and Harold Fisher, University of Rhode Island; **Fig. 2.1c–d:** Linda Stannard/Photo Researchers; **Fig. 2.1h:** Courtesy of Jean-Yves Sgro, University of Wisconsin, Madison; **Fig. 2.1i:** Courtesy of Richard Kuhn, Purdue University; **Fig. 2.1j:** Courtesy of Jan Orenstein, George Washington University Medical Center; **Fig. 2.2:** Figs. 2–3 reprinted from S.C. Harrison, "Principles of Virus Structure", Chapter 3 in *Fundamental Virology* 4th ed. (2001), D.M. Knipe and P.M. Howley (eds), Lippincott Williams & Wilkins (Baltimore); **Fig. 2.3a:** from VIPERdb2: an enhanced and web API enabled relational database for structural virology, Mauricio Carrillo-Tripp, Craig M. Shepherd, Ian A. Borelli,

Sangita Venkataraman, Gabriel Lander, Padmaja Natarajan, John E. Johnson, Charles L. Brooks, III and Vijay S. Reddy. *Nucleic Acids Research* 37, D436-D442 (2009); doi: 10.1093/nar/gkn840; **Fig. 2.3b:** Courtesy of Stephen C. Harrison; **Fig. 2.4:** from VIPERdb2;, *op. cit.*; **Fig. 2.5a:** Reprinted from S.C. Harrison, "Principles of Virus Structure," *op. cit.*, Fig. 6; **Fig. 2.5b:** from VIPERdb2, *op. cit;* **Fig. 2.6a:** VMD image courtesy of Dr. Jean-Yves Sgro, University of Wisconsin-Madison; **Fig. 2.6b:** Reprinted from Liddington, R.C. et al., "Structure of Simian Virus 40 at 3.8 A Resolution," *Nature* 354 (1991): 278–284; **Fig. 2.7:** Reprinted from Burnett R.M., et al. (eds) *Biological Macromolecules and Assemblies*, John Wiley and Sons (1984); **Fig. 2.8a:** Courtesy of Nicholas Acheson; **Fig. 2.8b:** PhotoDisc, Inc.; **Fig. 2.8c:** Reproduced from http://en.wikipedia.org/wiki/File:Eight_Allotropes_of_Carbon.png, Wikipedia Commons; **Fig. 2.9:** Reprinted from S.C. Harrison, "Principles of Virus Structure", *op. cit.*, Fig. 19; **Fig. 2.10:** Reprinted from Fig. 3 in Andrei Fokine, Anthony J. Battisti, Valorie D. Bowman, Andrei V. Efimov, Lidia P. Kurochkina, Paul R. Chipman, Vadim V. Mesyanzhinov, and Michael G. Rossmann, "Cryo-EM Study of the Pseudomonas Bacteriophage PhiKZ," *Structure* 15 (2007), 1099-1104, with permission from Elsevier and Michael G. Rossmann; **Fig. 2.11a:** Adapted from Fig. 24.2, p. 902, in James D. Watson, Nancy H. Hopkins, Jeffrey W. Roberts, Joan Argetsinger Steitz, and Alan M. Weiner, *Molecular Biology of the Gene*, 4th ed., The Benjamin/Cummings Publishing Company, Inc. Copyright 1965, 1970, 1976, 1987, reprinted by permission of Pearson Education, Inc (Atlanta); **Fig. 2.12:** Courtesy of Stephen Harrison.

Chapter 4

Fig. 4.1, 4.2: Courtesy of Ari Helenius. **Fig. 4.3a–c:** Courtesy of Jurgen Kartenbeck, German Cancer Research Center; **Fig. 4.4:** Reproduced from the Ph.D. thesis of Jennifer Gruenke, University of Virginia, with permission from Jennifer Greunke and Judith M. White; **Fig. 4.5:** Courtesy of Ari Helenius, modified from a figure from Whittaker, G.R. et al., "Viral entry into the nucleus," *Annual Review of Cell and Developmental Biology* 16 (2000): 627–651.

Chapter 5

Fig. 5.3: Courtesy of A.B. Jacobson, Reprinted from Fauquet, C. et al. (eds.), *Virus Taxonomy*, VIIIth Report of the International Committee on Taxonomy of Viruses, (2004) p. 741, with permission from Elsevier.

Chapter 6

Fig. 6.2, 6.5, 6.6: Courtesy of Bentley Fane.

Chapter 7

Fig. 7.1: Redrawn from Fig. 3 in F.W. Studier and J.J. Dunn, Cold Spring Harbor Symposia on Quantitative Biology 47 (1983): 999, with permission from Cold Spring Harbor Laboratory Press (Cold Spring Harbor, NY); Fig. 7.2: Redrawn from Fig. 1 in Molineux, I.J., "No Springs Please, Ejection of Phage T7 DNA from the Viron is Enzyme Driven," *Molecular Microbiology* 40 (2001):1–8, with permission of Blackwell Publishing Co.

Chapter 9

Fig. 9.1: Reproduced from http://en.wikipedia.org/wiki/File:Phylogenetic_tree.svg, Wikipedia Commons; Fig. 9.2a,c: Courtesy of Wolfram Zillig; Fig. 9.2b: Reproduced from Figure 1A in Redder, P., X. Peng, K. Brügger, S. A. Shah, F. Roesch, B. Greve, Q. She, C. Schleper, P. Forterre, R. A. Garrett, and D. Prangishvili, "Four newly isolated fuselloviruses from extreme geothermal environments reveal unusual morphologies and a possible interviral recombination mechanism," *Environmental Microbiology* 11 (2009), 2849–2862, with permission from John Wiley and Sons Inc.; Fig. 9.2d: Courtesy of Philippe Lemercier; Fig. 9.3: Figure 9 in Sabrina Fröls, Paul M.K. Gordon, Mayi Arcellana Panlilio, Christa Schleper, Christoph W. Sensen, "Elucidating the transcription cycle of the UV-inducible hyperthermophilic archaeal virus SSV1 by DNA microarrays," *Virology* 365 (2007), 48-59, with permission from Elsevier; Fig. 9.4a: Courtesy of Wolfram Zillig; Fig. 9.4b: Courtesy of Philippe Lemercier; Fig. 9.5a,b: Modified from Figure 1 in: Bettstetter M., X. Peng, R. Rachel, R. A. Garrett, and D. Prangishvili "AFV1, a novel virus of the hyperthermophilic crenarchaeon Acidianus," *Virology* 315 (2003), 68-79, with permission from Elsevier; Fig. 9.5c: Courtesy of Philippe Lemercier; Fig. 9.6a,b: Courtesy of Wolfram Zillig; Fig. 9.6c: Courtesy of Philippe Lemercier; Fig. 9.7a-c: Reprinted from Fig 4 A, Fig 5 A and G in: Monika Haring, Reinhard Rachel, Xu Peng, Roger A. Garrett, and David Prangishvili, "Viral Diversity in Hot Springs of Pozzuoli, Italy, and Characterization of a Unique Archaeal Virus, Acidianus Bottle-Shaped Virus, from a New Family, the Ampullaviridae," *Journal of Virology* 79 (2005), 9904–9911, with permission from American Society for Microbiology; Fig. 9.7d: Courtesy of Philippe Lemercier; Fig. 9.8a,b: Modified from Fig 1 b and c in: Häring M., X. Peng, K. Brügger, R. Rachel, K. O. Stetter, R. A. Garrett, and D. Prangishvili, "Morphology and genome organisation of the virus PSV of the hyperthermophilic archaeal genera Pyrobaculum and Thermoproteus: a novel virus family, the Globuloviridae," *Virology* 323 (2004), 233-242, with permission from Elsevier; Fig. 9.8c: Courtesy of Philippe Lemercier; Fig. 9.9a,b: Courtesy of Wolfram Zillig, reprinted by permission from Macmillan Publishers Ltd: figure 3a in David Prangishvili, Patrick Forterre and Roger A. Garrett, "Viruses of the Archaea: a unifying view," *Nature Reviews Microbiology* 4 (2006), 837-848; Fig. 9.9c: Courtesy of Phillipe Lemercier; Fig. 9.10a1-c2: Adapted From Figure 4 in: Bize A, E. A. Karlsson, K. Ekefjärd, T. E. Quax, M. Pina, M. C. Prevost, P. Forterre, O. Tenaillon, R. Bernander, and D. Prangishvili, A unique virus release

mechanism in the Archaea, *Proc. Natl. Acad. Sci. USA* 106 (2009), 11306-11311, with permission from PNAS; Fig. 9.11a-h: Reprinted by permission from Macmillan Publishers Ltd: Figure 1 in Häring M., G. Vestergaard, K. Brügger, R. Rachel, R. A. Garrett, and D. Prangishvili, "Independent virus development outside a host," *Nature* 436 (2005), 1101-1102; Fig. 9.11i: Courtesy of Philippe Lemercier; Fig. 9.12: Figure 4 reprinted from: Prangishvili, D., G. Vestergaard, M. Häring, R. Aramayo, T. Basta, R. Rachel, and R. A. Garrett, R. A., "Structural and genomic properties of the hyperthermophilic archaeal virus ATV with an extracellular stage of the reproductive cycle," *J. Mol. Biol.* 359 (2006), 1203-1216, with permission from Elsevier; Fig. 9.13: Part of Fig 6 reprinted from: Bath C, Cukalac T, Porter K, Dyall-Smith ML, "His1 and His2 are distantly related, spindle-shaped haloviruses belonging to the novel virus group, Salterprovirus," *Virology* 350 (2006), 228-239, with permission from Elsevier; Fig. 9.14a: Reprinted by permission from Macmillan Publishers Ltd: Fig 3 in Ortmann A.C., B. Wiedenheft, T. Douglas, and M.J. Young, "Hot crenarchaeal viruses reveal deep evolutionary connections," *Nat. Rev. Microbiol.* 4 (2006), 520-528; Fig. 9.14b: Fig 4D reprinted from: Jäälinoja HT, Roine E, Laurinmäki P, Kivelä HM, Bamford DH, & Butcher SJ, "Structure and host-cell interaction of SH1, a membrane-containing, halophilic euryarchaeal virus," *Proc. Natl. Acad. Sci. USA* 105 (2008), 8008-8013, with permission from PNAS, Copyright (2008) National Academy of Sciences, U.S.A.

Chapter 10

Fig. 10.1: Courtesy of Tom Zitter; Fig. 10.2: Courtesy of Robert Milne; Fig. 10.3: Courtesy of Keith Perry, from VIPERdb2, *op. cit.*; Fig. 10.6: Courtesy of Nicholas Acheson, with thanks to Cheng Kao; Fig. 10.9: Reproduced from DPV400 Figure 09 at http://www.dpvweb.net/dpv/showfig.php?dpvno=400&figno=09, Courtesy of Peter Palukaitis, with permission from the The Association of Applied Biologists.

Chapter 11

Fig. 11.1: Courtesy of Ny Carlsberg Glyptotek, Copenhagen; Fig. 11.2: Reproduced from http://www.virology.wisc.edu/virusworld/ictv8/p1m-human-poliovirus-1-ictv8.jpg, courtesy of Jean-Yves Sgro, University of Wisconsin; Fig. 11.3: Adapted from J.M. Hogle and V.R. Racaniello, "Poliovirus receptors and cell entry," *Molecular Biology of Picornaviruses*, Semler, B.L. and Wimmer, E. (eds.) (2002), with permission of ASM Press, Washington, D.C.; Fig. 11.4: Adapted from Leong, L.E., Cornell, C.T., and Semler, B.L., "Processing determinants and functions of the cleavage products of picornavirus polyproteins," *Molecular Biology of Picornaviruses*, Semler, B.L. and Wimmer, E. (eds.) (2002), with permission of ASM Press, Washington, D.C.; Fig. 11.6: Reprinted from Stewart, S.R. and Semler, B.L., "RNA determinants of picornavirus and cap-dependent translation initiation," *Seminars in Virology 8* (1997), 242–255, with permission from Elsevier; Fig. 11.8: Fig. 2 reprinted from Cornell, C.T. and Semler, B.L., "Gene expression and replication of the picornaviruses," *Encyclopedia of Molecular Cell Biology & Molecular Medicine*, 2nd ed. (2004), 93–117, Meyers, R.A. (ed.), Weinheim: Wiley-VCH.

Chapter 12

Fig. 12.1–12.6: Courtesy Richard Kuhn, Purdue University.

Chapter 13

Fig. 13.1: Courtesy Richard Kuhn, Purdue University; Fig. 13.3: Modified from Fig. 7 in J.A. Lemm et al., "Polypeptide requirements for assembly of functional Sindbis virus replication complexes: a model for temporal regulation of minus and plus strand RNA synthesis" *European Molecular Biology Organization Journal* 13 (1994): 2925–2934, by permission of Oxford University Press; Fig. 13.4: Adapted from Figs. 6 and 7 in Schlesinger, S. and Schlesinger, M.J., "Togaviridae," *Fundamental Virology, 3rd ed.*, Fields, B.N., Knipe, D.M., and Howley, P.M., (eds.). Philadelphia: Lippincott-Raven (1996) pp. 523–540.

Chapter 14

Fig. 14.2a: Courtesy of Dr. Guy Schoehn, CNBS, France.; Fig. 14.3: Reprinted from Fig 2A and B in: Kevin Knoops, Marjolein Kikkert, Sjoerd H. E. van den Worm, Jessika C. Zevenhoven-Dobbe, Yvonne van der Meer, Abraham J. Koster, A. Mieke Mommaas, Eric J. Snijder, "SARS-Coronavirus Replication Is Supported by a Reticulovesicular Network of Modified Endoplasmic Reticulum," *PLoS Biology* (doi:10.1371/journal.pbio.0060226), September 2008, Volume 6, Issue 9,e226, pp 1957-1974.

Chapter 15

Fig. 15.1a: Courtesy of George P. Leser, Northwestern University; Fig. 15.1b: Courtesy of Professor Rob Ruigrok, Universite Joseph Fournier, Grenoble, France; Fig. 15.2a: Courtesy of Guy Schoehn, CNBS, France; Fig. 15.2b: Courtesy of Professor Rob Ruigrok, Université Joseph Fournier, Grenoble, France; Fig. 15.5a–b: Reprinted from Fig. 1 in: Daniel Kolakofsky, Laurent Roux, Dominique Garcin, and Rob W. H. Ruigrok, "Paramyxovirus mRNA editing, the 'rule of six' and error catastrophe: a hypothesis," *Journal of General Virology* 86 (2005), 1869–1877.

Chapter 16

Fig. 16.1a–c: Courtesy of Dr. Yoshihiro Kawaoka and Dr. Takeshi Noda. Reprinted from Fig. 2a,b in Watanabe, S., et al., "Production of novel Ebola virus-like particles from cDNAs: an alternative to Ebola virus generation by reverse genetics." *Journal of Virology* 78 (2004): 999–1005.

Chapter 17

Fig. 17.1a: Courtesy of B. V. Venkataram Prasad, Baylor College of Medicine; Fig. 17.4: Courtesy of Esa Kuismanen,

University of Helsinki, Finland. Fig. 1A reprinted from Jantti, J. et al., *Journal of Virology* 7 (1997): 1162–1172.

Chapter 18

Fig. 18.1: Courtesy of Robert Dourmashkin, London Metropolitan University; Fig. 18.2: Courtesy of P.W. Choppin. Figs. 1–3 reprinted from Choppin, P.W., "On the emergence of influenza virus filaments from host cells." *Virology* 21 (1963): 278–281; Fig. 18.5: Redrawn from Portela, A. and Digard, P., "The influenza virus nucleoprotein: a multifunctional RNA-binding protein pivotal to virus replication," *Journal of General Virology* 83 (2002): 723–734, with permission from Society for General Microbiology; adapted from Compans, R.W., Content, J., and Duesberg, P.H., "Structure of the ribonucleoprotein of influenza virus," *Journal of Virology* 10 (1972): 795–800, with permission from American Society for Microbiology; Fig. 18.9: Reprinted by permission from Macmillan Publishers Ltd: Figures 1 a and c in Takeshi Noda, Hiroshi Sagara, Albert Yen, Ayato Takada, Hiroshi Kida, R. Holland Cheng & Yoshihiro Kawaoka, "Architecture of ribonucleoprotein complexes in influenza A virus particles," *Nature* 439 (2006), 490-492; Fig. 18.10: Figure 1 from Rebecca J. Garten et. al., "Antigenic and Genetic Characteristics of Swine-Origin 2009 A(H1N1) Influenza Viruses Circulating in Humans," *Science* 325 (2009), 197–201, reprinted with permission from AAAS.

Chapter 19

Fig. 19.5a–c: Courtesy Max L. Nibert, Harvard Medical School. Reprinted from Centonze, V.E., et al., "Structure of reovirus virions and subvirion particles," *Journal of Structural Biology* 115 (1995): 215–225. Fig. 19.7a, b: Reprinted from Becker, M.M., et al., "Reovirus sigma NS protein is required for nucleation of viral assembly complexes and formation of viral inclusions." *Journal of Virology* (2001) 75: 1459-1475, Figs 4a, c.

Chapter 20

Fig. 20.1: Reproduced from Fig. 3, p. 1373, in Mavis Agbandje-Mckenna, Antonio L. Liamas-Saiz, Feng Wang, Peter Tattersall, and Michael G. Rossmann, "Functional implications of the structure of the murine parvovirus, minute virus of mice," *Structure* (1998) Nov 15, vol 6 (11), 1369–1381, with permission from Elsevier. Fig. 20.2: Reproduced by permission of Edward Arnold from Fig. 14.1 in Siegl, G. and Cassinotti, P., "Parvoviruses," 261–279, *Topley & Wilson's Microbiology and Microbial Infections*, Vol. 1, 9th ed., Mahy, B.W.J., and Collier, L. (eds.). London: Edward Arnold Publishers (1998); Fig. 20.4: Courtesy of N. Acheson. Adapted from Fig. 4 in Berns, K.I. and Linden, R.M., "The Cryptic Lifestyle of Adeno-Associated Virus," *Bioessays* 17 (1995): 237–245, with permission from John Wiley and Sons; Fig. 20.5: Reprinted from Fig. 5 in N. Muzyczka and K.I. Berns, "Parvoviruses," Chapter 32 in *Fundamental Virology*, D.M. Knipe and P.M. Howley (eds), Lippincott Williams & Wilkins (Baltimore) (2001); Fig. 20.7: Reprinted with permission from Saudan, P., Ph.D. Thesis, University of Lausanne (2000).

Chapter 21

Fig. 21.1: Courtesy of Jean-Yves Sgro, Institute for Molecular Virology, University of Wisconsin-Madison. Fig. 21.11: Reprinted from Fig. 6a–c in Gai, D., et al., "Mechanisms of conformational change for a replicative hexameric helicase of SV40 large tumor antigen," *Cell* 119 (2004), 47–60, with permission from Elsevier.

Chapter 22

Fig. 22.1a–b: Courtesy of Norman H. Olson, University of California, San Diego; Fig. 22.3: Modified from Fig. 2 in Longworth, M.S. and Laimins, L.A., "Pathogenisis of human papilomaviruses in differentiating epithelia," *Microbiology and Molecular Biology Reviews* 68 (2004): 362–372, with permission from the American Society for Microbiology.

Chapter 23

Fig. 23.1, 23.2: Reprinted from Fig 1a in Rux, J.J, and Burnett, R.M., "Review: Adenovirus Structure," *Human Gene Therapy* (2004) 15: 1167–1176, with permission from Mary Ann Liebert Publishers. Fig. 23.9: Fig. 24.23, p. 932 from *Molecular Biology of the Gene, 4th ed.* By James D. Watson, Nancy H. Hopkins, Jeffrey W. Roberts, Joan Argetsinger Steitz, and Alan M. Weiner. Copyright 1965, 1970, 1976, 1987 by The Benjamin/Cummings Publishing Company, Inc. Reprinted by permission of Pearson Education, Inc. (Atlanta).

Chapter 24

Fig. 24.3b: Courtesy of Wah Chiu, Baylor College of Medicine, from Zhou, A.H., et al., "Visualization of tegument/capsid interactions and DNA in intact herpes simplex virus type 1 virions," *Journal of Virology* 73 (1999): 3210–3218; Fig. 24.4: Courtesy of Beate Sodeik, Hannover Medical School, reprinted from Fig. 1a–h in Sodeik, B. et al., "Microtubule-mediated transport of incoming herpes simplex virus 1 capsids to the nucleus," *The Journal of Cell Biology* 136 (1997): 1007–1021, with permission from Rockefeller University Press; Fig. 24.6: Reprinted from Fig. 6 in B. Roizman, "Herpesviruses", Ch. 33 in *Fundamental Virology*, D.M. Knipe and P.M. Howley (eds), Lippincott Williams & Wilkins (Baltimore) (2001); Fig. 24.9: Copyright Bernard Roizman. Figs. 24.11, 24.12: Courtesy of Richard Longnecker.

Chapter 25

Fig. 25.1, 25.2, 25.3: Courtesy Dr. Eric B. Carstens, Queen's University, Kingston, Ontario; Fig. 25.4: Reprinted from A.L. Passarelli and L.K. Miller, "Identification and characterization of Lef-1, a baculovirus gene involved in late and very late gene expression," *Journal of Virology* 67 (1993): 3481–3488, with permission from the American Society for Microbiology.

Chapter 26

Fig. 26.1a,b: Reprinted from Moss, B., et al., in *Mechanics of DNA Replication and Recombination*, Cozarelli, N. (ed.).

New York: A. Liss (1983), pp. 449–461; Fig. 26.1c: Reprinted from Baroudy BM, Venkatesan S, Moss B, "Incompletely base-paired flip-flop terminal loops link the two DNA strands of the vaccinia virus genome into one uninterrupted polynucleotide chain," *Cell* 28 (1982), 315–324, with permission from Elsevier; Fig. 26.2: Courtesy of Richard Condit, modified from Fig. 3 in Condit, R.C. and Niles, E.G., "Orthopoxvirus genetics," *Current Topics in Microbiology and Immunology* 163 (1990): 1–39, with kind permission of Springer Science and Business Media (Heidelberg); Fig. 26.3: Reprinted from Fig 2A in Wilton, S., Mohandas, A.R., and Dales, S., "Organization of the Vaccinia Envelope and Relationship to the Structure of the Intracellular Mature Virions," *Virology* 214 (1995):503–511, with permission from Elsevier; Fig. 26.4a: Reprinted from Fig 4C in Condit, R.C., Moussatche, N., Traktman, P., "In a Nutshell: Structure and Assembly of the Vaccinia Virion," *Adv Virus Res.* 66 (2006):31-124. Fig. 26.4b: Copyright The Rockefeller University Press. 2005. Originally published in *The Journal of Cell Biology* 169:269-83. Fig. 26.5: Courtesy of Richard Condit. Reprinted from Moyer, R.W. and Condit, R.C., "Poxviruses," *Topley & Wilson's Microbiology and Microbial Infections*, London: Edward Arnold Publishers (2005), reproduced by permission of Edward Arnold. Fig. 26.7: Courtesy Mariano Esteban, National Center of Biotechnology, Madrid, Spain. Fig. 5a from Juan Ramón Rodríguez, Cristina Risco, José L. Carrascosa, Mariano Esteban, and Dolores Rodríguez, "Vaccinia virus 15-kilodalton (A14L) protein is essential for assembly and attachment of viral crescents to virosomes," *Journal of Virology* (1998), vol 72 (2):1287–96.

Chapter 27

Fig. 27.1: Courtesy of James Van Etten; Fig. 27.3: Courtesy of Michael Rossman, reprinted from Figure 1A in Cherrier, M.V. et. al., "An icosahedral algal virus has a complex unique vertex decorated by a spike," *Proc Natl Acad Sci USA* 106 (2009):11085-9, with permission from PNAS; Fig. 27.4: Courtesy of Michael Rossman, reprinted from Figure 2 in Nandhagopal, N., et. al., "The structure and evolution of the major capsid protein of a large, lipid-containing DNA virus," *Proc Natl Acad Sci USA*, 99 (2002):14758-63, with permission from PNAS, Copyright (2002) National Academy of Sciences, U.S.A.; Fig. 27.5a,b: courtesy of James Van Etten, reprinted from Figure 2c and f in Meints, R.H. et al., "Infection of a Chlorella-like alga with the virus, PBCV-1: ultrastructural studies," *Virology* 138 (1984):341-6, with permission from Elsevier; Fig. 27.5c: Reprinted from Figure 2d in Meints, R. H., K. Lee, and J. L. Van Etten, "Assembly site of the virus PBCV-1 in a chlorella-like green alga: ultrastructural studies," *Virology* 154 (1986): 240–245, with permission from Elsevier; Fig. 27.5d: Reprinted from Figure 1d in Yanai-Balser, G.M., et. al., "Microarray analysis of Paramecium bursaria chlorella virus 1 transcription," *Journal of Virology* 84 (2010):532-42, PMID: 19828609, with permission from American Society for Microbiology; Fig. 27.6: Courtesy of Willie Wilson; Fig. 27.7: Courtesy of The Remote Sensing Group, Plymouth Marine Laboratory, Plymouth PL1 3DH, United Kingdom; Fig 27.8: Courtesy of Willie Wilson; Fig. 27.9: Courtesy of Mike Allen; Fig. 27.10: Courtesy of Karen Weynberg, republished from

Figure 4.6 d in Ph.D. thesis of Karen Weynberg; **Fig. 27.11:** Courtesy of Nathan Zauberman; **Fig. 27.12a-d:** Reprinted from Fig 5 A, B, D and E in Xiao, C. et. al., "Structural studies of the giant mimivirus," *PLoS Biol.* (2009) Apr 28;7(4):e92.PMID: 19402750; **Fig. 27.13, 14a-c:** Images kindly provided courtesy of Avi Minsky, reprinted from Figures 3A, 6A, B, C in Zauberman, N. et al., "Distinct DNA exit and packaging portals in the virus Acanthamoeba polyphaga mimivirus," *PLoS Biol.* (2008) May 13;6(5):e114. PMID: 18479185; **Fig. 27.15:** Courtesy of Jean-Michel Claverie, reprinted from Figure 4 (p. 97) in Claverie, J.M., Abergel, C., Ogata, H., "Mimivirus," *Curr Top Microbiol Immunol.* 328 (2009):89-121. PMID: 19216436, with kind permission of Springer Science+ Business Media.

Chapter 29

Fig. 29.1: Reprinted from Panteleo, G. and Fauci, A.S., "New concepts in the immunopathogenesis of HIV infection," *Annual Review of Immunology* (1995) 13: 487–512. **Fig. 29.2:** Courtesy Liwie Rong, McGill University.

Chapter 32

Fig. 32.1: Peter Morrison/AP/Wide World Photos; **Fig. 32.2:** Nobel Lecture, December 13, 1976 by D.C. Gajdusek. Copyright The Nobel Foundation, 1976; **Fig. 32.3:** Adapted from Fig. 6 in R. Moore et al., "Ataxia in prion protein (PrP)-deficient mice is associated with upregulation of the novel PrP-like protein doppel," *Journal of Molecular Biology* 292 (1999), 797–817, with permission from Elsevier; **Fig. 32.4a:** Courtesy of Holger Wille. **Fig. 32.4b** and **Fig. 32.6a–b:** Fig. 32.4b and Fig. 32.6a, b: Fig. ID and 4A,B reprinted from Holger Wille, Melissa D. Michelitsch, Vincent Guénebaut, Surachai Supattapone, Ana Serban, Fred E. Cohen, David A. Agard, and Stanley B. Prusiner, "Structural studies of the scrapie prion protein by electron crystallography," *Proceedings of the National Academy of Sciences, USA* 99 (2001): 3563–3568; **Fig. 32.5:** Courtesy of Katherine Sutliffe and P. Morrighan, Science. Reprinted with permission

from Priola, S.A., Chesebro, B., and Caughey, B., "A view from the top: Prion diseases from 10,000 feet," *Science* 300: 917–919, copyright 2003, AAAS. **Fig. 32.7:** Reprinted from Fig. 2 in Jackson, G.S. and Clarke, A.R., "Mammalian Prion Proteins," in *Current Opinion in Structural Biology* 10 (2000), 69–74, with permission from Elsevier.

Chapter 33

Fig. 33.11: Reprinted by permission from Macmillan Publishers Ltd: from Figure 2 in Gunter Meister & Thomas Tuschl, "Mechanisms of gene silencing by double-stranded RNA," *Nature* 431 (2004), 343-349.

Chapter 34

Fig. 34.3: From STARR, Biology: Concepts and Applications, 2E. © 1994 Brooks/Cole, a part of Cengage Learning, Inc. Reproduced by permission. www.cengage.com/permissions.

Chapter 36

Fig. 36.2a: Adapted from Fig. 3b, Smith, T., et al., "Neutralizing antibody to human rhinovirus 14 penetrates the receptor-binding canyon," *Nature* 383 (1996): 350–354; **Fig. 36.2b:** Fig. 2, reprinted with permission from Badger, J. et al., "Structural analysis of a series of antiviral agents complexed with human rhinovirus 14," *Proc. Natl. Acad. Sci. USA* 85 (1988): 3304–3308; **Fig. 36.10a:** Reprinted from figure 36-10 in *Principles of Pharmacology: The Pathophysiologic Basis of Drug Therapy*, 2nd edition (2007), p.668, David E. Golan, Armen H. Tashjian, Jr., Ehrin A. Armstrong, April W. Armstrong, Editors, with the permission of Lippincott Williams and Wilkins; **Fig. 36.10b:** Courtesy of Elspeth Garman, reprinted with permission from Graeme, L. and Garman, E., "The origin and control of pandemic influenza," *Science* 293: 1776–1777, copyright 2001 AAAS.